Hussaini · Van Leer · Van Rosendale (Eds.)
Upwind and High-Resolution Schemes

Springer
Berlin
Heidelberg
New York
Barcelona
Budapest
Hong Kong
London
Milan
Paris
Santa Clara
Singapore
Tokyo

M. Yousuff Hussaini
Bram van Leer
John Van Rosendale (Eds.)

Upwind and High-Resolution Schemes

Springer

Professor M. Yousuff Hussaini
Program in Computational Science & Engineering
Florida State University
Tallahassee, FL 32306-3075, USA

Professor Bram van Leer
Department of Aerospace Engineering
University of Michigan
Ann Arbor, MI 48109-2118, USA

Dr. John Van Rosendale
ICASE, NASA Langley Research Center
Mail Stop 132C
Hampton, VA 23681-0001, USA

Library of Congress Cataloging-in-Publication Data.

Upwind and high-resolution schemes / M. Yousuff Hussaini, Bram van Leer, John Van Rosendale (eds.). p. cm. Includes bibliographical references. ISBN-13: 978-3-642-64452-8 (Berlin: alk. paper)1.Fluid dynamics – Mathematical models. 2. Compressibility. I. Hussaini, M. Yousuff. II. Leer, Bram van. III. Van Rosendale, John R. QA911.U68 1997 532'.0535'0013–dc21 96-39965

ISBN-13:978-3-642-64452-8 e-ISBN-13: 978-3-642-60543-7
DOI: 10.1007/978-3-642-60543-7

Softcover reprint of the hardcover 1st edition 1997

Typesetting of the Preface, the two introductory articles, the annotations to the four chapters, and the bibliography: Data conversion by Springer-Verlag
Cover design: *design & production* GmbH, Heidelberg

SPIN 10548466 55/3144 – 5 4 3 2 1 0 – Printed on acid-free paper

This volume is dedicated to

AMI HARTEN

September 22, 1946 – August 5, 1994

Preface

One of the major achievements in computational fluid dynamics has been the development of numerical methods for simulating compressible flows. A problematic feature of compressible flows is the possible occurrence of shocks, structures much too narrow to be resolved on any practical grid. John von Neumann's solution to this problem is almost as old as his concept of the electronic computer. During World War II, in order to compute the effect of a strong explosion in the atmosphere, he introduced the concept of artificial viscosity, spreading a shock over a few computational cells and thereby "capturing" it without identifying it as a discontinuity. However, it was not until the 1970s that computational scientists succeeded in formulating methods that combine higher-order accuracy in smooth regions with a sharp, oscillation-free representation of embedded shocks – methods now known as "high-resolution schemes".

During the early 1980s knowledge of high-resolution schemes permeated aeronautical science; from then on this discipline took the lead in further advancing these methods. In this chain of events, ICASE played a crucial role. ICASE, where computational scientists and aerospace engineers meet without effort, proved to be the right place at the right time for interdisciplinary endeavors, the breeding of new ideas, and technology transfer. The list of ICASE reports from 1980 onward gives testimony to this historic development.

In 1989, one of us (Hussaini) had the idea of capturing the spirit of this exciting period by reprinting a collection of the most significant papers on high-resolution schemes. Ami Harten, one of the principal contributors to the subject, was asked to write a technical introduction to this collection. He did begin such an introduction, but with the many projects he was involved in, never finished it, and the manuscript was eventually lost. The task was later completed by another key player in the field, Phil Roe.

The present volume collects in one place many of the most significant papers in the development of high-resolution schemes, as this occurred at ICASE. Some of these stand alone, representing a milestone in the development of high-resolution schemes; others are selected as being typical of a particular thread of research, without being comprehensive. The papers follow a historical perspective, and have been grouped by research theme into four chapters, each beginning with notes on the relevance of the papers included.

The volume begins with a historical introduction by one of us (Van Leer). Following this comes the technical introduction by Phil Roe to which we have alluded. While the papers in this volume are self contained, they were written from the point of view of their day, when knowledge was incomplete. Roe's introduction describes high-resolution schemes from the modern vantage point, placing the papers in perspective. At the same time, it connects the historical development of high-resolution schemes to the modern push towards truly multidimensional schemes, an effort being actively pursued at many places around the world.

The remainder of the volume consists of four chapters. Chapter I contains papers highlighting the early development of approximate Riemann solvers in the context of monotonic schemes, which are necessarily of first-order accuracy. To obtain second-order accuracy, one has to relax the monotonicity requirement to the weaker condition of diminishing of total-variation, and Chapter II consists of papers dealing with second-order total-variation-diminishing (TVD) schemes. To obtain arbitrarily high-order accuracy, the total-variation-diminishing requirement has to be relaxed to boundedness of total-variation. Chapter III consists of papers which provide a historical development of total-variation-bounded (TVB) schemes at ICASE. Finally, Chapter IV collects together papers pertaining to inherently multidimensional schemes, which preserve the favorable features of their one-dimensional predecessors.

We hope this volume will find its way onto the bookshelves of those who share with us a fascination for computational science in general, and computational fluid dynamics in particular.

Finally, we wish to express our appreciation to Shelly M. Johnson and Shannon L. Keeter, members of ICASE's excellent secretarial staff, whose patient hard work and organizational skills were invaluable in putting this volume together. We would also like to thank Professor W. Beiglbock for encouraging this effort and for seeing the volume through to completion.

M. Yousuff Hussaini
Bram van Leer
John Van Rosendale

Contents

ICASE and the History of High-Resolution Schemes
B. van Leer 1
A Brief Introduction to High-Resolution Schemes
P. Roe 9

Chapter I. Riemann Solvers and Upwind Schemes

Annotation 31
1. On the Relation Between the Upwind-Differencing Schemes of Godunov, Engquist–Osher and Roe
B. van Leer 33
2. On Upstream Differencing and Godunov-Type Schemes for Hyperbolic Conservation Laws
A. Harten, P.D. Lax, and B. van Leer 53
3. Flux-Vector Splitting for the Euler Equation
B. van Leer 80

Chapter II. Second-Order Total-Variation Diminishing Schemes

Annotation 93
1. A Comparative Study of Computational Methods in Cosmic Gas Dynamics
G.D. van Albada, B. van Leer, and W.W. Roberts, Jr. 95
2. High Resolution Schemes and the Entropy Condition
S. Osher and S. Chakravarthy 104
3. Convergence of Generalized MUSCL Schemes
S. Osher 134
4. A Simplified TVD Finite Difference Scheme Via Artificial Viscosity
S.F. Davis 149
5. On Central-Difference and Upwind Schemes
R.C. Swanson and E. Turkel 167

Chapter III. Essentially Non-oscillatory Schemes

Annotation ... 185
1. Uniformly High-Order Accurate Nonoscillatory Schemes. I
A. Harten and S. Osher ... 187
2. Uniformly High Order Accurate Essentially Non-oscillatory Schemes, III
A. Harten, B. Engquist, S. Osher, and S.R. Chakravarthy ... 218
3. ENO Schemes with Subcell Resolution
A. Harten ... 291
4. Efficient Implementation of Essentially Non-oscillatory Shock-Capturing Schemes, II
C.-W. Shu and S. Osher ... 328
5. Essentially Nonoscillatory Spectral Fourier Methods for Shock Wave Calculations
W. Cai, D. Gottlieb, and C.-W. Shu ... 375
6. Cell Averaging Chebyshev Methods for Hyperbolic Problems
W. Cai, D. Gottlieb, and A. Harten ... 397

Chapter IV. Multidimensional Schemes

Annotation ... 421
1. A Rotationally Biased Upwind Difference Scheme for the Euler Equations
S.F. Davis ... 423
2. Discrete Models for the Numerical Analysis of Time-Dependent Multidimensional Gas Dynamics
P.L. Roe ... 451
3. Discontinuous Solutions to Hyperbolic Systems Under Operator Splitting
P.L. Roe ... 470
4. High Resolution Finite Volume Methods on Arbitrary Grids Via Wave Propagation
R.J. LeVeque ... 491
5. Design of an Essentially Nonoscillatory Reconstruction Procedure on Finite-Element-Type Meshes
R. Abgrall ... 519
6. A Contribution to the Great Riemann Solver Debate
J.J. Quirk ... 550
7. An Approximate Riemann Solver for Magnetohydrodynamics (That Works in More Than One Dimension)
K.G. Powell ... 570

Papers Reprinted ... 585

ICASE and the History of High-Resolution Schemes

Bram van Leer

The need for high-resolution schemes is a direct consequence of the nonlinear properties of hyperbolic systems of conservation laws such as the Euler equations of inviscid compressible flow. Due to the nonlinearity of these equations, any compression wave, no matter how smooth initially, will evolve into a shock discontinuity, while the interaction of shock waves in turn may create shear and entropy discontinuities. When attempting to compute a solution with such features on a grid of discrete meshes, we have the choice of representing these as actual discontinuities – as in "shock fitting" [1, 2], "jump recovery" [3], or "subcell resolution" [4], or spreading them artificially, as in "shock capturing" [5, 6]. In the latter technique the goal is to automatically represent any discontinuity by a sharp but oscillation-free transition, with most of the jump realized within one to three meshes, without compromising accuracy away from such jumps. The term "high resolution" appeared first in an article by A. Harten [7], who attributed it to P. R. Woodward.

For an appreciation of the problem of designing a high-resolution scheme for the Euler equations it is useful to first consider modeling the linear advection of a step function. Here we immediately run into a famous theorem presented by S. K. Godunov [8] in his influential 1959 paper: if an advection scheme preserves the monotonicity of the solution it is at most first-order accurate. Thus, if our step representation is to be monotone, it will necessarily be quite smeared. This result could discourage anyone attempting to improve advection schems; fortunately, there is a way to circumvent it. In the proof of this theorem it is tacitly assumed that the linear advection equation is approximated by a linear discretization; once *nonlinear* discretizations are admitted the theorem no longer stands and high-resolution schemes become possible.

In the same publication Godunov points out that, among the simplest advection schemes, the first-order upwind scheme is in some sense "best": it is the most accurate monotonicity-preserving scheme. He then shows how to extend the concept of upwind differencing to a nonlinear system of conservation laws, viz, the one-dimensional Lagrangean flow equations, by using the solution of Riemann's initial-value problem. This is an evolution problem for a hyperbolic conservation law, with initial values

$$U_0(x) = \begin{cases} U_l, & x < 0, \\ U_r, & x > 0. \end{cases}$$

In Godunov's method the numerical solution at each time level is regarded to be constant within each computational cell. One can then compute its evolution exactly, for a short time, by solving a Riemann problem at each cell interface. At the end of the time step the solutions are averaged in each cell, yielding a new piecewise uniform distribution. The overall method is very robust and produces the steepest monotone shock profiles obtainable with a first-order method.

Godunov's finite-volume scheme initially was of little influence in Western numerical circles, although it is mentioned in the book by R. D. Richtmyer and K. W. Morton [5]. By contrast, the second-order finite-difference scheme devised by P. D. Lax and B. Wendroff [9] at about the same time had an immediate impact. It produces shocks that are sharp but can be highly oscillatory, even nonlinearly unstable. The scheme's built-in dissipation is too small for most uses; additional dissipative terms are needed for oscillation control and robustness. The Lax-Wendroff scheme and its various predictor-corrector variants [5, 10, 11], and similar third-order schemes [12, 13], dominated research in hyperbolic methods for more than a decade. The most useful of these schemes was R. W. MacCormack's [11] efficient method, which became a real work-horse in industrial applications.

The realization that Godunov's theorem could be circumvented came at the beginning of the 1970s, when, less than a year apart, two approaches were launched for the construction of oscillation-free higher-order advection schemes. J. P. Boris (Naval Research Laboratory) presented the first non-oscillatory second-order scheme SHASTA – Sharp and Smooth Transport Algorithm – at a seminar course in Trieste in August 1971 [14], followed by B. van Leer (Leiden Observatory, Netherlands), who presented a non-oscillatory modification of the Lax-Wendroff scheme at the 3rd International Conference on Numerical Methods in Fluid Dynamics (ICNMFD) in Paris in July 1972 [15]. These approaches, while quite distinct, are similar in their use of nonlinearity to prevent numerical oscillations in regions of strongly varying solution gradients, a theme that has persisted in hyperbolic method development ever since. It is useful to discuss the two approaches separately in some detail.

The algorithms developed by Boris, D. L. Book et al. are known as Flux-Corrected Transport (FCT) methods and have a predictor-corrector structure. A first-order, non-oscillatory scheme is used to estimate the solution at the advanced time level; a correction step then removes the large dissipative error made in the first step, uncovering a solution with second- or third-order accuracy. During the second step the corrective fluxes are compared to the provisional solution values and limited where necessary, in order to ensure that no new extrema will arise, nor existing extrema grow. The comparison step makes the overall method nonlinear: the coefficients in the scheme depend on the solution itself, even when applied to a linear equation.

The development of FCT methods is described in four key papers published in the period 1973–79 in the Journal of Computational Physics (JCP), viz a sequence of three by Boris et al. [16–18] and a fourth by S. Zalesak [19], also at NRL. FCT methods are widely used, especially for simulating violent time-dependent flows, but have not had as much influence on the broader numerical community as the Godunov-type high-resolution methods. Harten's early work, though, in particular his Artificial-Compression Method (ACM) [20], was inspired by the FCT approach.

The methodology of Van Leer was published in a series of five papers, starting with the above conference paper; the other four appeared in JCP in the period 1974–79 [21–24]. In this work, oscillations are regarded as the result of oscillatory interpolation of the discrete intitial values; the remedy therefore consists of introducing non-oscillatory or monotonicity-preserving interpolation. Whereas Godunov's method is based on a uniform value of each state variable throughout a cell, the simplest higher-order schemes reconstructing a linear or quadratic distribution in a cell using three contiguous cell averages. Monotonicity is preserved by limiting the values of the first and second derivatives of the distribution.

Following Godunov, Van Leer's schemes include fluxes derived from the solution of Riemann problems. When combined with higher-order reconstruction this leads to upwind-biased differencing, hence the name MUSCL – Monotone Upstream Scheme for Conservation Laws – given to the first computer code (written by Woodward) based on the above principles. It uses piecewise linear interpolation and achieves second-order accuracy for smooth flow. The MUSCL approach was further advanced and popularized in the early 1980s by Woodward and P. Colella at Lawrence Livermore Laboratory [25, 26]. Their Piecewise Parabolic Method (PPM) is based on quadratic reconstruction; although formally only second-order accurate it has a much lower truncation error than MUSCL, and is currently the most accurate method for simulating violent time-dependent flows. A landmark paper is their 1984 JCP review [27], an elaborate comparative study in which Godunov-type schemes, FCT and more traditional methods are pitted against one another.

It is worth noticing that incorporating the Riemann problem was not unique to MUSCL. A. J. Chorin [28] had used it as early as 1976 when applying J. Glimm's [29] "random-choice method" for scalar advection to the Euler equations. In this method the Riemann solutions are not averaged to ensure conservation, as in Godunov's method, but sampled at random in each cell. Later Colella [30] greatly improved its accuracy by introducing pseudo-random sampling, but also showed that it breaks down in two dimensions.

The emergence of Godunov-type methods created several research trends in the 1980s, for instance, designing effective limiters for use in one-dimensional higher-order reconstruction [31–33]. Knowledge of non-oscillatory interpolation greatly increased with the development of Essentially Non-Oscillatory (ENO) schemes by Harten, S. Osher, B. Engquist, S. Chakravarthy, C.-W. Shu [34–37], and others, from 1985 onward. In these schemes a systematic procedure selects the discrete stencil whose data will give the smoothest interpolant, i.e., the function with the lowest values of its derivatives. ENO schemes sacrifice the TVD property, but are higher-order accurate, particularly near extrema.

A parallel trend was the development of algorithms for the approximate solution of Riemann problems, or "approximate Riemann solvers"; these were to replace the computationally expensive "exact Riemann solver". The justification was that conservative schemes suppress most of the information in Riemann solutions anyway, by averaging over them in the final update. Most prominent became P. L. Roe's [38] approximate Riemann solver, based on a local linearization of the conservation laws, and Osher's [39] solver based on simple waves. Both solvers can be used to split flux differences across a mesh into parts attributable to either forward- or backward-moving waves; this technique is called "fluctuation splitting" after Roe, or "flux-difference splitting". Less accurate are the solvers based on the Boltzmann approach [40], in which the distribution functions of adjacent cells are merged; this leads to formulas commonly known as "flux-vector splitting" or simply "flux splitting".

Meanwhile, upwind differencing for the calculation of steady Euler solutions was gaining ground at both NASA Ames and NASA Langley Research Centers. On the West coast J. L. Steger and R. F. Warming [41] in 1978 presented a flux splitting that could be used for implementing upwind differencing, although it was mainly intended for approximate LU factorization of implicit operators. Furthermore, the interaction of P. Goorjian with Engquist and Osher [42] (both at UCLA) in 1979 gave birth to the upwind Engquist-Osher flux for the small-disturbance transonic potential equation. In 1980 Osher extended this to the approximate Riemann solver for the full Euler equations mentioned earlier; this became a key ingredient of the high-resolution schemes he later developed with Chakravarthy (Rockwell) [43]. The year 1980 also marks the first public presentation by Roe [44], at the 7th ICNMFD in Stanford/NASA Ames, of his ideas on fluctuation splitting, which were kept quiet for four years by the Royal Aircraft Establishment, for reasons of security.

On the East coast Harten had already become a regular visitor to ICASE, when Van Leer came for a two year appointment (1979–81), this marks the introduction of Godunov-type methods to NASA Langley. Harten himself had been stimulated by Roe's presentation to work on upwind differencing [40], but did not embrace the concept of non-oscillatory interpolation, because it seemed restricted to one space dimension. Instead he was inspired by Glimm and Lax [45], who had shown that the total variation of the solution of a scalar 1-D conservation law cannot increase, and actually decreases in a shock. By analogy, Harten introduced the total variation of a discrete function as a measure of its oscillatory nature [7, 46], a quantity with the promise of multidimensional applicability. This led to the formulation of Total-Variation-Diminishing (TVD) schemes; the term TVD quickly became synonymous with high-resolution, usually upwind, schemes. Harten derived local conditions sufficient for ensuring the TVD property; for a linear equation these reduce to the constraint of monotone interpolation. It later was shown by J. B. Goodman and R. J. LeVeque [47] that the total variation is too blunt a tool to be of use in constraining multidimensional discrete functions: a multidimensional TVD advection scheme can be no better than first-order accurate. Harten then reexamined non-oscillatory interpolation theory, and developed the ENO schemes. These are only Total-Variation-Bounded (TVB) and do generalize to multidimensional equations.

From 1980 onwards ICASE became a center of knowledge exchange and collaboration in high-resolution schemes for aeronautical problems. Two streams of activity in this area could be distinguished. On the one hand ICASE was the meeting place *par excellence* of numerical analysts and computational fluid dynamicists such as S. F. Davis, Engquist, Harten, Lax, LeVeque, Osher, Roe, E. Tadmor, E. Turkel, and Van Leer, its relaxed research setting stimulating a rapid dissemination and discussion of new ideas, and general brain-storming. On the other hand ICASE was the place where these computational scientists could meet NASA engineers with formidable, realistic flow problems to solve, often leading to mutually beneficial collaboration. One such collaboration, started in 1983 among J. L. Thomas, W. K. Anderson, R. W. Walters (all at NASA), and Van Leer [48, 49], gave birth to the upwind high-resolution code CFL3D for compressible flow, which became a standard of industry.

To describe the above activities would mean condensing a long list of research papers. Instead, we shall let the papers speak for themselves. In the further chapters of this volume the development of high-resolution schemes and their key ingredients is highlighted by a sequence of papers produced at ICASE in the years 1980–1994, some of which may be regarded as the key papers in the field. Each chapter focuses on a particular research theme and starts with historical background information; the papers then follow in chronological order. The themes are also arranged in the order in which they became important through the years.

The first selection of papers we made would easily have filled two volumes; a page limit forced us to narrow down the choice. If some papers are dearly missed, the editors are painfully aware of this. In almost all cases, we have obtained a reprint of the final journal or book version of the ICASE Report and have reproduced this version.

References

[1] G. Moretti, "Efficient Euler solver with many applications", *AIAA Journal*, vol. 26, pp. 655–660, 1988.

[2] K. W. Morton and M. F. Paisley, "A finite volume scheme with shock-fitting for the Euler equations", *Journal of Computational Physics*, vol. 80, pp. 168–203, 1989.

[3] K. W. Morton and M. A. Rudgyard, "Shock recovery and the cell-vertex scheme for the steady Euler equations", *Lecture Notes in Physics*, vol. 323, pp. 424–428, 1989.

[4] A. Harten, "ENO schemes with subcell resolution", *Journal of Computational Physics*, vol. 83, pp. 148–184, 1989.

[5] R. D. Richtmyer and K. W. Morton, *Difference Methods for Initial-Value Problems.* Interscience, 1967.

[6] P. D. Lax, "Weak solutions of nonlinear hyperbolic equations and their numerical computation", *Communications on Pure and Applied Mathematics*, vol. 7, pp. 159–193, 1954.

[7] A. Harten, "High-resolution schemes for hyperbolic conservation laws", *Journal of Computational Physics*, vol. 49, pp. 357–393, 1983.

[8] S. K. Godunov, "A finite-difference method for the numerical computation and discontinuous solutions of the equations of fluid dynamics", *Matematicheskii Sbornik*, vol. 47, pp. 271–306, 1959.

[9] P. D. Lax and B. Wendroff, "Systems of conservation laws", *Communications in Pure and Applied Mathematics*, vol. 13, pp. 217–237, 1960.
[10] E. L. Rubin and S. Z. Burstein, "Difference methods for the inviscid and viscous equations of a compressible gas", *Journal of Computational Physics*, vol. 2, pp. 178–196, 1967.
[11] R. W. MacCormack, "The effect of viscosity in hypervelocity impact cratering", AIAA Paper 69-354, 1969.
[12] V. V. Rusanov, "On difference schemes of third-order accuracy for nonlinear hyperbolic systems", *Journal of Computational Physics*, vol. 5, pp. 507–516, 1970.
[13] S. Z. Burstein and A. A. Mirin, "Third order difference methods for hyperbolic systems", *Journal of Computational Physics*, vol. 5, pp. 547–571, 1970.
[14] J. P. Boris, "A fluid transport algorithm that works", in *Computing as a Language of Physics*, pp. 171–189, International Atomic Energy Commission, 1971.
[15] B. van Leer, "Towards the ultimate conservative difference scheme. I. The quest of monotonicity", *Lecture Notes in Physics*, vol. 18, pp. 163–168, 1973.
[16] J. P. Boris and D. L. Book, "Flux-Corrected Transport I: SHASTA, a fluid-transport algorithm that works", *Journal of Computational Physics*, vol. 11, pp. 38–69, 1973.
[17] J. P. Boris, D. L. Book, and K. H. Hain, "Flux-Corrected Transport II: Generalization of the method", *Journal of Computational Physics*, vol. 18, pp. 248–283, 1975.
[18] J. P. Boris and D. L. Book, "Flux-Corrected Transport III: Minimal error FCT methods", *Journal of Computational Physics*, vol. 20, pp. 397–431, 1976.
[19] S. Zalesak, "Fully multidimensional Flux-Corrected Transport algorithms for fluids", *Journal of Computational Physics*, vol. 31, pp. 335–362, 1979.
[20] A. Harten, "The Artificial Compression Method for computation of shocks and contact discontinuities: III. Self-adjusting hybrid schemes", *Mathematics of Computation*, vol. 32, pp. 363–389, 1978.
[21] B. van Leer, "Towards the ultimate conservative difference scheme. II. Monotonicity and conservation combined in a second-order scheme", *Journal of Computational Physics*, vol. 14, pp. 361–370, 1974.
[22] B. van Leer, "Towards the ultimate conservative difference scheme. III. Upstream-centered finite-difference schemes for ideal compressible flow", *Journal of Computational Physics*, vol. 23, pp. 263–275, 1977.
[23] B. van Leer, "Towards the ultimate conservative difference scheme. IV. A new approach to numerical convection", *Journal of Computational Physics*, vol. 23, pp. 276–299, 1977.
[24] B. van Leer, "Towards the ultimate conservative difference scheme. V. A second-order sequel to Godunov's method", *Journal of Computational Physics*, vol. 32, pp. 101–136, 1979.
[25] P. Colella, "A direct Eulerian MUSCL scheme for gas dynamics", *SIAM Journal on Scientific and Statistical Computing*, vol. 6, pp. 104–117, 1985.
[26] P. Colella and P. R. Woodward, "The Piecewise-Parabolic Method (PPM) for gas-dynamical simulations", *Journal of Computational Physics*, vol. 54, pp. 174–201, 1984.
[27] P. R. Woodward and P. Colella, "The numerical simulation of two-dimensional fluid flow with strong shocks", *Journal of Computational Physics*, vol. 54, pp. 115–173, 1984.
[28] A. J. Chorin, "Random choice solution of hyperbolic systems", *Journal of Computional Physics*, vol. 22, pp. 517–533, 1976.

[29] J. Glimm, "Solution in the large for nonlinear hyperbolic systems of equations", *Comm. Pure Appl. Math.*, vol. 18, pp. 697–715, 1965.

[30] P. Colella, "Glimm's method for gas dynamics", *SIAM Journal on Numerical Analysis*, vol. 18, pp. 289–315, 1981.

[31] P. K. Sweby, "High resolution schemes using flux limiters for hyperbolic conservation laws", *SIAM Journal on Numerical Analysis*, vol. 21, pp. 995–1011, 1984.

[32] S. P. Spekreijse, *Multigrid Solution of the Steady Euler Equations.* PhD thesis, Technische Universiteit Delft, 1987. Published by the Centrum voor Wiskunde en Informatica (CWI), Amsterdam.

[33] B. Koren, *Multigrid and Defect Correction for the Steady Navier-Stokes Equations.* PhD thesis, Technische Universiteit Delft, 1989. Published by the Centrum voor Wiskunde en Informatica (CWI), Amsterdam.

[34] A. Harten, S. Osher, B. Engquist, and S. R. Chakravarthy, "Some results on uniformly high-order accurate essentially non-oscillatory schemes", *J. Appl. Num. Math.*, vol. 2, pp. 347–377, 1986.

[35] A. Harten and S. Osher, "Uniformly high-order accurate non-oscillatory schemes I", *SIAM J. Numer. Anal.*, vol. 24, pp. 279–309, 1987.

[36] A. Harten, B. Engquist, S. Osher, and S. R. Chakravarthy, "Uniformly high-order accurate non-oscillatory schemes III", *Journal of Computational Physics*, vol. 71, pp. 231–303, 1987.

[37] C. W. Shu and S. Osher, "Efficient implementation of Essentially Non-Oscillatory shock capturing schemes", *Journal of Computational Physics*, vol. 77, pp. 439–471, 1988.

[38] P. L. Roe, "Approximate Riemann solvers, parameter vectors and difference schemes", *Journal of Computational Physics*, vol. 43, pp. 357–372, 1981.

[39] S. Osher and F. Solomon, "Upwind schemes for hyperbolic systems of conservation laws", *Mathematics and Computation*, vol. 38, pp. 339–374, 1982.

[40] A. Harten, P. D. Lax, and B. van Leer, "Upstream differencing and Godunov-type schemes for hyperbolic conservation laws", *SIAM Review*, vol. 25, pp. 35–61, 1983.

[41] J. L. Steger and R. F. Warming, "Flux vector splitting of the inviscid gas-dynamic equations with applications to finite difference methods", *Journal of Computational Physics*, vol. 40, pp. 263–293, 1981.

[42] B. Engquist and S. Osher, "Stable and entropy satisfying approximations for transonic flow calculations", *Mathematics of Computation*, vol. 34, pp. 45–75, 1980.

[43] S. K. Chakravarthy and S. Osher, "A new class of high-accuracy TVD schemes for hyperbolic conservation laws", AIAA Paper 85-0363, 1985.

[44] P. L. Roe, "The use of the Riemann problem in finite-difference schemes", *Lecture Notes in Physics*, vol. 141, pp. 354–359, 1980.

[45] J. Glimm and P. D. Lax, "Decay of solutions of systems of nonlinear hyperbolic conservation laws", *Mem. Amer. Math. Soc.*, no. 101, 1970.

[46] A. Harten, "On a class of high resolution total variation stable finite difference schemes", *SIAM Journal on Numerical Analysis*, vol. 21, pp. 1–23, 1984.

[47] J. B. Goodman and R. J. LeVeque, "On the accuracy of stable schemes for 2D conservation laws", *Mathematics of Computation*, vol. 45, pp. 15–21, 1985.

[48] W. K. Anderson, J. L. Thomas, and B. van Leer, "Comparison of finite volume flux vector splittings for the Euler equations", *AIAA Journal*, vol. 24, pp. 1453–1460, 1985.

[49] J. L. Thomas, B. van Leer, and R. W. Walters, "Implicit flux-split schemes for the Euler equations", AIAA Paper 85-1680, 1985.

A Brief Introduction to High-Resolution Schemes

Philip Roe

High-resolution schemes are a class of algorithm for solving problems involving first-order partial differential equations in which wave propagation, especially nonlinear wave propagation, is an important feature. They are strongly rooted in the mathematical theory of characteristics, and it is one purpose of these introductory remarks to provide the reader with a brief overview of this idea and its numerical implementation. The material is not intended to be mathematically rigorous; indeed rigor is not yet to be found in all aspects of the schemes. It aims instead to convey the strategies involved. Readers interested in a more precise explanation of the elements that make up the strategy should consult the references provided. Textbooks that provide an overview of high-resolution schemes are [13, 18, 24].

1 Characteristic Equations

We will begin with the equations describing unsteady isentropic flow in a parallel pipe. The unknowns are pressure p and velocity v, obeying

$$\partial_t p + v\partial_x p + \gamma p \partial_x v = 0, \tag{1}$$

$$\partial_t v + v\partial_x v + \frac{1}{\rho}\partial_x p = 0. \tag{2}$$

Here γ is the ratio of specific heats and the density ρ is not a constant but can be found from the isentropic relationship for weak waves that p/ρ^γ is constant.

A condition for equations like these to admit solutions describing wavelike behavior is that some linear combination of them reduces to an ordinary differential equation [22, 51]. In this particular case, if we introduce the speed of sound a, such that $a^2 = \gamma p/\rho$ and then add ρa times the second equation to the first equation, we get

$$(\partial_t p + (v+a)\partial_x p) + \rho a(\partial_t v + (v+a)\partial_x v) = 0, \tag{3}$$

where each bracketed expression does indeed represent an ordinary derivative along the *characteristic line* whose speed is given by $dx/dt = v + a$. If we denote this derivative by $\mathcal{D}/\mathcal{D}t$, then (3) becomes

$$\frac{\mathcal{D}p}{\mathcal{D}t} + \rho a \frac{\mathcal{D}v}{\mathcal{D}t} = 0. \tag{4}$$

An ordinary differential equation derived from partial differential equations is called a *characteristic equation.* It can be interpreted as a statement that the characteristic lines are pathways for information, along which waves may propagate. For the pipe flow problem it is easy to show that there are just two such paths, the other one being obtained by replacing a with $-a$ in (3). The wavespeeds are $\pm a$ relative to the fluid, and $v \pm a$ in a laboratory frame. In subsonic flow, with $v < a$, one of these waves will propagate to the left, and one to the right, as in Fig 1(a). From one point of view the characteristic equations merely restate the original problem; from another they expose its inner essence.

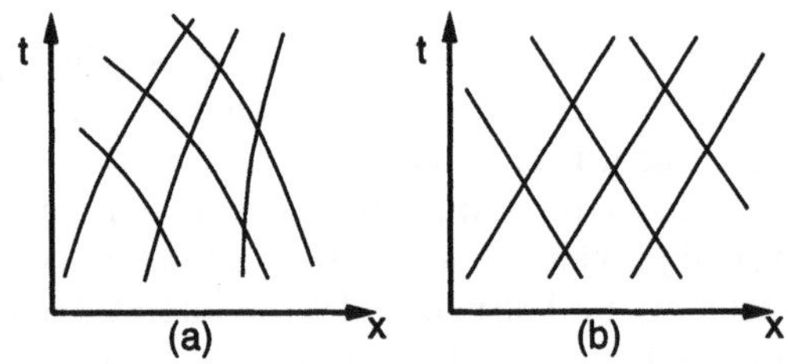

Figure 1: Characteristic curves in the (x, t)-plane, (a) in the nonlinear case, and (b) in the linear case.

Any set of equations that is completely equivalent to a set of characteristic equations is said to be *hyperbolic*, and this embraces many equations of practical importance, such as the Euler equations of compressible gas dynamics, the Saint-Venant equations of shallow-water flow, the equations of magnetohydrodynamics, and so on. In each of these situations, certain kinds of information are carried through the fluid, for example as acoustic waves, and others are carried with the fluid, at the local particle speed, like temperature and vorticity. For our present purposes, there is little need to distinguish the two types of propagation, although it is beginning to be stressed in some current research.

Anyone devising a scheme to solve hyperbolic problems numerically might be struck by the fact that they present themselves in (at least) two different forms, characteristic and non-characteristic. Which form would be the best basis for a numerical procedure? It could be (and has been) argued that since the possibility of wave propagation is a property of the physical system, whatever form the equations are written in, one should not have to take special care to see waves appearing in the solution of the discrete (digitized) equations. On the other hand, the waves that appear in a naively written code are often corrupted by noise that makes the solution unintelligible. Thus, one is led to consider the incorporation of wave propagation properties directly into the numerical procedure, even if this is likely to carry some overhead in the form of additional computation. In general this overhead proves not to be worthwhile if the waves involved are smooth and gentle, or if they decay to a smooth steady state which is the main item of interest. But if the waves are highly nonlinear and discontinuous the benefits are great, as the numerical analysis community has come to realise over the period covered by this collection of papers.

Equations (1,2) can be expressed in a compact *matrix form* as

$$\partial_t \begin{pmatrix} p \\ v \end{pmatrix} + \begin{bmatrix} v & \gamma p \\ \frac{1}{\rho} & v \end{bmatrix} \partial_x \begin{pmatrix} p \\ v \end{pmatrix} = 0 \tag{5}$$

and a general system of first-order equations can be written

$$\partial_t \mathbf{u} + A \partial_x \mathbf{u} = 0, \tag{6}$$

where $\mathbf{u}$ is a generic symbol for a vector of m unknowns, and A is a square $m \times m$ matrix. The advantage of this is that it makes available many powerful tools of matrix analysis, in particular the eigenvector, or singular value decomposition. It is a condition for the problem to be hyperbolic that A has as many real eigenvalues as there are unknowns, and that if any eigenvalues are repeated then the corresponding eigenvectors must be distinct [13, 18, 22, 24, 51]. Linear problems are characterized by the fact that A is a constant matrix; in particular the eigenvalues are constant, and so the wavespeeds are constant, as in Fig. 1(b). Usually, in significant applications such as (5), A is not constant, and the problem is nonlinear. In computations, it is often convenient to treat problems as locally linear by 'freezing' A at some local value.

If the eigenvector factorization of A is

$$A = R\Lambda L, \tag{7}$$

where R, L are respectively the matrices containing the right and left eigenvectors, consistently ordered, and Λ is the diagonal matrix containing the eigenvalues, then all elements of the decomposition are significant. An eigenvalue λ_i is a possible wavespeed, the right eigenvector $\mathbf{r}_i$ is the pattern of disturbances carried with that speed, and the left eigenvector $\boldsymbol{\ell}_i$ can be used to find the local strength of that wave. Specifically, if we have initial data $\mathbf{u}(x)$ the local strength of the ith wave is $\boldsymbol{\ell}_i \cdot \partial_x \mathbf{u}$. This accounts for the incessant references to eigenstructure in the papers that follow. An automatic way to find characteristic forms of hyperbolic equations is to premultiply (7) by L. The ith row of the resulting equation is

$$\boldsymbol{\ell}_i \partial_t \mathbf{u} + \boldsymbol{\ell}_i A \partial_x \mathbf{u} = 0,$$

or

$$\boldsymbol{\ell}_i (\partial_t \mathbf{u} + \lambda_i \partial_x \mathbf{u}) = 0,$$

implying that

$$\boldsymbol{\ell}_i \cdot d\mathbf{u} = 0 \qquad \text{on} \qquad dx/dt = \lambda_i$$

and this is the typical characteristic equation. You might like to verify this for the pipe flow equations.

For linear problems, the characteristic equation simplifies further to

$$d(\boldsymbol{\ell}_i \cdot \mathbf{u}) = 0 \qquad \text{on} \qquad dx/dt = \lambda_i,$$

so that the *Riemann invariant* $(\boldsymbol{\ell}_i \cdot \mathbf{u})$ is a constant along the characteristic path. The value taken by this variable in the initial data, or at one end of the domain, will be carried through the flow along the characteristic line, until the line leaves the domain, or (in the nonlinear case) enters a shockwave. This case makes particularly vivid the propagation of information along characteristics, and also shows that we have succeeded in decoupling our original problem, whatever it was, into a certain number of independent problems for the Riemann invariants. The decoupling is not in general perfect for nonlinear problems with more than two unknowns [22], but it has proved to be a very successful practical strategy to ignore this, and to proceed numerically as though the various waves were completely independent.

2 Scalar Advection

In the linear case where they are independent, each wave is governed by an equation like

$$u_t + au_x = 0, \tag{8}$$

where u is one of the Riemann invariants and a is its wavespeed. This is called the *scalar advection*[1] equation. This seemingly trivial equation (any function of $x - at$ solves it exactly) contains the essence of the difficulty encountered in numerical approaches to hyperbolic problems. A proper understanding of its correct treatment was for long elusive, and the correct generalization of that treatment to "real-world" hyperbolic problems forms much of the content of the following pages. Because no paper that appears here actually treats (8) in any detail, we discuss it briefly now.

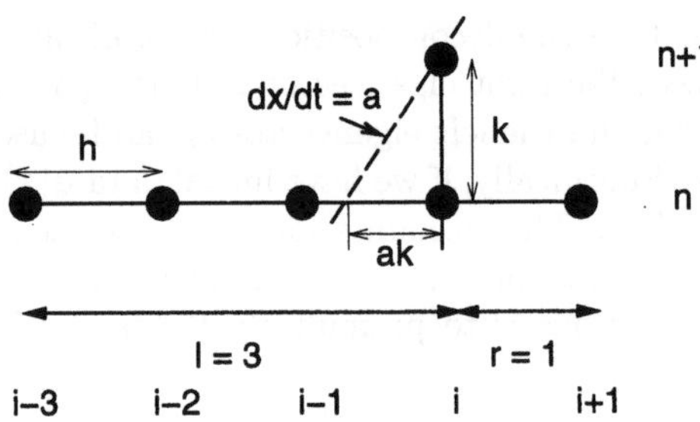

Figure 2: Options for the stencil employed to solve Eq. (8) numerically. In the exact solution, u is constant along the dashed line.

We consider discrete approximations u_i^n to $u(ih, nk)$ where h, k are uniform increments in space and time respectively, and note that (8), like any hyperbolic problem, is well-posed as an initial-value problem where $u(x,t)$ is defined through its initial value $u(x, 0)$ and then obtained by 'marching' in time. Numerically, the equivalent of this is to compute each value u_i^{n+1} in terms of

[1]The term originates with early studies of numerical modelling of transport phenomena by meteorologists. In their vocabulary, advection is the long-range transport of, say, moisture by the prevailing wind, and convection is the mixing of moist and dry air by smaller scale air currents

$u^n_{i-l}, u^n_{i-l+1}, \ldots u^n_{i+r-1}, u^n_{i+r}$, where the set of $(l + r + 1)$ previous values, consecutive in space, is called the *stencil* of the numerical method. A surprising amount of information about the numerical scheme can be deduced simply from its stencil. For example, if a is positive, a stencil with $l < ak/h$ will be unstable, because the numerical information available to it does not include the required information along the characteristic. Similarly, if a is negative, a scheme with $r < |a|k/h$ will be unstable. The parameter ak/h is the Courant-Friedrichs-Lewy number (or Courant number or CFL number), usually denoted by ν.

The accuracy that is normally associated with symmetry would suggest taking a symmetric stencil, with $r = l$, but in fact wave propagation is inherently an asymmetric process. We may note that taking $r = 0$ would lead to numerical information being propagated exclusively from left to right, and we might find this attractive for a problem with a positive. Certainly, if the problem is not an initial-value problem, but one in which in which information is fed in at the left of some domain, and propagates from this boundary to the right into the domain, such a property would be useful. Choosing the stencil to reflect the direction of wave propagation is called *upwinding*. For a positive a stencil with $r = 0$ is called an *upwind stencil* and a scheme with $r < l$ is called an *upwind-biassed stencil.* In a very fundamental contribution, Iserles (see [19]) showed that stable schemes are obtained only if $(l - r)$ has absolute value 0, 1, or 2, and the same sign as a. Zero is the symmetric choice, referred to as *central differencing.* Viable algorithms can of course be constructed from central diffferences; they usually have advantages of speed and simplicity because they treat all variables alike. As problems get more complicated, however, there comes a stage, not precisely definable, at which the more elaborate upwind schemes treated in this book become worthwhile.

Choosing the correct stencil is, however, only part of the solution. Since (8) is a linear equation, we naturally think of solving it by linear methods; specifically, by formulas of the type

$$u_i^{n+1} = \sum_{k=-l}^{k=r} c_k u_{i+k}^n \tag{9}$$

representing a weighted sum over the data contained in the stencil. All "classical" methods, that is, those developed prior to 1970 are in this class, and all share a disadvantage first pointed out by Godunov [14]. Many workers had observed when solving wave propagation problems numerically, even ones as simple as (8), that regions where the wave was steep (on the scale of the mesh) were either spread out unrealistically or else corrupted by meaningless oscillations (see Fig. 3), and no completely satisfactory cure for this had ever been discovered. It seemed to many that this was a natural punishment for the folly of attempting to discretize a partial differential equation (8) close to a discontinuity where the derivatives did not exist.

Although, as we will soon see, this objection is not a valid one, Godunov's result appeared to confirm it by showing that the dilemma was inevitable, in that all schemes of the form (9) would suffer one fault or the other. Specifically, no linear scheme with better than first-order truncation error could avoid the

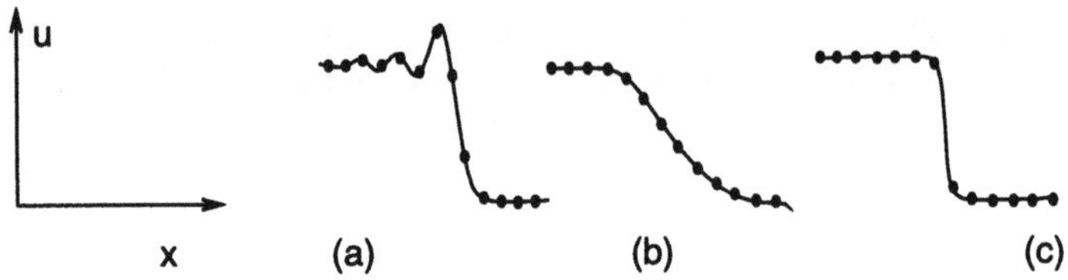

Figure 3: Representation of captured discontinuities. Linear methods of type (9) produce results like (a) if they are second-order accurate, or like (b) if they are first-order accurate. The crisp resolution shown in (c) can only be achieved by nonlinear methods.

oscillations. He also showed that the oscillation-free scheme with the smallest possible (necessarily first-order) error constant was the simple upwind scheme defined by

$$u_i^{n+1} = (1-\nu)u_i^n + \nu u_{i-1}^n \quad \text{if } a > 0, \qquad u_i^{n+1} = (1+\nu)u_i^n - \nu u_{i+1}^n \quad \text{if } a < 0. \tag{10}$$

Thus, the best linear scheme, in a rather natural sense, turns out to depend on the sense of wave propagation (or information flow), but is, even so, unduly diffusive in its handling of discontinuities.

The direction taken by researchers developing what came to be called *high-resolution schemes* was, in one way or another, to add terms to (10) that would improve its accuracy in smooth regions of the solution, but which could be switched off in non-smooth regions. The detection of, and response to, non-smooth regions, turned out to be an essentially nonlinear procedure, not expressible in the form (9), and thereby escaped Godunov's dilemma. The simplest, and still widely used, measure of smoothness is the ratio of two consecutive differences in the data,

$$r_i = \frac{u_{i+1} - u_i}{u_i - u_{i-1}}. \tag{11}$$

We will defer discussing exactly how this is employed until we have examined another aspect of wave propagation.

3 Conservation

Many wave propagation problems are governed by laws of *conservation*, such as those for mass, momentum, and energy. It is important that numerical treatments conserve the same quantities, since otherwise it is found that discontinuities (which are governed by *integral* relationships) may travel with the wrong speed. To handle conservation, almost all high-resolution schemes have been developed in the context of the *finite-volume* approach, where space (the reader should now be thinking three-dimensionally) is divided into non-overlapping control volumes, each containing a certain mass, momentum, energy, or whatever. The calculation proceeds by determining a certain *numerical flux* through each

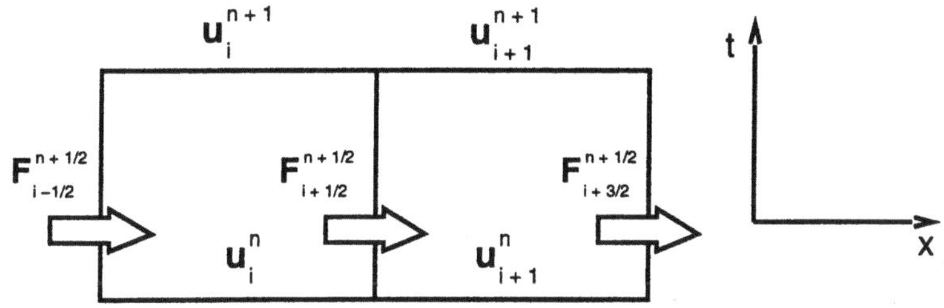

Figure 4: Illustrating a generic finite volume method.

interface between two adjacent volumes. This flux transfers a certain amount of mass, say, removing it from one volume and adding it to the other. Overall mass is therefore conserved, except to the extent that it is added or removed at boundaries of the domain. The goal of the research became concentrated on the design of a (necessarily nonlinear) *numerical flux function* to do this.

Many of the issues relating to general conservation laws can be illustrated by the simple example of the *inviscid Burgers' equation*

$$\partial_t u + \partial_x (u^2/2) = 0 \tag{12}$$

for which the *conserved quantity* is u, and the *flux function* $f(u)$ is $u^2/2$. A generic conservation law $\partial_t u + \partial_x f(u) = 0$ can always be written in characteristic form as $\partial_t u + \partial_u f \partial_x u = 0$ showing that the wave speed is $\partial_u f$ which in this case is simply u. Note that the wave speed may change sign. Burgers' equation is often presented as a purely abstract problem that allows the properties of conservation laws to be exemplified very simply, but it is the correct asymptotic description of most physical wave phenomena in the weakly nonlinear limit [26].

A conservative scheme for this problem has the general form

$$u_i^{n+1} = u_i^n - \frac{k}{h}\left[F_{i+\frac{1}{2}}^{n+\frac{1}{2}} - F_{i-\frac{1}{2}}^{n+\frac{1}{2}}\right]. \tag{13}$$

Here the quantities F are the numerical fluxes evaluated on the interfaces between two volumes. The flux at $i+\frac{1}{2}$ appears with a negative sign in the formula for u_i^{n+1} and with a positive sign in the formula for u_{i+1}^{n+1}; it therefore serves only to transfer the conserved quantity from one cell to another, and nothing is lost. Because of this *conservation form* the discrete description remains valid near discontinuities, and so there is no *a priori* reason why it cannot be used to compute shocks, even though the scheme may look like a finite-difference approximation to an (invalid) differential equation[2].

[2]It is natural still to feel some anxiety about replacing the discrete jump conditions (Rankine-Hugoniot relations), that hold across an infinitesimally thin shock, with a 'smeared-out' statement of conservation. It seems likely that such a strategy will lead to some sort of unavoidable error. Currently, rather delicate computations of the interaction between a strong shock and weak acoustic waves are not successful unless the shock is either represented as an explicit discontinuity or else the grid spacing is greatly reduced in its vicinity.

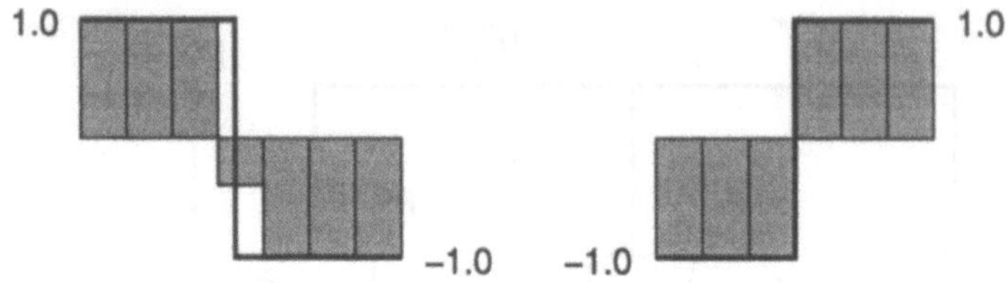

Figure 5: At left is the finite volume representation of a stationary shock solution to Burgers' equation. This is a physically realistic shock because characteristics converge into it with speed +1 from the left and −1 from the right. At right is a finite-volume representation of a rarefaction shock.

We create an explicit scheme by assuming that each F depends on some combination of the nearby states at time-level n. The simplest such combination would be

$$F_{i+\frac{1}{2}}^{n+\frac{1}{2}} = \frac{1}{2}\left[(u^2/2)_i^n + (u^2/2)_{i+1}^n\right]$$

leading to a central differencing approach, which is however unstable without the addition of numerical dissipation. Note, by the way, that since finite-volume methods can be thought of simply as finite-difference methods of a special kind, any general considerations of accuracy and stability apply equally to both types of method.

A simple upwind flux formula could be written as

$$F_{i+\frac{1}{2}}^{n+\frac{1}{2}} = \frac{1}{2}\left[(u^2/2)_i^n + (u^2/2)_{i+1}^n\right] - \frac{1}{4}\left|u_i^n + u_{i+1}^n\right|\left(u_{i+1}^n - u_i^n\right). \qquad (14)$$

The effect of the modulus bars in the extra term (which could be regarded as a form of numerical dissipation) in this formula is to select the flux from one side or other of the interface, according to the local direction of information flow. It is entirely equivalent to (dropping the superscripts)

$$F_{i+\frac{1}{2}} = \text{IF } \ u_i + u_{i+1} > 0 \ \text{ THEN } \ (u^2/2)_i \ \text{ ELSE } \ (u^2/2)_{i+1} \ \text{ FI}. \qquad (15)$$

and the choice between (14) and (15) is machine-dependent, depending on which compiles most efficiently.

It is important that the upwind scheme admits steady discrete solutions consisting of two constant states, i.e., $u_i = u_L$, $(i \leq I)$, $u_i = u_R$ $(i > I)$. For example, it admits the solution with $u_L = 1$, $u_R = -1$, which is the discrete equivalent of a steady shock located midway between grid points. It continues to be a steady discrete solution if a state u_M with $-1 \leq u_M \leq 1$ is inserted between the constant states, thereby allowing representation of a shock in a general location[3]. Unfortunately there is also a steady solution with $u_L = -1$, $u_R = 1$. This corresponds to an *entropy-violating* shock, such that the characteristics on

[3] It is easy to see that at least one such intermediate state is needed if the shock is supposed to move continuously across the grid. Representing shocks with *minimum* width is not neccessarily the aim of the game, especially now that adaptive grid procedures have been developed. A compact representation is, however, still the main objective of a high resolution scheme.

either side do not converge into it. In the physical world such situations are unstable, and break up into rarefaction waves. This has been a source of difficulty in the process of creating schemes which try to keep numerical dissipation small; they need to be provided with some other mechanism that can, when necessary, break symmetry. Such an additional mechanism is often called an *entropy fix.*

Theoretically, no entropy fix is needed if one follows Godunov's original suggestion of computing the flux between two cells as though the state in each cell were locally constant at the beginning of the time step. The breakup of an initial discontinuity between two constant states is a classical problem for hyperbolic partial differential equations, called the *Riemann problem* [14, 18, 22, 24, 51]. For Burgers' equation the solution is a shock moving with speed $\frac{1}{2}(u_i + u_{i+1})$ if $u_i > u_{i+1}$, otherwise it is a rarefaction wave occupying the region

$$u_i < (x - x_{i+\frac{1}{2}})/t < u_{i+1}.$$

If u_i, u_{i+1} are of the same sign the flux through the interface is correctly given by (14) or (15); also if they are of different sign and form a shock. If they are of different sign and form a rarefaction (the *sonic case*) the flux is zero, which makes a rarefaction shock break up. The formula is

$$\begin{aligned} F_{i+\frac{1}{2}} &= \text{IF } u_i < 0 \text{ AND } u_{i+1} > 0 \text{ THEN } 0 \text{ ELSE} \\ &\quad \text{IF } u_i + u_{i+1} > 0 \text{ THEN } (u^2/2)_i \text{ ELSE } (u^2/2)_{i+1} \text{ FI.} \end{aligned} \tag{16}$$

The flux that is found in this way by solving the Riemann problem exactly is sometimes called the *Godunov flux* (and the resulting scheme is called the *Godunov method*). It is seldom used in practice because for most physical problems it is relatively expensive computationally; it has been more usual to solve the Riemann problem approximately and add fixes as needed. In fact, it was shown by Osher [31] that for scalar conservation laws, the Godunov flux is the weakest possible way to break symmetry; it has been found that stronger methods work better in practice. Therefore, it is not in fact correct to think of the Godunov flux as an ideal to which all other flux formulas are merely cheap approximations. The perfect flux function has many tasks to perform, and subtle pitfalls to avoid, as the paper by Quirk in Chapter IV shows.

Nevertheless, the style of argument leading to the Godunov flux has been very influential. We solve exactly (at least for one time step) an approximation of the real problem in which the real initial data is replaced by piecewise constant data. The effect of the conservation formula (13) is to compute the new average contents of each cell, which may be taken as the new (piecewise constant) approximate data for the next timestep. The question that one is asking is "How would the data evolve if they really were piecewise constant?" This is a quite different method of building in the essential physics, compared with more traditional numerical methods based on making Taylor expansions of an unknown function. The Godunov procedure leads almost inevitably to flux functions containing some logical components[4], such as the modulus bars

[4]These may disappear due to cancellation of terms if the governing equations are simple and highly symmetric.

in (14), or the IF..THEN..ELSE in (15),(16). This is a computational cost to be paid for faithful imitation of the physics.

4 Nonlinear Reconstruction

Returning to the original, linear, problem (8), it is easy to verify that the Godunov scheme reduces to the simple upwind method (10), which is only first-order accurate. We recall that any higher-order correction terms must be added in a nonlinear way, even for this linear problem. There are several ways to do this, as will be realized from the papers in the second section, but the one that has become most popular is based on the idea of reconstructing the data in progressively more sophisticated detail. For example, given cell averages $\overline{u}_i$, we could try to find a linear, quadratic, or higher-order function $u_i(x)$ within each cell, such that its average value coincides with the given one, and its derivatives reflect information available from neighboring cells. The simplest case, which leads to a second-order method, and is due to Van Leer [23], is to make piecewise linear reconstructions such that in the ith cell

$$u_i(x) = \overline{u}_i + s_i(x - \overline{x}_i), \tag{17}$$

where the local gradient s_i needs to be determined from surrounding data. Closest at hand are the gradients

$$\begin{aligned} s_i^- &= (\overline{u}_i - \overline{u}_{i-1})/h \\ s_i^+ &= (\overline{u}_{i+1} - \overline{u}_i)/h. \end{aligned}$$

Since reconstruction of data, unlike wave propagation, has no inherent asymmetry, these two estimates should be used symmetrically, but simply taking their arithmetic mean

$$s_i = \frac{1}{2}(s_i^- + s_i^+)$$

leads to a linear formula and hence to oscillations. One needs to take some form of nonlinear average (geometric, harmonic, etc.) and a powerful idea is simply to choose the averaging procedure so that no new extrema arise in the

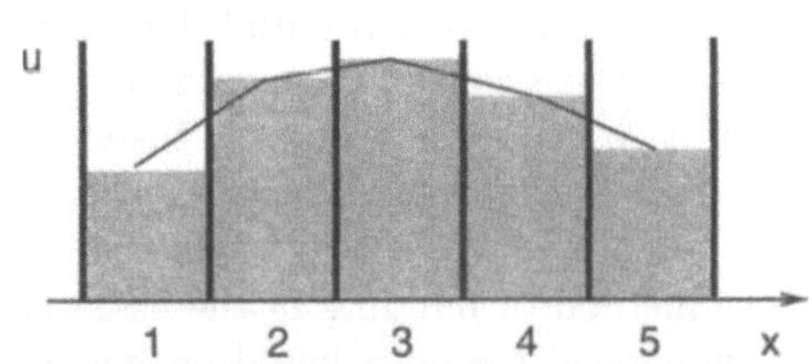

Figure 6: (a) A typical problem in nonlinear data reconstruction. Given are the average values in each cell of some conserved quantity, and the gradients s^-, s^+ connecting the cell to its left and right neighbors.

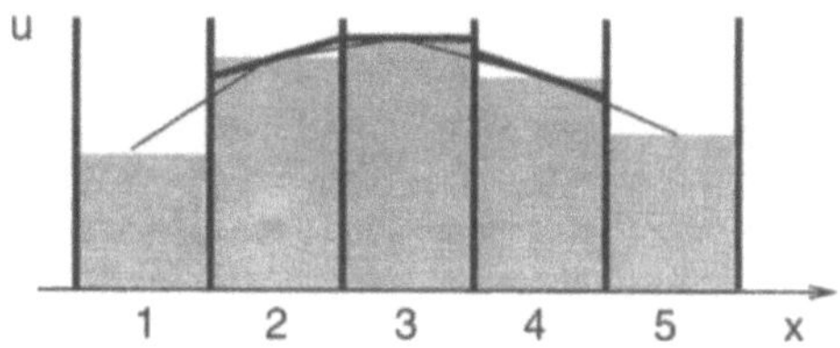

Figure 6: (b) The reconstructed gradients are denoted by heavy lines. In cell 3 no gradient is allowed because that would accentuate the existing maximum. In cell 4, s^-, s^+ are quite similar, and a slope close to their arithmetic average is allowed. In cell 2, s^-, s^+ are rather different and the average must be restricted.

reconstructed data that were not present in the cell averages themselves, and so that existing extrema are not accentuated.

Such a constraint is sometimes misinterpreted, because of course wave propagation can cause new extrema to arise. For example, in the pipe acoustic problem with which we began, if compression waves are started from each end of the pipe, the pressure generated as they pass through each other will exceed the pressure generated by either one alone. Nevertheless, the individual waves do not increase in strength (at least not in linear problems) and it is that fact that is modelled by studying (8), and requiring that its extrema (that is, the extrema of the Riemann invariant being carried) do not increase.

It is easy to show that a necessary and sufficient condition for this is that if s_i^-, s_i^+ are of opposite sign then $s_i = 0$ (the cell i was already a local extremum, and we must do nothing to exaggerate it), and if s_i^-, s_i^+ are of the same sign then s_i should share that sign, but its magnitude should not exceed twice the larger of $|s_i^-|$, $|s_i^+|$. The geometric mean does not have that property, but the harmonic mean does, and we can neatly modify the usual definition to cover the case of opposite signs[5],

$$s_i = \frac{|s_i^+|s_i^- + |s_i^-|s_i^+}{|s_i^-| + |s_i^+|}.$$

Now we ask Godunov's question: what would happen if the data really had this piecewise linear form? At least for small times, the flux would be determined by the Riemann problem for the two local states either side of the interface, that is

$$u^L_{i+\frac{1}{2}} = \overline{u}_i + \frac{h}{2}s_i, \qquad u^R_{i+\frac{1}{2}} = \overline{u}_{i+1} - \frac{h}{2}s_{i+1}.$$

The flux function for Burgers' equation would be one of (14), (15),(16), with u_i replaced by $u^L_{i+\frac{1}{2}}$, and u_{i+1} by $u^R_{i+\frac{1}{2}}$. High-resolution schemes derived like this are often referred to as *higher-order Godunov-type schemes*.

[5]There is a practical problem that this expression becomes indeterminate whenever $s_i^- = s_i^+ = 0$ as happens, for example, in a region of uniform flow. To avoid this we may define a small positive parameter ϵ, and add ϵ to the denominator, but ϵ^2 to the numerator.

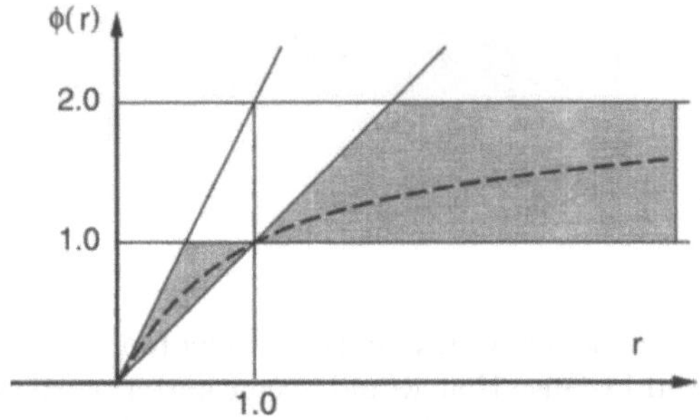

Figure 7: The limiter diagram, with the harmonic mean limiter displayed.

Because the flux found in this way is only appropriate for small times, the accuracy of the resulting scheme is second-order in space, but only first-order in time. That is, it has a truncation error of order (h^2+k). A two-step modification that is second-order in time, due to Hancock, is described in the paper by Van Albada, Van Leer, and Roberts. For calculating simply a final steady state, no such modification is needed. Quite commonly, some form of multistep time integration is employed, either as an alternative way to recover time-accuracy, or for the opposite purpose of removing transients as quickly as possible en route to the steady state.

The harmonic mean is a perfectly viable form of nonlinear averaging, but many workers have experimented with other averages to see how they influence the quality of the results, in particular how well they resolve captured discontinuities. The standard way of presenting these averages is to exploit the fact that any reasonable average $M(x, y)$ has the homogeneous property,

$$\forall k, \;\; M(kx, ky) = kM(x, y) \;\Rightarrow\; \frac{M(x, y)}{y} = M\left(\frac{x}{y}, 1\right),$$

so that nonlinear averages can be characterized by plotting the one-dimensional function $M(r, 1) = \phi(r)$ versus r. It is this mathematical point that justifies the earlier claim, in connection with Eq. (11), that taking a ratio of consecutive differences in the data would be a useful way to monitor solution smoothness.

The constraints that the average must satisfy to avoid overshoots are that

$$0 \le \phi(r) \le \min(2, r + |r|). \tag{18}$$

and the natural requirement that an average lies between its two inputs (if these are of the same sign) can be translated into

$$\min(1, r) \le \phi(r) \le \max(1, r). \tag{19}$$

A plot that shows these constraints and a function that satisfies them is often called a *limiter diagram* [36, 47]. It is sometimes held that high-resolution schemes are not as free from user input as they claim, because the user must still choose a *limiter function*. To some extent this is true, but the sensitivity to this choice is not great, and fairly intuitive. The lower border of the allowable region is safe, but leads to rather diffusive schemes, whereas near the upper border, where more weight is given in the reconstruction to the larger of the two slopes, we obtain sharper discontinuities, with more risk of oscillations or instabilities arising from causes not allowed for in the analysis.

It has to be realized that a strategy of reconstruction based only on comparing pairs of consecutive gradients cannot accomplish all desirable objectives. For example, it would be good to relax the constraint that existing extrema must not be accentuated because, although this prevents oscillations arising near discontinuities, it does cause a loss of accuracy at genuine extrema. Think of advecting a parabolic profile containing three consecutive values 0.9, 1.0, 0.9. When the peak value falls halfway between grid points the parabola would be represented as $0.775, 0.975, 0.975, 0.775$. An inflexible constraint on the maximum value would never allow 1.0 to reappear. Such reappearance is allowed by more advanced forms of reconstruction, known as Total Variation Bounded (TVB) schemes and Essentially Non-Oscillatory (ENO) schemes, which are discussed in some detail in the third section of this book.

At the present stage of their evolution, high-resolution schemes can be thought of as a modular set of tools, of which the nonlinear reconstruction is probably the most important. A technique for solving the Riemann problems that arise is also needed; as already mentioned an exact solution, although possible, is costly and not really essential, or even preferable. High-resolution schemes usually employ some approximate solution to compute the flux function, and the development of these is rather thoroughly covered in the first section of the book.

5 Higher Dimensions

All of the preceding discussion has referred to one space dimension, but is of course driven by the need to solve 'real' problems in two- and three-dimensional space. Early high-resolution methods often employed a form of *operator splitting*, where the real problem is replaced by a sequence of one-dimensional problems. A theoretical criticism of this technique is delivered by Roe in Chapter IV. Later on, especially in conjunction with the unstructured grids that are often thought necessary to handle complex geometrical shapes, a more multidimensional approach was adopted. To date, this has been chiefly restricted to the reconstruction step: the paper in Chapter IV by Abgrall deals with this. Once the data is reconstructed within each cell, however, the cells are still presumed to interact in a one-dimensional manner by exchanging waves normal to their interfaces. Codes that are based on that assumption have in fact been extremely sucessful, easily outperforming codes that lack even that much physical basis. However, the paper by Quirk shows that when embarking on very ambitious computations requiring much highly resolved detail over long periods of time, anomalies can appear that can only be avoided by making careful choices within the available modules.

6 Current Research

Research into high-resolution schemes remains very active, and it is fitting to close this short survey with a summary of themes that are current at the time of publication.

6.1 New Riemann Solvers

Further refinement of the modules making up the basic quasi-one-dimensional approach continues, often in the original context of the Euler equations. The object is usually to reduce computational cost without relaxing quality. Jameson [20] has shown by an elegant analysis how this may be possible if only steady shocks are of interest. Liou, in a series of papers (see [27] and references therein) has pursued an interesting approach that appears in some respects more robust than earlier choices. Several researchers have attempted to derive flux functions by considering the place of the Euler equations in the hierarchy of approximations to kinetic theory. Xu, Martinelli, and Jameson [52] present an interesting implementation, with some references to earlier work.

Extremely high-speed flows present special difficulty, due to the fact that a conservative scheme, although it predicts total energy directly, computes the kinetic energy indirectly from the mass and momentum. The internal energy is then the small difference of two large numbers, and may not even be positive. When this happens the code will crash. Einfeldt et al. [11] defined a scheme to be *positively conservative* if it could be guaranteed to predict only positive densities and internal energies. They showed that the original Godunov scheme was positively conservative, but that Roe's scheme was not, although they obtained a 'fix'. This issue has been interestingly revisited by Buffard and Herard [5].

6.2 New Application Areas

A topic that has been well-studied but continues to be of interest is the extension of the methods to fluids not obeying the ideal gas relationship $p = (\gamma - 1)\rho i$, but some more *general equation of state*. In that case, the solution either of the exact or approximate Riemann problem becomes more intricate, and there is debate over the most effective procedures. See for example Vinokur and Montagné [50], and Saurel et al. [41]. Flows involving two or more gas species seem straightforward in theory, but in practice one must take care near the interfaces between materials. Specific *front tracking* algorithms have been developed [7] but much recent work has concentrated on capturing the fronts, with impressive results [16, 35]. When the fluids interact, for example chemically, the equations acquire source terms representing species conversion. The incorporation of these is straightforward, except when the reaction times are small compared with the residence time in a typical computational cell. Techniques for such *stiff problems* are getting a lot of attention [2, 21, 33] and effective techniques are beginning to emerge.

Application of the basic methods to problems governed by hyperbolic equations other than Eulers is proving very profitable. Some equations, such as the Saint-Venant equations, are so similar in structure that little modification is required. Gasdynamics in a relativistic setting is rather more challenging. Computations of flows in which the velocities approach the speed of light within one part in 10^5 are an extreme test of numerical technique. Nevertheless, Eulderink and Mellema [12] describe the translation of high-resolution methods from non-relativistic to general relativistic hydrodynamics as 'laborious but straightforward'. Their treatment is perhaps the most comprehensive available, but significant contributions also appear in [3, 28, 29, 42].

Magnetohydrodynamic flow (MHD) is another difficult area, complicated by the fact that much of the basic physics is still poorly understood. A rather straightforward extension of high-resolution schemes to one-dimensional MHD was made by Brio and Wu [6]. Their results highlighted the *nonconvexity* of the MHD equations. This means that in a magnetoacoustic wave the characteristic speed may not change monotonically, and waves may form that are partly shocks and partly centered compressions. Such waves always reverse the sign of the magnetic field component tangential to the wave front, and there are alternative solutions involving Alfvèn waves. The solution to the Riemann problem in MHD is therefore not unique, and the theoretical issues raised by this are far from being resolved at the time of writing. Nevertheless, several authors have achieved impressive results by following the most direct lines of attack. Simplifications to the Riemann solver were presented in [10, 39, 53] and multidimensional applications in [15, 54]. Reference [15] exploits a neat observation by Powell, included in Chapter IV, that works around the instabilities often observed in MHD codes if the divergence of the magnetic field is not constrained to stay small (analytically it should vanish, but this is hard to enforce numerically).

Yet another application area having a rich wave structure to which high-resolution methods have been extended is that of elastic wave propagation [49]. A spin-off of upwinding is its extension to incompressible flow. In this limit acoustic waves propagate infinitely fast compared with the fluid speed and the velocity remains divergence-free. A high-resolution method can be used that does not strictly enforce this condition, but its timesteps are interleaved with a projection method that removes any divergence that may have been created [1].

6.3 Application to Adaptive Grids

It is becoming apparent that upwind methods have many advantages in the context of grid adaptivity, a technique that is very neccessary for the economical computation of flows involving disparate length scales. In [15] a calculation is made of the interaction between the solar wind and the ionized particles subliming from a comet. A particular feature of that interaction is that it creates a bow shock whose stand-off distance is some five orders of magnitude greater than the diameter of the nucleus. This calculation is only feasible because it employs a dynamically adaptive grid that relates the local scale of the mesh to the scale of the features needing to be resolved. The cells of a coarse Cartesian grid are progressively divided into four smaller cells (eight in three dimensions) whenever the grid is incapable of resolving the local detail. For the comet calculation this happens twenty-four times, producing cells that differ in size by $2^{24} \approx 1.6 \times 10^7$.

For this particular adaptation strategy, which permits efficient searching of the grid to find neighbor cells but produces discontinuous variations in cell size, upwind methods are much to be preferred. Central differencing schemes tend to generate reflected waves at these grid interfaces, but this is much reduced by building the directionality of the waves into the scheme. This was first recognised by Berger and Colella [4], and a series of papers have confirmed its value. In this volume the paper by Quirk contains additional information, and several of the papers cited above [16, 35, 49] make use of the idea. In each of these the original prescription of [4] is followed; cells flagged as needing refinement are formed into groups and each group is covered with a patch. All the cells in a given patch are refined, whether they were originally flagged or not. In [15] a variation is employed that links flagged cells into a tree structure. See also [34, 55].

6.4 Radical Alternatives

Finally, it needs to be mentioned that several researchers, including some of those who helped to build the current generation of high-resolution upwind schemes, have questioned whether there might not be methods that create an even more effective link between numerics and physics[6]. A known drawback of current methods is that the solution of one-dimensional Riemann problems fails to recognise certain common multidimensional situations. The paper in Chapter IV by Davis was perhaps the first to note the limitations of an essentially one-dimensional analysis. He pointed out that by analysing a more extensive region of data one could determine a probable orientation for any shockwave present and use such information to compute a more accurate flux. He showed much improved resolution of captured shocks in a first-order scheme, but the extra cost did not seem justifiable in a second-order method. Conventional high-resolution schemes actually capture oblique shocks rather well, aided by the fact that hydrodynamic shocks are naturally compressive due to the focussing of characteristics.

There are however other situations that cause more serious difficulty. An often-quoted example is the appearance of a shear wave oblique to the grid, as when a boundary layer separates from the surface. This case is discussed by, among others, [9, 38, 40]. For shear waves there is no natural focussing to counteract superfluous numerical dissipation. For such reasons, a continuing effort has been made to base the numerical scheme on a more correct analysis of local fluid behavior.

[6] It has become fashionable recently to describe algorithms as "multidimensional", or "genuinely/truly multidimensional" if they contain ingredients not found in the one-dimensional case. The writer feels that this is too inclusive, and that the term should be reserved for methods that go beyond the use of one-dimensional physics. That is, they do not solve Riemann problems. Nevertheless, there have been interesting attempts to extract the most mileage from Riemann solutions. Colella [8] stresses chiefly accuracy, but finds, for his examples, only marginal improvements over operator splitting. LeVeque [25] is more concerned with efficiency, devising ways to extract all the information needed for a time-accurate fully-discrete "monotone" method from the minimum number of Riemann problems.

Early attempts [37, 46] seem in retrospect to have followed the one-dimesional case too slavishly, trying to reduce the multidimensional problem also to a set of scalar advection problems by representing data within a small region as a summation of oblique simple-wave solutions. For cases where the flow is in fact dominated by such waves the strategy is rather successful, but there are other instances where the new approach is no better than the established one, and indeed can be somewhat worse. For example, close to stagnation points, the flow is compressed normal to the surface but expanded along the surface. A standard upwind code interprets the resulting Riemann problems as evidence for unsteady, mutually cancelling, acoustic waves, compressive in one direction but expansive in the other. Dissipation added in response to these meaningless waves damages the accuracy, whether the waves are added in the grid direction or not.

One response to this situation is to refuse to assign any particular meaning to analysis of the individual gradients in the coordinate directions, but to combine them into a time-derivative before attempting any analysis. Sidilkover [44, 45] has presented very interesting work along these lines, making what is essentially a formal extension to the Euler equations of a successful nonlinear scheme for linear advection on a minimal stencil [43].

An even more fundamental approach is to try and reveal more explicitly the distinction between the elliptic and hyperbolic modes of behavior that coexist in supersonic flow. This work is really too recent to make a firm pronouncement at this time, but a flavor of the general idea can be given. Recent research has revealed, for the steady Euler equations in two dimensions, *canonical variables* which reduce the description of the flow as completely as possible into independent subsystems[7]. In supersonic flow there are four scalar subproblems, one associated with each characteristic direction [17]. The canonical variables are entropy and enthalpy, which are both constant along streamlines, and pressure and flow direction, which interact along Mach lines. In subsonic flow, the canonical variables are the same, but the subsystem comprising pressure and direction is now elliptic and resembles the Cauchy-Riemann system. Ta'asan [48] has proposed, for subsonic flow, the use of velocity components to describe the elliptic system, but this creates some coupling between the elliptic and hyperbolic parts.

For the hyperbolic parts of the problem, very effective and efficient schemes are available on structured or unstructured grids [43] that resolve shocks just as well as the usual schemes but resolve shear and contact discontinuities much better because they can recognize them more clearly. For the elliptic parts of the problem it has proved possible to somewhat modify the scalar advection schemes [32], or to take a more 'truly elliptic' approach [30]. In either case one can obtain excellent accuracy. There remain however, some unclear problems related to speed of convergence (Ta'asan [48] avoids this by keeping closer to a traditional elliptic methodology), the transition between subsonic and supersonic regions, and the extension to three-dimensional and/or unsteady flow.

[7] The analysis is not completely new (see for example [17], pp. 494-496) but its significance, especially for computation, seems to have been missed.

If these remaining problems can be overcome, then the work described in this collection will turn out to have been merely a step along the path to the goal of efficient, reliable, discrete simulation of nonlinear wave phenomena. It will however have been a step that broke decisively with past traditions, and which by itself enlarged greatly our capacity to simulate the physical world.

References

[1] A. S. Almgren, J. B. Bell, P. Colella, T. Marthaler, "A cell-centred Cartesian grid projection method for the incompressible Euler equations in complex geometries", AIAA paper 95-1743-CP in *Proceedings 12th AIAA CFD Conference (San Diego, June, 1995)*, AIAA, 1995.

[2] M. Arora, "Explicit characteristic-based solution algorithms for hyperbolic conservation laws with stiff source terms", Ph.D. thesis, Department of Aerospace Engineering, University of Michigan, 1996.

[3] D. Balsara, "Riemann solver for relativistic hydrodynamics", *J. Comput. Phys.* **114**, p 284, 1994.

[4] M. J. Berger, P. Colella, "Adaptive mesh refinement for hyperbolic differential equations", *J. Comput. Phys.* **53**, p 484, 1989.

[5] T. Buffard, J. M. Herard, "A conservative fractional step method to solve nonisentropic Euler equations", *Computer Methods in Applied Mechanics and Engineering*, to appear.

[6] M. Brio, C. C. Wu, "An upwind differencing scheme for the equations of ideal magnetohydrodynamics", *J. Comput. Phys.* **75**, p 400, 1988.

[7] I. Chern, P. Collela, "A conservative front-tracking method for hyperbolic conservation laws", UCRL JC-97200, Lawrence Livermore National Laboratory, July 1987.

[8] P. Colella, "Multidimensional upwind methods for hyperbolic conservation laws", *J. Comput. Phys.* **87**, p 171, 1990.

[9] A. Dadone, B. Grossman, "A rotated upwind differencing scheme for the Euler equations", *AIAA J.* **30**, no 9, p 2219, 1992.

[10] W. Dai, P. R. Woodward, "A simple Riemann solver and high-order Godunov scheme for hyperbolic systems of conservation laws," *J. Comput. Phys.* **121**, p 51, 1995.

[11] B. Einfeldt, C. D. Munz, P. L. Roe, B. Sjögreen, "On Godunov-type methods near low densities", *J. Comput. Phys.* **92**, p 17, 1990.

[12] F. Eulderink, G. Mellema, "General relativistic hydrodynamics with a Roe solver," *Astron. Astrophys. Suppl. Ser.* **110**, p 587, 1995.

[13] E. Godlewski, P-A. Raviart, *Hyperbolic Systems of Conservation Laws,* Société de Mathématiques Appliquées et Industrielles, Ellipses, Paris, 1991.

[14] S. K. Godunov, "A difference method for the numerical calculation of discontinuous solutions of hydrodynamic equations", *Mat. Sbornik* **47**, (1959), translated by U.S Department of Commerce, JPRS 7225, 1960.

[15] T. I. Gombosi, K. G. Powell, D. De Zeeuw, "Axisymmetric modelling of cometary mass loading on an adaptively refined grid; MHD results", *J. Geophysical Res.-Space Phys.* **99**, No. A11, p 21,525, 1994.

[16] J. A. Greenough, V. Beckner, R. B. Pember, W. Y. Crutchfield, J. B. Bell, P. Colella, "An adaptive multifluid interface-capturing method for compressible

flow in complex geometries", AIAA paper 95-1718-CP in *Proceedings 12th AIAA CFD Conference (San Diego, June, 1995)*, AIAA, 1995.
[17] W. D. Hayes, R. Probstein, *Hypersonic Flow Theory*, 2nd edition, Vol 1, *Inviscid Flow*, Academic Press, 1966.
[18] Ch. Hirsch, *Numerical Computation of Internal and External Flows*, (two volumes), John Wiley 1988.
[19] A. Iserles, G. Strang, "The optimal accuracy of difference schemes", *Trans. Am. Math. Soc.* **277**, vol 2, p 779, 1983.
[20] A. Jameson, "Analysis and design of numerical schemes for gas dynamics 2; artificial diffusion and discrete shock structure", *Int. J. Comp. Fluid Dyn.*, submitted.
[21] S. Jin, "Runge-Kutta methods for hyperbolic conservation laws with stiff relaxation terms", *J. Comput. Phys.* **122**, p 51, 1995.
[22] P. D. Lax, *Hyperbolic Systems of Conservation Laws and the Mathematical Theory of Shock Waves*, SIAM, 1973.
[23] B. van Leer, "Towards the ultimate conservative differencing scheme IV, A new approach to numerical convection", *J. Comput. Phys.* **23**, p 276, 1977.
[24] R. J. LeVeque, *Numerical Methods for Conservation Laws*, Birkhäuser, Zurich, 1992.
[25] R. J. LeVeque, "Simplified multi-dimensional flux-limiter methods", in *Numerical Methods for Fluid Dynamics*, eds M. J. Baines and K. W. Morton, Oxford University Press, 1993.
[26] M. J. Lighthill, *Waves in Fluids*, Cambridge University Press, 1978.
[27] M-S. Liou, "Progress toward an improved CFD method", AUSM$^+$, AIAA paper 95-1701-CP, in *Proceedings 12th AIAA CFD Conference (San Diego, June, 1995)*, AIAA, 1995.
[28] J. M. Martí, J. M. Ibáñez, J. A. Miralles, *Phys. Rev. D.* **43**, p 3794, 1991.
[29] A. Marquine et al., *Astron. Astrophys.* **258**, p 566, 1992.
[30] L. Mesaros, P. L. Roe, "Multidimensional fluctuation splitting schemes based on decomposition methods," AIAA paper 95-1699-CP in *Proceedings 12th AIAA CFD Conference (San Diego, June, 1995)*, AIAA, 1995.
[31] S. Osher, "Riemann solvers, the entropy condition, and difference approximations", *SIAM J. Num. Anal.* **21**, p 217, 1984.
[32] H. Paillière. H. Deconinck, P. L. Roe, "Conservative upwind residual-distribution schemes based on the steady characteristics of the Euler equations", AIAA paper 95-1700-CP in *Proceedings 12th AIAA CFD Conference (San Diego, June, 1995)*, AIAA, 1995.
[33] R. Pember, "Numerical methods for hyperbolic conservation laws with stiff relaxation, II higher-order Gonudov methods", *SIAM J. Sci. Stat. Comput.* **14**, no 4, p 824, 1993.
[34] K. G. Powell, P. L. Roe, J. J. Quirk, "Adaptive-mesh algorithms for computational fluid dynamics", in *Algorithmic Trends in Computational fluid dynamics*, eds M.Y Hussaini, A. Kumar, M.D. Salas, Springer, 1993.
[35] J. J. Quirk, S. Karni, "On the dynamics of a shock-bubble interaction", ICASE Report 94-75, and *J. Fluid Mech.*, to appear.
[36] P. L. Roe, "Some contributions to the modelling of discontinuous flow", in *Large-Scale Computations in Fluid Mechanics*, eds B. E. Engquist, S. Osher, R. C. J. Somerville, AMS Lectures in Applied Mathematics **22**, 1985.
[37] P. L Roe, "Discrete models for the numerical analysis of time-dependent multidimensional gas dynamics", *J. Comput. Phys.* **63**, no.2, p 458, 1986.
[38] P. L. Roe, "Beyond the Riemann problem I", in *Algorithmic Trends in Compu-*

tational Fluid Dynamics, eds M. Y. Hussaini, A. Kumar, M. D. Salas, Springer, 1993.

[39] P. L. Roe, D. Balsara, "Notes on the eigensystem of magnetohydrodynamics", *SIAM J. App. Math.* to appear, 1996.

[40] C. R. Rumsey, B. van Leer, P. L. Roe, "A multidimensional flux function with applications to the Euler and Navier-Stokes equations", *J. Comput. Phys.* **105**, p 306, 1993.

[41] R. Saurel, M. Larini, J.C. Loraud, "Exact and approximate Riemann solvers for real gases," *J. Comput. Phys.* **126**, p 276, 1994.

[42] V. Schneider et al., "New algorithms for ultrarelativistic hydrodynamics", *J. Comput. Phys.* **105**, p 92-107, 1993.

[43] D. Sidilkover, P. L. Roe, "Unification of some advection schemes in two dimensions", *ICASE Report* 95-10, 1995, and *Math Comp*, to appear.

[44] D. Sidilkover, "A genuinely multidimensional upwind scheme and efficient multigrid solver for the compressible Euler equations", ICASE Report 94-84, submitted.

[45] D. Sidilkover, "Multidimensional upwinding and multigrid", AIAA paper 95-1759-CP in *Proceedings 12th AIAA CFD Conference (San Diego, June, 1995)*, AIAA, 1995.

[46] R. Struijs, H. Deconinck, P. de Palma, P. L. Roe, K G. Powell, "Progress on multidimensional upwind Euler solvers for unstructured grids", AIAA paper 91-1550, in *Proceedings, AIAA 10th Computational Fluid Dynamics Conference, Hawaii, 1991*, AIAA, 1991.

[47] P. K. Sweby, "High-resolution schemes using flux limiters for hyperbolic conservation laws", *SIAM J. Num. Anal.* **21**, p 995, 1984.

[48] S. Ta'asan, "Canonical forms of multidimensional steady inviscid flows", *ICASE Report* 93-34, 1993.

[49] J. R. Trangenstein, "Adaptive mesh refinement for wave propagation in nonlinear solids", *SIAM J. Sci. Comput.* **16**, p 809, 1995.

[50] M. Vinokur, J-L. Montagné, "Generalized flux-vector splitting and Roe average for an equilibrium real gas", *J. Comput. Phys.* **89**, p 276, 1990.

[51] G.B. Whitham, *Linear and Nonlinear Waves*, Wiley-Interscience, 1974.

[52] K. Xu, L. Martinelli, A. Jameson, "Gas-kinetic finite volume methods, flux-vector splitting, and artificial diffusion", *J. Comput. Phys.* **120**, p 48, 1995.

[53] A. Zachary, P. Collela, "A higher-order Godunov method for the equations of ideal magnetohydrodynamics", *J. Comput. Phys.* **99**, p 341, 1992.

[54] A. Zachary, A. Malagoli, P. Collela, *SIAM J. Scientific Computing* **15**, p 263, 1994.

[55] D. De Zeeuw, K. G. Powell, "An adaptively-refined Cartesian mesh solver for the Euler equations", *J. Comput. Phys.* **89**, p 276, 1990.

Chapter I

Riemann Solvers and Upwind Schemes

Annotation

In the early 1980's, when upwind discretizations were beginning to receive full appreciation, there was strong interest in improving the computational efficiency of such schemes. The path to achieving this was clearly to replace the exact Riemann solver in Godunov's method, by a less expensive, yet sufficiently accurate, approximate Riemann solver. There were some pitfalls, though, such as the discovery that Roe's solver in its original form would admit shocks that violate the entropy condition. After that, designers of hyperbolic discretization paid due attention to the implementation of this inequality.

This chapter looks at the development of Riemann solvers in the context of first-order upwind schemes.

1. **On the Relation Between the Upwind-Differencing Schemes of Godunov, Engquist–Osher and Roe** (B. van Leer)
   Highlights the differences among upwind schemes resulting from different Riemann solvers, on the basis of the inviscid form of Burgers' equation. This is one of the first papers showing a solution with an entropy-violating shock. Another feature of interest is the consideration of source terms.

2. **On Upstream Differencing and Godunov-Type Schemes for Hyperbolic Conservation Laws** (A. Harten, P. D. Lax and B. van Leer)
   A review paper summarizing the state of the art at its time. It offers a unified treatment of a wide class of approximate Riemann solvers for hyperbolic systems, including flux splitting, which is related to the collisionless Boltzmann equation ("Boltzmann solvers").

3. **Flux-Vector Splitting for the Euler Equation** (B. van Leer)
   Development of a flux splitting with the goal of overcoming the lack of smoothness of the earlier Steger-Warming split fluxes.

ON THE RELATION BETWEEN THE UPWIND-DIFFERENCING SCHEMES OF GODUNOV, ENGQUIST–OSHER AND ROE*

BRAM VAN LEER†

Abstract. The upwind-differencing first-order schemes of Godunov, Engquist–Osher and Roe are discussed on the basis of the inviscid Burgers equations. The differences between the schemes are interpreted as differences between the approximate Riemann solutions on which their numerical flux-functions are based. Special attention is given to the proper formulation of these schemes when a source term is present. Second-order two-step schemes, based on the numerical flux-functions of the first-order schemes are also described. The schemes are compared in a numerical experiment and recommendations on their use are included.

Key words. upwind differencing, approximate Riemann solution, conservation laws

1. Introduction. Upwind differencing, while trivial for a diagonalized hyperbolic system, is difficult to achieve when the difference scheme has to be written in conservation form. The oldest and most complicated version is due to Godunov [1], [2]; the increasing popularity of upwind differencing recently has led to a variety of simpler implementation techniques. A review of these is given by Harten, Lax and van Leer [3].

Among the recent additions to the family of upwind conservative schemes the method of Engquist and Osher [4], [5] is closest to the original Godunov scheme, while the method of Roe [6], [7] offers the greatest simplification. In the present paper the differences between these schemes are discussed on the basis of the inviscid Burgers equation. Part of the discussion covers known, but not well-known, aspects of the schemes, thus rendering the paper to some extent a review paper. For maximum clarity the presentation leans heavily on geometrical insights.

The first-order accurate schemes are explained in §§ 2, 3 and 4 for the homogeneous equation; § 6 describes how to include a source term and § 7 how to achieve second-order accuracy in a two-step format. Their rendition of a stationary shock and a transonic expansion is discussed in § 5 and illustrated in the numerical experiments of § 8. Section 9 rounds off with recommendations regarding the application of these schemes to single conservation laws and systems of conservation laws.

2. Godunov's method. Godunov's [1], [2] method for integrating a hyperbolic system of conservation laws

$$u_t+[f(u)]_x=0 \tag{1}$$

is a scheme in conservation form:

$$(u_i^{n+1}-u_i^n)/\Delta t+\{F(u_i^n, u_{i+1}^n)-F(u_{i-1}^n, u_i^n)\}/\Delta x=0; \tag{2}$$

here u_i^n represents the average value at time $t^n=n\Delta t$ in the computational zone centered on $x_i=i\Delta x$. The numerical flux-function $F_G(u_i^n, u_{i+1}^n)$ in the Godunov scheme

* Received by the editors June 5, 1981, and in revised form June 2, 1982. This work was supported under NASA contract NAS1-14472 while the author was in residence at ICASE, NASA Langley Research Center, Hampton, Virginia 23665.

† Leiden State University, Leiden, The Netherlands. Present address: Department of Mathematics and Computer Science, Delft University of Technology, 2600 AJ Delft, the Netherlands.

is taken to be the flux value arising at $x_{i+1/2}$ in the exact solution of the initial-value problem with a piecewise uniform initial distribution

$$u^n(x) = u_i^n, \qquad x_i - \Delta x/2 < x < x_i + \Delta x/2. \tag{3.1}$$

That is, if

$$u(x,t) = v(x/t; u_L, u_R) \tag{3.2}$$

is the (weak) similarity solution of the Riemann problem with initial values

$$u = \begin{cases} u_L, & x<0, \\ u_R, & x>0, \end{cases} \tag{3.3}$$

then

$$F_G(u_i^n, u_{i+1}^n) = f[v(0; u_i^n, u_{i+1}^n)]. \tag{4}$$

For Burgers' equation in the inviscid limit,

$$u_t + (\tfrac{1}{2}u^2)_x = 0, \tag{5}$$

the similarity solution to the initial-value problem (3.2) is

$$\begin{matrix} u_L \leqq u_R \\ \text{(expansion)} \end{matrix} \begin{cases} v = u_L, & x/t \leqq u_L, \\ v = x/t, & u_L < x/t < u_R, \\ v = u_R, & x/t \geqq u_R, \end{cases} \tag{6.1}$$

$$\begin{matrix} u_L > u_R \\ \text{(shock)} \end{matrix} \begin{cases} v = u_L, & x/t < u_S = \tfrac{1}{2}(u_L + u_R), \\ v = u_R, & x/t > u_S. \end{cases} \tag{6.2}$$

For numerical reasons it is useful to distinguish in the formula for $F_G(u_L, u_R)$ three cases:

(i) *fully supersonic/subsonic*

$$F_G(u_L, u_R) = \begin{cases} \tfrac{1}{2}u_L^2, & u_L, u_R > 0, \\ \tfrac{1}{2}u_R^2, & u_L, u_R < 0; \end{cases} \tag{7.1}$$

(ii) *transonic expansion*

$$F_G(u_L, u_R) = 0, \qquad u_L \leqq 0 \leqq u_R; \tag{7.2}$$

(iii) *transonic shock*

$$F_G(u_L, u_R) = \begin{cases} \tfrac{1}{2}u_L^2, & u_L > u_S \geqq 0 \geqq u_R, \\ \tfrac{1}{2}u_R^2, & u_L \geqq 0 > u_S > u_R. \end{cases} \tag{7.3}$$

These cases are illustrated in Figs. 1, 2 and 3. The reference to a sonic speed (=0) arises from the use of (5) in transonic aerodynamics.

In order to combine the formulas (7) in one compact algorithm, we follow Engquist and Osher by introducing

$$u^+ \equiv \max(u, 0), \tag{8.1}$$

$$u^- \equiv \min(u, 0), \tag{8.2}$$

$$u = u^+ + u^-, \tag{8.3}$$

$$|u| = u^+ - u^-; \tag{8.4}$$

we then have

$$F_G(u_L, u_R) = \max[\tfrac{1}{2}(u_L^+)^2, \tfrac{1}{2}(u_R^-)^2]. \tag{9}$$

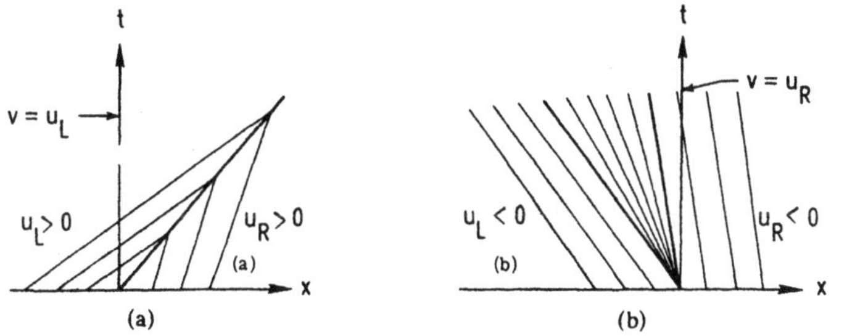

FIG. 1. *Riemann solution*: *(x, t) diagrams for case* (i), *with* $u_L > u_R > 0$ (a); $u_L < u_R < 0$ (b).

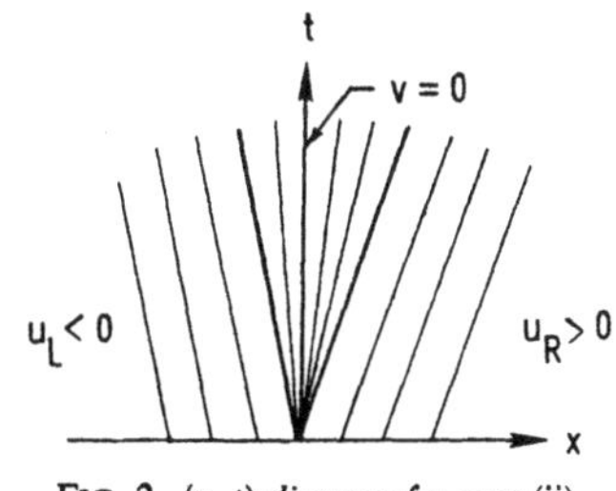

FIG. 2. *(x, t) diagram for case* (ii).

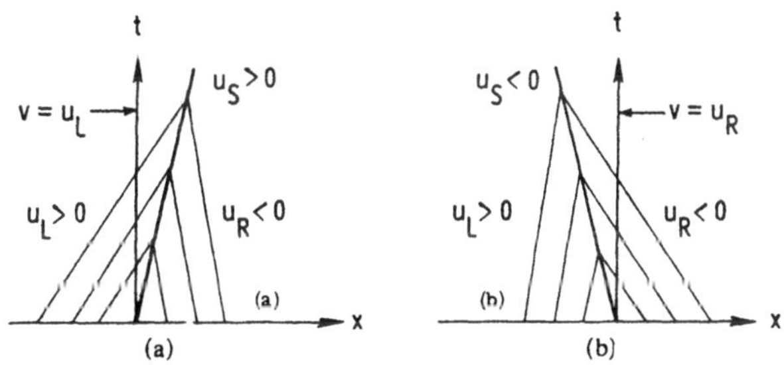

FIG. 3. *(x, t) diagram for case* (iii), *with* $u_S > 0$ (a); $u_S < 0$ (b).

3. The Engquist–Osher scheme. The Engquist–Osher [4], [5] scheme for integrating (1) also has the conservation form (2) and tacitly assumes the initial-value distribution to be (3.1); its numerical flux-function is

$$(10)\qquad \begin{aligned} F_{EO}(u_L, u_R) &= f(u_L) + \int_{u_L}^{u_R} A^-(u)\,du = f(u_R) - \int_{u_L}^{u_R} A^+(u)\,du \\ &= \frac{1}{2}[f(u_L)+f(u_R)] - \frac{1}{2}\int_{u_L}^{u_R} |A(u)|\,du, \end{aligned}$$

where the matrices $A^+(u)$, $A^-(u)$ and $|A(u)|$ are related to

$$(11)\qquad A(u) \equiv df/du$$

by an extension of (8). For precise definitions and a description of the integration path used for the integrals in (10), see [5] or the review [3].

For Burgers' equation (5) the above recipe boils down to

$$
(12)\qquad \begin{aligned} F_{EO}(u_L, u_R) &= \frac{1}{2}u_L^2 + \int_{u_L}^{u_R} u^- \,du = \frac{1}{2}u_R^2 - \int_{u_L}^{u_R} u^+\,du \\ &= \frac{1}{2}\left(\frac{1}{2}u_L^2 + \frac{1}{2}u_R^2\right) - \frac{1}{2}\int_{u_L}^{u_R} |u|\,du. \end{aligned}
$$

The integrals over u in (12) are defined in phase space, without reference to any mapping onto the (x, t)-plane. We may, however, introduce a mapping in the style of (3.2), namely,

$$
(13)\qquad u(x, t) = w(x/t; u_L, u_R), \qquad \min(u_L, u_R) \leqq u \leqq \max(u_L, u_R)
$$

and consider $w(x/t; u_L, u_R)$ an approximation to the exact solution $v(x/t; u_L, u_R)$ of the Riemann problem (3.3). It follows that w is the following function of x/t:

$$
(14)\qquad w = \begin{cases} u_L, & x/t \leqq u_L, \\ x/t, & \min(u_L, u_R) < x/t < \max(u_L, u_R), \\ u_R, & x/t \geqq u_R. \end{cases}
$$

In relating $F_{EO}(u_L, u_R)$ to $w(0; u_L, u_R)$ after the example of (4), caution is required, since, for $u_L > u_R$, w becomes multivalued in the domain $u_L \geqq x/t \geqq u_R$. Specifically, we have three branches

$$
(15)\qquad w^{(1)} = u_L, \quad w^{(2)} = x/t, \quad w^{(3)} = u_R, \qquad u_L \geqq x/t \geqq u_R.
$$

The picture associated with this case is that of an overturned centered compression wave or folded characteristic field (see Figs. 4 and 5); in the exact Rieman solution (6) such a wave would be replaced by a shock discontinuity.

The proper formula for $F_{EO}(u_L, u_R)$ in terms of $w(x/t; u_L, u_R)$, equivalent to (12), is

$$
(16)\qquad F_{EO}(u_L, u_R) = \sum_k (-1)^{k-1}\frac{1}{2}\{w^{(k)}(0; u_L, u_R)\}^2,
$$

where the sum is taken over all branches present at $x/t = 0$.

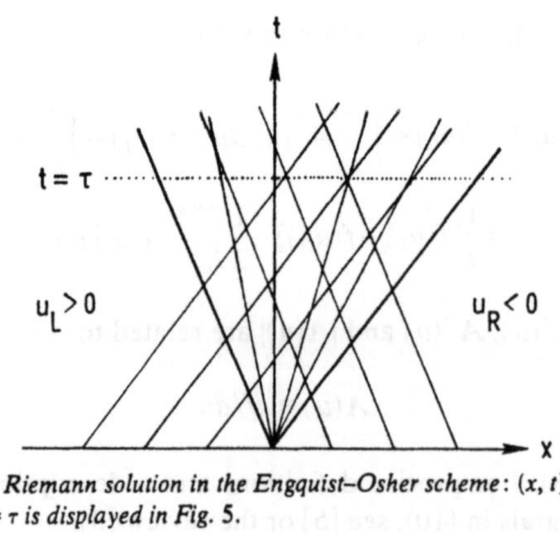

FIG. 4. *Approximate Riemann solution in the Engquist–Osher scheme*: (x, t) *diagram for case* (iii). *The multivalued solution at* $t = \tau$ *is displayed in Fig.* 5.

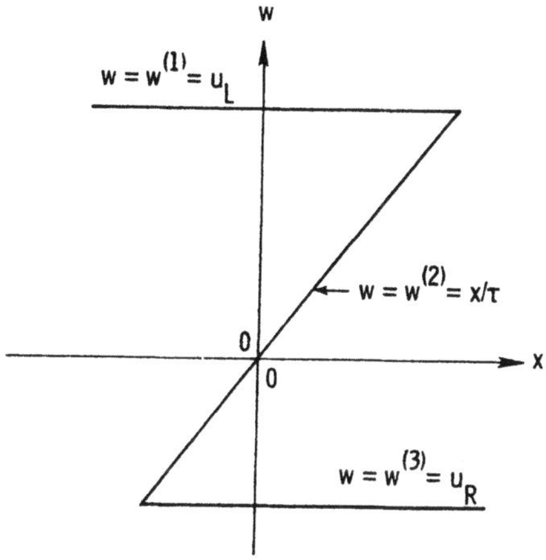

FIG. 5. *Approximate Riemann solution at $t=\tau$.*

In the cases (i), (ii) and (iii), distinguished earlier in (7) for $F_G(u_L, u_R)$, (12) or (16) yields

$$(17)\quad \begin{array}{lll} \text{(i)} & F_{EO}(u_L, u_R) = F_G(u_L, u_R) = \begin{cases} \frac{1}{2}u_L^2, \\ \frac{1}{2}u_R^2, \end{cases} & \begin{array}{l} u_L, u_R > 0, \\ u_L, u_R < 0, \end{array} \\ \text{(ii)} & F_{EO}(u_L, u_R) = F_G(u_L, u_R) = 0, & u_L \leqq 0 \leqq u_R, \\ \text{(iii)} & F_{EO}(u_L, u_R) = \frac{1}{2}u_L^2 + \frac{1}{2}u_R^2 \neq F_G(u_L, u_R), & u_L \geqq 0 \geqq u_R. \end{array}$$

As indicated in [4], these formulas combine into

$$(18)\qquad F_{EO}(u_L, u_R) = \tfrac{1}{2}(u_L^+)^2 + \tfrac{1}{2}(u_R^-)^2.$$

The difference between the Godunov and Engquist–Osher schemes lies entirely in the treatment of a transonic compression (iii). The latter scheme is the simpler one, since it does away with one test, namely, a test for the sign of u_S or for the maximum of $\frac{1}{2}(u_L^+)^2$ and $\frac{1}{2}(u_R^-)^2$.

4. Roe's method. Roe's [6], [7] method for integrating (1) again has the conservation form (2) and employs the initial-value distribution (3.1). Its numerical flux-function is defined as

$$(19)\qquad F_R(u_i^n, u_{i+1}^n) = f[w(0; u_i^n, u_{i+1}^n)],$$

where $w(x/t; u_L, u_R)$ is the exact solution of the Riemann problem (3.3) for the locally linearized system

$$(20.1)\qquad w_t + A(u_L, u_R)w_x = 0.$$

The approximate Jacobian $A(u_L, u_R)$ is constructed such as to satisfy the discrete version of (11)

$$(20.2)\qquad f(u_R) - f(u_L) = A(u_L, u_R)(u_R - u_L).$$

For a scalar equation like (5), (20.2) uniquely determines $A(u_L, u_R)$:

(21) $$A(u_L, u_R) = \tfrac{1}{2}(u_L + u_R) = u_S;$$

thus, the Riemann problem for the linear equation (20.1) with (21) has the similarity solution

(22) $$w(x/t, u_L, u_R) = \begin{cases} u_L, & x/t < u_S, \\ u_R, & x/t > u_S. \end{cases}$$

In all cases (i), (ii) and (iii) this yields

(23) $$F_R(u_L, u_R) = \begin{cases} \tfrac{1}{2}u_L^2, & u_S > 0, \\ \tfrac{1}{2}u_R^2, & u_S < 0. \end{cases}$$

The difference with Godunov's method lies in the treatment of an expansion: where the exact Riemann solution, used in Godunov's method, would include an expansion fan, Roe's method puts in a so-called expansion shock (see Fig. 6).

The numerical flux-function (23) deviates from the Godunov flux (7) only in case (ii), when the expansion is transonic. The computational simplification is, again, the elimination of one test.

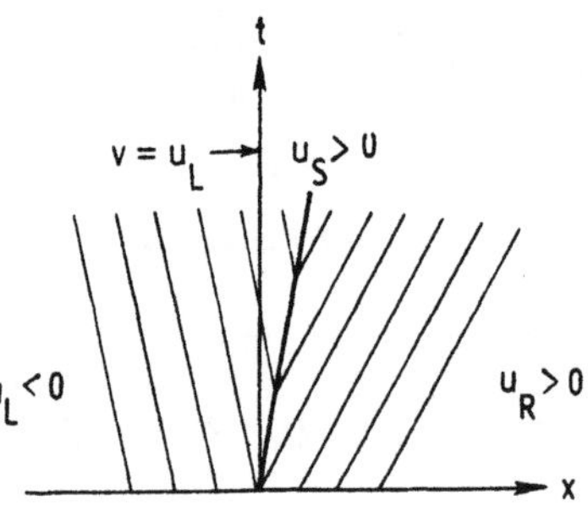

FIG. 6. *Approximate Riemann solution in Roe's scheme*: (x, t) *diagram for case* (iii), *with* $u_S > 0$, *showing the expansion shock.*

Rewriting (23) as an approximation to the Engquist–Osher flux (12),

(24) $$F_R(u_L, u_R) = \tfrac{1}{2}(\tfrac{1}{2}u_L^2 + \tfrac{1}{2}u_R^2) - \tfrac{1}{2}|\tfrac{1}{2}(u_L + u_R)|(u_R - u_L),$$

we see that, for Burgers' equation, Roe's scheme is identical to the scheme of Murman and Cole [8], used in transonic aerodynamics. The latter is known to occasionally yield numerical results that include a (physically inadmissible) expansion shock. This is a direct consequence of the admission of an expansion shock in the underlying approximate Riemann solution.

Roe [14] has proposed a modification in the numerical flux-function for a transonic expansion that not only does away with expansion shocks, but may be an improvement over the flux-function in the Godunov and Engquist–Osher schemes. This is discussed at the end of the next section.

The problem of preventing inadmissible discontinuities in the use of upwind schemes has been addressed in general by Harten and Lax [15] and for gas dynamics by Colella [16].

5. Steady-shock and sonic-point representation. The fluxes in the schemes of Godunov and Engquist–Osher differ only on meshes where the data constitute a transonic compression; therefore, numerical results from these schemes differ only if a transonic shock, in particular, a stationary shock, is present. Likewise, the results of Roe's scheme will differ from those of Godunov's scheme only if a transonic expansion is present.

For Burgers' equation (5), Godunov's scheme admits the following stationary discrete representation of a stationary shock connecting the states $u_L > 0$ and $u_R = -u_L < 0$:

$$\begin{aligned} u_i &= u_L, & i &\leqq -1, \\ u_0 &= u_M, & u_L &\geqq u_M \geqq u_R, \\ u_i &= -u_L, & i &\geqq 1. \end{aligned} \tag{25}$$

This is the only stationary distribution admitted by the scheme.

The interior value u_M indicates the subgrid position x_S of the shock; by conservation we have (see Fig. 7a),

$$u_L(x_S + \tfrac{1}{2}\Delta x) + u_R(\tfrac{1}{2}\Delta x - x_S) = u_M \Delta x \tag{26.1}$$

or

$$x_S = \tfrac{1}{2}\Delta x\, u_M / u_L. \tag{26.2}$$

A shock standing exactly on the boundary of a zone will be represented without any interior value.

The Engquist–Osher scheme admits discrete steady shock-profiles with, in general, two interior states:

$$\begin{aligned} u_i &= u_L, & i &\leqq -1, \\ u_0 &= u_M, & u_L &\geqq u_M \geqq 0, \\ u_i &= u_N, & 0 &\geqq u_N \geqq -u_L, \\ u_i &= -u_L, & i &\geqq 2, \end{aligned} \tag{27.1}$$

where u_M and u_N are constrained by

$$\tfrac{1}{2}u_M^2 + \tfrac{1}{2}u_N^2 = \tfrac{1}{2}u_L^2. \tag{27.2}$$

Again, this is the only stationary distribution admitted by the scheme.

The pair of interior states indicates the subgrid position of the shock; by conservation we have (see Fig. 7b):

$$u_L(x_S + \tfrac{1}{2}\Delta x) + u_R(\tfrac{3}{2}\Delta x - x_S) = (u_M + u_N)\Delta x \tag{28.1}$$

or

$$x_S = \tfrac{1}{2}\Delta x(u_M + u_N + u_L)/u_L. \tag{28.2}$$

A shock standing exactly in the middle of a zone will be represented with only one interior value.

The representation of a stationary shock by Roe's scheme is the same as by Godunov's scheme, at least for Burgers' equation. However, Roe's scheme also admits as a stationary solution the expansion shock

$$\begin{aligned} u_i &= u_L < 0, & i &\leqq -1, \\ u_i &= -u_L > 0, & i &\geqq 0. \end{aligned} \tag{29.1}$$

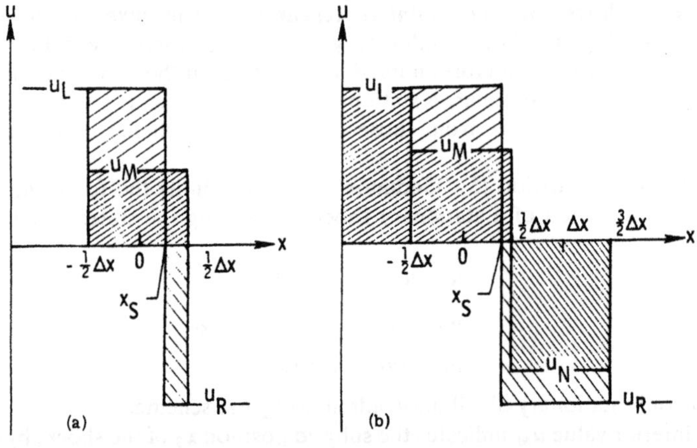

FIG. 7. *Fitting a shock discontinuity to stationary discrete shocks yielded by the schemes of Godunov or Roe* (a) *and Engquist–Osher* (b). *The numerical values u_M and u_N represent the average value of u in the intervals $(-\frac{1}{2}\Delta x, \frac{1}{2}\Delta x)$ and $(\frac{1}{2}\Delta x, \frac{3}{2}\Delta x)$. A subgrid distribution representing a shock transition from u_L to u_R at x_S must have the same integral as the numerical solution. In case* (a) *the integrals from $-\frac{1}{2}\Delta x$ to $\frac{1}{2}\Delta x$ are compared, in case* (b) *from $-\frac{1}{2}\Delta x$ to $\frac{3}{2}\Delta x$.*

A transonic shock profile, steady or not, obtained with any of the upwind schemes, has the property that the interior zones cannot influence the exterior solution (see Fig. 8a). Inversely, in a transonic expansion computed with the Roe–Murman–Cole scheme, the zone with the value closest to the sonic value cannot be influenced by the rest of the grid (see Fig. 8b). Once established by transient waves, the value in this zone changes no more, and it fixes the amplitude of the steady expansion shock that will remain.

Colella [16] has pointed out that, in the approximate solution to the Riemann problem, two steps can be distinguished. The first step is to determine the speeds and amplitudes of the finite-amplitude waves; the second step is to compute the full solution as a function of x/t. It is only in the second step that the method of Roe errs. A rarefaction wave must always be given a finite spread; for Burgers' equation this boils down to replacing Roe's scheme by Godunov's.

The remedy that Roe [14] proposes cannot be formulated as a change in the approximate Riemann solution. It is based on regarding the initial values as nodal values of a smooth distribution, rather than zone averages of a piecewise uniform distribution. For neighboring zones L and R enclosing a sonic point we may write down the following upwind formulas:

(29.2)
$$\left(\frac{\partial u}{\partial t}\right)_L = -u_L(u_R - u_L)/\Delta x, \qquad \left(\frac{\partial u}{\partial t}\right)_R = -u_R(u_R - u_L)/\Delta x, \qquad u_L \leqq 0 \leqq u_R,$$

from which follows

(29.3)
$$(\partial u/\partial t)_L/(\partial u/\partial t)_R = u_L/u_R.$$

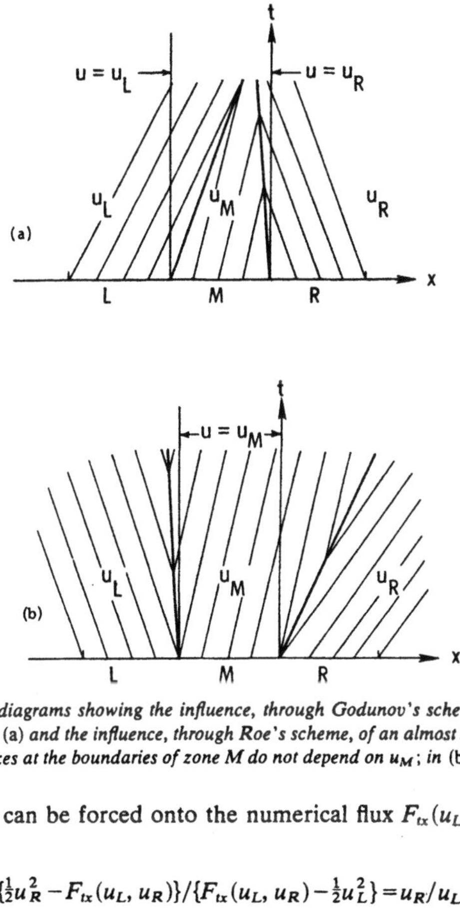

FIG. 8. *Two (x, t) diagrams showing the influence, through Godunov's scheme, of the adjacent zones on a zone inside a shock* (a) *and the influence, through Roe's scheme, of an almost sonic zone on the adjacent zones* (b). *In* (a), *the fluxes at the boundaries of zone M do not depend on* u_M; *in* (b) *they depend only on* u_M.

The latter relation can be forced onto the numerical flux $F_{tx}(u_L, u_R)$ for a transonic expansion:

$$\{\tfrac{1}{2}u_R^2 - F_{tx}(u_L, u_R)\}/\{F_{tx}(u_L, u_R) - \tfrac{1}{2}u_L^2\} = u_R/u_L, \tag{29.4}$$

leading to

$$F_{tx}(u_L, u_R) = \tfrac{1}{2}u_L u_R, \qquad u_L \leqq 0 \leqq u_R. \tag{29.5}$$

This unphysical flux (it is negative!) breaks down an expansion shock *faster* than the zero flux dictated by the exact Riemann solution. This is a direct consequence of the underlying assumption of smooth initial values.

While not fitting into the general framework of approximate Riemann solutions, the above approach has its appeal. The piecewise uniform values associated with a Riemann problem are not necessarily a good representation of the solution near a sonic point. In particular, the errors in the signal speed $df(u)/du$ $(=u)$ locally are of the order of the speed itself. The sonic point is always moved to a zone boundary and, in consequence, the flux gradient computed for each bordering zone becomes independent of the true gradient of the solution. In numerical solutions obtained with first-order schemes this shows up as a transonic expansion shock with amplitude $O(\Delta x)$. For a second-order scheme the effect disappears or reduces to $O[(\Delta x)^3]$.

In contrast, first-order accurate solutions obtained with the transonic flux (29.5) do not show these weak expansion shocks.

6. Inclusion of a source term. When approximating the equation

(30) $$u_t+[f(u)]_x=s(x)$$

with any upwind scheme, it is not sufficient to add the source term to the right-hand side of (2); we must also change the initial-value distribution implied in our numerical model. Instead of assuming that it is piecewise uniform, as in (3.1), we rather take it to be *piecewise stationary*, that is,

(31.1) $$[f(u^n(x))]_x=s(x),\qquad x_i-\Delta x/2<x<x_i+\Delta x/2,$$

with

(31.2) $$\frac{1}{\Delta x}\int_{x_i-\Delta x/2}^{x_i+\Delta x/2}u^n(x)\,dx=u_i^n.$$

This was first proposed by Liu [17] in constructing a random-choice method for the inhomogeneous conservation law (30).

The extension preserves a fundamental property of the homogeneous scheme, namely, that a zone-average can change only through the finite-amplitude waves entering from the zone boundaries. Moreover, if the numerical solution globally tends to a steady state, it will be the zone-averaged exact steady state almost everywhere (that is, provided that the scheme, like the three considered here, can render a shock transition in a finite number of zones). We shall come back to this further below.

As before, the initial values on either side of a zone boundary become the arguments of the numerical flux-function. Transient effects caused by a shock wave returning to the zone boundary under the influence of the source term, or by a curved transonic rarefaction fan, are ignored in order to keep the flux-function independent of Δt. The full scheme reads

(32.1) $$(u_i^{n+1}-u_i^n)/\Delta t+\{F(u^n_{(i+1/2)-},u^n_{(i+1/2)+})-F(u^n_{(i-1/2)-},u^n_{(i-1/2)+})\}/\Delta x=s_i^n,$$

with

(32.2) $$s_i\equiv\frac{1}{\Delta x}\int_{x_{i-1/2}}^{x_{i+1/2}}s(x)\,dx.$$

For the inhomogeneous Burgers equation

(33) $$u_t+(\tfrac{1}{2}u^2)_x=s(x)$$

the initial-value representation must satisfy

(34) $$\frac{1}{2}[u^n(x)]^2=\frac{1}{2}[u^n(x_{(i-1/2)+})]^2+\int_{x_{i-1/2}}^{x}s(x')\,dx',\qquad x_i-\Delta/2<x<x_i+\Delta/2,$$

under the constraint (31.2). This constraint, however, is not strong enough to define a unique distribution in each zone, since the distribution may (and, in some zones, must) contain a discontinuity. Uniqueness can be achieved by selecting the distribution with, say, the weakest possible shock.

If we take Δx small enough to ensure that $s(x)$ does not change sign more than once in any zone, the following algorithm for the piecewise construction of $u^n(x)$ is adequate. First, determine for the zone under consideration the smallest continuous solution (see Fig. 9). Let ω_i^n be the average value on the positive branch of this solution; then the stationary distribution inside zone i will be continuous if

(35) $$|u_i^n|\geqq\omega_i^n.$$

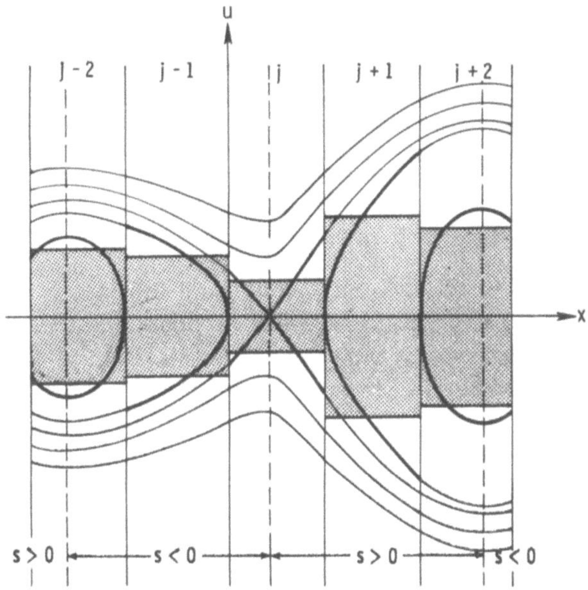

FIG. 9. *Stationary solutions of* (30) *for some source term s(x). In each zone the heavy line traces the positive and the negative branch of the smallest continuous steady solution. The average value on these branches is indicated by the upper and lower boundary of the shaded areas. If the average value in a zone falls in the range of the shaded area, a continuous stationary solution cannot be realized.*

If this condition is satisfied, we search, iteratively, for the continuous distribution with the proper zone average. If it is violated, the stationary distribution will include parts of the upper and lower branch of the smallest continuous distribution, connected by a shock positioned so as to achieve the proper average value (see Fig. 10).

When using the upwind scheme to approach a globally stationary solution of (33), any zone containing a sonic point must be treated with special care, since the chance of numerically realizing the exact transonic structure without a shock is zero.

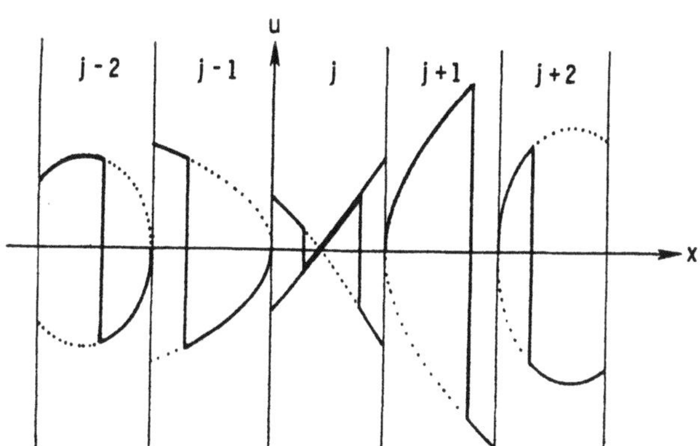

FIG. 10. *Examples of steady structure with a discontinuity, in the zones shown in Fig. 9. A (steady) shock connects the upper and lower branch of the smallest continuous steady solution for the zone considered. In zone j two different possibilities are shown.*

Specifically, in order to prevent the zone-boundary values from flipping sign, making global convergence impossible, smoothing may be introduced (see Fig. 11).

In a zone containing a stationary shock, no particular interior structure is needed to ensure global convergence, since the zone cannot influence its neighbors (see earlier Fig. 8a). After convergence one may insert the proper structure by enforcing continuity across the zone boundaries (see Fig. 12).

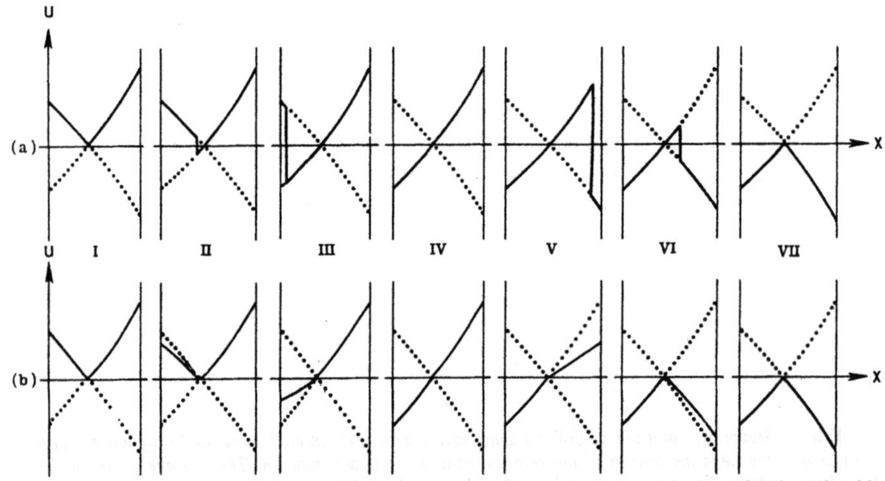

FIG. 11. (a) *A sequence of stationary structures close to the transonic expansion (graph* IV), *ordered by decreasing zone-average. Going from* III *to* IV, *and from* IV *to* V, *the slight change in zone-average causes one boundary value to flip its sign.* (b) *The solutions have been smoothed on the side where the shock occurs, so that the boundary values now vary continuously with the zone-average. The structures are not stationary but will allow convergence to the smooth transonic solution* IV.

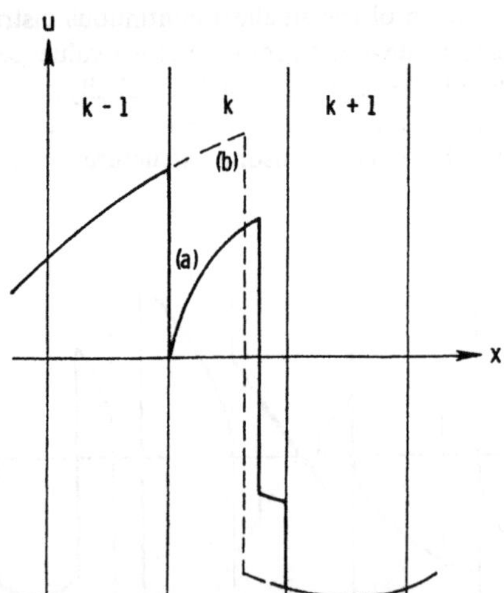

FIG. 12. *Shock structure in a globally stationary solution. Global convergence of the zone averages to a steady solution containing a shock has been obtained, say, with Godunov's scheme. For zone k containing the shock the scheme had adopted the structure* (a), *based on the smallest continuous steady solution for that zone. This may now be replaced by the stationary solution* (b) *that smoothly connects to the solution in zones* $k-1$ *and* $k+1$.

The scheme described above must be considered a research tool from which more practical schemes may be derived. In the numerical experiment of § 8 we simplified the procedure by allowing only s_i to enter the zone-boundary values. Away from sonic points or shocks the boundary values $u_{(i\pm 1/2)\mp}$ can be approximated by

(36.1) $$u^n_{(i\pm 1/2)\mp} = u^n_i \pm \tfrac{1}{2} s_i \Delta x / u^n_i + O\{(\Delta x)^2\}$$

or

(36.2) $$u^n_{(i\pm 1/2)\mp} = \sqrt{(u^n_i)^2 \pm s_i \Delta x} + O\{(\Delta x)^2\}.$$

We used (36.2), extending it, for use in sonic and shock zones, to

(36.3) $$u^n(x_{(i\pm 1/2)\mp}) = \operatorname{sgn}(u^n_i)\sqrt{(u^n_i)^2 \pm \min\{|s_i|\Delta x, (u^n_i)^2\}\operatorname{sgn}(s_i)}.$$

Note that (34), with $x = x_{i+1/2}$, is satisfied everywhere, but transonic structure in a zone is avoided. For zones that should include a sonic point the error in the boundary values is $O(\Delta x)$.

The technique of computing the boundary values in a zone from a locally stationary solution can easily be extended to systems of conservation laws of the form

(37) $$u_t + [f(u, x)]_x = s(x).$$

7. Second-order upwind schemes. Any numerical flux-function used in a first-order upwind scheme for (30) can be used in a second-order upwind scheme. For second-order accuracy in space we must introduce structure inside the zones by interpolation, while second-order accuracy in time may be achieved by advancing the cell-boundary values, to be used in the flux-function, to the intermediate time level $t^{n+1/2} = t^n + \frac{1}{2}\Delta t$. In predicting these time-centered values, *the interaction between cells can be fully ignored.* This observation, due to Hancock [9], has led to a drastic simplification of second-order upwind schemes since these first were formulated for systems of conservation laws by van Leer [10]. As in [10] we assume that the initial values form a *piecewise linear* distribution

(38) $$u^n(x) = u^n_i + (x - x_i)\frac{(\delta u)^n_i}{\Delta x}, \qquad x_{i-1/2} < x < x_{i+1/2},$$

with

(39) $$(\delta u)^n_i = \operatorname{ave}(u^n_{i+1} - u^n_i, u^n_i - u^n_{i-1});$$

ave (a, b) is an averaging procedure to be specified later. We particularly need the initial boundary values inside cell i

(40.1) $$u^n_{(i\pm 1/2)\mp} = u^n_i \pm \tfrac{1}{2}(\delta u)^n_i.$$

These boundary values are now advanced to $t^{n+1/2}$ by

(40.2) $$u^{n+1/2}_{(i\pm 1/2)\mp} = u^n_{(i\pm 1/2)\mp} - \frac{\Delta t}{2\Delta x}\{f(u^n_{(i+1/2)-}) - f(u^n_{(i-1/2)+})\} + \frac{\Delta t}{2} s_i.$$

The full scheme becomes

(41) $$(u^{n+1}_i - u^n_i)/\Delta t + \{F(u^{n+1/2}_{(i+1/2)-}, u^{n+1/2}_{(i+1/2)+}) - F(u^{n+1/2}_{(i-1/2)-}, u^{n+1/2}_{(i-1/2)+})\}/\Delta x = s_i.$$

The function ave (a, b) is chosen such that it tends to $\frac{1}{2}(a+b)$ if a and b are subsequent finite differences of a smooth solution, but tends to the smallest value where the solution is not smooth. Examples can be found in [11], [10]; we chose a refined formula due to Van Albada [12]

$$\text{ave}\,(a, b) = \frac{(b^2+c^2)a+(a^2+c^2)b}{a^2+b^2+2c^2}, \tag{42.1}$$

where c^2 is a small bias of the order $O((\Delta x)^3)$.

The weighted averaging prevents central differencing across a discontinuity in the solution or in its first derivative which would lead to numerical oscillations. It is an effective way of administering artificial dissipation wherever needed and nowhere else. By rewriting (42.1) as

$$\text{ave}\,(a, b) = \frac{a+b}{2}\left\{1 - \frac{(a-b)^2}{a^2+b^2+2c^2}\right\}, \tag{42.2}$$

we see that, wherever the solution is smooth, the artificial-viscosity coefficient generally is of the order $O((\Delta x)^2)$. In a smooth extremum the coefficient grows to $O(\Delta x)$; the bias $2c^2$ in the denominator prevents a further increase of the viscosity that could lead to an undesirable clipping phenomenon (see [13, § 3(e)]).

With regard to Burgers' equation, an acceptable expression for the bias is

$$c^2 = (u_{\max} - u_{\min})^2(\Delta x)^3/(x_{\max} - x_{\min})^3, \tag{43}$$

where $u_{\max}$ and $u_{\min}$ are certain upper and lower bounds of the numerical solution, fixed a priori or determined at each time level and $x_{\max} - x_{\min}$ is the length-scale of the problem.

8. Numerical comparison. The performance of the three schemes was tested on the basis of the periodic intial-value problem

$$u_t + (\tfrac{1}{2}u^2)_x = (\pi/2)\sin[2\pi(x-\xi)], \qquad 0 \leqq x \leqq 1, \quad 0 \leqq \xi \ll 1, \tag{44.1}$$

$$u(x, 0) = 0, \tag{44.2}$$

$$u(0, t) = u(1, t). \tag{44.3}$$

The solution tends to the steady state

$$u(x, \infty) = \begin{cases} +\sin \pi(x-\xi), & 0 \leqq x < \xi + \frac{1}{2}, \\ -\sin \pi(x-\xi), & \xi + \frac{1}{2} < x \leqq 1, \end{cases} \tag{45}$$

including a sonic point at $x = \xi$ and a shock at $x = x_S = \xi + \frac{1}{2}$. We tested the ability of the schemes to approximate this steady state, and the accuracy of the approximation. For Δx we chose a value of $\frac{1}{16}$; the zones were centered on $x_i = (i - \frac{1}{2})\Delta x$, $i = 1, \cdots, 16$. For ξ we chose 0, $\frac{1}{4}\Delta x$ and $\frac{1}{2}\Delta x$, in order to achieve different subgrid shock positions. The Courant–Friedrichs–Lewy condition on the timestep, based on the maximum characteristic speed (= 1) in the steady solution is

$$\frac{\Delta t}{\Delta x} \leqq 1; \tag{46}$$

accordingly we used $\Delta t = \frac{1}{2}\Delta x$ or $\Delta t = \Delta x$.

Table 1 shows the number of steps N_ε it took the schemes to converge according to the L_1-criterion

$$\sum_{i=1}^{16} |u_i^n - u_i^{n-1}| < \varepsilon, \tag{47}$$

with $\varepsilon = 1\times10^{-3}$ or 1×10^{-6} and $\Delta t = \frac{1}{2}\Delta x$ or (for Godunov's scheme only) $\Delta t = \Delta x$. When the time-step is doubled, N_ε appears to be halved. The L_1-error E_ε was evaluated with respect to the zone-averaged exact solution, for the smallest value of ε. The subgrid shock positions in Table 1 were calculated with aid of (26.1) and (28.1), approximately valid because the source term is small in the shock region.

TABLE 1

Number of steps N_ε till convergence, L_1-error E_ε and steady-shock position x_S, for the initial-value problem (44), *solved with the first-order schemes of Godunov* (G), *Engquist–Osher* (EO) *and Roe* (R). *Mesh*: $\Delta x = \frac{1}{16}$, $\Delta t = \frac{1}{2}\Delta x$ *or* Δx (G *only, numbers between brackets*).

$\xi/\Delta x$	N_ε				E_ε		$(x_S - \frac{1}{2})/\Delta x$	
	$\varepsilon = 1\times10^{-3}$		$\varepsilon = 1\times10^{-6}$		$\varepsilon = 1\times10^{-6}$		$\varepsilon = 1\times10^{-6}$	
	G, R	EO	G, R	EO	G, R	EO	G, R	EO
0	62	61	112 [55]	111	8.8×10^{-3}	4.6×10^{-2}	0	0
$\frac{1}{4}$	68	66	138 [70]	136	9.6×10^{-3}	1.8×10^{-2}	.23	.26
$\frac{1}{2}$	52	52	88 [42]	88	4.6×10^{-3}	4.6×10^{-3}	$\frac{1}{2}$	$\frac{1}{2}$

The converged solutions for $\xi = \frac{1}{4}\Delta x$, with $\varepsilon = 1\times10^{-6}$, are listed in Table 5; these are independent of Δt. The zone-averaged exact solution is given for comparison. The error in the zone with the sonic point is comparable to the value in the zone, as anticipated with the use of (36.3).

The results of Roe's scheme happen to be identical to those of Godunov's scheme; in particular, no expansion shock shows up. The reason is that, with the use of the boundary values (36.3), a sonic point in a zone is always moved to a boundary; in this case Roe's flux function (23) yields the same value ($=0$) as Godunov's (7.2). Furthermore, the initial average value in the zone ultimately containing the sonic point was close to the steady value to begin with.

The results of the Engquist–Osher scheme are identical to the Godunov–Roe results except for the expected spread in the shock profile for $\xi = 0$ and $\frac{1}{4}\Delta x$. The subgrid shock positions are slightly, but not significantly, more accurate than for the other schemes. Likewise, convergence with the Engquist–Osher scheme is faster, but not significantly so.

Better results were obtained with the second-order versions of the schemes. Convergence was achieved in 10–35% fewer steps than with the first-order schemes (Table 2) and the local error near the sonic point is now of the same order as elsewhere outside the shock (Table 5).

TABLE 2

The quantities N_ε, E_ε and x_S ($\varepsilon = 1\times 10^{-6}$) for the initial-value problem (44), *solved with the second-order two-step schemes based on the numerical flux-functions of Godunov* (G2, G2a), *Engquist–Osher* (EO2) *and Roe* (R2). *In* G2a *the algebraic average* (48) *is used instead of* (42.1). *Mesh*: $\Delta x = \frac{1}{16}$, $\Delta t = \frac{1}{2}\Delta x$.

$\xi/\Delta x$	N_ε			E_ε			$(x_S - \frac{1}{2})/\Delta x$		
	G2, R2	G2a	EO2	G2, R2	G2a	EO2	G2, R2	G2a	EO2
0	75	76	75	1.4×10^{-3}	3.1×10^{-2}	2.9×10^{-2}	0	0	0
$\frac{1}{4}$	89	92	89	1.3×10^{-3}	2.2×10^{-3}	4.7×10^{-3}	.24	.19	.26
$\frac{1}{2}$	79	79	79	1.3×10^{-3}	1.7×10^{-3}	1.3×10^{-3}	$\frac{1}{2}$	$\frac{1}{2}$	$\frac{1}{2}$

The Engquist–Osher scheme still yields a shock structure with two interior zones. For the second-order Godunov scheme we checked that the use of algebraic averaging of differences,

$$\text{ave}\,(a, b) = \tfrac{1}{2}(a + b), \tag{48}$$

causes the numerical shock to overshoot and undershoot (Table 5). The accuracy of the shock position also suffers (Table 2).

The results of Roe's scheme are practically, but not exactly, identical to the Godunov results, indicating a slightly different transient behavior. Again, the internal structure (36.3) in the zones, in combination with the particular choice of initial values, provides a safeguard against the occurrence of an expansion shock.

In order to make the problem more challenging, we changed the initial values (44.2), for $\xi = 0$, into

$$u_i^0 = \begin{cases} 1, & i = 1, \cdots, 8, \\ -1, & i = 9, \cdots, 16, \end{cases} \tag{49}$$

including an expansion shock at $x = 0$. The results are listed in Tables 3 and 6. Among the second-order schemes Roe's scheme now requires 30% more steps than the other schemes: apparently, the expansion shock is not so easily dissipated by Roe's scheme. Among the first-order schemes the discrepancy gets worse: the results of Roe's scheme quickly converge to the wrong solution, including the full initial expansion shock. This

TABLE 3

The quantities N_ε and E_ε ($\varepsilon = 1\times 10^{-6}$) for the initial-value problem (44.1), (49), (44.3), *solved with both first- and second-order upwind schemes. Scheme* R *converges to the wrong weak solution. The schemes* R* *and* R2* *incorporate the flux-function* (29.5) *for a transonic expansion. Mesh*: $\Delta x = \frac{1}{16}$, $\Delta t = \frac{1}{2}\Delta x$.

$\xi/\Delta x$	N_ε								E_ε	
	G	R	EO	G2	R2	EO2	R*	R2*	R	R2
0	170	30	169	77	99	76	103	71	5.7×10^{-1}	1.4×10^{-3}

is because the zones on either side of the expansion shock cannot be influenced by their neighbors (see Fig. 8b), and they cannot change by themselves because of (36c).

We then replaced the transonic value of the flux-function in Roe's scheme by Roe's new formula (29.5), with dramatic results. Not only was the proper solution obtained, but in considerably fewer steps than required by the other schemes (Table 3). Even for the second-order scheme convergence was significantly faster.

Finally, Table 4 shows the same quantities as Table 1 for experiments in which, in the first-order schemes, the source-term dependence of zone-boundary values was dropped. It took the schemes of Engquist–Osher and Godunov 15–25% more steps than previously to converge to a much less accurate solution (Table 5). Roe's scheme is now unstable: it yields an expansion shock where a smooth transonic transition should be; the amplitude of this shock grows linearly with time. The reason is that

TABLE 4

The quantities N_ε, E_ε and x_S ($\varepsilon = 1 \times 10^{-6}$) for the initial-value problem (44), *solved with the first-order upwind schemes under the assumption of piecewise uniform initial values* (3.1). (*The label* "u" *in* Gu, Ru, EOu *stands for* "*uniform*".) *Mesh*: $\Delta x = \frac{1}{16}$, $\Delta t = \frac{1}{2}\Delta x$.

$\xi/\Delta x$	N_ε			E_ε			$(x_S - \frac{1}{2})/\Delta x$		
	Gu	Ru	EOu	Gu	Ru	EOu	Gu	Ru	EOu
0	135	-	135	6.0×10^{-2}	unstable	9.5×10^{-2}	0	-	0
$\frac{1}{4}$	174	-	172	6.1×10^{-2}	unstable	6.7×10^{-2}	.22	-	.24
$\frac{1}{2}$	103	103	103	4.7×10^{-2}	4.7×10^{-2}	4.7×10^{-2}	$\frac{1}{2}$	$\frac{1}{2}$	$\frac{1}{2}$

TABLE 5

Converged numerical solutions and zone-averaged asymptotic solutions for the experiments of Tables 1, 2 *and* 4, *with* $\xi = \frac{1}{4}\Delta x$, $\varepsilon = 1 \times 10^{-6}$, $\Delta t = \frac{1}{2}\Delta x$.

i	$u_i^{N_\varepsilon}$, u_i^∞							
	$\xi/\Delta x = \frac{1}{4}$							
	G, R	EO	G2, R2	G2a	EO2	Gu	EOu	exact
1	.09778	.09778	.04916	.04917	.04916	.13828	.13828	.04899
2	.25516	.25516	.24249	.24244	.24248	.33330	.33330	.24259
3	.43185	.43184	.42635	.42610	.42639	.51176	.51175	.42687
4	.59602	.59602	.59379	.59344	.59364	.66976	.66976	.59474
5	.73869	.73869	.73871	.73726	.73933	.80171	.80171	.73976
6	.85367	.85367	.85462	.85711	.85231	.90266	.90266	.85635
7	.93631	.93631	.94080	.91445	.95106	.96879	.96879	.94003
8	.98329	.88033	.98172	1.15231	.95622	.99759	.89488	.98759
9	-.52996	-.42699	-.49587	-.60164	-.47889	-.54359	-.44088	-.49739
10	-.96441	-.96441	-.96449	-1.01358	-.96440	-.98796	-.98796	-.96847
11	-.89927	-.89927	-.90232	-.89318	-.90235	-.94026	-.94026	-.90254
12	-.79997	-.79997	-.80042	-.80067	-.80041	-.85632	-.85632	-.80192
13	-.67047	-.67047	-.66948	-.66861	-.66948	-.73932	-.73932	-.67048
14	-.51616	-.51616	-.51252	-.51223	-.51252	-.59368	-.59368	-.51328
15	-.34426	-.34426	-.33603	-.33589	-.33603	-.42473	-.42473	-.33635
16	-.16827	-.16827	-.14650	-.14649	-.14650	-.23797	-.23797	-.14649

TABLE 6

Converged numerical solutions and zone-averaged asymptotic solution for the experiments of Table 3 with schemes R *and* R2. *Note the expansion shock in the results of* R. *Parameters*: $\xi = 0$, $\varepsilon = 1\times10^{-6}$, $\Delta t = \frac{1}{2}\Delta x$.

i	$u_i^{N_\varepsilon}$, u_i^∞		
	R	R2	exact
1	1.00000	.09815	.09786
2	1.03596	.28976	.28982
3	1.09933	.47012	.47064
4	1.17699	.63242	.63337
5	1.25564	.77076	.77177
6	1.32417	.87880	.88051
7	1.37431	.95648	.95540
8	1.40069	.98762	.99359
9	-1.40069	-.98762	-.99359
10	-1.37431	-.95648	-.95540
11	-1.32417	-.87880	-.88051
12	-1.25564	-.77076	-.77177
13	-1.17699	-.63242	-.63337
14	-1.09933	-.47012	-.47064
15	-1.03596	-.28976	-.28982
16	-1.00000	-.09815	-.09786

the zone containing the sonic point (say, zone j), not influenced by its neighbors, no longer has a steady internal structure, so that the scheme locally reduces to

$$u_j^n = u_j^0 + n\Delta t s_j. \tag{50}$$

9. Recommendations. For a scalar conservation law like Burgers' equation there seems to be little reason to abandon Godunov's first-order scheme in favor of the first-order Engquist–Osher scheme. The slight simplification in the flux calculation is accompanied by a degradation of steady-shock representation and no significant acceleration of the convergence to a steady state. The simplification achieved in Roe's first-order scheme is too drastic: the scheme cannot be used "as is" near a sonic point. With the modification (29.5), however, the scheme surpasses Godunov's scheme.

Both Godunov and Engquist–Osher schemes become appreciably more elaborate when applied to a nonlinear hyperbolic system like the one-dimensional Euler equations. The interpretation of the Engquist–Osher scheme as a Godunov-type scheme in which any shock in the solution of a local Riemann problem is replaced by an overturned centered compression wave remains valid. This modification makes it possible to compute $F_{EO}(u_L, u_R)$ explicitly, while $F_G(u_L, u_R)$ must be determined iteratively.

The Engquist–Osher scheme, on the other hand, requires two interior points in a discrete stationary shock [5], while Godunov's scheme probably requires only one (this has not yet been proven in general).

For the Euler equations, however, Roe's more drastic simplification of Godunov's method will pay off, even though the scheme must be somewhat modified in order to reject (almost) stationary expansion shocks. This can be achieved by spreading of rarefaction waves, as in [16], or by generalizing (29.3). Numerical experience with the latter technique is still lacking.

The first-order schemes, when formulated as in § 6, have the potential of achieving any desired order of spatial accuracy in a steady numerical solution. To what degree this potential can be realized for the one-dimensional Euler equations is at present not clear. Meanwhile, the second-order two-step schemes, intended primarily for solving transient problems, seem to outperform the first-order schemes based on (36.3) in obtaining a steady solution for Burgers' equation. Experiments for the Euler equations with both kinds of schemes, conducted for [13], but not fully reported therein, indicate a similar performance.

All schemes discussed in this paper are explicit. When using their numerical flux-functions in an implicit configuration, which may be desirable in approaching a steady state, the above recommendations do not automatically carry over. As noted by Engquist and Osher [4], the nonsmooth dependence of $F_G(u_L, u_R)$ on its arguments in the case (iii) of a transonic shock, that is: $F_G(u_L, u_R) \in C^{(0)}$, invalidates the linearization needed to invert the implicit difference equations. In contrast, $F_{EO}(u_L, u_R) \in C^{(1)}$ in all cases (i), (ii) and (iii). Whether this makes a difference in practice remains to be shown.

Acknowledgments. The ideas of Steve Hancock (Physics International, San Leandro, California) and Dick van Albada (University of Groningen, Netherlands), embodied in (40) and (42), respectively, are greatly appreciated, while discussions with Stanley Osher (University of California at Los Angeles) strengthened my belief in the treatment of source terms discussed in § 6. Phil Colella (Lawrence Berkeley Laboratory) pointed out the weak transonic expansion shocks arising in the use of the exact Riemann solution, while Phil Roe (Royal Aircraft Establishment, United Kingdom) kindly gave me permission to discuss his hitherto unpublished sonic flux-function.

REFERENCES

[1] S. K. GODUNOV, *Finite-difference method for numerical computation of discontinuous solutions of the equations of fluid dynamics*, Mat. Sbornik, 47 (1959), pp. 271–306. (In Russian.)

[2] S. K. GODUNOV, A. W. ZABRODYN AND G. P. PROKOPOV, *A difference scheme for two-dimensional unsteady problems of gas dynamics and computation of flow with a detached shock wave*, Z. Vyčisl. Matem. Mat. Fiz., 1 (1961), pp. 1020–1050. (In Russian.)

[3] A. HARTEN, P. D. LAX AND B. VAN LEER, *On upstream differencing and Godunov-type schemes for hyperbolic conservation laws*, ICASE Report No. 82-5, 1982; SIAM Rev., 25 (1983), pp. 35–62.

[4] B. ENGQUIST AND S. OSHER, *Stable and entropy satisfying approximations for transonic flow calculations*, Math. Comp., 34 (1980), pp. 45–75.

[5] S. OSHER AND F. SOLOMON, *Upwind difference schemes for hyperbolic systems of conservation laws*, Math. Comp. (1983), to appear.

[6] P. L. ROE, *The use of the Riemann problem in finite-difference schemes*, Lecture Notes in Physics, 141, Springer-Verlag, New York, 1981, pp. 354–559.

[7] ———, *Approximate Riemann solvers, parameter vectors and difference schemes*, J. Comp. Phys., 43 (1981), pp. 357–372.

[8] E. M. MURMAN, *Analysis of embedded shock waves calculated by relaxation methods*, AIAA J., 12 (1974), pp. 626–633.

[9] S. L. HANCOCK, private communication (1980).

[10] B. VAN LEER, *Towards the ultimate conservative difference scheme. V. A second order sequel to Godunov's method*, J. Comp. Phys., 32 (1979), pp. 101–136.

[11] ———, *Towards the ultimate conservative difference scheme. IV. A new approach to numerical convection*, J. Comp. Phys. 23 (1977), pp. 276–299.

[12] G. D. VAN ALBADA, private communication (1981).

[13] G. D. VAN ALBADA, B. VAN LEER AND W. W. ROBERTS, JR., *A comparative study of computational methods in cosmic gas dynamics*, Astron. Astrophys., 108 (1982), pp. 76–84.

[14] P. L. ROE, private communication (1981).
[15] A. HARTEN AND P. D. LAX, *A random-choice finite-difference scheme for hyperbolic conservation laws*, SIAM J. Numer. Anal., 18 (1981), pp. 289–315.
[16] P. COLELLA, *Glimm's method for gas dynamics*, this Journal, 3 (1982), pp. 76–110.
[17] T.-P. LIU, *Quasilinear hyperbolic systems*, Comm. Math. Phys., 68 (1979), pp. 141–172.

ON UPSTREAM DIFFERENCING AND GODUNOV-TYPE SCHEMES FOR HYPERBOLIC CONSERVATION LAWS*

AMIRAM HARTEN†, PETER D. LAX‡, AND BRAM VAN LEER§

Abstract. This paper reviews some of the recent developments in upstream difference schemes through a unified representation, in order to enable comparison between the various schemes. Special attention is given to the Godunov-type schemes that result from using an approximate solution of the Riemann problem. For schemes based on flux splitting, the approximate Riemann solution can be interpreted as a solution of the collisionless Boltzmann equation.

Introduction. Upstream-differencing schemes attempt to discretize hyperbolic partial differential equations by using differences biased in the direction determined by the sign of the characteristic speed.

In recent years upstream differencing has become very popular, and a multitude of new techniques of implementing directionally biased differencing have been suggested. This popularity is primarily due to the robustness of upstream-differencing schemes, the availability of an underlying physical model, and the possibility of achieving high resolution of stationary discontinuities. Most of these schemes are an extension of the Courant–Isaacson–Rees scheme [1] to nonlinear conservation laws, and therefore a unified description may be given.

The present paper concentrates on reviewing basic concepts and deriving design principles; numerical experimentation is not presented. The paper is built up as follows. Section 1 reviews some properties of the equations essential to their proper numerical approximation. In §2 we discuss a straightforward extension of linear upstream differencing to nonlinear systems. Section 3 introduces the physical picture due to Godunov, useful to interpret certain schemes from §2 and to construct new schemes. Finally, in §4 we comment on flux splitting, another form of upstream differencing, and relate it to a class of schemes motivated by the Boltzmann equation.

1. Weak solutions and their numerical approximation. In this paper we consider numerical solutions of the initial-value problem for hyperbolic systems of conservation laws

$$u_t + f(u)_x = 0, \quad u(x,0) = u_0(x), \qquad -\infty < x < \infty. \tag{1.1a}$$

Here $u(x,t)$ is a column vector of m unknowns and $f(u)$, the flux, is a vector-valued function of m components. We can write (1.1a) in matrix form

$$u_t + Au_x = 0, \qquad A(u) = f_u. \tag{1.1b}$$

(1.1) is called *hyperbolic* if all eigenvalues of the Jacobian matrix A are real. We assume that the eigenvalues $a_1(u), \cdots, a_m(u)$ are distinct and arranged in an increasing order.

*Received by the editors March 29, 1982, and in revised form May 5, 1982. This research was sponsored by the National Aeronautics and Space Administration under contract NAS1-15810 while the first and third authors were in residence at ICASE, NASA Langley Research Center, Hampton, Virginia.

†Tel Aviv University, Ramat Aviv, Israel. The work of this author was supported in part by the National Aeronautics and Space Administration under contract NCA2-OR525-001 at NASA Ames Research Center, Moffett Field, California, and by the U.S. Department of Energy under contract DE-AC02-76ER03077.

‡Courant Institute of Mathematical Sciences, New York University, New York, New York 10012. The work of this author was supported in part by the U.S. Department of Energy under contract DE-AC02-76ER03077.

§Institute for Computer Applications in Science and Engineering, Leiden State University, Leiden, the Netherlands.

To allow for discontinuous solutions we admit weak solutions that satisfy (1.1) in the sense of distribution theory, i.e.,

$$\int_0^\infty \int_{-\infty}^\infty [w_t u + w_x f(u)]\, dx dt + \int_{-\infty}^\infty w(x,0)\, u_0(x)\, dx = 0, \tag{1.2a}$$

for all C^∞ test functions $w(x, t)$ that vanish for $|x| + t$ large.

Condition (1.2a) is equivalent to requiring that for all rectangles $(a, b) \times (t_1, t_2)$ the relation obtained by integrating (1.1a) over the rectangle should hold:

$$\int_a^b u(x, t_2)\, dx - \int_a^b u(x, t_1)\, dx + \int_{t_1}^{t_2} f(u(b, t))\, dt - \int_{t_1}^{t_2} f(u(a, t))\, dt = 0. \tag{1.2b}$$

Clearly, a piecewise-smooth weak solution of (1.1) satisfies (1.1) pointwise in each smooth region; across each curve of discontinuity the Rankine–Hugoniot relation

$$f(u_R) - f(u_L) = S(u_R - u_L) \tag{1.3}$$

holds, where S is the speed of propagation of the discontinuity, and u_L and u_R are the states on the left and the right, respectively.

Since weak solutions of (1.1) are not uniquely determined by their initial data, we select physically relevant solutions, defined as those solutions that are limits as $\varepsilon \to 0$ of solutions $u(\varepsilon)$ of the viscous equations

$$u_t + f(u)_x = \varepsilon u_{xx}, \qquad \varepsilon > 0. \tag{1.4}$$

In this paper we consider systems of conservation laws (1.1) that possess an *entropy function* $U(u)$, defined as follows:

(i) U is a convex function of u, i.e., $U_{uu} > 0$.

(ii) U satisfies

$$U_u f_u = F_u \tag{1.5a}$$

where F is some other function called *entropy flux;* it follows from (1.5a) that every smooth solution of (1.1) also satisfies

$$U(u)_t + F(u)_x = 0. \tag{1.5b}$$

Limit solutions of (1.4) satisfy, in the weak sense, the following inequality:

$$U(u)_t + F(u)_x \leqq 0; \tag{1.6a}$$

i.e., for all nonnegative smooth test functions $w(x, t)$ of compact support

$$-\int_0^\infty \int_{-\infty}^\infty (w_t U + w_x F)\, dx dt - \int_{-\infty}^\infty w(x, 0) U(u_0(x))\, dx \leqq 0. \tag{1.6b}$$

Condition (1.6b) is equivalent to requiring that for all rectangles $(a, b) \times (t_1, t_2)$ the inequality obtained by integrating (1.6a) over the rectangle should hold:

$$\begin{aligned} &\int_a^b U(u(x, t_2))\, dx - \int_a^b U(u(x, t_1))\, dx \\ &\qquad + \int_{t_1}^{t_2} F(u(b, t))\, dt - \int_{t_1}^{t_2} F(u(a, t))\, dt \leqq 0. \end{aligned} \tag{1.6c}$$

If u is piecewise smooth with discontinuities, then (1.5b) holds pointwise in the smooth regions, while across a discontinuity

$$F(u_R) - F(u_L) - S[U(u_R) - U(u_L)] \leqq 0. \tag{1.6d}$$

Relations (1.6) are called *entropy conditions* (see [12]).

In the following we shall describe numerical approximations to weak solutions of (1.1) that are obtained by 3-point explicit schemes in conservation form:

(1.7a) $$v_j^{n+1} = v_j^n - \lambda f_{j+1/2}^n + \lambda f_{j-1/2}^n, \qquad \lambda = \tau/\Delta,$$

where

(1.7b) $$f_{j+1/2}^n = f(v_j^n, v_{j+1}^n).$$

Here $v_j^n = v(j\Delta, n\tau)$, and $f(u, v)$ is a *numerical flux*. We require the numerical flux to be *consistent* with the physical flux in the following sense:

(1.7c) $$f(u, u) = f(u).$$

We say that the difference scheme (1.7) is *consistent* with the entropy condition (1.6a) if an inequality of the following kind is satisfied:

(1.8a) $$U_j^{n+1} \leqq U_j^n - \lambda F_{j+1/2}^n + \lambda F_{j-1/2}^n,$$

where the following abbreviations are used:

(1.8b) $$U_j^{n+1} = U(v_j^{n+1}), \qquad U_j^n = U(v_j^n),$$

(1.8c) $$F_{j+1/2}^n = F(v_j^n, v_{j+1}^n);$$

here $F(u, v)$ is a *numerical entropy flux*, consistent with the entropy flux:

(1.8d) $$F(u, u) = F(u).$$

The following is an easy (but useful) extension of an easy (but useful) theorem of Lax and Wendroff [12]:

THEOREM 1.1. *Suppose the difference scheme* (1.7) *is consistent with the conservation law* (1.1a), *and with the entropy condition* (1.6a). *Let v_j^n be a solution of* (1.7), *with initial values $v_j^0 = \phi(j\Delta)$. Extend the lattice function v_j^n to continuous values of x, t by setting, as usual*

(1.9) $$v(x, t) = v_j^n, \quad j = [x/\Delta], \quad n = [t/\tau].$$

Suppose that for some sequence $\Delta_k \to 0$, $\tau/\Delta = \lambda$, the limit

$$\lim_{\Delta_k \to 0} v(x, t) = u(x, t)$$

exists in the sense of bounded, L_1^{loc} convergence. Then the limit u satisfies the weak form (1.2) *of the conservation law, and the weak form* (1.6b) *of the entropy condition.*

The proof consists, just as in [12], of multiplying (1.7a) by a test function, summing by parts over n and j, writing the sum as an integral, and passing to the limit $\Delta_k \to 0$.

Theorem 1.1 remains true, and its proof the same, when the fluxes f and F are allowed to be functions of $2l$ arguments:

(1.10) $$f_{j+1/2} = f(u_{j-l+1}, u_{j-l+2}, \cdots, u_{j+l}),$$

and similarly for $F_{j+1/2}$.

Assume that $u_0(x)$ is equal to some reference state u_* for $|x|$ large:

(1.11a) $$u_0(x) = u_* \quad \text{for } |x| > M.$$

Then

(1.11b) $$v_j^n = u_* \quad \text{for } \Delta|j| > M + n\Delta.$$

The entropy U may be altered by adding to it an arbitrary inhomogeneous linear function; this follows from definition (1.5a). Adding such a linear function to U will not alter its convexity, but achieves the following:

$$U(u_*) = 0, \qquad U_u(u_*) = 0. \tag{1.12}$$

Since U is convex, it follows from (1.12) that $U(u) > 0$ for $u \neq u_*$; in fact if U is strictly convex,

$$U(u) \geqq c|u - u^*|^2. \tag{1.13}$$

Now sum (1.8a) with respect to j over all integers j; we obtain

$$\sum_j U_j^{n+1} \leqq \sum_j U_j^n. \tag{1.14a}$$

In other words: total entropy is a decreasing function of time. In particular

$$\sum_j U_j^n \leqq \sum_j U_j^0. \tag{1.14b}$$

This is an a priori inequality for solutions of the difference scheme (1.7), analogous to the energy inequality for linear symmetric hyperbolic differential and difference equations. Since by (1.13) U is positive for $u \neq u_*$, this is an a priori estimate for solutions of the difference scheme (1.7), and indicates that the scheme is stable. However (1.14b) is not strong enough to prove the pointwise boundedness of solutions of (1.7), nor the existence of convergent subsequences.

Theorem 1.1 holds in any number of space variables. Furthermore multidimensional schemes that are composites of one-dimensional fractional steps satisfy the multidimensional analogue of the entropy condition (1.8a) if each individual one-dimensional step satisfies an entropy inequality of the form (1.8a); see Crandall–Majda [2].

A word of caution: when dealing with equations of mathematical physics, in particular the equations of compressible flow, we must make sure that the difference scheme we are using keeps the variables within their physical range, i.e., that density and pressure are always positive quantities.

2. Upstream-differencing schemes. We start our review with the description of the first-order-accurate Courant–Isaacson–Rees scheme [1], the simplest upstream differencing scheme for the constant-coefficient scalar equation

$$u_t + a u_x = 0, \qquad a = \text{constant}, \tag{2.1}$$

$$v_j^{n+1} = v_j^n - \lambda a \cdot \begin{cases} v_{j+1}^n - v_j^n & \text{for } a < 0, \\ v_j^n - v_{j-1}^n & \text{for } a > 0. \end{cases} \tag{2.2a}$$

Introducing the notation

$$a^- \equiv \min(a, 0) = \tfrac{1}{2}(a - |a|),$$

$$a^+ \equiv \max(a, 0) = \tfrac{1}{2}(a + |a|),$$

we rewrite (2.2a) as

$$v_j^{n+1} = v_j^n - \lambda[a^+(v_j^n - v_{j-1}^n) + a^-(v_{j+1}^n - v_j^n)], \tag{2.2b}$$

which can be rewritten as

$$v_j^{n+1} = \lambda a^+ v_{j-1}^n + \lambda(1 - |a|)\, v_j^n - \lambda a^- v_{j+1}^n. \tag{2.2c}$$

Under the Courant–Friedrichs–Lewy condition

$$\lambda|a| \leq 1 \tag{2.2d}$$

all coefficients of v_i^n on the right in (2.2b) are positive. Such a scheme is called *monotone*, and is stable in the maximum norm; that is, for a monotone scheme

$$\max_j |v_j^{n+1}| \leq \max_j |v_j^n|. \tag{2.2e}$$

Equation (2.2b) can be rewritten as

$$v_j^{n+1} = v_j^n - \frac{\lambda}{2} a(v_{j+1}^n - v_{j-1}^n) + \frac{\lambda}{2}|a|(v_{j+1}^n - 2v_j^n + v_{j-1}^n). \tag{2.2f}$$

This shows that solutions of (2.2) can be thought of as approximating solutions of

$$w_t + aw_x = \tfrac{1}{2}\Delta x|a|(1 - \lambda|a|)w_{xx} \tag{2.3}$$

to second-order accuracy. We observe that the viscosity term in (2.3) vanishes for $a = 0$; this fact later will allow perfectly resolved stationary shocks but may also result in admitting entropy violating discontinuities.

We describe now the extension of (2.2) to systems of equations with constant coefficients:

$$u_t + Au_x = 0, \qquad A = \text{constant}. \tag{2.4}$$

Because of the hyperbolicity assumption, the system (2.4) can be diagonalized by a similarity transformation

$$w = T^{-1}u, \quad T^{-1}AT = \Lambda, \quad \Lambda_{ij} = a_i\delta_{ij}, \tag{2.5a}$$

$$w_t + \Lambda w_x = 0. \tag{2.5b}$$

The components of w are called characteristic variables and (2.5b) is a system of decoupled characteristic equations.

We extend the Courant–Isaacson–Rees scheme to systems by applying the scalar scheme (2.2) to each of the decoupled scalar characteristic equations. In matrix form this can be written as

$$w_j^{n+1} = w_j^n - \frac{\lambda}{2}\Lambda(w_{j+1}^n - w_{j-1}^n) + \frac{\lambda}{2}|\Lambda|(w_{j+1}^n - 2w_j^n + w_{j-1}^n), \tag{2.6a}$$

where the diagonal matrix $|\Lambda|$ is defined by $|\Lambda|_{ij} = |a_i|\delta_{ij}$. In the original variables, the scheme (2.6a) takes the form

$$v_j^{n+1} = v_j^n - \frac{\lambda}{2}A(v_{j+1}^n - v_{j-1}^n) + \frac{\lambda}{2}|A|(v_{j+1}^n - 2v_j^n + v_{j-1}^n), \tag{2.6b}$$

where $|A| = T|\Lambda|T^{-1}$. Clearly, the stability condition for (2.6a) and (2.6b) is

$$\frac{\Delta t}{\Delta x}\max_k |a_k| \leq 1. \tag{2.6c}$$

In general we define the matrix $\chi(A)$ by

$$\chi(A) = T\chi(\Lambda)T^{-1}, \qquad (\chi(\Lambda))_{ij} = \chi(a_i)\delta_{ij}. \tag{2.7}$$

We remark that under our assumption of a full set of eigenvectors we can compute $\chi(A)$ by $\chi(A) = P(A)$, where $P(x)$ is the Lagrangian interpolation polynomial such that $P(a_i) = \chi(a_i), i = 1, \cdots, m$.

In the following we shall describe various techniques to extend the upstream-differencing scheme (2.6b) to nonlinear systems of conservation laws (1.1).

The linear system (2.4) can be regarded as a system of conservation laws (1.1a), where the flux depends linearly on u:

$$f(u) = Au. \tag{2.8a}$$

The upstream-differencing scheme (2.6b) is in conservation form (1.7), with numerical flux given by

$$f(u, v) = A^+u + A^-v \tag{2.8b}$$

where A^+ and A^- are the positive and negative parts of A, defined by the functional calculus (2.7) as

$$A^+ = \chi^+(A), \qquad A^- = \chi^-(A) \tag{2.8c}$$

where we set

$$\chi^+(a) = a^+, \qquad \chi^-(a) = a^-. \tag{2.8d}$$

Note that since $\chi^+(a) + \chi^-(a) = a$, and $\chi^+(a) - \chi^-(a) = |a|$, we can write

$$A^+ = ½(A + |A|), \qquad A^- = ½(A - |A|). \tag{2.9}$$

DEFINITION. A difference scheme in conservation form (1.7) is said to be an *upstream* scheme if:

(i) For u and v nearby states, (2.8b) is a linear approximation to the numerical flux $f(u, v)$.

(ii) When all signal speeds associated with the numerical flux $f(u, v)$ are > 0,

$$f(u, v) = f(u).$$

When all signal speeds are < 0,

$$f(u, v) = f(v).$$

The relevant signal speeds generally differ from the characteristic speeds of the states u and v.

We restate (i) in analytic terms: Suppose u and v are near some reference state u_*; then we require that

$$\begin{aligned} f(u, v) = f(u_*) &+ A^+(u_*)(u - u_*) + A^-(u_*)(v - u_*) \\ &+ o(|u - u_*| + |v - u_*|). \end{aligned} \tag{2.10a}$$

A natural choice for u_* is $(u + v)/2$; setting this into (2.10a), making use of (2.9) to set

$$A^+ - A^- = |A|$$

and noting that

$$f\left(\frac{u+v}{2}\right) = \frac{f(u) + f(v)}{2} + o(|u - v|)$$

we get

$$f(u, v) = \frac{f(u) + f(v)}{2} - \frac{1}{2}\left|A\left(\frac{u+v}{2}\right)\right|(v - u) + o(|u - v|). \tag{2.10b}$$

We can write any numerical flux in the form

$$f(u, v) = \frac{f(u) + f(v)}{2} - \frac{1}{2} d(u, v); \tag{2.11a}$$

for the sake of consistency (1.7c) we need

$$d(u, u) = 0. \tag{2.11b}$$

The upstream condition (2.10b) can be expressed then as

$$d(u, v) = \left| A\left(\frac{u+v}{2}\right) \right| (v - u) + o(|u - v|). \tag{2.11c}$$

Formula (2.6a) shows that linear upstream difference schemes contain a large dose of artificial viscosity, except for those components where a_k is small, in particular where $a_k = 0$. The same appears to be true for all upstream difference schemes for nonlinear conservation laws: when all characteristic speeds are not zero, each component is treated like a scheme with a hefty amount of artificial viscosity, smearing discontinuities. There is however quite a distinction among the schemes when one of the characteristic speeds is zero; this shows up in the way each scheme resolves a stationary shock, centered transonic rarefaction wave, and stationary contact discontinuity. We turn now to examining these matters.

The most critical difference in performance occurs in resolving a stationary shock (see (1.6d)):

$$u_0(x) = \begin{cases} u, & x < 0, \\ v, & x > 0, \end{cases} \qquad f(u) = f(v), \qquad F(v) < F(u). \tag{2.12a}$$

The lack of numerical dissipation allows the design of schemes that perfectly resolve stationary shocks, i.e., (2.12a) is a steady solution of the numerical scheme. The condition for that is

$$d(u, v) = 0 \quad \text{if } f(u) = f(v), \qquad F(v) < F(u) \tag{2.12b}$$

On the other hand

$$u_0(x) = \begin{cases} u, & x < 0, \\ v, & x > 0, \end{cases} \qquad F(v) > F(u) \tag{2.12c}$$

is not an admissible discontinuity and should not be a steady solution of the finite difference scheme, i.e., we require that

$$d(u, v) \neq 0 \quad \text{if } f(u) = f(v), \qquad F(v) > F(u). \tag{2.12d}$$

We remark that the danger that a given upstream scheme selects a nonphysical solution will occur only for stationary or near-stationary discontinuities; otherwise there is enough numerical viscosity in (2.3) to enforce the selection of a physically relevant solution. Hence there are two options in designing an upstream-differencing scheme for solving problems with discontinuous solution:

(1) to switch direction of differencing in a way that will effectively introduce nonlinear dissipation at the expense of slightly spreading the shock;

(2) to satisfy (2.12) and thus get perfect resolution of a stationary shock, but to add a mechanism for checking the admissibility of the discontinuity.

We turn now to describing various forms of $d(u, v)$ in (2.11a). The most straightforward way to generate such functions is by

$$d(u,v) = |A|(u,v)(v-u), \tag{2.13a}$$

where $|A|(u, v)$ is a matrix function of u and v that has nonnegative eigenvalues, and such that

$$|A|(u,u) = |A(u)|; \tag{2.13b}$$

$|A(u)|$ is defined by (2.7).

The simplest forms of (2.13) are

$$|A|(u,v) = \left| A\left(\frac{u+v}{2}\right) \right|, \tag{2.14a}$$

or

$$|A|(u,v) = \tfrac{1}{2}[|A(u)| + |A(v)|]. \tag{2.14b}$$

The latter was used by van Leer in [13], and introduces some nonlinear numerical dissipation that somewhat smears stationary shocks but on the other hand excludes nonphysical discontinuities.

Another form of (2.13a) with (2.14a) has been suggested by Huang [10]:

$$d(u,v) = \operatorname{sgn}\left(A\left(\frac{u+v}{2}\right)\right)[f(v) - f(u)]; \tag{2.15}$$

here sgn (x) is the sign of x, and sgn (A) is defined by (2.7).

Yet another type of scheme has been designed by Roe [19]. His scheme is of the form (2.13a), where the matrix function $A(u, v)$ is required to have these properties:

(i)

$$f(v) - f(u) = A(u,v)(v-u). \tag{2.16a}$$

(ii) $A(u, v)$ has real eigenvalues and a complete set of eigenvectors.

(iii)

$$A(u,u) = A(u). \tag{2.16b}$$

For the Euler equations of compressible flow Roe [19] has constructed a linearization of form (2.16a) having these properties. We show now that such a linearization exists quite generally:

THEOREM 2.1 (Harten-Lax). *Suppose* (1.1a) *has an entropy function. Then* (1.1a) *has a Roe-type linearization.*

Proof. We shall construct an A satisfying (2.16a) which is of form $A = BP$, B symmetric, P positive definite. Clearly, such an A is similar to the symmetric matrix $P^{1/2}BP^{1/2}$ and so has property (ii). In our construction we use the entropy function $U(u)$; since U is convex, the mapping $u \rightarrow w = U_u$ is one-to-one; we introduce w as new variable in place of u. Let u_1 and u_2 be two arbitrary states, $w_1 = w(u_1)$, $w_2 = w(u_2)$, $f_1 = f(u_1)$, $f_2 = f(u_2)$. Then

$$\begin{aligned} f_2 - f_1 &= \int_0^1 \frac{d}{d\theta} f(\theta w_2 + (1-\theta)w_1)\, d\theta \\ &= \int_0^1 f_w d\theta (w_2 - w_1) = B(w_2 - w_1). \end{aligned} \tag{2.17}$$

We claim that f_w is a symmetric matrix; this implies that B is symmetric. To see about f_w, we use relation (1.5a); differentiating with respect to u we get

$$U_{uu} f_u + U_u f_{uu} = F_{uu}.$$

The second term on the left is a linear combination of symmetric matrices f^i_{uu}; the right side, F_{uu}, also is symmetric. Therefore so is the first term

$$U_{uu} f_u.$$

It follows then that also

$$f_u U_{uu}^{-1} \tag{2.18}$$

is symmetric.

Differentiating

$$w = U_u$$

with respect to u shows that $w_u = U_{uu}$; therefore

$$U_{uu}^{-1} = u_w.$$

Substituting this into (2.18) shows that

$$f_{uu} u_w = f_w$$

is symmetric, as asserted.

Next we express

$$\begin{aligned} w_2 - w_1 &= \int_0^1 \frac{d}{d\theta} w(\theta u_2 + (1-\theta) u_1)\, d\theta \\ &= \int_0^1 w_u d\theta (u_2 - u_1) = P(u_2 - u_1). \end{aligned} \tag{2.19}$$

Using $w_u = U_{uu}$ and the convexity of U we conclude that P as defined by (2.19) is positive definite.

Combining (2.17) and (2.19) gives

$$f_2 - f_1 = BP(u_2 - u_1) = A(u_2 - u_1).$$

Thus the A we have constructed can be factored as BP, as asserted. Condition (2.16b) is clearly satisfied. □

Note that B depends symmetrically on w_1 and w_2, and P symmetrically on u_1 and u_2. This shows that A is a symmetric function of u_1 and u_2.

A similar result holds for systems of conservation laws in any number of space variables as long as there is an entropy (see Harten [8]).

Having constructed A, we can define its absolute value by (2.7); then we set

$$d(u, v) = |A(u, v)|(v - u). \tag{2.20}$$

When u and v correspond to a stationary discontinuity (2.12), then it follows from $f(v) - f(u) = 0$ and (2.16a) that $v - u$ is a null vector of $A(u, v)$, and consequently in the null space of $|A(u, v)|$; thus $d(u, v) = 0$, whether or not the entropy condition $F(v) - F(u) < 0$ is satisfied. Therefore the corresponding upstream differencing may admit nonphysical solutions. In an appendix to [9], Harten and Hyman describe a viscosity-like term, i.e., of the form (3.33a), that can be added to (2.15) and (2.16) to reject inadmissible discontinuities without affecting the perfect resolution of the physical ones.

Yet another way to construct d in (2.11c) is

$$d(u, v) = \int_u^v |A(w)|\, dw, \tag{2.21}$$

where the integration in (2.21) is carried out on a path in state-space connecting u and v. Osher in [16] suggests a path of integration Γ that is piecewise parallel to the right eigenvectors R_k of A:

$$\Gamma = \bigcup_{k=1}^{m} \Gamma_k, \tag{2.22a}$$

$$\Gamma_k: \begin{cases} \dfrac{du^k}{dl} = R_k(u^k), & 0 < l < l_k, \quad k = 1, \cdots, m, \\ u^{k+1}(l_{k+1}) = u^k(0), \end{cases} \tag{2.22b}$$

$$u^m(0) = u, \qquad u^1(l_1) = v. \tag{2.22c}$$

Existence of a unique solution to (2.22b)–(2.22c) is guaranteed if $\|u - v\|$ is sufficiently small [16]. A consequence of this choice of path is that d in (2.21) decouples into characteristic contributions

$$d(u, v) = \sum_{k=1}^{m} \int_0^{l_k} |a_k(u(l))|\, R_k(u(l))\, dl. \tag{2.23}$$

Osher shows that limit solutions of (2.11) with (2.23) satisfy the entropy condition and that a stationary contact discontinuity is perfectly resolved; the representation of a stationary shock requires at least one intermediate state.

3. Godunov-type schemes.

A. The general theory. Godunov, in his construction of the "best" monotone scheme [4], has used the exact solutions of local Riemann problems to obtain an upstream-differencing scheme.

The solution of the Riemann problem

$$u_t + f(u)_x = 0, \qquad u(x, 0) = \begin{cases} u_L, & x < 0, \\ u_R, & x > 0, \end{cases} \tag{3.1}$$

depends only on the states u_L and u_R and the ratio x/t; it will be denoted by $u(x/t; u_L, u_R)$. Since signals propagate with finite velocity,

$$u(x/t; u_L, u_R) = u_L \quad \text{for } x/t \leq a_L, \tag{3.2a}$$

$$u(x/t; u_L, u_R) = u_R \quad \text{for } x/t \geq a_R; \tag{3.2b}$$

a_L and a_R are the smallest and largest signal velocity.

Godunov derives his scheme by considering the numerical approximation $v(x, t_n)$ of the discrete time levels t_n, $n = 0, 1, \cdots$, to be a piecewise constant function in x, i.e.,

$$v(x, t_n) = v_j^n \quad \text{for } x \text{ in } I_j = ((j - \tfrac{1}{2})\Delta, (j + \tfrac{1}{2})\Delta). \tag{3.3a}$$

To calculate the numerical approximation at the next time level $t_{n+1} = t_n + \tau$, we first solve exactly the initial-value problem

$$u_t + f(u)_x = 0, \qquad u(x, t_n) = v(x, t_n), \qquad -\infty < x < \infty, \tag{3.3b}$$

for $t_n \leqq t \leqq t_n + \tau$, and denote its solution by $u_n(x, t)$. Each discontinuity in $v(x, t_n)$ constitutes locally a Riemann problem. If we keep $\lambda |a_{max}| < 1/2$ where $|a_{max}|$ is the largest signal speed, then because of (3.2) there is no interaction between neighboring Riemann problems, and $u_n(x, t)$ can be expressed exactly in terms of the solutions of local Riemann problems:

$$u_n(x, t) = u\left(\frac{x - (j + 1/2)\Delta}{t - t_n}; v_j^n, v_{j+1}^n\right) \quad \text{for } j\Delta < x < (j + 1)\Delta, \quad t_n \leqq t \leqq t_{n+1}. \tag{3.3c}$$

Godunov obtains a piecewise-constant approximation $v(x, t_{n+1})$ by averaging $u_n(x, t_{n+1})$, i.e., he sets

$$v_j^{n+1} = \frac{1}{\Delta} \int_{I_j} u_n(x, t_n + \tau)\, dx. \tag{3.3d}$$

We can rewrite (3.3d) in terms of the solutions to the local Riemann problem as

$$v_j^{n+1} = \frac{1}{\Delta} \int_0^{\Delta/2} u(x/\tau; v_{j-1}^n, v_j^n)\, dx + \frac{1}{\Delta} \int_{-\Delta/2}^{0} u(x/\tau; v_j^n, v_{j+1}^n)\, dx. \tag{3.4a}$$

Since u_n is an exact solution of the conservation laws (3.1), we can evaluate the integral defining v_j^{n+1} in (3.3d) by applying (1.2b) over $I_j \times (t_n, t_{n+1})$; we get

$$v_j^{n+1} = v_j^n - \lambda[f(\hat{v}_{j+1/2}) - f(\hat{v}_{j-1/2})], \tag{3.4b}$$

where

$$\hat{v}_{j+1/2} = u(0; v_j^n, v_{j+1}^n). \tag{3.4c}$$

This shows that (3.4b) is in conservation form, with

$$f(v, w) = f(u(0; v, w)). \tag{3.4d}$$

The exact solution $u_n(x, t)$ of the Riemann problem satisfies the entropy condition (1.6c):

$$\int_{I_j} U(u_n(x, t_{n+1}))\, dx \leqq \Delta U(v_j^n) - \tau F(\hat{v}_{j+1/2}) + \tau F(\hat{v}_{j-1/2}).$$

Since U is a convex function, Jensen's inequality holds:

$$U\left(\frac{1}{\Delta} \int_{I_j} u(x, t)\, dx\right) \leqq \frac{1}{\Delta} \int_{I_j} U(u(x, t))\, dx.$$

Combining the last two inequalities, we deduce that Godunov's scheme satisfies the entropy inequality (1.8a).

The description (3.3c) makes sense only if the local Riemann problems don't interact, i.e., if

$$\lambda |a_{max}| < 1/2.$$

On the other hand, (3.4b) remains consistent with (3.3d) as long as the waves issuing from $(j \pm 1/2)\Delta$ do not reach $(j \mp 1/2)\Delta$ during the time interval $t_n \leqq t \leqq t_{n+1}$. This will be the case as long as

$$\lambda |a_{max}| \leqq 1.$$

It follows from the Rankine–Hugoniot relation (1.3) that $f(u(s; v, w)) - su(s; v, w)$

is a continuous function; however it is only piecewise differentiable. It follows that the Godunov flux function $f(v, w)$ defined by (3.4d) is only piecewise differentiable.

Godunov's scheme satisfies criterion (ii) for upstream schemes because of (3.2). To

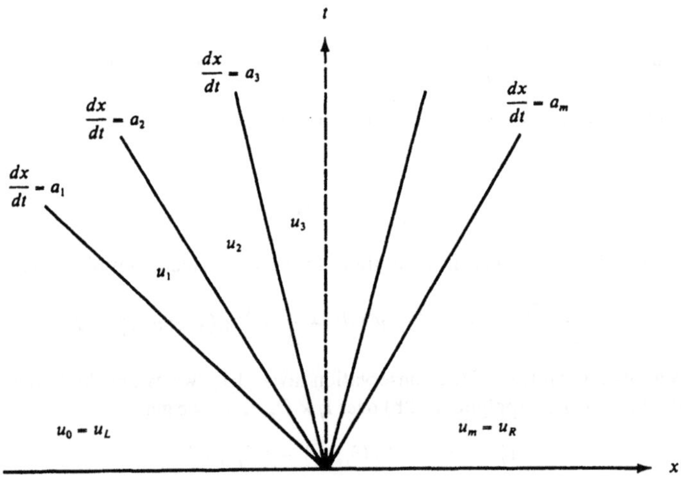

FIG. 1.

verify that it also satisfies criterion (i) we shall show that Godunov's scheme, when applied to linear equations, reduces to (2.6b).

Consider

$$u_t + Au_x = 0, \tag{3.5a}$$

A a constant matrix. Here the solution of the Riemann problem is composed of constant states separated by a fan of m characteristic lines (see Fig. 1).

$$u(x/t; u_L, u_R) = u_k \quad \text{for } a_k < x/t < a_{k+1}, \quad k = 0, \cdots, m, \tag{3.5b}$$

where we have defined

$$u_0 = u_L, \quad u_m = u_R, \quad a_0 = -\infty, \quad a_{m+1} = +\infty.$$

The intermediate states u_R can be calculated from the representation of $u_R - u_L$ in terms of the right eigenvectors R_k of A in the following way:

$$u_R - u_L = \sum_{i=1}^{m} J_i R_i, \tag{3.5c}$$

$$u_k = u_L + \sum_{i=1}^{k} J_i R_i. \tag{3.5d}$$

We can write this by (2.7) as

$$u_k = u_L + \sigma_k(A)(u_R - u_L) \tag{3.5e}$$

where σ_k is the function

$$\sigma_k(a) = \begin{cases} 1 & \text{for } a < a_k, \\ 0 & \text{for } a > a_k. \end{cases} \tag{3.5f}$$

Let N be an integer such that

$$a_N < 0 < a_{N+1}.$$

Then by (3.5e)

$$(3.6)\qquad \begin{aligned} u(0; u_L, u_R) = u_N &= u_L + \sigma_N(A)(u_R - u_L) \\ &= (I - \sigma_N(A))u_L + \sigma_N(A)u_R. \end{aligned}$$

Since by (2.8a), in the linear case $f(u) = Au$, substituting (3.6) into (3.4d) gives

$$f(u, v) = A(I - \sigma_N(A))u + A\sigma_N(A)v.$$

From the functional calculus (2.7) and (2.8c) we deduce that

$$A(I - \sigma_N(A)) = A^+, \qquad A\sigma_N(A) = A^-,$$

so that

$$f(u, v) = A^+u + A^-v,$$

in full agreement with (2.8b). Thus in the linear case Godunov's scheme reduces to the upstream scheme (2.6b).

The solution to the Riemann problem (3.1) has a rather complicated structure: as in the constant coefficient case (3.5), the solution to (3.1) depends on x/t and consists of constant states u_k, $k = 0, \cdots, m$; $u_0 = u_L$, $u_m = u_R$, separated by a fan of waves. Unlike in the constant coefficient case, the k-wave separating u_{k-1} and u_k is not necessarily a single line having a characteristic speed a_k. If the kth characteristic field is genuinely nonlinear then the k-wave is either a rarefaction wave $(a_k(u_{k-1}) < a_k(u_k))$ or a shock propagating with speed S, $(a_k(u_{k-1}) > S > a_k(u_k))$. If the kth characteristic field is linearly degenerate then the k-wave is a contact discontinuity propagating with speed $a_k(u_{k-1}) = a_k(u_k)$ (see [12]).

It is evident from (3.3d) that, due to averaging, the Godunov scheme does not make use of all the information contained in the exact solution of the Riemann problem. We therefore consider replacing the exact solution to the Riemann problem $u(x/t; u_L, u_R)$ in (3.4a) by an approximation $w(x/t; u_L, u_R)$; the latter can have a much simpler structure as long as it does not violate the essential properties of conservation and entropy inequality. The following theorem due to Harten and Lax [7, Thm. 2.1] shows that this type of approximation is consistent:

THEOREM 3.1 (Harten-Lax). *Let $w(x/t; u_L, u_R)$ be an approximation to the solution of the Riemann problem that satisfies the following conditions:*

(i) *consistency with the integral form of the conservation law in the sense that*

$$(3.7a)\qquad \int_{-\Delta/2}^{\Delta/2} w(x/t; u_L, u_R)\,dx = \frac{\Delta}{2}(u_L + u_R) - \tau f_R + \tau f_L$$

for $\Delta/2 > \tau \max|a_k|$, where

$$f_R = f(u_R), \qquad f_L = f(u_L);$$

(ii) *consistency with the integral form of the entropy condition in the sense that*

$$(3.7b)\qquad \int_{-\Delta/2}^{\Delta/2} U(w(x/t; u_L, u_R))\,dx \le \frac{\Delta}{2}(U_L + U_R) - \tau F_R + \tau F_L$$

for $\Delta/2 > \tau \max|a_k|$, where

$$F_R = F(u_R), \qquad F_L = F(u_L).$$

Using the approximation w to the Riemann problem, we can define a Godunov-type scheme as follows:

$$v_j^{n+1} = \frac{1}{\Delta}\int_0^{\Delta/2} w(x/t; v_{j-1}^n, v_j^n)\,dx + \frac{1}{\Delta}\int_{-\Delta/2}^{0} w(x/t; v_j^n, v_{j+1}^n)\,dx. \tag{3.8}$$

Assertion. If conditions (3.7a) *and* (3.7b) *are satisfied, the scheme* (3.8) *is in conservation form consistent with* (3.1), *and satisfies the entropy inequality* (1.8a).

For proof, see [7, after Thm. 2.1]. It is shown there that the Godunov-type averaging can be replaced by Glimm-type sampling. [7] contains a good account of this technique and an analysis of its accuracy.

Theorem 3.1 shows that Godunov-type schemes that satisfy conditions (i) and (ii) above satisfy the hypotheses of Theorem 1.1; this shows that if such a scheme converges, the limit satisfies the conservation law and the entropy condition in the weak sense.

We note that Godunov's scheme is of Godunov type.

We have shown at the beginning of this section that Godunov's scheme (3.3) can also be written as a scheme (3.4) in conservation form. The appropriate numerical flux was obtained from the integral conservation laws (1.2b). We show now that all schemes of Godunov type can be expressed in conservation form; we can obtain the appropriate numerical flux by applying the integral conservation law (1.2b) to the approximate solution of the Riemann problem over the rectangle $(-\Delta/2, 0) \times (0, \tau)$:

$$\int_{-\Delta/2}^{0} w(x/\tau; u_L, u_R)\,dx - \frac{\Delta}{2} u_L + \tau[f_{LR} - f_L] = 0, \tag{3.9a}$$

where

$$f_{LR} = f(u_L, u_R).$$

This gives

$$f_{LR} = f_L - \tau^{-1}\int_{-\Delta/2}^{0} w(x/\tau; u_L, u_R)\,dx + \frac{\Delta}{2\tau} u_L. \tag{3.9b}$$

If we apply the integral conservation law (1.2b) over the rectangle $(0, \Delta/2) \times (0, \tau)$, we obtain

$$f'_{LR} = f_R + \tau^{-1}\int_0^{\Delta/2} w(x/\tau; u_L, u_R)\,dx - \frac{\Delta}{2\tau} u_R. \tag{3.9c}$$

The equality of (3.9b) and (3.9c) is just the content of the consistency relation (3.7a).

Using formula (3.9b) for $f_{j+1/2}$ in (1.7b) and (3.9c) for $f_{j-1/2}$ in (1.7b) and setting the resulting expressions into (1.7a) gives (3.8), i.e., it puts the Godunov-type scheme in conservation form (1.7a):

$$v_j^{n+1} = v_j^n - \frac{\tau}{\Delta}[f(v_j, v_{j+1}) - f(v_{j-1}, v_j)]. \tag{3.10}$$

If all signal speeds are positive, then $w(s; u_L, u_R) = u_L$ *for* $s < 0$; according to (3.9b) in this case $f_{LR} = f_L$. Similarly, if all signal speeds are negative, then $w(s; u_L, u_R) = u_R$ for $s > 0$; it follows from (3.9c) that in this case $f_{LR} = f_R$. This is property (ii) of upstream schemes, which is thus satisfied by all Godunov-type schemes.

In a scheme of Godunov type we can incorporate into the numerical flux all physical insights that we can put into the approximate solution of the Riemann problem. Also, as Harten and Lax pointed out in [7], a Godunov-type scheme (3.8) can be used just as easily on a grid that varies in time, by adjusting the intervals of integration on the right in (3.8). This makes these schemes the natural choice for adaptive grids; further development of such algorithms and numerical experiments are described in Harten and Hyman [9].

B. Construction of particular Riemann solvers. We turn now to describing two different approximate Riemann solvers, and the Godunov-type schemes corresponding to them. The first, due to Roe, is based on a linearization notion of type (2.16). Roe approximates solutions of the Riemann problem for (3.1) by exact solutions of the Riemann problem for the following linear hyperbolic equation with constant coefficients:

$$w_t + A_{LR} w_x = 0, \qquad w(x,0) = \begin{cases} u_L, & x < 0, \\ u_R, & x > 0. \end{cases} \tag{3.11a}$$

Here A_{LR} is a matrix that satisfies (2.16a)–(2.16b) and has properties (i)–(iii) listed there. Combining (2.16a) with (3.5c) yields

$$f_R - f_L = A_{LR}(u_R - u_L) = \sum a_i J_i R_i, \tag{3.11b}$$

where a_i are the eigenvalues of A_{LR}, R_i the corresponding right eigenvectors, and J_i the coefficients in the resolution (3.5c):

$$u_R - u_L = \sum J_i R_i. \tag{3.11c}$$

The approximate Riemann solver is given by (3.5b), with u_k defined by (3.5d).

The numerical flux associated with an approximate Riemann solver is given by (3.9b); substituting (3.5d) into (3.9b) we get

$$f_{LR} = f_L + \sum a_i^- J_i R_i \tag{3.12a}$$

where

$$a^- = \min(a, 0) = \tfrac{1}{2}(a - |a|).$$

Substituting this into (3.12) and using (3.11b) gives

$$\begin{aligned} f_{LR} &= \tfrac{1}{2}(f_L + f_R) - \tfrac{1}{2} \sum |a_i| J_i R_i \\ &= \tfrac{1}{2}(f_L + f_R) - \tfrac{1}{2}|A_{LR}|(u_R - u_L); \end{aligned} \tag{3.12b}$$

in the last step we have used the definition of $|A|$ as given by (2.7). Indeed, (3.12b) is Roe's scheme defined in (2.11a), (2.20).

As already pointed out in §2, Roe's scheme admits nonphysical, i.e. entropy-violating, stationary discontinuities. In an appendix to [9], Harten and Hyman show how to modify Roe's scheme to eliminate such entropy-violating discontinuities while retaining those that satisfy the entropy law.

We note that the numerical flux (3.12b) of Roe's scheme resembles Osher's scheme (2.23). There the jumps J_k in the characteristic state variables are represented by the path length l_k. Osher's scheme, however, is not of Godunov type in the sense of (3.9) since the integration path Γ in state-space does not correspond to a univalued approximate Riemann solution $w(x/t, u_L, u_R)$ as in (3.8).

Roe's Riemann solver contains a great amount of detail: $m - 1$ intermediate states.

We describe next a hierarchy of Riemann solvers where much of this detail is lumped together. The simplest of these schemes contains only one intermediate state.

i) Denote by a_L and a_R lower and upper bounds, respectively, for the smallest and largest signal velocity, calculated according to some algorithm. Define the approximate Riemann solver by

$$u(x/t; u_L, u_R) = \begin{cases} u_L & \text{for } x/t < a_L, \\ u_{LR} & \text{for } a_L < x/t < a_R, \\ u_R & \text{for } a_R < x/t, \end{cases} \tag{3.13}$$

where the state u_{LR} is determined from the conservation law (3.7a):

$$(\tau a_L + \Delta/2)u_L + \tau(a_R - a_L)u_{LR} + (\Delta/2 - \tau a_R)u_R = \frac{\Delta}{2}(u_L + u_R) - \tau[f_R - f_L].$$

This gives

$$u_{LR} = \frac{a_R u_R - a_L u_L}{a_R - a_L} - \frac{f_R - f_L}{a_R - a_L}. \tag{3.14}$$

We turn now to the determination of the associated numerical flux. We substitute (3.14) into (3.13), and then into (3.9b):

$$f_{LR} = \begin{cases} f_L & \text{when } 0 < a_L, \\ \dfrac{-a_L}{a_R - a_L} f_R + \dfrac{a_R}{a_R - a_L} f_L + \dfrac{a_L a_R}{a_R - a_L}(u_R - u_L) & \text{when } a_L < 0 < u_R, \\ f_R & \text{when } a_R < 0. \end{cases} \tag{3.15a}$$

This can be combined into a single formula

$$f_{LR} = \frac{a_R^- - a_L^-}{a_R - a_L} f_R + \frac{a_R^+ - a_L^+}{a_R - a_L} f_L - \frac{1}{2}\frac{a_R|a_L| - a_L|a_R|}{a_R - a_L}(u_R - u_L). \tag{3.15b}$$

Since u_{LR} was chosen to satisfy the conservation law (3.7a), we conclude that u_{LR} is the mean value of the exact solution over the interval $(\tau a_L, \tau a_R)$. It follows therefore from Jensen's inequality that (3.15) satisfies the entropy inequality (3.7b).

Suppose that u_L and u_R can be connected with a shock of the first or the mth family. In these cases the exact solution is

$$u(x, t) = \begin{cases} u_L & \text{for } x/t < S, \\ u_R & \text{for } S < x/t, \end{cases} \tag{3.16}$$

where S is the speed of propagation of the shock. Suppose the algorithm for calculating a_L and a_R is such that it furnishes $a_L = S$ or $a_R = S$, depending on whether the shock belongs to the first or the mth family. Then it follows from the equality of u_{LR} with the mean value of the exact solution that (3.15) is the exact solution.

ii) We describe next a class of approximate Riemann solvers, where u_L and u_R are linked through *two* intermediate states. These states are so chosen that:

a) The conservation laws are satisfied.

b) If the exact solution of the Riemann problem links u_L and u_R through a single shock (or contact discontinuity) of any of the m families of waves, then so does the approximate Riemann solver.

c) The entropy law is satisfied.

Such an approximate Riemann solver was constructed in [7]; the one presented here differs from it in some important details. We are grateful to Paul Woodward for a suggestion which has been incorporated in the scheme.

Let the velocities a_L and a_R be defined as in approximation i) described above. We define a velocity V as follows: Let U be an entropy function defined by (1.5); denote its gradient by $w = U_u$, and introduce the abbreviation

(3.17a) $$w(u_R) - w(u_L) = l_{LR}.$$

We set

(3.17b) $$V = l_{LR} \cdot (f_R - f_L)/l_{LR} \cdot (u_R - u_L)$$

where the dot denotes the Euclidean scalar product.

Next we show that V is well defined, and derive its salient properties:

LEMMA 3.2: i) *The denominator in* (3.17b) *is positive for* $u_L \neq u_R$.

ii) *V is uniformly bounded.*

iii) *If u_L and u_R satisfy the Rankine–Hugoniot condition* (1.3),

(3.18) $$f_R - f_L = S(u_R - u_L),$$

then $V = S$.

Proof. i) Combining (3.17a) and (2.19),

(3.19) $$l_{LR} \cdot (u_R - u_L) = P(u_R - u_L) \cdot (u_R - u_L);$$

since P is positive definite, the above quantity is positive for $u_R \neq u_L$.

ii) Use (2.17), (3.17a) and (2.19) to express the numerator of (3.17b), and (3.19) for the denominator; we get

$$V = \frac{P(u_R - u_L) \cdot BP(u_R - u_L)}{(u_R - u_L) \cdot P(u_R - u_L)}.$$

This is a ratio of two quadratic forms and therefore lies between the smallest and largest eigenvalue of $P^{1/2}BP^{1/2}$. These are equal to the eigenvalues a_j of $A = BP$ constructed for Theorem 2.1. In fact, V can be represented as a weighted average of the eigenvalues of A.

iii) Setting (3.18) into (3.17b) gives

$$V = S.$$

This completes the proof of Lemma 3.2.

We now outline two methods for constructing approximate Riemann solvers $w(x/t; u_L, u_R)$ with two intermediate states u_L^* and u_R^* separated by the line $dx/dt = V$; i.e., w is of the form

(3.20) $$w(x/t; u_L, u_R) = \begin{cases} u_L, & x/t < a_L, \\ u_L^*, & a_L < x/t < V, \\ u_R^*, & V < x/t < a_R, \\ u_R, & a_R < x/t. \end{cases}$$

(See Fig. 2.)

The flux across the line $x = st$ for equation (1.1) is defined as

(3.21) $$f_s(u) = f(u) - su.$$

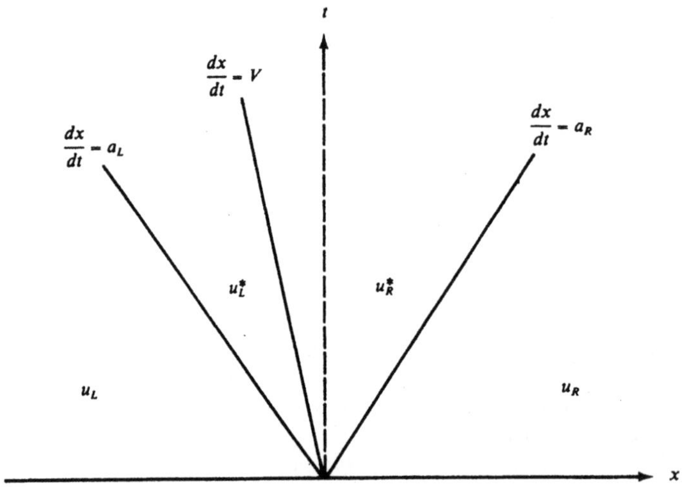

FIG. 2.

We introduce a numerical flux across the line $x = Vt$, denoted as

$$f_V(u_L, u_R).$$

We require consistency with the exact flux:

$$f_V(u, u) = f_V(u) = f(u) - Vu. \tag{3.22}$$

Having introduced flux (3.21) across lines, we can write approximate conservation laws for the triangular regions bounded by $t = \tau$, $x = Vt$, and $x = a_L t$ or $x = a_R t$ respectively:

$$(V - a_L)u_L^* + f_V(u_L, u_R) - f_{a_L}(u_L) = 0 \tag{3.23a}$$

and

$$(a_R - V)u_R^* + f_{a_R}(u_R) - f_V(u_L, u_R) = 0; \tag{3.23b}$$

u_L^* and u_R^* can be determined from (3.23). Clearly, since (3.23a,b) are conservation laws, the resulting scheme (3.20) satisfies the consistency relation (3.7a). Thus requirement a) on conservation is fulfilled.

We turn now to requirement b), the exact resolution of single shocks and contact discontinuities. A shock or contact discontinuity is characterized by the Rankine–Hugoniot condition (3.18) and the entropy condition (1.6d). Using the notation (3.21), these can be written as follows:

$$f_S(u) = f_S(u_R), \qquad F_S(u_L) \geq F_S(u_R) \tag{3.24}$$

where

$$F_S(u) = F(u) - SV. \tag{3.25}$$

We denote by $\hat{f}_V$ a vector function that has the following property:

If $f_V(u_L) = f_V(u_R)$, then

$$\hat{f}_V(u_L, u_R) = f_V(u_L) = f_V(u_R). \tag{3.26}$$

Below we shall specify two distinct ways of constructing such an $\hat{f}_V$. We now set

$$f_V(u_L, u_R) = \hat{f}_V(u_L, u_R) - \beta(u_L, u_R)(u_R - u_L), \tag{3.27}$$

where β has the following property:

$$\beta(u_L, u_R) = 0 \quad \text{when (3.24) holds.} \tag{3.28}$$

We take (3.27) as our numerical flux. It follows from (3.26) that this flux satisfies the consistency condition (3.22), and from (3.26), (3.28) and part iii) of Lemma 3.2 that single shocks and contact discontinuities are resolved exactly.

Here are our choices for $\hat{f}_V$ and β: we define δ_R and δ_L by

$$\delta_R = \frac{a_R - V}{a_R - a_L}, \qquad \delta_L = \frac{V - a_L}{a_R - a_L}. \tag{3.29a}$$

Note that

$$0 \leq \delta_R, \quad 0 \leq \delta_L, \quad \delta_R + \delta_L = 1. \tag{3.29b}$$

Then we define

$$\hat{f}^a_V(u_L, u_R) = \delta_R f_V(u_L) + \delta_L f_V(u_R) \tag{3.30a}$$

and

$$\hat{f}^b_V(u_L, u_R) = f_V(u_{LR}) - f(\delta_L u_L + \delta_R u_R) + \delta_L f_L + \delta_R f_R, \tag{3.30b}$$

where u_{LR} is defined by (3.14).

It can be verified immediately that f^a_V satisfies (3.26). To verify it for f^b_V we note that if $f_V(u_L) = f_V(u_R)$, then $u_{LR} = \delta_R u_R + \delta_L u_1$; setting this into the right side of (3.30b) we see that it equals the right side of (3.30a).

We define β as follows:

$$\beta = C_1\beta_1 + C_2\beta_2 \tag{3.31a}$$

where

(3.31b)
$$\beta_1(u_L, u_R) = [F_V(u_R) - F_V(u_L) - \tfrac{1}{2}(u_L + u_R)\cdot(f_V(u_R) - f_V(u_L))]^+ \|u_R - u_L\|^{-2}$$

where p^+ denotes max $(0, p)$, and

$$\beta_2(u_L, u_R) = (a_R - a_L)^{-1} \|f_V(u_R) - f_V(u_L)\|^2 \|u_R - u_L\|^{-2}. \tag{3.31c}$$

The analysis in [7, §4] shows β_1 is a bounded function.

By construction, $\beta_1 = 0$ when the shock condition (3.24) is satisfied; $\beta_2 = 0$ when the Rankine–Hugoniot condition (3.18) alone is satisfied. Thus our choice of β satisfies (3.28), and so requirement b) is fulfilled.

An analysis similar to that carried out in [7, §4] shows that the positive constants C_1 and C_2 in (3.31a) can be so chosen that the entropy condition (3.7b) is satisfied. Thus is requirement c) fulfilled.

To derive the numerical flux associated with the above Godunov-type scheme we substitute (3.20) into (3.9b), using (3.23) to express u^*_L and u^*_R:

$$\begin{aligned} f_{LR} = \tfrac{1}{2}\{f_L + f_R &+ \gamma_L[f_V(u_L) - f_V(u_L, u_R)] \\ &+ \gamma_R[f_V(u_L, u_R) - f_V(u_R)] - |V|(u_R - u_L)\} \end{aligned} \tag{3.32a}$$

where

$$\gamma_L = \frac{|V| - |a_L|}{V - a_L}, \qquad \gamma_R = \frac{|a_R| - |V|}{a_R - V}. \tag{3.32b}$$

One can easily verify that when $a_L \gtrless 0, f_{LR} = f_R$, and that when $V = 0, f_{LR} = f_V(u_L, u_R)$.

If on the right side of (3.32a) we substitute (3.27) for $f_V(u_L, u_R)$, the following term containing β appears:

$$-\tfrac{1}{2}(\gamma_R - \gamma_L)\beta(u_R - u_L). \tag{3.33a}$$

From (3.32b)

$$\tfrac{1}{2}(\gamma_R - \gamma_L) = \begin{cases} 0 & \text{if } a_L > 0 \text{ or } a_R < 0, \\ \dfrac{|a_L|}{|a_L| + |V|} & \text{if } a_L < 0 < V < a_R, \\ \dfrac{|a_R|}{|a_R| + |V|} & \text{if } a_L < V < 0 < a_R, \end{cases} \tag{3.33b}$$

a nonnegative quantity. This shows that β enters the difference scheme as an artificial viscosity. Note that, unlike classical artificial viscosity, our β is zero across a shock and is positive across an incipient rarefaction wave.

Unlike the previous schemes described in this review, the schemes (3.32) are nonlinear even when applied to linear equations. Thus it is not upstream in the sense of our definition in §2. The decrease of entropy guarantees the L_2 stability of the scheme.

For systems with many components, (3.32) requires less computational effort than either Godunov's or Roe's scheme; more experimentation is needed to see whether its accuracy is comparable.

Schemes of type (3.32) are especially suitable for computation on a moving mesh: we move each meshpoint with velocity V. Such mesh algorithms have been studied in [9].

We remark that any scheme in conservation form (1.7) with a numerical flux $f(u, v)$ that yields perfect resolution of discontinuities but also admits entropy-violating ones may be corrected by modifying its numerical flux to be

$$f(u, v) - C_1\beta(v - u).$$

4. Flux-splitting. In this section we discuss generalizations of the upstream-differencing scheme (2.6b) to nonlinear systems of conservation laws that are based on the flux-splitting

$$f(w) = f^+(w) + f^-(w). \tag{4.1a}$$

We consider schemes in conservation form (1.7a) with the numerical flux

$$f(u, v) = f^+(u) + f^-(v); \tag{4.1b}$$

clearly (4.1a) implies the consistency relation (1.7c). Let us define

$$f^a(w) = f^+(w) - f^-(w), \tag{4.2a}$$

and rewrite (4.1b):

$$f(u, v) = \tfrac{1}{2}[f(u) + f(v) - (f^a(v) - f^a(u))]. \tag{4.2b}$$

Recalling the notation (2.11a) we write

$$d(u, v) = f^a(v) - f^a(u). \tag{4.2c}$$

It is easy to see that (4.1)–(4.2) reduces to (2.6b) in the constant-coefficient case if and only if $f^a(w)$ becomes $|A|w$.

Steger and Warming [22] introduced the notion of flux splitting for the equations of gas dynamics. They took advantage of the fact that in gas dynamics f is a homogeneous function of w of degree one. Then Euler's identity holds:

(4.3a) $$f(w) = A(w)w, \qquad A = f_w.$$

Steger and Warming define

(4.3b) $$f^+(w) = A^+(w)w, \qquad f^-(w) = A^-(w)w,$$

where A^+ and A^- are defined by (2.7), (2.9). Clearly (4.1a) is satisfied; (4.2a) becomes

(4.3c) $$f^a(w) = |A(w)|w,$$

where $|A(w)|$ is defined by (2.7).

Substituting (4.3c) into (4.2c) gives, after rearrangement,

(4.3c′) $$d(u,v) = \tfrac{1}{2}(|A(u)| + |A(v)|)(v-u) + \tfrac{1}{2}(|A(v)| - |A(u)|)(u+v).$$

The right-hand side is of the upstream form (2.11c), except for those nearby values of u and v for which $\operatorname{sgn} a_k(u) \neq \operatorname{sgn} a_k(v)$ for some k. The consequence of this nonsmoothness is a kink in computed solutions near such transitions, e.g. near sonic points. This can be rectified to some extent if one replaces $|A|$ by $\chi(A)$, where $\chi(s)$ is a smooth approximation to $|s|$; see Steger [23] or Harten [5]. Van Leer [15] has derived a smooth flux splitting for the Euler equations without reference to (4.3a).

We describe now a way of splitting when $f(u)$ is not a homogeneous function of u, by introducing a reference state u_0 and applying the mean-value linearization (2.16a) to u and u_0:

(4.4a) $$f(u) - f(u_0) = A(u_0, u)(u - u_0).$$

We consider $u - u_0$ to be a new state variable, and $f(u) - f(u_0)$ to be a new flux function; denote them again by u and $f(u)$, respectively. Thus any solution of the conservation law (1.1) satisfies

(4.4b) $$u_t + [A(u_0, u)u]_x = 0.$$

The new flux

(4.5a) $$f(u) = A(u_0, u)u$$

can be split as

(4.5b) $$f(u) = A^+(u_0, u)u + A^-(u_0 u)u = f^+(u) + f^-(u),$$

exactly as in the homogeneous case (4.3a).

We turn now to a class of upstream schemes that are a natural generalization of Steger and Warming's flux splitting. These schemes are obtained by approximating each conservation law in (1.1a) by a collisionless Boltzmann equation.

Let $\phi(x,t,q)$ be a vector function whose ith component denotes the density of a particle of ith kind at position x and time t, travelling with velocity q. We assume that the particles stream freely, i.e. that ϕ satisfies the collisionless Boltzmann equation:

(4.6a) $$\phi_t + q\phi_x = 0.$$

Note that the densities of the different kinds of particles are completely decoupled.

Using equation (4.6a) we can determine the value of ϕ for $t > 0$ in terms of $\phi_0(x, q) = \phi(x, 0, q)$:

$$\phi(x, t, q) = \phi_0(x - qt, q). \tag{4.6b}$$

We denote by z and g the total density and flux associated with the density ϕ:

$$z(x, t) = \int \phi(x, t, q)\, dq, \tag{4.7a}$$

$$g(x, t) = \int q\phi(x, t, q)\, dq. \tag{4.7b}$$

These satisfy the conservation law obtained by integrating (4.6a) with respect to q:

$$z_t + g_x = 0. \tag{4.8}$$

The initial values of z and g can be obtained by setting $t = 0$ in (4.7):

$$z_0(x) = \int \phi_0(x, q)\, dq, \qquad g_0(x) = \int q\phi_0(x, q)\, dq. \tag{4.9}$$

If the initial values of z and g are equal to those of u and $f(u)$:

$$z_0(x) = u_0(x), \qquad g_0(x) = f(u_0(x)); \tag{4.10}$$

then for t small enough the solution z of (4.8) would be a reasonable approximation to the solution u of (1.1a). We show now how to choose ϕ_0 so that (4.10) holds: we introduce a vector distribution $\mu(q, u)$, depending on a vector parameter u, satisfying

$$\int \mu(q, u)\, dq = u, \tag{4.11a}$$

$$\int q\mu(q, u)\, dq = f(u). \tag{4.11b}$$

Then we simply set

$$\phi_0(x, q) = \mu(q, u_0(x)). \tag{4.12a}$$

Substituting (4.12a) into (4.9) and using (4.11) shows that (4.10) is satisfied.

Equation (4.8) can be solved explicitly; as in the Godunov-type schemes, averages of these explicit solutions will be used to approximate solutions of (1.1a).

Let $v(x)$ be an approximation at t_n to the solution u of (1.1a). We define the initial value ϕ_0 by (4.12a):

$$\phi_0(x, q) = \mu(q, v(x)). \tag{4.12b}$$

Using this in formula (4.6b) for the exact solution of equation (4.6a) gives

$$\phi(x, t, q) = \mu(q, v(x - qt)). \tag{4.12c}$$

Substituting this into (4.7a) results in

$$z(x, t) = \int \mu(q, v(x - qt))\, dq. \tag{4.13}$$

We assume that $v = v^n$ is piecewise constant. We define an approximation v^{n+1} to u at $t_{n+1} = t_n + \tau$ that is piecewise constant on each interval $I_j = (x_{j-1/2}, x_{j+1/2})$ by defining the value v_j^{n+1} of v^{n+1} on I_j to be the average of $z(x, \tau)$ over I_j. Using (4.13) we get

$$v_j^{n+1} = |I_j|^{-1} \int_{I_j} \int \mu(q, v^n(x - q\tau)\, dq\, dx. \tag{4.14}$$

We show next how to express (4.14) as a scheme in conservation form. We integrate (4.6a) over the rectangle $I_j \times (0, \tau)$; we get

$$\int_{I_j} \phi \, dx \Big|_0^\tau + \int_0^\tau q\phi \, dt \Big|_{x_{j-1/2}}^{x_{j-1/2}} = 0.$$

We integrate this with respect to q, obtaining the conservation form

(4.15a) $$|I_j|(v_j^{n+1} - v_j^n) + \tau(f_{j+1/2}^n - f_{j-1/2}^n) = 0,$$

where v_j^{n+1}, v_j^n are the values of v^{n+1}, v^n on I_j, and $f_{j+1/2}^n$ is defined by

$$\tau f_{j+1/2}^n = \int \int_0^\tau q\phi(x_{j+1/2}, t, q) \, dt \, dq,$$

with $f_{j-1/2}^n$ defined similarly. Using (4.12c) to express ϕ on the right-hand side we get

(4.15b) $$\tau f_{j+1/2}^n = \int \int_0^\tau q\mu(q, v^n(x_{j+1/2} - qt)) \, dt \, dq.$$

This is the numerical flux associated with the scheme (4.14).

We show now that this numerical flux is consistent with the flux f in (1.1a). Let's take the case, sure to be satisfied in any scheme of practical significance, that μ has bounded q-support. Then it follows from (4.15b), and the fact that v^n is piecewise constant, that there is an integer N, whose value depends on τ, such that

$$f_{j+1/2}^n = f(v_{j-N+1}^n, \cdots, v_{j+N}^n).$$

Suppose $v_{j+k}^n = v$, $k = -N + 1, \cdots, N$; then $v^n(x_{j+1/2} - qt)$ on the right-hand side of (4.15b) equals v on the q-support of μ; using (4.11b) we see that the right-hand side of (4.15b) equals $\tau f(v)$. This proves consistency.

We turn now to the task of determining the distribution μ. Clearly, since only the first two moments of μ are specified by conditions (4.11), there is a great deal of leeway. For guidance we turn to the linear case,

$$f(u) = Au, \qquad A \text{ constant}.$$

The solution of the linear equation

(4.16a) $$u_t + Au_x = 0,$$

with initial values $u(x, 0) = u_0(x)$, has the form

(4.16b) $$u(x, t) = \sum P_i u_0(x - a_i t).$$

Here a_j are the eigenvalues of A, and P_i is projection onto the line spanned by the right eigenvector R_i. On the other hand, substituting (4.6b) into (4.7a) gives the following expression for z:

(4.17) $$z(x, t) = \int \phi_0(x - qt, q) \, dq.$$

Clearly, comparing (4.16b) and (4.17) we see that

(4.18a) $$z(x, t) \equiv u(x, t)$$

if

(4.18b) $$\phi_0(x, q) = \sum P_i u_0(x) \delta(q - a_i).$$

It is not hard to show that (4.18a) holds for no other choice of ϕ_0 whose support is bounded in x and q. Setting (4.18b) for ϕ_0 into (4.12a) we conclude that μ must be of the form

$$\hat{\mu}(q, u) = \sum \delta(q - a_i) P_i u. \tag{4.19}$$

We turn now to the nonlinear case. As we have shown earlier in this section, the flux can be put in form (4.5a):

$$f(u) = A(u)u$$

where the matrix $A(u)$ has real eigenvalues $a_i(u)$ and a complete set of eigenvectors. According to the spectral theory of matrices,

$$\sum P_i = I, \qquad \sum a_i P_i = A. \tag{4.20}$$

It follows from this that the distribution $\hat{\mu}$ defined by (4.19) satisfies relations (4.11) even when a_1 and P_i are functions of u.

Since relations (4.11) are linear, the most general μ that satisfies (4.11) is of the form

$$\mu(q, u) = \hat{\mu}(q, u) + \eta(q, u), \tag{4.21}$$

where η is any distribution whose first two q-moments are zero for all values of u.

We call schemes of form (4.14) Boltzmann-type; we list some of their properties.

(a) Suppose the support of the distribution μ is contained in $|q| \leqq Q$. Suppose for simplicity that the mesh over which we discretize is uniform, i.e., that each interval I_j has the same length Δ. Then it follows easily that (4.14) is a three-point scheme if τ is chosen so that

$$\tau Q \leqq \Delta. \tag{4.22}$$

For Boltzmann-type schemes, the flux (4.15b) splits naturally into two parts. We define

$$\mu_+(q, u) = \begin{cases} \mu(q, u) & \text{for } q \geqq 0, \\ 0 & \text{for } q < 0, \end{cases}$$

$$\mu_-(q, u) = \begin{cases} 0 & \text{for } q > 0, \\ \mu(q, u) & \text{for } q < 0. \end{cases} \tag{4.23}$$

Clearly $\mu = \mu_+ + \mu_-$; substituting this into (4.15b) we obtain a splitting

$$f_{j+1/2} = f^+_{j+1/2} + f^-_{j+1/2}.$$

If (4.22) holds, $f^+_{j+1/2}$ depends only on v_j, and $f^-_{j+1/2}$ only on v_{j+1}, as in (4.1b).

(b) Suppose f is of the form (4.5a) and μ is chosen to be of form (4.19). The support of $\hat{\mu}$ extends from $a_{\min}$ to $a_{\max}$, so $Q = \max_i |a_i|$, and the restriction (4.22) is the Courant–Friedrichs–Lewy condition. The decomposition of $\tilde{\mu}$ is

$$\hat{\mu}_+ = \sum_{a_i \geqq 0} \delta(q - a_i) P_i u, \qquad \hat{\mu}_- = \sum_{a_i < 0} \delta(q - a_i) P_i u.$$

Substituting this into (4.15b) gives

$$f^+_{j+1/2} = A^+(v_j) v_j \qquad f^-_{j+1/2} = A^-(v_{j+1}) v_{j+1}$$

where A^+ and A^- are the positive and negative parts of the matrix A defined by (2.7) and (2.9). This is the Steger–Warming scheme (4.3).

(c) Consider the equations of compressible flow in Euler coordinates. In this case the three components of ϕ describe the transport of mass, momentum and energy. Since momentum is mass $\times$ velocity, and q is velocity, it is reasonable to stipulate that the second component of ϕ be q times its first component. In view of (4.12a), this would be the case if and only if the same relation holds for the components of μ:

$$\mu^{(2)} = q\mu^{(1)}.$$

We claim that this is true for $\hat{\mu}$ as given by (4.19). For

$$P_i u = w_i R_i$$

and so, by (4.19),

$$\hat{\mu} = \sum \delta(q - a_i) w_i R_i, \tag{4.24a}$$

$$q\hat{\mu} = \sum \delta(q - a_i) a_i w_i R_i. \tag{4.24b}$$

For the equations of compressible flow, the first component $f^{(1)}$ of flux and the second conserved quantity $u^{(2)}$ are equal. This implies that the first row of $A = f_u$ is $(0, 1, 0)$. It follows then from the eigenvalue equation

$$AR_i = a_i R_i$$

that the second component of R_i is a_i times its first component. This shows that the second component of (4.24a) equals the first component of (4.24b), as asserted.

The use of $\hat{\mu}$ as given in (4.19) in a Boltzmann-like difference scheme for the Euler equations goes back to Sanders and Prendergast [21]. Other choices of μ are found in [11], [17], and [18].

(d) We analyze now the stability of Boltzmann-type schemes for linear equations with constant coefficients. In this case we take μ to depend linearly on u,

$$\mu(q, u) = M(q)u, \tag{4.25}$$

M a matrix-valued function. The consistency conditions (4.11) become

$$\int M(q)\, dq = I, \qquad \int qM(q)\, dq = A. \tag{4.26}$$

The scheme (4.13) is

$$z(x, t) = \int M(q) v(x - qt)\, dq.$$

Taking the Fourier transform we get

$$\tilde{z}(\xi, t) = \tilde{M}(t\xi)\tilde{v}(\xi).$$

A condition for stability is that $\tilde{M}(\xi)$ should be power bounded, uniformly for all ξ. Note that there is no stability restriction of the time step t; the reason for this is that Boltzmann-type schemes automatically adjust their domain of dependence.

Note that we have analyzed here the stability of scheme (4.13). The full scheme (4.14) is a combination of (4.13) and projection onto the space of piecewise-constant functions. The latter decreases every weighted L_2 norm. Therefore if (4.13) decreases some weighted L_2 norm, the combined scheme (4.14) is L_2-stable.

For the case (4.19), we have

$$M(q) = \sum \delta(q - a_i)P_i \quad \text{and} \quad \tilde{M}(\xi) = \sum e^{ia_j\xi}P_j.$$

If A is symmetric, the P_j are orthogonal projections, and $\|\tilde{M}(\xi)\| = 1$; so in this case the scheme is stable. For A nonsymmetric, stability can be proved by replacing the Euclidean norm by some matrix-weighted norm.

(e) We have not carried out any stability analysis in the nonlinear case, nor studied the interesting question of how to assure the entropy condition.

(f) We conclude by observing that flux-splitting schemes cannot exactly resolve stationary discontinuities. For suppose that the stationary Rankine–Hugoniot condition $f(u) = f(v)$ is satisfied. It does not follow from this that also

$$f^a(u) = f^a(v). \tag{4.27}$$

But for a split-flux scheme, (4.2b) shows that (4.27) is necessary for the exact resolution of stationary discontinuities.

The split-flux scheme of van Leer [15] for the Euler equations can represent a stationary shock with, at best, one intermediate state, just as the Osher [16] scheme. A contact discontinuity, however, will keep on spreading with the use of any split flux scheme. This follows immediately from the diffusive nature of the underlying Boltzmann model.

REFERENCES

[1] R. Courant, E. Isaacson and M. Rees, *On the solution of nonlinear hyperbolic differential equations*, Comm. Pure Appl. Math., 5 (1952), pp. 243–255.

[2] M. G. Crandall and A. Majda, *Monotone difference approximations for scalar conservation laws*, Math. Comp., 34 (1980), pp. 1–21.

[3] J. Glimm, *Solutions in the large for nonlinear hyperbolic systems of equations*, Comm. Pure Appl. Math., 18 (1965), pp. 697–715.

[4] S. K. Godunov, *A difference scheme for numerical computation of discontinuous solutions of equations of fluid dynamics*, Math. Sbornik, 47 (1959), pp. 271–306. (In Russian.)

[5] A. Harten, *High resolution schemes for hyperbolic conservation laws*, Report DOE/ER/03077-.67, New York Univ., March 1982.

[6] A. Harten, J. M. Hyman and P. D. Lax, *On finite-difference approximations and entropy conditions for shocks*, Comm. Pure Appl. Math., 29 (1976), pp. 297–322.

[7] A. Harten and P. D. Lax, *A random choice finite-difference scheme for hyperbolic conservation laws*, SIAM J. Numer. Anal., 18 (1981), pp. 289–315.

[8] A. Harten, *On the symmetric form of systems of conservation laws with entropy*, ICASE Report 81-34, ICASE, Hampton, VA, 1981.

[9] A. Harten and J. M. Hyman, *A self-adjusting grid for the computation of weak solutions of hyperbolic conservation laws*, Report LA9105, Center for Nonlinear Studies, Theoretical Division, Los Alamos National Lab., Los Alamos, NM, 1981.

[10] L. C. Huang, *Pseudo-unsteady difference schemes for discontinuous solutions of steady-state one-dimensional fluid dynamics problems*, J. Comp. Phys., 42 (1981), pp. 195–211.

[11] S. Kaniel and J. Falcovitz, *Transport approach for compressible flow*, Proc. 4th International IRIA Symposium on Computing Methods in Science and Engineering, Versailles, Dec. 1979, R. Glowinski and P. L. Lions, eds., North Holland, Amsterdam, 1981.

[12] P. D. Lax, *Hyperbolic Systems of Conservation Laws and the Mathematical Theory of Shock Waves*, CBMS Regional Conference Series in Applied Mathematics 11, Society for Industrial and Applied Mathematics, Philadelphia, 1972.

[13] B. VAN LEER, *Towards the ultimate conservative difference scheme.* III, J. Comp. Phys., 23 (1977) pp. 263–275.

[14] ———, *Upwind differencing for hyperbolic systems of conservation laws,* in Numerical Methods for Engineering, Vol. 1, Dunod, Paris, 1980, pp. 137–149.

[15] ———, *Flux-vector splitting for the Euler equations,* presented at the 8th Internat. Conference on Numerical Methods for Engineering, Aachen, June, 1982.

[16] S. OSHER, *Numerical solution of singular perturbation problems and hyperbolic systems of conservation laws,* in Mathematics Studies, 47, North-Holland, Amsterdam, 1981, pp. 179–205.

[17] D. J. PULLIN, *Direct simulation methods for compressible inviscid ideal gas flow,* J. Comp. Phys., 34 (1980), pp. 231–244.

[18] R. D. RIETZ, *One-dimensional compressible gas dynamics calculations using the Boltzmann equation,* J. Comp. Phys., 42 (1981), pp. 108–123.

[19] P. L. ROE, *The use of the Riemann problem in finite-difference schemes,* in Proc. 7th International Conference on Numerical Methods in Fluid Dynamics, Stanford/NASA Ames, June 1980, Lecture Notes in Physics, 141, Springer-Verlag, New York, 1981, pp. 354–359.

[20] ———, *Approximate Riemann solvers, parameter vectors, and difference schemes,* J. Comp. Phys., 43 (1981), pp. 357–372.

[21] R. H. SANDERS AND K. H. PRENDERGAST, *On the origin of the 3-kiloparsec arm,* Ap. J., 188 (1974), pp. 489–500.

[22] J. L. STEGER AND R. F. WARMING, *Flux-vector splitting of the inviscid gas dynamic equations with applications to finite-difference methods,* J. Comp. Phys., 40 (1981), pp. 263–293.

[23] J. L. STEGER, *A preliminary study of relaxation methods for the inviscid conservative gas dynamics equations using flux-vector splitting,* Report 80-4, Flow Simulations, Inc., Sunnyvale, CA, August 1980.

Flux-Vector Splitting for the Euler Equation

Bram van Leer*
Leiden University Observatory
P.O. Box 9513, 2300 RA Leiden, The Netherlands

Introduction. When approximating a hyperbolic system of conservation laws $w_t+\{f(w)\}_x = 0$ with so-called upwind differences, we must, in the first place, establish which way the wind blows. More precisely, we must determine in which direction each of a variety of signals moves through the computational grid. For this purpose, a physical model of the interaction between computational cells is needed; at present two such models are in use.

In one model, neighboring cells interact through discrete, finite-amplitude *waves*. The nature, propagation speed and amplitude of these waves are found by solving, exactly or approximately, Riemann's initial-value problem for the discontinuity at the cell interface. We may call this the Riemann approach (Fig. 1a). The numerical technique of distinguishing between the influence of the forward- and the backward-moving waves is called *flux-difference splitting*; examples are the methods of Roe [4] and of Osher [3].

In the other model, the interaction of neighboring cells is accomplished through mixing of *pseudo-particles* that move in and out of each cell according to a given velocity distribution. We may call this the Boltzmann approach (Fig. 1b). The numerical technique of dis-

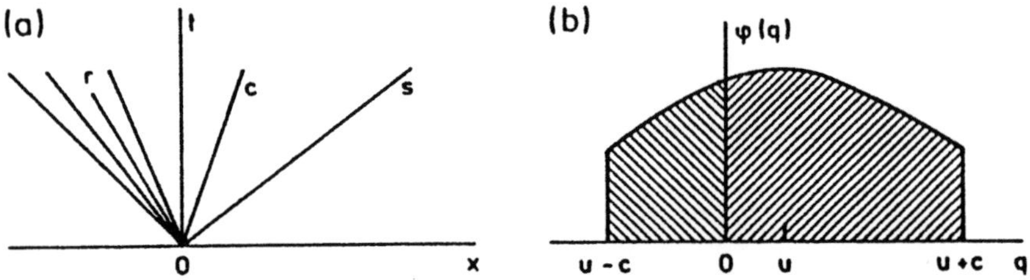

FIGURE 1. Two physical models for upwind differencing. (a) Space-time diagram showing waves in the solution of a Riemann problem: forward shock s and discontinuity c, backward rarefaction r. (b) Velocity-distribution function used in the Boltzmann model. The densities of the forward-moving and backward-moving particles are represented by the differently shaded areas.

*Research was supported under NASA Contract No. NAS1-15810 while the author was in residence at ICASE, NASA Langley Research Center, Hampton, VA 23665.

tinguishing between the influence of the forward- and the backward-moving particles is called *flux-vector splitting* or simply *flux-splitting*; an example is the "beam scheme" of Prendergast [6], rediscovered by Steger and Warming [8].

Both kinds of splitting are discussed by Harten, Lax, and Van Leer [2].

The present paper is restricted to flux-vector splitting for the Euler equations of compressible flow, with the ideal-gas law used as equation of state.

1. Goal. We wish to split the flux $f(w)$ in a forward flux $f^+(w)$ and a backward flux $f^-(w)$, that is,

(1) $f(w) = f^+(w) + f^-(w)$,
(2) df^+/dw must have all eigenvalues ≥ 0,
df^-/dw must have all eigenvalues ≤ 0,

under the following restrictions:

(3) $f^{\pm}(w)$ must be continuous, with
$f^+(w) \equiv f(w)$ for Mach-numbers $M \geq 1$,
$f^-(w) \equiv f(w)$ for Mach-numbers $M \leq -1$,
(4) the components of f^+ and f^- together must mimic the symmetry of f with respect to M (all other state quantities held constant), that is,
$f_k^+(M) = \pm f_k^-(-M)$ if $f_k(M) = \pm f_k(-M)$,
(5) $df^{\pm}/dw$ must be continuous,
(6) $df^{\pm}/dw$ must have one eigenvalue vanish for $|M| < 1$,
(7) $f^{\pm}(M)$, like $f(M)$, must be a polynomial in M, and of the lowest possible degree.

Restriction (3) ensures that in supersonic regions flux-vector splitting leads to standard upwind differencing. Restrictions (4) and (5) are self-evident, although (5) was not satisfied in [8], with negative consequences for the smoothness of numerical results. Restriction (6) is crucial, greatly narrowing down the choice of functions. The degeneracy of $f^{\pm}(w)$ in subsonic cells makes it possible to build stationary shock structures with no more than two interior zones, just as does the flux-differencing splitting technique in [3]. Finally, requirement (7) makes the splitting unique.

2. Derivation. Consider the one-dimensional Euler equations. We shall regard the full, forward, and backward fluxes as functions of density, ρ, sound speed c and Mach number M. The full flux reads

$$(8) \quad f(\rho, c, M) = \begin{pmatrix} \rho c M \\ \rho c^2 (M^2 + \frac{1}{\gamma}) \\ \rho c^3 M (\frac{1}{2} M^2 + \frac{1}{\gamma - 1}) \end{pmatrix},$$

where γ is the specific-heat ratio.

From conditions (3) and (5) it follows that $f^+(\rho, c, M)$ as well as $\vartheta f^+(\rho, c, M)/\vartheta M$ must vanish for $M \downarrow -1$, while $f^-(\rho, c, M)$ and $\vartheta f^-(\rho, c, M)/\vartheta M$ must vanish for $M \uparrow 1$. Condition (7) then leads to the restriction that f^+ includes a factor $(M+1)^2$ and f^- a factor $(-M+1)^2$, for $|M| < 1$. Without introducing further factors depending on M we can now achieve the splitting of the mass flux:

$$(9) \quad \rho c M \equiv \rho c \{\tfrac{1}{2}(M+1)\}^2 - \rho c \{\tfrac{1}{2}(-M+1)\}^2, \quad |M| < 1,$$

satisfying (1), (3), (4), (5) and (7). In order to split the momentum flux in agreement with (3) and (5) we need cubic polynomials in M:

$$(10) \quad \begin{aligned} \rho c^2 (M^2 + \tfrac{1}{\gamma}) \equiv {} & \rho c^2 \{\tfrac{1}{2}(M+1)\}^2 (\tfrac{\gamma-1}{\gamma} M + \tfrac{2}{\gamma}) \\ & + \rho c^2 \{\tfrac{1}{2}(-M+1)\}^2 (-\tfrac{\gamma-1}{\gamma} M + \tfrac{2}{\gamma}), \qquad |M| < 1; \end{aligned}$$

again, (1), (3), (4), (5) and (7) are satisfied.

The splitting of the energy flux can now be achieved by combining the split mass- and momentum-fluxes:

$$(11) \quad f^{\pm}_{energy} = \frac{\gamma^2}{2(\gamma^2 - 1)} (f^{\pm}_{momentum})^2 / f^{\pm}_{mass}, \quad |M| < 1.$$

The scale factor is needed to satisfy (3). The relation (11) between the components of $f^{\pm}$ makes these fluxes degenerate: $df^{\pm}(w)/dw$ has a zero eigenvalue for $|M| < 1$. Thus condition (6) is fulfilled. Moreover, since the vanishing eigenvalue is continuous for $M = \pm 1$, we have sufficient information about the smoothness of $f^{\pm}$ to conclude that condition (5) is fully satisfied. Testing the fulfillment of (7), (4) and (1) by (11) is trivial.

We still must determine if condition (2) is satisfied by the splitting (9), (10), (11). There is a good reason to believe this is indeed the case: the eigenvalue $\mu_1^{\pm}$ of $df^{\pm}/dw$ that is most likely to violate condition (2), i.e. to have the wrong sign, has been forced to vanish for $|M| < 1$. We find that the non-zero eigenvalues $\mu_{2,3}^{+}$ of df^{+}/dw are the roots of the quadratic equation:

conserved quantity	*x-flux f*	*forward x-flux f^+, $\lvert M\rvert < 1$*
mass	ρu	$\rho c\{\frac{1}{2}(M+1)\}^2$
x-momentum	$\rho(u^2+c^2/\gamma)$	$f^+_{mass}\cdot\{(\gamma-1)u+2c\}/\gamma$
y-momentum	ρuv	$f^+_{mass}\cdot v$
z-momentum	ρuw	$f^+_{mass}\cdot w$
total energy	$\rho u\{\frac{1}{2}(u^2+v^2+w^2)$ $+c^2/(\gamma-1)\}$	$f^+_{mass}\cdot[\{(\gamma-1)u+2c\}^2/\{2(\gamma^2-1)\}$ $+\frac{1}{2}(v^2+w^2)]$

Table I. Flux-splitting for the Euler equations. $M = u/c$.

conserved quantity	*x-flux f*	*forward x-flux f^2, $\lvert M_*\rvert < 1$*
mass	ρu	$\rho c_*\{\frac{1}{2}(M_*+1)\}^2$
x-momentum	$\rho\{(\gamma+1)/(2\gamma)\}\cdot$ $(u^2+c_*^2)$	$f^+_{mass}\cdot\{(\gamma+1)/\gamma\}c_*$
y-momentum	ρuv	$f^+_{mass}\cdot v$
z-momentum	ρuw	$f^+_{mass}\cdot w$

Table II. Flux-splitting for the isenthalpic Euler equations.
$H \equiv$ enthalpy, $c_* = \sqrt{[\{2(\gamma-1)/(\gamma+1)\}\{H-\frac{1}{2}(v^2+w^2)\}]}$, $M_* = u/c_*$.

$$(12).\quad (\mu^+)^2 - c\mu^+\cdot\tfrac{3}{2}(1+M)[1-\tfrac{\gamma-1}{12\gamma(\gamma+1)}(M-1)\{\gamma(M-1)^2+ 2\gamma(M-1)-2(\gamma+3)\}] + c^2\cdot\tfrac{1}{4}(1+M)^3[1-\tfrac{M-1}{8\gamma(\gamma+1)}\cdot \{4\gamma(\gamma-1)(M-1)+(\gamma+1)(3-\gamma)\}] = 0, \quad |M|<1.$$

In the relevant range $1 \le \gamma \le 3$, the roots of (12) are positive. The negativity of $\mu^-_{2,3}$ follows by symmetry.

This completes the derivation of the split fluxes for the one-dimensional Euler equations. The formulas for the three-dimensional equations are given in Table I (full equations) and Table II (constant enthalpy assumed).

3. Stability. The stability analysis for the first-order upwind scheme based on the above split fluxes is complicated by the fact that df^+/dw and df^-/dw commute neither with each other nor with df/dw, for $|M| < 1$. This leads to a reduction of the CFL limit; in the worst case, $M = 0$, we find for the shortest waves:

$$(13)\quad \tfrac{\Delta t}{\Delta x}c \le 2\gamma/(\gamma+3).$$

A practical local stability criterion is

$$(14)\quad \tfrac{\Delta t}{\Delta x}(|u|+c) \le \{2\gamma+|M|(3-\gamma)\}/(\gamma+3), \quad |M|<1.$$

4. Steady discontinuities. The degeneracy of $f^{\pm}$ for $|M| < 1$ makes it possible to build numerical shock structures that satisfy the steady upwind-difference equations using only two interior zones.

Consider the one-dimensional Euler equations. We denote the supersonic pre-shock state by L, the subsonic post-shock state by R and the interior states by P and Q, as in Fig. 2. To require stationarity means to require constancy of net flux:

$$(15) \quad \begin{aligned} f_R &= f_L \\ &= f_L^+ + f_P^- \\ &= f_P^+ + f_Q^- \\ &= f_Q^+ + f_R^-. \end{aligned}$$

Assume that the first equality, i.e., the jump condition across the full shock structure, holds. The second equality is automatically satisfied if zone P, like L, is supersonic, with $f_P^- = 0, f_L^+ = f_L$. The fourth equality boils down to

$$(16) \quad f_R^+ = f_Q^+,$$

which, if zone Q is subsonic, implies only 2 independent equations. The third equality stands for 3 equations, so that we end up with 5 equations for 6 unknowns, the components of w_P and w_Q. The

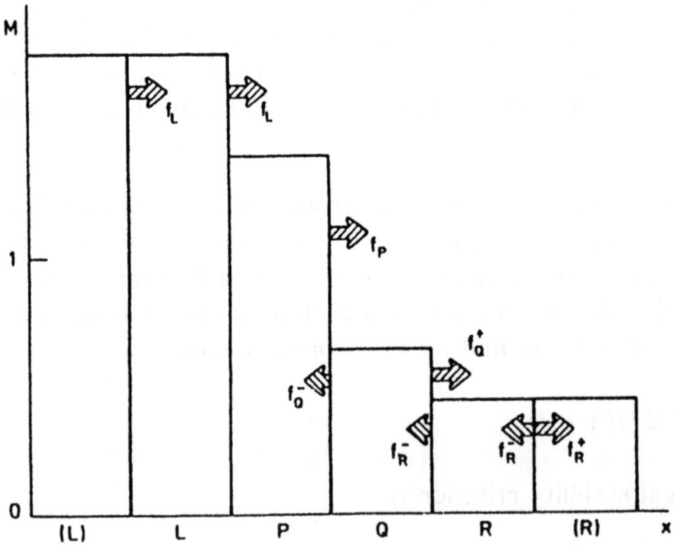

FIGURE 2. Steady shock profile from a flux-split upwind scheme.

solutions form a one-parameter family of steady shock profiles, the parameter relating to the sub-grid shock position. There exists one profile with only one interior zone that is precisely sonic. In contrast, Godunov's and Roe's schemes yield steady shocks with one or no interior zone.

Another scheme that yields steady shock structures with two interior states is Osher's flux-differencing scheme [3]. For a scalar conservation law it boils down to a flux-splitting method.

Unlike Osher's scheme, the present flux-vector splitting cannot preserve a stationary contact discontinuity; in fact, no flux-vector splitting scheme can (see [2]). This is readily understood from the underlying physical model: cell-interactions are achieved through mixing, a diffusive process. Present research is aimed at the development of a computationally simple "collision term" that could prevent the diffusion across a steady contact discontinuity.

5. A comparison. For one-dimensional isothermal flow ($\gamma = 1$, constant c) the flux-vector splitting of Steger and Warming, alias the beam scheme, can be derived from the assumption that there are two kinds of gas particles, equally abundant but moving at different rates: the beams. Per unit volume, $\frac{1}{2}\rho$ moves with velocity $u - c$ and $\frac{1}{2}p$ with velocity $u + c$, yielding the correct average flow speed u. For $|M| < 1$ there is a forward and a backward beam, with associated fluxes satisfying (1), (3) and (4):

$$(17) \quad f^{\pm}_{B/SW} = \begin{pmatrix} \frac{1}{2}\rho(u \pm c) \\ \frac{1}{2}\rho(u \pm c)^2 \end{pmatrix}, \quad |u| < c.$$

In Fig. 3a the dependence of the components of $f_{B/SW}$ on M is shown. The momentum flux is continuously differentiable at $M = \pm 1$; the mass flux is not. The splitting given by Eqs. (9) and (10) for $\gamma = 1$ reads

$$(18) \quad f^{\pm}_{VL} = \begin{pmatrix} \pm\frac{1}{4}\rho(u \pm c)^2/c \\ \frac{1}{2}\rho(u \pm c)^2 \end{pmatrix}, \quad |u| < c,$$

with the same momentum-flux splitting but improved mass-flux splitting. The dependence on M is shown in Fig. 3b. The Figs. 4a and 4b are graphs of the eigenvalues of $df^{\pm}_{B/SW}/dw$ and $df^{\pm}_{VL}/dw$. Their values are given by

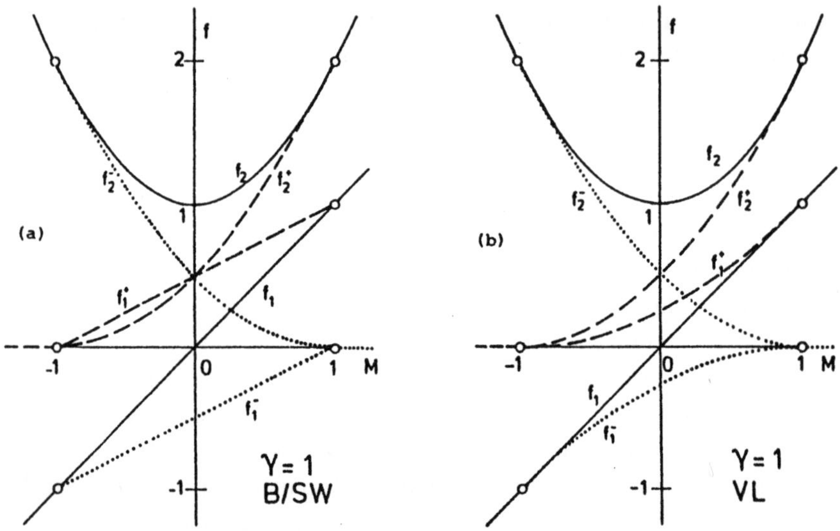

FIGURE 3. Flux-splitting for the isothermal Euler equations ($\gamma = 1$). Plotted against Mach number are mass fluxes (subscript 1) and momentum fluxes (subscript 2), normalized by ρc and ρc^2, respectively. (a) Beam scheme/Steger and Warming; (b) Van Leer.

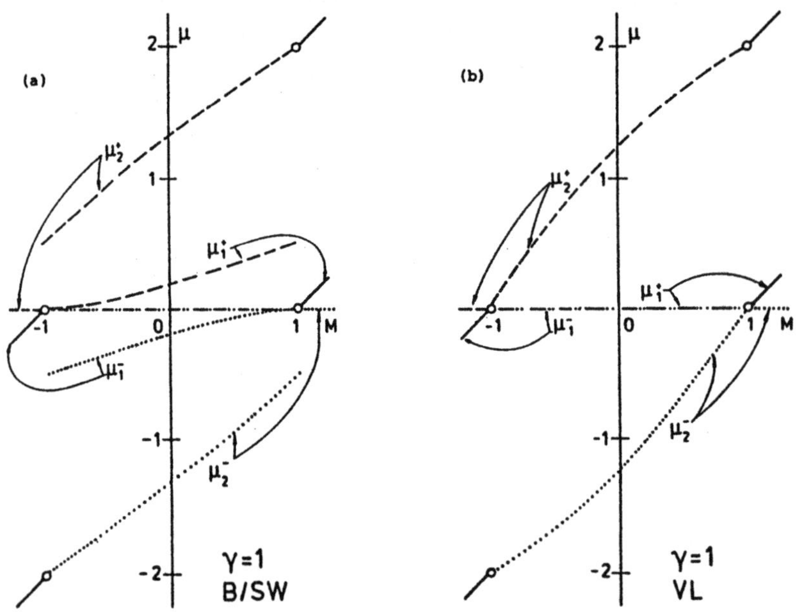

FIGURE 4. Eigenvalues of split-flux derivatives ($\gamma = 1$). Plotted against Mach number are the eigenvalues of df^+/dw and df^-/dw, normalized by c. (a) Beam scheme/Steger and Warming; (b) Van Leer.

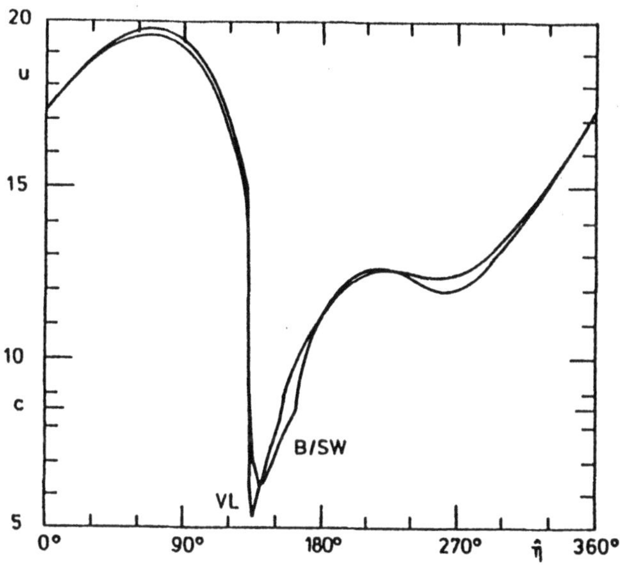

FIGURE 5. Numerical solutions of the periodic, nozzle-type, cosmic flow problem ($\gamma = 1$) from [9] by flux-split upwind schemes, on a 128-zone grid. Plotted is velocity against coordinate angle. $B/SW \equiv$ steady solution by Beam scheme/Steger and Warming; $VL \equiv$ steady solution by Van Leer.

$$
(19) \quad \begin{aligned} (\mu_{1,2}^{+})_{B/SW} &= \tfrac{1}{4}c\{3 + 2M \mp \sqrt{(5 + 4M)}\}, \\ (\mu_1^{+})_{VL} &= 0, \\ (\mu_2^{+})_{VL} &= \tfrac{1}{4}c(5 - M)(1 + M), \end{aligned} \qquad |M| < 1,
$$

and the symmetry condition (4). The eigenvalues shown in Fig. 4a, while satisfying (2), have a discontinuity at $M = +1$ or -1. In a stationary numerical solution this causes a discontinuity in the gradient at the sonic point, as seen from Fig. 5, curve B/SW. Correcting the mass-flux splitting performs a small miracle, as seen from curve VL. The gradient-discontinuity disappears, and the numerical diffusion in the subsonic region is substantially reduced. This reduction corresponds to the reduction of the eigenvalues $\mu_1^{\pm}$ to zero.

6. Conclusions. For the full or isenthalpic Euler equations combined with the ideal-gas law, the flux-vector splitting presented here is, by a great margin, the simplest means to implement upwind differencing. For a polytropic gas law, with $\gamma > 1$, closed formulas have not yet been derived.

The scheme produces steady shock profiles with two interior zones. There is evidence [1] that, among *implicit* versions of upwind methods, those with a two-zone steady-shock representation give faster convergence to a steady solution than those with a one-zone representation.

A disadvantage in using any flux-vector splitting is that it leads to numerical diffusion of a contact discontinuity at rest. This diffusion can be removed; present research is aimed at achieving this with minimal computational effort.

Numerical solutions by first- and second-order schemes including the above split fluxes can be found in Refs. [7], [9] (one-dimensional) and [10], [5] (two-dimensional).

REFERENCES

1. Goorjian, P.M. and R. van Buskirk, *Implicit calculations of transonic flow using monotone methods*, AIAA Paper 81-331, (1981).

2. Harten, A., P.D. Lax and B. van Leer, *Upstream differencing and Godunov-type schemes*, ICASE Report No. 82-5; to appear in SIAM Review.

3. Osher, S., *Numerical solution of singular perturbation problems and hyperbolic systems of conservation laws*, North-Holland Mathematical Studies, 47 (1981), p. 179.

4. Roe, P.L., J. Computational Phys., 43 (1981), p. 357.

5. Salas, M.D., *Recent developments in transonic flow over a circular cylinder*, NASA Technical Memorandum 83282, (April 1982).

6. Sanders, R.H. and K.H. Prendergast, Astrophys. J., 188 (1974), p. 489.

7. Steger, J.L., *A preliminary study of relaxation methods for the inviscid conservative gasdynamics equations using flux-vector splitting*, Report No. 80-4, Flow Simulations, Inc. (August 1980).

8. Van Albada, G.D., B. van Leer and W.W. Roberts, Jr., *A comparative study of computational methods in cosmic gas dynamics*, Astron. Astrophys., 108 (1982), p. 76.

9. Van Albada, G.D. and W.W. Roberts, Jr., Astrophys. J., 246 91981), p. 740.

Chapter II

Second-Order TVD Schemes

Annotation

Once the basic ingredients of upwind schemes were understood, attention turned to achieving second- and higher-order accuracy in these schemes. How to satisfy the entropy condition remained a favorite topic.

The construction of higher-order entropy-satisfying TVD schemes was the work of many people, both at ICASE and elsewhere. The papers collected in this chapter give the flavor of this research, which was important both in itself and as groundwork for the further developments covered by Chapters III and IV.

1. **A Comparative Study of Computational Methods in Cosmic Gas Dynamics** (G. D. van Albada, B. van Leer and W. W. Roberts, Jr.)
   Offers a numerical comparison of first- and second-order schemes based on a quasi-one-dimensional galactic flow problem. The much-used Van Albada limiter is introduced here; the paper also is the only published reference to S. Hancock's predictor-corrector scheme.

2. **High Resolution Schemes and the Entropy Condition** (S. Osher and S. Chakravarthy)
   A thorough treatment of the construction of high-resolution schemes satisfying the entropy condition, based on Osher's flux formula.

3. **Convergence of Generalized MUSCL Schemes** (S. Osher)
   Gives a convergence proof for second-order MUSCL-type schemes, as applied to scalar convex conservation laws.

4. **A Simplified TVD Finite Difference Scheme Via Artificial Viscosity** (S. F. Davis)
   Separates the TVD concept from upwind differencing by showing that one can construct TVD schemes using central differences and appropriate artificial viscosities. The motivation is computational economy.

5. **On Central-Difference and Upwind Schemes** (R. C. Swanson and E. Turkel)
   This paper represents the opposite of the paper by Davis, namely, a trend toward greater accuracy achieved by greater computational expense. It shows that, by introducing matrix viscosities in central difference schemes, one can reap most of the benefits of upwind schemes, without having to make structural changes to the code.

A Comparative Study of Computational Methods in Cosmic Gas Dynamics

G. D. van Albada[1,*], B. van Leer[2,**], and W. W. Roberts, Jr.[3,***]

[1] The National Radio Astronomy Observatory, Charlottesville, VA, USA and Kapteyn Laboratory, Groningen, The Netherlands
[2] Institute for Computer Applications in Science and Engineering NASA Langley Research Center, Hampton, VA, and Leiden State University, Leiden, The Netherlands
[3] Institute for Computer Applications in Science and Engineering NASA Langley Research Center, Hampton, VA, and The University of Virginia, Charlottesville, VA

Received August 25, accepted December 11, 1981

Summary. In search of reliable computational methods for cosmic flow problems, we apply several commonly used algorithms and one new algorithm, to a representative problem in galactic gas dynamics. A careful choice of the algorithm used in a calculation is found to be of the utmost importance in obtaining reliable results. Two methods most commonly employed in astronomy (the Beam scheme and FCT methods) prove to be highly unsuitable for our test problem. The penalty in programming effort and computer time per grid point required for the best second-order accurate codes tested is more than offset by the improvement in accuracy obtained and the possibility to reduce the number of points in a grid.

Key words: hydrodynamics – hydromagnetics – interstellar matter – spiral galaxies

I. Introduction

Many theoretical investigations of astrophysical fluid flows require extensive numerical calculations. The techniques of computational fluid dynamics have been employed in virtually every area from stellar structure to cosmology. The use of numerical calculations in astrophysics can only increase as the capacity to perform substantial calculations becomes more widespread. Indeed, the computation of two-dimensional flows may be considered almost routine. Thus, increasing numbers of astronomers will be asking that basic question of scientific computing: What reliable, accurate, efficient, and easy-to-program method should be used for this calculation?

The average astronomer with only a casual knowledge of numerical methods is likely to confine his choices to methods that have been used before on problems in his particular field of research and to methods described in standard textbooks on numerical analysis. In the former case he is likely to be exposed to a one-sided defense of a single method. In the latter case he is unlikely to encounter any comparison between methods on a physical problem that embodies some of the special difficulties posed by astrophysical flows.

The objective of this article is to supply a comparison of a variety of numerical methods on a representative astrophysical flow problem. The methods considered include some that are widely used in astronomy, some that are widely used in other fields such as aerodynamics, and one that is rather new. All the methods are explicit and are especially suited for transonic and supersonic flows. This comparison aims to acquaint astronomers with the virtues and failings of typical numerical methods.

II. The Problem

A simple, one-dimensional model of the gas flow in a spiral galaxy will serve as the test problem. Numerical calculations of the nonlinear response of the gas to a mild spiral structure in the more massive stellar component of a disk galaxy have played a major role in establishing the density-wave theory of Lin and Shu (1964, 1966) as a viable explanation for the coherent, large-scale spiral patterns observed in many galaxies. [Recent reviews of this topic have been given by Toomre (1977) and by Lin and Lau (1979).] Although the dynamics of spiral galaxies are dominated by the far more massive (and much hotter) stellar component, their appearance is largely a result of the fierce response of the (cold) gas to the stellar gravitational field. The ease with which a mild stellar spiral structure can induce shock waves in the gas, and its implications for the observed features of spiral structure, was demonstrated by Roberts (1969) using one-dimensional, steady-state gas equations which included a forcing term due to the spiral field of the stars. The actual evolution of such flows was studied by Woodward (1975) using a simplified time-dependent version of Roberts' equations. We use Woodward's equations and also a set of his parameter values.

The nonlinear response of the gas to an imposed spiral gravitational field has several of the distinguishing characteristics of astrophysical flows. A major role is played by source terms, strong shocks are apt to develop, and rotational effects are significant. The presence of source terms can lead to unexpected behavior in numerical methods which are usually analyzed and tested in the absence of such terms. Strong shocks demand reliability; the proper treatment of angular momentum requires accuracy. Many methods cope with the shock only at the price of

---

Send offprint requests to: G. D. van Albada (Groningen)

* The National Radio Astronomy Observatory is operated by Associated Universities, Inc., under contract with the National Science Foundation

** This work was supported under NASA Contract No. NAS1-14472 while the author was in residence at ICASE, NASA Langley Research Center, Hampton, VA 23665

*** This work was supported in part by the National Science Foundation under Grant AST-7909935; partial support was also received under NASA Contract No. NAS1-15810 while the author was in residence at ICASE, NASA Langley Research Center, Hampton, VA artificially redistributing the angular momentum. This can be a serious problem if one is investigating the dynamics of the gas, since the physical system can be very sensitive to redistribution of angular momentum.

The equations in an inertial frame for an isothermal gas are

$$\frac{\partial \varrho}{\partial t}+\nabla\cdot(\varrho \boldsymbol{q})=0, \tag{1}$$

$$\frac{\partial \boldsymbol{q}}{\partial t}+\boldsymbol{q}\cdot\nabla\boldsymbol{q}=-\frac{c^2}{\varrho}\nabla\varrho-\nabla\Phi, \tag{2}$$

where ϱ is the density, $\boldsymbol{q}$ is the velocity, c is the (constant) sound speed and Φ is the gravitational potential. The isothermal assumption is used since interstellar gas cools by radiative processes on a much shorter time-scale than that of any dynamical processes. It is customary to use for c an equivalent dispersive speed which is partly thermal, partly due to turbulence, and partly due to cosmic ray particles.

In the absence of the spiral forcing the gas flow is circular with angular velocity $\Omega(r)$ at radius r. A steady spiral field with small pitch angle α is assumed to rotate rigidly with pattern speed Ω_p. A convenient coordinate system is one which rotates at this speed and is aligned with the equipotential contours of the spiral. The coordinates parallel and perpendicular to the equipotential contours are denoted by ξ and η, respectively. The velocity components in this frame are written as

$$v=q_\xi,\quad u=q_\eta. \tag{3}$$

If we assume that the spiral has a pitch angle $\alpha \ll 1$, the equilibrium velocities are approximately

$$v_0=r(\Omega-\Omega_p),\quad u_0=\alpha r(\Omega-\Omega_p). \tag{4}$$

In this approximation derivatives with respect to η (normal to the spiral arms) are retained, but derivatives with respect to ξ (along the spiral arms) are discarded. For a two-armed spiral the resulting equations can be written as the system of conservation laws

$$\frac{\partial U}{\partial t}+\frac{\partial F}{\partial \eta}=H, \tag{5}$$

where the vector of conserved quantities is

$$U=\begin{pmatrix}\varrho\\ \varrho u\\ \varrho v\end{pmatrix}, \tag{6}$$

the vector of fluxes is

$$F=\begin{pmatrix}\varrho u\\ \varrho(u^2+c^2)\\ \varrho uv\end{pmatrix}, \tag{7}$$

and the vector of source terms is

$$H=\begin{pmatrix}0\\ 2\Omega(v-v_0)\varrho+\dfrac{2}{\alpha r}\varrho A\sin\hat{\eta}\\ -\dfrac{\kappa^2}{2\Omega}(u-u_0)\varrho\end{pmatrix} \tag{8}$$

The spiral phase $\hat{\eta}$ is defined by

$$\hat{\eta}=\frac{2\eta}{\alpha r}, \tag{9}$$

and the epicyclic frequency κ by

$$\kappa^2=\frac{2\Omega}{r}\frac{d}{dr}(r^2\Omega). \tag{10}$$

In this approximation the flow is periodic; in terms of the spiral phase the periodicity condition reads

$$U(\hat{\eta},t)=U(\hat{\eta}+2\pi,t). \tag{11}$$

The driving term $(2/\alpha r)\varrho A\sin\hat{\eta}$ arises from the assumed gravitational field of the stellar component. For the test problem we adopt the parameters thought to be appropriate for the neighborhood of the Sun in our own galaxy: $\Omega=25$ km s^{-1}/kpc, $\kappa=31.3$ km s^{-1}/kpc, $\Omega_p=13.5$ km s^{-1}/kpc, $c=8.56$ km s^{-1}, $r=10$ kpc, and $\alpha=\sin(6\overset{\circ}{.}7)\approx 0.11667$. For the amplitude A we choose $A=72.92$ (km s^{-1})^2, which makes the amplitude of the spiral force 2% of the equilibrium force $r\Omega^2$.

In the steady state the Eqs. (5) become

$$\frac{du}{d\eta}=\frac{u}{u^2-c^2}\left[2\Omega(v-v_0)+\frac{2A}{\alpha r}\sin\frac{2\eta}{\alpha r}\right], \tag{12}$$

$$\frac{dv}{d\eta}=-\frac{\kappa^2}{2\Omega}\frac{u-u_0}{u}. \tag{13}$$

The particular case studied here becomes supersonic with a sonic point at spiral phase $\hat{\eta}=155\overset{\circ}{.}53$ and a shock at $\hat{\eta}=131\overset{\circ}{.}68$. A procedure for solving the steady state Eqs. (12) and (13) plus the periodicity condition (11) is described in Roberts (1969); see also Shu et al. (1973). The noteworthy features of this flow are the rapid decompression after the shock and the secondary structure near a spiral phase of 270° which is caused by resonance effects. The time-dependent version of this problem challenges a numerical method to cope with the shock, while also resolving the remaining structure of the flow.

III. The Methods

A typical numerical method for the system (5) divides the spatial region $(0, \pi\alpha r)$ into N zones centered at the grid points $=(i-\frac{1}{2})\Delta\eta$, where $i=1,2,\ldots,N$ and $\Delta\eta=\pi\alpha r/N$ (or $\Delta\hat{\eta}=2\pi/N$), and advances the approximate solution from time t_n to time t_{n+1} (where $t_{n+1}=t_n+\Delta t$) by means of a discretized version of the partial differential equations. The approximate value of U at the point (η_i, t_n) is denoted by U_i^n; F_i^n and H_i^n are defined as $F(U_i^n)$ and $H(U_i^n,\eta_i)$ or, as explained below, $H(U_i^n,\eta_i,\Delta t)$. In all methods discussed here, U_i^n actually approximates the average value of the solution over zone i. A subscript $i+\frac{1}{2}$ denotes an interpolated numerical value at the zone boundary $\eta=i\Delta\eta$, or a finite difference across this boundary; a superscript $n+\frac{1}{2}$ denotes an approximate value at $t_{n+1/2}=(n+\frac{1}{2})\Delta t$.

Our sample of numerical methods includes

a) The beam scheme (B) [see Sanders and Prendergast (1974)].

b) Godunov's (1959) method (G).

c) Second-order flux-splitting method (FS2) [see Van Leer (1981a)].

d) MacCormack's (1969) method (MC2).

e) Flux Corrected Transport (FCT) methods of Boris and Book (1973).

Before discussing these methods at length, some general remarks are due. The methods (a) and (b) are first-order accurate, that is, these approximate Eq. (5) with an error $O(\Delta x)$. The remaining methods are of the second order of accuracy. All first-order methods and also method (c) are based on upwind differencing. This means that, in approximating $\partial F/\partial\eta$ in (5), a distinction is made between contributions from wave motion or material motion in the positive direction and in the negative direction [see the review by Harten et al. (1981)]. The methods (d) and (e) are based on central differencing, in which the above distinction is not made.

The schemes can be written in the form

$$\frac{U_i^{n+1}-U_i^n}{\Delta t}+\frac{\Phi_{i+1/2}^{\nu}-\Phi_{i-1/2}^{\nu}}{\Delta\eta}=H_i^{n+1/2}, \tag{14}$$

with $\nu=n$ for the first-order methods and $\nu=n+\frac{1}{2}$ for the higher order methods. The latter all are two-step algorithms in which time centering is achieved using the first-order accurate results of the first step at $t_{n+1/2}$ or t_{n+1}.

The numerical flux vector

$$\Phi_{i+1/2}^{\nu}\equiv\Phi(U_{i-k+1}^n,\ldots,U_{i+k}^n), \tag{15}$$

is a function of $2k$ arguments (initial values); for the time-step in question it is a representative value of F at $\eta_{i+1/2}$ according to some model of the interaction of the gas cells. It is the particular choice of Φ that distinguishes one scheme from another.

Note that (14) approximates (5) in the so-called "conservation form", that is, total differentials are approximated by perfect differences. This allows the use of the approximation in regions where the flow is discontinuous; see Lax and Wendroff (1960).

The methods usually are stable under the Courant-Friedrichs-Lewy condition, which says that the largest radial wave or material speed in a cell must not exceed the numerical signal speed $\Delta\eta/\Delta t$.

In programming methods (a), (b), and (c), the effect of the source term was accounted for in separate steps, that is, Eq. (5) was approximated by starting out with integrating the equation

$$\frac{\partial U}{\partial t}=H, \tag{16}$$

over a half time-step, then continuing by integrating

$$\frac{\partial U}{\partial t}+\frac{\partial F}{\partial\eta}=0, \tag{17}$$

over a full time-step, and finishing off with integrating Eq. (16) over a half time-step. This procedure is convenient in that it allows a very flexible program structure.

Another advantage of this method is that Eq. (16) can often be integrated accurately by itself. For the current problem the analytic solution of (16) is

$$\varrho(t_n+\tau,\eta_i)=\varrho_i^n,$$

$$u(t_n+\tau,\eta_i)-u_0=(u_i^n-u_0)\cos(\kappa\tau)+\frac{2\Omega}{\kappa}\left(v_i^n-v_0+\frac{A}{\alpha r\Omega}\sin\hat{\eta}\right)\sin(\kappa\tau), \tag{18}$$

$$v(t_n+\tau,\eta_i)-v_0=\frac{-A}{\alpha r\Omega}\sin\hat{\eta}+\left(v_i^n-v_0+\frac{A}{\alpha r\Omega}\sin\hat{\eta}\right)\cos(\kappa\tau)-\frac{\kappa}{2\Omega}(u_i^n-u_0)\sin(\kappa\tau).$$

Use of such an accurate solution proved essential for maintaining stability in very long runs (over 2000 time-steps). The cause of the instability that arises otherwise is the same that would make a linear first-order algorithm for integrating (16) unstable: instead of choosing $(u(t_n+\tau), v(t_n+\tau))$ on the ellipse given by (18), the linearized version will put it on the tangent to that ellipse, thus always leading to an amplification of the disturbance.

The exact solution to the test problem is known reliably only in the steady-state limit. Thus, the numerical methods cannot be evaluated precisely on how well they calculate the time evolution of the flow. We therefore restrict ourselves to testing the methods on the accuracy of the steady state they produce. Note that evaluating the methods on the time required for them to reach the steady-state starting, say, from uniform initial values $(\varrho,u,v)_i^0=(1,u_0,v_0)$, is unfair because the better methods will fare the worst. Because of the periodicity of the flow any transients will persist until they are damped out by the numerical viscosity, which is highest for the least accurate schemes. It seems to us that the fairest test is to use the exact steady state solution itself as the initial-value distribution, and compare how well the various methods preserve it. This constitutes the first test performed. In keeping with the spirit of the methods, the grid values U_i^0 employed are not the point-values of the steady-state but rather the zone-averaged values. Two methods that performed well in this test were applied to the problem with uniform initial values, mainly to determine their "robustness". This constitutes the second test.

The results presented are based on a computational grid of 64 zones; the numerical solutions for all methods are advanced by 1200 time-steps from the exact steady-state with a constant time-step corresponding initially to a global Courant number of 0.5. In this span the fastest moving signals can traverse the computational domain about 10 times. This is a reasonable amount of time to allow the computed solution to adjust towards the steady-state of the difference equations.

Many additional experiments were run with grid sizes of 16–128 zones, a variety of Courant numbers and a maximum number of time steps well over 10,000. These mainly served to check the consistency of our findings or to examine the long-term stability of the numerical solution.

The computed solutions of the first test are displayed in Figs. 1–6, while root-mean-square (rms) errors are listed in Table 1 for three consecutive output times. Since no method is expected to be equally accurate at the shock and away from the shock, 8 points straddling the shock, including 5 points in the decompression region, are excluded from the rms error calculation. The entries in Table 1 therefore mainly indicate the accuracy in the smooth part of the solution; for shock rendition we rely on visual inspection of the figures.

a) The Beam Scheme (B)

The beam scheme, used only in astrophysics, is due to Prendergast (see Sanders and Prendergast, 1974). It exemplifies the "Boltzmann approach," that is, mass and momentum are transported by pseudo-particles with a velocity distribution $f(w)$ designed for numerical convenience. In the beam scheme the velocity distribution is the sum of a number of delta functions (the beams); for the present calculations we used three beams:

$$f(w)=\tfrac{1}{6}\varrho\delta(w-[u-c\sqrt{3}])+\tfrac{2}{3}\varrho\delta(w-u)+\tfrac{1}{6}\varrho\delta(w-[u+c\sqrt{3}]), \tag{19}$$

although the middle beam is not really needed for this isothermal problem. Assuming that the velocity distribution is uniform and constant in each cell during the time-step, we can compute the flux of mass and momentum out of a cell in the positive direction,

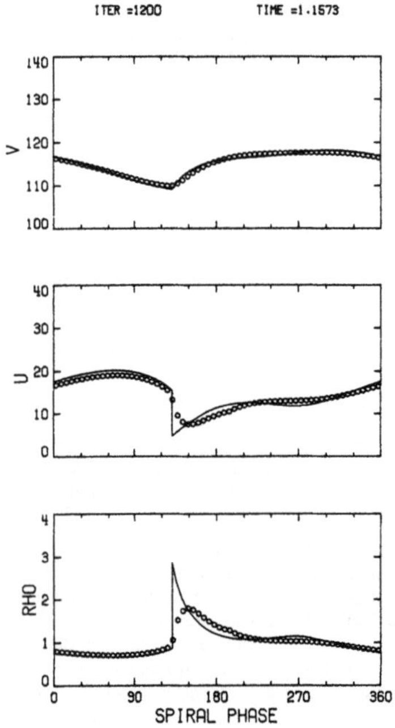

Fig. 1. Numerical results (circles) obtained with the Beam scheme (B) after 1200 time steps, starting from the exact solution (line), zone-averaged

Table 1. Test 1. Rms errors in the smooth region of the numerical solution, obtained with four schemes after 1200±150 time steps with a Courant number of 0.5, using the zone-averaged exact solution as initial-value distribution. Differences in the definition of the Courant number cause differences in the output times t_n. The results at $n=1200$ correspond to those of Figs. 1–6

			rms error (% of equilibrium value) in			
Scheme	n	t_n	ϱ	u	ϱu	v
	1050		7.8	7.6	0.70	0.37
B	1200	1.1573	8.4	7.7	0.77	0.36
	1350		8.0	7.8	0.74	0.37
	1050		3.5	4.0	0.57	0.18
G	1200	1.2266	3.6	4.1	0.40	0.19
	1350		3.6	4.5	0.48	0.19
	1050		0.54	0.42	0.25	0.026
FS2	1200	1.1983	0.62	0.53	0.25	0.020
	1350		0.48	0.46	0.11	0.026
	1050		1.0	0.71	0.58	0.052
MC2	1200	1.1962	1.0	1.0	0.51	0.037
	1350		0.79	0.80	0.32	0.049

$F^+(U)$, and in the negative direction, $F^-(U)$:

$$F^+(U)=\begin{cases} F(U)=\begin{pmatrix}\varrho u\\ \varrho(u^2+c^2)\\ \varrho uv\end{pmatrix} & u\geqq c\sqrt{3}\\[2ex] \begin{pmatrix}\frac{1}{6}\varrho(5u+c\sqrt{3})\\ \frac{1}{6}\varrho[4u^2+(u+c\sqrt{3})^2]\\ \frac{1}{6}\varrho v(5u+c\sqrt{3})\end{pmatrix} & 0\leqq u<c\sqrt{3}\\[2ex] \begin{pmatrix}\frac{1}{6}\varrho(u+c\sqrt{3})\\ \frac{1}{6}\varrho(u+c\sqrt{3})^2\\ \frac{1}{6}\varrho v(u+c\sqrt{3})\end{pmatrix} & -c\sqrt{3}<u<0\\[2ex] 0 & u\leqq -c\sqrt{3},\end{cases} \tag{20.1}$$

and $F^-(U)$ is obtained from

$$F^-(U)+F^+(U)=F(U). \tag{20.2}$$

The net flux across the cell interface at $\eta_{i+1/2}$, to be used in the scheme (14), is

$$\Phi^n_{i+1/2}=F^+(U^n_i)+F^-(U^n_{i+1}). \tag{21}$$

That Eq. (21) leads to upwind differencing becomes clear when we write down the central difference of Φ needed in scheme (14):

$$\Phi^n_{i+1/2}-\Phi^n_{i-1/2}=(F^+)^n_i-(F^+)^n_{i-1}+(F^-)^n_{i+1}-(F^-)^n_i. \tag{22}$$

Thus, the flux difference of the forward/backward moving material is biased in the backward/forward direction.

The local stability condition of a beam scheme with velocity dispersion $\pm\mu c$, $\mu\geqq 1$, is

$$\frac{\Delta t}{\Delta\eta}(|u|+\mu c)_i\leqq 1. \tag{23}$$

The beam scheme turns out to coincide, for $\mu=1$, with the "flux-vector splitting" method of Steger and Warming (1978), derived without reference to a velocity distribution. The term "splitting" refers to the splitting of F into F^+ and F^-. Figure 1 displays the results obtained with the beam scheme.

b) Godunov's Method (G)

In this method, designed by Godunov (1959), the interaction of a pair of cells at their interface is assumed to take place through discrete finite-amplitude waves rather than material beams. This yields a more accurate scheme. It has less diffusion than the beam scheme, yet still avoids non-physical oscillations in the vicinity of shock waves.

Assuming that at $t=t_n$ the distributions in each cell are uniform, we find discontinuities at the cell interfaces; these will be resolved instantaneously through shock waves and/or rarefaction waves, and (not for an isothermal gas) a contact discontinuity. Let $\hat{U}^n_{i+1/2}$ be the state remaining at $\eta_{i+1/2}$ after the resolution; Godunov's method incorporates

$$\Phi^n_{i+1/2}=F(\hat{U}^n_{i+1/2}). \tag{24}$$

It is easily seen that

$$(\delta F^-)^n_{i+1/2}\equiv\Phi^n_{i+1/2}-F^n_i, \tag{25}$$

and

$$(\delta F^+)^n_{i+1/2}\equiv F^n_{i+1}-\Phi^n_{i+1/2} \tag{26}$$

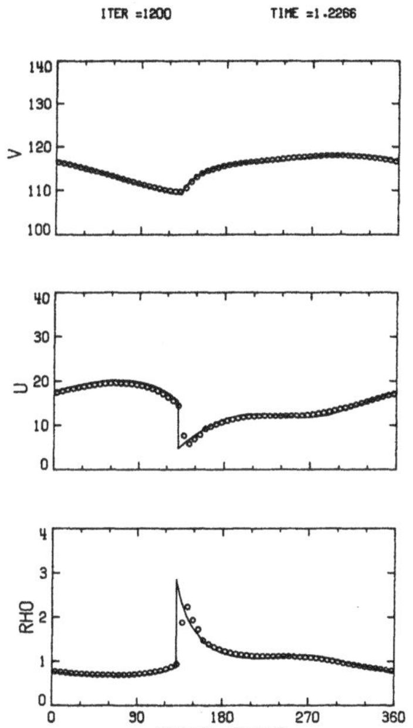

Fig. 2. Results of Godunov's method (G) after 1200 time steps

---

are the flux differences across the waves moving backward and forward from $\eta_{i+1/2}$, respectively. We have

$$\Phi_{i+1/2}^n - \Phi_{i-1/2}^n = \Phi_{i+1/2}^n - F_i^n + F_i^n - \Phi_{i-1/2}^n = (\delta F^-)_{i+1/2}^n + (\delta F^+)_{i-1/2}^n, \tag{27}$$

showing the upwind character of this method. The splitting evident in (25), (26), that is

$$\Delta_{i+1/2}F^n \equiv F_{i+1}^n - F_i^n = (\delta F^-)_{i+1/2}^n + (\delta F^+)_{i+1/2}^n, \tag{28}$$

is called flux-difference splitting, as opposed to the flux-vector splitting of Eq. (20.2).

The computation of $\hat{U}_{i+1/2}^n$ is the bottle-neck of Godunov's method, and a multitude of variations have been proposed based on employing a suitable approximation of $\hat{U}_{i+1/2}^n$; see Harten et al. (1981). In our computations we used an iterative procedure to find $\hat{U}_{i+1/2}^n$, which we stopped just short of convergence. For a compilation of the formulas and iteration procedures to resolve an arbitrary discontinuity (Riemann's initial-value problem) the reader may consult, e.g., Chorin (1976) or Van Leer (1979).

Note that in the supersonic region of the flow $U_{i+1/2}^n$ equals either U_i^n or U_{i+1}^n (the scheme becomes one-sided), so no full Riemann solution is needed. It therefore pays to test whether the flow is sub- or supersonic. The local stability condition for Godunov's method is

$$\frac{\Delta t}{\Delta\eta} W_{i+1/2} \leqq 1, \tag{29}$$

when $W_{i+1/2}$ is the largest wave speed occurring in the solution of the Riemann problem at $\eta_{i+1/2}$. The results obtained with the Godunov method are displayed in Fig. 2.

c) Second-order Flux-splitting Method (FS2)

Any first-order upwind-differencing method can be changed into a second-order method by first advancing the cell-boundary values, to be used in the numerical flux function, and the source term, to the intermediate time level $t_{n+1/2}$. In obtaining these values, *the interaction between cells can be fully ignored.* This observation, due to Hancock (1980), has led to a drastic simplification of second-order upwind schemes since these first were formulated by Van Leer (1979).

We choose Q to be a vector of (not necessarily conserved) quantities describing the state of the gas, in particular:

$$Q = \begin{pmatrix} \varrho \\ u \\ v \end{pmatrix} = \begin{pmatrix} \varrho \\ q \end{pmatrix}. \tag{30}$$

We then assume that the initial values for Q form a piecewise linear distribution:

$$Q^n(\eta) = Q_i^n + (\eta - \eta_i)\frac{(\delta Q)_i^n}{\Delta\eta} \qquad \eta_{i-1/2} < \eta < \eta_{i+1/2} \tag{31.1}$$

with

$$(\delta q)_i^n = c\cdot\text{ave}\left(\frac{q_{i+1}^n - q_i^n}{c}, \frac{q_i^n - q_{i-1}^n}{c}\right) \tag{31.2}$$

and

$$(\delta\varrho)_i^n = \varrho_i^n\cdot\text{ave}\left(2\frac{(\varrho_{i+1}^n - \varrho_i^n)}{(\varrho_{i+1}^n + \varrho_i^n)}, 2\frac{(\varrho_i^n - \varrho_{i-1}^n)}{(\varrho_i^n + \varrho_{i+1}^n)}\right), \tag{31.3}$$

where ave (a, b) is an averaging procedure to be specified later. The formulation in Eq. (31.3) guarantees positivity for ϱ when substituted in Eq. (31.1). Thus we have

$$\left(\frac{\partial Q}{\partial\eta}\right)_i^n = \frac{(\delta Q)_i^n}{\Delta\eta}, \tag{32}$$

allowing us to calculate $(\partial Q/\partial t)_i^n$ from the appropriate modification of Eq. (5). The cell averages are now advanced to $t_{n+1/2}$, and boundary values are calculated [the source terms have already been advanced, by Eq. (16)]:

$$Q_i^{n+1/2} = Q_i^n + \frac{\Delta t}{2}\left(\frac{\partial Q}{\partial t}\right)_i^n, \tag{33.1}$$

$$Q_{(i\pm1/2)\mp}^{n+1/2} = Q_i^{n+1/2} \pm \frac{1}{2}(\delta Q)_i^n, \tag{33.2}$$

$$U_{(i\pm1/2)\mp}^{n+1/2} = U(Q_{(i\pm1/2)\mp}^{n+1/2}). \tag{33.3}$$

The time-centered fluxes at cell boundary $i\pm\frac{1}{2}$ can now be computed from $U_{(i\pm1/2)-}^{n+1/2}$ and $U_{(i\pm1/2)+}^{n+1/2}$ by any upwind-biased numerical flux formula, such as used in B or G. Here we use a formula, due to Van Leer (1981b), based on flux-vector splitting and therefore related to the flux in B; however, no particular

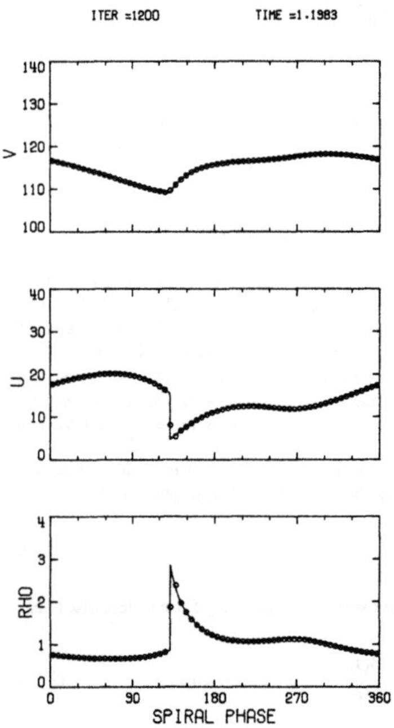

Fig. 3. Results of the second-order flux-splitting method (FS2) after 1200 time steps

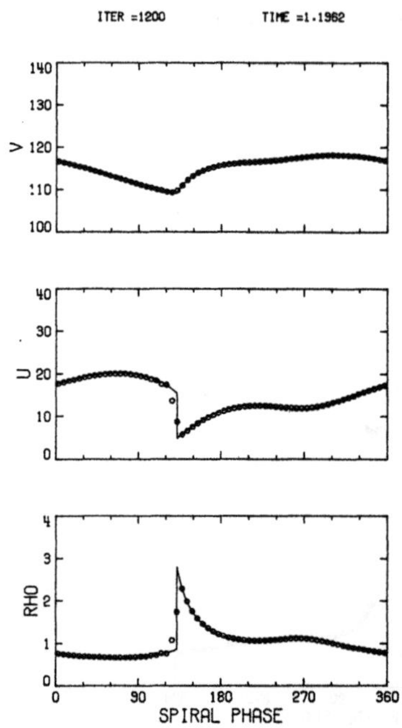

Fig. 4. Results of the second-order MacCormack method (MC2) after 1200 time steps

velocity distribution is used in its derivation. The forward and backward fluxes of mass and momentum are defined according to

$$F^+(U) = \begin{cases} F(U) & u \geqq c \\ \begin{pmatrix} \frac{\varrho}{4c}(u+c)^2 \\ \frac{\varrho}{2}(u+c)^2 \\ \frac{\varrho}{4c}(u+c)^2 v \end{pmatrix} & |u| < c, \\ 0 & u \leqq -c \end{cases} \tag{34}$$

and (20.2), while Φ again is given by (21). The split flux (34) is smoother than (20), having continuous first derivatives. Furthermore, the reduced mass flux in (34) relative to (20.1) results in a reduced numerical diffusion, making the numerical results of (34) alone come very close to the results of Godunov's flux (24) for the present test problem. The local stability condition is a combination of (23) with $\mu = 1$, and

$$\frac{\Delta t}{\Delta \eta} c_i \leqq \frac{1}{2}. \tag{35}$$

In FS2 we then use

$$\Phi^{n+1/2}_{i+1/2} = F^+(U^{n+1/2}_{(i+1/2)-}) + F^-(U^{n+1/2}_{(i+1/2)+}), \tag{36}$$

with $F^+(U)$ and $F^-(U)$ given by (34) and (20.2).

The function ave(a,b) is chosen such that it tends to $\frac{1}{2}(a+b)$ if a and b are subsequent finite differences of a smooth solution, but tends to the smallest value where the solution is not smooth (see Van Leer, 1977). We specifically chose

$$\text{ave}(a,b) = \frac{(b^2+\varepsilon^2)a + (a^2+\varepsilon^2)b}{a^2+b^2+2\varepsilon^2}, \tag{37}$$

where ε^2 is a small non-vanishing bias of the order $O((\Delta\eta)^3)$. This type of averaging prevents central differencing across a discontinuity in the solution or in its first derivative, which would lead to numerical oscillations. The bias prevents the undesirable clipping of a smooth extremum (see Sect. e) but otherwise has negligible influence. In the actual computations we used $\varepsilon^2 = 0.008$, but the results are not very sensitive to the precise value of ε^2. The results obtained with the second-order flux-splitting method are displayed in Fig. 3.

d) *MacCormack Method* (MC2)

This finite-difference method developed by MacCormack (1969) has been used widely in aerodynamics. On even time-steps it uses a forward predictor step which determines provisional values at t_{n+1}

$$\bar{U}_i^{n+1} = U_i^n - \frac{\Delta t}{\Delta y}(F^n_{i+1} - F^n_i) + \Delta t H^n_i, \tag{38}$$

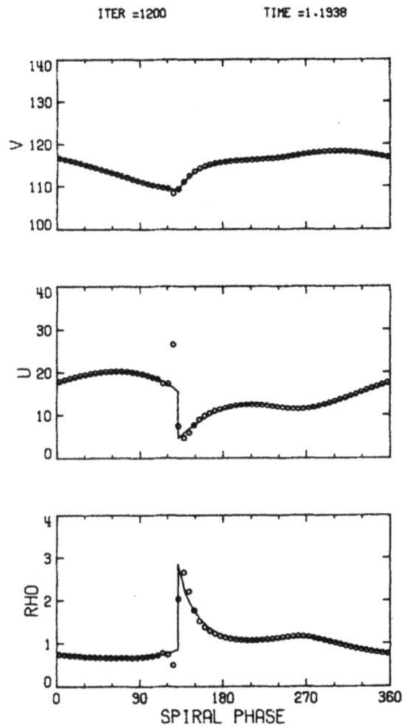

Fig. 5. Results of SHASTA after 1200 time steps as given by Eqs. (43)–(45), with the limiter *off*

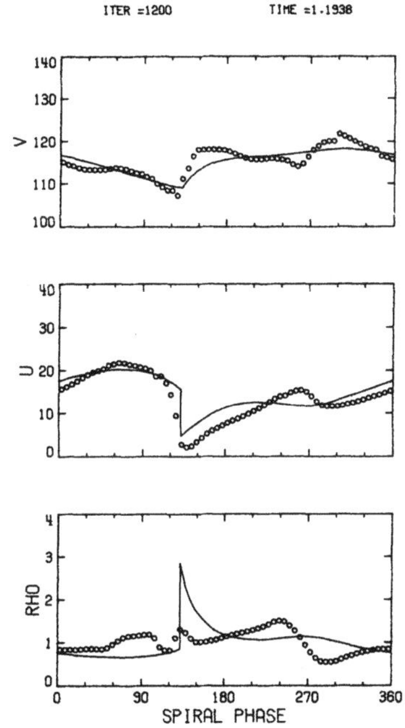

Fig. 6. Results of SHASTA after 1200 time steps, with the limiter *on*

followed by a backward corrector step which determines the final values at t_{n+1}

$$U_i^{n+1} = \tfrac{1}{2}\left[U_i^n + \bar{U}_i^{n+1} - \frac{\Delta t}{\Delta\eta}(\bar{F}_i^{n+1} - \bar{F}_{i+1}^{n+1}) + \Delta t\,\bar{H}_i^{n+1}\right]. \tag{39}$$

The corrector step corresponds to inserting

$$\Phi_{i+1/2}^{n+1/2} \equiv \tfrac{1}{2}(F_{i+1}^n + \bar{F}_i^{n+1}), \tag{40.1}$$

$$H_i^{n+1/2} \equiv \tfrac{1}{2}(H_i^n + \bar{H}_i^{n+1}), \tag{40.2}$$

into (14). On odd time-steps a backward predictor is employed with a forward corrector. This method is formally second-order accurate in both space and time.

Although MacCormack's method is slightly dissipative, an explicit smoothing term had to be added in order to control nonlinear instabilities in the test problem. This was implemented by adding the term

$$D_i^n = b\frac{\Delta t}{\Delta\eta}[\nu_{i+1/2}^n(U_{i+1}^n - U_i^n) - \nu_{i-1/2}^n(U_i^n - U_{i-1}^n)] \tag{41}$$

to the right-hand side of Eq. (39). The coefficient b is an adjustable constant of order unity; the choice

$$\nu_{i+1/2}^n = |u_{i+1}^n - u_i^n|, \tag{42}$$

for the artificial-viscosity coefficient was made on the basis of Liebovitch's (1978) experience with MacCormack's method in two-dimensional calculations. Figure 4 displays the results obtained with the second order MacCormack method [provided by T. A. Zang following the calculations of Zang and Hussaini (1980)].

e) Flux-corrected Transport Methods

We tested the following Flux-Corrected Transport (FCT) methods: SHASTA, Boris and Book (1973), and Phoenical SHASTA, Book et al. (1975) in the formulation given below, and also the low-phase-error routine of Boris (1976). These methods are known to perform well for flows with strong transient waves; we verified this with our programs.

The simplest FCT methods can be implemented by adding to the scheme of Lax and Wendroff (1960) or MacCormack (38, 39) the strong artificial diffusion

$$D_i^n = \tfrac{1}{8}(U_{i+1}^n - 2U_i^n + U_{i-1}^n), \tag{43}$$

thus suppressing numerical oscillations for a local Courant number $< \frac{1}{2}\sqrt{3}$, and then taking the diffusion away in the so-called anti-diffusion step, without creating new oscillations:

$$\tilde{U}_i^{n+1} = \bar{U}_i^{n+1} - \tfrac{1}{8}\{(\bar{U}_{i+1}^{n+1} - \bar{U}_i^{n+1})_{\lim} - (\bar{U}_i^{n+1} - \bar{U}_{i-1}^{n+1})_{\lim}\} \qquad \text{(SHASTA)}, \tag{44}$$

$$\tilde{U}_i^{n+1} = \bar{U}_i^{n+1} - \tfrac{1}{8}(U_{i+1}^{n+1} - U_i^{n+1})_{\lim} - (U_i^{n+1} - U_{i-1}^{n+1})_{\lim} \qquad \text{(Phoenical SHASTA)}. \tag{45}$$

Table 2. Test 2. Rms errors in the smooth region of the numerical solution, obtained with two schemes (FS2 and MC2) after 2400 time steps with a Courant number of 0.5, using uniform initial values

			rms errors (% of equilibrium value) in			
Scheme	n	t_n	ϱ	u	ϱu	v
FS2	2400	2.3946	2.4	4.8	7.0	0.20
MC2	2400	2.3934	3.1	3.9	7.0	0.19

Here U^{n+1} is the updated solution without the extra diffusion, $\bar{U}^{n+1}$ is the diffused updated solution and $\bar{\bar{U}}^{n+1}$ is the final result. The limiter works as follows:

$$(U_{i+1}-U_i)_{\lim}=\begin{cases}\min\{8|U_i-U_{i-1}|,|U_{i+1}-U_i|,8|U_{i+2}-U_{i+1}|\}\operatorname{sgn}(U_{i+1}-U_i)\\ \quad \text{if} \quad \operatorname{sgn}(U_i-U_{i-1})=\operatorname{sgn}(U_{i+1}-U_i)=\operatorname{sgn}(U_{i+2}-U_{i+1}),\\ 0 \text{ otherwise.}\end{cases} \tag{46}$$

In integrating our periodic test-problem starting from the stationary solution, we find that the limiter tries to give each extremum in ϱ, ϱu, and ϱv a flat top. This generates waves that keep running around, pile up and ultimately destroy any resemblance of the numerical results to the stationary solution.

Figures 5 and 6 show the results obtained with the Lax-Wendroff-based SHASTA. In order to illustrate the effect of the limiter, we first display the results generated with the limiter switched off in Fig. 5; these have reached a steady-state to the same degree as methods a)–d). Note the pre-shock spatial oscillations; without the limiter the scheme does not preserve monotonicity. Figure 6 shows the results generated with the limiter switched on. These are far from steady, and there is no numerical steady state to which they can converge. The other FCT methods tested behaved similarly.

A recent investigation of FCT methods by Zalesak (1981) suggests that it is not the form of the limiter that causes the problems, but the fact that it is placed in the final step of these methods, and that it acts independently on each of the conserved quantities, without synchronization.

It must be mentioned that the method FS2 also contains a limiter, albeit much more gently than (46), in the form of the averaging function (37). To our knowledge this does not hamper the convergence of the numerical solution in any way. However, divergence was observed with an earlier version of the formula, used by Van Leer (1977, 1979):

$$\text{ave}(a,b)=\frac{|b|a+|a|b}{|a|+|b|}. \tag{47}$$

This causes clipping of an extremum, since

$$\text{ave}(a,b)=0 \quad \text{if} \quad \operatorname{sgn} a=-\operatorname{sgn} b. \tag{A6}$$

In the present test problem, use of (47) in an FCT-like formulation of scheme FS2 caused a slowly growing travelling disturbance, invisible after 1200 time steps but a clear lump after 3600 time steps.

IV. Results

Method B a) is clearly singled out for its strong numerical diffusion. The computed shock appears as a shadow of the real one. The density maximum is severely underestimated and is displaced downstream by several zones. The corresponding values of u are barely subsonic. In the smooth region of the flow only the most general features are represented.

The displacement of the shock in the downstream direction is typical of upstream-differencing methods; the supersonic zone immediately before the shock is not influenced by the downstream subsonic region.

Method G b), although of the same order of accuracy, produces a substantial improvement. The density peak is better represented, the shock is much narrower, although still displaced downstream, and the smooth part of the solution is also closer to the truth.

Note the change in the numerical derivative of u and ϱ across the sonic point near a spiral phase of 160°. The downstream subsonic region cannot numerically influence the upstream supersonic region by sound waves moving upstream, while the numerical diffusion across those waves vanishes in the sonic point. The solutions on either side therefore are not strongly coupled. We may also say that the rapid variation of the scheme's truncation error shows up on this relatively coarse grid.

The very best results are produced by FS2 c). Its sharp, oscillation-free rendition of the shock shows the efficiency of applying dissipation through the averaging function (37). The rms entries in Table 1 are uniformly the lowest in all columns.

The virtues of method MC2 d) are evident in the smooth region of the flow. The shock is reasonably narrow and in the right place, with minor oscillations in front. The free parameter b in the numerical dissipation (41) was given the value 1.0; values in the range 0.2–2.0 produce acceptable results. No "fine-tuning" was attempted. Nonetheless, the fact that MC2 needs an adjustable parameter must be considered as a disadvantage of this method.

For methods a)–d) the rms error in the product ϱu, a quantity that becomes uniform in the steady state, is considerably smaller than the error in each of its factors. The errors in v, which by itself varies only a few percent in the solution, are again much smaller. The numbers in Table 1 are representative in that they do not change significantly when the iterations in time are continued far beyond 1200. Because of their poor quality, the FCT results have not been included in Table 1.

The methods, FS2 and MC2 were selected for the second test, to approch the steady state from uniform initial values. Neither of the schemes developed stability problems. The rms errors after 2400 time steps are listed in Table 2. They do not differ much from scheme to scheme and are substantially larger than the steady-state errors.

V. Conclusions and Recommendations

From the results of the first test a number of conclusions can be drawn about the spatial accuracy of the methods considered. The following three are quite firm.

1. The best second-order method (FS2) outperforms the best first-order method (G) by a huge margin, on a grid with resolution comparable to what is feasible in two-dimensional calculations. The second-order method, which includes a predictor step and a monotonicity algorithm, requires less than twice the computation time of the first-order method. A comparison of the rms errors suggests that the first-order method would require a six times finer

grid to match the accuracy of the second-order method. Thus, the extra programming effort reduces the total computational cost by an order of magnitude for the present mesh size. This holds even more strongly for computations on finer grids or in more dimensions.

2. Monotonicity algorithms for second-order methods have really come of age. The averaging procedure in FS2 that replaces central differencing is a simple and effective implementation of the ideas of artificial dissipation, filtering and limiting that have been explored during the last decade. The performance of FS2 derives to a great extent from the availability of this algorithm.

3. The second-order central-differencing scheme (MC2), while sufficiently accurate in rendering the smooth part of a solution, cannot compete with the second-order upwind-differencing method (FS2) in shock representation. The culprit seems to be the kind of artificial-dissipation term commonly used with this scheme. Better results can be expected from the use of terms derived from the scheme's truncation error (Klopfer and McRae, 1981); the derivation, however, is cumbersome.

The results of the second test show that the schemes selected could handle the more severe transients without heavily damping them. Firm conclusions about the accuracy of these schemes applied to transient phenomena should not be given, since the solutions obtained cannot be calibrated with a time-dependent exact solution.

One transient problem with known exact solution is the Riemann problem, describing the break-up of a discontinuity between two arbitrary uniform states. Comparative tests of twelve schemes on the basis of such a problem were carried out by Sod (1978). Among the correctly programmed schemes are central-difference schemes, including MC2. The results obtained for the same Riemann problem by Van Leer (1979) with an earlier version of FS2 are more accurate than any of the results of Sod, indicating the same hierarchy of accuracy as established in the present paper.

Our test results do not indicate how to proceed in order to reach a steady state that is not known a priori, with the minimum number of time steps. This was not the goal of the present paper. One possible strategy might be to start out with a first-order scheme while slowly turning on the source terms, and then finish off with a higher-order scheme.

In summary, on the basis of the present test results we may recommend the second-order upwind-differencing method FS2 for solving flow problems involving a shock of importance. For smoother problems, MC2 may do as good a job. One advantage of MC2 is that it has been extensively experimented within multi-dimensional computations, while the multi-dimensional experience with FS2 is still limited.

Acknowledgements. We are much indebted to T. A. Zang and M. Y. Hussaini for proposing the specific test problem, providing the MacCormack method, and for contributing to the manuscript.

References

Book, D.L., Boris, J.P., Hain, K.H.: 1975, *J. Computational Phys.* **18**, 248

Boris, J.P.: 1976, U. S. Naval Research Laboratory Memorandum Report 3237

Boris, J.P., Book, D.L.: 1973, *J. Computational Phys.* **11**, 38

Chorin, A.J.: 1976, *J. Computational Phys.* **22**, 517

Godunov, S.K.: 1959, *Mat. Sb.* **47**, 271 (also see Cornell Aeronautical Lab. Transl.)

Hancock, S.L.: 1980 (private communication)

Harten, A., Lax, R.D., Van Leer, B.: 1981, ICASE Report (in preparation)

Klopfer, G.H., McRae, D.S.: 1981, Proc. of AIAA Computational Fluid Dynamics Conference, Palo Alto, CA, 317

Lax, P.D., Wendroff, B.: 1960, *Comm. Pure Appl. Math.* **13**, 217

Liebovitch, L.S.: 1978, Ph. D. Thesis, Harvard University

Lin, C.C., Lau, Y.Y.: 1979, *Studies in Appl. Math.* **60**, 97

Lin, C.C., Shu, F.H.: 1964, *Astrophys. J.* **140**, 646

Lin, C.C., Shu, F.H.: 1966, *Proc. Natl. Acad. Sci. USA* **55**, 229

MacCormack, R.W.: 1969, AIAA Paper No. 69-354

Roberts, W.W.: 1969, *Astrophys. J.* **158**, 123

Sanders, R.H., Prendergast, K.H.: 1974, *Astrophys. J.* **188**, 489

Shu, F.H., Milione, V., Roberts, W.W.: 1973, *Astrophys. J.* **183**, 819

Sod, G.A.: 1978, *J. Computational Phys.* **27**, 1

Steger, J., Warming, R.: 1978, *J. Computational Phys.* **40**, 263

Toomre, A.: 1977, *Ann. Rev. Astron. Astrophys.* **15**, 437

Van Leer, B.: 1977, *J. Computational Phys.* **23**, 276

Van Leer, B.: 1979, *J. Computational Phys.* **32**, 101

Van Leer, B.: 1981a, ICASE Report No. 81-11

Van Leer, B.: 1981b, ICASE Report (in preparation)

Woodward, P.R.: 1975, *Astrophys. J.* **195**, 61

Zalesak, S.T.: 1981 (private communication)

Zang, T.A., Hussaini, M.Y.: 1980, Proc. 7th Intl. Conf. on Numerical Methods in Fluid Dynamics: *Lecture Notes in Physics*, Springer, Berlin, Heidelberg, New York

HIGH RESOLUTION SCHEMES AND THE ENTROPY CONDITION*

STANLEY OSHER† AND SUKUMAR CHAKRAVARTHY‡

Abstract. A systematic procedure for constructing semidiscrete, second order accurate, variation diminishing, five-point band width, approximations to scalar conservation laws, is presented. These schemes are constructed to also satisfy a single discrete entropy inequality. Thus, in the convex flux case, we prove convergence to the unique physically correct solution. For hyperbolic systems of conservation laws, we formally use this construction to extend the first author's first order accurate scheme, and show (under some minor technical hypotheses) that limit solutions satisfy an entropy inequality. Results concerning discrete shocks, a maximum principle, and maximal order of accuracy are obtained. Numerical applications are also presented.

Introduction. Recently a number of new shock capturing finite difference approximations have been constructed and found to be very useful in shock calculations, e.g. [3], [4], [16]. In addition to conservation form, these schemes are usually constructed to have as many as possible of the following properties:

(1) Stable and sharp discrete shock solutions.

(2) Limit solutions which satisfy a geometric and/or analytic entropy condition.

(3) A bound on the variation of the approximate solutions, at least in both the scalar, and linear systems, case.

(4) Second order accuracy in regions of smoothness, (with certain isolated exceptional points, as described in § 2 below).

In this paper we shall present a general procedure for constructing schemes with a five-point band width satisfying all of the above. We shall then prove a convergence theorem for a wide class of approximations to any scalar convex conservation laws in one space dimension, using recent uniqueness results of DiPerna [5].

Our convergence proof involves the following simple steps:

(a) We first obtain a variation bound for a wide class of second order accurate approximations. This implies that, for any sequence of mesh widths approaching zero, a subsequence of approximate solutions converges to a weak solution of the Cauchy problem.

(b) Next, we obtain a single discrete entropy inequality for a large subclass of the approximations mentioned above. Then all limit solutions in this subclass satisfy the inequality.

(c) We finally invoke the uniqueness results of DiPerna to prove convergence to the solution to the Cauchy problem, as the mesh width goes to zero, for the subclass mentioned in (b).

Earlier, Majda and Osher [20], modified the Lax–Wendroff scheme, keeping its essential properties, so that an entropy inequality was obtained. DiPerna (private communication) has proven L^2 convergence of this modification (again approximating convex conservation laws) using the theory of compensated compactness. No variation bound has been obtained for this second order accurate approximation.

From a practical point of view, this apparent lack of variation bound means that conventional schemes such as Lax–Wendroff, when approximating hyperbolic systems

* The research of this author was supported by the National Science Foundation under grant MCS 82-00788, the Army Research Office under grant DAAG 29-82-K-0090, and the National Aeronautics and Space Administration under grant NAG 1-270.

† Department of Mathematics, University of California, Los Angeles, California 90024.

‡ Rockwell Science Center, Thousand Oaks, California 91360.

of conservation laws, even with an entropy modification, suffer from a lack of robustness in computing complex flows with shock waves and steep gradients. While such schemes have been widely used in a variety of problems, (see [2] for references), that list of solved problems does not include flows with strong shocks (say Mach 5 upstream, normal to the shock), when the shocks are captured.

A main drawback of most finite-difference schemes is that discontinuities are approximated by discrete transitions, that when narrow, usually overshoot or undershoot, or when monotone, usually spread the discontinuity over many grid points.

Upwind schemes have been designed and used over the years, largely because of their success in treating this difficulty. Those based on solving the Riemann problem either exactly (Godunov's method [10]) or approximately e.g. (Osher's [24], or Roe's [28] with an entropy fix [34], [22]), have been extremely successful, especially when made second order accurate, [3], [4], or [16].

We should particularly mention the early investigations of van Leer [32], [33]. There he introduced the concepts of flux limiters and higher order Riemann solvers. Recently, Harten [16], using an argument also used in [1], [31] and [29], obtained simple sufficient conditions compatible with second order accuracy, which guarantee that a scalar one-dimensional approximation is TVD—total variation diminishing. He constructed a scheme having that property, and formally extended it to systems, using a field-by-field limiter and Roe's decomposition.

Using any of the three nonlinear decompositions (Godunov's Osher's or Roe's) one can obtain field-by-field limiters for systems, as described in § 6 below. Peter Sweby, [30], has investigated the properties of various limiters, and clarified their application considerably. The notation used in §§ 3–6 below is due to him.

We shall use the now introduced term "high resolution scheme" to mean a formal extension to systems via a field-by-field decomposition, of a scalar, second order accurate, variation diminishing scheme as in [16]. These schemes do not, in general, satisfy the entropy condition—e.g. expansion shocks exist as stable solutions of high resolution schemes based on Roe's (unmodified) scheme. In § 6, we use Osher's decomposition and certain limiters, to prove that under technically reasonable hypotheses, limit solutions of certain high resolution schemes do satisfy the entropy condition for hyperbolic systems of conservation laws.

Fundamental to our work is a formula measuring the discrete entropy in a cell for any scheme, obtained in [22]. See equations (4.2), (4.3), and (4.4) below. In [22], a class of approximations called E schemes was shown to be convergent, even for nonconvex, but scalar flux functions. This seems to be the widest known class of convergent schemes in this general case, but unfortunately the approximations are, at most, first order accurate.

All the results in this paper are for semi-discrete (continuous in time) approximations, and thus can serve as guidelines for a wide variety of time discretizations, both implicit and explicit. One of the principal applications of this theory has been in supersonic and transonic aerodynamics, where phenomena close to steady state are often studied. As an important example, one might mention the enormous advantage found when algorithms used to solve the transonic small disturbance equation, based on traditional Murman [21] (Roe's unfixed for scalar equations) differencing were replaced by using the E schemes of Engquist and Osher [12], [7], and later Godunov [13]. This minimal coding change, based on an entropy fix, increased the robustness of production codes by an order of magnitude, or more.

Goodman and LeVeque [11] have recently obtained a rather depressing result—two space dimensional, scalar approximations, cannot be TVD, and still be more than first order accurate, if the associated flux functions are reasonably smooth. Nevertheless, two-dimensional schemes based on dimension by dimension high resolution differencing, do perform extremely well, even for complex configurations, with very strong shocks. See [4], [16], [3], and § 7 below. This phenomenon remains to be justified analytically.

The format of this paper is as follows. In § 1, we review the relevant theory of weak solutions for hyperbolic systems of conservation laws. In § 2, we do the same for the theory of approximate solutions, and rederive results about bounding the variation in Lemma 2.1. We then obtain a maximum (minimum) principle and a nonlinear saturation result—Lemmas 2.2 and 2.3. In § 3, we develop a systematic procedure used to obtain high resolution schemes, based on any three-point first order accurate TVD scheme. We use second order "upwinding," together with flux limiters to do this. In § 4, we show how to obtain an entropy inequality, and hence convergence, which is our main result, Theorem 4.1. We use artificial compression or rarefaction operators (ACR). Harten introduced the notion of artificial compression in [15]. Here we use the compression part to firm up shocks, and contacts, and the rarefaction part to enforce an entropy condition. Rather precise admissible bounds on the ACR term are given. Section 5 is concerned with existence of steady discrete shocks for certain high resolution schemes. In § 6, we construct our high resolution scheme for systems, and then prove an entropy inequality: Theorem 6.2.

Finally, § 7 gives numerical evidence demonstrating the utility of these schemes, as well as some results illustrating a few theoretical points made throughout this paper. See also [3].

1. Review of theory of weak solutions. We shall consider numerical approximations to the initial value problem for nonlinear hyperbolic systems of conservation laws

$$\frac{\partial w}{\partial t}+\frac{\partial}{\partial x}f(w)=0, \qquad t>0, \quad -1\leqq x<1, \tag{1.1}$$

with periodic boundary conditions:

$$w(x+1,t)\equiv w(x,t),$$

and initial conditions $w(x,0)=w_0(x)$.

Here $w(x,t)$ is an m-vector of unknowns, and the flux function $f(w)$ is vector valued, having m components. The system is hyperbolic when the Jacobian matrix has real eigenvalues.

It is well known that solutions of (1.1) may develop discontinuities in finite time, even when the initial data are smooth. Because of this, we seek a weak solution of (1.1), i.e., a bounded measurable function w, such that for all $\varphi\in C_0^\infty(R\times R^+)$,

$$\text{(1.2)}\qquad \begin{aligned} &\text{(a)} && \iint_{R\times R^+}(w\varphi_t+f(x)\varphi_x)\,dx\,dt=0,\\ &\text{(b)} && \lim_{t\to 0}\|w(x,t)-w_0(x)\|_{L^1}=0.\end{aligned}$$

Solutions of (1.2) are not necessarily unique. For physical reasons, the limit solution of the viscous equation, as viscosity tends to zero, is sought. In the scalar case, this solution must satisfy, for all $\varphi\in C_0^\infty(R\times R^+)$, $\varphi\geqq 0$, and all real constants c.

$$\text{(1.3)}\quad \text{(a)}\qquad -\iint(|w-c|\varphi_t+\operatorname{sgn}(w-c)(f(w)-f(c))\varphi_x)\,dx\,dt\leqq 0.$$

This is equivalent to the statement:

$$\text{(1.3)} \quad \text{(b)} \qquad \frac{\partial}{\partial t}|w-c|+\frac{\partial}{\partial x}((f(w)-f(c))\operatorname{sgn}(w-c))\leqq 0,$$

in the sense of distributions.

Such solutions are called entropy solutions. Kruzkov has shown in [18] that two entropy solutions satisfy

$$\text{(1.4)} \qquad \|w(x,t_1)-v(x,t_1)\|_{L^1}\leqq \|w(x,t_0)-v(x,t_0)\|_{L^1},$$

for all $t_1 \geqq t_0$. Hence, condition (1.3) guarantees the uniqueness of solutions to the scalar version of (1.2). Existence was also obtained in [18].

For systems of equations, Lax has defined an entropy inequality [19], with the help of an entropy function $V(w)$ for (1.1), defined to have the following properties:

(i) V satisfies

$$\text{(1.5)} \qquad V_w f_w = F_w$$

where F is some other function, called the entropy flux.

(ii) V is a convex function of w.

It follows from (1.1), upon multiplication by V_w, that every smooth solution of (1.1) also satisfies

$$\text{(1.6)} \qquad V_t + F_x = 0.$$

It was also shown in [19], that if w is the bounded a.e. limit of solutions to the regularized equation, then the limit satisfies, in the weak sense, the following inequality:

$$\text{(1.7)} \qquad V_t + F_x \leqq 0.$$

Inequality (1.3b), for scalar equations, is just (1.7), with $V(w)=|w-c|$.

Inequalities (1.3) and (1.7) have important geometric consequences for piecewise continuous solutions. Suppose $w(x,t)$ is such a solution having a jump discontinuity, $w_L(t)$, $w_R(t)$ moving with speed $s(t)$. Then (1.2) implies the well known jump conditions

$$\text{(1.8)} \qquad f(w_L)-f(w_R)=s(w_L-w_R).$$

In the scalar case, (1.3) is equivalent to Oleinik's condition E across the shock

$$\text{(1.9)} \qquad \frac{f(w)-f(w_R)}{w-w_R}\leqq\frac{f(w_L)-f(w_R)}{w_L-w_R},$$

for all w between w_L and w_R.

If f is convex, then (1.9) is equivalent to the statement that characteristics flow *into* the discontinuity, as t increases. Also, for scalar convex f, inequality (1.7) for a *single fixed* convex V is equivalent to (1.3) for *all constants* c, if the solution is of bounded variation. This is a consequence of the recent results of DiPerna [5]. Thus uniqueness in this case is a consequence of Kruzkov's results [18]. This fact is crucial to our convergence proof in § 4.

For hyperbolic systems of equations, the Jacobian of f, denoted by ∂f, has real eigenvalues, which are usually assumed to be distinct:

$$\lambda_1<\lambda_2<\cdots<\lambda_m,$$

corresponding to right eigenvectors: $r_1, r_2, \cdots, r_m$. Lax [19] then defines the kth field to be linearly degenerate if

$$\nabla_w \lambda_k \cdot r_k \equiv 0.$$

He also defines the field to be genuinely nonlinear if

$$\nabla_w \lambda_k \cdot r_k \neq 0.$$

For genuinely nonlinear fields, a k shock moving with speed s is defined to be a discontinuity of w, such that $m+1$ characteristics flow into the shocks, and

$$\lambda_k(w_L) - s > 0 > \lambda_k(w_R) - s.$$

This geometric condition is equivalent to (1.7) for weak shocks [19].

2. Preliminary theory of approximate solutions. We consider a semi-discrete, method of lines approximation to (1.1). We break the interval $[-1,1)$ into subintervals

$$\mathcal{I}_j = \{x \mid (j-\tfrac{1}{2})\Delta \leqq x < (j+\tfrac{1}{2})\Delta\},$$

$j = 0, \pm 1, \cdots, \pm N$, with $(2N+1)\Delta = 2$.

Let $x_j = j\Delta$, be the center of each interval $\mathcal{I}_j$, with end points $x_{j-1/2}, x_{j+1/2}$. Define the step function for each $t > 0$, as

$$\mathcal{V}_\Delta(x, t) \equiv u_j(t)$$

for $x \in \mathcal{I}_j$.

The initial data is discretized via the averaging operator T_Δ,

$$T_\Delta w_0(x) = \frac{1}{\Delta} \int_{\mathcal{I}_j} w(s)\, ds = u_j(0) \quad \text{for } x \in \mathcal{I}_j. \tag{2.1}$$

For any step function, we define the difference operators:

$$\Delta_\pm u_j = \pm(u_{j\pm 1}, -u_j), \qquad D_\pm u_j = \frac{1}{\Delta} \Delta_\pm u_j.$$

A method of lines, conservation form discretization of (1.1), is a system of differential equations

$$\begin{gathered} \frac{\partial}{\partial t} u_j + D_+ h_{j-1/2} = 0, \qquad j = 0, \pm 1, \cdots, \pm N, \\ \mathcal{V}_\Delta(x, 0) = T_\Delta w_0(x) \quad \text{for } x \in \mathcal{I}_j. \end{gathered} \tag{2.2}$$

Here the numerical flux defined by

$$h_{j-1/2} = h(u_{j+k-1}, \cdots, u_{j-k}), \tag{2.3}$$

for $k \geqq 1$, is a Lipschitz continuous function of its arguments, satisfying the consistency condition:

$$h(w, w, \cdots, w) = f(w).$$

It is well known that bounded a.e. limits, as $\Delta \to 0$, of approximate solutions, converge to weak solutions of (1.1), i.e. (1.2a) is satisfied. However, this does not also imply that the limit solutions will satisfy any of the entropy conditions mentioned above, e.g. [8], [9]. Some restrictions on h are required.

A simple class of flux functions h, for which (2.2) converges to the unique entropy solution in $L^\infty(L^1(R);[0,T])$, as $\Delta\to 0$, for any $T>0$, are scalar "E" schemes, [22]. Such schemes satisfy the following:

DEFINITION 2.1. A consistent scheme whose numerical flux satisfies

$$\operatorname{sgn}(u_j-u_{j-1})[h_{j-1/2}-f(u)]\leqq 0, \tag{2.4}$$

for all u between u_{j-1} and u_j, is said to be an E scheme.

It is easy to see that this class includes the widely known class of three-point monotone schemes, i.e. those for which

$$h_{j-1/2}=h(u_j,u_{j-1})$$

with $h_{j-1/2}$ nonincreasing in its first argument, nondecreasing in its second.

Partial derivatives of a numerical flux, when needed, will be denoted via

$$\frac{\partial}{\partial u_{j+\gamma}}h(u_{j+k},\cdots,u_j,\cdots,u_{j-k+1})=h_\gamma.$$

Thus a C^1 three-point scheme is monotone iff

$$h_1\leqq 0\leqq h_0.$$

One particular three-point monotone scheme is due to Godunov [10], and has a special significance to this theory. The flux for Godunov's scalar scheme can be defined by

$$h^G_{j-1/2}=h_G(u_j,u_{j-1})=\begin{cases}\min\limits_{u_{j-1}\leqq u\leqq u_j} f(u) & \text{if } u_{j-1}\leqq u_j,\\ \max\limits_{u_{j-1}\geqq u\geqq u_j} f(u) & \text{if } u_{j-1}>u_j\end{cases} \tag{2.5}$$

Thus one can characterize E schemes as precisely those for which

$$\begin{aligned}&\text{(a)} && h_{j-1/2}\leqq h^G_{j-1/2} \quad \text{if } u_{j-1}<u_j,\\ &\text{(b)} && h_{j-1/2}\geqq h^G_{j-1/2} \quad \text{if } u_j>u_{j-1}.\end{aligned} \tag{2.6}$$

It follows from [22, Lemma 2.1] that these approximations are at most first order accurate.

Together with an entropy inequality, a key estimate involved in many convergence proofs, is a bound on the variation. We present an argument originally due to Sanders for monotone schemes [29], used to obtain this bound.

For any fixed $t>0$, the x variation of $\mathscr{V}_\Delta(x,t)$ is

$$B(\mathscr{V}_\Delta)=\sum_j|\Delta_+u_j(t)|.$$

Let

$$\chi_{j+1/2}=\begin{cases}1 & \text{if } \Delta_+u_j\geqq 0,\\ -1 & \text{if } \Delta_+u_j<0.\end{cases}$$

Then:

$$\frac{d}{dt}B(\mathcal{V}_\Delta)=\frac{1}{\Delta}\sum_j\frac{d}{dt}(\chi_{j+1/2}\Delta_+u_j)$$

$$=\frac{1}{\Delta}\sum_j\chi_{j+1/2}\frac{d}{dt}\Delta_+u_j=-\frac{1}{\Delta}\sum_j\chi_{j+1/2}\Delta_+\Delta_+h_{j-1/2}$$

$$=\frac{1}{\Delta}\sum_j(\chi_{j+1/2}-\chi_{j-1/2})\Delta_+h_{j-1/2}\leqq 0,$$

if we can write

(2.7)
(a) $\Delta_+h_{j-1/2}=-[C_{j+1/2}\Delta_+u_j-D_{j-1/2}\Delta_-u_j]$,
(b) $C_{j+1/2}\geqq 0$,
(c) $D_{j-1/2}\geqq 0$.

In the case of E (or three-point monotone) schemes, this follows by defining:

$$C_{j+1/2}=-\left(\frac{h_{j+1/2}-f(u_j)}{\Delta_+u_j}\right),\qquad D_{j-1/2}=\left(\frac{f(u_j)-h_{j-1/2}}{\Delta_-u_j}\right)$$

as in [22].

Harten in [16] pointed out for explicit methods that a variation bound could be obtained for schemes which are higher order accurate. In our present method of lines context, it involves a five-point consistent approximation

(2.8) (a) $$\frac{\partial u_j}{\partial t}=-\frac{1}{\Delta}\Delta_+h_{j-1/2}=(C_{j+1/2}\Delta_+u_j-D_{j-1/2}\Delta_-u_j)/\Delta,$$

with (b) $C_{j+1/2}=C(u_{j+2},u_{j+1},u_j,u_{j-1})\geqq 0$,

and (c) $D_{j-1/2}=D(u_{j+1},u_j,u_{j-1},u_{j-2})\geqq 0$,

both Lipschitz continuous functions of their arguments. (See also van Leer [32].)

We have thus proven for schemes of type (2.8):

LEMMA 2.1. *For any* $t_1>t_2\geqq 0$

$$B(\mathcal{V}_\Delta(\cdot,t_1))\leqq B(\mathcal{V}_\Delta(\cdot,t_2)).$$

We also have the maximum and minimum principles:

LEMMA 2.2. *Let* $\max u_j(0)=M$, $\min u_j(0)=m$. *Then, for* $0<t$ *and each* j;

$$m\leqq u_j(t)\leqq M.$$

Moreover, if $C_{j+1/2}>0$, *and* $u_j(t)=M$, *then* $u_{j+1}(t)=M$. *If* $D_{j-1/2}<0$ *and* $u_j(t)=m$, *then* $u_{j-1}(t)=m$.

Proof. The proof is trivial if the initial data is constant. Otherwise, we let $\{u_j^\varepsilon(t)\}_{j=-N}^N$, satisfy

(2.9) $$\frac{\partial}{\partial t}u_j^\varepsilon=(C_{j+1/2}+\varepsilon)\Delta_+u_j^\varepsilon-(D_{j-1/2}+\varepsilon)\Delta_-u_j^\varepsilon$$

for $\varepsilon>0$, with the initial data $u_j^\varepsilon(0)\equiv u_j(0)$.

Suppose the maximum of $u_j^\varepsilon(t)$, for $0 \leqq t \leqq T$, occurs for $u_{j_0}^\varepsilon(t_0)$, with $t_0 > 0$. Then (2.8), (2.9), imply that $u_j^\varepsilon(t_0) \equiv u_{j_0}^\varepsilon(t_0)$. Hence $u_j^\varepsilon(t) \equiv$ constant, for $0 \leqq t \leqq T$. This is a contradiction. Thus:

$$\max_{j, 0 \leqq t \leqq T} \{u_j^\varepsilon(t)\}_{j=-N}^N < M.$$

Let $\varepsilon \downarrow 0$. It follows that $u_j(t) \leqq M$ for each j, t. The remainder of the proof follows analogously.

Next, we prove a limit on the possible accuracy of approximations (2.8).

LEMMA 2.3. *Approximation* (2.8) *is at most first order accurate at nonsonic critical points of* u. (*A sonic point* $\bar{u}$ *is one such that* $f'(\bar{u}) = 0$.)

Proof. For any C^2 function $u(x)$, we let $C_{j+1/2}(u(x+2\Delta), u(x+\Delta), u(x), u(x-\Delta))$ be denoted by $C_{j+1/2}$, $C_{j+1/2}(u(x), u(x), u(x), u(x))$ be denoted by $C_{j+1/2}(u)$, and similarly for $D_{j-1/2}$.

Then

$$\begin{aligned} &-\frac{1}{\Delta}\Delta_- h_j(u(x+2\Delta), u(x+\Delta), u(x), u(x-\Delta)) \\ &\quad = (C_{j+1/2}(u) - D_{j-1/2}(u))u_x + \frac{\Delta}{2}u_{xx}(C_{j+1/2}(u) + D_{j-1/2}(u)) \\ &\qquad + u_x(C_{j+1/2} - C_{j+1/2}(u) - D_{j-1/2} - D_{j-1/2}(u)) + o(\Delta). \end{aligned} \tag{2.10}$$

Consistency implies

$$C_{j+1/2}(u) - D_{j-1/2}(u) = -f'(u), \tag{2.11}$$

while second order accuracy at critical points means

$$C_{j+1/2}(u) + D_{j-1/2}(u) = 0. \tag{2.12}$$

Solving (2.11), (2.12), gives us

$$C_{j+1/2}(u) = -\tfrac{1}{2}f'(u), \qquad D_{j-1/2}(u) = \tfrac{1}{2}f'(u).$$

Thus one of inequalities (2.8b, c) must fail at nonsonic u.

Thus, although schemes of the type (2.8) can be made to be as high as third order accurate, Lipschitz continuity of $C_{j+1/2}$, $D_{j-1/2}$ implies a local degeneracy to first order accuracy at smooth maxima and minima. This local degeneracy, together with some results on initial boundary value problems in [14], indicate strongly that overall second order accuracy is the best possible.

3. Total variation diminishing, second order accurate, scalar approximations. We now describe a systematic procedure used to construct second order methods of the form (2.8) from three-point, first order methods of the same type.

We begin with a three-point scheme

$$\begin{aligned} \frac{\partial u_j}{\partial t} &= -\frac{1}{\Delta}(h(u_{j+1}, u_j) - h(u_j, u_{j-1})) \\ &= -\frac{1}{\Delta}(h(u_{j+1}, u_j) - f(u_j)) - \frac{1}{\Delta}(f(u_j) - h(u_j, u_{j-1})) \\ &= \frac{1}{\Delta}C_{j+1/2}^{(1)}\Delta_+ u_j - \frac{1}{\Delta}D_{j-1/2}^{(1)}\Delta_- u_j \end{aligned} \tag{3.1}$$

for

$$C^{(1)}_{j+1/2} = -\left(\frac{h(u_{j+1}, u_j) - f(u_j)}{\Delta_+ u_j}\right) \geqq 0,$$

$$D^{(1)}_{j-1/2} = \left(\frac{f(u_j) - h(u_j, u_{j-1})}{\Delta_- u_j}\right) \geqq 0.$$

These quantities are both nonnegative for E schemes. Moreover, the entropy inequality was shown to be valid in [22] for this class. Thus, using Lemmas 2.1, 2.2 and a standard argument, e.g. [29], we have convergence to the unique entropy solution, as $\Delta \to 0$; for this class.

However, the nonnegativity of these functions, by itself, need not imply convergence, since the entropy condition may fail, as it does for Murman's (Roe's) scheme. See e.g. [34]. We do assume nonnegativity of $C^{(1)}_{j+1/2}$, $D^{(1)}_{j-1/2}$ in what follows, as well as Lipschitz continuity of $h(u, v)$.

Our first attempt at constructing a higher order accurate scheme comes from a simple upwind type of hybridization:

$$\frac{\partial u_j}{\partial t} = \frac{1}{\Delta}\left(1 - \frac{1}{2}\Delta_+\right) C^{(1)}_{j+1/2}\Delta_+ u_j - \frac{1}{\Delta}\left(1 + \frac{1}{2}\Delta_-\right) D^{(1)}_{j-1/2}\Delta_- u_j = -\frac{1}{\Delta}\Delta_+ \mathcal{H}_{j-1/2}, \tag{3.2}$$

with the numerical flux satisfying

$$\begin{aligned} \mathcal{H}_{j-1/2} &= h(u_j, u_{j-1}) - \tfrac{1}{2}(h(u_{j+1}, u_j) - f(u_j)) + \tfrac{1}{2}(f(u_{j-1}) - h(u_{j-1}, u_{j-2})) \\ &= \tfrac{1}{2}(f(u_j) + f(u_{j-1})) - \tfrac{1}{2}\Delta_+\Delta_- h(u_j, u_{j-1}). \end{aligned} \tag{3.3}$$

(Throughout this work we shall often use the fact that consistency is equivalent to the statement:

$$h(u, u) = f(u).\ )$$

This scheme is, within second order accuracy, just the central difference algorithm, and is thus always second order accurate. It cannot, by Lemma 2.3, obey (2.8).

We can define the associated quantities so that

$$-\Delta_+ \mathcal{H}_{j-1/2} = C^{(2)}_{j+1/2}\Delta_+ u_j - D^{(2)}_{j-1/2}\Delta_- u_j$$

where, in this case:

$$\begin{aligned} C^{(2)}_{j+1/2} &= C^{(1)}_{j+1/2}\left(1 - \frac{1}{2}\frac{(h(u_{j+2}, u_{j+1}) - h(u_{j+1}, u_{j+1}))}{(h(u_{j+1}, u_j) - h(u_j, u_j))} + \frac{1}{2}\right), \\ D^{(2)}_{j-1/2} &= D^{(1)}_{j-1/2}\left(1 + \frac{1}{2} - \frac{1}{2}\left(\frac{h(u_{j-1}, u_{j-1}) - h(u_{j-1}, u_{j-2})}{h(u_j, u_j) - h(u_j, u_{j-1})}\right)\right). \end{aligned} \tag{3.4}$$

It is clear that the two quantities on the right above may become negative if the nontrivial ratios above become sufficiently large.

One possible remedy is to employ a flux limiter, using the notation of Sweby [30] (this issue, pp. 995–1011). (See also van Leer [32] for very original related work.)

Let:

$$R_j^+ = \frac{f(u_j) - h(u_j, u_{j-1})}{f(u_{j+1}) - h(u_{j+1}, u_j)}, \qquad R_j^- = \frac{h(u_{j+1}, u_j) - f(u_j)}{h(u_j, u_{j-1}) - f(u_{j-1})}. \tag{3.5}$$

Our TVD approximation will be of the form

$$\frac{\partial u_j}{\partial t} = -\frac{1}{\Delta}\Big[h(u_{j+1}, u_j) - h(u_j, u_j) - \frac{1}{2}\Delta_-(\psi(R_{j+1}^-)(h(u_{j+1}, u_j) - h(u_j, u_j))) \tag{3.6}$$
$$+ h(u_j, u_j) - h(u_j, u_{j-1}) + \frac{1}{2}\Delta_+(\psi(R_{j-1}^+)(h(u_j, u_j) - h(u_j, u_{j-1}))\Big].$$

This scheme is easily seen to be second order accurate away from critical points, (i.e. points where the denominators in (3.5) approach zero), if the Lipschitz continuous function, $\psi(R)$, satisfies $\psi(1) = 1$.

The resulting quantities $C_{j+1/2}$, $D_{j-1/2}$ are required to be positive:

$$C_{j+1/2} = -\left(\frac{h(u_{j+1}, u_j) - h(u_j, u_j)}{\Delta_+ u_j}\right)\left[1 + \frac{1}{2}\left[\frac{\psi(R_j^-)}{R_j^-} - \psi(R_{j+1}^-)\right]\right] \geqq 0, \tag{3.7}$$
$$D_{j-1/2} = \left(\frac{h(u_j, u_j) - h(u_j, u_{j-1})}{\Delta_- u_j}\right)\left[1 + \frac{1}{2}\left[\frac{\psi(R_j^+)}{R_j^+} - \psi(R_{j-1}^+)\right]\right] \geqq 0.$$

We thus have the required inequalities for $\psi(R^\pm)$:

$$\text{(a)} \quad 1 + \frac{1}{2}\left[\frac{\psi(R_j^-)}{R_j^-} - \psi(R_{j+1}^-)\right] \geqq 0, \tag{3.8}$$
$$\text{(b)} \quad 1 + \frac{1}{2}\left[\frac{\psi(R_j^+)}{R_j^+} - \psi(R_{j-1}^+)\right] \geqq 0.$$

Various slope limiters have been developed. See Sweby [30], for a numerical and theoretical analysis of their properties. As an example, we may take

$$\psi(R) \equiv \begin{cases} 0 & \text{if } R < 0, \\ R, & \text{if } 0 \leqq R \leqq k \text{ for } 1 \leqq k \leqq 2, \\ k & \text{if } R > k. \end{cases} \tag{3.9}$$

Let the numerical flux defined from (3.1)–(3.8) be called

$$\mathscr{H}_{j-1/2} = \mathscr{H}(u_{j+1}, u_j, u_{j-1}, u_{j-2}).$$

Its precise definition is

$$\mathscr{H}(u_{j+1}, u_j, u_{j-1}, u_{j-2}) = h(u_j, u_{j-1}) - \tfrac{1}{2}\psi(R_j^-)(h(u_j, u_{j-1}) - h(u_{j-1}, u_{j-1})) \tag{3.10}$$
$$+ \tfrac{1}{2}\psi(R_{j-1}^+)(h(u_j, u_j) - h(u_j, u_{j-1})).$$

We thus have the following:

THEOREM 3.1. *The approximation defined through*

$$\frac{\partial u_j}{\partial t} = -\frac{1}{\Delta}\Delta_+ \tilde{H}_{j-1/2}, \qquad j = 0, \pm 1, \cdots, \pm N, \tag{3.11}$$
$$u_j(0) = \frac{1}{\Delta}\int_{\mathscr{I}_j} w_0(s)\, ds,$$

with $w_0 \in L^1 \cap L^\infty \cap BV$, *defines* $u_\Delta(x, t)$ *having the property that as* $\Delta \to 0$, *there exists a converging subsequence* $\mathscr{V}_{\Delta'}(x, t)$ *which converges in* $L^\infty(L^1(R), [0, T])$, *to a weak solution of* (1.1).

The proof is a routine consequence of Lemmas 2.1 and (2.2). See, e.g. [29].

THEOREM (3.2). *The approximation* (3.10) *is second order accurate, except at isolated zeros of* $h_1(w, w)w_x$, *or* $h_0(w, w_,)w_x$.

See Sweby [30] for a related result.

4. High resolution schemes and the entropy condition for scalar approximations. Following recent tradition, we christen the schemes constructed in the last section high resolution schemes—namely, they are second order accurate (with the usual exceptional points), variation diminishing, and have a five-point bandwidth. The question remains—is it true that limit solutions satisfy the entropy condition? For first order non-E schemes, variation diminishing is not enough. The perennial example is Murman's (Roe's) scheme with $f''(u)>0$ and

$$h^M(u_{j+1}, u_j)=\frac{1}{2}(f(u_{j+1})+f(u_j))-\frac{1}{2}\left|\frac{\Delta_+f(u_j)}{\Delta_+u_j}\right|\Delta_+u_j. \tag{4.1}$$

Given an entropy violating shock $u^L<u^R$, with $f(u^L)=f(u^R)$, it is well known that

$$u_j\equiv\begin{cases} u^L & \text{for } j\leqq 0,\\ u^R & \text{for } j>0\end{cases}$$

is a steady solution to (3.1). Moreover, since $h(u_{j+1}, u_j)\equiv f(u^L)\equiv f(u^R)\equiv h(u_j, u_{j-1})$, in this case, then any of our high resolution schemes (3.11) will have entropy violating solutions, if $h(u_{j+1}, u_j)=h^M(u_{j+1}, u_j)$.

Let $V(w)$ be any convex function. In [22, § 3] it was shown for any solution of (2.2) that

$$\Delta\left(\frac{d}{dt}V(u_j)+D_+F_A(u_j)\right)=\int_{u_j}^{u_{j+1}}dw\, V''(w)[h_{j+1/2}-f(w)], \tag{4.2}$$

where the approximate entropy flux is defined through

$$F_A(u_j)=F(u_j)+V'(u_j)[h_{j-1/2}-f(u_j)]. \tag{4.3}$$

Thus a sufficient condition that any limit solution satisfy (1.7), for a fixed V, is that

$$\int_{u_j}^{u_{j+1}}V''(w)[h_{j+1/2}-f(w)]\,dw\leqq 0. \tag{4.4}$$

In order that the above inequality be valid for all convex V, it is necessary and sufficient that $h_{j+1/2}$ correspond to an E scheme, and hence that the approximation be at most first order accurate, [22]. Thus we shall only obtain our entropy inequality for a single $V(w)$, say $V(w)=\frac{1}{2}w^2$. This is sufficient for convergence, if f is convex, as will be shown in § 4.

We now proceed to modify the flux difference quantities, to ensure that (4.4) is valid, and, moreover, to sharpen discrete shock profiles. We shall do this by using the notion of artificial compression introduced by Harten [15], cf. also [16]. We also add negative artificial compression (=artificial rarefaction), if $f'(u_j)<f'(u_{j+1})$, and a certain amount of (positive) artificial compression if $f'(u_j)>f'(u_{j+1})$. The high resolution properties are preserved, and (4.4) is shown to be valid.

It is known that too much artificial compression can cause expansion shocks to develop. We shall obtain fairly precise bounds on the amount allowed.

Once again, we define

$$h_1(u, v)=\frac{\partial}{\partial u}h(u, v),\qquad h_0(u, v)=\frac{\partial}{\partial v}h(u, v).$$

We shall also use two Lipschitz continuous functions of u_{j+1}, u_j, denoted by $a^+_{j+1/2}$, $a^-_{j-1/2}$ and defined below. Our scheme is an artificially compressed or rarefied version of (3.6).

We first modify our flux differences via the following:

$$(4.5)\quad \begin{aligned} &\text{(a)}\quad (h(u_{j+1},u_j)-h(u_j,u_j))^m\\ &\qquad=(h(u_{j+1},u_j)-h(u_j,u_j))\\ &\qquad\cdot\left[1+a^-_{j+1/2}\frac{(h_1(u_{j+1},u_{j+1})-h_1(u_j,u_j))\Delta_+u_j}{(h(u_{j+1},u_j)-h(u_j,u_j))}\right]\\ &\text{(b)}\quad (h(u_{j+1},u_{j+1})-h(u_{j+1},u_j))^m\\ &\qquad=(h(u_{j+1},u_{j+1})-h(u_{j+1},u_j))\\ &\qquad\cdot\left[1-a^+_{j+1/2}\frac{(h_0(u_{j+1},u_{j+1})-h_0(u_j,u_j))\Delta_+u_j}{(h(u_{j+1},u_{j+1})-h(u_{j+1},u_j))}\right].\end{aligned}$$

Here $a^{\mp}_{j+1/2}$ are both positive numbers, chosen first so the quantities in brackets in (4.5a, b) are both always between 0 and 2, i.e., we must have:

$$(4.6)\quad \begin{aligned} &\text{(a)}\quad |a^-_{j+1/2}(h_1(u_{j+1},u_{j+1})-h_1(u_j,u_j))|\leqq-\frac{(h(u_{j+1},u_j)-h(u_j,u_j))}{\Delta_+u_j},\\ &\text{(b)}\quad |a^+_{j+1/2}(h_0(u_{j+1},u_{j+1})-h_0(u_j,u_j))|\leqq\frac{h(u_{j+1},u_{j+1})-h(u_{j+1},u_j)}{\Delta_+u_j}.\end{aligned}$$

We now let

$$R^{m_+}_j=\frac{(h(u_j,u_j)-h(u_j,u_{j-1}))^m}{(h(u_{j+1},u_{j+1})-h(u_{j+1},u_j))^m},$$

$$R^{m_-}_j=\frac{(h(u_{j+1},u_j)-h(u_j,u_j))^m}{(h(u_j,u_{j-1})-h(u_{j-1},u_{j-1}))^m}.$$

Our scheme, which uses ACR (artificial compression-rarefaction) is

$$(4.7)\quad \begin{aligned}\frac{\partial u_j}{\partial t}=-\frac{1}{\Delta}\Big[&h(u_{j+1},u_j)-h(u_j,u_j)\\ &-\frac{1}{2}\Delta_-(\psi(R^{m_-}_{j+1})(h(u_{j+1},u_j)-h(u_j,u_j))^m)\\ &+h(u_j,u_j)-h(u_j,u_{j-1})\\ &+\frac{1}{2}\Delta_+(\psi(R^{m_+}_{j-1})(h(u_j,u_j)-h(u_j,u_{j-1}))^m)\Big]\end{aligned}$$

where we take $\psi(R)$ to be as defined in (3.4), with $k=1$.

We let the numerical flux defined through (4.5), (4.6), (4.7) be denoted by $\tilde{H}^{ac}_{j-1/2}$. It is precisely defined by (3.10), with the superscript m attached to both $R^{\pm}_j$ and the two flux differences.

Now, for TVD, we must have

$$0 \leqq -\left(\frac{h(u_{j+1}, u_j) - h(u_j, u_j)}{\Delta_+ u_j}\right)$$
$$\cdot\left[1 + \frac{1}{2}\frac{(h(u_{j+1}, u_j) - h(u_j, u_j))^m}{(h(u_{j+1}, u_j) - h(u_j, u_j))}\left[\frac{\psi(R_j^{m-})}{R_j^{m-}} - \psi(R_{j+1}^{m-})\right]\right],$$
$$0 \leqq \left(\frac{h(u_j, u_j) - h(u_j, u_{j-1})}{\Delta_- u_j}\right)$$
$$\cdot\left[1 + \frac{1}{2}\frac{(h(u_j, u_j) - h(u_j, u_{j-1}))^m}{(h(u_{j+1}, u_j) - h(u_j, u_j))}\left[\frac{\psi(R_j^{m+})}{R_j^{m+}} - \psi(R_{j-1}^{m+})\right]\right]$$

which is valid, because h satisfies (2.7), and because of (4.5), (4.6) and (3.10). Theorems (3.1) and (3.2) are also valid for this ACR version of our scheme, but our real goal is to choose the $a_{j+1/2}^{\pm}$ so that the inequality

$$\int_{u_j}^{u_{j+1}} [\mathscr{H}_{j+1/2}^{ac} - f(w)]\, dw \leqq 0 \tag{4.8}$$

is valid, i.e. to enforce inequality (4.4), with $V(w) = \frac{1}{2}w^2$.

It is easy to see that

$$\begin{aligned} -\int_{u_j}^{u_{j+1}} f(w)\, dw &= -\frac{1}{2}\Delta_+ u_j[f(u_{j+1}) + f(u_j)] \\ &\quad + \frac{1}{2}\int_{u_j}^{u_{j+1}} \left[\left(\frac{1}{2}(\Delta_+ u_j)\right)^2 - \left(w - \frac{1}{2}(u_{j+1} + u_j)\right)^2\right] f''(w)\, dw. \end{aligned} \tag{4.9}$$

Substituting this into (4.8) gives us the equivalent, desired inequality

$$\begin{aligned} &\frac{1}{2}\int_{u_j}^{u_{j+1}} \left[\left(\frac{1}{2}\Delta_+ u_j\right)^2 - \left(w - \frac{1}{2}(u_{j+1} + u_j)\right)^2\right] f''(w)\, dw \\ &\quad + \frac{1}{2}\Delta_+ u_j[(h(u_{j+1}, u_j) - h(u_j, u_j))^m (1 - \psi(R_{j+1}^{m-})) \\ &\qquad - a_{j+1/2}^-(h_1(u_{j+1}, u_{j+2}) - h_1(u_j, u_j))\Delta_+ u_j] \\ &\quad + \frac{1}{2}(\Delta_+ u_j)[(h(u_{j+1}, u_j) - h(u_{j+1}, u_{j+1}))^m (1 - \psi(R_j^{m+})) \\ &\qquad - a_{j+1/2}^+(h_0(u_{j+1}, u_{j+1}) - h_0(u_j, u_j))\Delta_+ u_j] \leqq 0. \end{aligned} \tag{4.10}$$

From (4.5), (4.6), and the fact that h is an E flux, it suffices to prove (after cancelling the factor $\frac{1}{2}$)

$$\begin{aligned} &\int_{u_j}^{u_{j+1}} f''(w)\, dw\left[\left(\frac{1}{2}\Delta_+ u_j\right)^2 - \left(w - \frac{1}{2}(u_{j+1} + u_j)\right)^2\right] \\ &\quad - a_{j+1/2}^-(\Delta_+ u_j)^2[h_1(u_{j+1}, u_{j+1}) - h_1(u_j, u_j)] \\ &\quad - a_{j+1/2}^+(\Delta_+ u_j)^2[h_0(u_{j+1}, u_{j+1}) - h_0(u_j, u_j)] \leqq 0. \end{aligned} \tag{4.11}$$

We thus have:

LEMMA 4.1. *Any solution to* (1.1) *which is the limit of a subsequence of approximate solutions* $\mathcal{V}_{\Delta'}$ as $\Delta' \to 0$ *of*

$$\begin{aligned} \frac{\partial u_j}{\partial t} &= -\frac{1}{\Delta}\Delta_+ \mathcal{H}^{ac}_{j-1/2}, \qquad j=0, \pm 1, \cdots, \pm N, \\ u_j(0) &= \frac{1}{\Delta}\int_{\mathcal{I}_j} w_0(s)\, ds, \end{aligned} \tag{4.12}$$

satisfies the entropy inequality (1.6), *provided that inequalities* (4.6) *and* (4.11) *are valid.*

As mentioned in the introduction, for scalar *convex* conservation laws where solutions lie in BV, a single entropy inequality implies uniqueness of solutions. This follows from the recent results of DiPerna [5]. We know that if $w(x, 0) \in BV$, then the function $\mathcal{V}_\Delta(x, t)$ has x variation bounded by that of $w(x, 0)$, for any fixed t. In particular

$$\frac{1}{h}\int_{-1}^{1} |\mathcal{V}_\Delta(x+h, t) - \mathcal{V}_\Delta(x, t)|\, dx \leqq \operatorname{var}(w(x, 0)) \tag{4.13}$$

for any $t>0$, $\Delta>0$, $h>0$.

The dominated convergence theorem then guarantees that any limit solution $w(x, t)$ will have x variation bounded by $w(x, 0)$, uniformly in t. Consider the interval $0 \leqq t \leqq T$, for any $T>0$. Then

$$\begin{aligned} &\frac{1}{h}\int_{-1}^{1} dx \int_0^T dt |\mathcal{V}_\Delta(x, t+h) - \mathcal{V}_\Delta(x, t)| \\ &= \frac{1}{h}\int_{-1}^{1} dx \int_0^T dt \left|\int_t^{t+h} \frac{\partial \mathcal{V}_\Delta}{\partial s}(x, s)\, ds\right| \\ &= \frac{1}{h}\int_0^T dt \left|\int_t^{t+h} \Delta_- \mathcal{H}_{j+1/2}(s)\, ds\right| \leqq 4MT \operatorname{var} w(x, 0), \end{aligned} \tag{4.14}$$

where M is the Lipschitz constant in

$$|\mathcal{H}(u_{j+1}, u_j, u_{j-1}, u_{j-2}) - \mathcal{H}(v_{j+1}, v_j, v_{j-1}, v_{j-2})| \leqq M \sum_{r=-2}^{1} |u_{j+r} - v_{j+r}|$$

for any pair of vectors $\{u_j\}$, $\{v_j\}$.

Thus the dominated convergence theorem implies that any limit solution $w(x, t)$ will have bounded variation in x and t. This together with Lemma 4.1 and the above mentioned uniqueness result gives our

THEOREM 4.1 (convergence). *The sequence of approximate solutions* $\mathcal{V}_\Delta$ *converges a.e. as* $\Delta \to 0$ *to the unique solution of the scalar convex conservation law* (1.1) *provided that the initial data is in BV and that inequalities* (4.6) *and* (4.11) *are valid.*

For simplicity, we exemplify this theory by considering semi-discrete versions of the following three monotone schemes.

(a) Lax–Friedrichs:

$$h^{LF}(u_{j+1}, u_j) = \tfrac{1}{2}f(u_{j+1}) + \tfrac{1}{2}f(u_j) - \tfrac{1}{2}k(u_{j+1} - u_j)$$

with $0<k$ chosen so that $|f'(u)| \leqq k\theta$, for some θ with $0<\theta<1$.

(b) Engquist–Osher:

$$h^{EO}(u_{j+1}, u_j) = \int_0^{u_{j+1}} \min(f'(s), 0)\,ds + \int_0^{u_j} \max(f'(s), 0)\,ds - f(0) \tag{4.15}$$
$$= f_-(u_{j+1}) + f_+(u_j) - f(0).$$

(c) Godunov:

$$h^G(u_{j+1}, u_j) = \begin{cases} \min\limits_{u_j \leqq u \leqq u_{j+1}} f(u) & \text{if } u_j < u_{j+1}, \\ \max\limits_{u_j \geqq u \geqq u_{j+1}} f(u) & \text{if } u_j > u_{j+1}. \end{cases}$$

For Lax–Friedrichs, (4.6a, b) become

$$|a^{\pm}_{j+1/2}(f'(u_{j+1}) - f'(u_j))| \leqq k \pm \frac{(f(u_{j+1}) - f(u_j))}{\Delta_+ u_j}, \tag{4.16}$$

while (4.11) is valid if

$$\int_{u_j}^{u_{j+1}} f''(w)\,dw\left[\left(\frac{\Delta_+ u_j}{2}\right)^2 - \left(w - \frac{1}{2}(u_{j+1} + u_j)\right)^2\right] \tag{4.17}$$
$$\leqq \min_{+,-} a^{\mp}_{j+1/2}\,(\Delta_+ u_j)^2 (f'(u_{j+1}) - f'(u_j)).$$

Clearly these inequalities are compatible if θ is sufficiently small. Moreover, if f is convex, then (4.16) becomes

(a) if $u_j < u_{j+1}$ (rarefaction)

$$a^{\mp}_{j+1/2} \geqq \frac{\int_{u_j}^{u_{j+1}} f''(w)\,dw\left[\left(\frac{\Delta_+ u_j}{2}\right)^2 - \left(w - \frac{1}{2}(u_{j+1} + u_j)\right)^2\right]}{(\Delta_+ u_j)^2 (f'(u_{j+1}) - f'(u_j))} \tag{4.18}$$

(b) if $u_j > u_{j+1}$ (shock)

$$a^{\mp}_{j+1/2} \leqq \frac{\int_{u_j}^{u_{j+1}} f''(w)\,dw\left[\left(\frac{\Delta_+ u_j}{2}\right)^2 - \left(w - \frac{1}{2}(u_{j+1} + u_j)\right)^2\right]}{(\Delta_+ u_j)^2 (f'(u_{j+1}) - f'(u_j))}$$

The right side of (4.18) is always positive for convex f. For example, for $f(w) = \frac{1}{2}w^2$, the right side is $\frac{1}{6}$.

The precise restrictions for this case become for Burgers' equation:

(4.19) (a)
$$|a^{\pm}_{j+1/2}\Delta_+ u_j| + \tfrac{1}{2}|u_{j+1} + u_j| \leqq k$$

and

(b)
$$\operatorname{sgn}(u_{j+1} - u_j)(a^{\pm}_{j+1/2} - \tfrac{1}{6}) \geqq 0.$$

Thus the high resolution scheme based on LF is a convergent approximation to Burgers' equation if θ is small enough and the $a^{\pm}_{j+1/2}$ are then chosen as above.

For the Engquist–Osher scheme, (4.6a,b) become

$$|a^{\mp}_{j+1/2}(f'_{\mp}(u_{j+1}) - f'_{\mp}(u_j))| \leqq \frac{\mp(f_{\mp}(u_{j+1}) - f_{\mp}(u_j))}{\Delta_+ u_j}, \tag{4.20}$$

while (4.11) is valid if

$$\int_{u_j}^{u_{j-1}} f''_{\mp}(w)\,dw\left[\left(\frac{\Delta_+ u_j}{2}\right)^2-\left(w-\frac{1}{2}(u_{j+1}+u_j)\right)^2\right] \tag{4.21}$$
$$\leqq a^{\pm}_{j+1/2}(\Delta_+ u_j)^2[f'_{\mp}(u_{j+1})-f'_{\mp}(u_j)].$$

Inequalities (4.20) and (4.21) are obviously compatible for convex f, for certain bounded $a^{\pm}_{j+1/2}$, if $|\Delta_+ u_j|$ is sufficiently small (i.e. in a region of smoothness). They are also compatible for $|\Delta_+ u_j|$ bounded away from zero. This follows since we may take in (4.20)

$$a^{\pm}_{j+1/2}=\frac{\mp(\Delta_+ f_{\mp} u_j)}{(\Delta_+ u_j)(\Delta_+ f'_{\mp}(u_j))}. \tag{4.22}$$

It now remains to prove that (4.21) is valid, i.e.,

$$\int_{u_j}^{u_{j+1}} f''(w)\,dw\left[\left(\frac{\Delta_+ u_j}{2}\right)^2-\left(w-\frac{1}{2}(u_{j+1}+u_j)\right)^2\right]\pm(\Delta_+ f_{\mp}(u_j))\Delta_+ u_j\leqq 0, \tag{4.23}$$

or equivalently

$$2\int_{u_j}^{u_{j+1}}(f_-(u_{j+1})-f_-(w))\,dw\leqq 0,$$
$$2\int_{u_j}^{u_{j+1}}(f_+(u_j)-f_+(w))\,dw\leqq 0. \tag{4.24}$$

These last are obvious facts.

In the case $f(w)=\frac{1}{2}w^2$, then (4.21) becomes

(4.25)

(Burgers' equation)

(a) if $u_j<u_{j+1}$ (rarefaction), then:

(i) $a^-_{j+1/2}\geqq\frac{1}{6}$, $a^+_{j+1/2}\geqq 0$ if $u_{j+1}\leqq 0$,

(ii) $a^-_{j+1/2}\geqq 0$, $a^+_{j+1/2}\geqq\frac{1}{6}$ if $u_j\geqq 0$,

(iii) $a^-_{j+1/2}\geqq\frac{1}{4}-\frac{1}{12}\left(\frac{3u_{j+1}^2+u_j^2}{(\Delta_+ u_j)^2}\right)$, $a^+_{j+1/2}\geqq-\frac{1}{12}+\frac{1}{12}\left(\frac{3u_{j+1}^2+u_j^2}{(\Delta_+ u_j)^2}\right)$ if $u_{j+1}>0>u_j$;

(b) if $u_j\geqq u_{j+1}$ (shock), then

(i) $a^-_{j+1/2}\leqq\frac{1}{6}$, $a^+_{j+1/2}\leqq 0$ if $u_j\leqq 0$,

(ii) $a^-_{j+1/2}\leqq 0$, $a^+_{j+1/2}\leqq\frac{1}{6}$ if $u_{j+1}\geqq 0$,

(iii) $a^-_{j+1/2}\leqq\frac{1}{4}-\frac{1}{12}\left(\frac{3u_j^2+u_{j+1}^2}{(\Delta_+ u_j)^2}\right)$, $a^+_{j+1/2}\leqq-\frac{1}{12}+\frac{1}{12}\left(\frac{3u_j^2+u_{j+1}^2}{(\Delta_+ u_j)^2}\right)$ if $u_{j+1}<0<u_j$.

Moreover, if $|a^{\mp}_{j+1/2}| \leqq \frac{1}{2}$ (which is admissible from (4.25)), then (4.20) is no restriction at all.

Thus, the E–O scheme incorporated into $H^{ac}_{j+1/2}$ yields a convergent high resolution scheme for any convex f, and any initial datum having bounded variation, provided the $a^{\pm}_{j+1/2}$ are judiciously chosen.

A tedious calculation yields the analogous results for Godunov's scheme.

5. Steady discrete shocks. We now check for the existence of discrete, steady, shock solutions to the high resolution, entropy condition satisfying schemes constructed in the previous section, based on the Engquist–Osher flux. Since the scheme satisfies Lemma 2.1, the profiles must be monotone. We shall also show that they are sharp.

Let u^L, u^R, be the left and right states for a physically correct steady shock, i.e.

$$f(u^L) = f(u^R),$$
$$f(u) < f(u^L) \quad \text{for all } u \text{ between } u^L \text{ and } u^R.$$

A steady, discrete shock $\{u_j\}$ will satisfy

$$\lim_{j\to-\infty} u_j = u^L, \qquad \lim_{j\to-\infty} u_j = u^R, \tag{5.1}$$
$$\tilde{H}^{ac}_{j+1/2} \equiv f(u^L) \quad \text{for all } j.$$

For simplicity, we assume $f''(u) > 0$, with $f'(0) = 0$ and $f(0) = 0$. The numerical flux becomes

$$\begin{aligned}\mathcal{H}^{ac}_{j+1/2} = f_-(u_{j+1}) + f_+(u_j) - \tfrac{1}{2}\psi(R^{M_-}_{j+1})(f_-(u_{j+1}) - f_-(u_j))^M \\ + \tfrac{1}{2}\psi(R^{M_+}_j)(f_+(u_{j+1}) - f_+(u_j))^M.\end{aligned} \tag{5.2}$$

If we take $k = 1$ in (3.9), we have

$$\begin{aligned}
&\text{(a)} \quad \psi(R^{M_-}_{j+1})(f_-(u_{j+1}) - f_-(u_j))^M \\
&\quad = -\Delta_+ u_j \max\Big[0, \min_{r=0,1}([\Delta_+ f_-(u_{j+r}) \\
&\qquad\qquad + a^-_{j+1/2+r}(\Delta_+ f'_-(u_{j+r})\Delta_+ u_{j+r}](-\Delta_+ u_j)^{-1})\Big], \\
&\text{(b)} \quad \psi(R^{M_+}_j)(f_+(u_{j+1}) - f_+(u_j))^M \\
&\quad = \Delta_+ u_j \max\Big[0, \min_{r=0,1}([\Delta_+ f_+(u_{j+r-1}) \\
&\qquad\qquad - a^+_{j-1/2+r}(\Delta_+ f'_+(u_{j+r-1}))\Delta_+ u_{j+r-1}](\Delta_+ u_j)^{-1})\Big].
\end{aligned} \tag{5.3}$$

We shall seek steady discrete shocks of the same general form obtained in [8] for the E–O scheme.

These are

$$u_j \equiv \begin{cases} u^L, & j \leqq -1, \\ u^R, & j \geqq 2, \end{cases} \qquad u^L > u_0 \geqq 0 > u_1 \geqq u^R. \tag{5.4}$$

For the first order scheme, u_0 can be viewed as a smooth function of u_1, satisfying the above and

$$f(u_0) + f(u_1) = f(u^L).$$

For the present scheme, we shall get a different one-parameter family of intermediate states.

It follows for all $j \neq 0$, that (5.3) implies: $\tilde{H}^{ac}_{j+1/2} \equiv f(u^L)$.

For $j=0$, we have the following equation:

$$\begin{aligned} 0 &= f(u_1)+f(u_0)-f(u^L) \\ &+\frac{1}{2}(u_1-u_0)\min\left[\frac{f(u_1)+a^-_{1/2}f'(u_1)(u_1-u_0)}{u_0-u_1},\right. \\ &\qquad \left.\frac{f(u^R)-f(u_1)+a^-_{3/2}(f'(u^R)-f'(u_1))(u^R-u_1)}{u_0-u_1}\right] \\ &+\frac{1}{2}(u_1-u_0)\min\left[\frac{f(u_0)-f(u^L)-a^+_{-1/2}(f'(u_0)-f'(u^L))(u-u^L)}{u_1-u_0},\right. \\ &\qquad \left.\frac{-f(u_0)-a^+_{-1/2}(-f'(u_0))(u_1-u_0)}{u_1-u_0}\right] = G(u_1, u_0). \end{aligned} \tag{5.5}$$

One special solution is, again, $u_1 = u^R$, $u_0 = 0$. We have

$$\begin{aligned} &\frac{\partial G}{\partial u_1}(u^R, 0) = \frac{3}{2}f'(u^R) < 0, \\ &\frac{\partial G}{\partial u_0}(u^R, 0) = 0, \\ &\frac{\partial^2 G}{\partial u_1^2}(u^R, 0) = \left(\frac{3}{2}-\frac{1}{2}a^-_{3/2}\right)f''(u^R) > 0, \\ &\frac{\partial^2 G}{\partial u_0 \partial u_1}(u^R, 0) = 0, \\ &\frac{\partial^2 G}{\partial u_0^2} = \frac{1}{2}f''(0) > 0. \end{aligned} \tag{5.6}$$

A simple application of the implicit function theorem gives us:

THEOREM 5.1. *There exists a family of sharp, discrete, shock solutions to* (4.12) *of the type* (5.3), *with u_1 a smooth function of u_0 and $0 < u_0$ small enough. By symmetry, the same is true with u_0 a smooth function of u_1, and $-u_1$ small enough.*

6. Systems of conservation laws. We shall now build a second order accurate scheme approximating (1.1) for systems: $m > 1$ and verify some of the desirable properties mentioned above. Our basic three point first order scheme will be the one devised by the first author [24], with Solomon [26], and analyzed jointly by us in [2], [25].

The numerical flux for this scheme is constructed as follows. First, we define a piecewise smooth, continuous path in phase space, connecting w_j to w_{j+1}, made up of m subpaths. Along subpath k, we have

$$\frac{dw}{ds} = r_k(w(s)), \tag{6.1}$$

with r_k, the kth eigenvalue of $\partial f(w)$ corresponding to eigenvalue r_k. (Here $\lambda_1 < \lambda_2 < \cdots < \lambda_m$.)

On such a subpath, the $m-1$ independent Riemann invariants, corresponding to field k, are constants. Our ordering is as follows. We begin at w with $k=m$, stop at some endpoint, use $k=m-1$, etc., arriving at w_{j+1} with the subpath corresponding to $k=1$. We call each such subpath $\Gamma^{\gamma}_{j+1/2}$, $\gamma=m,m-1,\cdots,1$. There exists exactly one such path $\Gamma_{j+1/2}=\bigcup_{\gamma=m}^{1}\Gamma^{\gamma}_{j+1/2}$, for $|w_{j+1}-w_j|$ sufficiently small (actually, for physical systems such as Euler's equations, the construction works in the large, as long as cavitation does not take place [26]).

We define the numerical flux to be

$$h^0_{j+1/2}=\frac{1}{2}\left[f(w_{j+1})+f(w_j)-\int_{\Gamma_{j+1/2}}|\partial f(w)|\,dw\right]. \tag{6.2}$$

Here $|\cdot|$ denotes the absolute value of a diagonalizable matrix.

This algorithm yields a closed form expression for $h^0_{j+1/2}$ for various physical systems, including Euler's equations for compressible, inviscid, gas dynamics, [26], [3], because their Riemann invariants can be easily tabulated, and because all fields are either linearly degenerate, or genuinely nonlinear. We assume this property throughout the remainder of this paper, for (1.1).

On such a k subpath, the integral in (6.2) is given by a simple expression involving $\pm f(w)$ at endpoints of the subinterval, and, in the genuinely nonlinear case, it may also involve $\pm 2f(\bar{w})$, where $\bar{w}$ is the unique sonic point for which $\lambda_k(\bar{w})=0$.

The full algorithm can be interpreted, geometrically, as solving the incoming Riemann problem in phase space, using only rarefaction, compression, or contact waves, then averaging the resulting multivalued solution, in physical space, as in Godunov's method [34].

The first order scheme may be written so as to yield a field by field decomposition:

$$\frac{\partial}{\partial t}w_j=-\frac{1}{\Delta}(h^0(w_{j+1},w_j)-h^0(w_j,w_j))-\frac{1}{\Delta}(h^0(w_j,w_j)-h^0(w_j,w_{j-1})), \tag{6.3}$$

with

$$\text{(a)}\quad h^0(w_{j+1},w_j)-h^0(w_j,w_j)=\int_{\Gamma_{j+1/2}}(\partial f(w))^+\,dw, \tag{6.4}$$
$$\text{(b)}\quad h^0(w_j,w_j)-h^0(w_j,w_{j-1})=\int_{\Gamma_{j-1/2}}(\partial f(w))^-\,dw.$$

Here

$$(\partial f)^+=\tfrac{1}{2}((\partial f)+|\partial f|),\qquad (\partial f)^-=\tfrac{1}{2}((\partial f)-|\partial f|).$$

For a linearly degenerate k field, we have λ_k constant on $\Gamma^k_{j+1/2}$, and hence:

$$\int_{\Gamma^k_{j+1/2}}(\partial f)^+\,dw=\chi(\lambda_k^{j+1/2})[f(w^u)-f(w^L)], \tag{6.5}$$
$$\int_{\Gamma^k_{j+1/2}}(\partial f)^-dw=(1-\chi(\lambda_k^{j+1/2}))[f(w^u)-f(w^L)]$$

where $\chi(x)\equiv 1$ if $x>0$, $\chi(x)\equiv 0$ if $x\leqq 0$, and w^u, w^L are the upper and lower limits of integration.

For a genuinely nonlinear k field, we have λ_k monotone on $\Gamma^k_{j+1/2}$, and hence:

$$(6.6)\quad \begin{aligned} &\text{(a)} \quad \int \Gamma^k_{j-1/2} (\partial f(w))^+ \, dw = f(w) \left\{ \begin{array}{ll} w^u, & \text{if } \lambda_k(w^u) > \lambda_k(\bar{w}), \\ \bar{w}, & \text{if } \lambda_k(\bar{w}) \geqq \lambda_k(w^u), \\ w^L & \text{if } \lambda_k(w^L) > \lambda_k(\bar{w}), \\ \bar{w}, & \text{if } \lambda_k(\bar{w}) \geqq \lambda_k(w^L); \end{array} \right. \\ &\text{(b)} \quad \int_{\Gamma^k_{j-1/2}} (\partial f(w))^- \, dw = f(w) \left\{ \begin{array}{ll} w^u, & \text{if } \lambda_k(w^u) \leqq \lambda_k(\bar{w}), \\ \bar{w}, & \text{if } \lambda_k(\bar{w}) < \lambda_k(w^u), \\ w^L, & \text{if } \lambda_k(w^L) \leqq \lambda_k(\bar{w}), \\ \bar{w}, & \text{if } \lambda_k(\bar{w}) < \lambda_k(w^L). \end{array} \right. \end{aligned}$$

As mentioned above, $\bar{w}$ is the unique sonic point (if it exists) for which $\lambda_k(\bar{w}) = 0$ on $\Gamma^k_{j+1/2}$.

This nonlinear field by field decomposition enables us to construct second order accurate schemes while working separately in each field. See, [16] for a similar treatment (without an entropy inequality), using Roe's nonlinear decomposition [28].

Let the end points of $\Gamma^k_{j+1/2}$ be denoted by $w^L = w_{j+1-k/m}$, $w^U = w_{j+1-(k-1)/m}$, and let the associated

$$(6.7)\quad \begin{aligned} &\text{(a)} \quad \int_{\Gamma^k_{j-1/2}} (\partial f)^- \, dw - N^-_{k,j+1/2}, \\ &\text{(b)} \quad \int_{\Gamma^k_{j-1/2}} (\partial f)^+ \, dw = N^+_{k,j-1/2}. \end{aligned}$$

As in the scalar case, a first attempt at a second order accurate extension comes via

$$(6.8)\quad \frac{\partial w}{\partial t} = -\frac{1}{\Delta}\left(\sum_{k=m}^{1} \left(1 - \frac{1}{2}\Delta_+\right) N^-_{k,j+1/2} + \sum_{k=m}^{1} \left(1 + \frac{1}{2}\Delta_-\right) N^+_{k,j-1/2} \right).$$

This differencing, although formally second order accurate, again leads to overshoot, undershoot, and occasionally nonlinear instability.

To remedy this, we limit the flux differencing, as in the scalar case, except that we do this separately in each field. We shall use the entropy function, discussed in the first section, in the construction of the scheme. We remark, that although the existence of a convex $V(x)$ follows only if the overdetermined system

$$(\nabla_w V)^T \left(\frac{\partial f}{\partial w} \right) = \nabla_w F$$

has a convex solution, the fact is that most of the equations of physics are endowed with such an entropy. See [6] for further details.

The entropy gradient will be used to construct a linear functional applied to vectors w. This is correct dimensionally, as pointed out by Harten–Lax in [17].

We first define

$$(6.9)\quad V_w(w_{j+1-(k-1/m)}) - V_w(w_{j+1-(k/m)}) = \Delta_+^{(j,k)} V_w.$$

The algorithm in each field will involve the quantities

$$(6.10)\qquad R^+_{k,j}=\frac{(N^+_{k,j-1/2},\Delta^{(j,k)}_+V_w)}{(N^+_{k,j+1/2},\Delta^{(j,k)}_+V_w)},\qquad R^-_{k,j}=\frac{(N^-_{k,j+1/2},\Delta^{(j,k)}_+V_w)}{(N^+_{k,j-1/2},\Delta^{(j,k)}_+V_w)}$$

for each field $k=1,\cdots,m$.

Here $(\cdot,\cdot)$ denotes the usual inner product in R^m.

Using any flux limiter of the type described in § 3 (i.e. $\psi(R)$ satisfies (3.8), with $\psi(1)=1$), we define a second order accurate, TVD type discretization:

$$(6.11)\qquad \begin{aligned}\frac{\partial w_j}{\partial t}=-\frac{1}{\Delta}\Bigg(&\sum_{k=1}^m\left(N^-_{k,j+1/2}-\frac{1}{2}\Delta_-(\psi(R^-_{k,j+1})N^-_{k,j+1/2})\right)\\&+\sum_{k=1}^m\left(N^+_{k,j-1/2}+\frac{1}{2}\Delta_+(\psi(R^+_{k,j-1})N^+_{k,j-1/2})\right)\Bigg).\end{aligned}$$

We now show that for linear systems, this nonlinear scheme decouples to a second order accurate variation diminishing algorithm, and hence approximate solutions converge, as $\Delta\to 0$, to the unique solution of (1.1). (See also [16], for the first proof of such a result.)

We have

$$\partial f(w)=A=R\Lambda L,$$

with:

$$R=[r_1,r_2,\cdots,r_m]\quad\text{and}\quad L=\begin{bmatrix}l_1\\l_2\\\vdots\\l_m\end{bmatrix},$$

the constant matrices made up of left and right eigenvalues of A, suitably normalized so that

$$(l_i,r_j)=\begin{cases}0 & \text{if } i\neq j,\\1 & \text{if } i=j,\end{cases}$$

and

$$\Lambda=\begin{bmatrix}\lambda_1 & & & 0\\ & \lambda_2 & & \\ & & \ddots & \\ 0 & & & \lambda_m\end{bmatrix}.$$

Then

$$N^+_{k,j-1/2}=(l_k,\Delta_-w_j)\max(\lambda_k,0)r_k,$$

$$N^-_{k,j+1/2}=(l_k,\Delta_+w_j)\min(\lambda_k,0)r_k.$$

Moreover, since the entropy is just defined by $V(w)=\frac{1}{2}(w,w)$ for linear systems, then

$$\Delta^{(j,k)}_+V_w=(l_k,\Delta_+w)r_k.$$

Thus, if we take the inner product of (6.11) with l_k, and denote (l_k, w_j) by $w_j^{(k)}$, we obtain

$$\frac{\partial}{\partial t} w_j^{(k)} = -\frac{1}{\Delta}\Big[\min(\lambda_k, 0)\Big(\Delta_+ w_j^{(k)} - \frac{1}{2}\Delta_-(\psi(R^-_{k,j+1})(\Delta_+ w_j^{(k)}))\Big)$$

$$+\max(\lambda_k, 0)\Big(\Delta_- w_j^{(k)} + \frac{1}{2}\Delta_+(\psi(R^+_{k,j-1})(\Delta_- w_j^{(k)}))\Big)\Big].$$

This scheme is of the type (3.6)–(3.8), and hence is TVD. Thus we have the variation bound for each $w_j^{(k)}$, as in § 2, as well as the maximum (minimum) principles stated there. We thus have the simple:

THEOREM 6.1. *The approximation to the linear hyperbolic system* (1.1), *defined by* (6.11), *with initial data in* $L^1 \cap L^\infty \cap BV$, *converges as* $\Delta \to 0$ *in* $L^\infty(L^1(R), [0, T])$, *to the unique solution of* (1.1).

Next, in order to sharpen shocks and contacts, and to enforce a discrete entropy condition for systems, we use ACR, as in the scalar case. We again (as in the scalar case) require that $\psi(R)$ be defined by (3.4), with $k = 1$.

In order to add ACR, we use the k eigenvalue difference along a k subpath. This is defined by

$$\begin{aligned} &\max(\lambda_k(w_{j+1-(k-1/m)}), 0) - \max(\lambda_k(w_{j+1-(k/m)}), 0) = \Delta_+^{(j,k)}\lambda_+, \\ &\min(\lambda_k(w_{j+1-(k-1/m)}), 0) - \min(\lambda_k(w_{j+1-(k/m)}), 0) = \Delta_+^{(j,k)}\lambda_-. \end{aligned} \tag{6.12}$$

Next we define

$$\text{(6.13)}\quad \text{(a)}\quad N^{M+}_{k,j+1/2} = N^+_{k,j+1/2}\left(1 - a^+_{k,j+1/2}(\Delta_+^{(j,k)}\lambda_+)\frac{(\Delta_+^{(j,k)}w, \Delta_+^{(j,k)}V_w)}{(N^+_{k,j+1/2}, \Delta_+^{(j,k)}V_w)}\right)$$

for genuinely nonlinear k fields, and

$$\text{(6.13)}\quad \text{(b)}\quad N^{M+}_{k,j+1/2} = N^+_{k,j+1/2}(1 + a^+_{k,j+1/2}(\Delta_+^{(j,k)}w, \Delta_+^{(j,k)}V_w)^{1/2})$$

for linearly degenerate fields.

Analogously, we define

$$\text{(6.14)}\quad \text{(a)}\quad N^{M-}_{k,j+1/2} = N^-_{k,j+1/2}\left(1 + a^-_{k,j+1/2}(\Delta_+^{(j,k)}\lambda_-)\frac{(\Delta_+^{(j,k)}w, \Delta_+^{(j,k)}V_w)}{(N^-_{k,j+1/2}, \Delta_+^{(j,k)}V_w)}\right)$$

for genuinely nonlinear fields, and

$$\text{(6.14)}\quad \text{(b)}\quad N^{M-}_{k,j+1/2} = N^-_{k,j+1/2}(1 + a^-_{k,j+1/2}(\Delta_+^{(j,k)}w, \Delta_+^{(j,k)}V_w)^{1/2})$$

for linearly degenerate fields.

We make the following restriction on the Lipschitz continuous functions $a^{\mp}_{k,j+1/2}$:

Restriction 6.1. The values of the coefficients multiplying $N^{\mp}_{k,j+1/2}$ in (6.13) and (6.14) are always strictly between 0 and 2.

We now define $R^{M_\pm}_{k,j}$ as in (6.10), replacing $N^\pm_{k,j+1/2}$ by these compressed or rarefied flux differences. The TVD and ACR scheme is constructed as follows:

$$\begin{aligned}\frac{\partial w_j}{\partial t} = -\frac{1}{\Delta}\Bigg(\sum_{k=1}^{m}\left(N^-_{k,j+1/2}-\frac{1}{2}\Delta_-(\psi(R^{M_-}_{k,j+1})N^{M_-}_{k,j+1/2})\right)\\ +\sum_{k=1}^{m}\left(N^+_{k,j-1/2}+\frac{1}{2}\Delta_+(\psi(R^{M_+}_{k,j-1})N^{M_+}_{k,j-1/2})\right)\Bigg)\\ = -\frac{1}{\Delta}\Delta_-\mathcal{H}^{ac}_{j+1/2}.\end{aligned} \tag{6.15}$$

We shall now prove our last result.

THEOREM 6.2. *There exist values of $a^\pm_{k,j+1/2}$, for each j, k such that if the solutions to* (6.15) *with the usual initial data have, for Δ small enough,* (a) *sufficiently small oscillation, and* (b) *the property that eigenvalues λ_k corresponding to linearly degenerate fields stay bounded away from zero, then bounded a.e. limits of the approximate solution satisfy the entropy inequality* (1.7).

Proof. It was shown in [22] that it suffices to obtain the inequality

$$\int_{\Gamma_{j+1/2}}(dw)^T V_{ww}(w)[\mathcal{H}^{ac}_{j+1/2}-f(w)]\leqq 0, \tag{6.16}$$

for all j, under the given hypotheses.

An integration by parts and rearrangement of terms gives us an equivalent problem:

$$\begin{aligned}\int_{\Gamma_{j+1/2}}(dw)^T(\partial f(w))\left[V_w(w)-\frac{1}{2}(V_w(w_{j+1})+V_w(w_j))\right]\\ -\frac{1}{2}\left((\Delta_+V_w(w_j)),\sum_{k=1}^{m}N^{M_+}_{k,j+1/2}(1-\psi(R^{M_+}_{k,j}))\right)\\ +\frac{1}{2}\left((\Delta_+V_w(w_j)),\sum_{k=1}^{m}N^{M_-}_{k,j+1/2}(1-\psi(R^{M_-}_{k,j+1}))\right)\\ -\frac{1}{2}\left((\Delta_+V_w(w_j)),\sum_{k=1}^{m}(N^+_{k,j+1/2}-N^{M_+}_{k,j+1/2})\right)\\ +\frac{1}{2}\left((\Delta_+V_w(w_j)),\sum_{k=1}^{m}(N^-_{k,j+1/2}-N^{M_-}_{k,j+1/2})\right)\leqq 0.\end{aligned} \tag{6.17}$$

Let k' denote indices corresponding to genuinely nonlinear fields, and k'' denote the remaining fields (which are all linearly degenerate).

Then

$$\begin{aligned}&\text{(a)}\quad N^{M_\pm}_{k',j+1/2}-N^\pm_{k,j+1/2}=\mp a^\pm_{k',j+1/2}(\Delta_+^{(j,k')}\lambda_\pm)\frac{(\Delta_+^{(j,k')}w,\Delta_+^{(j,k')}V_w)}{(N^\pm_{k',j+1/2},\Delta_+^{(j,k')}V_w)}N^\pm_{k',j+1/2},\\ &\text{(b)}\quad N^{M_\pm}_{k'',j+1/2}-N^\pm_{k'',j+1/2}=a^\pm_{k'',j+1/2}(\Delta_+^{(j,k'')}w,\Delta_+^{(j,k'')}V_w)^{1/2}N^\pm_{k'',j+1/2}.\end{aligned} \tag{6.18}$$

(Notice that $a^\pm_{k'',j+1/2}$ are not dimensionless quantities—they have dimension inverse to that of $(V)^{1/2}$. The $a^\pm_{k',j+1/2}$ are normalized to be dimensionless.)

Using the argument (and notation) in [22, § 3], it is easy to show that

$$\begin{aligned}(6.19)\quad &\frac{1}{2}\left(\Delta_+ V_w(w_j), \int_{\Gamma^k_{j+1/2}} (\partial f(w))\, dw\right)\\ &= \frac{1}{2} s_k \int_0^{s_k} dt (r_k(w(t))^T V_{ww} |\lambda_k(w(t))| r_k(w(t)))\\ &+ \frac{1}{2}\sum_{\nu=1}^m \int_0^{s_\nu} ds \int_0^{s_k} dt (V_{ww}(w(s)) r_\nu(w(s)) - V_{ww}(w(t) r_\nu(w(t)))^T)\\ &\qquad\cdot |\lambda_k(w(t))| r_\nu(w(t)).\end{aligned}$$

Let $\varepsilon_j = |\Delta_+ w_j|$ be sufficiently small. Then our assumption about the fields, the argument in [22], together with (6.20) and Restriction 6.1 guarantees that the second and third terms in (6.17) are nonpositive. Moreover, regarding the last two terms, we have, using (6.18), and letting $C>0$ be a universal constant:

$$\begin{aligned}&(\Delta_+ V_w(w_j))^T\left(\sum_{k=1}^m (N^+_{k,j+1/2} - N^{M_+}_{k,j-1/2}) + \sum_{k=1}^m (N^{M_-}_{k,j+1/2} - N^-_{k,j+1/2})\right)\\ (6.20)\quad &\geqq (1-C\varepsilon_j)\left(\sum_{k'} (a^+_{k',j+1/2}(\Delta^{(j,k')}_+ \lambda_+) + a^-_{k',j+1/2}(\Delta^{(j,k')}_+ \lambda_-))(\Delta^{(j,k')}_+ w, \Delta^{(j,k')}_+ V_w)\right)\\ &+(1-C\varepsilon_j)\left(\sum_{k''} (-a^+_{k'',j+1/2}(\Delta^{(j,k'')}_+ V_w, N^+_{k'',j+1/2})\right.\\ &\qquad \left. + a^-_{k'',j+1/2}(\Delta^{(j,k'')}_+ V_w, N^-_{k'',j+1/2}))(\Delta^{(j,k'')}_+ w, \Delta^{(j,k'')}_+ V_w)^{1/2}\right),\end{aligned}$$

recalling again, that the k' are the genuinely nonlinear, and the k'' are the linearly degenerate, fields.

Here we have taken

$$(6.21)\quad \begin{aligned}&\text{(a)} \qquad a^\pm_{k',j+1/2}(\Delta^{(j,k')}_+ \lambda_\pm) \geqq 0,\\ &\text{(b)} \qquad a^\pm_{k'',j+1/2} \leqq 0.\end{aligned}$$

This means that we have added only expansion, not compression, to our scheme. The freedom to add compression at shocks or contacts is lacking, only because of technical points in our proof.

The integral in (6.17) can be rewritten as

$$(6.22)\quad \begin{aligned}&-\frac{1}{2}\int_{\Gamma_{j+1/2}} \left(\int_{\Gamma_{j+1/2,w,U}} ((dv)^T V_{ww}(v))\partial f(w)\right) dw\\ &+\frac{1}{2}\int_{\Gamma_{j+1/2}} \left(\int_{\Gamma_{j+1/2,w,L}} ((dv)^T V_{ww}(v))\partial f(w)\right) dw.\end{aligned}$$

Here, $\Gamma_{j+1/2,w,U}$ is that part of $\Gamma_{j+1/2}$ which starts at w_j and ends at w_{j+1}, $\Gamma_{j+1/2,w,L}$ starts at w, and ends at w. Thus $\Gamma_{j+1/2} = \Gamma_{j+1/2,w,L} \cup \Gamma_{j+1/2,w,U}$. Using the arguments of [22, § 5], we can show that the sum of all the integrals along subsegments is

$$O\left(\sum_{k=1}^m |s_k|^2 \left| \int_0^{s_k} |\lambda_k(w(s))|\, ds \right|\right),$$

except, perhaps for a term

$$(6.23)\quad \frac{1}{2}\sum_k \int_0^{s_k} ds \left(\int_0^s dt - \int_s^{s_k} dt\right)((r_k(w(t)))^T V_{ww}(w(t)) r_k(w(t)) \lambda_k(w(t))).$$

Let $t = s_k - t'$ in the second integral. Using our hypotheses about the fields, easily gives us an upper bound

$$\sum_{k=1}^{M} D_k(s_k)^2 \left| \int_0^{s_k} |\lambda_k(w(s))| \, ds \right| \tag{6.24}$$

for this integral, where the D_k are certain universal positive constants.

Comparing (6.23), (6.24) to the right side of (6.20) shows that we may choose each constant $a^{\pm}_{k,j+1/2}$ subject to Restriction 6.1, and sufficiently large in magnitude, (if ε_j is sufficiently small) so that (6.17) is valid.

7. Results of numerical experiments. Since our preceding theory concerned only semidiscrete, continuous in time, approximations, we begin by describing the time

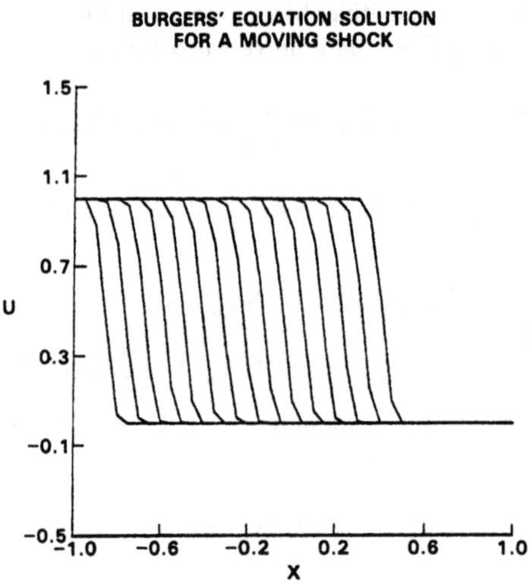

FIG. 1.

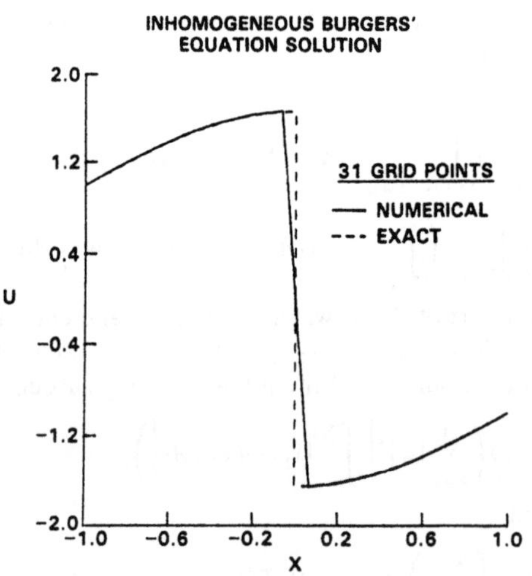

FIG. 2.

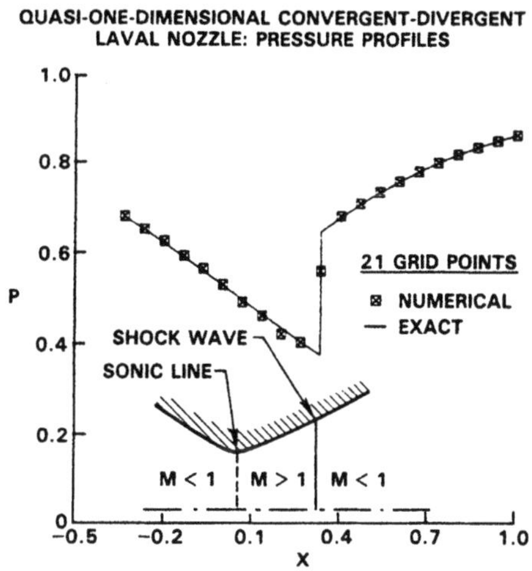

FIG. 3.

discretizations used for all calculations presented here. That was a two-step time differencing found in Richtmyer and Morton [27]. In one space dimension it can be written as

$$\text{(a)}\quad w^{n+1/2} = w^n - \frac{\Delta t}{2}\{\text{first order approximation to } f_x^n\},$$

$$(7.1)\quad \text{(b)}\quad w^{n+1} = w^n - \Delta t\,\{\text{first order approximation to } f_x^{n+1/2}\}$$

$$-\frac{\Delta t\,\Delta x}{2}\{\text{``upwind'' approximation to } \pm f_{xx}^n \text{ using flux limiters}\}.$$

The resulting explicit algorithm is not in general, truly variation diminishing, but still works well, even in multidimensional calculations.

We begin by computing a solution to inviscid Burger's equation which is a shock moving with speed $\frac{1}{2}$. The results are shown in Fig. 1. Next we compute a solution to the steady state Burger's equation having an inhomogeneous right-hand side, and compare the result with the known exact solution—Fig. 2.

Both of these scalar experiments were done using the flux limiter in (3.9), with $k=1$, and with the high resolution scheme based on Engquist–Osher first order differencing.

Next we compute solutions to Euler's equations of compressible gas dynamics, using the schemes constructed in the previous section. See [2], [3], for the precise equations, and algorithm used here. Figure 3 shows the results of quasi-one-dimensional Laval nozzle flow. Reference [2] has the equation with precise source term.

We next consider the case of an oblique shock reflecting from a flat plate, in two space dimensions. Figure 4a shows a cartesian grid with 61×21 points. Figure 4b shows a first order accurate solution, 4c was obtained using a non-TVD second order extension, and 4d shows the contours obtained using the TVD second order accurate scheme. Figure 4b has highly smeared incident, and reflected shocks, 4c has thick profiles because of undershoots and overshoots near the shock, and 4d has relatively sharp, nonoscillatory profiles.

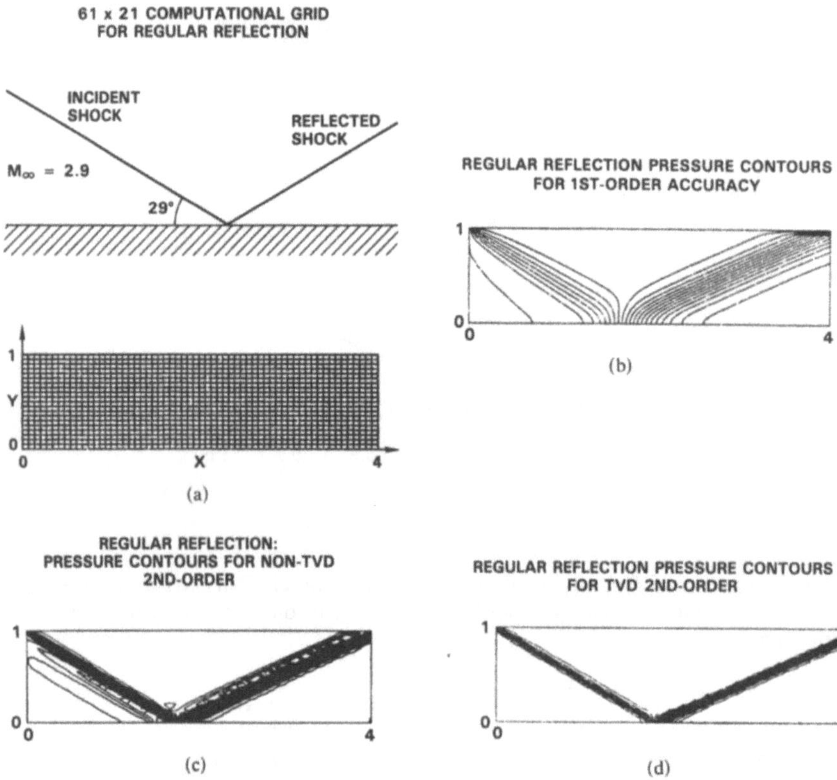
61 x 21 COMPUTATIONAL GRID
FOR REGULAR REFLECTION
INCIDENT
SHOCK
REFLECTED
SHOCK
M∞ = 2.9
29°
Y
X
(a)
REGULAR REFLECTION PRESSURE CONTOURS
FOR 1ST-ORDER ACCURACY
(b)
REGULAR REFLECTION:
PRESSURE CONTOURS FOR NON-TVD
2ND-ORDER
(c)
REGULAR REFLECTION PRESSURE CONTOURS
FOR TVD 2ND-ORDER
(d)

FIG. 4.

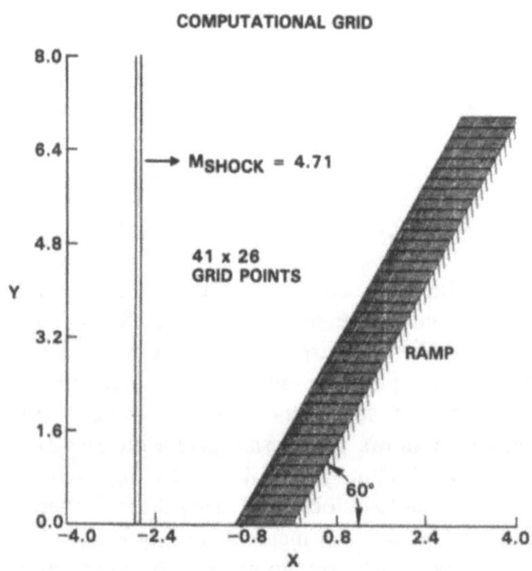
COMPUTATIONAL GRID
MSHOCK = 4.71
41 x 26
GRID POINTS
RAMP
60°
Y
X

FIG. 5.

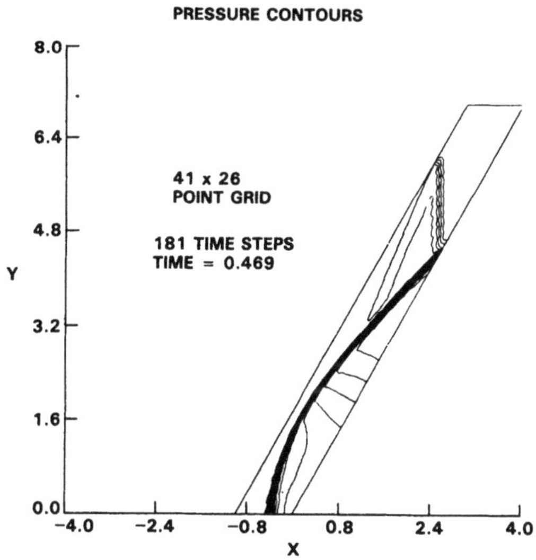

FIG. 6.

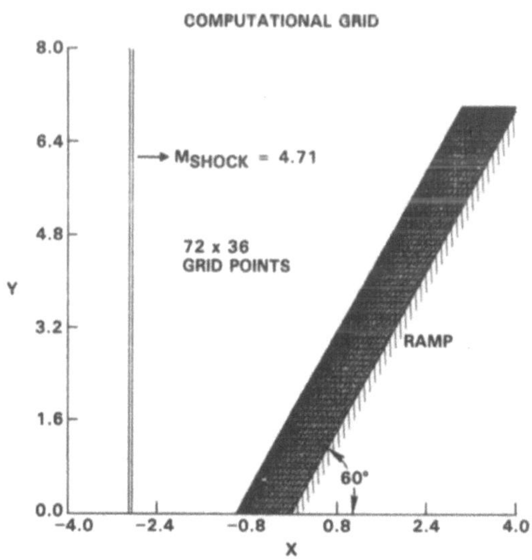

FIG. 7.

In Fig. 5, we present the computational grid to be used for a blast wave problem solved as an unsteady problem. In Fig. 6, we give the computed pressure contours using the TVD, second order scheme, at time $T=0.469$. In Fig. 7, we show a finer computational mesh, and in Fig. 8, we present the pressure contours using the TVD, second order scheme at time $T=0.459$.

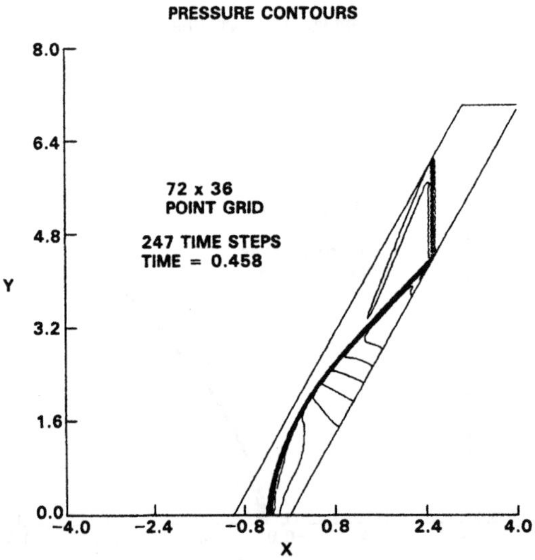

FIG. 8.

Acknowledgment. The authors would like to thank Peter Sweby and Bram van Leer for some very valuable conversations concerning flux limiters.

REFERENCES

[1] L. ABRAHAMSSON AND S. OSHER, *Monotone difference schemes for singular perturbation problems*, this Journal, 19 (1982), pp. 979–991.

[2] S. CHAKRAVARTHY AND S. OSHER, *Numerical experiments with the Osher upwind scheme for the Euler equations*, AIAA paper #82-0975, AIAA J., to appear.

[3] ———, *High resolution applications of the Osher upwind scheme for the Euler equations*, AIAA paper #83-1943, Danvers, MA, 1983.

[4] P. COLELLA AND P. R. WOODWARD, *The piecewise-parabolic method* (PPM) *for gas-dynamical simulations*, LBL report #14661, (July 1982).

[5] R. J. DIPERNA, *Convergence of approximate solutions to conservation laws*, Arch. Rat. Mech. Anal., 82 (1983), pp. 27–70.

[6] ———, *Decay of solutions of hyperbolic systems of conservation laws with a convex extension*, Arch. Rat. Mech. Anal., 64 (1977), pp. 1–92.

[7] J. W. EWARDS, R. M. BENNETT, W. WHITLOW, JR. AND D. A. SEIDEL, *Time marching transonic flutter solutions including angle-of-attack effects*, AIAA paper #82-0685, New Orleans, LA, 1982.

[8] B. ENGQUIST AND S. OSHER, *Stable and entropy condition satisfying approximations for transonic flow calculations*, Math. Comp., 34 (1980), pp. 45–75.

[9] ———, *One sided difference approximations for nonlinear conservation laws*, Math. Comp., 36 (1981), pp. 321–351.

[10] S. K. GODUNOV, *A finite difference method for the numerical computation of discontinuous solutions of the equations of fluid dynamics*, Mat. Sb., 47(1959), pp. 271–290.

[11] J. B. GOODMAN AND R. J. LEVEQUE, *On the accuracy of stable schemes for two dimensional conservation laws*, preprint.

[12] P. M. GOORJIAN AND R. VAN BUSKIRK, *Implicit calculations of transonic flow using monotone methods*, AIAA paper #81-0331, St. Louis, MO, 1981.

[13] P. M. GOORJIAN, M. MEAGHER, AND R. VAN BUSKIRK, *Monotone implicit algorithms for the small disturbance and full potential equation applied to transonic flow*, AIAA paper #83-0371, Reno, NV, 1983.

[14] B. GUSTAFSSON, *The convergence rate for difference approximations to mixed initial boundary value problems*, Math. Comp., 29 (1975), pp. 396–406.

[15] A. HARTEN, *The artificial compression method for computing shocks and discontinuities. I. Single conservation laws*, Comm. Pure Appl. Math., 30 (1977), pp. 611–638.

[16] ———, *High resolution schemes for hyperbolic conservation laws*, J. Comp. Physics, to appear.

[17] A. HARTEN AND P. D. LAX, *A random choice finite-difference scheme for hyperbolic conservation laws*, this Journal, 18 (1981), pp. 289–315.

[18] S. N. KRUZKOV, *First order quasi-linear equations in several independent variables*, Math. USSR Sb., 10 (1970), pp. 217–243.

[19] P. D. LAX, *Hyperbolic Systems of Conservation Laws and the Mathematical Theory of Shock Waves*, CBMS Regional Conference Series in Applied Mathematics 11, Society for Industrial and Applied Mathematics, Philadelphia, 1973.

[20] A. MAJDA AND S. OSHER, *Numerical viscosity and the entropy condition*, Comm. Pure Appl. Math., 32 (1979), pp. 797–838.

[21] E. M. MURMAN, *Analysis of embedded shock waves calculated by relaxation methods*, AIAA J., 12 (1974), pp. 626–633.

[22] S. OSHER, *Riemann solvers, the entropy condition, and difference approximations*, this Journal, 21 (1984), pp. 217–235.

[23] ———, *Shock modelling in transonic and supersonic flow*, Recent Advances in Numerical Methods in Fluids, Vol. 4, Advances in Computational Transonics, W. G. Habashi, ed., Pineridge Press, Swansea, UK, 1984.

[24] ———, *Numerical solution of singular perturbation problems and hyperbolic systems of conservation laws*, Mathematical Studies 47, S. Axelsson, L. S. Frank, and A. van der Sluis, eds., North-Holland, Amsterdam, 1981, pp. 179–205.

[25] S. OSHER AND S. CHAKRAVARTHY, *Upwind schemes and boundary conditions with applications to Euler equations in general geometries*, J. Comp. Phys., to appear.

[26] S. OSHER AND F. SOLOMON, *Upwind schemes for hyperbolic systems of conservation laws*, Math. Comp., 38 (1982), pp. 339–377.

[27] R. D. RICHTMYER AND K. W. MORTON, *Difference Methods for Initial-Value Problems*, 2nd ed., Interscience, New York, 1967.

[28] P. L. ROE, *Approximate Riemann solvers, parameter vectors, and difference schemes*, J. Comp. Phys., 43 (1981), pp. 357–372.

[29] R. SANDERS, *On convergence of monotone finite difference schemes with variable spatial differencing*, Math. Comp., 40 (1983), pp. 91–106.

[30] P. K. SWEBY, *High resolution schemes using flux limiters for hyperbolic conservation laws*, this Journal, this issue, pp. 995–1011.

[31] P. K. SWEBY AND M. J. BAINES, *Convergence of Roe's scheme for the general nonlinear scalar wave equation*, Numerical Analysis Report 8/81, Univ. Reading, 1981.

[32] B. VAN LEER, *Towards the ultimate conservative difference scheme, II. Monotonicity and conservation combined in a second-order scheme*, J. Comp. Phys., 14 (1974), pp. 361–370.

[33] ———, *Towards the ultimate conservative difference scheme, V. A second-order sequel to Godunov's method*, J. Comp. Phys., 32 (1979), pp. 101–136.

[34] ———, *On the relation between the upwind differencing schemes of Godunov, Engquist–Osher, and Roe*, SIAM J. Sci. Statist. Comput., 5 (1984), pp. 1–20.

CONVERGENCE OF GENERALIZED MUSCL SCHEMES*

STANLEY OSHER†

Abstract. Semidiscrete generalizations of the second order extension of Godunov's scheme, known as the MUSCL scheme, are constructed, starting with any three point "E" scheme. They are used to approximate scalar conservation laws in one space dimension. For convex conservation laws, each member of a wide class is proven to be a convergent approximation to the correct physical solution. Comparison with another class of high resolution convergent schemes is made.

1. Introduction and preliminaries. Recently, there has been an enormous amount of activity related to the construction and analysis of "high resolution" schemes approximating hyperbolic systems of conservation laws. Some examples of the successful consequences of this activity can be found in the proceedings of the latest (sixth) AIAA Computational Fluid Dynamics Conference, [1], [10], [18]. Extensive bibliographies can also be found in these papers.

Our aim here is merely to construct and prove the convergence of a subclass of these schemes approximating scalar convex conservation laws. This subclass is based on an idea of van Leer [17]. He christened his algorithms MUSCL (Monotonic Upstream Centered Schemes for Conservation Laws) schemes, and with mixed emotions, we shall use his acronym here.

In future work with S. Chakravarthy, we shall extend this construction to systems in multidimensions, using triangle-based algorithms. This work in progress will stress the computational aspects of the algorithms, especially as they relate to the Euler equations of compressible gas dynamics. In earlier work with the same author [12], we constructed, and proved convergence of, a class of high resolution schemes approximating scalar convex conservation laws. We also showed for certain high resolution approximations to systems that limit solutions satisfy an entropy inequality.

We shall consider numerical approximations to the initial value problem for a single conservation law in one space dimension

$$w_t + f(w)_x = 0, \quad t > 0, \quad -1 \leqq x < 1, \tag{1.1a}$$

with a periodic boundary condition

$$w(x+1, t) = w(x, t), \tag{1.1b}$$

and initial condition

$$w(x, 0) = w_0(x). \tag{1.1c}$$

It is well known that solutions of (1.1) may develop discontinuities in finite time, even when the initial data are smooth. Because of this, we seek a weak solution of

* Received by the editors February 14, 1984. This research was supported in part by the National Science Foundation under grant MCS 82-00788, the Army Research Office under grant DAAG 29-82-k-0090, and the National Aeronautics and Space Administration under grant NAG-1-270. Part of the research was carried out while the author was a visitor at the Institute for Computer Applications in Science and Engineering (ICASE), NASA Langley Research Center, Hampton, Virginia, which is operated under NASA contract NASI-17070.

† Department of Mathematics, University of California at Los Angeles, Los Angeles, California 90024.

(1.1), i.e. a bounded measurable function w, such that, for all $\phi \in C_0(R \times R^+)$,

$$\iint_{R\times R^+} (w\phi_t + f(w)\phi_x)\, dx\, dt = 0, \tag{1.2a}$$

$$\lim_{t\to 0} \|w(x,t) - w_0(x)\|_{L^1} = 0. \tag{1.2b}$$

Solutions of (1.2) are not necessarily unique. For physical reasons, the limit solution of the viscous equation, as viscosity tends to zero, is sought. In the scalar case, this solution must satisfy, for $\phi \in C_0^\infty(R \times R^+)$, $\phi \geqq 0$ and all real constants c

$$-\iint (|w-c|\phi_t + \operatorname{sgn}(w-c)(f(w)-f(c))\phi_x)\, dx\, dt \leqq 0. \tag{1.3a}$$

This is equivalent to the statement,

$$\frac{\partial}{\partial t}|w-c| + \frac{\partial}{\partial x}((f(w)-f(c))\operatorname{sgn}(w-c)) \leqq 0, \tag{1.3b}$$

in the sense of distributions.

Such solutions are called entropy solutions. Kruzkov has shown in [9] that two entropy solutions satisfy

$$\|w(x,t_1) - u(x,t_1)\|_{L^1} \leqq \|w(x,t_0) - u(x,t_0)\|_{L^1} \tag{1.4}$$

for all $t_1 \geqq t_0$. Hence, (1.3) guarantees the uniqueness of solutions to the scalar version of (1.2). Existence was also obtained there.

Equations (1.3) may be viewed as the statement

$$\frac{\partial}{\partial t} V(w) + \frac{\partial}{\partial x} F(w) \leqq 0 \tag{1.5}$$

for each of the convex "entropy" functions of solutions w to (1.1),

$$V(w) = |w-c|,$$

and their associated flux functions

$$F(w) = (f(w)-f(c))\operatorname{sgn}(w-c).$$

For scalar *convex* conservation laws whose solutions lie in the space BV, a *single* entropy inequality (1.5) for any strictly convex $V(w)$ and its associated entropy flux $F(w)$, satisfying

$$F'(w) = V'(w)f'(w), \tag{1.6}$$

are enough to imply existence and uniqueness of solutions to (1.1). This follows from the results of DiPerna [3].

Next we consider a semidiscrete, method of lines, approximation to (1.1). We break the interval $(-1, 1)$ into subintervals,

$$I_j = \{x \mid (j-\tfrac12)\Delta \leqq x < (j+\tfrac12)\Delta\}$$

$j = 0, \pm 1, \cdots, \pm N$, with $(2N+1)\Delta = 2$.

Let $x_j = j\Delta$, be the center of each interval I_j, with end points $x_{j-1/2}, x_{j+1/2}$.

Define the step function for each $t > 0$ as

$$U_\Delta(x,t) = u_j(t),$$

for $x \in I_j$.

The initial data is discretized via the averaging operator T_Δ,

$$T_\Delta w_0(x) = \frac{1}{\Delta}\int_{I_j} w_0(s)\,ds = u_j(0) \quad \text{for } x \in I_j. \tag{1.7}$$

For any step function, we define the difference operators

$$\Delta_\pm u_j = \pm(u_{j\pm1} - u_j), \qquad D_\pm u_j = \frac{1}{\Delta}\Delta_\pm u_j.$$

A method of lines, conservation form, discretization of (1.1), is a system of differential equations

$$\frac{\partial}{\partial t}u_j + D_+ h_{j-1/2} = 0, \qquad j = 0, \pm1, \cdots, \pm N,$$
(1.8)
$$U_\Delta(x,0) = T_\Delta w_0(x) \quad \text{for } x \in I_j.$$

Here, the numerical flux, defined by

$$h_{j-1/2} = h(u_{j+k-1}, \cdots, u_{j-k}), \tag{1.9}$$

for $k \geqq 1$, is a Lipschitz continuous function of its arguments and satisfies the consistency condition

$$h(w, w, \cdots, w) = f(w).$$

It is well known that bounded a.e. limits, as $\Delta \to 0$, of approximate solutions converge to weak solutions of (1.1), i.e. (1.2a) is satisfied. However, this does not also imply that the limit solutions will satisfy any of the entropy conditions (1.5), let alone the general condition (1.3). Some restrictions on h are required.

A simple class of flux functions h, for which (1.8) converges, for all f, to the unique entropy solution in $L^\infty(L^1(R);[0,T])$, as $\Delta \to 0$, for any $T > 0$, is the class of "E" schemes introduced in [11]. Such schemes satisfy the following.

A consistent scheme whose numerical flux satisfies

$$\operatorname{sgn}(u_j - u_{j-1})[h_{j-1/2} - f(u)] \leqq 0 \tag{1.10}$$

for all u between u_{j-1} and u_j is said to be an E scheme. This is at present the most general class of schemes known to converge in the nonconvex case.

It is clear that this class includes the widely known class of three-point monotone schemes, i.e. those for which

$$h_{j-1/2} = h(u_j, u_{j-1}),$$

with $h_{j-1/2}$ nonincreasing in its first argument, nondecreasing in its second.

We denote partial derivatives of a numerical flux via

$$\frac{\partial}{\partial u_{j+\gamma}} h(u_{j+k}, \cdots, u_{j-k+1}) = h_\gamma.$$

Thus a three-point scheme with a differentiable flux function is monotone iff

$$h_1 \leqq 0 \leqq h_0.$$

Any numerical flux can be written

$$h_{j+1/2} = \tfrac{1}{2}(f(u_j) + f(u_{j+1})) - \tfrac{1}{2}Q_{j+1/2}(\Delta_+ u_j) \tag{1.11}$$

where $Q_{j+1/2}$ can be viewed as the viscosity of the scheme [15].

One particular three point scheme is due to Godunov [5] and has a special significance in this theory. The flux for Godunov's scalar scheme can be defined by

$$h^G_{j-1/2} = h_G(u_j, u_{j-1}) = \begin{cases} \displaystyle\min_{u_{j-1} \leqq u \leqq u_j} f(u) & \text{if } u_{j-1} \leqq u_j, \\ \displaystyle\max_{u_{j-1} \geqq u \geqq u_j} f(u) & \text{if } u_{j-1} > u_j. \end{cases} \tag{1.12}$$

One can thus, [11], characterize E schemes as precisely those for which

$$h_{j-1/2} \leqq h^G_{j-1/2} \quad \text{if } u_j < u_{j-1}, \tag{1.13a}$$

$$h_{j-1/2} \geqq h^G_{j-1/2} \quad \text{if } u_j > u_{j-1}; \tag{1.13b}$$

or, as Tadmor [15] pointed out, those which have at least as much viscosity as Godunov's scheme

$$Q^G_{j-1/2} \leqq Q_{j-1/2}. \tag{1.14}$$

It follows from [11, Lemma 2.1], that these approximations are, at most, first order accurate.

Together with an entropy inequality, a key estimate involved in many convergence proofs, is a bound on the variation. For any fixed $t \geqq 0$, the x variation of $U_\Delta(x, t)$ is

$$B(U_\Delta) = \sum_j |\Delta_+ u_j(t)|.$$

If we can write

$$\Delta_+ h_{j-1/2} = C_{j+1/2}\Delta_+ u_j + D_{j-1/2}\Delta_- u_j, \tag{1.15a}$$

$$C_{j+1/2} \geqq 0, \tag{1.15b}$$

$$D_{j-1/2} \geqq 0, \tag{1.15c}$$

then it is easy to show [12] using an argument of [19], that, for $t_1 \geqq t_2 \geqq 0$,

$$B(U_\Delta(\cdot, t_1)) \leqq B(U_\Delta(\cdot, t_2)). \tag{1.16}$$

Harten in [7], pointed out for explicit methods, that this decomposition could be obtained for schemes which are higher order accurate. In our present method of lines context, it involves a five point, consistent, approximation

$$\frac{\partial u_j}{\partial t} = -\frac{1}{\Delta}\Delta_+ h_{j-1/2} = C_{j+1/2}\Delta_+ u_j - D_{j-1/2}\Delta_- u_j \tag{1.17a}$$

with

$$C_{j+1/2} = C(u_{j+2}, u_{j+1}, u_j, u_{j-1}) \geqq 0, \tag{1.17b}$$

$$D_{j-1/2} = D(u_{j+1}, u_j, u_{j-1}, u_{j-2}) \geqq 0, \tag{1.17c}$$

both Lipschitz continuous functions of their arguments. (See also van Leer [16].)

In addition to (1.16) we have a maximum principle for (1.17), [12],

$$\min_k u_k(0) \leqq u_j(t) \leqq \max_k u_k(0) \tag{1.18}$$

for each j and all $t \geqq 0$.

Moreover, in [12] we also showed a limit on the possible accuracy of approximations of the type (1.17). Any scheme of the type (1.17) is at most first order accurate at nonsonic critical points of u. (A sonic point $\bar{u}$ is one such that $f'(\bar{u}) = 0$). Thus, although schemes of this type can be made to be as high as third order accurate, Lipschitz continuity implies a local degeneracy to first order accuracy at smooth maxima and minima. This local degeneracy, together with some results on initial boundary value problems in [6], indicate strongly that overall second order accuracy is the best possible.

Following Harten [7], we call algorithms of the type (1.17), total variation diminishing, or TVD schemes.

It is known, that although (1.17) generates a compact family of solutions, $U_\Delta(x, t)$, in $L^\infty(L^1(R), [0, T])$, (if $w_0 \in L^1 \cap L^\infty \cap BV$)), limit solutions may not satisfy any of the entropy conditions (1.15), and hence may not even be unique (1.12). See e.g. [4], [11] and [15]. One way of avoiding this difficulty is to use an inequality obtained in [11].

Let $V(w)$ be any convex function. We showed in [11, § III], that, for any solution of any scheme (1.8),

$$\Delta\left(\frac{d}{dt} V(u_j) + D_+ F_A(u_j)\right) = \int_{u_j}^{u_{j+1}} dw\, V''(w)[h_{j+1/2} - f(w)], \tag{1.19}$$

where the approximate entropy flux is defined through

$$F_A(u_j) = F(u_j) + V'(u_j)[h_{j-1/2} - f(u_j)]. \tag{1.20}$$

Thus, a sufficient condition that any limit solution satisfy (1.5) for a fixed convex V, is that

$$\int_{u_j}^{u_{j+1}} dw\, V''(w)[h_{j+1/2} - f(w)] \leqq 0. \tag{1.21}$$

In order that the above inequality be valid for all convex V, it is necessary and sufficient that $h_{j+1/2}$ correspond to an E scheme, which implies again that the approximation be at most first order accurate. Thus, for the schemes to be constructed in the following section we shall only obtain our entropy inequality (1.21) for a single $V(w)$, say $V(w) = \frac{1}{2}w^2$. The following theorem summarizes the technical hypotheses needed for convergence.

THEOREM 1.1. *The sequence of approximate solutions converges a.e. as $\Delta \to 0$, to the unique solution of the scalar, convex conservation law* (1.1) *provided that the initial data is in BV and that inequalities* (1.17) *and* (1.21) *are valid for a single convex* $V(w)$.

The proof is analogous to that of [12, Thm. 4.1].

2. Construction of TVD MUSCL schemes. The MUSCL scheme as discussed, e.g. in [2] and [8], is a second order accurate extension of Godunov's method [5] that is based on ideas expressed by van Leer in [16].

We assume that $u_j(t)$ is known. The first step in MUSCL is the reconstruction of a piecewise linear description of the solution. In the interval I_j the result of this operation is

$$z(x, t) = u_j(t) + (x - x_j)s_j(t) \quad \text{for } (x - x_j) < \Delta/2. \tag{2.1}$$

Here $s_j(t)$ are slopes that satisfy

$$s_j = w_x(x_j, t) + O(\Delta), \tag{2.2}$$

subject to some monotonicity constraints discussed below.

Godunov's scheme, in our semidiscrete context, is the following. The numerical flux $h^G(u_{j+1}, u_j)$ is computed by solving the Riemann problem, i.e. the initial value problem (1.1) with initial data

$$w=\begin{cases} u_j(t) & \text{for } x<x_{j+1/2}, \\ u_{j+1}(t) & \text{for } x\geqq x_{j+1/2}. \end{cases}$$

The resulting unique entropy condition satisfying solution is a function of the type

$$w\left(\frac{x-x_{j+1/2}}{t-t_{j+1/2}}\right)=w(\zeta).$$

A closed form for w was recently obtained in [11].

Then Godunov's flux is defined through

$$h^G_{j+1/2}=f(w(0)) \tag{2.3}$$

which is the same as (1.12).

A fully discrete, explicit in time, Godunov scheme has the same numerical flux, after we impose a CFL restriction which prevents the interaction of solutions of adjacent Riemann problems. The literal second order accurate extension to this fully discrete approach would be to compute the solution to the initial value problem (1.1) with initial data

$$w=\begin{cases} u_j(t)+(x-x_j)s_j(t) & \text{for } x<x_{j+1/2}, \\ u_{j+1}(t)+(x-x_j)s_{j+1}(t) & \text{for } x\geqq x_{j+1/2}. \end{cases}$$

Find $f(w(x_{j+1/2}, s))$ for $t+\lambda\Delta\geqq s\geqq t$, with $\lambda>0$ satisfying a CFL restriction, then compute

$$\frac{1}{\lambda\Delta}\int_t^{t+\lambda\Delta} f(w(x_{j+1/2}, s))\,ds. \tag{2.4}$$

This is the MUSCL numerical flux.

Unfortunately, obtaining the exact solution to this nonlinear initial value problem with piecewise linear initial data is a nontrivial business, even in the scalar case. However, at $s=t^+$, $x=x_{j+1/2}$, the solution is the same as for the Riemann problem with initial data

$$w=\begin{cases} u_j(t)+\frac{\Delta}{2}s_j(t) & \text{for } x<x_{j+1/2}, \\ u_{j+1}-\frac{\Delta}{2}s_{j+1}(t) & \text{for } t\geqq x_{j+1/2}, \end{cases}$$

which is easily calculated, e.g. from the formula in [11]. Thus, the semidiscrete MUSCL extension of Godunov's algorithm comes through

$$\frac{\partial u_j}{\partial t}=-\frac{1}{\Delta}\Delta_- h^G\left(u_{j+1}-\frac{\Delta}{2}s_{j+1}, u_j+\frac{\Delta}{2}s_j\right). \tag{2.5}$$

We generalize this still further as follows. Let $h(u_{j+1}, u_j)=h_{j+1/2}$ be an arbitrary first order accurate numerical flux function. Then our generalized MUSCL algorithm is merely

$$\frac{\partial u_j}{\partial t}=-\frac{1}{\Delta}\Delta_- h\left(u_{j+1}-\frac{\Delta}{2}s_{j+1}, u_j+\frac{\Delta}{2}s_j\right). \tag{2.6}$$

We have the following lemma.

LEMMA 2.1. *At points $w(x_j, t) = u_j(t)$ in a neighborhood of which $h(u_{j+1}, u_j)$ is C^2 with Lipschitz continuous partial derivatives, and $\Delta s_{j+1}/\Delta_+ u_j = 1 + O(\Delta) = \Delta s_j/\Delta_+ u_j$, the algorithm* (2.6) *is at least second order accurate for smooth functions w.*

Proof. Let

$$u_{j+1/2} = \tfrac{1}{2}(u_{j+1} + u_j). \tag{2.7}$$

We shall show

$$\Delta_-\left(h\left(u_{j+1} - \frac{\Delta}{2} s_{j+1}, u_j + \frac{\Delta}{2} s_j\right) - f(u_{j+1/2})\right) = O(\Delta^2) \tag{2.8}$$

at these points.

Using the consistency of $h(u_{j+1}, u_j)$, the left side of (2.8) can be written as

$$\Delta_-\left[\int_0^1 h_1\left(u_{j+1/2} + \gamma\frac{\Delta_+ u_j}{2}\left(1 - \frac{\Delta s_{j+1}}{\Delta_+ u_j}\right), u_j + \frac{\Delta}{2} s_j\right) d\gamma \frac{\Delta_+ u_j}{2}\left(1 - \frac{\Delta s_{j+1}}{\Delta_+ u_j}\right)\right.$$
$$\left. + \int_0^1 h_0\left(u_j + \frac{\Delta}{2} s_j, u_{j+1/2} + \gamma\frac{\Delta_+ u_j}{2}\left(\frac{\Delta s_j}{\Delta_+ u_j} - 1\right)\right) d\gamma \frac{\Delta_+ u_j}{2}\left(\frac{\Delta s_j}{\Delta_+ u_j} - 1\right)\right].$$

The result is immediate.

Next we make an observation about the viscosity of this scheme, $Q_{j+1/2}$, as defined in (1.11).

LEMMA 2.2. *If h is monotone, then the viscosity of algorithm* (2.6) *is a decreasing function of $s_{j+1}/\Delta_+ u_j$ and $s_j/\Delta_+ u_j$.*

The proof comes directly from (1.11).

Thus, as expected, if $s_j/\Delta_+ u_j$ and $s_{j+1}/\Delta_+ u_j$ are both restricted to be positive, the most viscous, hence the least accurate, case occurs when these values are both zero. The scheme then degenerates to first order accuracy. We can compress or smear out the solution locally by increasing or decreasing these ratios.

We wish this scheme to be TVD, which will give a natural restriction on these above ratios.

LEMMA 2.3. *If h is a flux corresponding to an E scheme, then the scheme is* TVD *if*

$$0 \leqq \frac{\Delta s_j}{\Delta_+ u_j}, \frac{\Delta s_{j+1}}{\Delta_+ u_j} \leqq 1 \tag{2.9}$$

for each j.

Proof. We have

$$\begin{aligned}
&\frac{1}{\Delta}\Delta_+ h\left(u_j - \frac{\Delta}{2} s_j, u_{j-1} + \frac{\Delta}{2} s_{j-1}\right) \\
&\quad = \frac{1}{\Delta}\left[h\left(u_{j+1} - \frac{\Delta s_{j+1}}{2}, u_j + \frac{\Delta}{2} s_j\right) - f\left(u_j + \frac{\Delta}{2} s_j\right)\right. \\
&\qquad \left. + f\left(u_j + \frac{\Delta}{2} s_j\right) - h\left(u_j - \frac{\Delta s_j}{2}, u_j + \frac{\Delta}{2} s_j\right)\right] \\
&\quad + \frac{1}{\Delta}\left[h\left(u_j - \frac{\Delta s}{2}, u_j + \frac{\Delta}{2} s_j\right) - f\left(u_j - \frac{\Delta s_j}{2}\right)\right. \\
&\qquad \left. + f\left(u_j - \frac{\Delta s_j}{2}\right) - h\left(u_j - \frac{\Delta s_j}{2}, u_{j-1} + \frac{\Delta}{2} s_{j-1}\right)\right].
\end{aligned} \tag{2.10}$$

Thus, in (1.15) we can write

$$C_{j+1/2}=-\frac{1}{\Delta(\Delta_+u_j)}\left[h\left(u_{j+1}-\frac{s_{j+1}}{2}\Delta,\,u_j+\frac{\Delta}{2}s_j\right)-f\left(u_j+\frac{\Delta}{2}s_j\right)+f\left(u_j+\frac{\Delta}{2}s_j\right)-h\left(u_j-\frac{\Delta}{2}s_j,\,u_j+\frac{\Delta}{2}s_j\right)\right], \tag{2.11a}$$

$$D_{j-1/2}=\frac{1}{\Delta(\Delta_-u_j)}\left[h\left(u_j-\frac{\Delta s_j}{2},\,u_j+\frac{\Delta}{2}s_j\right)-f\left(u_j-\frac{\Delta}{2}s_j\right)+f\left(u_j-\frac{\Delta}{2}s_j\right)-h\left(u_j-\frac{\Delta}{2}s_j,\,u_{j-1}+\frac{\Delta}{2}s_{j-1}\right)\right]. \tag{2.11b}$$

Both $C_{j+1/2}$, $D_{j-1/2}$ are nonnegative because of (1.10) and (2.9).

The restrictions (2.9) can be relaxed somewhat if h is monotone; we have the next lemma.

LEMMA 2.4. *If h is a flux corresponding to a monotone scheme, then the scheme is TVD if*

$$1\geqq\frac{(s_{j+1}-s_j)\Delta}{2\Delta_+u_j}\quad \textit{near points where } h_1\neq 0, \tag{2.12a}$$

$$1\geqq\frac{(-s_{j+1}+s_j)\Delta}{2\Delta_+u_j}\quad \textit{near points where } h_0\neq 0 \tag{2.12b}$$

for each j.

Proof. We have

$$\begin{aligned}\frac{1}{\Delta}\Delta_+h\left(u_j-\frac{\Delta}{2}s_j,\,u_{j-1}+\frac{\Delta}{2}s_{j-1}\right)&=\frac{1}{\Delta}\left[h\left(u_{j+1}-\frac{\Delta s_{j+1}}{2},\,u_j+\frac{\Delta s_j}{2}\right)-h\left(u_j-\frac{\Delta s_j}{2},\,u_j+\frac{\Delta s_j}{2}\right)\right]\\&\quad+\frac{1}{\Delta}\left[h\left(u_j-\frac{\Delta s_j}{2},\,u_j+\frac{\Delta s_j}{2}\right)-h\left(u_j-\frac{\Delta s_j}{2},\,u_{j-1}+\frac{\Delta s_{j-1}}{2}\right)\right].\end{aligned} \tag{2.13}$$

Thus, in (1.15), we can write

$$C_{j+1/2}=-\frac{1}{\Delta(\Delta_+u_j)}h_1\left(\tilde{u}_j,\,u_j+\frac{\Delta s_j}{2}\right)\left[\Delta_+u_j-\frac{\Delta}{2}(s_{j+1}-s_j)\right], \tag{2.14a}$$

$$D_{j-1/2}=\frac{1}{\Delta(\Delta_-u_j)}h_0\left(u_j-\frac{\Delta s_j}{2},\,\tilde{u}_j\right)\left[\Delta_-u_j+\frac{\Delta}{2}(s_j-s_{j-1})\right]. \tag{2.14b}$$

Both $C_{j-1/2}$ and $D_{j-1/2}$ are nonnegative because of monotonicity and (2.12).

Remark 2.5. These last two lemmas can be made "local", i.e. in regions of monotonicity we may relax the restrictions from (2.9) to (2.12).

3. Convergence of MUSCL schemes. In order to prove convergence for convex f, we need only verify the discrete entropy inequality (1.21) for the schemes constructed in the previous section (Theorem 1.1). We shall do this for the entropy

$$V(w)=\tfrac{1}{2}w^2$$

and for schemes satisfying the TVD hypothesis of Lemma 2.3, in addition to some other restrictions near sonic shock points.

Suppose $u_j < u_{j+1}$, i.e. we are at a rarefaction. We then have, letting $u_{j+1/2} = \frac{1}{2}(u_j + u_{j+1})$ as above,

$$\begin{aligned}\int_{u_j}^{u_{j+1}} dw \left[h\left(u_{j+1} - \frac{\Delta s_{j+1}}{2}, u_j + \frac{\Delta s_j}{2} \right) - f(w) \right] \\ = (\Delta_+ u_j)\left(h\left(u_{j+1} - \frac{\Delta s_{j+1}}{2}, u_j + \frac{\Delta s_j}{2} \right) - f(u_{j+1/2}) \right) \\ + \int_{u_j}^{u_{j+1}} dw[f(u_{j+1/2}) - f(w)] \\ = [\mathrm{I}] + [\mathrm{II}].\end{aligned} \tag{3.1}$$

Now, using the hypotheses of Lemma 2.3, we have

$$[\mathrm{I}] \leqq 0. \tag{3.2}$$

Next we have

$$\begin{aligned}[\mathrm{II}] &= \int_{u_j}^{u_{j+1}} dw \int_w^{u_{j+1/2}} \left(\frac{\partial}{\partial s}(s-w) \right) f'(s)\, ds \\ &= \int_{u_j}^{u_{j+1}} dw (u_{j+1/2} - w) f'(u_{j+1/2}) \\ &\quad - \int_{u_j}^{u_{j+1}} dw \int_w^{u_{j+1/2}} (s-w) f''(s)\, ds \\ &= -\int_{u_j}^{u_{j+1}} dw \int_w^{u_{j+1/2}} (s-w) f''(s)\, ds \\ &= -\frac{1}{2} \int_{u_j}^{u_{j+1}} ds\, f''(s)(\min|s-u_{j+1}|, |s-u_j|)^2 < 0.\end{aligned} \tag{3.3}$$

Suppose $u_j > u_{j+1}$, i.e. we are at a shock. Let $\tilde{u}_{j+1/2}$ be chosen so that

$$\int_{u_j}^{u_{j+1}} f'(w)(w - \tilde{u}_{j+1/2})\, dw = 0,$$

i.e.

$$\tilde{u}_{j+1/2} = \frac{\int_{u_j}^{u_{j+1}} w f'(w)\, dw}{\int_{u_j}^{u_{j+1}} f'(w)} = \frac{\Delta_+ F(u_j)}{\Delta_+ f(u_j)}, \tag{3.4}$$

where $F(u)$ is the entropy flux corresponding to our entropy, $V(u) = \frac{1}{2}u^2$. Of course we need the denominator in (3.4) to be nonvanishing. In fact we restrict ourselves to the situation where

$$u_{j+1} \leqq \tilde{u}_{j+1/2} \leqq u_j. \tag{3.5}$$

We note

$$\begin{aligned}\tilde{u}_{j+1/2} - u_{j+1/2} &= \frac{\int_{u_j}^{u_{j+1}} (w - u_{j+1/2}) f'(w)\, dw}{\Delta_+ f(u_j)} \\ &= \frac{\int_{u_j}^{u_{j+1}} (w - u_{j+1/2})(f'(w) - f'(u_{j+1/2}))\, dw}{\Delta_+ f(u_j)}\end{aligned} \tag{3.6}$$

Thus, we have

$$|\tilde{u}_{j+1/2}-u_{j+1/2}|<\tfrac{1}{6}(\Delta_+u_j)^2\frac{\sup_{u_{j+1}\leqq u\leqq u_j} f''(u)}{|s_{j+1/2}|}, \tag{3.7}$$

where $s_{j+1/2}=\Delta_+ f(u_j)/\Delta_+u_j$.

We further restrict the s_j so that

$$0\leqq \Delta s_j \leqq 2\max(\min((u_j-\tilde{u}_{j+1/2}),(\tilde{u}_{j-1/2}-u_j)),0). \tag{3.8}$$

In view of (3.7), this restriction does not affect the second order accuracy of the scheme, except at sonic points.

We now write

$$\begin{aligned}\int_{u_j}^{u_{j+1}} dw\left[h\left(u_{j+1}-\frac{\Delta s_{j+1}}{2}, u_j+\frac{\Delta s_j}{2}\right)-f(w)\right] &= \Delta_+u_j\left[h\left(u_{j+1}-\frac{\Delta s_{j+1}}{2}, u_j+\frac{\Delta s_j}{2}\right)-f(\tilde{u}_{j+1/2})\right]\\ &\quad+\int_{u_{j+1}}^{u_j} dw[f(w)-f(\tilde{u}_{j+1/2})]\\ &=[\mathrm{III}]+[\mathrm{IV}].\end{aligned} \tag{3.9}$$

Now, the assumptions (3.5), (3.8) imply

$$[\mathrm{III}]\leqq 0. \tag{3.10}$$

An argument analogous to that of (3.3) suffices to show that $[\mathrm{IV}]\leqq 0$.

Finally, suppose $u_{j+1}<u_j$ and (3.5) is violated. The inequalities of Lemmas 2.1 and 3.8 imply that s_j and s_{j+1} vanish. Then inequality (1.21) is immediate, since h is an E flux.

We may summarize all this in the following theorem.

THEOREM 3.1. *The sequence of approximate solutions satisfying (2.6) converges a.e. to the unique solution of the scalar convex conservation law (1.1) provided that the initial data is in BV and that for each j*

$$0\leqq\frac{\Delta s_j}{\Delta_+u_j}\leqq 1,\qquad 0\leqq\frac{\Delta s_j}{\Delta_-u_j}\leqq 1,$$

and, in addition if $u_j>u_{j+1}$,

$$-\Delta s_j\leqq 2\max(\min((u_j-\tilde{u}_{j+1/2}),(\tilde{u}_{j-1/2}-u_j)),0).$$

4. Steady discrete shocks. We now check for the existence of discrete, steady, shock solutions to the convergent MUSCL schemes constructed in the previous section, based on the Engquist-Osher flux [4]. Since the scheme satisfies a maximum principle ([12, Lemma 2.1]), any profile must be monotone.

Let $f''(u)>0$, and for simplicity, we take $f'(0)=0=f(0)$. In this case, the monotone E-O scheme becomes

$$h^{\mathrm{EO}}(u_{j+1},u_j)=f_-(u_{j+1})+f_+(u_j), \tag{4.1a}$$

where

$$f_-(u)=\begin{cases}f(u) & \text{if } u<0,\\ 0 & \text{if } u\geqq 0,\end{cases} \tag{4.1b}$$

(4.1c) $$f_+(u)=\begin{cases} f(u) & \text{if } u\geqq 0,\\ 0 & \text{if } u<0.\end{cases}$$

Let u^L, u^R be the left and right states for a physically correct, steady shock, i.e.

$$f(u^L)=f(u^R), \qquad u^L>0>u^R.$$

A steady discrete shock $\{u_j\}$ will satisfy

(4.2a) $$\lim_{j\to-\infty} u_j=u^L,$$

(4.2b) $$\lim_{j\to+\infty} u_j=u^R,$$

(4.2c) $$h^{\mathrm{EO}}\left(u_{j+1}-\frac{\Delta s_{j+1}}{2},\, u_j+\frac{\Delta s_j}{2}\right)=f(u^L)=f_-\left(u_{j+1}-\frac{\Delta s_{j+1}}{2}\right)+f_+\left(u_j+\frac{\Delta s_j}{2}\right),$$

for all j.

We shall seek steady, discrete, shocks of the same general form obtained in [4] for the E-O scheme, c.f. also [12, § V].

These are

(4.3a) $$u_j=u^L, \qquad j\leqq -1,$$

(4.3b) $$u_j=u^R, \qquad j\geqq 2,$$

(4.3c) $$u^L\geqq u_0\geqq 0>u_1\geqq u^R.$$

For the first order scheme, u_0 can be viewed as a smooth function of u_1 satisfying the above and, in addition,

$$f(u_0)+f(u_1)=f(u_L).$$

For the present scheme, we shall get a different one parameter family of intermediate states.

It follows for all $j\neq 0$, that (4.3) implies (4.2c).

For $j=0$, we have the following equation,

(4.4) $$0=f_-\left(u_1-\frac{\Delta s_1}{2}\right)+f_+\left(u_0+\frac{\Delta s_0}{2}\right)-f(u^L)=G(u_1,u_0).$$

One special solution is, again, $u_1=u^R$, $u_0=0$.

For (u_0, u_1) close to $(0, u^R)$ and satisfying (4.3c), we have from the hypotheses of Theorem (3.1).

(4.5) $$s_1=a_{1/2}\frac{(u^R-u_1)}{\Delta}, \qquad s_0=0.$$

Here $0\leqq a_{1/2}$ is assumed, in addition, to be a C^2 function of u_1 near $u_1=u^R$, with $\partial a_{1/2}/\partial u_1=0$.

A straightforward calculation gives us

(4.6a) $$\frac{\partial G}{\partial u_1}(u^R,0)=f'(u_R)\left(1+\frac{a_{1/2}}{2}\right)<0,$$

(4.6b) $$\frac{\partial G}{\partial u_0}(u^R,0)=0,$$

(4.6c) $$\frac{\partial^2 G}{\partial u_1^2}(u^R,0)=f''(u_R)\left(1+\frac{a_{1/2}}{2}\right)^2>0,$$

(4.6d) $$\frac{\partial^2 G}{\partial u_0 \partial u_1}(u^R, 0) = 0,$$

(4.6e) $$\frac{\partial^2 G}{\partial u_0^2} = f''(0) > 0.$$

A simple application of the implicit function theorem gives us the next theorem.

THEOREM 4.1. *Given the hypotheses of Theorem* 3.1 *and the mild technical assumption concerning* $a_{1/2}$, *there exists a family of sharp, discrete, shock solutions to* (2.6) *of the type* (4.3), *with* u_1 *a smooth function of* u_0, *and* $0 < u_0$, *small enough. By symmetry, the same is true, with* u_0 *a smooth function of* $u_1 < 0$, *and* $-u_1$ *small enough.*

5. Comparison with other convergent high resolution schemes. In [12] we constructed a class of convergent high resolution schemes, using flux limiters and second order accurate upwinding. Our first step was to use a first order E scheme flux, $h(u_{j+1}, u_j)$, to construct a second order accurate TVD scheme. The resulting scheme is

(5.1) $$\frac{\partial u_j}{\partial t} = -\frac{1}{\Delta}\Delta_- H(u_{j+2}, u_{j+1}, u_j, u_{j-1})$$

with

(5.2) $$\begin{aligned} H(u_{j+2}, u_{j+1}, u_j, u_{j-1}) &= h(u_{j+1}, u_j) - \tfrac{1}{2}\psi(R_{j+1}^-)(h(u_{j+1}, u_j) - f(u_j)) \\ &\quad + \tfrac{1}{2}\psi(R_j^+)(f(u_{j+1}) - h(u_{j+1}, u_j)). \end{aligned}$$

Here

(5.3) $$R_j^+ = \frac{f(u_j) - h(u_j, u_{j-1})}{f(u_{j+1}) - h(u_{j+1}, u_j)}, \qquad R_j^- = \frac{h(u_{j+1}, u_j) - f(u_j)}{h(u_j, u_{j-1}) - f(u_{j-1})}.$$

The slope limiter $\psi(R)$ satisfies $\psi(1) = 1$ and the following inequalities

(5.4a) $$1 + \frac{1}{2}\left[\frac{\psi(R_j^-)}{R_j^-} - \psi(R_{j+1}^-)\right] \geqq 0,$$

(5.4b) $$1 + \frac{1}{2}\left[\frac{\psi(R_j^+)}{R_j^+} - \psi(R_{j-1}^+)\right] \geqq 0.$$

Various slope limiters have been developed. See Sweby [13] for a numerical and theoretical analysis of their properties. Perhaps the simplest example is the so-called min-mod [14],

(5.5) $$\psi(R) = \max(0, \min(R, 1)).$$

We have the following analogue of Lemma 2.2.

LEMMA 5.1. *If h is an E flux, then the viscosity of* (5.1) *is a decreasing function of* ψ.

Next we compare the schemes (2.6) and (5.1). If $f'(w) \neq 0$ and w is smooth at x_j, then it is easy to see that

$$H(u_{j+2}, u_{j+1}, u_j, u_{j-1}) = h\left(u_{j+1} - \frac{\Delta s_{j+1}}{2}, u_j + \frac{\Delta s_j}{2}\right) + O((\Delta_- u_j)^2 + (\Delta_+ u_j)^2).$$

If

(5.6) $$\Delta s_j = \psi(R_j^-)\Delta_- u_j + O((\Delta_- u_j)^2 + (\Delta_+ u_j)^2) \quad \text{if } h_1(w, w) \neq 0$$

and/or

$$\Delta s_j = \psi(R_j^+)\Delta_+ u_j + O((\Delta_- u_j)^2 + (\Delta_+ u_j)^2) \quad \text{if } h_0(w, w) \neq 0.$$

Dropping the quadratic terms, inequality (2.9) gives

(5.7a) $$O \leqq \frac{\psi(R_j^-)}{R_j^-}, \; \psi(R_j^-) \leqq 1 \quad \text{if } h_1(w, w) \neq 0,$$

(5.7b) $$O \leqq \frac{\psi(R_j^+)}{R_j^+}, \; \psi(R_j^+) \leqq 1 \quad \text{if } h_0(w, w) \neq 0.$$

These inequalities are a bit more restrictive than those needed in the monotone case. There we have

(5.8a) $$1 \geqq \frac{1}{2}\left[-\frac{\psi(R_j^-)}{R_j^-} + \psi(R_{j+1}^-)\right] \quad \text{if } h_1(w, w) \neq 0,$$

(5.8b) $$1 \geqq \frac{1}{2}\left[-\frac{\psi(R_j^+)}{R_j^+} + \psi(R_{j-1}^+)\right] \quad \text{if } h_0(w, w) \neq 0,$$

which are the same as inequalities (5.4a), (5.4b).

For the entropy condition to be proven valid, the flux was modified. We replaced H by H^{ac} (where a.c. stands for artificial compression) as follows.

First we modify the flux differences in (5.2),

(5.9a) $$\begin{aligned}(h(u_{j+1}, u_j) - f(u_j))^M \\ = (h(u_{j+1}, u_j) - f(u_j))\left[1 + \frac{a_{j+1/2}^-(\Delta_+ h_1(u_j, u_j))(\Delta_+ u_j)}{h(u_{j+1}, u_j) - f(u_j)}\right],\end{aligned}$$

(5.9b) $$\begin{aligned}(f(u_{j+1}) - h(u_{j+1}, u_j))^M \\ = (f(u_{j+1}) - h(u_{j+1}, u_j))\left[1 - \frac{a_{j+1/2}^+(\Delta_+ h_0(u_j, u_j))\Delta_+ u_j}{f(u_{j+1}) - h(u_{j+1}, u_j)}\right].\end{aligned}$$

Here $a_{j+1/2}^{\pm}$ are both positive, chosen first so that the quantities in the brackets in (5.9) are between 0 and 2, (inequalities (5.13) below).

We next let

$$R_j^{M_+} = \frac{(f(u_j) - h(u_j, u_{j-1}))^M}{(f(u_{j+1}) - h(u_{j+1}, u_j))^M},$$

$$R_j^{M_-} = \frac{(h(u_{j+1}, u_j) - f(u_j))^M}{(h(u_j, u_{j-1}) - f(u_{j-1}))^M}.$$

Then we define our scheme:

(5.10) $$\frac{\partial u_j}{\partial t} = -\frac{1}{\Delta}\Delta_- H^{ac}(u_{j+2}, u_{j+1}, u_j, u_{j-1})$$

with

(5.11) $$\begin{aligned}H^{ac}(u_{j+2}, u_{j+1}, u_j, u_{j-1}) \\ = h(u_{j+1}, u_j) - \tfrac{1}{2}\psi(R_{j+1}^{M_-})(h(u_{j+1}, u_j) - f(u_j))^M \\ + \tfrac{1}{2}\psi(R_j^{M_+})(f(u_{j+1}) - h(u_{j+1}, u_j))^M.\end{aligned}$$

The resulting scheme is convergent if the following inequalities are valid:

$$\begin{aligned}
&\int_{u_j}^{u_{j+1}} f''(w)\, dw((\tfrac{1}{2}(\Delta_+ u_j))^2-(w-\tfrac{1}{2}(u_{j+1}+u_j))^2) \\
(5.12)\qquad &-a^-_{j+1/2}(\Delta_+u_j)^2[h_1(u_{j+1}, u_{j+1})-h_1(u_j, u_j)] \\
&-a^+_{j+1/2}(\Delta_+u_j)^2[h_0(u_{j+1}, u_{j+1})-h_0(u_j, u_j)]\leqq 0,
\end{aligned}$$

$$(5.13a)\qquad |a^-_{j+1/2}(h_1(u_{j+1}, u_{j+1})-h_1(u_j, u_j))|\leqq -\frac{(h(u_{j+1}, u_j)-f(u_j))}{\Delta_+u_j},$$

$$(5.13b)\qquad |a^+_{j+1/2}(h_0(u_{j+1}, u_{j+1})-h_0(u_j, u_j))|\leqq \frac{f(u_{j+1})-h(u_{j+1}, u_j)}{\Delta_+u_j},$$

and if $\psi(R)$ is defined by (5.5).

Examples are given in [12].

It is interesting that this case allows compression to be added at shocks, i.e. the restrictions on the $a^{\pm}_{j+1/2}$ allow negative viscosity to be added to (5.2) if $u_j > u_{j+1}$, while, again for the proof of convergence, positive viscosity must be added for rarefactions, $u_j < u_{j+1}$. In the present MUSCL case the situation is (unnaturally) reversed. A comparison of the convergent MUSCL scheme with this one yields two main points:

(1) Away from sonic points, the fluxes differ only in $O((\Delta_\pm u_j)^3)$ terms

(2) At sonic points, the fluxes differ in $O((\Delta_\pm u_j)^2)$ terms.

REFERENCES

[1] S. R. CHAKRAVARTHY AND S. OSHER, *High resolution applications of the Osher upwind scheme for the Euler equations*, Proc. AIAA Computational Fluid Dynamics Conference, Danvers, MA, 1983, pp. 363-372.

[2] P. COLELLA, *A direct Eulerian* MUSCL *scheme for gas dynamics*, Lawrence Berkeley Lab. Report #LBL-14104, 1982.

[3] R. J. DIPERNA, *Convergence of approximate solutions to conservation laws*, Arch. Rational Mech. Anal., 82 (1983), pp. 27-70.

[4] B. ENGQUIST AND S. OSHER, *Stable and entropy condition satisfying approximations for transonic flow calculations*, Math. Comp., 34 (1980), pp. 45-75.

[5] S. K. GODUNOV, *A finite difference method for the numerical computation of discontinuous solutions of the equations of fluid dynamics*, Mat. Sb., 47 (1959), pp. 271-290.

[6] B. GUSTAFSSON, *The convergence rate for difference approximations to mixed initial boundary value problems*, Math. Comp., 29 (1975), pp. 396-406.

[7] A. HARTEN, *High resolution schemes for hyperbolic conservation laws*, J. Comp. Phys., 49 (1983), pp. 357-393.

[8] ———, *On second order accurate Godunov-type schemes*, NASA AMES Report #NCA2-ORS25-201.

[9] S. N. KRUZKOV, *First order quasi-linear equations in several independent variables*, Math. USSR Sb., 10 (1970), pp. 217-243.

[10] W. A. MULDER AND B. VAN LEER, *Implicit upwind computations for the Euler equations*, AIAA Computational Fluid Dynamics Conference, Danvers, MA, 1983, pp. 303-310.

[11] S. OSHER, *Riemann solvers, the entropy condition, and difference approximations*, this Journal, 21 (1984), pp. 217-235.

[12] S. OSHER AND S. R. CHAKRAVARTHY, *High resolution schemes and the entropy condition*, this Journal, 21 (1984), pp. 955-984.

[13] P. K. SWEBY, *High resolution schemes using flux limiters for hyperbolic conservation laws*, this Journal, 21 (1984), pp. 995-1011.

[14] P. K. SWEBY AND M. J. BAINES, *Convergence of Roe's scheme for the general non-linear scalar wave equation*, Numerical Analysis Report, 8/81, Univ. Reading, 1981.

[15] E. TADMOR, *Numerical viscosity and the entropy condition for conservative difference schemes*, ICASE NASA Contractor Report 172141, (1983), NASA Langley Research Center, Hampton, VA.

[16] B. van Leer, *Towards the ultimate conservative difference scheme, II. Monotonicity and conservation combined in a second order scheme*, J. Comp. Phys., 14 (1974), pp. 361-376.

[17] ———, *Towards the ultimate conservative difference scheme. V. A. second-order sequel to Godunov's method*, J. Comp. Phys., 32 (1979), pp. 101-136.

[18] H. C. Yee, R. F. Warming and A. Harten, *Implicit total variation diminishing (TVD) schemes for steady state calculations*, Proc. AIAA Computational Fluid Dynamics Conference, Danvers, MA, 1983, pp. 110-127.

[19] R. Sanders, *On convergence of monotone finite difference schemes with variable spatial differencing*, Math. Comp., 40 (1983), pp. 91-106.

A SIMPLIFIED TVD FINITE DIFFERENCE SCHEME VIA ARTIFICIAL VISCOSITY*

STEPHEN F. DAVIS†

Abstract. In this paper we show that the total variation diminishing (TVD) finite difference scheme which was analysed by Sweby [8] can be interpreted as a Lax-Wendroff scheme plus an upwind weighted artificial dissipation term. We then show that if we choose a particular flux limiter and remove the requirement for upwind weighting, we obtain an artificial dissipation term which is based on the theory of TVD schemes, which does not contain any problem dependent parameters and which can be added to existing MacCormack method codes. Finally, we conduct numerical experiments to examine the performance of this new method.

Key words. finite difference methods, hyperbolic partial differential equations

AMS(MOS) subject classifications. 65M05, 76-08

1. Introduction. A major result of recent research into numerical methods for the solution of systems of hyperbolic conservation laws has been the development of second order accurate total variation diminishing (TVD) finite difference schemes. These schemes have a number of attractive properties including the fact that they resolve discontinuous solutions well, they do not exhibit spurious oscillations and, under certain circumstances, they can be proven to converge.

Unfortunately, the apparent complexity of these schemes has thus far discouraged their widespread adoption. Instead, most applications programs use standard methods such as the MacCormack variant of the Lax-Wendroff method and add additional artificial dissipation to damp the spurious wiggles that occur near discontinuities. These dissipation terms are usually chosen in an ad hoc fashion and usually contain problem-dependent parameters which must be fine tuned before the method will work.

In this paper we attempt to remedy this situation. In particular, we interpret the TVD finite difference scheme which was analysed by Sweby [8] as a Lax-Wendroff scheme plus an upwind weighted artificial dissipation term. We then attempt to simplify this artificial dissipation term by removing the requirement that it be upwind weighted. The result is a dissipation term which is based on the theory of TVD schemes and which does not contain any free parameters.

An outline of this paper is as follows. In § 2 we briefly review the theory of TVD finite difference schemes. Using this theory, in § 3, we show that a TVD scheme can be constructed by adding an artificial viscosity-like term to a standard finite difference scheme. Numerical experiments are performed to examine the performance of this new scheme. These results are extended to systems of hyperbolic equations in § 4 and additional numerical experiments are performed. Section 5 contains some closing remarks.

2. Total variation diminishing finite difference schemes. We consider the initial value problem for a scalar conservation law. That is

$$\begin{aligned} &u_t+f(u)_x=u_t+a(u)u_x=0, \quad a(u)=\frac{df}{du}(u), \quad t>0 \\ &u(x,0)=u_0(x), \qquad -\infty<x<\infty \end{aligned} \tag{2.1}$$

* Received by the editors July 10, 1984, and in revised form April 22, 1985. This research was supported by the National Aeronautics and Space Administration under NASA contract NAS1-17070 while the author was in residence at ICASE, NASA Langley Research Center, Hampton, Virginia 23665.

† Applied Mathematics Branch, Naval Surface Weapons Center, White Oak, Silver Spring, Maryland 20903.

where $u_0(x)$ is assumed to have bounded total variation. A weak solution to this problem has the following monotonicity properties:

(1) No new extrema in $u(x)$ may be created.

(2) The value of a local minimum is nondecreasing and the value of a local maximum is nonincreasing.

The total variation of the solution to (2.1) at time t is defined by the formula

$$TV(u(x,t)) = \sup \sum_k |u(x_{k+1},t) - u(x_k,t)|, \tag{2.2}$$

where the supremum is taken over all partitions of the real line.

It follows from this monotonicity property that the total variation in x of $u(x,t)$ does not increase in t. That is,

$$TV(u(t_2)) \leqq TV(u(t_1)) \quad \text{for all } t_2 \geqq t_1. \tag{2.3}$$

Much recent research has been devoted to the construction of finite difference schemes that satisfy a discrete version of equation (2.2). We briefly describe this work below.

Consider explicit finite difference schemes in conservation form which approximate (2.1) and which we denote by

$$U^{n+1} = L \cdot U^n. \tag{2.4}$$

A scheme is called *total variation diminishing* if

$$TV(U^{n+1}) = TV(L \cdot U^n) \leqq TV(U^n). \tag{2.5}$$

In addition, a scheme is called *monotonicity preserving* if the finite difference operator L is monotonicity preserving; that is, U^n a monotone mesh function implies that $L \cdot U^n$ is also a monotone mesh function. The following result is due to Harten [3].

THEOREM 2.1 (Harten). *A total variation diminishing scheme is monotonicity preserving.*

This says that TVD schemes will not produce spurious oscillations. This is their chief attraction.

Another reason why TVD schemes are attractive is that it is very useful to have a bound on the total variation of the solution when proving convergence of nonlinear difference schemes (cf. [4], [7]). Equation (2.5) provides such a bound but convergence proofs are beyond the scope of this paper.

The scheme (2.4) can be written in the form

$$U_k^{n+1} = U_k^n - C_{k-1/2}\Delta U_{k-1/2}^n + D_{k+1/2}\Delta U_{k+1/2}^n \tag{2.6}$$

where

$$\Delta U_{k+1/2}^n = U_{k+1}^n - U_k^n \tag{2.7}$$

and $C_{k-1/2}$ and $D_{k+1/2}$ are functions of U^n. The choice of $C_{k-1/2}$ and $D_{k+1/2}$ is not unique. If a particular scheme can be rearranged so that C and D satisfy the hypotheses of the following lemma, the scheme can be proved to be TVD.

LEMMA 2.2 (Harten). *If the coefficients C and D of* (2.6) *satisfy the inequalities*

$$\begin{gathered} 0 \leqq C_{k+1/2}, \quad 0 \leqq D_{k+1/2}, \\ 0 \leqq C_{k+1/2} + D_{k+1/2} \leqq 1, \end{gathered} \tag{2.8}$$

then the scheme (2.6) *is total variation diminishing.*

In the next section we derive a scheme and use this lemma to prove that it is TVD.

3. Scalar equations. For ease of presentation, we first consider the scalar linear equation

$$u_t + au_x = 0, \qquad a = \text{const} > 0. \tag{3.1}$$

Sweby [8] considers the solution to this problem using a scheme of the form

$$U_k^{n+1} = U_k^n - \nu\{1 + \tfrac{1}{2}(1-\nu)[\phi(r_k^+)/r_k^+ - \phi(r_{k-1}^+)]\}\Delta U_{k-1/2}^n \tag{3.2}$$

where

$$\nu = \frac{a\Delta t}{\Delta x}, \qquad r_k^+ = \frac{\Delta U_{k-1/2}^n}{\Delta U_{k+1/2}^n} \tag{3.3}$$

and $\phi(r_k)$ is the flux limiter. This is a scheme of the form (2.6) with

$$\begin{aligned} C_{k-1/2} &= \nu\{1 + \tfrac{1}{2}(1-\nu)[\phi(r_k^+)/r_k^+ - \phi(r_{k-1}^+)]\}, \\ D_{k+1/2} &= 0. \end{aligned} \tag{3.4}$$

A sufficient condition for the scheme (3.2) to be TVD is that $\nu \leqq 1$ and

$$|\phi(r_k^+)/r_k^+ - \phi(r_{k-1}^+)| \leqq 2. \tag{3.5}$$

Sweby also specifies that $\phi(r) > 0$ and that $\phi(r) = 0$ for $r \leqq 0$. Under these additional restrictions the bound (3.5) becomes

$$0 \leq \frac{\phi(r)}{r} \leq 2, \qquad 0 \leq \phi(r) \leq 2. \tag{3.6}$$

If $\phi(r) = 1$, scheme (3.2) reduces to the centered difference Lax-Wendroff method. If $\phi(r) = r$, scheme (3.2) reduces to the second order upwind Warming and Beam [10] method.

The region defined by (3.6) is shown in Fig. 1 along with the limiters corresponding to the Lax-Wendroff and Warming-Beam methods. Since these schemes are known to produce spurious wiggles in solutions with strongly varying gradients, it is not surprising that these schemes are not uniformly within the TVD region.

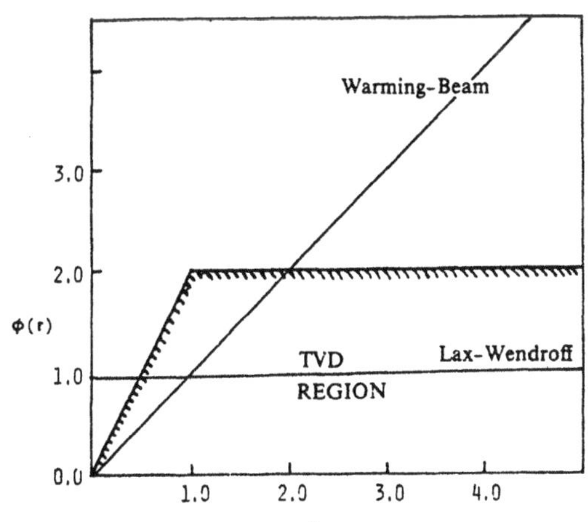

FIG. 1. TVD *region.*

Since the Lax-Wendroff method does not require one to determine an upwind direction and since many production computer codes are based on the Lax-Wendroff method or its variants, we wish to examine the possibility of adding terms to these codes to obtain a TVD scheme.

If we add a term of the form

$$K^+_{k+1/2}(r^+_k)\Delta U^n_{k+1/2} - K^+_{k-1/2}(r^+_{k-1})\Delta U^n_{k-1/2} \tag{3.7}$$

to the Lax-Wendroff scheme

$$U^{n+1}_k = U^n_k - \frac{\nu}{2}(\Delta U^n_{k+1/2} + \Delta U^n_{k-1/2}) + \frac{\nu^2}{2}(\Delta U^n_{k+1/2} - \Delta U^n_{k-1/2}) \tag{3.8}$$

and rearrange this to the form (3.2), we get

$$U^{n+1}_k = U^n_k - \left[\nu\left\{1 + \frac{1}{2}(1-\nu)[1/r^+_k - 1]\right\} - \left\{\frac{K^+_{k+1/2}}{r^+_k} - K^+_{k-1/2}\right\}\right]\Delta U^n_{k-1/2}. \tag{3.9}$$

That is,

$$\begin{aligned} C_{k-1/2} &= \nu\left\{1 + \frac{1}{2}(1-\nu)[1/r^+_k - 1]\right\} - \left\{\frac{K^+_{k+1/2}}{r^+_k} - K^+_{k-1/2}\right\}, \\ D_{k+1/2} &= 0, \end{aligned} \tag{3.10}$$

where the $^+$ superscript denotes terms which are added when $a > 0$. Note that the form (3.7) preserves conservation.

A comparison of (3.9) with (3.2) shows that we can obtain Sweby's scheme if we choose

$$K^+_{k+1/2} = \frac{\nu}{2}(1-\nu)[1 - \phi(r^+_k)]. \tag{3.11}$$

Next we consider the equation

$$u_t + au_x = 0, \qquad a = \text{const} < 0. \tag{3.12}$$

For this problem Sweby's scheme takes the form

$$U^{n+1}_k = U^n_k + \nu\left\{-1 + \frac{1}{2}(1+\nu)\left[\phi(r^-_{k+1}) - \frac{\phi(r^-_k)}{r^-_k}\right]\right\}\Delta U^n_{k+1/2} \tag{3.13}$$

where

$$r^-_k = \frac{\Delta U^n_{k+1/2}}{\Delta U^n_{k-1/2}}, \qquad \nu = \frac{a\Delta t}{\Delta x} < 0$$

and our modified Lax-Wendroff scheme takes the form

$$U^{n+1}_k = U^n_k + \nu\left[\left\{-1 + \frac{1}{2}(1+\nu)\left[1 - \frac{1}{r^-_k}\right]\right\} + \left\{K^-_{k+1/2} - \frac{K^-_{k-1/2}}{r^-_k}\right\}\right]\Delta U^n_{k+1/2}. \tag{3.14}$$

This reduces to Sweby's scheme if we choose

$$K^-_{k+1/2} = \frac{\nu(1+\nu)}{2}[\phi(r^-_{k+1}) - 1]. \tag{3.15}$$

A comparison of equations (3.11) and (3.15) shows how the sign of the coefficient a changes the method. In particular, we see that as a changes sign, the definition of

r changes and the form of the term involving the Courant number ν changes. We can combine these two cases into a Lax-Wendroff method with an upstream weighted artificial dissipation term as follows. Put the scheme into the form

$$\begin{aligned} U_k^{n+1} = U_k^n - \frac{\nu}{2}(U_{k+1}^n - U_{k-1}^n) + \frac{\nu^2}{2}(U_{k+1}^n - 2U_k^n + U_{k-1}^n) \\ + [K_{k+1/2}^+(r_k^+) + K_{k+1/2}^-(r_{k+1}^-)](U_{k+1}^n - U_k^n) \\ - [K_{k-1/2}^+(r_{k-1}^+) + K_{k-1/2}^-(r_k^-)](U_k^n - U_{k-1}^n) \end{aligned} \tag{3.16}$$

where

$$K_{k+1/2}^+ = \begin{cases} \frac{\nu}{2}(1-\nu)[1-\phi(r_k^+)] & \text{if } a>0, \\ 0 & \text{if } a \leqq 0, \end{cases}$$
$$K_{k+1/2}^- = \begin{cases} 0 & \text{if } a \geqq 0, \\ \frac{\nu}{2}(1+\nu)[\phi(r_{k+1}^-)-1] & \text{if } a<0. \end{cases} \tag{3.17}$$

This method still requires that we know which direction is upwind. The determination of upwind directions for hyperbolic systems is a complicated procedure. Therefore we redefine the K's as follows and attempt to construct a method without upstream weighting.

$$\begin{aligned} K_{k+1/2}^+ = \frac{|\nu|}{2}(1-|\nu|)[1-\phi(r_k^+)], \\ K_{k+1/2}^- = \frac{|\nu|}{2}(1-|\nu|)[1-\phi(r_{k+1}^-)]. \end{aligned} \tag{3.18}$$

That is, we always include terms involving both sets of K's.

Next we need to consider how the added terms affect the accuracy in smooth regions. To do this we need to define what we mean by a smooth region. For the purposes of this paper, we define a smooth region as a region in which $\Delta U = O(h)$ and $(1-r) = O(h)$ where h is the discretization parameter. With some changes of notations, this definition is consistent with the work of Harten [3] and Roe [5]. In [3] Harten shows that a scheme is second-order accurate if its numerical flux differs from the Lax-Wendroff numerical flux by $O(h^2)$ in smooth regions. In the following the limiter ϕ is chosen so that $(1-\phi(r)) = O(h)$ when $(1-r) = O(h)$ and the Harten condition is satisfied.

Since the terms involving the K's are second difference terms with positive coefficients, it is obvious that they are dissipative. The question that needs to be addressed is how much more dissipative (3.18) is than (3.17). To obtain some idea about this we compute the solution to the equation

$$u_t + u_x = 0 \tag{3.19}$$

with square wave initial data. We use the Lax-Wendroff method and the limiter defined by

$$\phi(r) = \begin{cases} \min(2r, 1) & \text{if } r>0, \\ 0 & \text{if } r \leqq 0. \end{cases} \tag{3.20}$$

Note that we are using the same limiter for both the upwind and nonupwind schemes and are thus not taking advantage of the wider choice of limiters allowed by upwind schemes. This is done to compare upwind and nonupwind versions of the same scheme and isolate the effect of upwinding. In a future paper we examine other choices of limiter functions for both methods.

Fig. 2a shows the result of this computation after 100 steps at a Courant number, $\nu=.9$ using the Lax-Wendroff method without additional dissipation. Note that the solution exhibits severe oscillations in those regions where r^+ is small as predicted by Fig. 1. Figure 2b is the result of the same computation using the Lax-Wendroff scheme with the upwind dissipation (3.17) and the flux limiter (3.20). Notice that the spurious oscillations have been removed. Finally, Fig. 2c shows the result of the same computation using the Lax-Wendroff scheme with the simplified dissipation (3.18) and the flux limiter (3.20). These results are almost indistinguishable from those of Fig. 2b. In addition we can prove the following theorem.

THEOREM 3.1. *The method* (3.16), (3.18) *with flux limiter* (3.20) *is* TVD *under the restriction that the Courant number* $|\nu|<1$.

The proof appears in the Appendix.

It is trivial to extend these results to scalar nonlinear problems. We simply define a local wave speed by

$$a_{k+1/2}=\begin{cases}\dfrac{\Delta f_{k+1/2}}{\Delta U_{k+1/2}} & \text{if } \Delta U_{k+1/2}\neq 0,\\[2ex] \dfrac{df(U_k)}{dU} & \text{if } \Delta U_{k+1/2}=0\end{cases} \tag{3.21}$$

and apply the schemes (3.16), (3.17) or (3.16), (3.18) as before. The wave speed definition (3.33) makes the resulting scheme conservative.

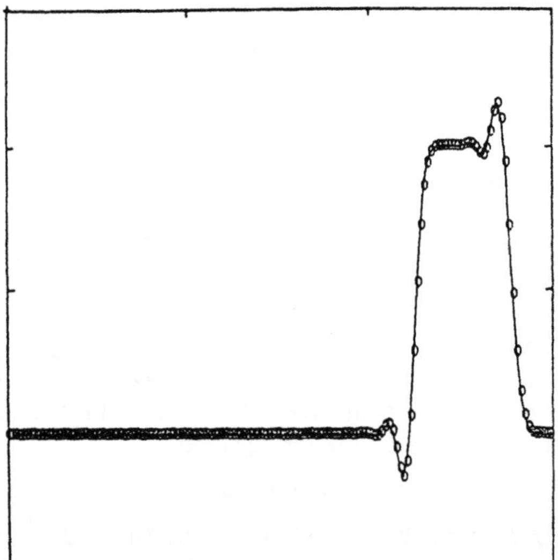

FIG. 2a. *Solution of* (3.19) *with square wave initial data after* 100 *steps at* $\nu=.9$ *using Lax-Wendroff method.*

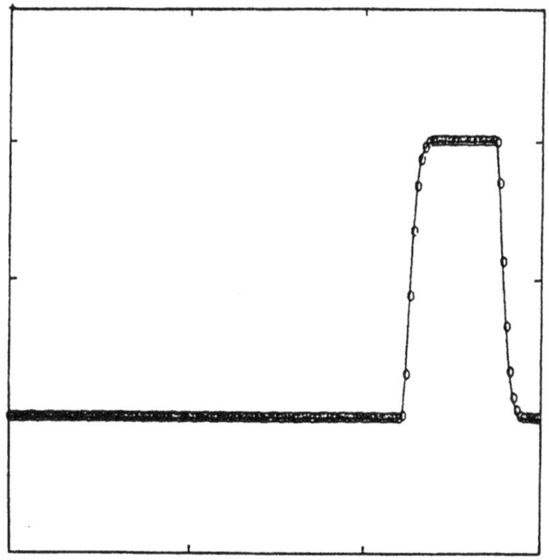

FIG. 2b. *Solution of* (3.19) *with square wave initial data after* 100 *steps at* $\nu = .9$ *using upwind scheme* (3.16), (3.17).

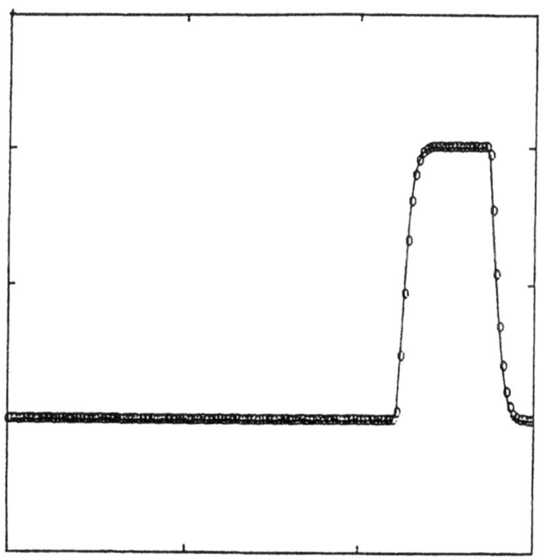

FIG. 2c. *Solution of* (3.19) *with square wave initial data after* 100 *time steps at* $\nu = .9$ *using* TVD *scheme* (3.16), (3.18).

Figures 3a, 3b and 3c show computed solutions of the inviscid Burgers' equation with square wave initial data and periodic boundary conditions. These results were obtained using the MacCormack two-step Lax-Wendroff scheme defined in the next section, the second order upwind scheme (3.16), (3.17) and the simplified scheme (3.16), (3.18), respectively. The MacCormack results exhibit severe oscillations in the vicinity of the shock and an entropy violating expansion shock. The upwind scheme

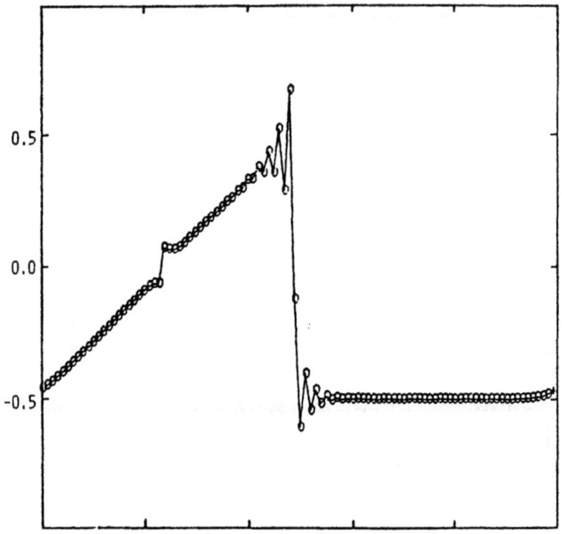

FIG. 3a. *Solution of Burgers' equation using MacCormack scheme.*

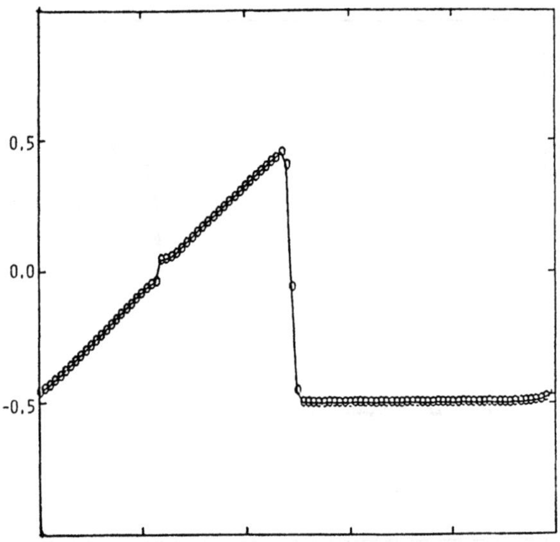

FIG. 3b. *Solution to Burgers' equation using upwind scheme* (3.16), (3.17).

eliminates the oscillations but not the expansion shock while the simplified scheme eliminates all but a small "entropy glitch" in the expansion region. There is no difference between the upwind method and the simplified method in their ability to resolve the shock.

The Burgers' equation results were highly dependent on the choice of initial conditions. Indeed, for certain initial conditions, the simplified method also computed solutions containing expansion shocks. To our knowledge, the only way to avoid

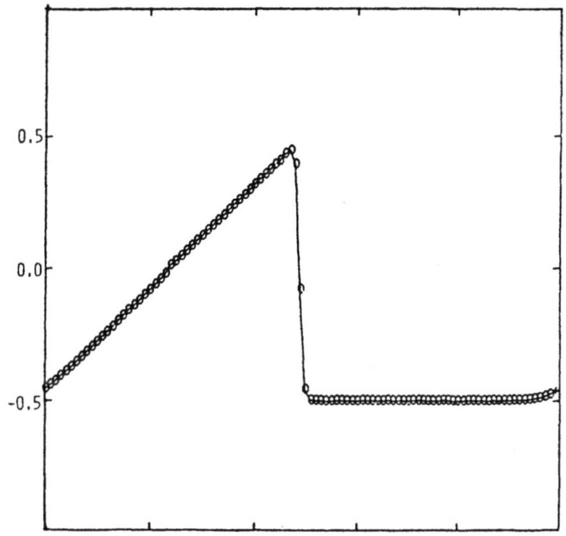

FIG. 3c. *Solution to Burgers' equation using TVD scheme* (3.16), (3.18).

expansion shocks in all cases is to explicitly add extra dissipation to the method when a sonic point occurs in an expansion region. Although we cannot guarantee that the simplified scheme will do this automatically, our computations indicate that this scheme is more robust than the unmodified upwind scheme.

4. Hyperbolic systems. In this section we extend the results of the previous section to hyperbolic systems. To that end, we consider first the linear, constant coefficient system

$$u_t + Au_x = 0, \qquad A = \text{const} \tag{4.1}$$

where u is an m vector and A is an $m \times m$ matrix.

If the system (4.1) is hyperbolic, the matrix A has real eigenvalues and a complete set of linearly independent right eigenvectors. If we let P denote the matrix whose columns are the right eigenvectors of A, then

$$P^{-1}AP = \Lambda \tag{4.2}$$

where

$$\Lambda = \begin{bmatrix} \lambda_1 & & & 0 \\ & \lambda_2 & & \\ & & \ddots & \\ 0 & & & \lambda_m \end{bmatrix} \tag{4.3}$$

and λ_j are the eigenvalues of A.

If we define a new set of dependent variables by the formula

$$v = P^{-1}u \tag{4.4}$$

and multiply (4.1) by P^{-1} we obtain

$$(P^{-1}u)_t + P^{-1}AP(P^{-1}u)_x = 0 \tag{4.5}$$

or

$$v_t + \Lambda v_x = 0. \tag{4.6}$$

This is an uncoupled set of scalar equations. We solve (4.6) using (3.16). That is,

$$\begin{aligned} V_k^{n+1} = V_k^n - \frac{\nu}{2}(V_{k+1}^n - V_{k-1}^n) + \frac{\nu^2}{2}(V_{k+1}^n - 2V_k^n + V_{k-1}^n) \\ + [K_{k+1/2}^+(r_k^+) + K_{k+1/2}^-(r_{k+1}^-)](V_{k+1}^n - V_k^n) \\ - [K_{k-1/2}^+(r_{k-1}^+) + K_{k-1/2}^-(r_k^-)](V_k^n - V_{k-1}^n) \end{aligned} \tag{4.7}$$

where the diagonal matrices $\nu = \Lambda \Delta t/\Delta x$ and $K^\pm$ and the vectors $r^\pm$ are defined below.

Multiply (4.7) by P to obtain an equation in terms of the original dependent variables. The result is

$$\begin{aligned} U_k^{n+1} = U_k^n - A\frac{\Delta t}{2\Delta x}(U_{k+1}^n - U_{k-1}^n) + A^2\frac{\Delta t}{2\Delta x}(U_{k+1}^n - 2U_k^n + U_{k-1}^n) \\ + P[K_{k+1/2}^+(r_k^+) + K_{k+1/2}^-(r_{k+1}^-)]P^{-1}(U_{k+1}^n - U_k^n) \\ - P[K_{k-1/2}^+(r_{k-1}^+) + K_{k-1/2}^-(r_k^-)]P^{-1}(U_k^n - U_{k-1}^n). \end{aligned} \tag{4.8}$$

The first three terms on the right of (4.8) comprise the well-known Lax-Wendroff scheme and need no further discussion. The last two terms require that we know the matrices P and P^{-1}. In the following we construct a simplified version of this scheme which removes this requirement.

We remove the requirement that P and P^{-1} be known by approximating the diagonal matrices $K^\pm$ by scalar matrices. That is, we let

$$K^\pm(r^\pm) \approx \bar{K}^\pm(r^\pm) I \tag{4.9}$$

where the $\bar{K}^\pm(r^\pm)$ are scalar functions of $r^\pm$.

We remove the local Courant number dependence by choosing

$$\bar{K}^\pm(r^\pm) = .5C(\nu)[1 - \phi(r^\pm)] \tag{4.10}$$

where the Courant number ν is defined as

$$\nu = \max_j |\lambda_j| \frac{\Delta t}{\Delta x} \tag{4.11}$$

and $C(\nu)$ is chosen as follows:

$$C(\nu) = \begin{cases} \nu(1-\nu) & \text{if } \nu \leqq .5, \\ .25 & \text{if } \nu > .5. \end{cases} \tag{4.12}$$

This definition of $C(\nu)$ is an upper bound to the Courant number-dependent coefficient in (3.18).

Thus far we have not defined r^+ and r^-. In light of the fact that we do not wish to compute P or P^{-1}, we have chosen the following definitions for r^+ and r^-:

$$r_k^+ = \frac{(\Delta U_{k-1/2}^n, \Delta U_{k+1/2}^n)}{(\Delta U_{k+1/2}^n, \Delta U_{k+1/2}^n)}, \tag{4.13a}$$

(4.13b) $$r_k^- = \frac{(\Delta U_{k-1/2}^n, \Delta U_{k+1/2}^n)}{(\Delta U_{k-1/2}^n, \Delta U_{k-1/2}^n)},$$

where $(\cdot,\cdot)$ denotes the usual inner product on R^m.

If P does not vary significantly over adjacent mesh intervals, these definitions can be interpreted as averages of the scalar definitions. These were the most simple definitions that we could construct. They worked so well in our numerical experiments that we saw no reason to investigate more sophisticated $r^\pm$ definitions. It might be advantageous to look at other definitions since we have not been able to prove rigorously that (4.13) always assures that (4.7) is TVD. Such a proof seems to require that we base $r^\pm$ on the largest and smallest jumps in the characteristic fields. We note in passing that other definitions of $r^\pm$ have been proposed by Sweby [8] and Chakravarthy and Osher [2].

With these simplifications, our numerical method takes the form

(4.14) $$\begin{aligned} U_k^{n+1} = U_k^n &- A\frac{\Delta t}{2\Delta x}(U_{k+1}^n - U_{k-1}^n) + A^2\frac{\Delta t}{2\Delta x}(U_{k+1}^n - 2U_k^n + U_{k-1}^n) \\ &+ [\bar{K}_{k+1/2}^+(r_k^+) + \bar{K}_{k+1/2}^-(r_{k+1}^-)](U_{k+1}^n - U_k^n) \\ &- [\bar{K}_{k-1/2}^+(r_{k-1}^+) + \bar{K}_{k-1/2}^-(r_k^-)](U_k^n - U_{k-1}^n), \end{aligned}$$

where $\bar{K}^\pm$ and $r^\pm$ are defined by equations (4.10) and (4.13) respectively. Note that the resulting scheme does not depend explicitly on the transformation (4.4). Therefore, we can use the scheme without modification when the matrix A is not a constant matrix. In a future paper we hope to examine the stability and TVD properties of this scheme rigorously. Here we present only numerical results.

For the computations which follow, we replace the Lax-Wendroff scheme (the first three terms on the right of (4.14)) by the conservative MacCormack scheme. These schemes are equivalent for linear problems.

The resulting algorithm is as follows:

1. For each grid interval $j+\frac{1}{2}$ compute

$$D_{j+1/2} \leftarrow [\bar{K}_{j+1/2}^+(r_k^+) + \bar{K}_{j+1/2}^-(r_{k+1}^-)](U_{j+1}^n - U_j^n)$$

and save.

2. Compute a provisional solution at each grid point using some variant of the MacCormack method, e.g.,

$$U_j^{(1)} = U_j^n - \lambda[F(U_j^n) - F(U_{j-1}^n)],$$
$$U_j^{(2)} = .5*\{U_j^n + U_j^{(1)} - \lambda[F(U_{j+1}^{(1)}) - F(U_j^{(1)})]\}$$

where $\lambda = \Delta t/\Delta x$. We note here that the computed results did not differ significantly when the other MacCormack variant was used.

3. Add the dissipative terms to the provisional solution, i.e.,

$$U_j^{n+1} = U_j^{(2)} + \{D_{j+1/2} - D_{j-1/2}\}.$$

As a first test, we demonstrate the performance of this method on the Riemann problem. That is, we solve the Euler equations

(4.15a) $$u_t + f(u)_x = 0, \quad -\infty < x < \infty, \quad t > 0$$

where

$$u = \begin{bmatrix} \rho \\ \rho u \\ E \end{bmatrix}, \qquad f(u) = \begin{bmatrix} \rho u \\ \rho u^2 + p \\ (E+p)u \end{bmatrix}, \tag{4.15b}$$

$$p = (\gamma - 1)(E - \tfrac{1}{2}\rho u^2) \tag{4.15c}$$

and $\gamma = 1.4$ with initial conditions

$$u(x, 0) = \begin{cases} u_L, & x < 0, \\ u_R, & x > 0. \end{cases} \tag{4.15d}$$

In this case the initial conditions were

$$u_L = \begin{bmatrix} .445 \\ .311 \\ 8.928 \end{bmatrix}, \qquad u_R = \begin{bmatrix} .5 \\ 0 \\ 1.4275 \end{bmatrix}. \tag{4.16}$$

Figure 4 shows the solution after 100 time steps computed on 140 grid points at a Courant number of .95. These conditions are the same as those used by Harten [3] and Chakravarthy and Osher [2] in their numerical experiments. The results shown in Fig. 4 cannot be distinguished from those shown in the cited references with one exception. Harten was able to obtain a dramatic improvement in the resolution of the contact discontinuity by selectively adding artificial compression to his second order upwind scheme. We intend to study this technique in the future.

Initially, we attempted to extend this algorithm to two dimensions in an obvious manner. That is:

1. For each x-direction grid interval compute

$$D_{i+1/2,j} = [\bar{K}^{+}_{i+1/2,j}(r^{+}_{xi,j}) + \bar{K}^{-}_{i+1/2,j}(r^{-}_{xi+1,j})](U^n_{i+1,j} - U^n_{i,j})$$

and save.

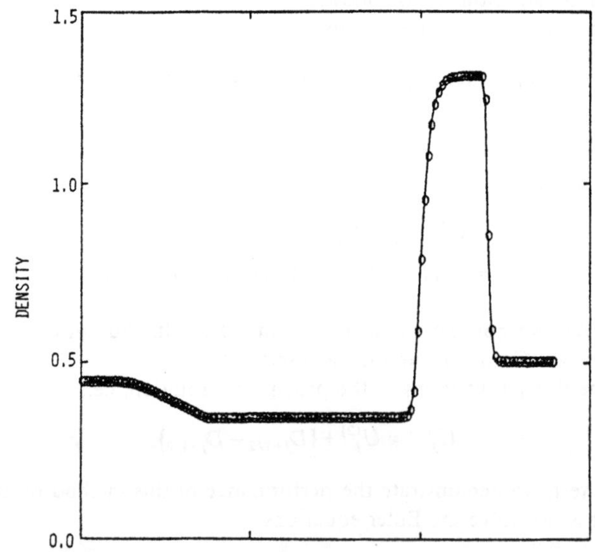

FIG. 4. *Solution to Riemann problem after* 100 *time steps at* $\nu = .95$ *using scheme* (4.14), (4.10).

2. For each y-direction grid interval compute

$$D_{i,j+1/2}=[\bar{K}^+_{i,j+1/2}(r^+_{yi,j})+\bar{K}^-_{i,j+1/2}(r^-_{yi,j+1})](U^n_{i,j+1}-U_{i,j})$$

and save.

3. Compute a provisional solution at each grid point using some variant of the MacCormack scheme, e.g.,

$$U^{(1)}_{i,j}=U^n_{i,j}-\lambda_x[F(U^n_{i,j})-F(U^n_{i-1,j})]-\lambda_y[G(U^n_{i,j+1})-G(U^n_{i,j})],$$

$$U^{(2)}_{i,j}=.5*\{U^n_{i,j}+U^{(1)}_{i,j}-\lambda_x[F(U^{(1)}_{i+1,j})-F(U^{(1)}_{i,j})]-\lambda_y[G(U^{(1)}_{i,j})-G(U^{(1)}_{i,j-1})]\}.$$

Results of computations made using different MacCormack variants did not differ significantly.

4. Add the dissipative terms to the provisional solutions, i.e.,

$$U^{n+1}_{i,j}=U^{(2)}_{i,j}+[D_{i+1/2,j}-D_{i-1/2,j}+D_{i,j+1/2}-D_{i,j-1/2}].$$

In the case that all of the r's are negative, each of the $\bar{K}$'s can be as large as .125. This would give a dissipative coefficient of .25 which can introduce into the method a weak checkerboard instability which is independent of the stability of the MacCormack method. To avoid this, we follow the suggestion of Roe [6] and combine the two coefficients $\bar{K}^+$ and $\bar{K}^-$ into a single coefficient $\bar{K}$ where

$$\bar{K}(r^+,r^-)=.5C(\nu)[1-\phi(r^+,r^-)],$$

where $C(\nu)$ is as before and

$$\phi(r^+,r^-)=\max[0,\min(2r^+,r^-,1),\min(2r^-,r^+,1)].$$

We are currently studying other limiters. This reduces the maximum dissipation coefficient to .125, which would be acceptable in three dimensions as well as two.

Next we demonstrate the performance of our method on a two-dimensional problem. Figure 5 shows a comparison of the present method with the second order upwind scheme of van Leer [9]. To obtain these results, we solve the two-dimensional Euler equations

$$u_t+f(u)_x+g(u)_y=0 \tag{4.17a}$$

where

$$u=\begin{bmatrix}\rho\\ \rho u\\ \rho v\\ E\end{bmatrix},\quad f(u)=\begin{bmatrix}\rho u\\ \rho u^2+p\\ \rho uv\\ (E+p)u\end{bmatrix},\quad g(u)=\begin{bmatrix}\rho v\\ \rho uv\\ \rho v^2+p\\ (E+p)v\end{bmatrix}, \tag{4.17b}$$

$$p=(\gamma-1)[E-\tfrac{1}{2}(\rho u^2+\rho v^2)] \tag{4.17c}$$

and $\gamma=1.4$ for the problem of the reflection of an oblique shock from a plane wall.

For the computations shown, we specify a uniform $M=2.9$ flow at the left boundary. At the top boundary, we specify the conditions behind a shock that would turn the flow 11° (cf. [1]). A flow tangency condition is specified at the wall and all variables are extrapolated at the right boundary. The computation is started with the upstream conditions specified everywhere except the top boundary. The results shown consist of a three-dimensional plot of the converged density solution and a longitudinal section of this plot taken at $y=.5$. Once again it is difficult to distinguish between the results computed using the two methods.

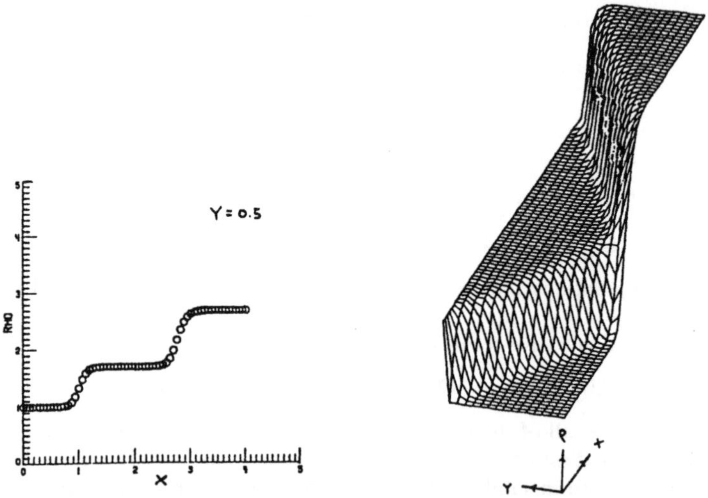

FIG. 5a. *Density profile for shock reflection problem using van Leer method.*

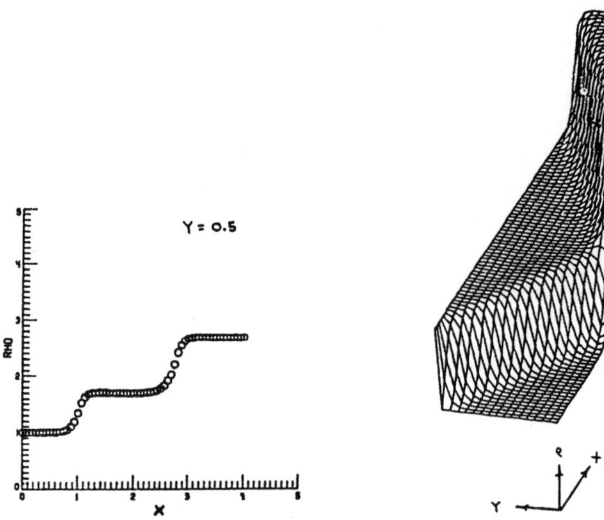

FIG. 5b. *Density profile for shock reflection problem using modified MacCormack scheme.*

Finally, we compare the present method to the second order upwind method of van Leer on a model transonic flow problem.

The two-dimensional Euler equations (4.17) are solved for the flow over a 10% thick parabolic arc bump in a channel. The flow is assumed to be uniform initially and then the condition that the flow be tangent to the bump is applied at the wall in the manner of small disturbance theory. Nonreflecting boundary conditions are applied at both upstream and downstream boundaries. Figures 6a and 6b show the converged pressure distribution on the wall for a flow with inlet Mach number .675 using the van Leer method and the present method, respectively.

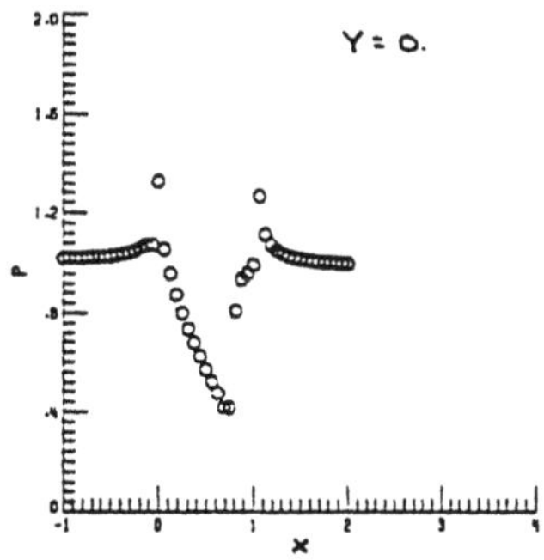

FIG. 6a. *Pressure distribution along wall for transonic problem using van Leer scheme.*

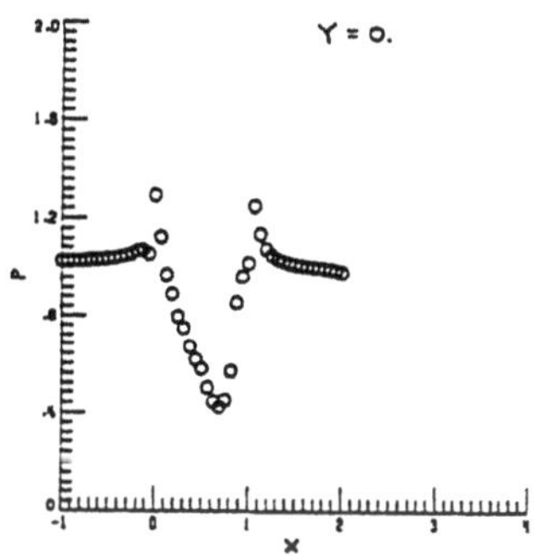

FIG. 6b. *Pressure distribution along wall for transonic problem using modified MacCormack scheme.*

The flow consists of an expansion region over the forward part of the bump, $0 \leqq x \leqq .5$, followed by a shock near $x = .75$. This is the type of problem that the van Leer method was designed to solve since the shock is steady and nearly aligned with the computing grid. Figure 6a shows that the van Leer scheme does an admirable job on this problem. In particular, the flow expands smoothly through the sonic point ($P \approx .717$) and there is only one grid point within the shock. By comparison, Fig. 6b shows that the flow computed using the present method is not quite as smooth in the expansion region and there are two points within the shock. Still, considering the fact that the current method is considerably easier to program and that it runs in $\frac{2}{3}$ the time of the van Leer scheme, the results shown are quite acceptable.

5. Concluding remarks. In this paper we construct a simple artificial viscosity term for Lax-Wendroff type methods which is based on the theory of Total Variation Diminishing upwind finite difference schemes. This method has advantages over both conventional artificial viscosity schemes and the TVD upwind schemes. In particular, the method does not contain the problem dependent parameters of conventional artificial viscosity schemes and it does not require the complex logic of upwind schemes.

The numerical experiments that have been performed thus far have been very encouraging but more numerical experimentation is needed. We intend to carry this out in the future and also to apply the ideas of this paper to other numerical schemes.

Appendix. Here we prove the following theorem.

THEOREM 3.1. *The method* (3.16), (3.18) *with flux limiter* (3.20) *is* TVD *under the restriction that the Courant number* $|\nu|<1$.

Proof. The proof is a direct application of Lemma 2.2. That is, we show that if the scheme is put into the form

$$U_k^{n+1}=U_k^n-C_{k-1/2}\Delta U_{k-1/2}^n+D_{k+1/2}\Delta U_{k+1/2}^n \tag{A.1}$$

then

$$C_{k+1/2}\geqq 0, \tag{A.2}$$

$$D_{k+1/2}\geqq 0, \tag{A.3}$$

$$0\leqq C_{k+1/2}+D_{k+1/2}<1. \tag{A.4}$$

Here we prove the result for $\nu>0$. The computations for $\nu<0$ are similar.

Rearrange (3.16) as follows

$$\begin{aligned}U_k^{n+1}=U_k^n-\left[\nu\left\{1+\frac{1}{2}(1-\nu)\left(\frac{1}{r_k^+}-1\right)\right\}-\left\{\frac{K_{k+1/2}^+}{r_k^+}-K_{k-1/2}^+-K_{k-1/2}^-\right\}\right]\Delta U_{k-1/2}^n\\ +K_{k+1/2}^-\Delta U_{k+1/2}^n.\end{aligned} \tag{A.5}$$

That is,

$$C_{k+1/2}=\nu\left\{1+\frac{1}{2}(1-\nu)\left(\frac{1}{r_{k+1}^+}-1\right)\right\}-\left\{\frac{K_{k+3/2}^+}{r_{k+1}^+}-K_{K+1/2}^+-K_{k+1/2}^-\right\}, \tag{A.6}$$

$$D_{k+1/2}=K_{k+1/2}^-, \tag{A.7}$$

$D_{K+1/2}>0$ since $K_{k+1/2}^->0$.

To show that $C_{k+1/2}>0$, substitute (3.18) into (A.6) and rearrange terms to obtain

$$C_{K+1/2}=\nu+\frac{\nu(1-\nu)}{2}\left[\frac{\phi(r_{k+1}^+)}{r_{k+1}^+}-\phi(r_k^+)+1-\phi(r_{k+1}^-)\right]. \tag{A.8}$$

Note that (3.20) implies

$$0<\frac{\phi(r)}{r}<2, \tag{A.9}$$

$$0<\phi(r)<1, \tag{A.10}$$

so

$$C_{k+1/2}\geqq\nu-\frac{\nu(1-\nu)}{2}=\frac{\nu}{2}+\frac{\nu^2}{2}>0.$$

Since $D_{k+1/2}>0$ and $C_{k+1/2}>0$, $C_{k+1/2}+D_{k+1/2}>0$. To prove that

$$C_{k+1/2}+D_{k+1/2}<1,$$

substitute (3.18) into (A.7) and add to (A.8). The result is

$$C_{k+1/2}+D_{k+1/2}=\nu+\frac{\nu(1-\nu)}{2}\left[\frac{\phi(r_{k+1}^+)}{r_{k+1}^+}-\phi(r_k^+)+2-2\phi(r_{k+1}^-)\right]. \tag{A.11}$$

(A.11) is $\leqq 1$, if the term in the square brackets is $\leqq 2$. To show this, we note that

(A.12)
$$r_{k+1}^{-}=\frac{1}{r_{k+1}^{+}}$$

and we consider four cases.

Case 1.

$$r_{k+1}^{+}<0\Rightarrow\frac{1}{r_{k+1}^{+}}<0\Rightarrow\phi(r_{k+1}^{+})=\phi\left(\frac{1}{r_{k+1}^{+}}\right)=0.$$

In this case the term in square brackets becomes

$$2-\phi(r_k^{+})\leqq 2.$$

Case 2.

$$2r_{k+1}^{+}<1\Rightarrow r_{k+1}^{+}<\frac{1}{2},\qquad\frac{1}{r_{k+1}^{+}}>2,$$

so

$$\phi(r_{k+1}^{+})=2r_{k+1}^{+},\qquad\phi\left(\frac{1}{r_{k+1}^{+}}\right)=1$$

and the term in square brackets is

$$2-\phi(r_k^{+})\leqq 2.$$

Case 3.

$$\frac{1}{2}<r_{k+1}^{+},\qquad\frac{1}{r_{k+1}^{+}}<2\Rightarrow\phi(r_{k+1}^{+})=\phi\left(\frac{1}{r_{k+1}^{+}}\right)=1.$$

Then the term in square brackets becomes

$$\frac{1}{r_{k+1}^{+}}-\phi(r_k^{+})\leqq 2-\phi(r_k^{+})\leqq 2.$$

Case 4.

$$2\leqq r_{k+1}^{+}\Rightarrow\frac{2}{r_{k+1}^{+}}<1\Rightarrow\phi(r_{k+1}^{+})=1,\qquad\phi\left(\frac{1}{r_{k+1}^{+}}\right)=\frac{2}{r_{k+1}^{+}}.$$

Then the term in square brackets becomes

$$\frac{1}{r_{k+1}^{+}}-\phi(r_k^{+})+2-\frac{4}{r_{k+1}^{+}}=2-\frac{3}{r_{k+1}^{+}}-\phi(r_k^{+})\leqq 2.$$

This completes the proof.

REFERENCES

[1] AMES RESEARCH STAFF, *Equations, tables and charts for compressible flow*, NACA Report 1135, 1953.

[2] S. CHAKRAVARTHY AND S. OSHER, *High resolution applications of the Osher upwind scheme for the Euler equations*, Proc. AIAA Computational Fluid Dynamics Conference, Danvers, MA, 1983, pp. 362-372.

[3] A. HARTEN, *High resolution schemes for hyperbolic conservation laws*, J. Comput. Phys., 49 (1983), pp. 357-393.

[4] A. Y. LE ROUX, *A numerical conception of entropy for quasi linear equations*, Math. Comput., 31 (1977), pp. 848–872.

[5] P. L. ROE, *Some contributions to the modelling of discontinuous flows*, Proc. AMS/SIAM Seminar on Large Scale Computations in Fluid Mechanics, San Diego, CA, June 26-July 8, 1983, to appear.

[6] ———, *Generalized formulations of TVD Lax-Wendroff schemes*, ICASE Report No. 84-53, NASA CR-172478, Hampton, VA, 1984.

[7] R. SANDERS, *On convergence of monotone finite difference schemes with variable space differencing*, Math. Comput., 40 (1983), pp. 91–106.

[8] P. K. SWEBY, *High resolution schemes using flux limiters for hyperbolic conservation laws*, SIAM J. Numer. Anal., 21 (1984), pp. 995–1011.

[9] B. VAN LEER, *Computational methods for ideal compressible flow*, ICASE Report No. 83-38, NASA CR 172180, Hampton, VA, 1983.

[10] R. F. WARMING AND R. M. BEAM, *Upwind second order difference schemes and applications in aerodynamics*, AIAA J., 14 (1976), pp. 1241–1249.

On Central-Difference and Upwind Schemes

R. C. SWANSON

NASA Langley Research Center, Hampton, Virginia 23665

AND

ELI TURKEL

Tel-Aviv University, Tel-Aviv 69978, Israel, and Institute for Computer Applications in Science and Engineering, Hampton, Virginia 23665

Received October 16, 1990; revised July 22, 1991

A class of numerical dissipation models for central-difference schemes constructed with second- and fourth-difference terms is considered. The notion of matrix dissipation associated with upwind schemes is used to establish improved shock capturing capability for these models. In addition, conditions are given that guarantee that such dissipation models produce a TVD scheme. Appropriate switches for this type of model to ensure satisfaction of the TVD property are presented. Significant improvements in the accuracy of a central-difference scheme are demonstrated by computing both inviscid and viscous transonic airfoil flows.

I. INTRODUCTION

Central-difference-type schemes are currently being applied on a regular basis in the solution of the Euler and Navier–Stokes equations. A numerical dissipation model is included in these schemes, and it plays a crucial role in the determination of their success. The form of the dissipation model is quite often a blending of second-difference and fourth-difference dissipation terms. The second-difference terms are used to prevent oscillations at shock waves, while the fourth-difference terms are important for stability and convergence to a steady state. There is a constant to be specified for each contribution. However, using the model in conjunction with appropriate numerical procedures, these constants can usually be selected and maintained for a fairly wide class of fluid dynamics problems. This dissipation model allows shock waves to be captured with smearing over three to four mesh cells.

Even though these central-difference schemes have proven to be reasonably effective in many cases, there are still strong motivations for reducing the numerical dissipation being produced. For example, by appropriate reduction of the artificial dissipation, shock wave representation and boundary-layer definition (especially the wall shear stresses) can be improved on coarse meshes. Such improvements in accuracy are especially beneficial for complex three-dimensional flows, which can demand extensive computational effort. In addition, better estimates of the limit of infinitely fine mesh values of aerodynamic coefficients for flows with shocks can be obtained. Also, the standard model has difficulties in hypersonic flow. Finally, for some problems, the influence of numerical dissipation needs to be severely limited in certain smooth regions of a flow field (i.e., near the trailing edge of an airfoil), while still maintaining stability near discontinuities. This difficulty cannot generally be resolved by simply reducing the global constants in the dissipation model.

One can appeal to ideas from upwind schemes to improve the dissipation model, especially in the vicinity of shock waves. Upwind algorithms utilize concepts from characteristic theory in order to determine the direction of spatial differencing. They have been extended to systems of conservation laws using such approaches as the flux-vector splitting of van Leer [1] and the approximate Riemann solver of Roe [2]. A fundamental feature of these schemes is the use of a matrix evaluation of the dissipation either in the implied or direct sense. In so doing, the dissipative terms of each discrete equation are scaled by the appropriate eigenvalues of the flux Jacobian matrices of the Euler equations, rather than the same eigenvalue as in the dissipation model employed with central-difference schemes. Also, upwind schemes can be designed to have the total variation diminishing (TVD) property, which prevents the occurrence of spurious oscillations. The disadvantage of these schemes is that, in general, they increase the operation count for processing mesh points by about a factor of two over that required by central-difference schemes. One would

292

certainly like to more closely imitate the highly desirable behavior of the upwind algorithms near flow discontinuities, and at the same time, retain the more efficient central-difference scheme over significant portions of a flow field. In addition, one would like to have the high degree of numerical efficiency that has been achieved by combining a central-difference scheme with a Runge–Kutta time-marching algorithm, which includes residual smoothing and multigrid acceleration techniques.

The primary purpose of this paper is to construct a numerical dissipation model for a central-difference scheme that has both the properties of matrix dissipation and of TVD. As a starting point, we consider the elements of a widely used scalar dissipation model. Modifications of this model that facilitate accurate viscous flow computations are also examined. In the next section of the paper, the intimate connection between the formulation for an upwind scheme and a centered-difference scheme is presented, so as to establish a foundation for a matrix dissipation model. Then a theorem is proved that provides a simple sufficient condition to determine when a central-difference scheme with dissipation terms comprised of second and fourth differences is TVD. In the following section, appropriate flux limiter functions consistent with the central-difference dissipation model are discussed. A multistage time-stepping scheme used in applications is next briefly described. Finally, numerical results are shown to demonstrate the benefits of using the matrix dissipation model. Both inviscid and viscous transonic airfoil flows are computed.

II. SCALAR DISSIPATION MODEL

The basic elements of the scalar dissipation model considered in this paper were first introduced by Jameson, Schmidt, and Turkel [3] in conjunction with Runge–Kutta explicit schemes. This dissipation model has been used by many investigators [4–6] to numerically solve the Euler equations for a wide range of fluid dynamic applications. The same type of dissipation model has been applied to alternating direction implicit (ADI) schemes [7] and LU factored implicit schemes [8]. Several modifications of the model have been investigated in [9] and [10] in order to improve it and make it suitable for obtaining accurate and efficient solutions of the Navier–Stokes equations. In this section, the basic model and important modifications are briefly reviewed.

Consider the Euler equations in the form

$$W_t + f_x + g_y = 0, \tag{2.1}$$

where W is the four-component vector of conserved variables, and f, g are the flux vectors. The independent variables are time t and Cartesian coordinates (x, y). If (2.1) is transformed to arbitrary curvilinear coordinates $\xi = \xi(x, y)$ and $\eta = \eta(x, y)$, then we obtain

$$(J^{-1}W)_t + F_\xi + G_\eta = 0, \tag{2.2}$$

where J^{-1} is the inverse transformation Jacobian, and

$$F = fy_\eta - gx_\eta, \qquad G = gx_\xi - fy_\xi.$$

In a cell-centered, finite-volume method, (2.2) is integrated over an elemental volume in the discretized computational domain, and J^{-1} is identified as the volume of the cell. Equation (2.2), assuming J^{-1} is independent of time, can also be written as

$$J^{-1}W_t + AW_\xi + BW_\eta = 0,$$

where A and B are the flux Jacobian matrices defined by $A = \partial F/\partial W$ and $B = \partial G/\partial W$.

To advance the scheme in time we use a multistage scheme. A typical step of a Runge–Kutta approximation to (2.2) is

$$W^{(k)} = W^{(0)} - \alpha_k \frac{\Delta t}{J^{-1}} [D_\xi F^{(k-1)} + D_\eta G^{(k-1)} - AD], \tag{2.3}$$

where D_ξ and D_η are spatial differencing operators, and AD represents the artificial dissipation terms. The dissipation terms are a blending of second and fourth differences. That is,

$$AD = (D_\xi^2 + D_\eta^2 - D_\xi^4 - D_\eta^4)W, \tag{2.4}$$

where

$$D_\xi^2 W = \nabla_\xi[(\lambda_{i+1/2,j}\varepsilon^{(2)}_{i+1/2,j})\Delta_\xi] W_{i,j}, \tag{2.5}$$

$$D_\xi^4 W = \nabla_\xi[(\lambda_{i+1/2,j}\varepsilon^{(4)}_{i+1/2,j})\Delta_\xi \nabla_\xi \Delta_\xi] W_{i,j}, \tag{2.6}$$

and Δ_ξ, ∇_ξ are the standard forward and backward difference operators, respectively, associated with the ξ direction. The variable scaling factor λ was originally chosen as

$$\lambda_{i+1/2,j} = \tfrac{1}{2}[(\lambda_\xi)_{i,j} + (\lambda_\xi)_{i+1,j} + (\lambda_\eta)_{i,j} + (\lambda_\eta)_{i+1,j}], \tag{2.7}$$

where λ_ξ and λ_η are proportional to the largest eigenvalues of the matrices A and B. The spectral radii λ_ξ and λ_η are given by

$$\lambda_\xi = |uy_\eta - vx_\eta| + c\sqrt{x_\eta^2 + y_\eta^2},$$

$$\lambda_\eta = |vx_\xi - uy_\xi| + c\sqrt{x_\xi^2 + y_\xi^2},$$

where u and v are the Cartesian velocity components, and c is the speed of sound. The coefficients $\varepsilon^{(2)}$ and $\varepsilon^{(4)}$ are adapted to the flow and are defined as

$$\varepsilon^{(2)}_{i+1/2,j} = \kappa^{(2)} \max(\nu_{i-1,j}, \nu_{i,j}, \nu_{i+1,j}, \nu_{i+2,j}), \qquad (2.8)$$

$$\nu_{i,j} = \left| \frac{p_{i+1,j} - 2p_{i,j} + p_{i-1,j}}{p_{i+1,j} + 2p_{i,j} + p_{i-1,j}} \right|, \qquad (2.9)$$

$$\varepsilon^{(4)}_{i+1/2,j} = \max[0, (\kappa^{(4)} - \varepsilon^{(2)}_{i+1/2,j})], \qquad (2.10)$$

where p is the pressure, and the quantities $\kappa^{(2)}$ and $\kappa^{(4)}$ are constants to be specified. The operators in (2.4) for the η direction are defined in a similar manner.

The second-difference dissipation term is nonlinear. Its purpose is to introduce an entropy-like condition and to suppress oscillations in the neighborhood of shocks. This term is small in the smooth portion of the flow field. The fourth-difference dissipation term is basically linear and is included to damp high-frequency modes and allow the scheme to approach a steady state. Only this term affects the linear stability of the scheme. Near shocks it is reduced to zero.

The isotropic scaling factor of the original dissipation model as given in (2.7) is generally satisfactory for inviscid flow problems when typical inviscid flow meshes (i.e., cell aspect ratio $O(1)$) are used. The factor can produce too much numerical dissipation in the cases of meshes with high aspect ratio cells. This is also an important consideration for high Reynolds number viscous flows, where a mesh providing appropriate spatial resolution can have cell aspect ratios $O(10^3)$. In [9] and [10] this difficulty is remedied by replacing the factor of (2.7) with the anisotropic one

$$\lambda_{i+1/2,j} = \tfrac{1}{2}[(\bar{\lambda}_\xi)_{i,j} + (\bar{\lambda}_\xi)_{i+1,j}],$$

where

$$(\bar{\lambda}_\xi)_{i,j} = \phi_{i,j}(r)(\lambda_\xi)_{i,j},$$

$$\phi_{i,j}(r) = 1 + r^{\zeta}_{i,j}, \qquad 0 < \zeta < 1,$$

and $r = \lambda_\eta / \lambda_\xi$. In the normal direction, one defines

$$(\bar{\lambda}_\eta)_{i,j} = \phi_{i,j}\left(\frac{1}{r}\right)(\lambda_\eta)_{i,j}.$$

Alternatives to the switching function presented in (2.9) have been investigated. Caution must be exercised in the selection of a switching variable. If a quantity with the same functional dependence as entropy (i.e., p/ρ^γ) is used, sharper shocks can be captured in viscous transonic flows. However, such a choice can result in a loss of accuracy for the surface shear stress, due to the significant variation in the entropy-type variable across the boundary layer. This difficulty can be removed by simply multiplying the scaling factor by a function of the local Mach number of the flow. An acceptable modifying function has proven to be $(M_L/M_{ref})^\beta$, for some $\beta \geqslant 1$, where M_L is the local Mach number and M_{ref} is a reference Mach number (i.e., free-stream value for external flows). Typically the parameter β is 1 or 2. It is important to note that the entropy-type function is generally not satisfactory for inviscid flows. In addition, one can consider the sum of two switches, one depending on pressure and the other on temperature, so that all thermodynamic changes are taken into account. When introducing the matrix-valued dissipation it will be possible to use separate switches for different characteristic variables.

The treatment of the artificial dissipation must be modified at the boundaries of the physical domain. In the case of the fourth-difference dissipation the standard five point difference stencil must be replaced at the first two interior mesh cells. This means that one-sided or one-sided biased stencils are used at these cells. The dissipative character of the artificial terms is important because it influences both stability and accuracy. For example, if the dissipation is too large at a solid boundary, an artificial boundary layer is created in an inviscid flow, and the effective Reynolds number of a viscous flow is altered. To improve accuracy at the wall boundaries of viscous flows, where gradients are steep due to physical boundary layers, the usual fourth-difference stencils are changed in this dissipation model.

Let the total dissipation for a mesh cell, in the direction represented by the index j, be denoted by d_j. For simplicity assume that $\lambda\varepsilon^{(4)} = 1$. Then,

$$d_j = df_{j+1/2} - df_{j-1/2},$$

where the dissipative flux

$$df_{j+1/2} = (\Delta W)_{j+3/2} - 2(\Delta W)_{j+1/2} + (\Delta W)_{j-1/2}$$

and thus

$$d_j = (\Delta W)_{j+3/2} - 3(\Delta W)_{j+1/2} + 3(\Delta W)_{j-1/2} - (\Delta W)_{j-3/2}, \qquad (2.11)$$

with the index i for ΔW suppressed for convenience. Consider the first two interior cells adjacent to a solid boundary, as depicted in Fig. 1. If

$$(\Delta W)_{1/2} = (\Delta W)_{3/2} = (\Delta W)_{5/2} \qquad (2.12)$$

then (2.11) gives

$$d_2 = W_4 - 2W_3 + W_2 \qquad (2.13)$$

$$d_3 = W_5 - 4W_4 + 5W_3 - 2W_2. \qquad (2.14)$$

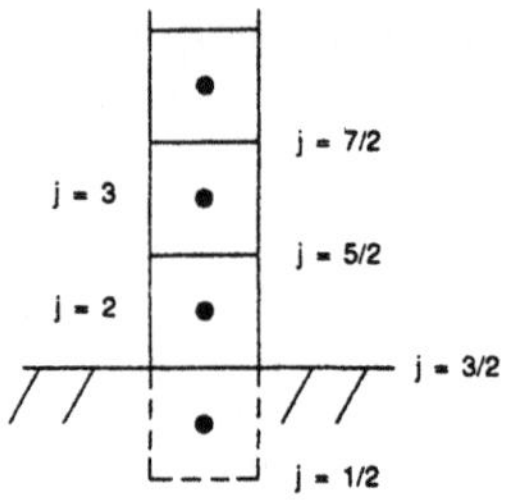

FIG. 1. Boundary point dissipation.

These boundary stencils are fairly standard ones, and they result in a nonpositive definite dissipation matrix for the system of difference equations [7]. An alternative form, which has reduced the sensitivity to solid surface normal mesh spacing for turbulent flow calculations without compromising stability or convergence, is given by

$$(\Delta W)_{1/2} = 2(\Delta W)_{3/2} - (\Delta W)_{5/2} \tag{2.15}$$

and

$$d_2 = W_4 - 3W_3 + 3W_2 - W_1, \tag{2.16}$$

$$d_3 = W_5 - 4W_4 + 6W_3 - 4W_2 + W_1. \tag{2.17}$$

This boundary condition is advantageous if the mesh is fine enough to adequately represent the laminar sublayer region of the boundary layer (i.e., at least two mesh points are inside the sublayer). For coarse meshes this treatment can be less accurate than the zeroth-order extrapolation of (2.12).

III. THE UPWIND CONNECTION

Upwind schemes for solving hyperbolic systems of conservation laws (i.e., Euler equations of gas dynamics) rely upon characteristic theory to determine the direction of propagation of information, and thus, the direction required for one-sided differencing approximations of the spatial derivatives. With such schemes shock waves can be captured without oscillations. Thus, a successful artificial dissipation model for a central-difference scheme should imitate an upwind scheme in the neighborhood of shocks. We now review the connection between these two types of schemes.

Consider the one-dimensional scalar wave equation

$$u_t + a u_x = 0$$

with a constant. The first-order upwind scheme can be written as

$$u_i^{n+1} = u_i - a\frac{\Delta t}{\Delta x}\begin{cases} u_{i+1} - u_i, & a < 0 \\ u_i - u_{i-1}, & a > 0, \end{cases} \tag{3.1}$$

where all discrete quantities are evaluated at time level $n\,\Delta t$ unless otherwise denoted. The scheme of (3.1) can be rewritten as

$$u_i^{n+1} = u_i - a\frac{\Delta t}{2\Delta x}(u_{i+1} - u_{i-1}) + |a|\frac{\Delta t}{2\Delta x}(u_{i+1} - 2u_i + u_{i-1}), \tag{3.2}$$

which now contains a central-difference term and a second-difference dissipation term. Now consider the system

$$u_t + A u_x = 0, \tag{3.3}$$

where u is a N-component vector. The system case can be converted to a scalar one by diagonalizing the $N \times N$ matrix A with a similarity transformation

$$\Lambda = T^{-1} A T,$$

where the columns of T are the right eigenvectors of A. After diagonalizing (3.3) and applying the scheme of (3.2), the first-order upwind scheme is given by

$$u_i^{n+1} = u_i - A\frac{\Delta t}{2\Delta x}(u_{i+1} - u_{i-1}) + |A|\frac{\Delta t}{2\Delta x}(u_{i+1} - 2u_i + u_{i-1}), \tag{3.4}$$

where

$$|A| = T\,|\Lambda|\,T^{-1}, \qquad \Lambda = \mathrm{Diag}[|\lambda_1| \cdots |\lambda_N|].$$

The generalization to a system of conservation laws is straightforward; namely,

$$u_t + f_x = 0,$$

with f being a N-component flux vector, and

$$u_i^{n+1} = u_i - \frac{\Delta t}{2\Delta x}(f_{i+1} - f_{i-1}) + \frac{\Delta t}{2\Delta x}\left[|A_{i+1/2}|\,(u_{i+1} - u_i) - |A_{i-1/2}|\,(u_i - u_{i-1})\right], \tag{3.5}$$

where the Jacobian matrix $A = \partial f/\partial u$, and $|A|$ is defined as for (3.4). The matrix $|A_{i+1/2}|$ can be computed as an arithmetic average or a Roe average. For transonic steady flows the differences are negligible; therefore, we use the simpler arithmetic average. For hypersonic flows Yee [11] found that the Roe average yields better results. For time-dependent problems the Roe average also seems to give slightly better results.

IV. MATRIX DISSIPATION MODEL

We now extend the scheme given in (3.5) to the two-dimensional equations of fluid dynamics. In particular, consider the transformed Euler equations of (2.2) with the Runge–Kutta scheme of (2.3). The necessary modification to the contributions for the ξ direction of the artificial dissipation term defined by (2.4) is to substitute $|A|$ for the eigenvalue scaling factor, λ, in (2.5) and (2.6). For the η direction, ξ and $|A|$ are replaced by η and $|B|$, respectively. We next define explicitly the form for the matrix $|A|$. Let

$$\Lambda = \mathrm{Diag}[\lambda_1 \lambda_2 \lambda_3 \lambda_3],$$

with

$$\lambda_1 = q + \sqrt{a_1^2 + a_2^2}\, c, \qquad \lambda_2 = q - \sqrt{a_1^2 + a_2^2}\, c,$$
$$\lambda_3 = q, \qquad a_1 = J^{-1}\xi_x,$$
$$a_2 = J^{-1}\xi_y, \qquad q = a_1 u + a_2 v.$$

Then

$$|A| = |\lambda_3|\, I + \left(\frac{|\lambda_1| + |\lambda_2|}{2} - |\lambda_3|\right) \times \left[\frac{\gamma - 1}{c^2} E_1 + \frac{1}{a_1^2 + a_2^2} E_2\right] + \frac{|\lambda_1| - |\lambda_2|}{2} \left(\frac{1}{\sqrt{a_1^2 + a_2^2}\, c}\right) \times [E_3 + (\gamma - 1) E_4],$$

where

$$E_1 = \begin{bmatrix} \phi & -u & -v & 1 \\ u\phi & -u^2 & -uv & u \\ v\phi & -uv & -v^2 & v \\ H\phi & -uH & -vH & H \end{bmatrix},$$

$$E_2 = \begin{bmatrix} 0 & 0 & 0 & 0 \\ -a_1 q & a_1^2 & a_1 a_2 & 0 \\ -a_2 q & a_2 a_1 & a_2^2 & 0 \\ -q^2 & q a_1 & q a_2 & 0 \end{bmatrix},$$

$$E_3 = \begin{bmatrix} -q & a_1 & a_2 & 0 \\ -uq & u a_1 & u a_2 & 0 \\ -vq & v a_1 & v a_2 & 0 \\ -Hq & H a_1 & H a_2 & 0 \end{bmatrix},$$

$$E_4 = \begin{bmatrix} 0 & 0 & 0 & 0 \\ a_1\phi & -a_1 u & -a_1 v & a_1 \\ a_2\phi & -a_2 u & -a_2 v & a_2 \\ q\phi & -qu & -qv & q \end{bmatrix},$$

H is the total enthalpy, and $\phi = (u^2 + v^2)/2$. Note the special form of $|A|$, where each row of E_i is a scalar times the first row or a scalar times the second row when the first row contains only zeros. Because of this special form for any λ_1, λ_2, and λ_3, an arbitrary vector x can be multiplied by $|A|$ very quickly. That is, we calculate $|A_{i+1/2}|\,(u_{i+1} - u_i)$ directly (see [12]) rather than calculate $|A_{i+1/2}|$ and multiply a matrix times a vector. The matrix $|B|$ is computed in the same way as $|A|$ by simply replacing ξ with η.

In practice one cannot choose λ_1, λ_2, λ_3 as given above. Near stagnation points λ_3 approaches zero while near sonic lines λ_1 or λ_2 approach zero. A zero artificial viscosity would create numerical difficulties, such as the appearance of nonlinear instabilities. Hence, we limit these values as

$$|\tilde{\lambda}_1| = \max(|\lambda_1|, V_n \rho(A)),$$
$$\rho(A) = |q| + c\sqrt{a_1^2 + a_2^2},$$
$$|\tilde{\lambda}_2| = \max(|\lambda_2|, V_n \rho(A)),$$
$$|\tilde{\lambda}_3| = \max(|\lambda_3|, V_l \rho(A)),$$

where the linear eigenvalue λ_3 can be limited differently than the nonlinear eigenvalues. The parameters V_n and V_l have been determined numerically. Various values have been evaluated by comparing their corresponding computed solutions on the basis of the following: (1) sharpness of shock waves captured (without producing oscillations) and (2) convergence rate of numerical scheme. A good choice for V_n and V_l is between 0.2 and 0.3.

We have thus far replaced $\lambda_{i+1/2,j}$ in (2.5) and (2.6) by a matrix while leaving the limiters $\varepsilon^{(2)}$ and $\varepsilon^{(4)}$ as scalars. One can also introduce $\varepsilon^{(2)}$ and $\varepsilon^{(4)}$ into the diagonal matrix Λ. This allows different limiters to be chosen for different characteristic variables. For example, the limiter may be based on pressure for the nonlinear waves. However, the pressure is smooth through a contact discontinuity. Hence, a switch based on temperature may be more appropriate for the linear wave. One could also use different mesh scalings, $\phi(r)$, for the linear and nonlinear waves. Also, a smoother cutoff [13] can be introduced.

V. THE TVD PROPERTY

Consider the one-dimensional scalar conservation law

$$\frac{\partial}{\partial t}[u(x,t)]+\frac{\partial}{\partial x}[f(u(x,t))]=0. \tag{5.1}$$

where

$$-\infty<x<\infty, \qquad t\geqslant 0.$$

Let $v(t)=\{v_i(t)\}$ be the approximate solution of (5.1) and consider the semidiscrete equation

$$\frac{d}{dt}v_i(t)+\frac{1}{2\Delta x}[f_{i+1}-f_{i-1}]$$
$$=\frac{1}{2\Delta x}[Q_{i+1/2}\,\Delta v_{i+1/2}-Q_{i-1/2}\,\Delta v_{i-1/2}]$$
$$-\frac{\kappa^{(4)}}{\Delta x}[R_{i+1/2}\,\Delta^3 v_{i+1/2}-R_{i-1/2}\,\Delta^3 v_{i-1/2}], \tag{5.2}$$

with

$$\Delta v_{i+1/2}=(\Delta v)_{i+1/2}=v_{i+1}(t)-v_i(t);$$

Δ^3 is a third-difference operator defined as

$$\Delta^3 v_{i+1/2}=(\Delta^3 v)_{i+1/2}$$
$$=v_{i+2}(t)-3v_{i+1}(t)+3v_i(t)-v_{i-1}(t).$$

The terms on the right hand side of (5.2) represent second- and fourth-difference numerical dissipation terms, with $\kappa^{(4)}$ a constant. Define

$$s_{i+1/2}=\operatorname{sgn}(\Delta v_{i+1/2}),$$

where sgn represents the signum function. Then, a modification of the theorem of Tadmor [14] is given by the following.

THEOREM. *The semidiscrete scheme of* (5.2) *is TVD if*

(a) $$[2-s_{i+1/2}(s_{i-1/2}+s_{i+3/2})]\,Q_{i+1/2} \geqslant s_{i+1/2}(s_{i+3/2}-s_{i-1/2})\frac{\Delta f_{i+1/2}}{\Delta v_{i+1/2}}$$

and

(b) $R_{i+1/2}=0$ *when* $s_{i+3/2}-2s_{i+1/2}+s_{i-1/2}\neq 0.$

Proof. Shifting the indices by one in (5.2) and subtracting (5.2) from the resulting equation, we obtain

$$\frac{d}{dt}\Delta v_{i+1/2}+\frac{1}{2\Delta x}[\Delta f_{i+3/2}-\Delta f_{i-1/2}]$$
$$=\frac{1}{2\Delta x}[Q_{i+3/2}\,\Delta v_{i+3/2}$$
$$-2Q_{i+1/2}\,\Delta v_{i+1/2}+Q_{i-1/2}\,\Delta v_{i-1/2}]$$
$$-\frac{\kappa^{(4)}}{\Delta x}[R_{i+3/2}\,\Delta^3 v_{i+3/2}$$
$$-2R_{i+1/2}\,\Delta^3 v_{i+1/2}+R_{i-1/2}\,\Delta^3 v_{i-1/2}]. \tag{5.3}$$

Multiply (5.3) by $s_{i+1/2}$ and sum over all i. Note that $s_{i+1/2}=\pm 1$, so $s^2_{i+1/2}=1$, and

$$s_{i+1/2}\,\Delta v_{i+1/2}=|\Delta v_{i+1/2}|.$$

We then obtain

$$\frac{d}{dt}\sum_i|\Delta v_{i+1/2}|=-\frac{1}{2\Delta x}\sum_i s_{i+1/2}(s_{i-1/2}-s_{i+3/2})$$
$$\times\frac{\Delta f_{i+1/2}}{\Delta v_{i+1/2}}|\Delta v_{i+1/2}|$$
$$+\frac{1}{2\Delta x}\sum_i s_{i+1/2}(s_{i+3/2}-2s_{i+1/2}+s_{i-1/2})$$
$$\times Q_{i+1/2}\,|\Delta v_{i+1/2}|$$
$$-\frac{\kappa^{(4)}}{\Delta x}\sum_i(s_{i+3/2}-2s_{i+1/2}+s_{i-1/2})$$
$$\times R_{i+1/2}\,\Delta^3 v_{i+1/2}. \tag{5.4}$$

We stress that the last term in (5.4) will not help for TVD. Its purpose is to eliminate high frequencies and accelerate convergence to a steady state. Hence, we want this contribution to be zero. This can be accomplished if we demand either

$$s_{i+3/2}-2s_{i+1/2}+s_{i-1/2}=0 \tag{5.5}$$

or

$$R_{i+1/2}=0, \tag{5.6}$$

i.e., condition (b). We are left with

$$\frac{d}{dt}(\mathrm{TV})=-\frac{1}{2\Delta x}\sum_i\Big[-s_{i+1/2}(s_{i+3/2}-s_{i-1/2})$$
$$\times\frac{\Delta f_{i+1/2}}{\Delta v_{i+1/2}}-s_{i+1/2}(s_{i+3/2}-2s_{i+1/2}+s_{i-1/2})$$
$$\times Q_{i+1/2}\Big]|\Delta v_{i+1/2}|, \tag{5.7}$$

where TV denotes the total variation as given by

$$\mathrm{TV} = \sum |\Delta v_{i+1/2}|.$$

Thus, a sufficient but not necessary condition that the total variation not increase is that the term of (5.7) in brackets must be positive. This means that

$$\begin{aligned} -s_{i+1/2}(s_{i+3/2} - 2s_{i+1/2} + s_{i-1/2})\, Q_{i+1/2} \\ \geqslant s_{i+1/2}(s_{i+3/2} - s_{i-1/2}) \frac{\Delta f_{i+1/2}}{\Delta v_{i+1/2}}. \end{aligned} \tag{5.8}$$

Since $s_{i+1/2}^2 = 1$, (5.8) is equivalent to condition (a) of the theorem. Defining

$$\chi_i = 1 - s_{i-1/2} s_{i+1/2},$$

(5.8) can be written in the form given by Tadmor; namely,

$$(\chi_i + \chi_{i+1})\, Q_{i+1/2} \geqslant (\chi_i - \chi_{i+1}) \frac{\Delta f_{i+1/2}}{\Delta v_{i+1/2}}.$$

Several remarks concerning this theorem are in order.

Remarks. (i) When no extrema are present locally ($s_{i-1/2} = s_{i+1/2} = s_{i+3/2}$), both conditions of the theorem are trivially satisfied for all $Q_{i+1/2}$ and $R_{i+1/2}$. For such regions we want $Q_{i+1/2} = O(\Delta x)$, for second-order accuracy and $R_{i+1/2}$ to be chosen so that high frequencies are damped.

(ii) If $s_{i-1/2} = s_{i+3/2} = -s_{i+1/2}$ (a local oscillation at $x_{i+1/2}$), we require

$$Q_{i+1/2} \geqslant 0, \qquad R_{i+1/2} = 0.$$

(iii) If $s_{i-1/2} = s_{i+1/2} = -s_{i+3/2}$ (a local extremum at x_{i+1}),

$$Q_{i+1/2} \geqslant \frac{-\Delta f_{i+1/2}}{\Delta v_{i+1/2}}, \qquad R_{i+1/2} = 0.$$

(iv) If $s_{i-1/2} = -s_{i+1/2} = -s_{i+3/2}$ (a local extremum at x_i),

$$Q_{i+1/2} \geqslant \frac{\Delta f_{i+1/2}}{\Delta v_{i+1/2}}, \qquad R_{i+1/2} = 0.$$

It follows from these inequalities that $Q_{i+1/2}$ can sometimes be negative. As fas as total variation is concerned, central differences are not nondissipative. That is, they sometimes increase and sometimes decrease the total variation. In cases where central differences decrease the total variation, $Q_{i+1/2}$ can be negative. For systems of equations, especially in multidimensions, this behavior can sometimes lead to difficulties. Hence, we demand the stronger condition that

$$Q_{i+1/2} \geqslant \left| \frac{\Delta f_{i+1/2}}{\Delta v_{i+1/2}} \right|.$$

For systems this condition is replaced by

$$Q_{i+1/2} \geqslant |A_{i+1/2}|,$$

where $A = \partial f/\partial u$, and the average at $i + 1/2$ is constructed by the technique of Roe [2]. In practice, we find that one can use the simpler arithmetic average rather than the Roe type average. The arithmetic average is used in all numerical results of this paper.

VI. FLUX LIMITERS

In this section switching functions are introduced that force the scheme to automatically satisfy the inequalities presented in Section V. These switches are required to be smooth so that limit cycles are not experienced when marching in time to obtain a steady-state solution. The discrete switching functions are defined as

$$\begin{aligned} Q_{i+1/2} &= \psi \cdot |a_{i+1/2}|, \qquad \text{with} \quad 0 \leqslant \psi \leqslant 1, \\ R_{i+1/2} &= \Gamma \cdot |a_{i+1/2}|, \end{aligned} \tag{6.1}$$

where $\psi = 1$ near extrema, so that the inequalities are satisfied, and $\psi = O(\Delta x)$ in smooth regions of the flow field. Conservely, $\Gamma = 0$ at extrema, and $\Gamma = 1$ in the smooth regions. Here we define $a_{i+1/2} = \Delta f_{i+1/2} / \Delta v_{i+1/2}$. In the case of the fluid dynamic equations, $a_{i+1/2}$ becomes a matrix, and the switch is usually based on the pressure, though sometimes the temperature can also be used.

The functions ψ and Γ of (6.1) can be defined in terms of a limiter function. Let

$$r = \frac{v_i - v_{i-1}}{v_{i+1} - v_i} = \frac{\Delta v_{i-1/2}}{\Delta v_{i+1/2}}.$$

The van Leer flux limiter is given by

$$\varphi_i(r) = \frac{r + |r|}{1 + |r|} = \begin{cases} \dfrac{2r}{1 + r}, & r > 0 \\ 0, & r \leqslant 0. \end{cases}$$

From Sweby [15] it is straightforward to see that for any flux limiter

$$\psi_i(r) = 1 - \varphi_i(r).$$

Since we want $\psi_i(r)$ to be positive, this relation is redefined as

$$\psi_i(r)=|1-\varphi_i(r)|.$$

Then, for the van Leer limiter,

$$\psi_i(r)=\left|\frac{1-r}{1+|r|}\right|=\begin{cases}\dfrac{|1-r|}{1+r}, & r>0\\ 1, & r\leqslant 0,\end{cases}$$

so $0\leqslant\psi_i\leqslant 1$ and define

$$\psi_{i+1/2}=\max(\psi_i,\psi_{i+1}).$$

We now show that the inequality (5.8) is satisfied with this ψ. By the first remark of Section V, the inequality is satisfied when $s_{i-1/2}=s_{i+1/2}=s_{i+3/2}$, and so in this case we only need $Q=O(\Delta x)$. Since $r>0$ in this case,

$$\psi_i(r)=\frac{|1-r|}{1+r}.$$

In addition, for smooth regions of the flow field,

$$r=1+O(\Delta x),$$

and thus

$$\psi_{i+1/2}=O(\Delta x),\qquad Q_{i+1/2}=O(\Delta x).$$

Next, consider the case $s_{i-1/2}=-s_{i+1/2}$. This implies $r_i<0$, and so, $\psi_i=1$. Moreover,

$$s_{i-1/2}=-s_{i+1/2}\Rightarrow\psi_i=1\Rightarrow\psi_{i+1/2}=1.$$

Similarly, if $s_{i+3/2}=-s_{i+1/2}$,

$$s_{i+1/2}=-s_{i+3/2}\Rightarrow\psi_{i+1}=1\Rightarrow\psi_{i+1/2}=1.$$

It also follows, using (6.1), that the inequalities in (ii) and (iii) of the remarks of Section V are satisfied. Also, for these two cases, setting $\Gamma_{i+1/2}=1-\psi_{i+1/2}$ guarantees that $\Gamma_{i+1/2}=0$. So

$$\Gamma_{i+1/2}=1-\max(\psi_i,\psi_{i+1}).$$

The flux limiters ψ and Γ can be connected to those used in upwind schemes. Sweby [15] considers an upwind Lax–Wendroff scheme. In particular, for the one-dimensional wave equation

$$u_t+au_x=0,$$

the numerical solution is obtained with

$$u_i^{n+1}=u_i-\nu(u_i-u_{i-1})-\Delta_-\left[\frac{\nu(1-\nu)}{2}\varphi_i(u_{i+1}-u_i)\right],\qquad(6.2)$$

where

$$\nu=a\frac{\Delta t}{\Delta x},\quad 0\leqslant\varphi(r)\leqslant 2,\quad\frac{\varphi(r)}{r}=\varphi\left(\frac{1}{r}\right),$$

and the backward difference operator Δ_- is defined by

$$\Delta_-u_i=u_i-u_{i-1}.$$

If (6.2) is rewritten as a central-difference scheme, then

$$\begin{aligned}u_i^{n+1}=u_i&-\frac{\nu}{2}(u_{i+1}-u_{i-1})\\&+\frac{\nu}{2}[(1-\varphi_i)(u_{i+1}-u_i)\\&-(1-\varphi_{i-1})(u_i-u_{i-1})]\\&+\frac{\nu^2}{2}[\varphi_i(u_{i+1}-u_i)-\varphi_{i-1}(u_i-u_{i-1})].\end{aligned}\qquad(6.3)$$

By dropping the ν^2 term in (6.3) and changing to a semidiscrete formulation, we get

$$\begin{aligned}\frac{d}{dt}u_i=&-\frac{\nu}{2\Delta t}(u_{i+1}-u_{i-1})\\&+\frac{\nu}{2\Delta t}[(1-\varphi_i)(u_{i+1}-u_i)\\&-(1-\varphi_{i-1})(u_i-u_{i-1})].\end{aligned}$$

If

$$\varphi_i=1-\psi_{i+1/2},$$

then the second-difference dissipation term has the same form as the one presented in Section V.

To complete the connection between the limiters for the central-difference and upwind schemes, we compare their behavior. For the central-difference case,

$$0\leqslant\psi\leqslant 1,\qquad 0\leqslant\varphi\leqslant 1,$$

whereas for the upwind case, $0\leqslant\varphi\leqslant 2$. Furthermore, for the central-difference limiter,

$$\psi\left(\frac{1}{r}\right)=\psi(r),\qquad\varphi\left(\frac{1}{r}\right)=\varphi(r)$$

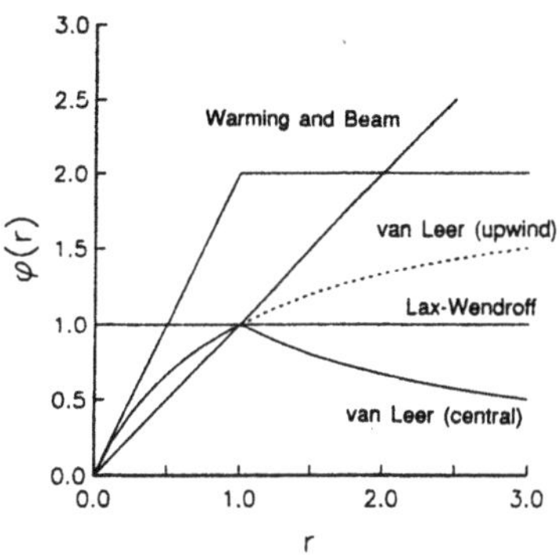

FIG. 2. Sweby diagram.

means that it does not matter if $\Delta u_{i+1/2} > \Delta u_{i-1/2}$ or $\Delta u_{i+1/2} < \Delta u_{i-1/2}$ (i.e., the sign of a does not affect the scheme), as opposed to the upwind limiter where

$$\frac{\varphi(r)}{r} = \varphi\left(\frac{1}{r}\right).$$

In Fig. 2 the Sweby diagram for the upwind van Leer flux limiter and the central-difference van Leer flux limiter is shown. For $r < 1$, the limiters are the same. For $r > 1$, the upwind version continues until $\varphi = 2$, while the central-difference version returns to zero. In Sweby's paper the flux limiters are not allowed to decrease when $r > 1$. However, there is no difficulty from such behavior. In both cases $\varphi(r) = 0$ when $r \leqslant 0$ so that the limiter is turned off for extrema. We note that other limiters besides that of van Leer can be used to get switches for central-difference schemes.

For systems of equations we use a scalar limiter (i.e., switching function). Using the matrix form of the dissipation it is easy to implement different limiters for different characteristic variables. In the computations discussed later, the switch is always based on pressure.

In terms of the pressure the switch becomes

$$\psi_i = \frac{1-r}{1+|r|}, \quad r = \frac{\Delta p_{i-1/2}}{\Delta p_{i+1/2}}, \quad a > 0$$

or

$$\psi_i = \frac{|p_{i+1} - p_i| - (p_i - p_{i-1})\operatorname{sgn}(p_{i+1} - p_i)}{|p_{i+1} - p_i| + |p_i - p_{i-1}| + \varepsilon}, \qquad a > 0,$$

with

$$\psi_{i+1/2} = \max(\psi_i, \psi_{i+1}),$$

and $\varepsilon = O(\Delta x^2)$ to prevent a zero denominator for constant pressure regions. Consider also the wave speed $a < 0$. Then,

$$\psi_i = \frac{p_{i+1} - 2p_i + p_{i-1}}{|p_{i+1} - p_i| + |p_i - p_{i-1}| + \varepsilon} \times \begin{cases} \operatorname{sgn}(p_{i+1} - p_i), & a > 0 \\ \operatorname{sgn}(p_{i-1} - p_i), & a < 0. \end{cases}$$

We use a conservative approach and take

$$\psi_i = \frac{|p_{i+1} - 2p_i + p_{i-1}|}{|p_{i+1} - p_i| + |p_i - p_{i-1}| + \varepsilon}. \tag{6.4}$$

There is difficulty in defining for this switching function a general ε, i.e., one that does not activate the switch when small oscillations occur in the discrete field. Some flexibility can be obtained by considering the switching function

$$\psi_i = \frac{|p_{i+1} - 2p_i + p_{i-1}|}{\begin{pmatrix} (1-\omega)(|p_{i+1} - p_i| + |p_i - p_{i-1}|) \\ +\,\omega(p_{i+1} + 2p_i + p_{i-1}) \end{pmatrix}}, \tag{6.5}$$

where $0 < \omega \leqslant 1$, and (6.4) is recovered when $\omega \ll 1$. In practice a reasonable value for ω is $\frac{1}{2}$. Larger values of ω do not prevent oscillations at high Mach numbers, and smaller values of ω can prevent a multigrid scheme from converging. Then the $\varepsilon^{(2)}_{i+1/2,j}$ appearing in (2.5) is replaced by $\psi_{i+1/2,j}$. Note that this switch is very similar to (2.9) for the original dissipation model of the Runge–Kutta scheme. There is only a change in the denominator. However, with this change, ω sufficiently small, and the factor $\frac{1}{2}$ in front of the second-difference dissipation term, the scalar equation becomes first-order upwind near shocks. In the case of the original ψ_i we find that $\psi_i \simeq 0.05$ near shock waves in transonic flows. One may require different parameters for the Runge–Kutta scheme to ensure stability. Extensions to other limiters and time-dependent problems are considered in [16].

We now no longer have a free parameter for the second-difference dissipation. The only free parameter is the coefficient $\kappa^{(4)}$ of the fourth-difference term. In the context of (5.2) and (6.1) one can define the function $\Gamma = \max[0, (1 - 2\psi)]$, resulting in the fourth-difference dissipation being cut off for $\psi \geqslant \frac{1}{2}$. In the case of the original dissipation formulation, one can simply replace $\varepsilon^{(2)}_{i+1/2,j}$ appearing in (2.10) with $\psi_{i+1/2,j}$.

VII. NUMERICAL ALGORITHM

The majority of the numerical results presented in this paper were obtained with a Navier–Stokes code developed by the authors, which is based on the explicit multistage time-stepping schemes of [3] and [17]. This class of schemes is currently in widespread use for solving the Euler equations. In references [18–21], these schemes were extended to allow solution of the compressible Navier–Stokes equations. Significant improvements in numerical efficiency were introduced in [9, 10, 22]. In the code of Swanson and Turkel a cell-centered, finite-volume method is employed to obtain centered type difference approximations for the flow equations. The scheme is second-order accurate in space for sufficiently smooth meshes (see [22, 23] for definition of sufficiently smooth). A modified five stage Runge–Kutta scheme is used for the time integration to obtain a steady-state solution. There is a weighted evaluation of the artificial dissipation terms on the first, third, and fifth stages, which provides a large parabolic stability limit. The absolute stability curve for this scheme is shown in [10], and the intersection of this curve with the negative real axis is -9. The physical viscous terms are computed on the first stage only and frozen for the remaining ones; this does not compromise the stability characteristics of the scheme. The spatial and temporal differencing are decoupled. Thus, the numerical algorithm is independent of time step and amenable to steady-state convergence acceleration techniques. These methods include local time stepping (a preconditioning for the system of difference equations), variable coefficient implicit residual smoothing, and multi-grid. Implicit residual smoothing is just a mathematical step, applied at each stage of the explicit scheme, to extend the local stability range. The multigrid method involves cycling through a sequence of successively coarser grids and relying upon effective high frequency damping for rapid removal of the errors in the fine grid solution. For effective high frequency damping the coefficient $\kappa^{(4)}$ of the fourth-difference dissipation terms is generally taken to be in the range from $\frac{1}{64}$ to $\frac{1}{32}$.

VIII. APPLICATIONS

Two airfoil flow problems are considered here to demonstrate the benefits of using a central-difference scheme with a matrix numerical dissipation model. The first problem concerns inviscid flow over an NACA 0012 airfoil. At a free-stream Mach number (M_∞) of 0.8 and angle of

TABLE I

Lift and Drag Coefficients for NACA 0012 Airfoil, $M_\infty = 0.8$, $\alpha = 1.25°$

Case	c_l	c_d
Scalar dissipation model (224 × 32 mesh)	0.3628	0.0231
Matrix dissipation model (224 × 32 mesh)	0.3591	0.0227
Scalar dissipation model (640 × 64 mesh)	0.3577	0.0228

Note. c_l, lift coefficient; c_d, drag coefficient.

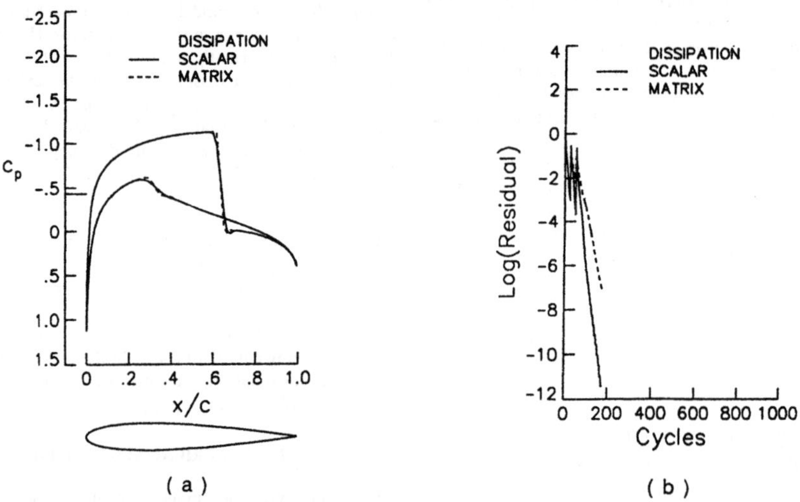

FIG. 3. Central-difference scheme results for inviscid flow over NACA 0012 airfoil ($M_\infty = 0.80$, $\alpha = 1.25°$): (a) surface pressure coefficient and (b) convergence history.

attack (α) of 1.25°, a fairly strong transonic shock wave occurs on the upper surface, while a much weaker one appears on the lower surface. On representative inviscid meshes, schemes based on central differencing capture the upper surface shock reasonably well, but they smear significantly the lower surface shock. The second problem involves transonic turbulent flow over an RAE 2822 airfoil. In this case the free-stream Mach number is 0.73, the angle of attack is 2.79°, and the Reynolds number based on chord (Re_{∞}) is 6.5×10^6. For such transonic viscous flows small differences in the shock strength can result in noticeable changes in the lift and drag of the airfoil. Thus, both of these problems can provide a reasonable measure of the performance of the artificial dissipation model, especially near shock waves.

A C-type mesh consisting of 224 cells around the airfoil

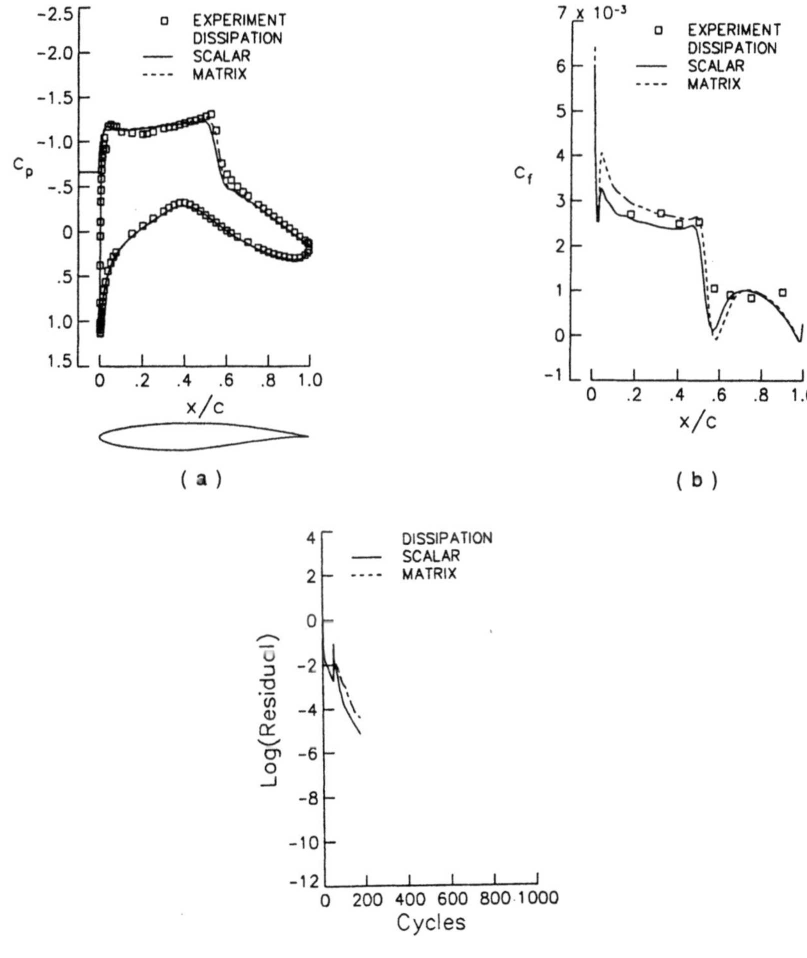

FIG. 4. Central-difference scheme results for turbulent flow over RAE 2822 airfoil (160 × 32 mesh, $M_{\infty} = 0.73$, $\alpha = 2.79°$, $Re_{\infty} = 6.5 \times 10^6$): (a) surface pressure coefficient, (b) upper surface skin-friction coefficient, and (c) convergence history.

(160 cells on the airfoil) and 32 cells normal to the airfoil was used for the first problem. The outer boundary of the finite domain was placed 20 chords away from the airfoil, so as to not produce significant effects on the solutions (see [24]). The normal mesh spacing at the surface was approximately 0.01 chords. The mesh was clustered at the leading and trailing edges of the airfoil in order to improve resolution of the steep gradients occurring in these regions. In Fig. 3 the surface pressure distributions computed with the scalar and matrix dissipation models are shown. There is a discernible improvement in the sharpness of the shock waves using the matrix model, especially for the one on the lower airfoil surface. The lift and drag coefficients obtained using the two models, along with the values from a high density mesh calculation, are compared in Table I. The values associated with the matrix model nearly match those corresponding to the fine mesh solution. The convergence histories in terms of number of multigrid cycles are also displayed in Fig. 3. Convergence is measured by the logarithm of the root mean square of the residual of the continuity equation. The convergence rates obtained with the scalar and matrix models are 0.819 and 0.888, respectively. The matrix model result required about 55 cpu sec on a Cray II computer for 125 multigrid cycles with the solution grid. This processing time represents a 15 percent increase in the time needed compared with the scalar dissipation model. It may be possible to improve the convergence rate with the matrix model by altering the coefficients of the time-stepping scheme and/or the type of implicit residual smoothing operator employed.

The following set of C-type meshes were used in solving the second problem with the two dissipation models: (1) 160×32 with 128 cells on the airfoil, (2) 320×64 with 256 cells on the airfoil, and (3) 640×128 with 512 cells on the airfoil. Each successively coarser mesh was generated by eliminating every other mesh line in both coordinate directions of the finer mesh. Again the outer boundary of the domain was located 20 chords from the airfoil. The normal spacing at the surface for the finest mesh was 7.5×10^{-6} chords. In Fig. 4a comparison is made between the experimental data of [25] and the predicted distributions of pressure and upper surface skin friction obtained on the 160×32 mesh. The skin-friction coefficient is the wall shear stress nondimensionalized by the dynamic pressure at the edge of the viscous layer. There is marked improvement in the results obtained with the matrix model, as verified by their reasonable agreement with the data. The convergence histories given in Fig. 4 indicate that the final residual level is somewhat higher with the matrix model. However, the terminal rate of convergence is nearly the same with both dissipation models.

Computations were also made on the coarse mesh with second-order upwind and "third-order" upwind biased forms of Roe's scheme [2], which has been shown to have low levels of dissipation. Figure 5 displays pressure and skin-friction distributions calculated with the somewhat

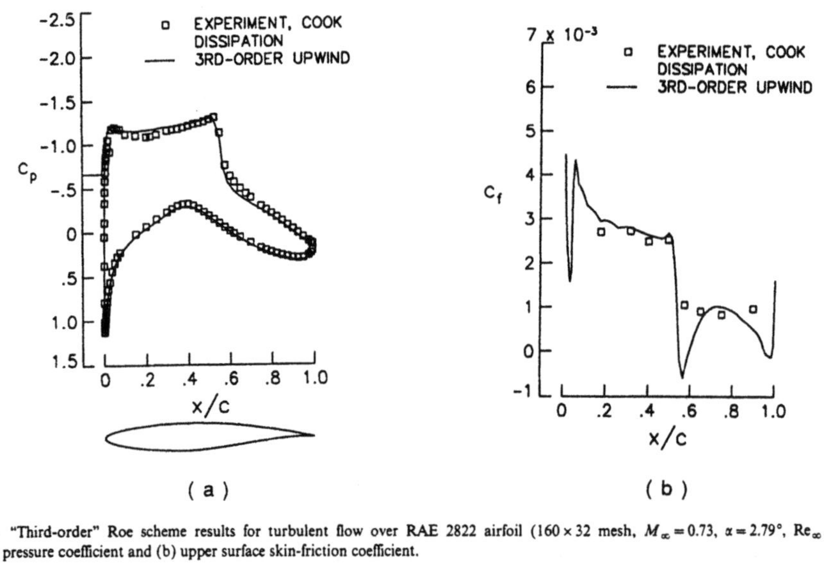

FIG. 5. "Third-order" Roe scheme results for turbulent flow over RAE 2822 airfoil (160×32 mesh, $M_\infty = 0.73$, $\alpha = 2.79°$, $Re_\infty = 6.5 \times 10^6$): (a) surface pressure coefficient and (b) upper surface skin-friction coefficient.

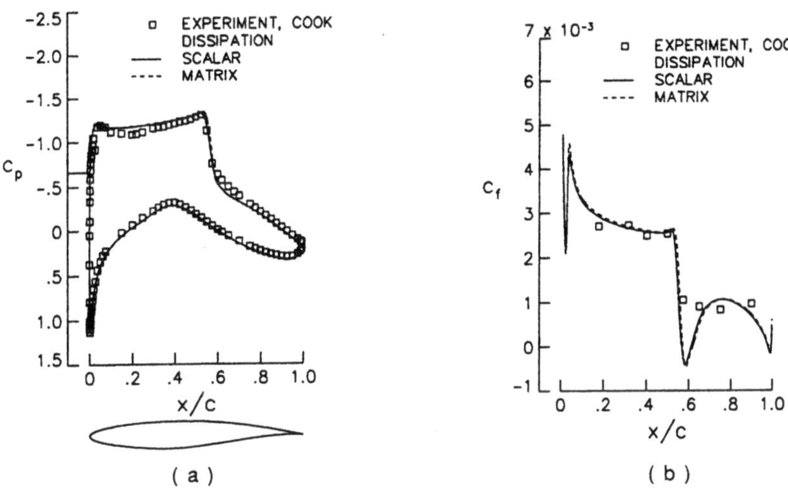

FIG. 6. Central-difference scheme results for turbulent flow over RAE 2822 airfoil (320 × 64 mesh, $M_\infty = 0.73$, $\alpha = 2.79°$, $Re_\infty = 6.5 \times 10^6$): (a) surface pressure coefficient and (b) upper surface skin-friction coefficient.

more accurate upwind biased form. The pressures are very close to those obtained with the matrix model. However, they do indicate a slightly higher flow acceleration just upstream of the shock wave. According to fine mesh results, this is the correct behavior. There is good agreement between the upwind skin-friction solution and the experimental data upstream of the shock. The skin-friction distribution computed with the matrix model also exhibits good agreement with the data in this region. As is revealed shortly, this part of the solution remains unchanged with mesh refinement. The skin-friction values of the upwind solution indicate a little larger separation region induced by the shock, which agrees fairly well with the matrix dissipation results on the finer grids. It is important to note that on the coarse mesh neither the upwind nor matrix dissipation solution gives the correct skin-friction values over the last 20% of the airfoil. This demonstrates the sensitivity of the solution to the shock position.

Numerical results for the 320 × 64 mesh are presented in Fig. 6. A slightly stronger shock is predicted using the

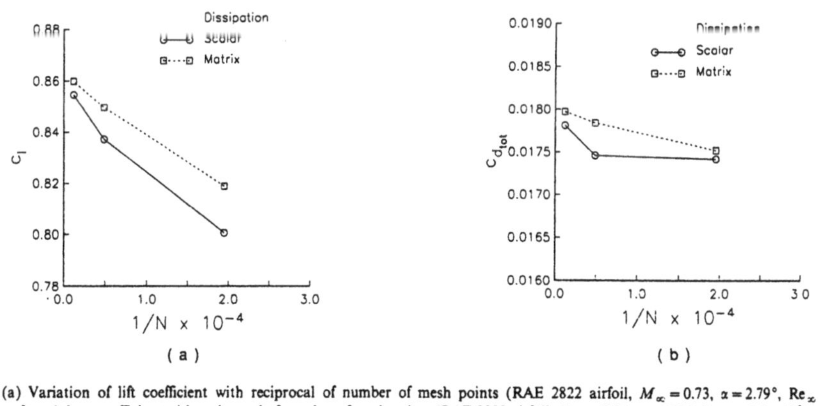

FIG. 7. (a) Variation of lift coefficient with reciprocal of number of mesh points (RAE 2822 airfoil, $M_\infty = 0.73$, $\alpha = 2.79°$, $Re_\infty = 6.5 \times 10^6$). (b) Variation of total drag coefficient with reciprocal of number of mesh points (RAE 2822 airfoil, $M_\infty = 0.73$, $\alpha = 2.79°$, $Re_\infty = 6.5 \times 10^6$).

matrix dissipation model. Otherwise, the solutions determined with the different dissipation models are nearly the same. Even so one must keep in mind that small changes in shock location can have significant effects on aerodynamics. The character of the convergence behavior with the two models is about the same as for the coarse mesh. On the 640×128 mesh, the application of the models gives essentially the same results.

In Fig. 7 the variation of the computed lift and total drag coefficients with the reciprocal of the number of mesh cells is plotted for each dissipation model. The curves corresponding to the matrix model are nearly linear. A linear curve indicates second-order accuracy for the full range of meshes being considered. The numerical values for the components of the drag coefficient (form and friction contributions) along with the lift coefficient (c_l) are given for all the viscous computations in Table II. All of the coarse mesh solutions give at least a four percent error in the c_l relative to the finest mesh solution. Notice that the values obtained with the matrix model on the 320×64 grid are very close to those calculated with the scalar model on the 640×128 grid. Such a reduction in the mesh required for acceptable accuracy is highly desirable, especially for three-dimensional flow problems. Some applications to three-dimensional viscous flows are presented in Turkel and Vatsa [26]. It should be emphasized that accurate prediction of lift and drag is very important in the design of aircraft, and thus, a good estimate of these quantities for an infinitely fine mesh is needed.

TABLE II

Lift and Drag Coefficients for RAE 2822 Airfoil, $M_\infty = 0.73$, $\alpha = 2.79°$, $\mathrm{Re}_\infty = 6.5 \times 10^6$

Mesh	c_l	c_{d_p}	c_{d_f}	$c_{d_{tot}}$
Scalar dissipation model				
160×32	0.8006	0.0127	0.0047	0.0174
320×64	0.8372	0.0120	0.0055	0.0175
640×128	0.8547	0.0123	0.0055	0.0178
Matrix dissipation model				
160×32	0.8189	0.0122	0.0053	0.0175
320×64	0.8497	0.0123	0.0055	0.0178
640×128	0.8599	0.0124	0.0055	0.0179
Second-order upwind				
160×32	0.8176	0.0119	0.0054	0.0173
"Third-order" upwind biased				
160×32	0.8220	0.0124	0.0054	0.0178

Note. c_l, lift coefficient; c_{d_p}, pressure drag coefficient; c_{d_f}, friction drag coefficient; $c_{d_{tot}}$, total drag coefficient.

In [27] it is shown that the TVD switch (6.4) allows converged solutions for hypersonic flow where the standard switch (2.9) does not.

IX. CONCLUDING REMARKS

We have thus shown that the second-difference artificial dissipation is equivalent to using a flux limiter, and hence, central-difference schemes are not any more "artificial" than upwind schemes. The central-difference scheme is slightly more dissipative for two reasons. First, $\psi_{i+1/2} = \max(\psi_i, \psi_{i+1})$, while for an upwind scheme ψ_{i+1} is equal to either ψ_i or ψ_{i+1}, depending on the direction of the wind. Second, we insist that ψ_i be positive (i.e., $\varphi_i \leqslant 1$) while upwind limiters allow $\varphi_i > 1$ (i.e., in some cases we can have negative viscosity but still be TVD). However, to compensate for this slight increase in dissipation central-difference schemes are simpler to program and require less computer time per time step, and work well with multigrid acceleration techniques.

In addition the central-difference schemes have a free parameter in conjunction with the fourth-difference dissipation. This dissipation is needed to approach a steady state and has noting to do with TVD properties. In fact the fourth-difference contribution is set equal to zero near local extrema. For time-dependent flows one can set this dissipation identically to zero. On the other hand TVD properties do not necessarily imply a rapid convergence to the steady state.

In summary, a formulation for a numerical dissipation model that makes a central-difference scheme closely resemble an upwind scheme near flow discontinuities has been described. A theorem has been proven that gives sufficient requirements for this type of dissipation model to satisfy the TVD property for a scalar equation. Flux limiter functions have been presented for this form of dissipation model. For a system of equations a matrix-valued dissipation is introduced. Solutions of the Euler and Navier–Stokes equations for airfoil flows have been obtained using the matrix dissipation model. These results have demonstrated noticeable improvements in accuracy in smooth regions of the flow field as well as near shock waves. There is a 15% increase in computational time for explicit multistage schemes when this model is used. However, the improved accuracy has permitted a significant reduction in the number of mesh points required. Such behavior of a scheme can have a dramatic effect on the necessary mesh size for three-dimensional flow calculations.

Finally, it is important to emphasize the different principal objectives associated with matrix-valued dissipation and the TVD switch. The purpose of the matrix form of the numerical dissipation model is to apply the appropriate scaling of the dissipation in each flow equation, yielding a reduction in the amount of dissipation being introduced and improved accuracy. The TVD switch plays a somewhat opposite role in that it causes more dissipation to be added in order to prevent overshoots, and thus, allows convergence and provides robustness in solving high speed flow problems. The combination of the two should give more dissipation near shocks and less dissipation in smooth regions, hence giving better accuracy in all regions.

ACKNOWLEDGMENT

The authors express their appreciation to Dr. Vankatakrishnan for providing numerical solutions obtained with the upwind scheme of Roe.

REFERENCES

1. B. van Leer, in *Proceedings of Eighth International Conference on Numerical Methods in Fluid Dynamics*, edited by E. Krause, *Lecture Notes in Physics*, Vol. 170 (Springer-Verlag, Berlin, 1982), p. 507.
2. P. L. Roe, *J. Comput. Phys.* **43**, 357 (1981).
3. A. Jameson, W. Schmidt, and E. Turkel. AIAA Paper 81-1259, 1981 (unpublished).
4. A. Jameson and T. J. Baker, AIAA Paper 84-0093, 1984 (unpublished).
5. J. E. Deese and R. K. Agarwal. AIAA Paper 83-1853, 1983 (unpublished).
6. S. V. Subramanian and R. Bozzola, AIAA Paper 85-1332, 1985 (unpublished).
7. T. H. Pulliam, *AIAA J.* **24**, 1931 (1986).
8. A. Jameson and S. Yoon, *AIAA J.* **25**, 929 (1987).
9. R. C. Swanson and E. Turkel. AIAA Paper 87-1107, 1987 (unpublished).
10. L. Martinelli and A. Jameson, AIAA Paper 88-0414, 1988 (unpublished).
11. H. C. Yee, NASA TM 89464, 1987 (unpublished).
12. E. Turkel, in *Proceedings of the 11th International Conference on Numerical Methods in Fluid Dynamics*, edited by D. L. Dwoyer, M. Y. Hussaini, and R. G. Voigt, *Lecture Notes in Physics*, Vol. 323 (Springer-Verlag, Berlin, 1989), p. 586.
13. A. Harten, *J. Comput. Phys.* **49**, 357 (1983).
14. E. Tadmor, *SIAM J. Numer. Anal.* **25**, 1002 (1988).
15. P. K. Sweby, *SIAM J. Numer. Anal.* **21**, 995 (1984).
16. P. C. E. Jorgenson and E. Turkel, AIAA Paper 92-0053, 1992 (unpublished).
17. A. Jameson, MAE Report 1743, Princeton University, 1985 (unpublished).
18. R. C. Swanson and E. Turkel, AIAA Paper 85-0035, 1985 (unpublished).
19. R. K. Agarwal and J. E. Deese, AIAA Paper 84-1551, 1984 (unpublished).
20. L. Martinelli, A. Jameson, and F. Grasso, AIAA Paper 86-0208, 1986 (unpublished).
21. V. N. Vatsa, *J. Aircraft* **24**, 377 (1987).
22. R. C. Swanson and R. Radespiel, *AIAA J.* **29**, 697 (1991).
23. E. Turkel, *Appl. Numer. Math.* **2**, 529 (1986).
24. J. L. Thomas and M. D. Salas, *AIAA J.* **24**, 1074 (1986).
25. P. H. Cook, M. A. McDonald, and M. C. P. Firmin, AGARD Advisory Report No. 138, 1979 (unpublished).
26. E. Turkel and V. N. Vatsa, AIAA Paper 90-1445, 1990 (unpublished).
27. N. Decker and E. Turkel, "Multigrid for Hypersonic Inviscid Flows," Proceedings of Third International Conference on Hyperbolic Problems, Uppsala, Sweden, June 1990 (unpublished).

Chapter III

Essentially Non-oscillatory Schemes

Annotation

From the outset it has been known that second-order Total-Variation-Diminishing (TVD) schemes, in spite of substantial merits, have one major defect: they are not able to distinguish between a shock profile and a smooth extremum, and therefore tend to clip the latter. Also, there appeared to be no natural generalization of the TVD concept to higher dimension. The cure was to step down from the TVD requirement to the Total-Variation-Bounded (TVB) requirement.

A systematic way of constructing TVB schemes is provided for in the concept of Essentially Non-Oscillatory (ENO) schemes. Here one achieves accurate representation of smooth extrema at the price of allowing, near shocks, small overshoots of magnitude $(O(h^p))$ for a p-th order method. ENO schems are based on solution-adapted stencils, and can be given arbitrarily high accuracy.

In this chapter we survey the historical development of TVB schemes at ICASE.

1. **Uniformly High-Order Accurate Nonoscillatory Schemes. I**
   (A. Harten and S. Osher)
   The first in a planned series of papers detailing the design of ENO schemes. This is the only paper in this book that did not appear first as an ICASE report, although the work was done in large part at ICASE.

2. **Uniformly High Order Accurate Essentially Non-oscillatory Schemes, III**
   (A. Harten, B. Engquist, S. Osher and S. Chakravarthy)
   Formally, the third paper of the series. The second paper, though, was never written.

3. **ENO Schemes with Subcell Resolution** (A. Harten)
   One of the few papers written on the subject of jump recovery, in this case using ENO reconstruction away from the jump. Harten did not succeed in extending the technique to multiple dimensions.

4. **Efficient Implementation of Essentially Non-oscillatory Shock-Capturing Schemes, II** (C.-W. Shu and S. Osher)
   A fundamental paper on preserving the non-oscillatory property of a spatial operator when combined with multi-stage time marching.

5. **Essentially Nonoscillatory Spectral Fourier Methods for Shock Wave Calculations** (W. Cai, D. Gottlieb and C.-W. Shu)
   Using ENO concepts to enhance spectral methods. A key idea here is to add to the pool of spectral basis functions a discontinuous basis function.

6. **Cell Averaging Chebyshev Methods for Hyperbolic Problems**
(W. Cai, D. Gottlieb and A. Harten)
Describes an efficient method for transforming between cell-averaged and pointwise values of higher-order polynomials, useful in all higher-order finite-volume schemes.

UNIFORMLY HIGH-ORDER ACCURATE NONOSCILLATORY SCHEMES. I*

AMI HARTEN† AND STANLEY OSHER‡

Abstract. We begin the construction and the analysis of nonoscillatory shock capturing methods for the approximation of hyperbolic conservation laws. These schemes share many desirable properties with total variation diminishing schemes, but TVD schemes have at most first-order accuracy, in the sense of truncation error, at extrema of the solution. In this paper we construct a uniformly second-order approximation, which is nonoscillatory in the sense that the number of extrema of the discrete solution is not increasing in time. This is achieved via a nonoscillatory piecewise-linear reconstruction of the solution from its cell averages, time evolution through an approximate solution of the resulting initial value problem and an average of this approximate solution over each cell.

Key words. conservation laws, finite difference scheme, nonoscillatory, TVD

AMS(MOS) subject classifications. Primary 65 M10; secondary 65 M05

1. Introduction. In this paper we consider numerical approximations to weak solutions of the scalar initial value problem (IVP)

$$u_t + f(u)_x = u_t + a(u)u_x = 0, \tag{1.1a}$$

$$u(x, 0) = u_0(x). \tag{1.1b}$$

The initial data $u_0(x)$ are assumed to be piecewise-smooth functions that are either periodic or of compact support.

Let $v_j^n = v_h(x_j, t_n)$, $x_j = jh$, $t_n = n\tau$, denote a numerical approximation in conservation form

$$v_j^{n+1} = v_j^n - \lambda(\hat{f}_{j+1/2} - \hat{f}_{j-1/2}) \equiv (E_h \cdot v^n)_j. \tag{1.2a}$$

Here E_h is the numerical solution operator, $\lambda = \tau/h$, and $\hat{f}_{j+1/2}$, the numerical flux is a function of $2k$ variables

$$\hat{f}_{j+1/2} = \hat{f}(v_{j-k+1}^n, \cdots, v_{j+k}^n), \tag{1.2b}$$

which is consistent with (1.1a) in the sense that

$$\hat{f}(u, u, \cdots, u) = f(u). \tag{1.2c}$$

We consider the numerical approximation $v_h(x, t)$ in (1.2) to be a piecewise-constant function

$$v_h(x, t) = v_j^n, \quad x_{j-1/2} < x < x_{j+1/2}, \quad n\tau < t \leqq (n+1)\tau. \tag{1.3}$$

Accordingly we define its total variation in x to be

$$TV(v^n) = TV(v_h(\cdot, t_n)) = \sum_j |v_{j+1}^n - v_j^n|. \tag{1.4}$$

---

* Received by the editors May 1, 1985; accepted for publication (in revised form) April 12, 1986.

† School of Mathematical Sciences, Tel-Aviv University, Tel-Aviv, Israel and Department of Mathematics, University of California, Los Angeles, California 90024. The research of this author was supported by Army Research Office contract DAAG-29-80 C-0041 while he was in residence at the Mathematics Research Center, University of Wisconsin, Madison, Wisconsin 53706, and by National Aeronautics and Space Administration Consortium Agreement NCA2-IR390-403 and Army Research Office grant DAAG 29-82-0090 while at the University of California, Los Angeles, California 90024.

‡ Department of Mathematics, University of California, Los Angeles, California 90024. The research of this author was supported by National Science Foundation grant 82-00788, Army Research Office grant DAAG 29-82-0090, National Aeronautics and Space Administration Consortium Agreement NCA2-IR390-403, and National Aeronautics and Space Administration Langley grants NAG 1-270 and NAGI-506.

If the total variation of the numerical solution is uniformly bounded in h for $0 \leqq t \leqq T$

$$TV(v_h(\cdot, t)) \leqq C \cdot TV(u_0) \tag{1.5}$$

then any refinement sequence $h \to 0$, $\tau = O(h)$ has a subsequence $h_j \to 0$ so that

$$v_{h_j} \xrightarrow{L_1} u \tag{1.6}$$

where u is a weak solution of (1.1).

If all limit solutions (1.6) of the numerical solution (1.2) satisfy an entropy condition that implies uniqueness of the IVP (1.1), then the numerical scheme is convergent (see e.g. [3], [12]).

Recently we have introduced the notion of total variation diminishing (TVD) schemes (see [3]), where the approximate solution operator is required to diminish the total variation (1.4) of the numerical solution at each time-step

$$TV(v^{n+1}) \leqq TV(v^n); \tag{1.7}$$

these schemes trivially satisfy (1.5) with $C = 1$. It follows from (1.7) that in the numerical solution of a TVD scheme the value of an isolated local maximum may only decrease in time, while that of a local minimum may only increase. Some early work along these lines was done by van Leer in [15].

We were able to construct TVD schemes that in the sense of local truncation error are high-order accurate everywhere except at local extrema where they necessarily degenerate into first-order accuracy (see [4], [13], [10], [11], [14]). The perpetual damping of local extrema determines the cumulative global error of the "high-order TVD schemes" to be $O(h)$ in the L_∞ norm, $O(h^{3/2})$ in the L_2 norm and $O(h^2)$ in the L_1 norm (see [17]).

In this paper we introduce a new class of nonoscillatory schemes, in which the solution operator is only required to diminish the *number* of local extrema in the numerical solution (as is customary we use "diminishing" loosely as short for "nonincreasing," throughout this paper). This property is satisfied by all the essentially 3-pt TVD schemes that can be described as an average of monotone Riemann solvers; most of the computationally interesting TVD schemes (with a more restrictive CFL condition) are of this type (see [12]). Unlike TVD schemes, nonoscillatory schemes are not required to damp the values of each local extremum at every single time-step, but are allowed to occasionally accentuate a local extremum.

In a sequence of papers, of which the present paper is the first, we show how to construct nonoscillatory schemes that are uniformly high-order accurate (in the sense of global error for smooth solutions of (1.1)). In this first paper we describe a second-order accurate scheme of this type.

The fact that the number of local extrema in the numerical solution may only diminish in time is sufficient by itself to guarantee that the application of the scheme to monotone data results in a monotone function. Thus nonoscillatory schemes, like TVD schemes, are monotonicity preserving. In particular, when applied to a step-function, they do not generate spurious oscillations.

We note that since the number of local extrema in the solution of nonoscillatory schemes is bounded by that of the initial data, uniform boundedness of its total variation (1.5) follows immediately if the maximum norm of the solution is shown to be uniformly bounded.

2. Design principle and overview. In this section we describe how to construct a nonoscillatory scheme that is uniformly second-order accurate.

Integrating the partial differential equation (1.1a) over the computational cell $(x_{j-1/2}, x_{j+1/2}) \times (t_n, t_{n+1})$ we get

$$\bar{u}_j^{n+1} = \bar{u}_j^n - \lambda[\hat{f}_{j+1/2}(u) - \hat{f}_{j-1/2}(u)] \tag{2.1a}$$

where

$$\hat{f}_{j+1/2}(u) = \frac{1}{\tau}\int_{t_n}^{t_{n+1}} f(u(x_{j+1/2}, t))\, dt, \tag{2.1b}$$

and

$$\bar{u}_j^n = \frac{1}{h}\int_{x_{j-1/2}}^{x_{j+1/2}} u(x, t_n)\, dx. \tag{2.1c}$$

We observe that although (2.1a) is a relation between the cell-averages $\bar{u}_j^n$ and $\bar{u}_j^{n+1}$, the evaluation of the fluxes $\hat{f}_{j+1/2}(u)$ in (2.1b) requires knowledge of the solution itself and not its cell-averages.

As in Godunov's scheme and its second-order extension by van Leer [16] and Colella and Woodward [2], we derive our scheme as a direct approximation to (2.1). We denote by v_j^n the numerical approximation to the cell-averages $\bar{u}_j^n$ of the exact solution in (2.1c), and set v_j^0 to be the cell-averages of the initial data. Given $v^n = \{v_j^n\}$ we compute v^{n+1} as follows.

First we reconstruct $u(x, t_n)$ out of its approximate cell-averages $\{v_j^n\}$ to the appropriate accuracy and denote the result by $L(x; v^n)$. Next we solve the IVP

$$v_t + f(v)_x = 0, \qquad v(x, 0) = L(x; v^n), \tag{2.2}$$

and denote its solution by $v(x, t)$. Finally we obtain v_j^{n+1} by taking cell-averages of $v(x, \tau)$

$$v_j^{n+1} = \frac{1}{h}\int_{x_{j-1/2}}^{x_{j+1/2}} v(x, \tau)\, dx. \tag{2.3}$$

The averaging operator in (2.3) is nonoscillatory, therefore the number of local extrema in v^{n+1} (interpreted as a mesh-function or the piecewise-constant function (1.3)) does not exceed that of $v(x, \tau)$. Assuming $v(x, t)$ to be the exact solution of (2.2) implies (since the exact solution operator is TVD) that the number of local extrema in $v(x, \tau)$ is less than or equal to that of $v(x, 0) = L(x; v^n)$. Therefore if the number of local extrema in $L(x; v^n)$ does not exceed that of v^n, then the resulting scheme is nonoscillatory.

We conclude that the design of nonoscillatory high-order accurate schemes essentially boils down to a problem on the level of approximation of functions: Given cell-averages $\bar{u}_j$ of a piecewise-smooth function $u(x)$, reconstruct $u(x)$ to a desired accuracy. Prior to studying this problem we tackle another related question in approximation of functions, that of constructing a nonoscillatory high-order accurate interpolation of piecewise-smooth functions.

In § 3 we construct a nonoscillatory piecewise-parabolic function $Q(x; u)$ that interpolates a piecewise-smooth function $u(x)$ at the mesh points

$$Q(x_j; u) = u(x_j) \tag{2.4a}$$

and satisfies, wherever $u(x)$ is smooth,

$$Q(x; u) = u(x) + O(h^3), \tag{2.4b}$$

$$\frac{d}{dx}Q(x \pm 0; u) = \frac{d}{dx}u(x) + O(h^2). \tag{2.4c}$$

In § 4 we make use of this nonoscillatory piecewise-parabolic interpolant to design a nonoscillatory reconstruction of a piecewise-smooth function from its cell-averages. As in [16], [2], [5] and [9] we take $L(x;\bar{u})$ to be the following piecewise-linear function:

$$L(x;\bar{u})=\bar{u}_j+S_j(x-x_j)/h \quad \text{for } |x-x_j|<\frac{h}{2}. \tag{2.5a}$$

Unlike the above references that present "second-order accurate" TVD schemes, we compute the slopes S_j/h from $Q(x;\bar{u})$ by

$$S_j/h=m\left(\frac{d}{dx}Q(x_j-0;\bar{u}),\frac{d}{dx}Q(x_j+0;\bar{u})\right). \tag{2.5b}$$

Here $m(x,y)$ is the min mod function

$$m(x,y)=\begin{cases} s\cdot\min(|x|,|y|) & \text{if } \operatorname{sgn}(x)=\operatorname{sgn}(y)=s,\\ 0 & \text{otherwise.}\end{cases} \tag{2.6}$$

We show in § 4 that $L(x;\bar{u})$ is a proper reconstruction of $u(x)$ in the sense that whenever $u(x)$ is smooth

$$L(x;\bar{u})=u(x)+O(h^2) \tag{2.7a}$$

and

$$\bar{L}(x;\bar{u})=\bar{u}(x)+O(h^3). \tag{2.7b}$$

Here

$$\bar{u}(x)=h^{-1}\int_{-h/2}^{h/2}u(x+y)\,dy \quad\text{and}\quad \bar{L}(x;\bar{u})=h^{-1}\int_{-h/2}^{h/2}L(x+y;\bar{u})\,dy;$$

like $Q(x;\bar{u})$, the latter is also a nonoscillatory piecewise-parabolic interpolant of $\bar{u}(x)$,

$$\bar{L}(x_j;\bar{u})=\bar{u}(x_j). \tag{2.7c}$$

We remark that the "second-order accurate" TVD schemes described in the above-mentioned references use a slope S_j/h in (2.5a) that approximates $(d/dx)u(x_j)$ to $O(h)$, and their loss of second-order accuracy at local extrema points is due to lack of smoothness of the coefficient in the $O(h)$ term at these points.[1] This problem is circumvented in the present scheme by taking S_j/h to be (2.5b) which is an $O(h^2)$ approximation to $(d/dx)u(x_j)$. Unfortunately there is a price to pay for this extra accuracy, namely the loss of the TVD property. As in TVD schemes

$$TV(v^{n+1})\leqq TV(L(\cdot;v^n)); \tag{2.8}$$

however, here

$$TV(L(\cdot;v^n))\geqq TV(v^n)$$

and indeed the scheme may occasionally increase the variation of the numerical solution. Although we prove that the scheme is nonoscillatory, we have not been able as yet to complete a proof of uniform boundedness of the total variation of the numerical solution; this is due to lack of techniques to verify uniform boundedness of the maximum norm of the numerical solution.

[1] We repeat that the results of [8] and [11] imply that TVD schemes, no matter how they are constructed, must have this loss of accuracy at local extrema.

In § 5 we study the proposed scheme in the constant coefficient case. We verify that it is uniformly second-order accurate, examine its behavior at local extrema points and get estimates for the possible increase in total variation per time-step.

In this paper where we consider numerical schemes of the form (1.2) that are derived from approximating the relation (2.1), it is only natural to consider truncation error in the sense of cell-averages. That is, we say that the scheme (1.2) is second-order accurate if

$$\bar{u}^{n+1} = E_h \cdot \bar{u}^n + O(h^3) \tag{2.9}$$

where $\bar{u}^n$ is the cell-average (2.1c) of the exact solution. Since

$$\bar{u}(x) = u(x) + O(h^2) \tag{2.10}$$

whenever $u(x)$ is smooth, (2.9) holds also for pointwise values of the solution. However, in the context of third and higher order accurate schemes, this difference in definitions of truncation error will be not only conceptual but of practical importance as well.

Up to this point we have assumed that $v(x, \tau)$ in (2.3) is the exact solution to (2.2). The resulting scheme

$$v_j^{n+1} = v_j^n - \lambda[\hat{f}_{j+1/2}(v) - \hat{f}_{j-1/2}(v)], \tag{2.11a}$$

where $\hat{f}_{j+1/2}(v)$ is (2.1b) applied to $v(x, t)$,

$$\hat{f}_{j+1/2}(v) = \frac{1}{\tau}\int_0^\tau f(v(x, t))\, dt, \tag{2.11b}$$

is certainly second-order accurate in the sense of (2.9). Starting with the exact cell-averages $v_j^n = \bar{u}_j^n$ in (2.11) we get from (2.7a) that

$$v(x, t) = u(x, t + t_n) + O(h^2) \quad \text{for } 0 \leqq t \leqq \tau, \tag{2.12a}$$

and consequently,

$$\hat{f}_{j+1/2}(v) = \hat{f}_{j+1/2}(u) + O(h^2), \tag{2.12b}$$

which implies (2.9) due to the sufficient smoothness of the coefficient in the $O(h^2)$ term in (2.12b).

In § 6 we replace the exact solution $v(x, t)$ in (2.3) by an approximate one, which we denote by $v_n(x, t)$. This approximate solution is conservative, TVD, and second-order accurate in the sense of (2.12a). Thus replacing $v(x, t)$ in (2.3) by this approximate solution results in a conservative scheme that is nonoscillatory and uniformly second-order accurate.

We remark that an alternative approach to the above is to approximate $\hat{f}_{j+1/2}(v)$ in (2.11b) by using a midpoint rule (or trapezoidal rule) for the integral and by replacing $v(x, t)$ with a nonoscillating second-order accurate approximate one $v_n(x, t)$ (see [16] and [2]). The resulting scheme

$$v_j^{n+1} = v_j^n - \lambda(\hat{f}_{j+1/2} - \hat{f}_{j-1/2}), \tag{2.13a}$$

$$\hat{f}_{j+1/2} = f(v_n(x_{j+1/2}, \tau/2)), \tag{2.13b}$$

is certainly second-order accurate, and it is nonoscillatory in the constant coefficient case. Since we have not used the cell-averaging (2.3) to derive this scheme, we cannot ascertain in general that the resulting scheme is nonoscillatory. Nevertheless, our numerical experiments as well as many other experiments in the context of TVD schemes (see e.g. [1], [2]) demonstrate that the numerical results are nonoscillatory in many (if not all) applications.

In § 7 we present some numerical experiments that compare the present scheme with a typical "second-order accurate" TVD scheme.

3. Nonoscillatory interpolation. The oscillatory nature of second-order accurate Lax-Wendroff type schemes results from a Gibbs phenomenon associated with high-order interpolation across discontinuities. In this section, as a preparatory step towards designing a nonoscillatory approximation to (1.1), we construct a nonoscillatory piecewise-parabolic interpolant $Q(x; u)$ to a piecewise-smooth function $u(x)$ such that

$$Q(x_i; u) = u(x_i), \tag{3.1a}$$

$$Q(x; u) \equiv q_{i+1/2}(x; u), \qquad x_i \leqq x \leqq x_{i+1}, \tag{3.1b}$$

where $q_{i+1/2}$ is a quadratic polynomial, and

$$Q(x; u) - u(x) = O(h^3), \qquad \frac{d}{dx}Q(x \pm 0; u) - \frac{d}{dx}u(x) = O(h^2) \tag{3.1c}$$

wherever $u(x)$ is smooth.

$Q(x; u)$ is nonoscillatory in the sense that the number of its local extrema does not exceed that of $u(x)$.

Since

$$q_{i+1/2}(x_i; u) = u_i, \qquad q_{i+1/2}(x_{i+1}; u) = u_{i+1},$$

it can be written in the form

$$q_{i+1/2}(x; u) = u_i + d_{i+1/2}u \cdot (x - x_i)/h + \tfrac{1}{2}D_{i+1/2}u \cdot (x - x_i)(x - x_{i+1})/h^2 \tag{3.2a}$$

where

$$d_{i+1/2}u = u_{i+1} - u_i \tag{3.2b}$$

and $D_{i+1/2}u$ is yet to be determined.

$$D_{i+1/2}u = h^2 q''_{i+1/2}(x; u), \qquad x_i \leqq x \leqq x_{i+1}.$$

We consider as candidates for $q_{i+1/2}$ the two quadratic polynomials $\bar{q}_i$ and $\bar{q}_{i+1}$, interpolating $u(x)$ at (x_{i-1}, x_i, x_{i+1}) and (x_i, x_{i+1}, x_{i+2}), respectively, and choose $q_{i+1/2}$ to be the one that is least oscillatory in $[x_i, x_{i+1}]$. Both $\bar{q}_j$, $j = i$ and $j = i+1$, can be written as (3.2a) with $D_{i+1/2}u = D_ju$ where

$$D_ju = d_{j+1/2}u - d_{j-1/2}u = u_{j+1} - 2u_j + u_{j-1}. \tag{3.2c}$$

Since the least oscillatory of $\bar{q}_i$ and $\bar{q}_{i+1}$ can be characterized as the one that deviates the least from the line connecting (x_i, u_i) with (x_{i+1}, u_{i+1}) we choose $D_{i+1/2}u$ in (3.2a) to be

$$D_{i+1/2}u = m(D_iu, D_{i+1}u), \tag{3.2d}$$

where $m(x, y)$ is the min mod function

$$m(x, y) = \begin{cases} s \cdot \min(|x|, |y|) & \text{if } \operatorname{sgn}(x) = \operatorname{sgn}(y) = s, \\ 0 & \text{otherwise.} \end{cases} \tag{3.3}$$

If $u(x)$ is smooth in $[x_{j-1}, x_{j+1}]$, then $\bar{q}_j$ as a quadratic interpolant of u satisfies

$$\bar{q}_j(x) - u(x) = O(h^3), \qquad \frac{d}{dx}\bar{q}_j(x) - \frac{d}{dx}u(x) = O(h^2), \qquad x_{j-1} \leqq x \leqq x_{j+1}. \tag{3.4}$$

If $D_i u \cdot D_{i+1} u \geqq 0$ then $q_{i+1/2}$ is either $\bar{q}_i$ or $\bar{q}_{i+1}$. Otherwise we set $D_{i+1/2} u = 0$, but then smoothness of u implies that $D_j u = O(h^3)$ and consequently $q_{i+1/2} - \bar{q}_j = O(h^3)$ for $j = i$, $i+1$. Thus (3.1c) follows from (3.4).

We turn now to prove that $Q(x; u)$ is a nonoscillatory interpolant of u, i.e., that the number of its local extrema does not exceed that of u. We do so by showing a one-to-one correspondence between local extrema of Q to those of the mesh function $\{u_j\}$, the number of which certainly does not exceed that of $u(x)$.

Q may have a local extremum in either the interior of some interval (x_i, x_{i+1}) or at a mesh point x_i. The first case, which will be referred to as interior-extremum, occurs when there is a point x^*, $x_i < x^* < x_{i+1}$, such that

$$\frac{d}{dx} Q(x^*; u) = 0 \quad \text{but} \quad \frac{d}{dx} q_{i+1/2} \not\equiv 0.$$

From (3.2a) it follows that Q has an interior-extremum in (x_i, x_{i+1}) if and only if

$$|D_{i+1/2} u| > 2|d_{i+1/2} u|. \tag{3.5}$$

$q^*_{i+1/2} = q_{i+1/2}(x^*)$, the value of the interior-extremum is then

$$q^*_{i+1/2} = u_i - \frac{1}{2} D_{i+1/2} u \cdot \left(\frac{d_{i+1/2} u}{D_{i+1/2} u} - \frac{1}{2} \right)^2; \tag{3.6}$$

if $D_{i+1/2} u < 0$ it is a local maximum; if $D_{i+1/2} u > 0$ it is a local minimum. Since $D_{i+1/2} u = m(D_i u, D_{i+1} u)$, (3.5) holds if and only if

$$D_i u \cdot D_{i+1} u > 0, \tag{3.7a}$$

$$|D_j u| > 2|d_{i+1/2} u|, \qquad j = i, \quad i+1. \tag{3.7b}$$

This implies that $q_{i+1/2}$ has a local extremum in (x_i, x_{i+1}) if and only if *both* $\bar{q}_i$ and $\bar{q}_{i+1}$ also have a local extremum in (x_i, x_{i+1}) and of the same kind. Since a parabola has at most one local extremum, it follows then that $\bar{q}_i$ does not have a local extremum in (x_{i-1}, x_i) and $\bar{q}_{i+1}$ does not have one in (x_{i+1}, x_{i+2}). Consequently Q is monotone in both (x_{i-1}, x_i) and (x_{i+1}, x_{1+2}), but in an opposite sense, i.e., $d_{i-1/2} u \cdot d_{i+3/2} u < 0$; the latter implies that u has a local extremum in $[x_i, x_{i+1}]$ and that either u_i or u_{i+1} is a local extremum of the mesh function $\{u_j\}$ (for obvious reasons the case $u_i = u_{i+1}$ is counted as a single-extremum). The above analysis also shows that interior-extrema are isolated, i.e., if Q has an interior-extremum in (x_i, x_{i+1}), then it is the only local extremum of Q in (x_{i-1}, x_{i+2}).

We turn now to examine the case that Q has a local extremum at a mesh point x_i; this will be referred to as a mesh-extremum. The above observation that interior-extrema are isolated excludes the possibility that Q has an interior-extremum in either (x_{i-1}, x_i) or (x_i, x_{i+1}) and consequently Q is monotone in these intervals. This implies that $d_{i-1/2} u \cdot d_{i+1/2} u < 0$ and therefore u_i is a local extremum of the mesh function $\{u_j\}$. This concludes the proof that $Q(x; u)$ is a nonoscillatory interpolant of u.

We next express the nonoscillatory nature of Q in terms of total variation. If $|D_{j+1/2} u| \leqq 2|d_{j+1/2} u|$ then (2.5) implies that Q is monotone in $[x_j, x_{j+1}]$. Thus

$$|D_{j+1/2} u| \leqq 2|d_{j+1/2} u| \Rightarrow TV_{[x_j, x_{j+1}]}(Q) = |d_{j+1/2} u|. \tag{3.8a}$$

If $|D_{j+1/2} u| > 2|d_{j+1/2} u|$ then Q has a local extremum in (x_j, x_{j+1}) and

$$TV_{[x_j, x_{j+1}]}(Q) = |q^*_{j+1/2} - u_j| + |u_{j+1} - q^*_{j+1/2}|.$$

Using (2.6) we get

$$|D_{j+1/2}u| > 2|d_{j+1/2}u| \Rightarrow TV_{[x_j,x_{j+1}]}(Q) = |d_{j+1/2}u| + |D_{j+1/2}u|\left(\left|\frac{d_{j+1/2}u}{D_{j+1/2}u}\right| - \frac{1}{2}\right)^2. \tag{3.8b}$$

We conclude that

$$0 \leqq TV(Q) - \sum_j |d_{j+1/2}u| \leqq \sum_{m\in M} |D_{m+1/2}u|\left(\left|\frac{d_{m+1/2}u}{D_{m+1/2}u}\right| - \frac{1}{2}\right)^2 \leqq \frac{1}{4}\sum_{m\in M} |D_{m+1/2}u|. \tag{3.8c}$$

The sum in the RHS of (3.8c) is taken over the set of indices M of intervals (x_m, x_{m+1}) in which $|D_{m+1/2}u| > 2|d_{m+1/2}u|$, i.e., where Q has interior-extrema.

Next we show that if $u(x)$ is a piecewise-smooth function of bounded variation, then

$$\lim_{h\to 0} TV(Q(\cdot\,; u)) = TV(u). \tag{3.9}$$

We observe that in this case the number of intervals in M is finite and is uniformly bounded by the number of local extrema in $u(x)$. Hence (3.9) will follow if we prove that $D_{m+1/2}u \to 0$ as $h \to 0$ for all $m \in M$. To accomplish this we show that for h sufficiently small M does not include intervals (x_m, x_{m+1}) in which $u(x)$ is discontinuous. To see that let us examine the case where $u(x)$ has a discontinuity at $\bar{x} \in (x_i, x_{i+1})$. Clearly $d_{i+1/2}u$ approaches the size of the jump in u while $d_{i-1/2}u$ approaches zero as $h \to 0$. Consequently

$$|D_i u/d_{i+1/2}u| = |1 - d_{i-1/2}u/d_{i+1/2}u| \to 1 \quad \text{as } h \to 0. \tag{3.10a}$$

Hence for h sufficiently small

$$2|d_{i+1/2}u| > |D_i u| \geqq |D_{i+1/2}u| \tag{3.10b}$$

which implies $i \notin M$.

4. Nonoscillatory reconstruction. Let $u(x)$ be a piecewise-smooth function and denote by $\bar{u}(x)$ its mean over $(x - h/2, x + h/2)$, i.e.,

$$\bar{u}(x) = \frac{1}{h}\int_{x-h/2}^{x+h/2} u(y)\,dy. \tag{4.1}$$

We denote $\bar{u}_j = \bar{u}(x_j)$ and refer to these values as cell-averages of $u(x)$. Given $\{\bar{u}_j\}$, the task at hand is to reconstruct $u(x)$ to $O(h^2)$ in a nonoscillatory way; denote the approximately reconstructed function by $L(x; \bar{u})$. To achieve $O(h^2)$ accuracy it is sufficient to consider $L(x; \bar{u})$ to be a piecewise-linear function. To make $L(x; \bar{u})$ a nonoscillatory approximation we use the nonoscillatory piecewise parabolic interpolation $Q(x; \bar{u})$ to compute its slopes as follows:

$$L(x; \bar{u}) = \bar{u}_j + S_j(x - x_j)/h \quad \text{for } |x - x_j| < \tfrac{1}{2}h, \tag{4.2a}$$

$$S_j = h \cdot m\left(\frac{d}{dx}Q(x_j - 0; \bar{u}), \frac{d}{dx}Q(x_j + 0; \bar{u})\right). \tag{4.2b}$$

Here m is the min mod function (3.3); $d_{i+1/2}\bar{u}$ and $D_{i+1/2}\bar{u}$ are (3.2b) and (3.2d), respectively.

We note that $L(x;\bar{u})$ may be discontinuous at $\{x_{j+1/2}\}$ and that

$$L(x_j;\bar{u})=\bar{u}_j. \tag{4.3a}$$

To see that wherever $u(x)$ is smooth

$$L(x,\bar{u})-u(x)=O(h^2) \tag{4.3b}$$

we observe that in this case

$$\bar{u}(x)=u(x)+O(h^2) \tag{4.4a}$$

and therefore it follows from (3.1c) that

$$\frac{1}{h}S_j=\frac{d}{dx}\bar{u}(x_j)+O(h^2)=\frac{d}{dx}u(x_j)+O(h^2). \tag{4.4b}$$

Consequently the RHS of (4.2a) can be expanded as

$$\begin{aligned} L(x;\bar{u}) &= u_j+(x-x_j)\frac{d}{dx}u(x_j)+O(h^2) \\ &= u(x)+O(h^2) \quad \text{for } |x-x_j|<\tfrac{1}{2}h, \end{aligned} \tag{4.4c}$$

and thus (4.3b) follows.

Denote by $\bar{L}(x;\bar{u})$ the mean value of $L(x;\bar{u})$ in $(x-h/2,x+h/2)$, i.e.,

$$\bar{L}(x;\bar{u})=\frac{1}{h}\int_{x-h/2}^{x+h/2}L(y;\bar{u})\,dy. \tag{4.5a}$$

Using (4.2a) to evaluate the integral in (4.5a) we find

$$\bar{L}(x;\bar{u})=\bar{u}_j+d_{j+1/2}\bar{u}(x-x_j)/h+(\tfrac{1}{2})(S_{j+1}-S_j)(x-x_j)(x-x_{j+1})/h^2 \quad \text{for } x_j\leqq x\leqq x_{j+1}, \tag{4.5b}$$

$$\bar{L}(x_j;\bar{u})=\bar{u}_j. \tag{4.5c}$$

Hence $\bar{L}(x;\bar{u})$, like $Q(x;\bar{u})$, is a piecewise-parabolic interpolant of $\bar{u}(x)$. Comparing (4.5b) with (3.2) we find that for $x_j\leqq x\leqq x_{j+1}$

$$\bar{L}(x;\bar{u})-Q(x;\bar{u})=\tfrac{1}{2}(S_{j+1}-S_j-D_{j+1/2}\bar{u})(x-x_j)(x-x_{j+1})/h^2. \tag{4.6a}$$

From (4.4b) we see that $S_i=h(d/dx)\bar{u}(x_i)+O(h^3)$ (note that this is true also at local extrema points) and therefore

$$S_{j+1}-S_j=h^2\frac{d^2}{dx^2}u(x_{j+1/2})+O(h^3).$$

On the other hand, (3.2) shows that

$$D_{j+1/2}\bar{u}=h^2\frac{d^2}{dx^2}u(x_{j+1/2})+O(h^3).$$

Therefore

$$S_{j+1}-S_j-D_{j+1/2}\bar{u}=O(h^3) \tag{4.6b}$$

which shows that RHS of (4.6a) is $O(h^3)$. Since (2.1c) shows that

$$Q(x;\bar{u})-\bar{u}(x)=O(h^3)$$

we conclude from (4.6a)-(4.6b) that

$$\bar{L}(x;\bar{u})-\bar{u}(x)=O(h^3). \tag{4.6c}$$

We turn now to prove that $L(x; \bar{u})$ is a nonoscillatory approximation to $\bar{u}(x)$; this certainly implies that $L(x; \bar{u})$ is a nonoscillatory approximation to $u(x)$. We shall do so by showing that $TV_{[x_j,x_{j+1}]}(L(\cdot\,; \bar{u}))$, the total-variation of $L(x; \bar{u})$ in $[x_j, x_{j+1}]$, which has the value

$$TV_{[x_j,x_{j+1}]}(L(\cdot\,; \bar{u})) = \tfrac{1}{2}(|S_j| + |S_{j+1}|) + |d_{j+1/2}\bar{u} - \tfrac{1}{2}(S_j + S_{j+1})|, \tag{4.7a}$$

can also be expressed as

$$TV_{[x_j,x_{j+1}]}(L(\cdot\,; \bar{u})) = \max(|d_{j+1/2}\bar{u}|, \tfrac{1}{2}|D_{j+1/2}\bar{u}|). \tag{4.7b}$$

Then it follows immediately from (4.7b), (4.3a) and (3.8) that L is monotone in $[x_j, x_{j+1}]$ if and only if Q is; consequently L is a nonoscillatory approximation to $u(x)$ in exactly the same sense as Q is to the interpolated function (see § 3).

Next let us denote

$$S_j^{\pm} = h \cdot \frac{d}{dx} Q(x_j \pm 0; \bar{u}), \tag{4.8a}$$

i.e.,

$$S_j^- = d_{j-1/2}\bar{u} + \tfrac{1}{2}D_{j-1/2}\bar{u}, \qquad S_j^+ = d_{j+1/2}\bar{u} - \tfrac{1}{2}D_{j+1/2}\bar{u}, \tag{4.8b}$$

and observe that (4.2b) implies that

$$\begin{aligned} \tfrac{1}{2}(|S_j| + |S_{j+1}|) &= \tfrac{1}{2}[|m(S_j^-, S_j^+)| + |m(S_{j+1}^-, S_{j+1}^+)|] \\ &\leqq \tfrac{1}{2}(|S_j^+| + |S_{j+1}^-|) = \tfrac{1}{2}(|d_{j+1/2}\bar{u} - \tfrac{1}{2}D_{j+1/2}\bar{u}| + |d_{j+1/2}\bar{u} + \tfrac{1}{2}D_{j+1/2}\bar{u}|) \\ &= \max(|d_{j+1/2}\bar{u}|, \tfrac{1}{2}|D_{j+1/2}\bar{u}|). \end{aligned} \tag{4.8c}$$

We note that if $|d_{j+1/2}\bar{u}| \geqq \tfrac{1}{2}|D_{j+1/2}\bar{u}|$ then

$$\operatorname{sgn}(d_{j+1/2}\bar{u} \pm \tfrac{1}{2}D_{j+1/2}\bar{u}) \cdot \operatorname{sgn}(d_{j+1/2}\bar{u}) \geqq 0$$

which in turn implies

$$\operatorname{sgn}(S_j) \cdot \operatorname{sgn}(d_{j+1/2}\bar{u}) \geqq 0, \qquad \operatorname{sgn}(S_{j+1}) \cdot \operatorname{sgn}(d_{j+1/2}\bar{u}) \geqq 0.$$

It follows then from (4.8c) that the RHS of (4.7a) is $|d_{j+1/2}\bar{u}|$. This shows that

$$|d_{j+1/2}\bar{u}| \geqq \tfrac{1}{2}|D_{j+1/2}\bar{u}| \Rightarrow TV_{[x_j,x_{j+1}]}(L(\cdot\,; \bar{u})) = |d_{j+1/2}\bar{u}|. \tag{4.9a}$$

To complete the verification of (4.7b) we still have to show that

$$|d_{j+1/2}\bar{u}| < \tfrac{1}{2}|D_{j+1/2}\bar{u}| \Rightarrow TV_{[x_j,x_{j+1}]}(L(\cdot\,; \bar{u})) = \tfrac{1}{2}|D_{j+1}\bar{u}|. \tag{4.9b}$$

First we observe that

$$\begin{aligned} S_i^+ - S_i^- &= (d_{i+1/2}\bar{u} - \tfrac{1}{2}D_{i+1/2}\bar{u}) - (d_{i-1/2}\bar{u} + \tfrac{1}{2}D_{i-1/2}\bar{u}) \\ &= D_i\bar{u} - \tfrac{1}{2}(D_{i+1/2}\bar{u} + D_{i-1/2}\bar{u}). \end{aligned} \tag{4.10}$$

Since (3.2d) implies

$$|D_i\bar{u}| \leqq \tfrac{1}{2}(|D_{i-1/2}\bar{u}| + |D_{i+1/2}\bar{u}|), \tag{4.11}$$

we conclude from (4.10) that

$$(S_i^+ - S_i^-) \cdot \operatorname{sgn}(D_i\bar{u}) \geqq 0. \tag{4.12}$$

We turn now to prove (4.9b). First let us consider the case that $Q(x; \bar{u})$ has a local maximum in (x_j, x_{j+1}), i.e., $D_j\bar{u} < 0$, $D_{j+1}\bar{u} < 0$, and $|d_{j+1/2}\bar{u}| < \tfrac{1}{2}|D_{j+1/2}\bar{u}|$.

It follows from (4.12) that

(4.13a) $$S_j^- \geqq S_j^+ = d_{j+1/2}\bar{u} - \tfrac{1}{2}D_{j+1/2}\bar{u} > 0,$$

(4.13b) $$0 > d_{j+1/2}\bar{u} + \tfrac{1}{2}D_{j+1/2}\bar{u} = S_{j+1}^- \geqq S_{j+1}^+.$$

The relations (4.13) and the definitions (4.8a), (4.2b) imply that

(4.14a) $$S_j = S_j^+ = d_{j+1/2}\bar{u} - \tfrac{1}{2}D_{j+1/2}\bar{u},$$

(4.14b) $$S_{j+1} = S_{j+1}^- = d_{j+1/2}\bar{u} + \tfrac{1}{2}D_{j+1/2}\bar{u}.$$

The same analysis shows that (4.14) holds also for the case that $Q(x;\bar{u})$ has a local minimum in (x_j, x_{j+1}). (4.9b) follows immediately from (4.14) and (4.7a).

We note that since $L(x;\bar{u})$ is continuous at x_j

(4.15)
$$\begin{aligned} TV(L(\cdot\,;\bar{u})) &= \sum_j TV_{[x_j,x_{j+1}]}(L(\cdot\,;\bar{u})) = \sum_j \max\left(|d_{j+1/2}\bar{u}|, \frac{1}{2}|D_{j+1/2}\bar{u}|\right) \\ &= \sum_j |d_{j+1/2}\bar{u}| + \sum_{m\in M}\left(\frac{1}{2}|D_{m+1/2}\bar{u}| - |d_{m+1/2}\bar{u}|\right). \end{aligned}$$

Here M is the set of indices of intervals (x_m, x_{m+1}) in the interior of which L (and also Q) has a local extremum. The number of these intervals is finite and is bounded by the number of local extrema of $\bar{u}(x)$. Comparing (4.9) with (3.8) we note that

(4.16) $$TV(L(\cdot\,;\bar{u})) \geqq TV(Q(\cdot\,;\bar{u})).$$

5. The constant coefficient case. In this section we study the constant coefficient case

(5.1) $$u_t + au_x = 0, \qquad a = \text{const.}$$

The exact solution of the IVP (2.2) is

(5.2) $$v(x,t) = L(x - at;\, v^n).$$

Hence our scheme (2.3) is

(5.3a) $$v_j^{n+1} = \frac{1}{h}\int_{x_{j-1/2}}^{x_{j+1/2}} L(x - a\tau;\, v^n)\,dx = \bar{L}(x_j - a\tau;\, v^n)$$

where $\bar{L}$ is (4.5a). We have shown in § 3 that the number of local extrema in $L(x;\, v^n)$ does not exceed that of v^n. Since v^{n+1} in (5.3a) is a cell-average of L, it follows that the number of local extrema in v^{n+1} does not exceed that of v^n, and consequently the scheme (5.3a) is nonoscillatory.

Using (4.5b) in (5.3a) we get the following expression for the scheme

(5.3b)
$$\begin{aligned} v_j^{n+1} &= (E_h\cdot v^n)_j \\ &= \begin{cases} v_j^n - \mu d_{j-1/2}v^n - \tfrac{1}{2}\mu(1-\mu)(S_j^n - S_{j-1}^n) & \text{if } a>0, \\ v_j^n - \mu d_{j+1/2}v^n + \tfrac{1}{2}\mu(1+\mu)(S_{j+1}^n - S_j^n) & \text{if } a<0, \end{cases} \end{aligned}$$

E_h denotes the operator form of the finite difference scheme; $\mu = \lambda a$, the CFL-number, is assumed to satisfy

(5.3c) $$|\mu| \leqq 1.$$

We turn now to prove that (5.3) is second-order accurate in the sense of (2.9), i.e., if $\bar{u}^n(x)$ denotes the mean value (4.1) of $u(x, t_n)$ then

(5.4a) $$\bar{u}_j^{n+1} - (E_h\cdot \bar{u}^n)_j = O(h^3).$$

To show that we observe that in the constant coefficient case (5.1) $\bar{u}_j^{n+1} = \bar{u}^n(x_j - a\tau)$, and by (5.3a) $(E_h \cdot \bar{u}^n)_j = \bar{L}(x_j - a\tau; \bar{u}^n)$. Hence the LHS of (5.4a) is nothing but

(5.4b) $$\bar{u}^n(x_j - a\tau) - \bar{L}(x_j - a\tau; \bar{u}^n),$$

which is $O(h^3)$ as a direct consequence of (4.6c).

Next we study the time-dependence of the total variation and the maximum norm of the numerical solution (5.3). In § 2 we have pointed out that

(5.5a) $$TV(\nu^{n+1}) \leqq TV(L(\cdot\,; \nu^n)).$$

Using (4.15) and (5.5a) we get the following upper bound on the possible growth of the total variation of the numerical solution per time-step:

(5.5b) $$TV(\nu^{n+1}) - TV(\nu^n) \leqq \sum_{m \in M_n} \left(\frac{1}{2}|D_{m+1/2}\nu^n| - |d_{m+1/2}\nu^n|\right).$$

Here M_n is the set of indices of intervals (x_m, x_{m+1}) in the interior of which $L(x; \nu^n)$ (and also $Q(x; \nu^n)$) has a local extremum. The number of these intervals is finite and remains uniformly bounded in time by the number of local extrema in the initial data.

Clearly the upper bound (5.5b) is overly pessimistic. It estimates the possible increase in variation in the reconstruction step due to replacing the cell-averages ν_j^n by the piecewise-linear function $L(x; \nu^n)$. It does not take into account the possible decrease in variation in the averaging step (2.3), resulting from doing just the opposite, i.e., replacing the piecewise-linear function $L(x - a\tau; \nu^n)$ in (5.2) by its cell-averages (5.3a).

In the following we shall examine the temporal behavior of the local extrema of the numerical solution and its total variation by analyzing the explicit values of the cell-averages ν_j^{n+1} given by (5.3b). To simplify our presentation let us assume that $a > 0$.

First we note that (4.8b) implies

(5.6a) $$|S_j - S_{j-1}| \leqq |S_j| + |S_{j+1}| \leqq 2 \max\left(|d_{j-1/2}\nu^n|, \tfrac{1}{2}|D_{j-1/2}\nu^n|\right).$$

Hence

(5.6b) $$|d_{j-1/2}\nu^n| \geqq \tfrac{1}{2}|D_{j-1/2}\nu^n| \Rightarrow |\gamma_{j-1/2}| \equiv |S_j^n - S_{j-1}^n| / |d_{j-1/2}\nu^n| \leqq 2.$$

Rewriting (5.3) in this case as

(5.7a) $$\begin{aligned} \nu_j^{n+1} &= \nu_j^n - \mu d_{j-1/2}\nu^n - \tfrac{1}{2}\mu(1-\mu)\gamma_{j-1/2} d_{j-1/2}\nu^n \\ &= (1 - \sigma_{j-1/2})\nu_j^n + \sigma_{j-1/2}\nu_{j-1}^n \end{aligned}$$

where

(5.7b) $$\sigma_{j-1/2} = \mu + \tfrac{1}{2}\mu(1-\mu)\gamma_{j-1/2},$$

we see that the CFL condition $0 < \mu \leqq 1$ and (5.6b) imply that

(5.7c) $$0 \leqq \sigma_{j-1/2} \leqq 1;$$

thus we conclude

(5.8) $$|d_{j-1/2}\nu^n| \geqq \tfrac{1}{2}|D_{j-1/2}\nu^n| \Rightarrow \nu_j^{n+1} \in [\nu_{j-1}^n, \nu_j^n].$$

Relation (5.8) shows that if ν^n is monotone for $J_L \leqq j \leqq J_R$, i.e. $\nu_{J_L} \leqq \nu_{J_{L+1}} \leqq \cdots \leqq \nu_{J_R}$, or $\nu_{J_L} \geqq \nu_{J_{L+1}} \geqq \cdots \geqq \nu_{J_R}$, then ν^{n+1} is monotone for $J_L + 1 \leqq j \leqq J_R$, and in the same sense. Relation (5.8) also shows that mesh-extrema of ν^n, i.e., those for which Q has its local extremum at a mesh point, are being damped at the nth time-step. Namely,

(5.9a) $|d_{j\pm1/2}\nu^n| \geqq \frac{1}{2}|D_{j\pm1/2}\nu^n|, \qquad \nu_{j-1}^n \leqq \nu_j^n \geqq \nu_{j+1}^n \Rightarrow \max(\nu_j^{n+1}, \nu_{j+1}^{n+1}) \leqq \nu_j^n,$

(5.9b) $|d_{j\pm1/2}\nu^n| \geqq \frac{1}{2}|D_{j\pm1/2}\nu^n|, \qquad \nu_{j+1}^n \geqq \nu_j^n \leqq \nu_{j+1}^n \Rightarrow \min(\nu_j^{n+1}, \nu_{j+1}^{n+1}) \geqq \nu_j^n.$

We turn now to consider interior local extrema of ν^n, i.e. those for which Q has its local extremum in the interior of some (x_i, x_{i+1}). We recall that such an extremum is characterized by $|d_{i+1/2}\nu^n| < \frac{1}{2}|D_{i+1/2}\nu^n|$ and that S_i^n and S_{i+1}^n in this case are given by (4.14); therefore $S_{i+1}^n - S_i^n = D_{i+1/2}\nu^n$. From (5.3a) and (4.6a) we see that in general

(5.10a) $$\nu_{i+1}^{n+1} - Q(x_{i+1} - a\tau; \nu^n) = \tfrac{1}{2}\mu(1-\mu)(D_{i+1/2}\nu^n - S_{i+1}^n + S_i^n).$$

Hence

(5.10b) $$|d_{i+1/2}\nu^n| < \tfrac{1}{2}|D_{i+1/2}\nu^n| \Rightarrow \nu_{i+1}^{n+1} = Q(x_{i+1} - a\tau; \nu^n).$$

Relation (5.10b) confirms the second-order accuracy of the scheme at local extrema. Although it does not necessitate accentuation of the extremal values, as ν_{i+1}^{n+1} in (5.10b) may still be in $[\nu_i^n, \nu_{i+1}^n]$, it does allow ν_{i+1}^{n+1} to deviate from this interval by as much as

(5.10c) $$\tfrac{1}{2}|D_{i+1/2}\nu^n|(|d_{i+1/2}\nu^n|/|D_{i+1/2}\nu^n| - \tfrac{1}{2})^2.$$

Thus (5.10b) is the essential difference between the present scheme and the "second-order" TVD schemes.

A similar analysis, which we do not present here, shows that if ν_i^n is a mesh-extremum then ν_j^{n+1}, $j = i, i+1$, relates to $Q(x_j - a\tau; \nu^n)$ in the following way:

(5.11a) $$\nu_j^{n+1} \geqq Q(x_j - a\tau; \nu^n), \quad j = i, i+1 \quad \text{if } \nu_i^n \text{ is a maximum},$$

(5.11b) $$\nu_j^{n+1} \leqq Q(x_j - a\tau; \nu^n), \quad j = i, i+1 \quad \text{if } \nu_i^n \text{ is a minimum}.$$

From (5.9)-(5.11) we deduce the following relation between the total variation of the numerical solution and that of its piecewise-parabolic interpolant Q:

(5.12) $$TV(\{Q(x_j - a\tau; \nu^n)\}) \leqq TV(\nu^{n+1}) \leqq TV(Q(\cdot; \nu^n)).$$

The LHS of (5.12) is the total variation of the mesh function $\{Q(x_j - a\tau; \nu^n)\}$. Relation (5.12) suggests to consider an equivalent definition $\overline{TV}$ of the total variation of the numerical approximation of the form

$$\overline{TV}(\nu^n) = TV(Q(\cdot; \nu^n))$$

with the hope that the scheme (5.3) is TVD with respect to this modified definition. Unfortunately our numerical experiments have shown that there are instances, although rather rare, that $\overline{TV}(\nu^n)$ is increasing with n; the same is true for $\overline{TV}(\nu^n) = TV(L(\cdot; \nu^n))$.

As we have mentioned in the Introduction, because of the nonoscillatory nature of the scheme, uniform total variation boundedness of the numerical solution is implied by uniform boundedness of its maximum norm. If we follow a particular local maximum of the initial data we see from (5.9)-(5.10a) that it actually decreases most of the time, and whenever it does increase (5.10c) and (3.10) suggest that it does so by a "small amount" that vanishes with $h \to 0$. Since the initial data is only piecewise-smooth we have not been able as yet to rigorize these arguments.

We remark that our numerical experiments clearly indicate that in a normal computational situation the maximum norm of the numerical solution is indeed uniformly bounded. We feel that our inability to prove this fact stems only from lack of theoretical tools to analyze pointwise regularity of the numerical solution.

6. The nonlinear case. In this section we describe an approximate solution $v_n(x, t)$ of [5] for the IVP (2.2)

$$v_t + f(v)_x = 0, \qquad v(x, 0) = L(x; v^n). \tag{6.1}$$

This approximate solution is consistent with the conservation form of the equation (6.1) in the sense that the cell-averaging (2.3) results in a scheme in conservation form, i.e.,

$$v_j^{n+1} = \frac{1}{h}\int_{x_{j-1/2}}^{x_{j+1/2}} v_n(x, \tau)\, dx = v_j^n - \lambda(\hat{f}_{j+1/2} - \hat{f}_{j-1/2}) \tag{6.2}$$

where the numerical flux $\hat{f}_{j+1/2}$ is consistent with $f(u)$ in the sense of (1.2c). Furthermore, the approximate solution operator is TVD

$$TV(v_n(\cdot\,; t)) \leqq TV(v_n(\cdot\,; 0)) = TV(L(\cdot\,; v^n)) \quad \text{for } 0 \leqq t \leqq \tau, \tag{6.3}$$

and thus by the reasoning presented in § 2, the resulting scheme (6.2) is nonoscillatory.

We turn now to outline the derivation of this approximate solution. To simplify our presentation we ignore entropy considerations and refer the reader to future papers for details of appropriate modifications. We assign to the point $x_{j+1/2}$ a characteristic speed that corresponds to the Rankine-Hugoniot speed $\bar{a}_{j+1/2}$ of the two neighboring cell-averages v_j^n and v_j^{n+1}

$$\bar{a}_{j+1/2} = \begin{cases} \dfrac{f(v_{j+1}^n) - f(v_j^n)}{v_{j+1}^n - v_j^n} & \text{if } v_j^n \neq v_{j+1}^n, \\ a(v_j^n) & \text{if } v_j^n = v_{j+1}^n, \end{cases} \tag{6.4a}$$

and denote by $\bar{a}(x)$ the piecewise-linear interpolant of $\{\bar{a}_{j+1/2}\}$, i.e.,

$$\bar{a}(x) = \bar{a}_{j-1/2} + (\bar{a}_{j+1/2} - \bar{a}_{j-1/2}) \cdot (x - x_{j-1/2})/h \quad \text{for } x_{j-1/2} \leqq x \leqq x_{j+1/2}. \tag{6.4b}$$

The approximate solution $v_n(x, t)$ is defined by specifying its constancy along the approximate characteristics

$$x(t) = x_0 + \bar{a}(x_0) \cdot t, \tag{6.5a}$$

i.e.,

$$v_n(x(t), t) \equiv v_n(x_0, 0) = L(x_0; v^n). \tag{6.5b}$$

Using (6.5a) and (6.4b) to express x_0 in terms of x and t

$$x_0(x, t) = x_{j-1/2} + h \cdot [x - x_{j-1/2}(t)]/[x_{j+1/2}(t) - x_{j-1/2}(t)] \quad \text{for } x_{j-1/2}(t) < x < x_{j+1/2}(t) \tag{6.5c}$$

we get from (6.5b) that

$$v_n(x, t) = L\left(x_{j-1/2} + h \cdot \frac{x - x_{j-1/2}(t)}{x_{j+1/2}(t) - x_{j-1/2}(t)}; v^n\right) \quad \text{for } x_{j-1/2}(t) < x < x_{j+1/2}(t) \tag{6.6a}$$

where

$$x_{i+1/2}(t) = x_{i+1/2} + t \cdot \bar{a}_{i+1/2}. \tag{6.6b}$$

Taking cell-averages of the approximate solution (6.6) we get the conservation form (6.2)

$$v_j^{n+1} = v_j^n - \lambda(\hat{f}_{j+1/2} - \hat{f}_{j-1/2}) \tag{6.7a}$$

with the numerical flux

$$\hat{f}_{j+1/2} = \begin{cases} f(\nu_j^n) + \frac{1}{2}\bar{a}_{j+1/2}(1-\lambda\bar{a}_{j-1/2}) \cdot \hat{S}_j & \text{if } \bar{a}_{j+1/2} \geqq 0, \\ f(\nu_{j+1}^n) - \frac{1}{2}\bar{a}_{j+1/2}(1+\lambda\bar{a}_{j+3/2}) \cdot \hat{S}_{j+1} & \text{if } \bar{a}_{j+1/2} \leqq 0, \end{cases} \tag{6.7b}$$

where

$$\hat{S}_j = S_j^n / [1 + \lambda(\bar{a}_{j+1/2} - \bar{a}_{j-1/2})]. \tag{6.7c}$$

Note that (6.7) is identical to (5.30) in the constant coefficient case.

We turn now to prove that the scheme (6.7) is uniformly second-order accurate in the sense of (2.9). We start with the exact cell-averages $\nu_j^n = \bar{u}_j^n$ in (6.7); this amounts to showing that

$$\hat{f}_{j+1/2} = \hat{f}_{j+1/2}(u) + O(h^2) \tag{6.8a}$$

with a sufficiently smooth coefficient in the $O(h^2)$ term; here $\hat{f}_{j+1/2}$ is the numerical flux (6.7b) computed with the exact cell-averages, and $\hat{f}_{j+1/2}(u)$ is (2.1b). We shall do this in two steps: first we shall show that

$$\hat{f}_{j+1/2}(u) = \tfrac{1}{2}[f(L(x_{j+1/2}; \bar{u}^n)) + f(L(x_0(x_{j+1/2}, \tau); \bar{u}^n))] + O(h^2) \tag{6.8b}$$

where $x_0(x_{j+1/2}, \tau)$ is (6.5c), and then we shall verify that

$$\tfrac{1}{2}[f(L(x_{j+1/2}; \bar{u}^n)) + f(L(x_0(x_{j+1/2}, \tau); \bar{u}^n))] = \hat{f}_{j+1/2} + O(h^2). \tag{6.8c}$$

Special attention will be given to the smoothness of the $O(h^2)$ coefficients.

To show (6.8b) we start by using the trapezoidal rule to approximate the integral in (2.1b); we get

$$\hat{f}_{j+1/2}(u) = \tfrac{1}{2}[f(u(x_{j+1/2}, t_n)) + f(u(x_{j+1/2}, t_n + \tau))] + O(h^2). \tag{6.9a}$$

The smoothness of the $O(h^2)$ term follows from that of $f(u)$ and $u(x, t)$. Next we observe that $\bar{a}(x)$ in (6.4b) approximates $a(u(x, t_n))$ to $O(h^2)$, and therefore we can use the approximate characteristic line (6.5c) to trace $u(x_{j+1/2}, t_n + \tau)$ to $u(x_0(x_{j+1/2}, \tau), t_n)$ with $O(h^3)$ accuracy; consequently

$$f(u(x_{j+1/2}, t_n + \tau)) = f(u(x_0(x_{j+1/2}, \tau), t_n)) + O(h^3). \tag{6.9b}$$

Finally we obtain (6.8b) by approximating $u(x, t_n)$ in (6.9a) and (6.9b) to $O(h^2)$ by $L(x; \bar{u}^n)$ (see (4.4)). The smoothness of the $O(h^2)$ term in this approximation is due to (4.4c):

$$S_j^n = h \cdot u_x(x, t_n) + O(h^3).$$

(We recall that the degeneracy to first-order accuracy at local extrema points of some "second-order accurate" TVD schemes is due to lack of smoothness there of the $O(h^2)$ term in (2.7a).)

We turn now to verify (6.8c). First let us consider the case $\bar{a}_{j+1/2} \geqq 0$:

$$L(x_{j+1/2}; \bar{u}^n) = \bar{u}_j^n + \tfrac{1}{2}S_j^n,$$

$$L(x_0(x_{j+1/2}, \tau); \bar{u}^n) = \bar{u}_j^n + \left[\frac{1}{2} - \frac{\lambda\bar{a}_{j+1/2}}{1+\lambda(\bar{a}_{j+1/2} - \bar{a}_{j-1/2})}\right] S_j^n,$$

and expand the LHS of (6.8c) around $\bar{u}_j^n$. We get

$$\begin{aligned} &f(\bar{u}_j^n) + \frac{1}{2}a(\bar{u}_j^n)(1-\lambda\bar{a}_{j+1/2})\hat{S}_j + \frac{1}{8}a'(\bar{u}_j^n)[(1-\lambda\bar{a}_{j-1/2})^2 + (\lambda\bar{a}_{j+1/2})^2](\hat{S}_j)^2 + O(h^3) \\ &= \hat{f}_{j+1/2} + \frac{h^2}{8}(2\lambda^2 a^2 - 1) \cdot a' \cdot (u_x)^2|_{j+1/2} + O(h^3). \end{aligned} \tag{6.10a}$$

Similarly in the case $\bar{a}_{j+1/2} \leqq 0$,

$$L(x_{j+1/2}; \bar{u}^n) = \bar{u}_{j+1} - \frac{1}{2} S^n_{j+1}, L(x_0(x_{j+1/2}, \tau); \bar{u}^n)$$

$$= \bar{u}^n_{j+1} - \left[\frac{1}{2} + \frac{\lambda \bar{a}_{j+1/2}}{1 + \lambda(\bar{a}_{j+3/2} - \bar{a}_{j+1/2})}\right] \cdot S^n_{j+1},$$

we expand the LHS of (6.8c) around $\bar{u}^n_{j+1}$ to get

$$f(\bar{u}^n_{j+1}) - \frac{1}{2} a(\bar{u}^n_{j+1})(1 + \lambda \bar{a}_{j+3/2}) \hat{S}_{j+1}$$

$$+ \frac{1}{8} a'(u^n_{j+1})[(1 + \lambda \bar{a}_{j+3/2})^2 + (\lambda \bar{a}_{j+1/2})^2] \cdot (\hat{S}_{j+1})^2 + O(h^3) \tag{6.10b}$$

$$= \hat{f}_{j+1/2} + \frac{h^2}{8}(2\lambda^2 a^2 - 1) \cdot a' \cdot (u_x)^2|_{j+1/2} + O(h^3).$$

We see from (6.10a) and (6.10b) that independently of the sign of $\bar{a}_{j+1/2}$, the $O(h^2)$ term in (6.8c) is the same, namely

$$\frac{h^2}{8}(2\lambda^2 a^2 - 1) \cdot a' \cdot (u_x)^2|_{j+1/2}.$$

This completes the proof that the scheme (6.7) is second-order accurate in the sense of (2.9) wherever $u(x, t)$ is smooth, including local extrema and sonic ($f' = 0$) points.

Remarks. (1) The numerical flux (6.7b) can be rewritten as

$$\hat{f}_{j+1/2} = \tfrac{1}{2}[f(v^n_j) + f(v^n_{j+1}) - |\bar{a}_{j+1/2}|(v^n_{j+1} - v^n_j) + \max(0, \bar{a}_{j+1/2}) \cdot (1 - \lambda \bar{a}_{j-1/2}) \cdot \hat{S}_j]$$

$$- [\min(0, \bar{a}_{j+1/2}) \cdot (1 + \lambda \bar{a}_{j+3/2}) \cdot \hat{S}_{j+1}]. \tag{6.11}$$

(2) Our proof that the scheme (6.7) is nonoscillatory is based on the representation of (6.7) as the cell-average (6.2) of the nonoscillatory approximate solution $v_n(x, t)$ in (6.6). To ensure that $v_n(x, t)$ remains univalued for $0 \leqq t \leqq \tau$ we have to restrict the time-step τ so that for all j

$$x_{j+1/2}(\tau) > x_{j-1/2}(\tau). \tag{6.12a}$$

This implies the condition

$$\tau \max_j (\bar{a}_{j-1/2} - \bar{a}_{j+1/2}) \leqq h. \tag{6.12b}$$

On the other hand, to derive the particular form of the numerical flux (6.7b) we have made the assumption

$$|x_{j+1/2}(\tau) - x_{j+1/2}| < h \quad \text{for all } j, \tag{6.13a}$$

which implies the condition

$$\tau \max_j |\bar{a}_{j+1/2}| < h. \tag{6.13b}$$

The two time-step restrictions (6.12b) and (6.13b) are characteristic to Godunov-type schemes. The practice of reconciling the two conditions by

$$\tau \max_j |\bar{a}_{j+1/2}| \leqq h/2$$

is completely unnecessary: The scheme (6.7), as the original Godunov scheme, remains nonoscillatory (or TVD in the case analyzed in [10]) for the full CFL-restriction (6.13b). The reasoning for this observation is as follows: The approximate solution (6.6) under the CFL restriction (6.13b) may fail to be univalued in the jth cell only if $\bar{a}_{j-1/2} > 0$ and $\bar{a}_{j+1/2} < 0$. In this case the numerical fluxes $\hat{f}_{j\pm1/2}$ as defined by cell-averaging in the neighboring cells $j \pm 1$, remain (6.7b). Thus the only thing that needs to be changed in this case is the definition of $v_n(x, \tau)$ in the jth cell.

(3) We observe that once $v_n(x, \tau)$ is defined globally in (6.6) there is no intrinsic need to average it on the original mesh. We may average it on different intervals and still conclude that the resulting approximation is nonoscillatory and conservative. Furthermore, the construction of the interpolant Q, the approximation L and the approximate characteristic field $\bar{a}(x)$ needed to define $v_n(x, t)$, does not depend on the uniformity of the mesh. Therefore the scheme (6.7) generalizes immediately to nonuniform moving meshes. Of particular computational interest are the self-adjusting moving grids of the type described in [12], which make it possible to obtain perfectly resolved shocks and contact discontinuities.

(4) We note that since the approximate solution $v_n(x, t)$ in (6.6) is conservative, it is possible to consider an associated random-choice method obtained by replacing the cell-averaging in (6.2) by a sampling with respect to a variable that is uniformly distributed in the cell, i.e.,

$$v_j^{n+1} = v_n(x_j + \theta_j^n h, \tau)$$

where θ_j^n is uniformly distributed in $[-\frac{1}{2}, \frac{1}{2}]$. Following the reasoning of [7] it is clear that the resulting random-choice method is nonoscillatory and that its limits are weak solutions of (1.1). Although the random-choice approach has many attractive computational features, it has been our experience that in many applications it is possible to accomplish the same computational goals with a self-adjusting moving grid. In this case the use of the latter is preferable as it offers gain in resolution without a loss in pointwise accuracy that is associated with sampling.

7. Numerical illustration. In this section we compare the new uniformly second-order nonoscillatory scheme of this paper (to be referred to as UNO2) to the typical second-order TVD scheme (to be referred to as TVD2). Both schemes can be written in the form (6.7), i.e.,

(7.1a)
$$v_j^{n+1} = v_j^n - \lambda(\hat{f}_{j+1/2} - \hat{f}_{j-1/2}),$$

(7.1b)
$$\hat{f}_{j+1/2} = \begin{cases} f(v_j^n) + \frac{1}{2}\bar{a}_{j+1/2}(1-\lambda\bar{a}_{j-1/2})S_j^n/[1+\lambda(\bar{a}_{j+1/2}-\bar{a}_{j-1/2})] & \text{if } \bar{a}_{j+1/2} \geqq 0, \\ f(v_{j+1}^n) - \frac{1}{2}\bar{a}_{j+1/2}(1+\lambda\bar{a}_{j+3/2})S_{j+1}^n/[1+\lambda(\bar{a}_{j+3/2}-\bar{a}_{j+1/2})] & \text{if } \bar{a}_{j+1/2} \leqq 0, \end{cases}$$

(7.1c)
$$S_j^n = m(S_j^+, S_j^-);$$

here $\bar{a}_{j+1/2}$ is (6.4a) and $m(x, y)$ is the min mod function (3.3). $S_j^\pm$ are different for TVD2 and UNO2:

$$\text{TVD2: } S_j^\pm = d_{j\pm1/2}v^n, \tag{7.2}$$

$$\text{UNO2: } S_j^\pm = d_{j\pm1/2}v^n \mp \tfrac{1}{2}D_{j\pm1/2}v^n, \tag{7.3}$$

where $d_{i+1/2}$ and $D_{i+1/2}$ are defined in (3.2).

UNO2 and TVD2 are both second-order accurate Godunov-type schemes that differ only in the reconstruction step (4.2a):

$$L(x; u) = u_j + S_j(x - x_j)/h \quad \text{for } |x - x_j| < \frac{h}{2}, \tag{7.4}$$

where the slopes of the lines are calculated by (7.3) and (7.2), respectively. Therefore we start our comparison on the approximation level.

In Table 1 and Fig. 1 we present approximations to $u(x) = \sin \pi x$, $-1 \leqq x \leqq 1$. We divide $[-1, 1]$ into N equal intervals and define

$$x_j = -1 + j \cdot \frac{2}{N}, \qquad 0 \leqq j \leqq N. \tag{7.5}$$

TABLE 1
Approximations to $u(x) = \sin(\pi x)$, $-1 \leq x \leq 1$, with periodic boundary conditions.

	L_∞-ERROR			L_1-ERROR		
N	Q	L_{UNO2}	L_{TVD2}	Q	L_{UNO2}	L_{TVD2}
10	1.545×10^{-2}	5.122×10^{-2}	1.420×10^{-1}	1.494×10^{-2}	2.467×10^{-2}	7.016×10^{-2}
20	1.971×10^{-3}	1.231×10^{-2}	3.558×10^{-2}	1.802×10^{-3}	5.576×10^{-3}	1.525×10^{-2}
40	2.476×10^{-4}	3.083×10^{-3}	9.163×10^{-3}	2.148×10^{-4}	1.355×10^{-3}	3.902×10^{-3}
80	3.104×10^{-5}	7.710×10^{-4}	2.308×10^{-3}	2.617×10^{-5}	3.351×10^{-4}	9.787×10^{-4}

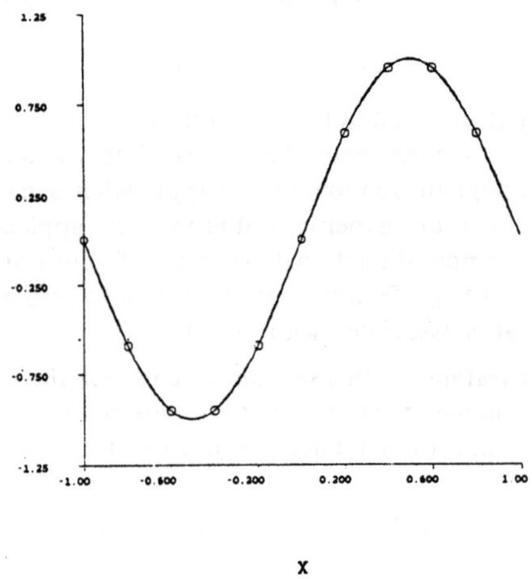

FIG. 1(a). *Approximations of $u = \sin \pi x$, $-1 \leq x \leq 1$, with $N = 10$. Piecewise-parabolic interpolant $Q(x, u)$.*

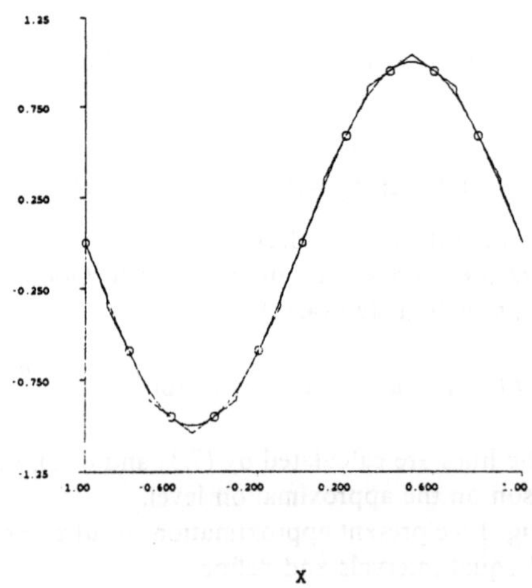

FIG. 1(b). *Approximations of $u = \sin \pi x$, $-1 \leq x \leq 1$, with $N = 10$. Piecewise-linear approximation $L^{UNO2}(x; u)$.*

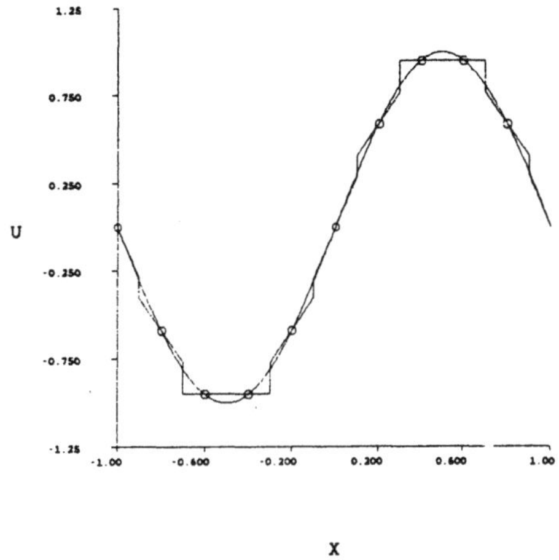

FIG. 1(c). *Approximations of* $u = \sin \pi x$, $-1 \leq x \leq 1$, *with* $N = 10$. *Piecewise-linear approximation* $L^{TVD2}(x; u)$.

The symbols in Fig. 1 denote values of $u_j = \sin \pi x_j$ for $N = 10$ in (7.5). In Fig. 1(a) we show the piecewise-parabolic interpolant $Q(x; u)$ (see § 3). In Fig. 1(b) we show the piecewise-linear approximation $L^{UNO2}(x; u)$ which is (7.4) with (7.1c) and (7.3). In Fig. 1(c) we show the piecewise-linear approximation $L^{TVD2}(x; u)$ which is (7.4) with (7.1c) and (7.2). We make the following observations regarding Fig. 1: (i) Q is a better approximation than L^{UNO2}; L^{UNO2} is a better approximation than L^{TVD2}. (ii) $TV(L^{UNO2}) > TV(u) > TV(L^{TVD2})$. In Table 1 we quantify the first observation; we list the L_∞-error and the L_1-error of these approximations to $\sin \pi x$ for a refinement sequence of $N = 10$, 20, 40, 80 in (7.5). Clearly Q is an $O(h^3)$ approximation, while L^{UNO2} and L^{TVD2} are $O(h^2)$. The error in L^{UNO2} is about a $\frac{1}{3}$ of the error in L^{TVD2}.

In Table 2 and Fig. 2 we present solutions of UNO2 and TVD2 for the constant coefficient case

$$u_t + u_x = 0, \quad u(x, 0) = \sin \pi x, \quad -1 \leq x \leq 1, \tag{7.6}$$

with periodic boundary conditions, at $t = 2$ with $\tau/h = 0.8$. Figures 3(a) and 3(b) show UNO2 and TVD2, respectively, with $N = 20$ in (7.5). In Table 2 we list the L_∞-error and L_1-error for a refinement sequence with $N = 20$, 40, 80, 160. Clearly UNO2 is second-order accurate in both L_∞ and L_1, while TVD2 is second-order accurate in L_1 but only first-order accurate in L_∞. Figure 3(b) demonstrates that the degeneracy to

TABLE 2
Numerical solutions of $u_t + u_x = 0$, $u(x, 0) = \sin \pi x$, $-1 \leq x \leq 1$ *at* $t = 2$ *with periodic boundary conditions and* $\tau/h = 0.8$.

	L_∞-ERROR		L_1-ERROR	
N	UNO2	TVD2	UNO2	TVD2
20	7.097×10^{-3}	8.119×10^{-2}	8.944×10^{-3}	6.778×10^{-2}
40	1.607×10^{-3}	3.477×10^{-2}	2.044×10^{-3}	2.033×10^{-2}
80	3.870×10^{-4}	1.453×10^{-2}	4.926×10^{-4}	5.626×10^{-3}
160	9.201×10^{-5}	5.975×10^{-3}	1.172×10^{-4}	1.528×10^{-3}

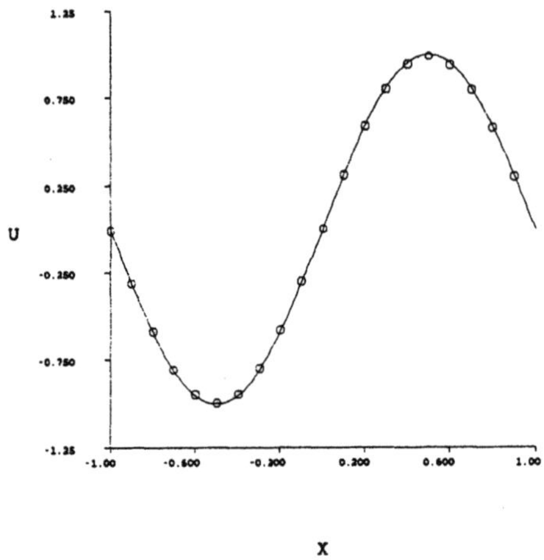

(a)

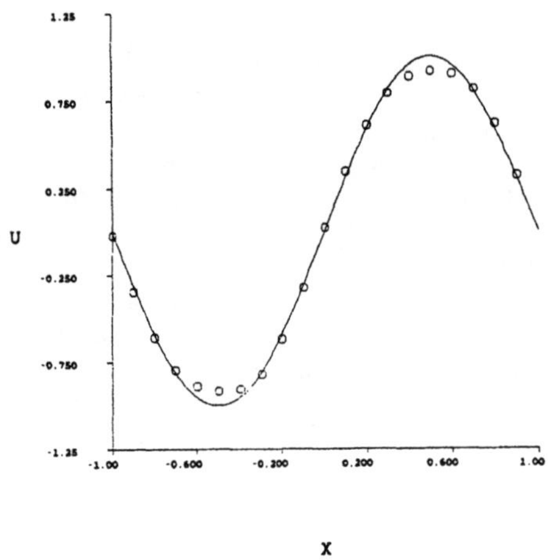

(b)

FIG. 2. *Numerical solutions of* $u_t + u_x = 0$, $u(x, 0) = \sin \pi x$ *at* $t = 2$, *with* $N = 20$ *and* $\tau/h = 0.8$. (a) UNO2. (b) TVD2.

first-order accuracy at local extrema in the TVD scheme adversely affects the accuracy everywhere. (Because the scheme is TVD it cannot switch abruptly to second-order accuracy as this would introduce oscillations; consequently it takes quite a while to recover the second-order accuracy.)

Next we approximate the discontinuous function

$$u(x) = \begin{cases} -x \sin(3\pi x^2/2), & -1 < x < -\frac{1}{3}, \\ |\sin(2\pi x)|, & |x| < \frac{1}{3}, \\ 2x - 1 - \frac{1}{6}\sin(3\pi x), & \frac{1}{3} < x < 1, \end{cases} \tag{7.7}$$

which we extend to have period 2 outside $[-1, 1]$.

In Fig. 3 we present approximations to $\phi(x)$, using $N = 20$. Figure 3(a) shows $Q(x; \bar{u})$, Fig. 3(b) shows $L^{\text{UNO2}}(x; \bar{u})$, and Fig. 3(c) shows $L^{\text{TVD2}}(x; \bar{u})$. We again observe that Q is a better approximation than L^{UNO2}, while L^{UNO2} is a better approximation than L^{TVD2}.

In Fig. 4 we present solutions of UNO2 and TVD2 for the constant coefficient problem (7.6), initial data given by (7.2), and periodic boundary conditions. We take $t = 2$ and $\tau/h = 0.8$. Figures 4(a) and 4(b) show UNO2 and TVD2, respectively, with $N = 40$. Figure 4(b) shows the damping effect that the TVD scheme imposes due to its degeneracy to first-order accuracy at local extrema.

In Fig. 5 we solve the same problems, except we impose boundary conditions. At $x = -1$ we impose the given function (7.7) evaluated at $-1 - t$. No boundary conditions are imposed at $x = 1$. We implement this numerically using UNO2 and TVD2 except

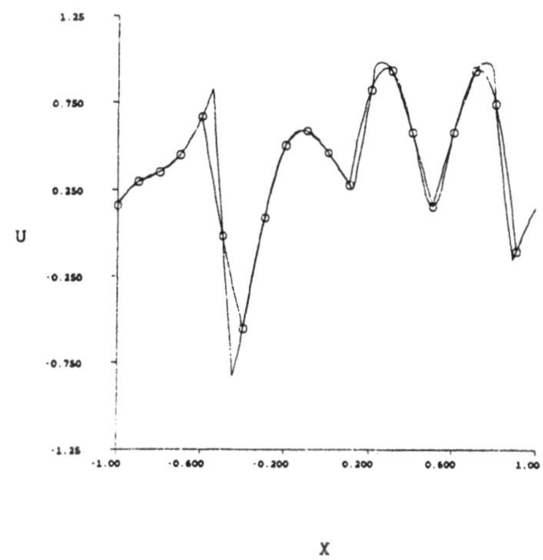

FIG. 3(a). $Q(x; \bar{u})$ *for u given* (7.7) *with* $N = 20$.

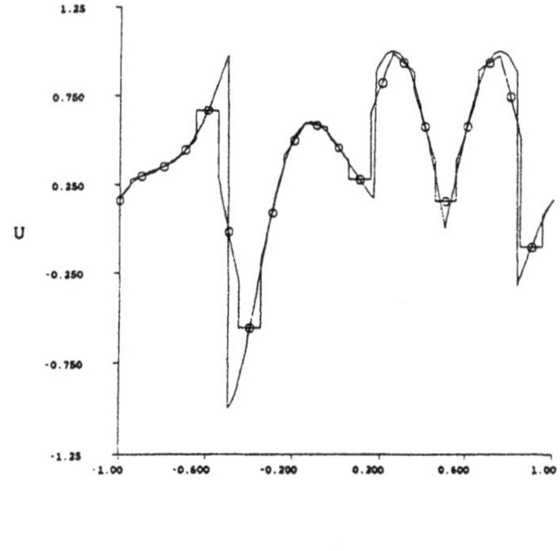

FIG. 3(b). $L^{\text{UNO2}}(x; \bar{u})$ *for u given by* (7.7) *with* $N = 20$.

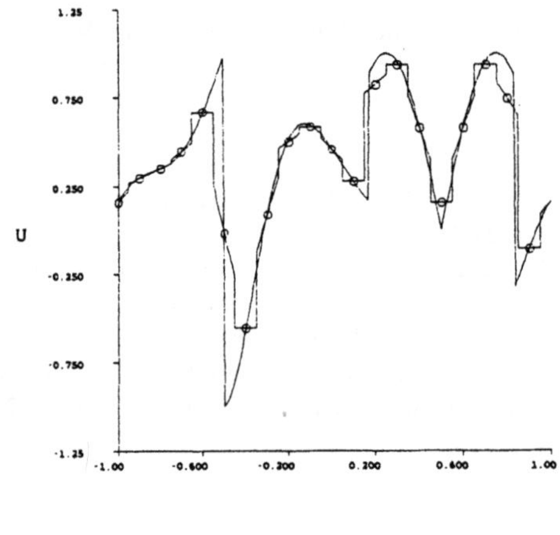

FIG. 3(c). $L^{TVD2}(x;\bar{u})$ *for* u *given by* (7.7) *with* $N=20$.

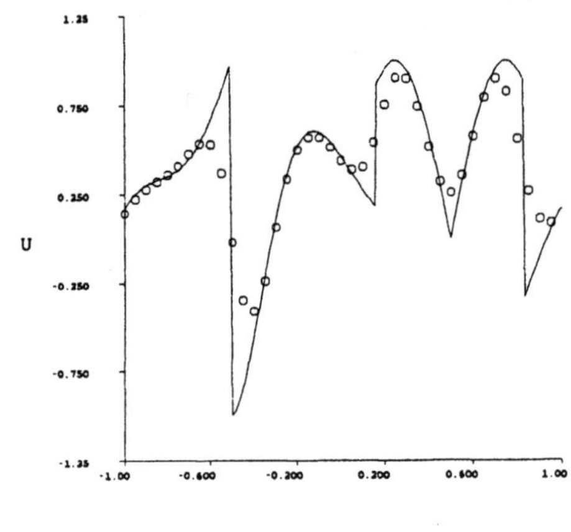

FIG. 4(a). *Numerical solution of* $u_t+u_x=0$, $u(x,0)$ *defined by* (7.7) *at* $t=2$ *with periodic boundary conditions* $N=40$ *and* $\tau/h=0.8$, *using* UNO2.

at the boundary points. There we are, in general, unable to construct nonoscillatory piecewise-parabolic interpolants $Q(x,\bar{u})$, so we construct the only possible parabolic interpolant through x_i, x_{i+1} and the point to either the left or right which lies in the region. The analogous procedure is carried out at the reconstruction stage. Figures 5(a) and 5(b) again show the results at $t=2$ with $\tau/h=0.8$.

The possible introduction of oscillations through the boundary conditions does not seem to have degraded the performance of either scheme (in fact the opposite is observed). Again the TVD2 scheme shows a damping effect.

In Table 3 and Fig. 6 we present results for Burgers' equation

$$u_t+uu_x=0,\quad u(x,0)=\alpha+\sin\pi(x+\beta),\quad -1\leqq x\leqq 1, \tag{7.8}$$

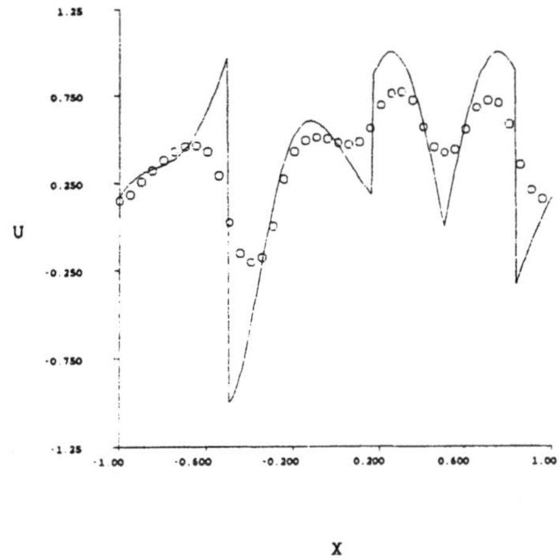

FIG. 4(b). *Same as Fig.* 4(a) *using* TVD2.

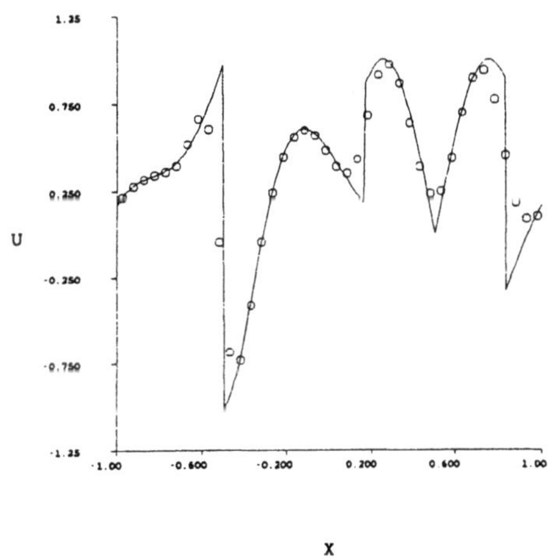

FIG. 5(a). *Same as Fig.* 4(a) *except with given data at* $x=-1$, *and outflow at* $x=1$.

with periodic boundary conditions and $\tau/h(1+|\alpha|)=0.5$. The solution to (7.7) is smooth for $t<1/\pi$; at $t=1/\pi$ it develops shocks. In Table 3(a) we list the L_∞-error and L_1-error of UNO2 and TVD2 at $t=0.15$ for $\alpha=\beta=0$ in (7.7). This table shows the same asymptotic behavior as Table 2, except that because of the large gradients it shows for a smaller h.

In Figs. 6(a) and 6(b) we show results of UNO2 and TVD2 for (7.8) with $\alpha=2$ and $\beta=1$ at $t=0.35$ with $N=20$. In this case the solution to (7.8) develops a shock moving with speed 2 beginning at time $t=1/\pi\approx 0.318$.

In Table 3(b) and Figs. 6(c) and 6(d) we repeat the previous calculations for the schemes (2.13):

(7.9a) $$v_j^{n+1}=v_j^n-\lambda(\hat{f}_{j+1/2}-\hat{f}_{j-1/2}),$$

(7.9b) $$\hat{f}_{j+1/2}=f(v_n(x_{j+1/2},\tau/2))=f(L(x_0(x_{j+1/2},\tau/2),v^n))$$

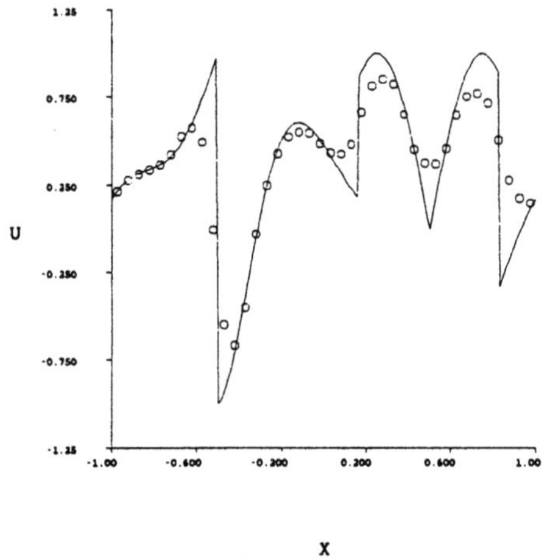

FIG. 5(b). *Same as Fig.* 4(b) *except with given data at* $x=-1$, *and outflow at* $x=1$.

TABLE 3(a)

Numerical solutions of $u_t + uu_x = 0$, $u(x,0) = \sin \pi x$, *at* $t = 0.15$ *and* $\tau/h = 0.5$ – UNO2 *and* TVD2.

	L_∞-ERROR		L_1-ERROR	
N	UNO2	TVD2	UNO2	TVD2
20	1.890×10^{-2}	2.238×10^{-2}	1.090×10^{-2}	1.854×10^{-2}
40	5.712×10^{-3}	1.054×10^{-2}	3.034×10^{-3}	5.051×10^{-3}
80	1.552×10^{-3}	4.422×10^{-3}	7.771×10^{-4}	1.340×10^{-3}
160	3.985×10^{-4}	1.837×10^{-3}	1.965×10^{-4}	3.621×10^{-4}

TABLE 3(b)

Same as Table 3(a) *for* FNO2 *and* FVD2.

	L_∞-ERROR		L_1-ERROR	
N	FNO2	FVD2	FNO2	FVD2
20	6.938×10^{-3}	2.091×10^{-2}	3.726×10^{-3}	1.322×10^{-2}
40	1.050×10^{-3}	1.054×10^{-2}	8.869×10^{-4}	3.835×10^{-3}
80	5.106×10^{-4}	4.424×10^{-3}	2.163×10^{-4}	1.072×10^{-3}
160	1.251×10^{-4}	1.837×10^{-3}	5.270×10^{-5}	2.946×10^{-4}

TABLE 3(c)

Same as Table 3(a) *for* ANO2.

N	L_∞-ERROR	L_1-ERROR
20	2.249×10^{-2}	1.221×10^{-2}
40	6.623×10^{-3}	3.243×10^{-3}
80	1.781×10^{-3}	8.259×10^{-4}
160	4.597×10^{-4}	2.079×10^{-4}

where $x_0(x_{j+1/2}, \tau/2)$ is (6.5c), i.e.,

$$\text{(7.9c)} \qquad \hat{f}_{j+1/2} = \begin{cases} f\left(v_j^n + \dfrac{1}{2} \cdot \dfrac{1-\lambda(\hat{a}_{j+1/2}+\hat{a}_{j-1/2})/2}{1+\lambda(\hat{a}_{j+1/2}-\hat{a}_{j-1/2})/2} \cdot S_j^n \right) & \text{if } \hat{a}_{j+1/2} \geqq 0, \\ f\left(v_{j+1}^n - \dfrac{1}{2} \cdot \dfrac{1+\lambda(\hat{a}_{j+3/2}+\hat{a}_{j+1/2})/2}{1+\lambda(\hat{a}_{j+3/2}-\hat{a}_{j+1/2})/2} \cdot S_{j+1}^n \right) & \text{if } \hat{a}_{j+1/2} \leqq 0. \end{cases}$$

As we have remarked in § 2, v_j^{n+1} in (7.9a) is not a cell-average of $v_n(x, \tau)$, but only an approximation to it. Therefore it is not necessary to take $\hat{a}_{j+1/2}$ in (7.8c) to be (6.4a). We choose $\hat{a}_{j+1/2}$ so that (7.9c) is continuous at $\hat{a}_{j+1/2} = 0$:

$$\text{(7.9d)} \quad \hat{a}_{j+1/2} = [f(v_{j+1}^n - \tfrac{1}{2}S_{j+1}^n) - f(v_j^n + \tfrac{1}{2}S_j^n)]/[(v_{j+1}^n - \tfrac{1}{2}S_{j+1}^n) - (v_j^n + \tfrac{1}{2}S_j^n)].$$

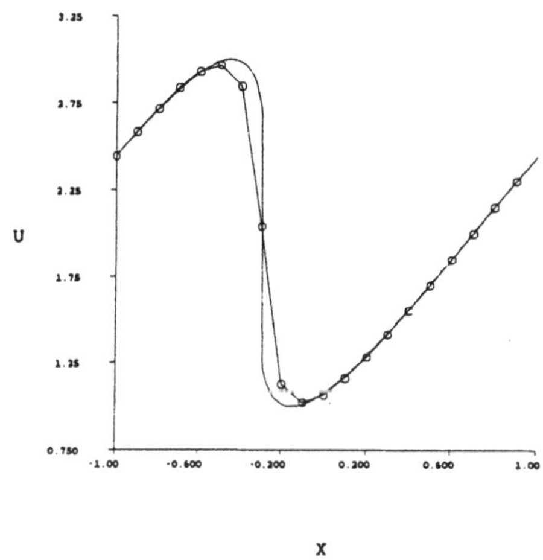

FIG. 6(a). *Solution to* (7.8) *with* $\alpha = 2$, $\beta = 1$, CFL = .5, $N = 20$, *at* $t = 0.35$ *using* UNO2

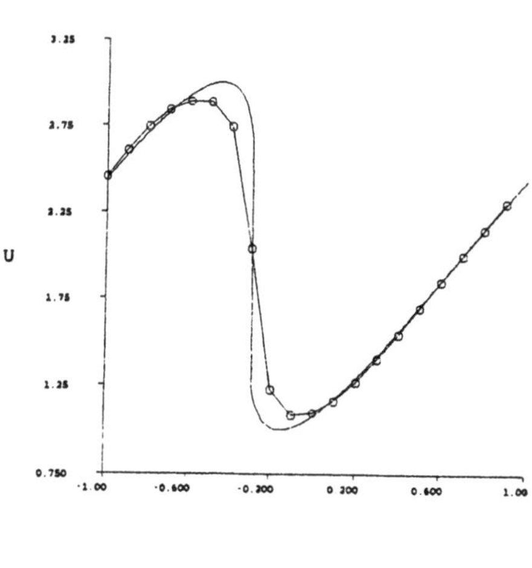

FIG. 6(b). *Same as Fig.* 6(a) *using* TVD2.

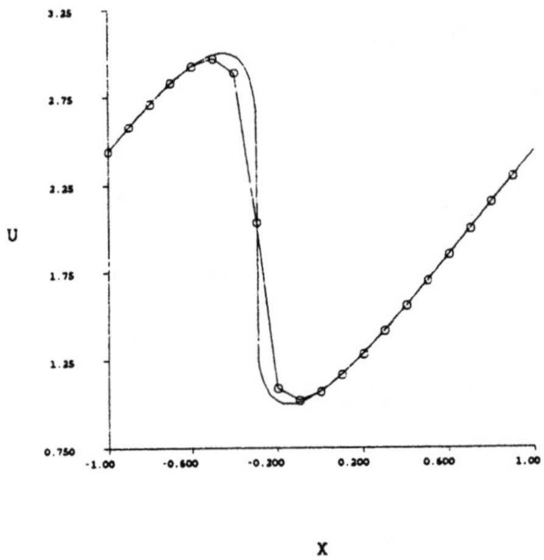

FIG. 6(c). *Same as Fig.* 6(a) *using* FNO2.

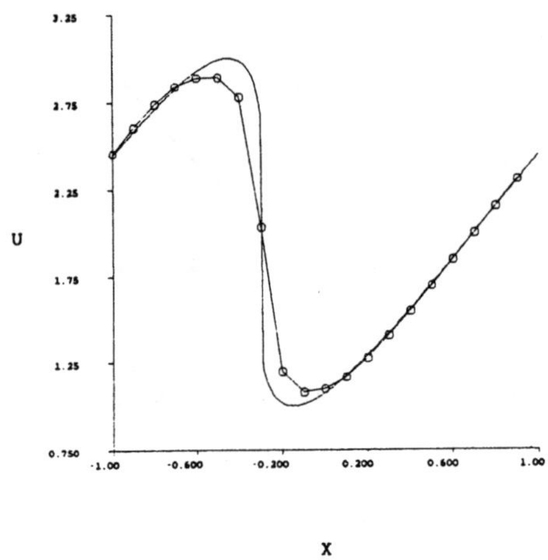

FIG. 6(d). *Same as Fig.* 6(a) *using* FVD2.

We denote the schemes (7.9) with S_j^n defined by (7.1c) and either (7.2) or (7.3) by FVD2 and FNO2, respectively. We note that (7.9) is identical to (7.1) in the constant coefficient case, and consequently FVD2 and FNO2 are nonoscillatory in the constant coefficient case. Figures 6(c) and 6(d) show that FNO2 and FVD2 are also nonoscillatory in the case (7.8). Furthermore, Table 3(b) shows that FNO2 is much more accurate than UNO2 (FVD2 is about the same as TVD2).

In all previous examples we have presented pointwise calculations; namely, we have initialized the numerical solution by taking v_j^0 to be the value of the initial data at x_j, and we have considered v_j^n to be an approximation to $u(x_j, t_n)$. (Surely this is an acceptable practice for second-order accurate schemes.) In Table 3(c) we repeat the calculation for UNO2 in Table 3(a), but now in a sense of cell-averages and denote it by ANO2. Now we initialize UNO2 for (7.8) with $\alpha=\beta=0$ by cell-averages of the initial data, i.e.,

$$\text{(7.10a)} \qquad \nu_j^0 = -\frac{1}{\pi h}[\cos(\pi x_{j+1/2}) - \cos(\pi x_{j-1/2})] = \frac{\sin(\pi h/2)}{(\pi h/2)} \cdot \sin(\pi x_j),$$

and regard ν_j^n to represent cell-averages of $u(x, t_n)$. To obtain a pointwise approximation to $u(x, t_n)$ we first compute point values $V_{j+1/2}^n$ of its indefinite integral $u^n(x_{j+1/2}) = \int_{x_0}^{x_{j+1/2}} u(y, t_n)\, dy$ by

$$\text{(7.10b)} \qquad V_{j+1/2}^n = h \sum_{i=i_0}^{j} \nu_i^n.$$

Next we obtain a global piecewise-linear approximation $\nu(x, t_n)$ to $u(x, t_n)$ by

$$\text{(7.10c)} \qquad \nu(x, t_n) = \frac{d}{dx} Q(x; V^n)$$

where Q is the piecewise-parabolic interpolant of § 3. Finally we get

$$\text{(7.10d)} \qquad \nu(x_j, t_n) = \frac{d}{dx} Q(x_j; V^n) = \frac{1}{h}(V_{j+1/2}^n - V_{j-1/2}^n) = \nu_j^n.$$

Thus the only difference between ANO2 in Table 3(c), and UNO2 in Table 3(a) is the initialization (7.10a), which itself differs only slightly from the mesh values of the initial data (since $\sin(\pi h/2)/(\pi h/2) = 1 - \frac{1}{6}(\pi h/2)^2 + O(h^4)$).

We remark that cell-averages do play a significant role when the initial data is discontinuous (since they provide information about the location of the discontinuity) and in higher-order Godunov-type schemes; this will be described elsewhere.

We turn now to present calculations with a formal extension of UNO2 and TVD2 for systems of conservation laws. We consider a Riemann problem for the Euler equations of gas dynamics

$$\text{(7.11a)} \qquad u_t + f(u)_x = 0, \qquad u(x, 0) = \begin{cases} u_L, & x < 0, \\ u_R, & x > 0, \end{cases}$$

$$\text{(7.11b)} \qquad u = (\rho, m, E)^T, \quad f(u) = (m, m^2/\rho + P, m(E+P)/\rho)^T,$$

$$\text{(7.11c)} \qquad P = (\gamma - 1)(E - \tfrac{1}{2} m^2/\rho).$$

Here ρ, m, E and P are the density, momentum, total energy and pressure, respectively; we take $\gamma = 1.4$.

In the following we apply the extension technique of [3] to UNO2 and TVD2. The idea is to extend UNO2 and TVD2 to systems in such a way that will be identical to (7.1) in the scalar case, and will decouple into (5.3) for each of the characteristic variables in the constant coefficient system case. To accomplish that we use Roe's averaging for (7.11) (see [13])

$$\text{(7.12a)} \qquad \nu_{j+1/2} = V(\nu_j^n, \nu_{j+1}^n)$$

for which

$$\text{(7.12b)} \qquad f(\nu_{j+1}^n) - f(\nu_j^n) = A(\nu_{j+1/2})(\nu_{j+1}^n - \nu_j^n), \qquad A(u) = \partial f/\partial u,$$

and define local characteristic variables with respect to the right-eigenvector system $\{R_{j+1/2}^k\}_{k=1}^3$ of $A(\nu_{j+1/2})$. We extend (6.11) to systems as follows:

$$\text{(7.13a)} \qquad \nu_j^{n+1} = \nu_j^n - \lambda(\hat{f}_{j+1/2} - \hat{f}_{j-1/2}),$$

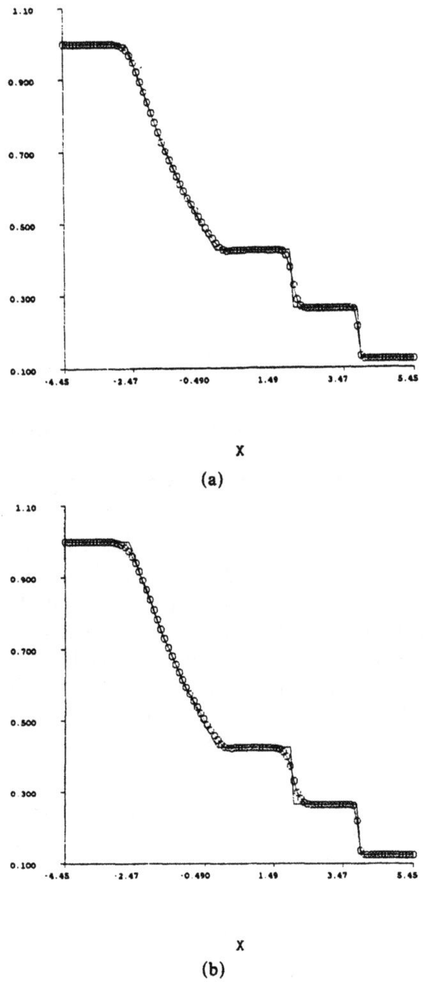

FIG. 7. *Numerical solution of density in a Riemann problem for Euler's equations.* (a) UNO2. (b) TVD2.

(7.13b) $$\hat{f}_{j+1/2}=\frac{1}{2}\left[f(v_j^n)+f(v_{j+1}^n)-\sum_{k=1}^{3} c_{j+1/2}^k R_{j+1/2}^k\right],$$

(7.13c) $$\begin{aligned} c_{j+1/2}^k=|\bar{a}_{j+1/2}^k| d_{j+1/2}^k w-\max(0,\bar{a}_{j+1/2}^k)(1-\lambda\bar{a}_{j-1/2}^k)\hat{S}_j^k \\ +\min(0,\bar{a}_{j+1/2}^k)(1+\lambda\bar{a}_{j+3/2}^k)\hat{S}_{j+1}^k. \end{aligned}$$

Here $\bar{a}_{j+1/2}^k$ is the kth eigenvalue of $A(v_{j+1/2})$ corresponding to $R_{j+1/2}^k$, and $d_{j+1/2}^k w$ denotes the component of $d_{j+1/2}v=v_{j+1}^n-v_j^n$ in the kth characteristic field, i.e.,

(7.13d) $$d_{j+1/2}v=\sum_{k=1}^{3}(d_{j+1/2}^k w)R_{j+1/2}^k.$$

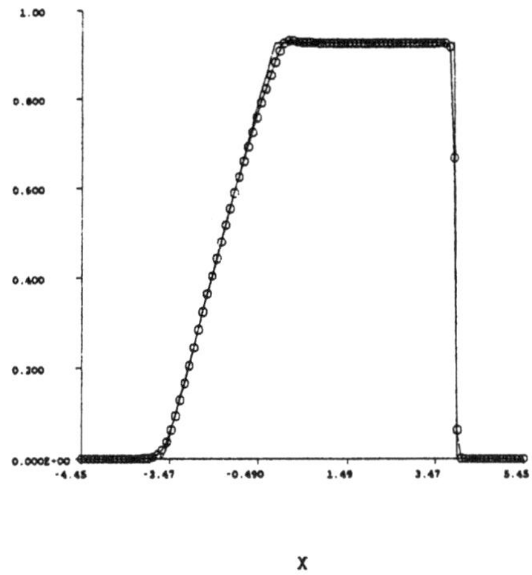

(a)

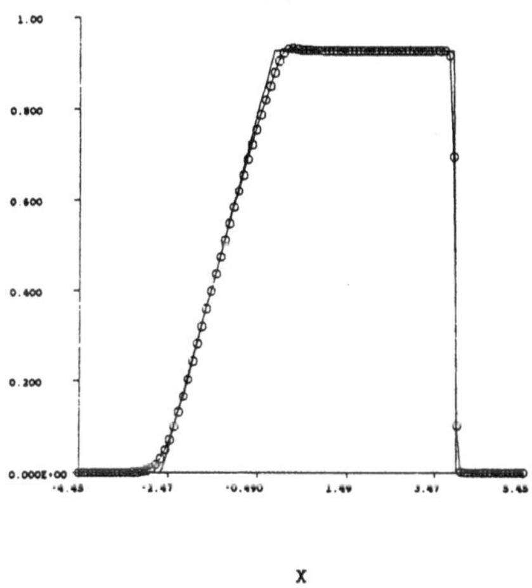

(b)

FIG. 8. *Numerical solution of velocity in a Riemann problem for Euler's equations.* (a) UNO2. (b) TVD2.

Likewise $\hat{S}_j^k$ denotes the component of the vector of slopes in the kth characteristic field, and is defined as follows:

$$\hat{S}_j^k = m(S_{-,j}^k, S_{+,j}^k)/[1+\lambda(\bar{a}_{j+1/2}^k - \bar{a}_{j-1/2}^k)]; \tag{7.13e}$$

$m(x, y)$ is the min mod function (3.3). $S_{\pm,j}^k$ are different for VD2 and NO2:

$$\text{TVD2:} \quad S_{\pm j}^k = d_{j\pm 1/2}^k w, \tag{7.14}$$

$$\text{UNO2:} \quad S_{\pm,j}^k = d_{j+1/2}^k w \mp \tfrac{1}{2} D_{j\pm 1/2}^k w; \tag{7.15}$$

$$D_{i\pm 1/2}^k w = m(d_{i+3/2}^k w - d_{i+1/2}^k w, d_{i+1/2}^k w - d_{i-1/2}^k w).$$

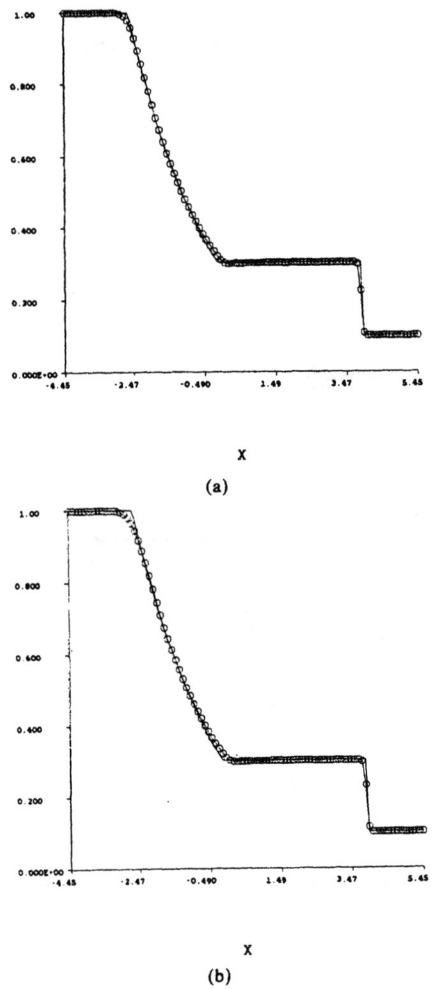

FIG. 9. *Numerical solution of pressure in a Riemann problem for Euler's equations.* (a) UNO2. (b) TVD2.

In Figs. 7, 8 and 9 we show numerical solutions of UNO2 and TVD2, respectively, for the Riemann problem (7.1b) with

$$U_L = (1, 0, 2.5)^T, \qquad U_R = (0.125, 0, 0.25).$$

These figures demonstrate that the formal extension to systems is nonoscillatory in this case. Since the solution to the Riemann problem is just constant states separated by waves we do not get to see here the extra resolution power of UNO2, except that its numerical solution is somewhat "crisper" than that of TVD2. In this calculation we have not employed any artificial compression in the linearly degenerate field and therefore the contact discontinuity smears like $n^{1/3}$, as expected. The interested reader is referred to [4], [5] and [10] for a detailed description of such compression techniques, as well as for details of entropy enforcement mechanisms.

REFERENCES

[1] S. R. CHAKRAVARTHY AND S. OSHER, *A new class of high accuracy* TVD *schemes for hyperbolic conservation laws*, AIAA paper 85-0363, Reno, NV, 1985.

[2] P. COLELLA AND P. R. WOODWARD, *The piecewise-parabolic method* (PPM) *for gas-dynamical simulations*, J. Comput. Phys., 54 (1984), pp. 174-201.

[3] A. HARTEN, *On a class of high resolution total-variation-stable finite-difference schemes*, this Journal, 21 (1984), pp. 1-23.

[4] ———, *High resolution schemes for hyperbolic conservation laws*, J. Comput. Phys., 49 (1983), pp. 357-393.

[5] ———, *On second-order accurate Godunov-type schemes*, to appear.

[6] A. HARTEN AND J. M. HYMAN, *A self-adjusting grid for the computation of weak solutions of hyperbolic conservation laws*, J. Comput. Phys., 50 (1983), pp. 235-269.

[7] A. HARTEN AND P. D. LAX, *A random choice finite difference scheme for hyperbolic conservation laws*, this Journal, 18 (1981), pp. 289-315.

[8] A. JAMESON AND P. D. LAX, *Conditions for the construction of multi-point total variation diminishing difference schemes*, Princeton Univ., MAE Report 1650, 1984.

[9] S. OSHER, *Convergence of generalized* MUSCL *schemes*, NASA Langley Contractor Report 172306 1984; this Journal, 22 (1985), pp. 947-961.

[10] S. OSHER AND S. R. CHAKRAVARTHY, *High-resolution schemes and the entropy condition*, this Journal, 21 (1984), pp. 955-984.

[11] ———, *Very high order accurate* TVD *schemes*, ICASE Report #84-44, 1984.

[12] S. OSHER AND E. TADMOR, *On the convergence of difference approximations to conservation laws*, Math. Comp., submitted.

[13] P. L. ROE, *Approximate Riemann solvers, parameter vectors, and difference schemes*, J. Comput. Phys., 43 (1981), pp. 357-372.

[14] P. K. SWEBY, *High resolution schemes using flux limiters for hyperbolic conservation laws*, this Journal, 21 (1984), pp. 995-1011.

[15] B. VAN LEER, *Towards the ultimate conservative finite difference scheme*, II. *Monotonicity and conservation combined in a second-order scheme*, J. Comput. Phys., 14 (1974), pp. 361-376.

[16] ———, *Towards the ultimate conservative finite difference scheme*, IV. *A new approach to numerical convection*, J. Comput. Phys., 23 (1977), pp. 276-298.

Uniformly High Order Accurate Essentially Non-oscillatory Schemes, III

AMI HARTEN*

Department of Mathematics, UCLA, and School of Mathematical Sciences, Tel-Aviv University, Israel

BJORN ENGQUIST* AND STANLEY OSHER*

Department of Mathematics, UCLA

AND

SUKUMAR R. CHAKRAVARTHY

Rockwell Science Center, Thousand Oaks, California

Received April 11, 1986; revised October 13, 1986

DEDICATED TO PETER LAX ON HIS 60TH BIRTHDAY

We continue the construction and the analysis of essentially non-oscillatory shock capturing methods for the approximation of hyperbolic conservation laws. We present an hierarchy of uniformly high-order accurate schemes which generalizes Godunov's scheme and its second-order accurate MUSCL extension to an arbitrary order of accuracy. The design involves an essentially non-oscillatory piecewise polynomial reconstruction of the solution from its cell averages, time evolution through an approximate solution of the resulting initial value problem, and averaging of this approximate solution over each cell. The reconstruction algorithm is derived from a new interpolation technique that, when applied to piecewise smooth data, gives high-order accuracy whenever the function is smooth but avoids a Gibbs phenomenon at discontinuities. Unlike standard finite difference methods this procedure uses an adaptive stencil of grid points and, consequently, the resulting schemes are highly non-linear.

1. INTRODUCTION

In this paper, the third in a series, we continue to study the use of essentially non-oscillatory, uniformly high-order accurate schemes for the numerical approximation of weak solutions of hyperbolic systems of conservation laws

$$u_t + f(u)_x = 0 \tag{1.1a}$$

$$u(x, 0) = u_0(x). \tag{1.1b}$$

* Researh supported by NSF Grant DMS85-03294, ARO Grant DAAG 29-85-K-0190, NASA Consortium Agreement NCA 2-IR390-403, and NASA Langley Grant NAG 1-270.

231

Here $u=(u_1,..., u_m)^T$ is a state vector and $f(u)$, the flux, is a vector valued function of m components. The system is hyperbolic in the sense that the $m\times m$ Jacobian matrix

$$A(u)=\partial f/\partial u$$

has m real eigenvalues

$$a_1(u)\leqslant a_2(u)\leqslant \cdots \leqslant a_m(u)$$

and a complete set of m linearly independent right-eigenvectors $\{r_k(u)\}_{k=1}^m$. We denote by $\{l_k(u)\}_{k=1}^m$ the left-eigenvectors of $A(u)$ and assume that $l_i r_k=\delta_{ik}$.

We assume that the initial value problem (IVP) (1.1) (embedded in an appropriate setting which includes entropy considerations) is well posed in the sense of Cauchy and that its weak solutions are generically piecewise smooth. We denote its evolution operator by $E(t)$, i.e.,

$$u(\cdot, t)=E(t)\cdot u_0. \tag{1.2}$$

Let $\bar{w}(x)$ denote the sliding average of $w(x)$

$$\bar{w}(x)\equiv\frac{1}{h}\int_{-h/2}^{h/2} w(x+y)\,dy\equiv(A_h\cdot w)(x) \tag{1.3a}$$

We note that $\bar{w}$ is smoother than w by one derivative, and that at the points of smoothness

$$\bar{w}(x)=w(x)+O(h^2). \tag{1.3b}$$

The sliding average in x of a weak solution of (1.1), $\bar{u}(x, t)$, satisfies

$$\frac{\partial}{\partial t}\bar{u}(x, t)+\frac{1}{h}\left[f(u(x+h/2, t))-f(u(x-h/2, t))\right]=0. \tag{1.4}$$

Integrating this relation from t to $t+\tau$, we get

$$\bar{u}(x, t+\tau)=\bar{u}(x, t)-\lambda[\hat{f}(x+h/2, t; u)-\hat{f}(x-h/2, t; u)], \tag{1.5a}$$

where $\lambda=\tau/h$ and

$$\hat{f}(x, t; w)=\frac{1}{\tau}\int_0^\tau f(w(x, t+\eta))\,d\eta. \tag{1.5b}$$

Let $\{I_j\times[t_n, t_{n+1}]\}$, where $I_j=[x_{j-1/2}, x_{j+1/2}]$, and $x_\alpha=\alpha h$, $t_n=n\tau$, be a partition of $\mathbf{R}\times\mathbf{R}^+$. Writing relation (1.5) at $x=x_j$, $t=t_n$ we get

$$\bar{u}_j^{n+1}=\bar{u}_j^n-\lambda[\hat{f}(x_{j+1/2}, t_n; u)-\hat{f}(x_{j-1/2}, t_n; u)]. \tag{1.6a}$$

Here

$$\bar{u}_j^n = \bar{u}(x_j, t_n) = \frac{1}{h} \int_{I_j} u(x, t_n)\, dx \tag{1.6b}$$

is the "cell-average" of u at time t_n.

In this paper we describe a class of numerical schemes that generalizes Godunov's scheme [5] and its second-order extensions [22, 4, 15] to any finite order of accuracy. These schemes can be written in standard conservation form

$$v_j^{n+1} = v_j^n - \lambda(\bar{f}_{j+1/2} - \bar{f}_{j-1/2}) \equiv (\bar{E}_h(\tau) \cdot v^n)_j. \tag{1.7a}$$

Here $\bar{E}_h(\tau)$ denotes the numerical solution operator and $\bar{f}_{j+1/2}$, the numerical flux, denotes a function of $2k$ variables

$$\bar{f}_{j+1/2} = \bar{f}(v_{j-k+1}^n, \ldots, v_{j+k}^n), \tag{1.7b}$$

which is consistent with the flux $f(u)$ in (1.1) in the sense that $\bar{f}(u, u, \ldots, u) = f(u)$. We design these schemes so that the conservation form (1.7a) will approximate (1.5) to a high order of accuracy. Setting $v_j^n \equiv \bar{u}_j^n$ in (1.7) and comparing it to (1.6) we see that if the numerical flux (1.7b) can be expanded as

$$\bar{f}_{j+1/2} = \hat{f}(x_{j+1/2}, t_n; u) + d(x_{j+1/2})\, h^r + O(h^{r+1}) \tag{1.8a}$$

then

$$v_j^{n+1} = \bar{u}_j^{n+1} - \lambda[d(x_{j+1/2}) - d(x_{j-1/2})]\, h^r + O(h^{r+1}).$$

This shows that if the numerical flux $\bar{f}_{j+1/2}$ satisfies (1.8a) then the truncation error in the sense of cell averages is

$$\bar{u}(x_j, t_n + \tau) - [\bar{E}_h(\tau)\, \bar{u}(\cdot, t_n)]_j = \lambda[d(x_{j+1/2}) - d(x_{j-1/2})]\, h^r + O(h^{r+1}), \tag{1.8b}$$

which is $O(h^{r+1})$ where $d(x)$ is Lipschitz continuous.

When $f(u)$ is a nonlinear function of u, the approximation of $\hat{f}(x_{j+1/2}, t_n; u)$ to $O(h^r)$ requires knowledge of pointwise values of the solution to the same order of accuracy. In order to design a numerical flux that satisfies (1.8a), we must extract high order accurate pointwise information from the given $\{v_j^n\}$, which are approximations to $\{\bar{u}_j^n\}$, the cell averages (1.6b) of the solution. Solving this reconstruction problem to an arbitrarily high order of accuracy r, without introducing $O(1)$ Gibbs-like spurious oscillations at points of discontinuity, is the most important step in the design of our new schemes.

Given $\bar{w}_j = \bar{w}(x_j)$, cell averages of a piecewise smooth function $w(x)$, we construct $R(x; \bar{w})$, a piecewise polynomial function of x of uniform polynomial degree $(r-1)$ that satisfies:

(i) At all points x for which there is a neighborhood where w is smooth

$$R(x; \bar{w}) = w(x) + e(x)\,h^r + O(h^{r+1}). \tag{1.9a}$$

(ii) Conservation in the sense of

$$\bar{R}(x_j; \bar{w}) = \bar{w}_j \tag{1.9b}$$

here $\bar{R}$ denotes the sliding average (1.3) of R.

(iii) It is essentially non-oscillatory

$$TV(R(\cdot; \bar{w})) \leqslant TV(w) + O(h^r), \tag{1.9c}$$

where TV denotes total variation in x.

The inequality (1.9c) implies that the reconstruction R is essentially non-oscillatory in the sense that it does not have a Gibbs-like phenomenon of generating $O(1)$ spurious oscillations at points of discontinuity that are proportional to the size of the jump there. In [16, 11, 17] we describe $R(x; \bar{w})$ in the scalar case. We show there that R may occasionally produce $O(h^r)$ spurious oscillations which are on the level of the truncation error. These small spurious oscillations may occur only in the smooth part of w and they usually disappear once $w(x)$ is adequately resolved on the computational mesh. For sake of completeness we review this reconstruction algorithm in Section 3; we shall extend it to vector functions $w(x)$ in Section 5 of this paper.

Using the reconstruction (1.9) we can express the abstract form of our new schemes by

$$\bar{E}_h(\tau) \cdot \bar{w} \equiv A_h \cdot E(\tau) \cdot R(\cdot; \bar{w}). \tag{1.10}$$

Here A_h is the cell-averaging operator on the RHS of (1.3), $E(t)$ is the exact evolution operator (1.2), and w is any piecewise smooth function of x. These schemes are a generalization of Godunov's scheme and its second-order extensions in the sense that (1.10) with the first-order accurate piecewise constant reconstruction

$$R(x; \bar{w}) = \bar{w}_j \qquad \text{for} \quad x_{j-1/2} \leqslant x < x_{j+1/2} \tag{1.11}$$

is exactly Godunov's schemes [5]; (1.10) with the second-order accurate piecewise linear reconstruction

$$R(x; \bar{w}) = \bar{w}_j + s_j(x - x_j) \qquad \text{for} \quad x_{j-1/2} \leqslant x < x_{j+1/2} \tag{1.12a}$$

such that

$$s_j = w_x(x_j) + O(h), \tag{1.12b}$$

is the abstract form of the second-order extensions to Godunov's scheme described in [22, 4, and 15].

We recall that the evolution operator $E(t)$ is monotone in the scalar case. Since A_h, the cell-averaging operator is also monotone we see that in the scalar case

$$TV(\bar{E}_h(\tau)\,\bar{w}) = TV(A_h \cdot E(\tau) \cdot R(\cdot\,; \bar{w})) \leqslant TV(R(\cdot\,; \bar{w})). \tag{1.13a}$$

If $\bar{w}$ in (1.13a) is the sliding average of a piecewise smooth function $w(x)$, it follows then from (1.9c) that

$$TV(\bar{E}_h(\tau)\,\bar{w}) \leqslant TV(w) + O(h^r). \tag{1.13b}$$

This shows that the schemes (1.10) in the scalar case are essentially non-oscillatory in exactly the same way as the reconstruction: They do not have a Gibbs-like phenomenon at discontinuities, yet they may occasionally produce small spurious oscillations on the level $O(h^r)$ of the trunction error (see Remark 1.3 at the end of this section).

Equation (1.10) is the abstract operator expression of a scheme in the conservation form (1.7). Although the scheme generates discrete values v_j^n, which are rth order accurate approximations to the cell-averages $\bar{u}_j^n$, its operation involves a globally defined pointwise approximation to $u(x, t)$ of the same order of accuracy, which we denote by $v_h(x, t)$. The latter is defined for all x in the time-strips $t_n \leqslant t < t_{n+1}$, with a possible discontinuity at $\{t_k\}$; we shall use the standard notation $v_h(x, t_n \pm 0)$ to distinguish between the two possibly different values.

We define $v_h(x, t)$ via the following algorithmic description of the scheme (1.10). We start by setting

$$v_j^0 = \bar{u}_0(x_j),$$

where u_0 is the given initial datum (1.1b), and $\bar{u}_0$ is its sliding average (1.3a). Having defined $v^n = \{v_j^n\}$, approximation to $\{\bar{u}_j^n\}$ in (1.6b), we proceed to evaluate v^{n+1} by the following three steps:

(i) *Reconstruction.* Define

$$v_h(x, t_n + 0) = R(x; v^n), \tag{1.14a}$$

Note that $v_h(x, t_n + 0)$ is a pointwise approximation to $u(x, t_n)$.

(ii) *Solution in the small.* For $t_n \leqslant t < t_n + \tau = t_{n+1}$, define

$$v_h(\cdot\,; t) = E(t - t_n) \cdot v_h(\cdot\,; t_n + 0). \tag{1.14b}$$

(iii) *Cell averaging.* Close the time loop of the algorithm by defining

$$v_j^{n+1} = \bar{v}_h(x_j; t_{n+1} - 0) = \frac{1}{h}\int_{x_{j-1/2}}^{x_{j+1/2}} v_h(x, t_{n+1} - O)\,dx. \tag{1.14c}$$

We note that v_h, being an exact solution of (1.1) in $t_n \leqslant t < t_{n+1}$, satisfies (1.5) in this strip. Using the conservation property (1.9b) of the reconstruction in (1.14a), i.e.,

$$\bar{v}_h(x_j, t_n + 0) = v_j^n, \tag{1.15}$$

we get from (1.5) that the schemes (1.10) and (1.14) satisfy the conservation form

$$v_j^{n+1} = v_j^n - \lambda(\bar{f}_{j+1/2} - \bar{f}_{j-1/2}) \tag{1.16a}$$

with the numerical flux

$$\bar{f}_{j+1/2} = \hat{f}(x_{j+1/2}, t_n; v_h) \equiv \frac{1}{\tau} \int_0^\tau f(v_h(x_{j+1/2}, t_n + \eta))\, d\eta. \tag{1.16b}$$

We turn now to examine the local truncation error of the scheme. For this purpose we consider a single application of (1.14) starting with $v_j^n \equiv \bar{u}_j^n$, the exact cell averages of the solution. It follows from (1.9a) and (1.14a) that

$$v_h(x, t_n + 0) = u(x, t_n) + e(x)\, h^r + O(h^{r+1}). \tag{1.17a}$$

The definition (1.14b) and our assumption of the well-posedness of the IVP (1.1) imply that

$$v_h(x, t) = u(x, t) + O(h^r) \qquad \text{for} \quad t_n \leqslant t < t_{n+1}. \tag{1.17b}$$

This in turn implies that the numerical flux (1.16b) of the scheme satisfies (1.8a), i.e.,

$$\bar{f}_{j+1/2} = \hat{f}(x_{j+1/2}, t_n; u) + d(x_{j+1/2})\, h^r + O(h^{r+1}). \tag{1.17c}$$

Clearly non-smoothness of $d(x)$ in (1.17c) can result only from non-smoothness of the coefficient $e(x)$ in (1.17a). It follows then from (1.8b) that away from points of discontinuity and points at which $e(x)$ fails to be Lipschitz continuous, the local truncation error in the sense of cell-averages is $O(h^{r+1})$.

Let $u(x, t)$ be a smooth solution of (1.1) and let us suppose that as $h \to 0$, $\tau = O(h)$, the numerical approximation converges pointwise to $u(x, t)$. If $e(x)$ is globally Lipschitz continuous then the local truncation error in the sense of cell averages is globally $O(h^{r+1})$. At time t, after performing $N = t/\tau$ time-steps, we expect the cumulative error to be $O(h^r)$, i.e.,

$$v_j^N = \bar{u}(x_j, t_N) + O(h^r) \tag{1.18a}$$

In this case we see from (1.9a) that

$$v_h(x, t_N + 0) = R(x; v^N) = u(x, t_N) + O(h^r). \tag{1.18b}$$

Thus at the end of the computation we have two sets of output data at our disposal: (i) discrete values $\{v_j^N\}$ that approximate $\{\bar{u}(x_j, t_N)\}$ to $O(h^r)$ and (ii) a piecewise polynomial function of x, $R(x; v^N)$, that approximates $u(x, t_N)$ to $O(h^r)$.

Remark. 1.1. Note that (1.8) is quite different from the truncation error in a pointwise sense which is used in formulating Lax–Wendroff-type schemes [20, 21]. There we take $v_j^n = u(x_j, t_n)$ and require $v_j^{n+1} = u(x_j, t_{n+1}) + O(h^{r+1})$. To accomplish that we need a numerical flux that satisfies

$$-\lambda(\bar{f}_{j+1/2} - \bar{f}_{j-1/2}) = \sum_{k=1}^{r} \frac{\tau^k}{k!} \frac{\partial^k u}{\partial t^k}\bigg|_{(x_j, t_n)} + O(h^{r+1}).$$

We shall see in the following that condition (1.8a) for the accuracy in a cell-average sense is more manageable in many respects.

Remark 1.2. When $e(x)$ fails to be Lipschitz continuous at a point, the local truncation error (1.8b) is only $O(h^r)$. In the MUSCL-type schemes [22, 4] this happens at local extremum points; in higher order accurate schemes this may occur at roots of higher derivatives of u (see [15, 11]). Due to local accumulation we expect the pointwise error at time t, after $N = t/\tau$ time-steps, to be only $O(h^{r-1})$ at such points. Away from these points we expect the pointwise cumulative error to remain $O(h^r)$. Consequently the scheme is $(r-1)$th order accurate in the maximum norm. Because of the non-oscillatory nature of the schemes, we expect the number of points where $e(x)$ fails to be Lipschitz-continuous to remain bounded as $h \to 0$. In this case the L_1-norm of the cumulative error is $O(h^r)$. To distinguish between schemes that are rth order accurate in the usual pointwise sense, and those that are rth order accurate in the L_1-norm but only $(r-1)$th accurate in the maximum norm, we shall use "rth order accurate" for the latter, thus qualifying the difference by the use of quotation marks.

Remark 1.3. It is well known that if the total variation of the numerical approximation is uniformly bounded, i.e.,

$$TV(v_h(\cdot, t)) \leqslant C \cdot TV(u_0), \tag{1.19}$$

where the constant C is independent of h for $0 \leqslant t \leqslant T$, then any refinement sequence $h \to 0$, $\tau = O(h)$ has a subsequence that converges in L_1^{loc} to a weak solution of (1.1). Therefore uniform boundedness of the total variation is an appropriate sense of stability for numerical approximations to discontinuous solutions of (1.1); see [9, 10] and the references cited there.

Inequality (1.13) shows that the total variation of our new schemes is dominated by that of reconstruction step

$$TV(v^{n+1}) \leqslant TV(R(\cdot; v^n)). \tag{1.20}$$

When R is the piecewise-constant function (1.11) or the piecewise-linear function (1.12) (where the slope s_j is that of the MUSCL scheme) then

$$TV(R(\cdot\,; v)) \leqslant TV(v) \tag{1.21a}$$

for any function v of bounded total variation. Consequently Godunov's scheme and the MUSCL scheme are total variation diminishing (TVD) in the scalar case

$$TV(v^{n+1}) \leqslant TV(v^n); \tag{1.21b}$$

this trivially implies (1.19) with $C = 1$.

In proving relation (1.9c) for higher order reconstructions we have used the assumption that for h sufficiently small there are at least $r+1$ points of smoothness between discontinuities. Consequently we cannot apply this result to the numerical solution v^n. Nevertheless, based on heuristic analysis and extensive numerical experimentation, we conjecture that in the scalar case

$$TV(v^{n+1}) \leqslant TV(v^n) + O(h^{p+1}) \tag{1.22}$$

for some $p > 0$.

2. Review and Overview

In [15], the first paper of this series, we present a second-order accurate scheme which is strictly non-oscillatory in the scalar case ($m = 1$), i.e.,

$$N_0(v^{n+1}) \leqslant N_0(v^n), \tag{2.1}$$

where $N_0(v)$ denotes the number of local extreme in v. This scheme is a modification of the "second-order accurate" MUSCL scheme [22, 4], which is total-variation-diminishing (TVD) in the scalar case, i.e.,

$$TV(v^{n+1}) \leqslant TV(v^n). \tag{2.2}$$

In order to enforce (2.2), the slope s_j (1.12a) in the MUSCL scheme is subjected to a so called "limiter." Due to the operation of this limiter, the coefficient in the $O(h)$ term in the Taylor expansion (1.12b) becomes discontinuous at local extrema: Consequently $e(x)$ in (1.9a) fails to be Lipschitz continuous at such points, which leads to a loss of accuracy at local extrema. In [15] this difficulty is circumvented by using a modified slope s_j in (1.12a) which satisfies

$$s_j = w_x(x_j) + O(h^2), \tag{2.3}$$

thus leading to a globally smooth $e(x)$ in (1.9a).

Although the end result is a simple technical modification of the formula for the slope s_j, the design of the scheme in [15] invokes major conceptual changes.

Realizing that TVD schemes, independent of their particular form, are necessarily only first-order accurate at local extrema, we seek a weaker notion of control over possible growth of the total variation of the numerical solution. For this purpose we introduce the notion of non-oscillatory schemes, which satisfy in the scalar case for piecewise smooth w,

$$TV(\bar{E}_h(\tau)\cdot\bar{w})\leqslant TV(w)+O(h^2) \tag{2.4}$$

rather than (2.2). In [16], the second paper in this series, we show that even the notion of (strictly) non-oscillatory schemes (2.1) is too restrictive in the sense that it limits the order of accuracy to 2. To enable the design of higher order accurate schemes we then introduce the notion of *essentially* non-oscillatory schemes (1.13), which excludes a Gibbs-like phenomenon but allows for the production of spurious oscillations on the level of the truncation error.

Another conceptual change is the removal of the "monotonicity limiters" which are an essential part of TVD schemes [30] and may cause a reduction of the order of accuracy at some points. Our new schemes are of uniform order of accuracy r. The control over possible growth of the total variation of the numerical solution is obtained by an adaptive stencil that at each point attempts to use the smoothest information available. This adaptive selection of stencil is introduced to the algorithm through the reconstruction step (1.14a). The number of points in the stencil, independent of its orientation, is always $(r+1)$.

In [16], the second paper in the series, we investigate the stability of our new schemes in the scalar constant coefficient case

$$u_t+au_x=0, \qquad a=\text{constant}. \tag{2.5a}$$

The exact evolution operator (1.2) in this case is just a translation with the constant speed a. Therefore our schemes (1.14) take the particularly simple form

$$v_j^{n+1}=\bar{R}(x_j-a\tau;v^n). \tag{2.5b}$$

Due to the adaptive selection of stencil in the reconstruction step, the scheme (1.23b) is highly nonlinear; consequently the use of the standard linear stability analysis is inappropriate. We demonstrate this point in [16] by choosing initial data for which the reconstruction algorithm selects a stencil that is biased in the "down-wind" direction (i.e., in the direction opposite to that of the wind); a constant choice of such a stencil is notoriously unstable. Such an instability usually exhibits itself by the production of increasing oscillations which start at the highest derivative and propagate to the function itself. The numerical experiment in [16] shows that once these oscillations begin to appear on the level of the highest derivative, the adaptive selection of stencil in (2.5b) reacts by changing the orientation of the stencil and thus avoids the buildup of instability.

In [16] we also investigate the initial-boundary value problem (IBVP) for (2.5a). Unlike the treatment of boundaries in standard finite-difference schemes we do not use "numerical boundary conditions." Instead we modify the scheme (2.5b) by restricting the selection of the stencil to available information. As a result the scheme is biased "against the wind" at one of the two boundaries. Nevertheless, numerical experiments show the scheme to be strongly stable.

In the present paper, the third in the series, we turn to consider the general nonlinear case. The abstract form of our schemes, (1.10) and (1.14), call for the evaluation of the exact *solution in the small* (i.e., for $0 \leqslant t < \tau$, τ small) of the IVP (1.1) with the initial data $R(x; v^n)$; the latter is a piecewise polynomial function of x with possible discontinuities at $\{x_{j+1/2}\}$.

When $R(x; v^n)$ is the piecewise-constant function (1.11) (i.e., Godunov's scheme), we can express this solution in terms of local solutions to the Riemann problem

$$u_t + f(u)_x = 0, \qquad u(x, 0) = \begin{cases} v_j^n & x < 0 \\ v_{j+1}^n & x > 0 \end{cases}. \tag{2.7}$$

When $R(x; v^n)$ is a piecewise polynomial function of higher degree we cannot in general express the solution of the IVP (1.1) in a simple closed form. Nevertheless, (see [1, 6]) we can obtain a local Taylor expansion of the solution to any desired order of accuracy.

We note, however, that the step of "solution-in-the-small" (1.14b) is followed by the step of "cell-averaging" (1.14c). Consequently many of the fine details of the exact solution, which may be very costly to compute, are later ignored in evaluating v_j^{n+1} by averaging the exact solution over $(x_{j-1/2}, x_{j+1/2})$. To economize on the cost of our schemes it makes sense to use simplified approximate "solvers" that carry only this information which determines the value of the cell average, namely the one needed to compute a numerical flux satisfying (1.17c). The study of such approximate solvers is a main issue of the present paper. In Section 4 we consider the scalar case; in Section 5 we extend the scheme to hyperbolic systems of conservation laws.

When we consider the reconstruction (1.9) in the context of approximation of functions, the assumption that $w(x)$ is piecewise smooth with a finite number of discontinuities implies that for h sufficiently small there are at least $(r+1)$ points of smoothness separating discontinuities on the computational grid. Therefore at any point of smoothness it is possible to select a stencil from the smooth part of the function. Although the x behavior of weak solutions of (1.1) is generically of this type, their time dependence allows for collision of discontinuities, as well as their collision with a boundary, e.g., solid walls. For points in a region between two discontinuities that are about to collide, no matter how small h is, there must come a time when there are not enough points to select a stencil of $(r+1)$ points from the region of smoothness. Consequently, a component-wise extension of the scalar reconstruction algorithm in [16] to vector functions may produce large spurious oscillations during this brief encounter.

The elimination of such spurious oscillations has been a major consideration in designing the extension of our scalar schemes to systems of conservation laws. In

Section 5 we show that this can be accomplished to a great extent by extending the scalar reconstruction algorithms to systems via the use of locally defined characteristic variables.

In Section 6 we describe in detail the algorithm for the solution of the Euler equations of gas dynamics. In Section 7 we present some numerical experiments.

In future papers we shall present the extension of these schemes to two-dimensional problems and study the dependence of the computational efficiency on the order of accuracy of the scheme.

3. Reconstruction

In this section we present a brief description of the reconstruction $R(x; w)$ to be used in (1.14a); we refer the reader to [16, 11, 17] for more details and analysis. For this purpose we introduce $H_m(x; w)$, a piecewise polynomial function of x that interpolates w at the points $\{x_j\}$, i.e.,

$$H_m(x_j; w) = w(x_j), \tag{3.1a}$$

$$H_m(x; w) \equiv q_{m,j+1/2}(x; w) \qquad \text{for} \quad x_j \leqslant x \leqslant x_{j+1}, \tag{3.1b}$$

where $q_{m,j+1/2}$ is a polynomial in x of degree m.

We take $q_{m,j+1/2}$ to be the (unique) mth degree polynomial that interpolates $w(x)$ at the $(m+1)$ successive points $\{x_i\}$, $i_m(j) \leqslant i \leqslant i_m(j) + m$, that include x_j and x_{j+1}, i.e.,

$$q_{m,j+1/2}(x_i; w) = w(x_i) \qquad \text{for} \quad i_m(j) \leqslant i \leqslant i_m(j) + m. \tag{3.2a}$$

$$1 - m \leqslant i_m(j) - j \leqslant 0. \tag{3.2b}$$

Clearly there are exactly m such polynomials corresponding to the m different choices of $i_m(j)$ subject to (3.2b). This freedom is used to assign to (x_j, x_{j+1}) a stencil of $(m+1)$ points (3.2) so that $w(x)$ is "smoothest" in $(x_{i_m(j)}, x_{i_m(j)+m})$ in some asymptotic sense.

The information about smoothness of $w(x)$ is extracted from a table of divided differences of w. The latter can be defined recursively by

$$w[x_i] = w(x_i) \tag{3.3a}$$

$$w[x_i, ..., x_{i+k}] = (w[x_{i+1}, ..., x_{i+k}] - w[x_i, ..., x_{i+k-1}])/(x_{i+k} - x_i). \tag{3.3b}$$

It is well known that if w is C_∞ in $[x_i, x_{i+k}]$ then

$$w[x_i, ..., x_{i+k}] = \frac{1}{k!} \frac{d^k}{dx^k} w(\xi_{i,k}), \qquad x_i \leqslant \xi_{i,k} \leqslant x_{i+k}. \tag{3.3c}$$

However if w has a jump discontinuity in the pth derivative in this interval, $0 \leqslant p \leqslant k$, then

$$w[x_i, ..., x_{i+k}] = O(h^{-k+p}[w^{(p)}]); \tag{3.3d}$$

here $[w^{(p)}]$ denotes the jump in the pth derivative. Equations (3.3c)–(3.3d) show that $|w[x_i, ..., x_{i+k}]|$ provides an asymptotic measure of the smoothness of w in (x_i, x_{i+k}), in the sense that if w is smooth in (x_{i_1}, x_{i_1+k}) but is discontinuous in (x_{i_2}, x_{i_2+k}), then for h sufficiently small $|w[x_{i_1}, ..., x_{i_1+k}]| < |w[x_{i_2}, ..., x_{i_2+k}]|$. Hence the problem of choosing a stencil of points for which w is "smoothest" is basically the same as that of finding an interval in which w has the "smallest divided differences" (see [16, 11] for more details).

In [11] we propose the following recursive algorithm to evaluate $i_m(j)$. We start by setting

$$i_1(j) = j, \tag{3.4a}$$

i.e., $q_{1,j+1/2}$ is the first-degree polynomial interpolating w at x_j and x_{j+1}. Let us assume that we have already defined $i_k(j)$, i.e., $q_{k,j+1/2}$ is the kth degree polynomial interpolating w at

$$x_{i_k(j)}, ..., x_{i_k(j)+k}.$$

We consider now as candidates for $q_{k+1,j+1/2}$ the two $(k+1)$th degree polynomials obtained by adding to the above stencil the neighboring point to the left or the one to the right; this corresponds to setting $i_{k+1}(j) = i_k(j) - 1$ or $i_{k+1}(j) = i_k(j)$, respectively. We choose the one that gives a $(k+1)$th order divided difference that is smaller in absolute value, i.e.,

$$i_{k+1}(j) = \begin{cases} i_k(j) - 1 & \text{if } |w[x_{i_k(j)-1}, ..., x_{i_k(j)+k}]| < |w[x_{i_k(j)}, ..., x_{i_k(j)+k+1}]| \\ i_k(j) & \text{otherwise.} \end{cases} \tag{3.4b}$$

In [16] we analyze this interpolation technique for a piecewise smooth function w and show that:

(i) wherever $w(x)$ is smooth

$$\frac{d^k}{dx^k} H_m(x; w) = \frac{d^k}{dx^k} w(x) + O(h^{m+1-k}), \qquad 0 \leqslant k \leqslant m; \tag{3.5a}$$

(ii) $H_m(x; w)$ is an essentially non-oscillatory interpolation of w in the sense that

$$TV(H_m(\cdot; w)) \leqslant TV(w) + O(h^{m+1}). \tag{3.5b}$$

We turn now to describe two different techniques to solve the reconstruction problem (1.9) in terms of interpolation. (See Appendix for an algorithmic description.)

(1) *Reconstruction via a Primitive Function*

Given cell averages $\bar{w}_j$ of a piecewise smooth function w

$$\bar{w}_j = \frac{1}{h_j} \int_{x_{j-1/2}}^{x_{j+1/2}} w(y)\, dy, \qquad h_j = x_{j+1/2} - x_{j-1/2}, \tag{3.6}$$

we can immediately evaluate the point values of the primitive function $W(x)$

$$W(x) = \int_{x_0}^{x} w(y)\, dy \tag{3.7a}$$

by

$$W(x_{j+1/2}) = \sum_{i=i_0}^{j} h_i \bar{w}_i. \tag{3.7b}$$

Since

$$w(x) \equiv \frac{d}{dx} W(x)$$

we apply interpolation to the point values (3.7b) of the primitive function $W(x)$ (3.7a) and then obtain an approximation to $u(x)$ by defining

$$R(x; \bar{w}) = \frac{d}{dx} H_r(x; W). \tag{3.8}$$

We note that this procedure does not require uniformity of the mesh.

The primitive function $W(x)$ is one derivative smoother than $w(x)$, therefore it follows from (3.5a) that wherever $W(x)$ is smooth

$$\frac{d^k}{dx^k} H_r(x; W) = \frac{d^k}{dx^k} W(x) + O(h^{r+1-k});$$

thus we get from the definition (3.8) that

$$\frac{d^l}{dx^l} R(x; \bar{w}) = \frac{d^l}{dx^l} w(x) + O(h^{r-l}), \tag{3.9}$$

which implies (1.9a) for $l = 0$.

The conservation property of the reconstruction (1.9b) follows immediately from the definition (3.8):

$$\frac{1}{h_j} \int_{x_{j-1/2}}^{x_{j+1/2}} R(x; \bar{w})\, dx = \frac{1}{h_j} [H_r(x_{j+1/2}, W) - H_r(x_{j-1/2}; W)]$$

$$= \frac{1}{h_j} [W(x_{j+1/2}) - W(x_{j-1/2})] = \bar{w}_j. \tag{3.10}$$

The non-oscillatory nature of the reconstruction (1.9c) follows primarily from the non-oscillatory nature of the interpolation (3.5b), see [16].

We denote the reconstruction via the primitive function (3.8) by RP.

(2) *Reconstruction via Deconvolution*

We assume that the mesh is uniform and consider the given cell averages $\bar{w}_j$ to be point values of $\bar{w}(x)$, the globally defined sliding-average function (1.3) of w, i.e.,

$$\bar{w}_j = \bar{w}(x_j), \tag{3.11a}$$

where

$$\bar{w}(x) = \frac{1}{h} \int_{-h/2}^{h/2} w(x+y)\, dy. \tag{3.11b}$$

Expanding $w(x+y)$ in (3.11b) around $y=0$, we get

$$\bar{w}(x) = \sum_{k=0}^{\infty} \frac{w^{(k)}(x)}{k!} \int_{-h/2}^{h/2} y^k\, dy = \sum_{k=0}^{\infty} \alpha_k h^k w^{(k)}(x), \tag{3.12a}$$

where

$$\alpha_k = \begin{cases} 0 & k \text{ odd} \\ 2^{-k}/(k+1)! & k \text{ even}. \end{cases} \tag{3.12b}$$

Multiplying both sides of (3.12a) by $h^l d^l/dx^l$ and then truncating the expansion in the RHS at $O(h^r)$, we get

$$h^l \bar{w}^{(l)}(x) = \sum_{k=0}^{r-l-1} \alpha_k h^{k+l} w^{(k+l)}(x) + O(h^r). \tag{3.13a}$$

Writing the relations (3.13a) for $l = 0, \ldots, r-1$ in a matrix form, we obtain

$$\begin{bmatrix} \bar{w}(x) \\ h\bar{w}'(x) \\ h^2\bar{w}''(x) \\ \vdots \\ h^{r-1}\bar{w}^{(r-1)}(x) \end{bmatrix} = \begin{pmatrix} 1 & 0 & \alpha_2 & 0 & \alpha_4 & \cdots & \alpha_{r-1} \\ & \ddots & \ddots & \ddots & \ddots & \ddots & \vdots \\ & & \ddots & \ddots & \ddots & \ddots & \alpha_4 \\ & & & \ddots & \ddots & \ddots & 0 \\ & & & & \ddots & \ddots & \alpha_2 \\ & & & & & \ddots & 0 \\ 0 & & & & & & 1 \end{pmatrix} \begin{bmatrix} w(x) \\ hw'(x) \\ h^2 w''(x) \\ \vdots \\ h^{r-1} w^{(r-1)}(x) \end{bmatrix} + O(h^r). \tag{3.13b}$$

Let us denote the coefficient matrix in the RHS of (3.13b) by C. This matrix is upper triangular and diagonally dominant. Multiplying both sides of (3.13b) by C^{-1} from the left we get

$$\begin{pmatrix} w(x) \\ hw'(x) \\ \vdots \\ h^{r-1}w^{(r-1)}(x) \end{pmatrix} = C^{-1} \begin{pmatrix} \bar{w}(x) \\ h\bar{w}'(x) \\ \vdots \\ h^{r-1}\bar{w}^{(r-1)}(x) \end{pmatrix} + O(h^r). \tag{3.13c}$$

Given $\bar{w}_j$ we interpolate $\bar{w}(x)$ by $H_m(x; \bar{w})$ with $m \geqslant r-1$. Since $\bar{w}(x)$ is smoother than $w(x)$ it follows from (3.5a) that

$$\frac{d^k}{dx^k} H_m(x; \bar{w}) = \frac{d^k}{dx^k} \bar{w}(x) + O(h^{m+1-k})$$

wherever $w(x)$ is smooth. We note that although H_m is only continuous at x_j, the one-sided derivatives at $x_j \pm 0$ do satisfy the above relations, i.e.,

$$\frac{d^k}{dx^k} H_m(x_j \pm 0; \bar{w}) = \frac{d^k}{dx^k} \bar{w}(x_j) + O(h^{m+1-k}). \tag{3.14}$$

Next we define

$$\bar{D}_{0,j} = \bar{w}_j \tag{3.15a}$$

$$\bar{D}_{l,j} = h^l M\left(\frac{d^l}{dx^l} H_m(x_j - 0; \bar{w}), \frac{d^l}{dx^l} H_m(x_j + 0; \bar{w})\right) \tag{3.15b}$$

for $1 \leqslant l \leqslant r-1$, (3.15b)

where $M(x, y)$ is the min mod function

$$M(x, y) = \begin{cases} s \cdot \min(|x|, |y|) & \text{if } \operatorname{sgn}(x) = \operatorname{sgn}(y) = s \\ 0 & \text{otherwise.} \end{cases} \tag{3.16}$$

Clearly

$$\bar{D}_{l,j} = h^l \bar{w}^{(l)}(x_j) + O(h^r); \tag{3.17a}$$

using $\bar{D}_j = (\bar{D}_{0,j}, ..., \bar{D}_{r-1,j})^T$ to approximate the vector, on the RHS of (3.13c) we get that

$$D_j = C^{-1} \bar{D}_j \tag{3.17b}$$

satisfies

$$D_j = (w(x_j), hw'(x_j), ..., h^{r-1} w^{(r-1)}(x_j))^T + O(h^r). \tag{3.17c}$$

Finally we define

$$R(x; \bar{w}) = \sum_{k=0}^{r-1} \frac{1}{k!} D_{k,j}[(x-x_j)/h]^k \qquad \text{for} \quad |x-x_j| < h/2. \tag{3.18}$$

We note that since C is upper triangular $D_{l,j}$ in (3.17b) can be computed by back-substitution, i.e., we set

$$D_{r-1,j} = \bar{D}_{r-1,j} \tag{3.19a}$$

and then compute for $k = r-2, ..., 0$

$$D_{k,j} = \bar{D}_{k,j} - \sum_{l=k+1}^{r-1} \alpha_l D_{l,j}. \tag{3.19b}$$

It follows immediately from the definition (3.18) and the relations (3.17a) that wherever $w(x)$ is smooth

$$\frac{d^l}{dx^l} R(x; \bar{w}) = \frac{d^l}{dx^l} w(x) + O(h^{r-l}); \tag{3.20}$$

this for $l = 0$ implies (1.9a). The conservation property of the reconstruction (1.9b) follows from

$$\frac{1}{h}\int_{-h/2}^{h/2} R(x+y; \bar{w})\, dy = \sum_{k=0}^{r-1} \frac{D_{k,j}}{k!} \frac{1}{h^{k+1}} \int_{-h/2}^{h/2} y^k\, dy = D_{0,j} + \sum_{k=1}^{r-1} \alpha_k D_{k,j}$$
$$= \bar{D}_{0,j} = \bar{w}_j. \tag{3.21}$$

The last two equalities in (3.21) follow from (3.19b) with $k = 0$ and (3.15a).

The non-oscillatory nature of the reconstruction (1.9c) follows primarily from the non-oscillatory nature of the interpolation $H_m(x; \bar{w})$; see [16] for more details.

We note that $\bar{w}(x)$ is the convolution of $w(x)$ with $\psi_h(x)$, the characteristic function of a cell, i.e.,

$$\bar{w}(x) = (w * \psi_h)(x) \tag{3.22a}$$

$$\psi_h(x) = \begin{cases} 1/h & \text{for} \quad |x| < h/2. \\ 0 & \text{for} \quad |x| > h/2 \end{cases} \tag{3.22b}$$

Hence (3.13c) is actually a deconvolution to $O(h^r)$. Therefore we refer to (3.18) as reconstruction via deconvolution and denote it by RD.

Remark 3.1. We note that for RP with $m = r$ and RD with $m = r-1$ the coefficient $e(x)$ of h^r in the reconstruction error (1.9a) is discontinuous at points where there is a change of orientation in the stencil of the associated interpolation; this may happen at critical points of the function and its derivatives. Hence the resulting schemes (1.14) are "rth order accurate" (See Remark 1.4). On the other hand, RD with $m=r$ yield $e(x)$ which is globally Lipschitz continuous, thus resulting in schemes that are rth order accurate in a pointwise sense. This follows from the fact that (3.17a) is upgraded to

$$\bar{D}_{l,j}=h^l\bar{w}^{(l)}(x)+O(h^{r+1}) \tag{3.23}$$

which has the effect of pushing the non-smoothness due to the change of stencil orientation in the associated interpolation to the $O(h^{r+1})$ level.

Remark 3.2. We note that both RD with $r=2$, $m=1$ and RP with $r=2$ are piecewise linear reconstructions of the form (1.12). The slope s_j for RD is identical to that of the "2nd-order accurate" TVD scheme in [5]. The slope for RP is the same as that of RD except at local extrema, where $s_j=0$ for RD while for RP

$$s_j=\begin{cases} w[x_j, x_{j+1}] & \text{if } |w[x_j, x_{j+1}]|<|w[x_{j-1}, x_j]| \\ w[x_{j-1}, x_j] & \text{otherwise.} \end{cases} \tag{3.24}$$

Although RP does not "chop" local extrema as RD, the lack of smoothness in (3.24) results in the same loss of accuracy at local extrema.

We note that RD with $m=r=2$ is essentially the same reconstruction that gives the non-oscillatory second-order accurate scheme of [15].

4. Scalar Conservation Laws

The abstract form of our scheme calls in (1.14b) for the evaluation of the exact solution in the small of the IVP (1.1) with the initial data $R(x; v^n)$. This step is followed by the cell-averaging operation in (1.14c) which results in the conservation form (1.16). Thus we are spared the task of having to compute a global solution. All we need to do is evaluate

$$\bar{f}_{j+1/2}=\frac{1}{\tau}\int_0^\tau f(v_h(x_{j+1/2}, t_n+\eta))\, d\eta. \tag{4.1}$$

To simplify our notation let us denote $v_h(x, t_n+t)$ by $v(x, t)$. Thus $v(x, t)$ is the solution of

$$v_t+f(v)_x=0 \tag{4.2a}$$

with the piecewise-polynomial initial data

$$v(x, 0)=R(x; v^n)=\sum_{l=0}^{r-1} b_{j,l}(x-x_j)^l/l! \qquad \text{for} \quad x_{j-1/2}\leqslant x<x_{j+1/2} \tag{4.2b}$$

in the time strip $-\infty<x<\infty$, $0\leqslant t\leqslant\tau$, τ small.

The solution $v(x, t)$, for sufficiently small τ, is composed of sections of smoothness separated by "fans" that emerge from the discontinuities at $\{x_{i+1/2}\}$. We use here the term "fan" loosely, allowing a "fan" with zero spread which is just a curve. In the linear case discontinuities propagate along characteristic curves; in this case all the "fans" are just curves. In the nonlinear case the "fans" with zero spread are shock curves, while "fans" with positive spread are rarefaction fans—or possibly a succession of rarefaction fans separated by contact shocks in the case of nonconvex flux. We denote by $v_j(x, t)$ the section of smoothness of $v(x, t)$ that is connected to the polynomial data in $(x_{j-1/2}, x_{j+1/2})$.

A global description of $v(x, t)$ can be quite complicated. Fortunately all we need is $v(x_{j+1/2}, t)$ for small t, which can be easily described in terms of $v_j(x, t)$, $v_{j+1}(x, t)$ and the "fan" eminating from $x = x_{j+1/2}$ at $t = 0$ as follows: If for $t > 0$ the "fan" stays to the right of $x = x_{j+1/2}$ then $v(x_{j+1/2}, t) = v_j(x_{j+1/2}, t)$; if this "fan" stays to the left of $x = x_{j+1/2}$ then $v(x_{j+1/2}, t) = v_{j+1}(x_{j+1/2}, t)$; if the "fan" covers $x = x_{j+1/2}$, then $v(x_{j+1/2}, t) = \text{constant} = V(0; v_j(x_{j+1/2}, 0), v_{j+1}(x_{j+1}, 0))$. Here $V(x/t; u_L, u_R)$ denotes the self-similar solution of the Riemann problem

$$u_t + f(u)_x = 0, \quad u(x, 0) = \begin{cases} u_L & x < 0 \\ u_R & x > 0, \end{cases} \tag{4.3}$$

with constant u_L and u_R. We note that the "fan" covers $x = x_{j+1/2}$ only when it contains a sonic centered rarefaction wave (i.e., one that includes a point for which $f' = 0$); this wave retains its self-similar form as long as it does not interact with shocks. Therefore if we choose τ sufficiently small so that no shock crosses $x = x_{j+1/2}$ for $0 < t \leqslant \tau$, we can express $f(v(x_{j+1/2}, t))$ by

$$f(v(x_{j+1/2}, t)) = \begin{cases} f(v_j(x_{j+1/2}, t)) & \text{"fan" stays to the right of } x = x_{j+1/2} \\ f^R(v_j(x_{j+1/2}, 0), v_{j+1}(x_{j+1/2}, 0)) & \text{"fan" covers } x = x_{j+1/2}. \\ f(v_{j+1}(x_{j+1/2}, t)) & \text{"fan" stays to the left of } x = x_{j+1/2} \end{cases} \tag{4.4}$$

Here f^R denotes the flux at $x = 0$ of the solution to the Riemann problem (4.3), i.e.,

$$f^R(u_1, u_2) = f(V(0; u_1, u_2)); \tag{4.5}$$

using the formula in [23] it can be expressed by

$$f^R(u_1, u_2) = \begin{cases} \min_{u_1 \leqslant u \leqslant u_2} f(u) & \text{if } u_1 \leqslant u_2 \\ \max_{u_1 \geqslant u \geqslant u_2} f(u) & \text{if } u_1 > u_2. \end{cases} \tag{4.6a}$$

When $f(u)$ is a convex function of u, i.e., $f''(u) > 0$, $f(u)$ may have only a single local extremum which is a minimum; let us denote its location by u_s. Using this fact in (4.6a) we can express $f^{R}(u_1, u_2)$ in the convex case by

$$f^{R}(u_1,u_2)=\begin{cases} f(u_1) & \text{if } u_s<u_1\leqslant u_2\\ f(u_s) & \text{if } u_1\leqslant u_s\leqslant u_2\\ f(u_2) & \text{if } u_1\leqslant u_2<u_s\\ f(u_1) & \text{if } u_1>u_2 \text{ and } \bar{a}(u_1,u_2)>0\\ f(u_2) & \text{if } u_1>u_2 \text{ and } \bar{a}(u_1,u_2)\leqslant 0.\end{cases} \tag{4.6b}$$

Here

$$\bar{a}(u_1,u_2)=[f(u_2)-f(u_1)]/(u_2-u_1) \tag{4.7}$$

is the speed of the shock with $u_L=u_1$ and $u_R=u_2$ in (4.3). We remark that (4.4) is deliberately formulated in terms of $f(v(x_{j+1/2},t))$ rather than $v(x_{j+1/2},t)$ in order to remove ambiguity in the definition when v is discontinuous at $x_{j+1/2}$. The continuity of $f(v)$ in this case follows from the Rankine–Hugoniot relation for a stationary shock.

We turn now to derive a simple but adequate approximation to the numerical flux (4.1), which is

$$\bar{f}_{j+1/2}=\frac{1}{\tau}\int_0^{\tau} f(v(x_{j+1/2},t))\,dt \tag{4.8}$$

with the integrand given by (4.4). Note that the integrand is a smooth function of t.

The first step is to discretize the integral in (4.8) by using a numerical quadrature

$$\frac{1}{\tau}\int_0^{\tau} g(t)\,dt=\sum_{k=0}^{K}\alpha_k\,g(\beta_k\tau)+O(\tau^r); \tag{4.9}$$

thus

$$\bar{f}_{j+1/2}\approx\sum_{k=0}^{K}\alpha_k\,f(v(x_{j+1/2},\beta_k\tau)). \tag{4.10}$$

The second step is to approximate $v_j(x,t)$ in (4.4) by its Taylor expansion which is obtained by the following local Cauchy–Kowalewski procedure. We start by expressing $\partial^l v(x,0)/\partial x^l$ at x_j by

$$\frac{\partial^l v(x_j,0)}{\partial x^l}=\begin{cases} b_{j,l} & \text{for } 0\leqslant l\leqslant r-1\\ 0 & \text{for } l\geqslant r.\end{cases} \tag{4.11a}$$

Next we use (4.11a) to evaluate

$$\frac{\partial^l v}{\partial x^k\,\partial t^{l-k}}(x_j,0) \qquad \text{for all } l \text{ and } 0\leqslant k\leqslant l \tag{4.11b}$$

by taking derivatives of the partial differential equation (1.1a) in the following ordered way

$$\begin{aligned} v_t &= -f'v_x \\ v_{xt} &= -[f''(v_x)^2 + f'v_{xx}] \\ v_{tt} &= -[f''v_t v_x + f'v_{xt}] \\ v_{xxt} &= -[f'''(v_x)^3 + 3f''v_x v_{xx} + f'v_{xxx}] \\ v_{xtt} &= -[f'''(v_x)^2 v_t + f''(2v_x v_{xt} + v_t v_{xx}) + f'v_{xxt}] \\ v_{ttt} &= -[f'''(v_t)^2 v_x + f''(2v_t v_{xt} + v_x v_{tt}) + f'v_{xtt}] \\ &\text{etc.,} \end{aligned} \tag{4.11c}$$

and then compute (4.11b) by successively evaluating the RHS of (4.11c); note that this procedure always uses known values which are either initially given by (4.11a) or previously computed in the algorithm (4.11c). We observe that

$$\tilde{v}_j(x, t) = \sum_{l=0}^{r-1} \sum_{k=0}^{l} \frac{\partial^l v(x_j, 0)}{\partial x^k \, \partial t^{l-k}} \cdot \frac{(x-x_j)^k}{k!} \frac{t^{l-k}}{(l-k)!} \tag{4.12a}$$

satisfies

$$\tilde{v}_j(x, 0) = v_j(x, 0) = v(x, 0) \qquad \text{for} \quad x_{j-1/2} < x < x_{j+1/2}, \tag{4.12b}$$

and that

$$\tilde{v}_j(x, t) = v_j(x, t) + O(h^r), \tag{4.12c}$$

wherever $v_j(x, t)$ is well defined.

The last step in our derivation of the numerical flux is to approximate $f(v(x_{j+1/2}, t))$ in (4.4) by

$$f(v(x_{j+1/2}, t)) \approx f^{R}(\tilde{v}_j(x_{j+1/2}, t), \tilde{v}_{j+1}(x_{j+1/2}, t)), \tag{4.13}$$

where f^R is (4.5)–(4.6). The resulting numerical scheme is

$$v_j^{n+1} = v_j^n - \lambda(\bar{f}_{j+1/2} - \bar{f}_{j-1/2}) \tag{4.14a}$$

$$\bar{f}_{j+1/2} = \sum_{k=0}^{K} \alpha_k f^{R}(\tilde{v}_j(x_{j+1/2}, \beta_k \tau), \tilde{v}_{j+1}(x_{j+1/2}, \beta_k \tau)). \tag{4.14b}$$

In the following we show that the numerical flux $\bar{f}_{j+1/2}$ in (4.14b) is an adequate approximation to the "abstract numerical flux" (4.1).

We start by proving that the scheme (4.14) is rth order accurate in the sense of

(1.8). To do so we take in (4.14) $v_j^n = \bar{u}(x_j, t_n)$, where $u(x, t)$ is a smooth (either globally or locally) solution of (1.1) and show that

$$\bar{f}_{j+1/2} = \frac{1}{\tau} \int_0^\tau f(u(x_{j+1/2}, t_n + \eta))\, d\eta + O(h^r). \tag{4.15}$$

When we apply the reconstruction R to $\bar{u}^n$ we get (3.9) and (3.20) that

$$\frac{d^k}{dx^k} R(x; \bar{u}^n) = \frac{\partial^k}{\partial x^k} u(x, t_n) + O(h^{r-k}) \qquad \text{for} \quad 0 \leqslant k \leqslant r-1. \tag{4.16a}$$

Consequently it follows from the Cauchy–Kowalewski procedure (4.11)–(4.12) and $\tau = O(h)$ that

$$\tilde{v}_i(x_{j+1/2}, t) = u(x_{j+1/2}, t_n + t) + O(h^r) \qquad \text{for} \quad i = j, j+1, \tag{4.16b}$$

$f^{\mathrm{R}}(u_1, u_2)$ is Lipschitz-continuous with respect to u_1 and u_2, and it is consistent with $f(u)$ in the sense that $f^{\mathrm{R}}(u, u) = f(u)$; therefore

$$f^{\mathrm{R}}(u_1, u_2) = f(u) + O(|u - u_1| + |u - u_2|). \tag{4.16c}$$

Applying (4.16c) to (4.16b) we get that

$$f^{\mathrm{R}}(\tilde{v}_j(x_{j+1/2}, t), \tilde{v}_{j+1}(x_{j+1/2}, t)) = f(u(x_{j+1/2}, t_n + t)) + O(h^r). \tag{4.16d}$$

Finally using the assumed smoothness of $u(x, t)$ and the order of accuracy of the numerical quadrature (4.9) we obtain (4.15).

Next we consider the constant coefficient case

$$v_t + a v_x = 0, \qquad a = \text{constant}. \tag{4.17a}$$

Here the "fan" in (4.4) is the characteristic line

$$x_{j+1/2}(t) = x_{j+1/2} + at \tag{4.17b}$$

and

$$v_j(x, t) = v(x - at, 0) = R(x - at; v^n) \qquad \text{for} \quad x_{j-1/2}(t) < x < x_{j+1/2}(t). \tag{4.17c}$$

Since $v_j(x, t)$ in (4.17c) is a polynomial of degree $r-1$ in $(x - at)$ we get that

$$\frac{\partial^l v}{\partial x^k\, \partial t^{l-k}} \equiv 0 \qquad \text{for} \quad l \geqslant r; \tag{4.17d}$$

this implies in (4.11)–(4.12) that

$$\tilde{v}_j(x, t) \equiv v_j(x, t). \tag{4.18a}$$

Hence

$$f^{R}(\tilde{v}_j(x_{j+1/2}, t), \tilde{v}_{j+1}(x_{j+1/2}, t)) \equiv f(v(x_{j+1/2}, t)). \tag{4.18b}$$

Since the numerical quadrature (4.9) is exact for polynomials of degree $r-1$ we get that the numerical flux $\bar{f}_{j+1/2}$ (4.14b) is identical to (4.8). It follows then that the numerical scheme (4.14) in the constant coefficient case is the "abstract scheme" (1.14), (1.16), i.e.,

$$v_j^{n+1} = \bar{R}(x_j - a\tau; v^n). \tag{4.18c}$$

We observe that since the "fans" in the solution $v(x, t)$ in the constant coefficient case have zero spread, the evaluation of $f(v(x_{j+1/2}, t))$ in (4.4) involves only the smooth parts of the solution $v_j(x, t)$. The "fans" in the numerical approximation mark the domain of validity of the Taylor expansions $\tilde{v}_j(x, t)$. Therefore the only role of the Riemann solver in the formulation of the numerical flux (4.14b)

$$f^{R}(\tilde{v}_j(x_{j+1/2}, t), \tilde{v}_{j+1}(x_{j+1/2}, t)) = \begin{cases} f(\tilde{v}_j(x_{j+1/2}, t)) & \text{if } a > 0 \\ f(\tilde{v}_{j+1}(x_{j+1/2}, t)) & \text{if } a < 0 \end{cases} \tag{4.19}$$

is to serve as a pointer, i.e., to identify whether $x = x_{j+1/2}$ falls into the domain of validity of $\tilde{v}_j$ or into that of $\tilde{v}_{j+1}$. Since $\tilde{v}_i(x_{j+1/2}, t) = v(x_{j+1/2} - at, 0)$, the use of the Cauchy–Kowalski procedure is equivalent to that of a characteristic method that traces the characteristic curve through $(x_{j+1/2}, t)$ to the initial data.

Next we consider the scalar IVP (1.1) with convex $f(u)$ and smooth initial data $u_0(x)$ and we show that the above interpretation of the numerical approximations applies to this nonlinear case as well. The numerical solution $v_j^n \approx \bar{u}(x_j, t_n)$ typically forms a monotone transition of 1–2 points across shocks and stays close to $\bar{u}(x_j, t_n)$ in the smooth parts of the solution (see the numerical experiments with $u_t + uu_x = 0$ and $u(x, 0) = \sin \pi x$ in Sect. 7). Let us now examine the discontinuities of $R(x; v^n)$ at $\{x_{j+1/2}\}$ and the nature of the "fans" emanating from these points. Relation (4.16b) with $t = 0$ shows that the jump at $x_{j+1/2}$ in the smooth part of the solution is of the order of the local error, say $O(h^p)$ with $0 < p \leqslant r$. Hence the "fan" emerging from $x_{j+1/2}$ in a region of smoothness is either a shock curve or a rarefaction fan with $O(h^p)$ spread. On the other hand, in the vicinity of shocks of $u(x, t_n)$ the side of this jump is $O(1)$; however, the "fan" is necessarily a shock curve. We see therefore that the global picture is very similar to that of the constant coefficient case, i.e., the "fans" separating $\{v_j(x, t)\}$ are either shock curves with zero spread or rarefaction fans with $O(h^p)$ spread (these can be thought of as "blurred" characteristic curves); these "fans" are the boundaries of the domains of validity of the Taylor expansions $\tilde{v}_i(x, t)$. We note that the value given by $\tilde{v}_i(x_{j+1/2}, t)$ differs of $O(\tau^r)$ from that obtained by solving the nonlinear characteristic relation for v

$$v = R(x_{j+1/2} - a(v)t; v^n). \tag{4.20}$$

Hence the use of the local Cauchy–Kowalewski procedure is again computationally equivalent to tracing the characteristic curve through $(x_{j+1/2}, t)$ to the initial data. Since the evaluation of $f(v(x_{j+1/2}, t))$ in (4.4) essentially involves only $\tilde{v}_j(x_{j+1/2}, t)$ and $\tilde{v}_{j+1}(x_{j+1/2}, t)$, the role of the Riemann solver in the numerical flux (4.14b) is again that of a pointer, i.e., to identify to which domain of validity $x = x_{j+1/2}$ belongs. This indicates that the monotonicity of the exact solution operator is approximately preserved by our procedure. See C1 in Section 7. This also suggests that f^{R} in (4.14b) can be adequately replaced by the simpler expression f^{ROE} which corresponds to Roe's approximate solution of the Riemann problem (see [25, 14]):

$$f^{\mathrm{ROE}}(u_1, u_2) = \tfrac{1}{2}[f(u_1) + f(u_2) - |\bar{a}(u_1, u_2)|\,(u_2 - u_1)]$$
$$= \begin{cases} f(u_1) & \text{if } \bar{a}(u_1, u_2) > 0 \\ f(u_2) & \text{if } \bar{a}(u_1, u_2) \leqslant 0, \end{cases} \tag{4.21}$$

where $\bar{a}(u_1, u_2)$ is defined by (4.7). Observe that f^{ROE} in (4.21) satisfies (4.16c) and therefore the modified scheme remains rth order accurate.

The heuristic analysis presented above is applicable only when all the discontinuities in the solution to the IVP (1.1) are shocks; discontinuities that are not shocks may be present in the solution either by being introduced through the initial data $u_0(x)$ or as a result of a shock–shock interaction in the non-convex case. Clearly f^{ROE} in its form (4.21a) should not be used when the solution contains a sonic rarefaction wave since it admits any discontinuity with $\bar{a}(u_{\mathrm{L}}, u_{\mathrm{R}}) = 0$ as a stationary solution. This problem is well known and there are many ways to overcome it (see [13, 26, 9], and Sect. 7).

In Section 7 we present numerical experiments testing the performance of the scheme (4.14) in the solution of the Riemann IVP (4.3), where $f(u)$ is non-convex and $\bar{a}(u_L, u_R) = 0$. In all these experiments, as well as in others not reported here, we have found the scheme to develop the correct structure of the solution.

5. Systems of Conservation Laws

In this section we extend the reconstruction algorithm of Section 3 and the solution-in-the-small procedure of Section 4 to the case of hyperbolic systems of conservation laws.

As always we are interested only in "computable" solutions and therefore assume that the initial data $u_0(x)$ in (1.1b) are such that $u(x, t)$, which is a vector function of m components $u = (u_1, ..., u_m)^T$, is, at any given t a piecewise smooth function of x with a finite number of discontinuities. Given cell averages $\bar{u}_j^n = \bar{u}(x_j, t_n)$, it seems natural from the point of view of approximation theory to reconstruct $u(x, t_n)$ by applying the scalar reconstruction R to each of the scalar component $\bar{u}_k^n$, i.e.,

$$\mathbf{R}(x; \bar{u}^n) = (R((x; \bar{u}_1^n), ..., R(x; \bar{u}_m^n))^T; \tag{5.1}$$

here **R** denotes vector reconstruction. However, componentwise reconstruction seems natural only if we disregard the time-dependence of $u(x, t)$ which allows discontinuities in the solution to collide with each other.

We recall that the scalar reconstruction is non-oscillatory only if discontinuities are separated by at least $r+1$ points of smoothness, where r is the order of accuracy. Consequently the component-by-component reconstruction (5.1) may cease to be non-oscillatory around the discrete set of points (x_c, t_c), where discontinuities of $u(x, t)$ interact. In the following we describe an algorithm to reconstruct $u(x, t_n)$ from $\bar{u}^n$ which avoids this difficulty by decomposing $\bar{u}^n$ which avoids this difficulty by decomposing $\bar{u}^n$ into m locally defined scalar characteristic variables.

We start by examining the constant coefficient case $f(u) = Au$, where A is a constant $m \times m$ matrix

$$u_t + Au_x = 0 \tag{5.2a}$$

$$u(x, 0) = u_0(x). \tag{5.2b}$$

We note that the eigenvalues $\{a_k\}$ as well as the eigenvectors $\{r_k\}$, $\{l_k\}$ are also constant. We assume that

$$a_1 < a_2 < \cdots < a_m \tag{5.3a}$$

$$l_i r_j = \delta_{ij}. \tag{5.3b}$$

We define the kth characteristic variable w^k by

$$w^k = l_k u. \tag{5.4a}$$

It follows then from (5.3b) that

$$u = \sum_{k=1}^{m} w^k r_k. \tag{5.4b}$$

Multiplying (5.2) from the left by l_k we see that $w^k(x, t)$ satisfies the following scalar IVP

$$(w^k)_t + a_k (w^k)_x = 0 \tag{5.5a}$$

$$w^k(w, 0) = l_k u_0(x) \equiv w_0^k(x), \tag{5.5b}$$

the solution to which is

$$w^l(x, t) = w_0^k(x - a_k t). \tag{5.5c}$$

Using (5.4b) and (5.5c) we can express the solution $u(x, t)$ of the constant coefficient IVP (5.2) by

$$u(x, t) = \sum w_0^k(x - a_k t)\, r_k.$$

Let us now consider the following initial data in (5.2b)

$$u_0(x) = \begin{cases} u_L & x < x_L \\ u_M & x_L \leqslant x \leqslant x_R \\ u_R & x_R < x. \end{cases} \tag{5.6a}$$

First let us consider the case $x_L = x_R = 0$ which is the Riemann IVP (4.3). The solution $u(x, t)$ is a self-similar solution $V(x/t; u_L, u_R)$ of the following form

$$u(x, t) = V(x/t; u_L, u_R) = \begin{cases} u_L & x/t < a_1 \\ u^k & a_k < x/t < a_{k+1},\ 1 \leqslant k \leqslant m-1 \\ u_R & a_m < x/t, \end{cases} \tag{5.6b}$$

where

$$u^k = u_L + \sum_{i=1}^{k} (w_R^i - w_L^i)\, r_i, \qquad 1 \leqslant k \leqslant m-1. \tag{5.6c}$$

In the case $x_R > x_L$ in (5.6a) the solution $u(x, t)$, for t small, is

$$u(x, t) = \begin{cases} V\left(\dfrac{x - x_L}{t}; u_L, u_M\right) & \text{for } x < x_L + a_m t \\ u_M & \text{for } x_L + a_m t \leqslant x \leqslant x_R + a_1 t \\ V\left(\dfrac{x - x_R}{t}; u_M, u_R\right) & \text{for } x_R + a_1 t < x. \end{cases} \tag{5.6d}$$

As t increases, the discontinuity in the kth characteristic field originating at $x = x_L$ will eventually collide with any discontinuity in the lth field, $l = 1, ..., k-1$ originating at $x = x_R$.

The example (5.6) demonstrates the difficulty encountered in using the componentwise reconstruction (5.1). We may get oscillations for small t in both (5.6b) and (5.6d) since the discontinuities are too close due to the self-similar nature of the solution to the Riemann problem. Later on we may get more oscillations in (5.6d) as discontinuities collide.

We observe that there are no such problems with $w^k(x, t) = w_0^k(x - a_k t)$. Therefore it makes sense to use the scalar reconstrution $R(x; \bar{w}^k)$ to define

$$\mathbf{R}(x: \bar{u}) = \sum_{k=1}^{m} R(x; \bar{w}^k)\, r_k, \tag{5.7a}$$

where

$$\bar{w}^k = l_k \bar{u}. \tag{5.7b}$$

We generalize (5.7) to the nonlinear system case by using locally defined characteristic variables. To reconstruct u from $\bar{u}$ in $(x_{j-1/2}, x_{j+1/2})$ we define

$$\mathbf{R}(x; \bar{u}) = \sum_{k=1}^{m} R(x; \bar{w}^k(\bar{u}_j))\, r_k(\bar{u}_j) \qquad \text{for} \quad x_{j-1/2} \leqslant x < x_{j+1/2}, \tag{5.8a}$$

where the mesh function $\bar{w}^k(\bar{u}_j) = \{\bar{w}_i^k(\bar{u}_j)\}$ is defined by

$$\bar{w}_i^k(\bar{u}_j) = l_k(\bar{u}_j)\, \bar{u}_i \qquad \text{for} \quad j-p \leqslant i \leqslant j+p; \tag{5.8b}$$

here p is the desired order of reconstruction.

In Section 7 we present calculations for the Euler equations of gas dynamics with the initial data (5.6a). The results of these calculations (as well as those of shocks reflecting from a wall) demonstrate that the reconstruction (5.8) works well also in the nonlinear case.

We turn now to describe our scheme in the case of hyperbolic systems of conservation laws. This scheme is identical in form to (4.14):

$$v_j^{n+1} = v_j^n - \lambda(\bar{f}_{j+1/2} - \bar{f}_{j-1/2}) \tag{5.9a}$$

$$\bar{f}_{j+1/2} = \sum_{k=0}^{K} \alpha_k f^{\mathrm{R}}(\tilde{v}_j(x_{j+1/2}, \beta_k \tau), \tilde{v}_{j+1}(x_{j+1/2}, \beta_k \tau)). \tag{5.9b}$$

The derivation of (5.9), although different in some details, is basically the same as the one presented in Section 4 for the scalar case. Rather than repeating ourselves we shall use the formulae of Section 4 (which are to be interpreted here in a vector sense), and point out the differences whenever they do exist.

The problem to be solved in the "solution-in-the-small" step of the algorithm (1.14b) is as before (4.2). The general structure of the solution $v(x, t)$ is similar to that of the scalar case, i.e., it is composed of sections of smoothness separated by "fans" emanating from the discontinuities at $\{x_{j+1/2}\}$. As in the scalar case we can use a local Cauchy–Kowalewski procedure to approximate $v_j(x, t)$, the section of smoothness of $v(x, t)$ that is connected to the polynomial initial data in $(x_{j-1/2}, x_{j+1/2})$, by $\tilde{v}_j(x, t)$ in (4.12) to any desired order of accuracy. Since $f(u)$ is now a vector, $f'(u)$ is a matrix, $f''(u)$ is a tensor, and so on; consequently, (4.11c) has to be replaced by a much more complicated expression. Rather than doing this we shall present an algorithm in Section 6 to carry out the Cauchy–Kowalewski procedure in the specific case of the Euler equations for gas dynamics.

Next we consider the "fan" that emanates from the discontinuity at $x_{j+1/2}$. As in the scalar case this "fan" starts at $t=0$ as a self-similar solution to the Riemann problem (4.3), which in the system case is a packet of m fans corresponding to the different characteristic fields. A major difference from the scalar case is that (except when the initial data in (4.2b) are piecewise constant) the "fan" emerging from $x_{j+1/2}$ at $t=0$ immediately loses its self-similar nature. Therefore it is no longer possible to express $v(x_{j+1/2}, t)$ in a simple closed form as we did in (4.4). However

$v(x_{j+1/2}, t)$ can be expressed to any desired order of accuracy via a local Taylor expansion of the various curves in the "fan" and the states in between (We refer the interested reader to [1] where Ben-Artzi and Falcowitz describe such an expansion for the Euler equations of gas dynamics.) Thus as in the scalar case, although at considerably more effort, it is possible to obtain an explicit expression that approximates the "abstract numerical flux" (4.8) to any desired order of accuracy.

We turn now to show that the numerical scheme (5.9) is an adequate approximation to the "abstract scheme" (1.16). First we observe that relations (4.16) hold also for the system case; therefore (4.15) follows in exactly the same way as in the scalar case and consequently the scheme (5.9) is likewise rth order accurate.

Next we consider the scheme in the constant coefficient case (5.2). Since both the PDE (1.1a) and the scheme (5.9) decouple into m scalar relations for the characteristic variables w^k in (5.4a), we can apply the analysis of the scalar constant coefficient case to systems in a characteristic-wise fashion. It follows then from (4.17)–(4.18) that the numerical flux (5.9b) is exact and that the numerical scheme (5.9) is identical to the "abstract scheme" (1.16). Let us examine now the structure of the solution $v(x, t)$: The "fan" emanating from $x_{j+1/2}$ has the same form as (5.6b) except that u_L, u^k, and u_R are now functions of x and t. The section of smoothness $x_{j-1/2} + a_m t < x < x_{j+1/2} + a_1 t$ is also the domain of validity of the Taylor expansion (x, t). We note, however, that $l_k \tilde{v}_j(x, t)$, which is the Taylor expansion of $w_j^k(x, t)$, is valid in the larger domain $x_{j-1/2} + a_k t < x < x_{j+1/2} + a_k t$. Next let us examine the role of f^R in formulating the numerical flux (5.9b):

$$f^R(\tilde{v}_j(x_{j+1/2}, t), \tilde{v}_{j+1}(x_{j+1/2}, t)) = \sum_k (a_k)^+ [l_k \tilde{v}_j(x_{j+1/2}, t)] r_k$$

$$+ \sum_k (a_k)^- [l_k \tilde{v}_{j+1}(x_{j+1/2}, t)] r_k; \qquad (5.10a)$$

where

$$(a_k)^+ = \max(0, a_k), \qquad (a_k)^- = \min(0, a_k). \qquad (5.10b)$$

We see from (5.10) that as in the scalar case the role of f^R is that of a pointer, i.e., to identify for each characteristic variable $w^k = l_k v$ to which domain of validity of $\{l_k \tilde{v}_i\}$ does $x = x_{j+1/2}$ belong. Since $l_k \tilde{v}_i(x_{j+1/2}, t) = l_k v(x_{j+1/2} - a_k t, 0)$, the use of Cauchy–Kowalewski procedure in this procedure in this fashion is again computationally equivalent to that of a characteristic method.

In the following we argue that except for the discrete set $\{(x_c, t_c)\}$ of interactions, the above interpretation can be applied to the nonlinear case as well. Unlike the scalar case we do not consider in this paper the "non-convex case" for systems and assume that each characteristic field is either genuinely nonlinear or linearly degenerate (see [19]). When we consider the IVP (4.2) in the context of the numerical scheme where $v(x, 0) = R(x; v^n)$ we see that the "fans" in the solution $v(x, t)$ are related to the global structure of $u(x, t_n)$ in the following way (see Figs. 14 and 16): When $u(x, t_n)$ is smooth, the "fan" has the basic structure of the constant coefficient case linearized around $u(x_{j+1/2}, t_n)$, except that the k waves may have a spread of $O(h^p)$. When $x_{j+1/2}$ is in the vicinity of a shock of $u(x, t_n)$, the "fan" is essentially a shock wave with small perturbations in the other fields. We see that typically (excluding interactions) the "fan" eminating from $x_{j+1/2}$ in the solution $v(x, t)$ is degenerate in the sense that except possibly for a single large shock (or a contact discontinuity) all the waves in it are weak. This heuristic analysis suggests that $f(v(x_{j+1/2}, t))$ can be adequately approximated by a local Roe's linearization; this linearization is exact for a single shock or a contact-discontinuity and amounts to a characteristic approximation for weak waves.

As in the scalar case, f^{ROE} is obtained by a local linearization with respect to a particular average $\hat{u} = \hat{u}(u_{\mathrm{L}}, u_{\mathrm{R}})$ for which

$$f(u_{\mathrm{R}}) - f(u_{\mathrm{L}}) = A(\hat{u})(u_{\mathrm{R}} - u_{\mathrm{L}}). \tag{5.11a}$$

f^{ROE} is defined as the flux at $x = 0$ of the solution to the constant coefficient Riemann IVP:

$$u_t + A(\hat{u})\, u_x = 0$$

$$u(x, 0) = \begin{cases} u_{\mathrm{L}} & x < 0 \\ u_{\mathrm{R}} & x > 0, \end{cases}$$

which can be expressed as

$$f^{\mathrm{ROE}}(u_{\mathrm{L}}, u_{\mathrm{R}}) = \frac{1}{2}\left[f(u_{\mathrm{L}}) + f(u_{\mathrm{R}}) - \sum_{k=1}^{m} \delta_k(u_{\mathrm{L}}, u_{\mathrm{R}}) |a_k(\hat{u})| r_k(\hat{u}) \right], \tag{5.11b}$$

where

$$\delta_k(u_{\mathrm{L}}, u_{\mathrm{R}}) = l_k(\hat{u})(u_{\mathrm{R}} - u_{\mathrm{L}}); \tag{5.11c}$$

here $a_k(\hat{u})$, $l_k(\hat{u})$, and $r_k(\hat{u})$ are evaluated with respect to the Jacobian matrix $A(\hat{u})$. The derivation of Roe's Riemann solver is well documented in the literature (see [25, 8, 9, 14]). In Section 6 we describe f^{ROE} for the Euler equations of gas dynamics.

Finally let us examine the performance of the scheme (5.9) during an interaction of discontinuities in the solution $u(x, t)$ of the IVP (1.1). We observe that it takes some time until the outcoming waves can be properly described on the computational grid. Till then $R(x; v^n)$, which is based on polynomial interpolation, can only be a crude approximation to $u(x, t_n)$. Under these circumstances we expect the "fans" in the solution $v(x, t)$ (4.12) that originate from discontinuities in the interaction zone of $u(x, t_n)$, to be adequately approximated by the self-similar solution to the local Riemann problem. We note that once the outcoming waves are properly resolved on the computational grid, the previous analysis applies.

In Section 7 we present numerical experiments where the scheme (5.9) with f^{R} replaced by f^{ROE} (5.11) is applied to an interaction problem for the Euler equations of polytropic gas. In all these experiments the scheme (5.9) has developed the correct structure of the solution.

We remark that the scheme (5.9) with f^{ROE} in its form (5.11b) admits a stationary "expansion shock" as its steady solution. This can be easily rectified by adding entropy viscosity terms for the genuinely nonlinear characteristic fields (see [13, 14, 9]).

6. Euler Equations of Gas Dynamics

In this section we describe how to apply the scheme (5.1) to the Euler equations of gas dynamics for a polytropic gas:

$$u_t + f(u)_x = 0 \tag{6.1a}$$

$$u = (\rho, m, E)^T \tag{6.1b}$$

$$f(u) = qu + (0, P, qP)^T \tag{6.1c}$$

$$P = (\gamma - 1)(E - \tfrac{1}{2}\rho q^2). \tag{6.1d}$$

Here ρ, q, P, and E are the density, velocity, pressure, and total energy, respectively; $m = \rho q$ is the momentum and γ is the ratio of specific heats.

The eigenvalues of the Jacobian matrix $A(u) = \partial f/\partial u$ are

$$a_1(u) = q - c, \qquad a_2(u) = q, \qquad a_3(u) = q + c, \tag{6.2a}$$

where $c = (\gamma P/\rho)^{1/2}$ is the speed of sound.

The corresponding right-eigenvector are

$$r_1(u) = \begin{pmatrix} 1 \\ q - c \\ H - qc \end{pmatrix}, \qquad r_2(u) = \begin{pmatrix} 1 \\ q \\ \frac{1}{2}q^2 \end{pmatrix}, \qquad r_3(u) = \begin{pmatrix} 1 \\ q + c \\ H + qc \end{pmatrix}; \tag{6.2b}$$

here

$$H = (E + P)/\rho = c^2/(\gamma - 1) + \tfrac{1}{2}q^2 \tag{6.2c}$$

is the enthalpy.

To compute $\{l_k(u)\}$ which is bi-orthonormal to $\{r_k(u)\}$ in (6.2b), we first form the matrix $\mathbf{T}(u)$, the columns of which are the right-eigenvectors in (6.2b)

$$\mathbf{T}(u) = (r_1(u), r_2(u), r_3(u))$$

and then define $l_k(u)$ to be the kth row in $\mathbf{T}^{-1}(u)$, the inverse of $\mathbf{T}(u)$. We get

$$\begin{aligned} l_1(u) &= \tfrac{1}{2}(b_2 + q/c,\ -b_1 q - 1/c,\ b_1) \\ l_2(u) &= (1 - b_2,\ b_1 q,\ -b_1) \\ l_3(u) &= \tfrac{1}{2}(b_2 - q/c,\ -b_1 q + 1/c,\ b_1), \end{aligned} \tag{6.2d}$$

where

$$b_1 = (\gamma - 1)/c^2 \tag{6.2e}$$

$$b_2 = \tfrac{1}{2} q^2 b_1. \tag{6.2f}$$

Given $\{v_j^n\}$, approximation to $\{\bar{u}(x_j, t_n)\}$, we use (6.2d)–(6.2f) to evaluate the locally defined characteristic variables (5.8b)

$$\bar{w}_i^k(v_j^n) = l_k(v_j^n)\, v_i^n \qquad \text{for } i = j - r, \ldots, j + r \text{ and } k = 1, 2, 3. \tag{6.3a}$$

Next we apply our scalar reconstruction algorithm to each of the locally defined characteristic variables in (6.3a) The scalar reconstruction $R(x, \bar{w})$ is described in an algorithmic form in an Appendix; the output of this algorithm is in the form of the finite Taylor series in (4.2b). Thus we get for each characteristic variable in $(x_{j-1/2}, x_{j+1/2})$

$$R(x; \bar{w}^k(v^n)) = \sum_{l=0}^{r-1} b_{j,l}^k (x - x_j)^l / l! \tag{6.3b}$$

Rearranging terms we can express the vector reconstruction (5.8a) by

$$\mathbf{R}(x; v^n) = \sum_{l=0}^{r-1} \mathbf{b}_{j,l} (x - x_j)^l / l!, \tag{6.3c}$$

where

$$\mathbf{b}_{j,l} = \sum_{k=1}^{3} b_{j,l}^k r_k. \tag{6.3d}$$

Note that wherever the solution is smooth

$$\mathbf{b}_{j,l} = \frac{\partial^l}{\partial x^l} (\rho, m, E)^T \bigg|_{x = x_j} + O(h^{r-1}) \qquad \text{for} \quad 0 \leqslant l \leqslant r - 1. \tag{6.3e}$$

We turn now to describe the Cauchy–Kowalewski procedure (4.4)–(4.5) for the Euler equations of gas dynamics. We start by using the reconstruction output (6.3c) to define (4.4a), i.e.,

$$\frac{\partial^l v(x_j, 0)}{\partial x^l} = \begin{cases} \mathbf{b}_{j,l} & \text{for } 0 \leqslant l \leqslant r - 1 \\ 0 & \text{for } l \geqslant r. \end{cases} \tag{6.4a}$$

We find it convenient to express $\partial^l v(x_j, 0)/\partial x^k \partial t^{l-k}$ in terms of derivatives of the 4-vector $Z=(\rho, m, P, q)^T$. For this purpose we use (6.4a) and the relations

$$m=\rho q$$

$$P=(\gamma-1)(E-\tfrac{1}{2}qm)$$

to find the x-derivatives of q and P, by

$$\begin{aligned} m_x &= q\rho_x+\rho q_x \Rightarrow q_x=(m_x-q\rho_x)/\rho \\ P_x &= (\gamma-1)[E_x-\tfrac{1}{2}(q_x m+qm_x)] \end{aligned} \tag{6.4b}$$

$$\begin{aligned} m_{xx} &= \rho q_{xx}+2\rho_x q_x+q\rho_{xx} \Rightarrow q_{xx}=(m_{xx}-q\rho_{xx}-2q_x\rho_x)/\rho \\ P_{xx} &= (\gamma-1)[E_{xx}-\tfrac{1}{2}(mq_{xx}+2q_x m_x+qm_{xx})] \end{aligned} \tag{6.4c}$$

and so on. Having evaluated $\partial^l Z(x_j, 0)/\partial x^l$ for $0 \leqslant l \leqslant r-1$, we proceed to obtain the rest of the derivatives $\partial^l Z(x_j, 0)/\partial x^k \partial t^{l-k}$, $0 \leqslant l \leqslant r-1$, $0 \leqslant k \leqslant l$ by differentiating the PDEs

$$\rho_t+m_x=0 \tag{6.5a}$$

$$m_t+(qm)_x+P_x=0 \tag{6.5b}$$

$$P_t+qP_x+\gamma P u_x=0 \tag{6.5c}$$

and the algebraic relation

$$m=q\rho \tag{6.5d}$$

in the following ordered way: Compute $Z_t(x_j, 0)$ from

$$\begin{aligned} \rho_t+m_x&=0 \\ m_t+q_x m+qm_x+P_x&=0 \\ P_t+qP_x+\gamma P q_x&=0 \\ \rho q_t+\rho_t q&=m_t; \end{aligned} \tag{6.6a}$$

compute $Z_{xt}(x_j, 0)$ from

$$\begin{aligned} \rho_{xt}+m_{xx}&=0 \\ m_{xt}+2q_x m_x+qm_{xx}+mq_{xx}+P_{xx}&=0 \\ P_{xt}+qP_{xx}+\gamma P q_{xx}+(1+\gamma)\,q_x P_x&=0 \\ \rho q_{xt}+\rho_x q_t+\rho_t q_x+q\rho_{xt}&=m_{xt}; \end{aligned} \tag{6.6b}$$

compute $Z_{tt}(x_j, 0)$ from

$$\begin{aligned} \rho_{tt} + m_{xt} &= 0 \\ m_{tt} + qm_{xt} + mq_{xt} + P_{xt} + (q_x m_t + q_t m_x) &= 0 \\ P_{tt} + qP_{xt} + \gamma P q_{xt} + (q_t P_x + \gamma P_t q_x) &= 0 \\ \rho q_{tt} + 2\rho_t q_t + q\rho_{tt} &= m_{tt}, \end{aligned} \tag{6.6c}$$

compute $Z_{xxt}(x_j, 0)$ from

$$\begin{aligned} \rho_{xxt} + m_{xxx} &= 0 \\ m_{xxt} + qm_{xxx} + mq_{xxx} + P_{xxx} + 3q_x m_{xx} + 3q_{xx} m_x &= 0 \\ P_{xxt} + qP_{xxx} + \gamma P q_{xxx} + (2+\gamma)\, q_x P_{xx} + (1+2\gamma)\, q_{xx} P_x &= 0 \\ \rho q_{xxt} + 2(\rho_x q_{xt} + \rho_{xt} q_x) + q\rho_{xxt} + \rho_{xx} q_t + \rho_t q_{xx} &= m_{xxt}; \end{aligned} \tag{6.6d}$$

compute $Z_{xtt}(x_j, 0)$ from

$$\begin{aligned} \rho_{xtt} + m_{xxt} &= 0 \\ m_{xtt} + qm_{xxt} + mq_{xxt} + P_{xxt} + 2(q_x m_{xt} + m_x q_{xt}) + (m_t q_{xx} + q_t m_{xx}) &= 0 \\ P_{xtt} + qP_{xxt} + \gamma P q_{xxt} + (\gamma+1)[q_{xt} P_x + q_x P_{xt}] + q_t P_{xx} + \gamma P_t q_{xx} &= 0 \\ \rho q_{xtt} + q\rho_{xtt} + \rho_x q_{tt} + q_x \rho_{tt} + 2(\rho_{xt} q_t + \rho_t q_{xt}) &= m_{xtt}; \end{aligned} \tag{6.6e}$$

compute $Z_{ttt}(x_j, 0)$ from

$$\begin{aligned} \rho_{ttt} + m_{xtt} &= 0 \\ m_{ttt} + qm_{xtt} + mq_{xtt} + P_{xtt} + 2(q_t m_{xt} + m_t q_{xt}) + (q_x m_{tt} + m_x q_{tt}) &= 0 \\ P_{ttt} + qP_{xtt} + \gamma P q_{xtt} + \gamma q_x P_{tt} + 2(q_t P_{xt} + \gamma P_t q_{xt}) + q_{tt} P_x &= 0 \\ \rho q_{ttt} + q\rho_{ttt} + 3(\rho_t q_{tt} + q_t \rho_{tt}) &= m_{ttt}; \end{aligned} \tag{6.6f}$$

and so on.

We note that one can differentiate the algebraic relation (6.1d) in order to obtain $\partial^l E(x_j, 0)/\partial x^k\, \partial t^{l-k}$ in terms of the already evaluated derivatives of P, q, and m and use the derivatives of the conserved quantities ρ, m, E to compute $\bar{v}_j(x, t)$ in (4.5). However, it is more convenient to evaluate the flux $f(u)$ and $f^{R}(u_1, u_2)$ in terms of p, q, and P; since $\bar{v}_j(x, t)$ is smooth and the scheme (5.9) is in conservation form we do not really to worry about relation (4.12b). See Remark 6.2. For this reason we use the first, third, and fourth components of $\tilde{Z}_j(x, t)$

$$\tilde{Z}_j(x, t) = \sum_{l=0}^{r-1} \sum_{k=0}^{l} \frac{\partial^l Z(x_j, 0)}{\partial x^k\, \partial t^{l-k}} \cdot \frac{(x-x_j)^k}{k!} \frac{t^{l-k}}{(l-k)!} \tag{6.7}$$

to define $\tilde{\rho}_j(x, t)$, $\tilde{P}_j(x, t)$, and $\tilde{q}_j(x, t)$, respectively.

Once we have computed (6.7) we can compute the numerical flux $\bar{f}_{j+1/2}$ in (5.9b).

An exact $f^{\mathrm{R}}(u_1, u_2)$, i.e.,

$$f^{\mathrm{R}}(u_1, u_2) = f(V(0; u_1, u_2)), \tag{6.8}$$

where $V(x/t; u_1, u_2)$ is the exact solution of the Riemann problem for the Euler equations of gas dynamics, can be computed through an iterative algorithm. This algorithm is rather complicated, and we refer the reader to [5, 3, and 28] for its details.

To compute $f^{\mathrm{R}} \equiv f^{\mathrm{ROE}}$ in (5.11b) and (5.9b) all we need is to describe the particular average $\hat{u}(u_1, u_2)$ for the Euler equations of gas dynamics (see [25, 9]). To do so we denote the arithmetic mean of $b(u)$ with respect to u_1 and u_2 by

$$\langle b \rangle = \tfrac{1}{2}[b(u_1) + b(u_2)] \tag{6.9a}$$

and define

$$\hat{q} = \langle \sqrt{\rho} q \rangle / \langle \sqrt{\rho} \rangle, \qquad \hat{H} = \langle \sqrt{\rho} H \rangle / \langle \sqrt{\rho} \rangle, \qquad \hat{c} = (\gamma - 1)^{1/2} \cdot \sqrt{\hat{H} - \tfrac{1}{2}\hat{q}^2};$$

here H is the enthalpy (6.2c). Having prescribed $\hat{q}$, $\hat{H}$, and $\hat{c}$, we have all the quantitites needed to define the eigenvalues and eigenvectors in (6.2).

Remark 6.1. The importance of using the particular average (6.9) rather than a simpler one is that when (u_1, u_2) corresponds to a single shock or a single contact discontinuity in the solution of the Riemann problem $V(x/t; u_1, u_2)$, then f^{ROE} is exact, i.e.

$$f^{\mathrm{ROE}}(u_1, u_2) = f^{\mathrm{R}}(V(0; u_1, u_2)). \tag{6.10}$$

Remark 6.2. No matter how we derive $\tilde{v}_j(x, t)$ the numerical approximation (1.16a) is in conservation form. However, in order to make v_j^{n+1} approximate the cell average (1.14c) (which is desirable for stability purposes), we need the reconstruction to be conservative, i.e., (1.14b) should be valid.

7. Numerical Experiments

In this section we present results of several computer experiments with the schemes (4.14) and (5.9). These schemes will be referred to as rth order ENO schemes (or "rth order" when applicable—see Remark 1.2); ENO stands for essentially non-oscillatory.

The ENO schemes are highly nonlinear and consequently do not easily lend themselves to rigorous analysis. At present we have completed the analysis of the non-oscillatory interpolation H_m (3.1)–(3.5) and have acquired a fairly good understanding of the reconstruction $R(x; \bar{w})$; these reconstruction results can be extended to a single application of the "abstract scheme" (1.10) to piecewise smooth data. Unfortunately we have not been able as yet to analyze rigorously the crucial question of accumulation of error. Under these circumstances, computer experiments have become our main tool of analysis. We have performed a large number of numerical experiments with initial data ranging from random noise to smooth functions. We have studied two notions of "stability": (i) boundedness of a refinement sequence $h \to 0$, $\tau = O(h)$ for $0 \leqslant t \leqslant T$; (ii) boundedness of the numerical solution as $n \to \infty$ with fixed h and τ. In all our experiments[1] we have found the ENO schemes to be stable under a CFL restriction of 1 and strongly so, in the sense that they strongly damp high frequency noise—this is probably due to the cell-averaging step (1.14c).

In [15], the first paper in this series, we have presented numerical results which compare the second-order ENO scheme based on RD with $r = 2$ to a "second-order accurate" MUSCL-type scheme, which is computationally equivalent to the "second-order" ENO scheme based on RP with $r = 2$.

In [16], the second paper in this series, we have presented numerical experiments that verify our statements about the accuracy and non-oscillatory nature of the reconstruction $R(x; \bar{w})$, and demonstrate the stability of the ENO schemes in the scalar constant coefficient case for both the pure IVP and the mixed initial-boundary value problem (IBVP).

In this paper we present a sample of our numerical experiments for the nonlinear scalar case and the Euler equations of gas dynamics in 1D. The purpose of this presentation is to address the open questions that we could not fully answer by analysis: The accumulation of error, the adequacy of the "solution in the small" procedure, consistency with entropy inequalities, and the effectiveness of the characteristic-wise reconstruction for systems. We have performed most of the numerical experiments for $r = 1, 2, 3, 4, 5, 6$. Since it is not practical to present six sets of data for each problem we usually compare $r = 2$, which is the current state of the art scheme, to $r = 4$ which seems to be optimal for smooth solutions. However, presentation of a comprehensive efficiency study is deferred to a future paper.

A. Scalar Conservation Laws

A1. *Convex $f(u)$ with smooth initial data*

In this subsection we show results of applying the ENO schemes to

$$u_t + (u^2/2)x = 0 \tag{7.1a}$$

$$u(x, 0) = \alpha + \beta \sin(\pi x + \gamma), \tag{7.1b}$$

[1] The only expection where we had to reduce the CFL number is for the initial data of the mesh oscillation function $v_j^0 = (-1)^j$. This choice of initial data forces the ENO scheme to become linear; for $v_j^0 = \theta_j(-1)^j$, where θ_j is a positive random number, the scheme is again stable under a CFL restriction of 1 (see [16] for more details).

TABLE IA

Solution of the Periodic IVP (7.1) at $t=0.3$ by ENO Schemes Based on RP

J	$R=1$	r_c	$r=2$	r_c	$r=3$	r_c	$r=4$	r_c	$r=5$	r_c
L_∞-Error										
8	1.582×10^{-1}		9.047×10^{-2}		3.804×10^{-2}		2.715×10^{-2}		2.117×10^{-2}	
		0.80		1.07		2.24		2.76		2.86
16	9.082×10^{-2}		4.300×10^{-2}		8.038×10^{-3}		3.998×10^{-3}		2.906×10^{-3}	
		0.87		1.24		2.10		3.18		3.76
32	4.964×10^{-2}		1.819×10^{-2}		1.876×10^{-3}		4.423×10^{-4}		2.151×10^{-4}	
		0.91		1.32		2.35		3.39		3.65
64	2.648×10^{-2}		7.296×10^{-3}		3.684×10^{-4}		4.211×10^{-5}		1.713×10^{-5}	
		0.92		1.33		2.14		3.02		4.50
128	1.404×10^{-2}		2.900×10^{-3}		8.356×10^{-5}		5.188×10^{-6}		7.562×10^{-7}	
L_1-Error										
8	8.440×10^{-2}		4.507×10^{-2}		1.669×10^{-2}		1.003×10^{-2}		7.000×10^{-3}	
		0.98		1.79		2.70		3.21		3.57
16	4.279×10^{-2}		1.304×10^{-2}		2.574×10^{-3}		1.086×10^{-3}		5.875×10^{-4}	
		0.97		1.81		2.69		3.53		4.08
32	2.186×10^{-2}		3.707×10^{-3}		3.992×10^{-4}		9.400×10^{-5}		3.462×10^{-5}	
		0.96		1.89		2.69		3.57		4.48
64	1.124×10^{-2}		9.980×10^{-4}		6.165×10^{-5}		7.905×10^{-6}		1.546×10^{-6}	
		0.99		1.83		2.56		3.53		4.62
128	5.675×10^{-3}		2.813×10^{-4}		1.042×10^{-5}		6.835×10^{-7}		6.270×10^{-8}	

TABLE IB

Solution of the Periodic IVP (7.1) at $t=0.3$ by ENO Schemes Based on RD

J	$r=1$	r_c	$r=2$	r_c	$r=3$	r_c	$r=4$	r_c	$r=5$	r_c
L_∞-Error										
8	1.582×10^{-1}		5.204×10^{-2}		4.484×10^{-2}		2.787×10^{-2}		2.115×10^{-2}	
		0.80		1.94		3.01		2.97		3.17
16	9.082×10^{-2}		1.352×10^{-2}		5.577×10^{-3}		3.565×10^{-3}		2.343×10^{-3}	
		0.87		1.92		2.67		3.24		3.61
32	4.964×10^{-2}		3.562×10^{-3}		8.791×10^{-4}		3.777×10^{-4}		1.191×10^{-4}	
		0.91		2.05		3.96		3.69		5.20
64	2.648×10^{-2}		8.610×10^{-4}		5.658×10^{-5}		2.927×10^{-5}		5.242×10^{-6}	
		0.92		2.30		3.22		4.76		5.36
128	1.404×10^{-2}		1.748×10^{-4}		6.081×10^{-6}		1.077×10^{-6}		1.276×10^{-7}	
L_1-Error										
8	8.440×10^{-2}		2.231×10^{-2}		1.333×10^{-2}		8.600×10^{-3}		6.045×10^{-3}	
		0.98		2.40		3.26		3.71		3.90
16	4.279×10^{-2}		4.228×10^{-3}		1.388×10^{-3}		6.575×10^{-4}		4.063×10^{-4}	
		0.97		2.30		3.31		4.02		4.40
32	2.186×10^{-2}		8.565×10^{-4}		1.399×10^{-4}		4.056×10^{-5}		1.919×10^{-5}	
		0.96		2.23		3.67		4.39		5.40
64	1.124×10^{-2}		1.826×10^{-4}		1.096×10^{-4}		1.096×10^{-5}		1.936×10^{-6}	
		0.99		2.18		3.35		4.53		5.53
128	5.675×10^{-3}		4.039×10^{-5}		1.071×10^{-6}		8.385×10^{-8}		9.810×10^{-9}	

TABLE IIA

Solution of the IBVP (7.1) and (7.5) at $t = 0.3$ by ENO Schemes Based on RP

J	$r=1$	r_c	$r=2$	r_c	$r=3$	r_c	$r=4$	r_c	$r=5$	r_c
L_∞-Error										
8	1.471×10^{-1}		9.047×10^{-2}		2.849×10^{-2}		2.936×10^{-2}		1.766×10^{-2}	
		0.68		1.07		0.77		1.56		0.69
16	9.184×10^{-2}		4.300×10^{-2}		1.672×10^{-2}		9.943×10^{-3}		1.096×10^{-2}	
		0.77		1.24		2.75		2.98		3.58
32	5.382×10^{-2}		1.819×10^{-2}		2.477×10^{-3}		1.275×10^{-3}		9.149×10^{-4}	
		0.75		1.32		2.75		5.14		5.56
64	3.198×10^{-2}		7.296×10^{-3}		3.684×10^{-4}		3.614×10^{-5}		1.944×10^{-5}	
		0.88		1.33		2.14		2.83		4.69
128	1.742×10^{-2}		2.900×10^{-3}		8.356×10^{-5}		5.089×10^{-6}		7.540×10^{-7}	
L_1-Error										
8	7.745×10^{-2}		3.355×10^{-2}		1.221×10^{-2}		8.065×10^{-3}		5.265×10^{-3}	
		0.95		1.33		1.60		2.40		1.92
16	4.015×10^{-2}		1.339×10^{-2}		4.032×10^{-3}		1.528×10^{-3}		1.389×10^{-3}	
		0.91		1.88		3.14		3.40		4.15
32	2.143×10^{-2}		3.643×10^{-3}		4.563×10^{-4}		1.445×10^{-4}		7.810×10^{-5}	
		0.96		1.86		2.89		4.27		5.04
64	1.102×10^{-2}		1.003×10^{-3}		6.175×10^{-5}		7.505×10^{-6}		2.371×10^{-6}	
		0.97		1.81		2.70		3.58		5.07
128	5.630×10^{-3}		2.855×10^{-4}		9.485×10^{-6}		6.285×10^{-7}		7.060×10^{-8}	

TABLE IIB

Solution of the IBVP (7.1) and (7.5) at $t = 0.3$ by ENO Schemes Based on RD

J	$r=1$	r_c	$r=2$	r_c	$r=3$	r_c	$r=4$	r_c	$r=5$	r_c
L_∞-Error										
8	1.471×10^{-2}		3.264×10^{-2}		3.122×10^{-2}		1.799×10^{-2}		9.192×10^{-3}	
		0.68		0.78		1.86		0.69		-.65
16	9.184×10^{-2}		1.907×10^{-2}		8.571×10^{-3}		1.114×10^{-2}		1.441×10^{-2}	
		0.77		2.46		2.65		3.58		3.54
32	5.382×10^{-2}		3.454×10^{-3}		1.366×10^{-3}		9.312×10^{-4}		1.240×10^{-3}	
		0.75		2.06		4.71		5.01		6.96
64	3.198×10^{-2}		8.302×10^{-4}		5.220×10^{-5}		2.897×10^{-5}		9.727×10^{-6}	
		0.88		2.24		3.12		4.75		6.07
128	1.742×10^{-2}		1.754×10^{-4}		5.996×10^{-6}		1.075×10^{-6}		1.473×10^{-7}	
L_1-Error										
8	7.745×10^{-2}		1.488×10^{-2}		8.425×10^{-3}		5.480×10^{-3}		2.240×10^{-3}	
		0.95		1.52		2.45		2.00		0.88
16	4.015×10^{-2}		5.190×10^{-3}		1.540×10^{-3}		1.372×10^{-3}		1.217×10^{-3}	
		0.91		2.57		3.18		4.07		3.83
32	2.143×10^{-2}		8.740×10^{-4}		1.694×10^{-4}		8.105×10^{-5}		8.570×10^{-5}	
		0.96		2.22		4.00		4.90		6.26
64	1.102×10^{-2}		1.879×10^{-4}		1.059×10^{-5}		2.747×10^{-6}		1.119×10^{-6}	
		0.97		2.15		3.39		4.90		6.68
128	5.630×10^{-3}		4.219×10^{-5}		1.012×10^{-6}		9.200×10^{-8}		1.092×10^{-8}	

for $-1 \leqslant x \leqslant 1$, $t \geqslant 0$. In these calculations we have used the ENO schemes with f^{R} replaced by f^{ROE} (4.21) (without any entropy correction).

Let $\mathscr{X}(x, t)$ denote the solution of (7.1a) with $\mathscr{X}(x, 0) = \sin \pi x$, i.e., $\beta = 1$, $\alpha = \gamma = 0$ in (7.1b). The solution $\mathscr{X}(x, t)$ is smooth for $0 \leqslant t < 1/\pi$; when $t = 1/\pi$ a shock develops at $x = \pm 1$ and stays there as a stationary shock for $t > 1/\pi$. Some time after its development, this shock starts interacting with the expansion wave in $-1 < x < 1$; this brings about a fast decay of the solution. The "exact" solution presented in the following is computed in $0 \leqslant x < 1$ by using Newton–Raphson iterations to solve the characteristic relation

$$\mathscr{X} = \sin \pi(x - \mathscr{X} t); \tag{7.2}$$

$\mathscr{X}$ in $(-1, 0)$ is obtained from $\mathscr{X}$ in $(0, 1)$ by $\mathscr{X}(-x, t) = -\mathscr{X}(x, t)$. The general solution of (7.1) is computed from $\mathscr{X}(x, t)$ in $(-1, 1)$ by

$$u(x, t) = \alpha + \beta \mathscr{X}(x - \alpha t + \gamma, \beta t). \tag{7.3}$$

In Tables I and II and Fig. 1 we present the computation of (7.1) with $\alpha = 1$, $\beta = \frac{1}{2}$, $\gamma = 0$, i.e., $u(x, t) = 1 + \frac{1}{2}\mathscr{X}(x - t, \frac{1}{2}t)$; thus the shock develops at $t = 2/\pi$. The results are presented at $t = 0.3$ when the solution is still smooth. We divide $(-1, 1)$ into J equal intervals and define

$$x_j = -1 + (j - 1/2)h, \qquad h = 2/J, \qquad 1 \leqslant j \leqslant J \tag{7.4}$$

First we consider the pure IVP for (7.1), i.e., periodic boundary conditions at $x = \pm 1$. In Figs. 1a and 1b we show the results of the ENO schemes with RD at $t = 0.3$; Fig. 1a shows the second-order ENO scheme, while Fig. 1b shows the fourth-order one. Both calculations were performed with $J = 10$ and CFL $= 0.6$. The

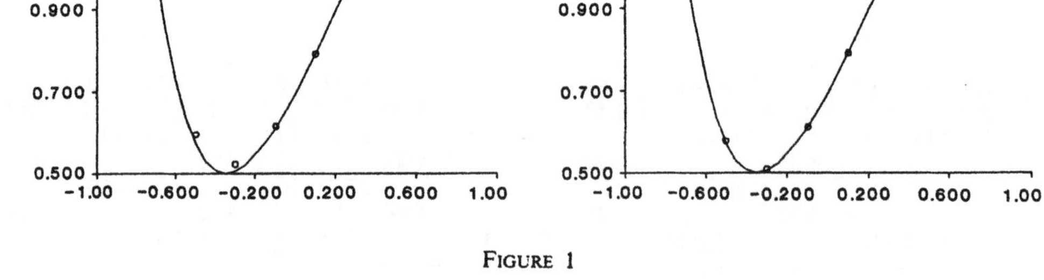

FIGURE 1

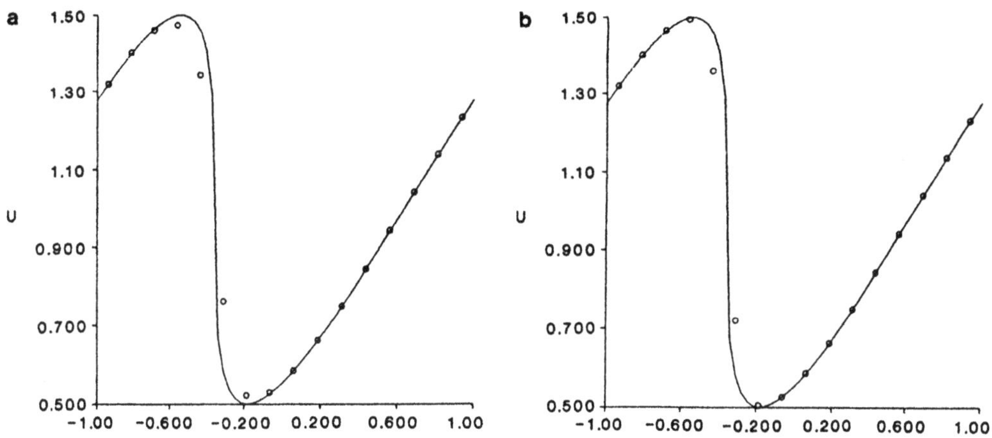

FIGURE 2

continuous line in these figures is the exact solution; the circles represent the values of $R(x_j; v^N)$. In Tables IA and IB we list the L_∞-error and the L_1-error at $t=0.3$ of a refinement sequence $J=8$, 16, 32, 64, 128 for $r=1, 2, 3, 4, 5$ with CFL $=0.6$. Table IA shows the results of the ENO schemes with RP while Table IB shows the ones with RD. The value of r_c in Tables I and II is the "computational order of accuracy" which is calculated by assuming the error to be a constant times h^{r_c}; this definition is meaningful only for h sufficiently small.

In Figs. 2a and 2b we use the same schemes as in Fig. 1, but with $J=16$, and show the results at $t \approx 2/\pi$ (after 17 time steps) which is the time of the formation of the shock. In Figs. 3a and 3b we show the reconstruction $R(x; v^n)$ corresponding to

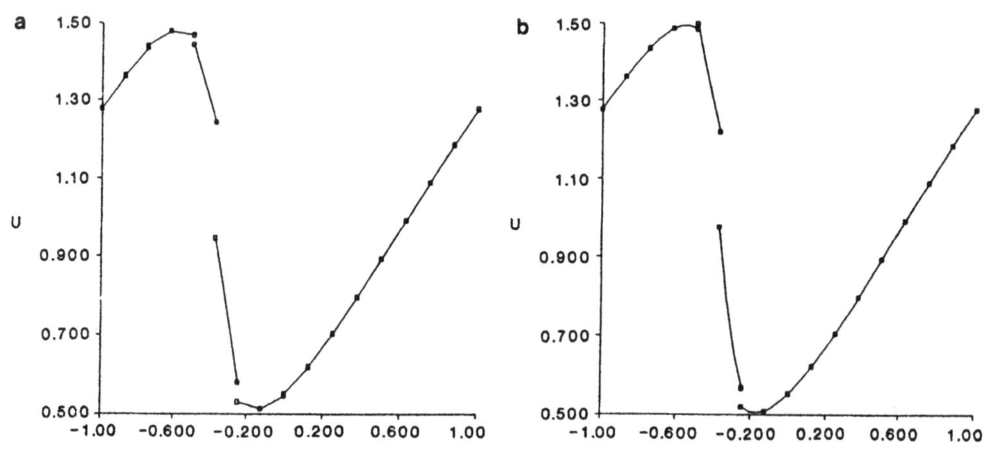

FIGURE 3

the numerical solutions of Fig. 2. The squares in Figs. 3a and 3b mark the values of $R(x_{j+1/2} \pm 0; v^n)$. $R(x; v^n)$ is a piecewise linear in Fig. 3a and piecewise cubic in Fig. 3b.

Next we consider the IBVP for (7.1); since the characteristic speed for (7.3) with $\alpha = 1$, $\beta = \frac{1}{2}$, $\gamma = 0$ is positive, we prescribe

$$u(-1, t) = g(t); \tag{7.5a}$$

$x = +1$ is an outflow boundary and no condition is prescribed there. To be able to compare with the periodic problem we take $g(t)$ in (7.5a) to be the value of the periodic solution at $x = -1$, i.e.,

$$g(t) = 1 + \tfrac{1}{2}\mathscr{X}(-1 - t, \tfrac{1}{2}t). \tag{7.5b}$$

The point of view that we have taken in treating boundary conditions is consistent with the presentation of the "abstract scheme" (1.10), (1.14) as a sequence of global operations. Thus in the reconstruction step, as in the pure IVP case, we use the given cell averages $\{v_j^n\}$, $1 \leqslant j \leqslant J$, to get $R(x; v^n)$ for $-1 < x < 1$; in the presence of boundaries we restrict the choice of stencil to the available information by imposing the condition

$$1 \leqslant i_k(j) \leqslant J - r \quad \text{for} \quad 1 \leqslant k \leqslant r \tag{7.6}$$

in the algorithm (3.4). Note that we do not use the given boundary data $g(t)$ (7.5a) in the reconstruction step. The boundary data is incorporated into the scheme on the PDE level by considering the solution-in-the-small step to be an IBVP. Obviously the resulting scheme is biased "against the wind" near $x = -1$; nevertheless, numerical experiments in the nonlinear case as well as in the constant coefficient case (see [16]), indicate that the ENO schemes are stable. We observe

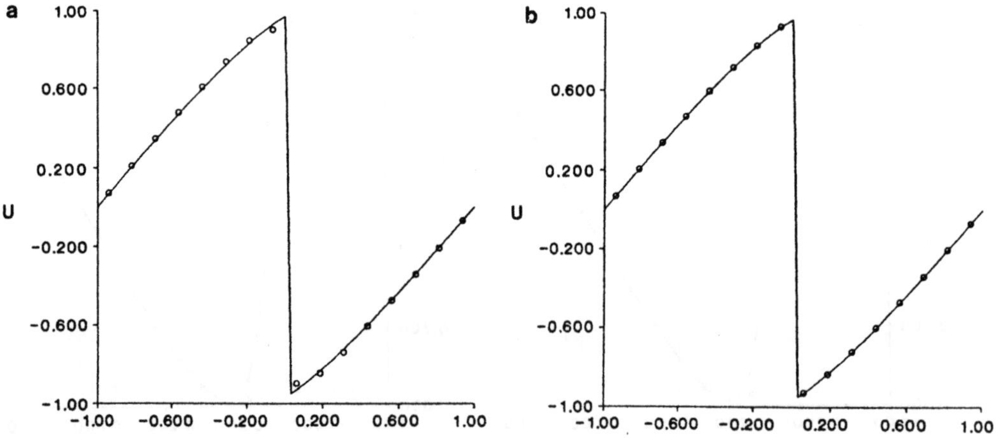

FIGURE 4

that a similar choice of stencil occurs near discontinuities in the interior of the computational grid.

In Tables IIA and IIB we repeat the calculation in Table I for the IBVP (7.1) with (7.5). In Figs. 4a and 4b we show the calculations of the ENO schemes based on RP with $r=2$ and $r=4$, respectively, for the IBVP (7.1) with $\alpha=0$, $\beta=1$, $\gamma=\pi$. Here the boundaries $x=\pm 1$ are characteristic, and a stationary shock develops at $x=0$ at $t=1/\pi$. In these calculations we have treated $x=-1$ as an inflow boundary and specified

$$u(-1, t)=0;$$

$x=+1$ was treated as an outflow boundary. The results show the numerical solution with $J=16$ and $\mathrm{CFL}=0.6$ at $t=0.6$, at which time the solution has already started to decay considerably due to the interaction of the shock with the expansion waves.

A2. *Riemann IVP for Nonconvex $f(u)$*

In this subsection we show results of applying the ENO schemes to the Riemann IVP

$$u_t+f(u)_x=0, \qquad u(x, 0)=\begin{cases} u_L & x<0 \\ u_R & x>0, \end{cases} \tag{7.7a}$$

where $f(u)$ is the nonconvex function

$$f(u)=\tfrac{1}{4}(u^2-1)(u^2-4). \tag{7.7b}$$

We recall that the main difficulty in justifying the approximation (4.13) is when the "fan" in (4.4) covers $x=x_{j+1/2}$; the same difficulty is encountered in justifying the use of f^{ROE} (4.21) instead of the exact flux of the Riemann problem (4.6). Therefore we present two cases in which $\bar{a}(u_L, u_R)=0$; in the first case, $x=0$ is covered by a centered sonic rarefaction fan while, in the second one, there is a stationary (sonic) shock at $x=0$.

In each case we present two sets of experiments. In the first set we use the ENO schemes with the exact f^{R} which is defined by (4.6a); these results, which we consider to be rather pleasing, are presented in Figs. 5 and 7. In the second set of experiments we use the ENO schemes with f^{R} replaced by the following modification of f^{ROE} in (4.21)

$$f^{\mathrm{ROE}}(u_1, u_2)=\tfrac{1}{2}[f(u_1)+f(u_2)-\max(|\bar{a}(u_1, u_2)|, \varepsilon)(u_2-u_1)]. \tag{7.8a}$$

The addition of the linear viscosity term $-\varepsilon(u_2-u_1)/2$ for $|\bar{a}|<\varepsilon$, is the simplest but crudest entropy correction of (4.21). We note that $\varepsilon=0$ in (7.8a) corresponds to (4.21), while $\varepsilon=1/\lambda$ ($\lambda=\tau/h$) corresponds to Lax's first order scheme [18]; since

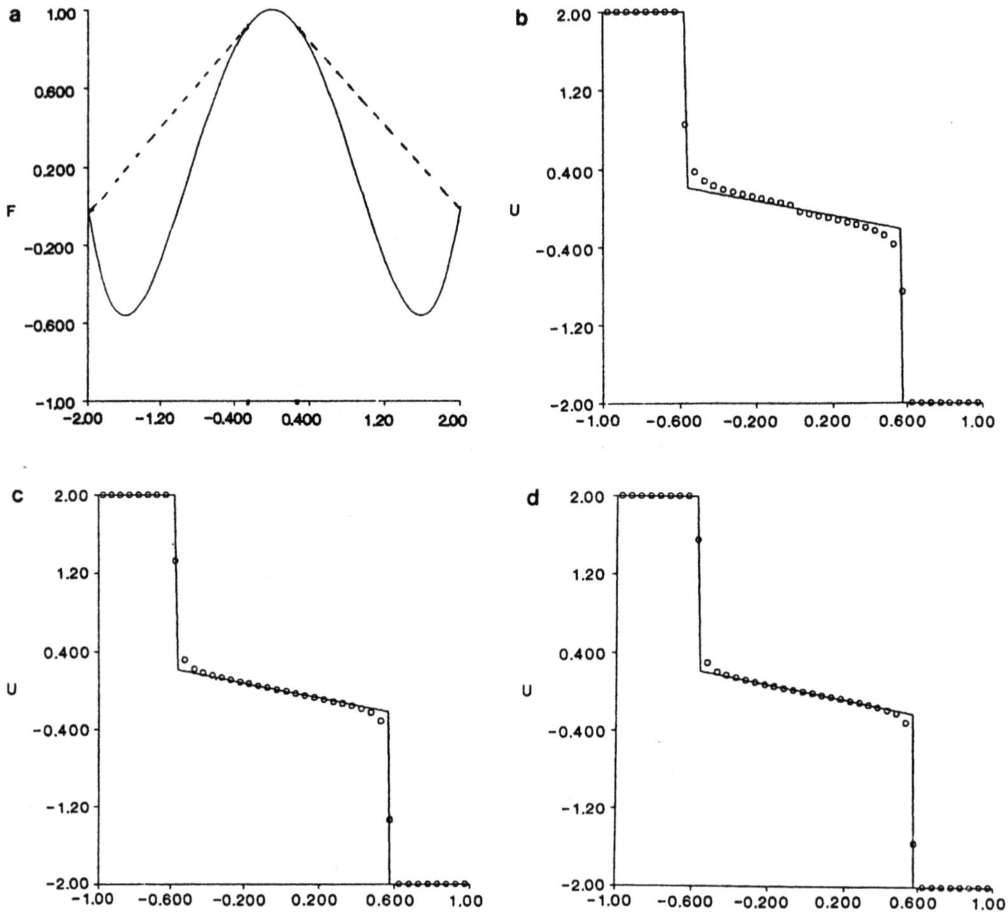

FIGURE 5

(7.8a) satisfies relation (4.16c) the modified scheme remains rth order accurate. In our calculations we take

$$\varepsilon = 0.1/\lambda. \tag{7.8b}$$

Analysis presented in [24] shows that using (7.8a)–(7.8b) in the "second-order" TVD scheme of [9] results in a scheme which converges to entropy correct solutions for convex $f(u)$, provided that λ is sufficiently small; numerical experiments in the convex case [9] and the non-convex case [32] seem to verify this statement even for a CFL number close to 1.

The numerical results of the ENO schemes using (7.8a)–(7.8b) are shown in Figs. 6 and 8. These results show that the ENO schemes converge to entropy

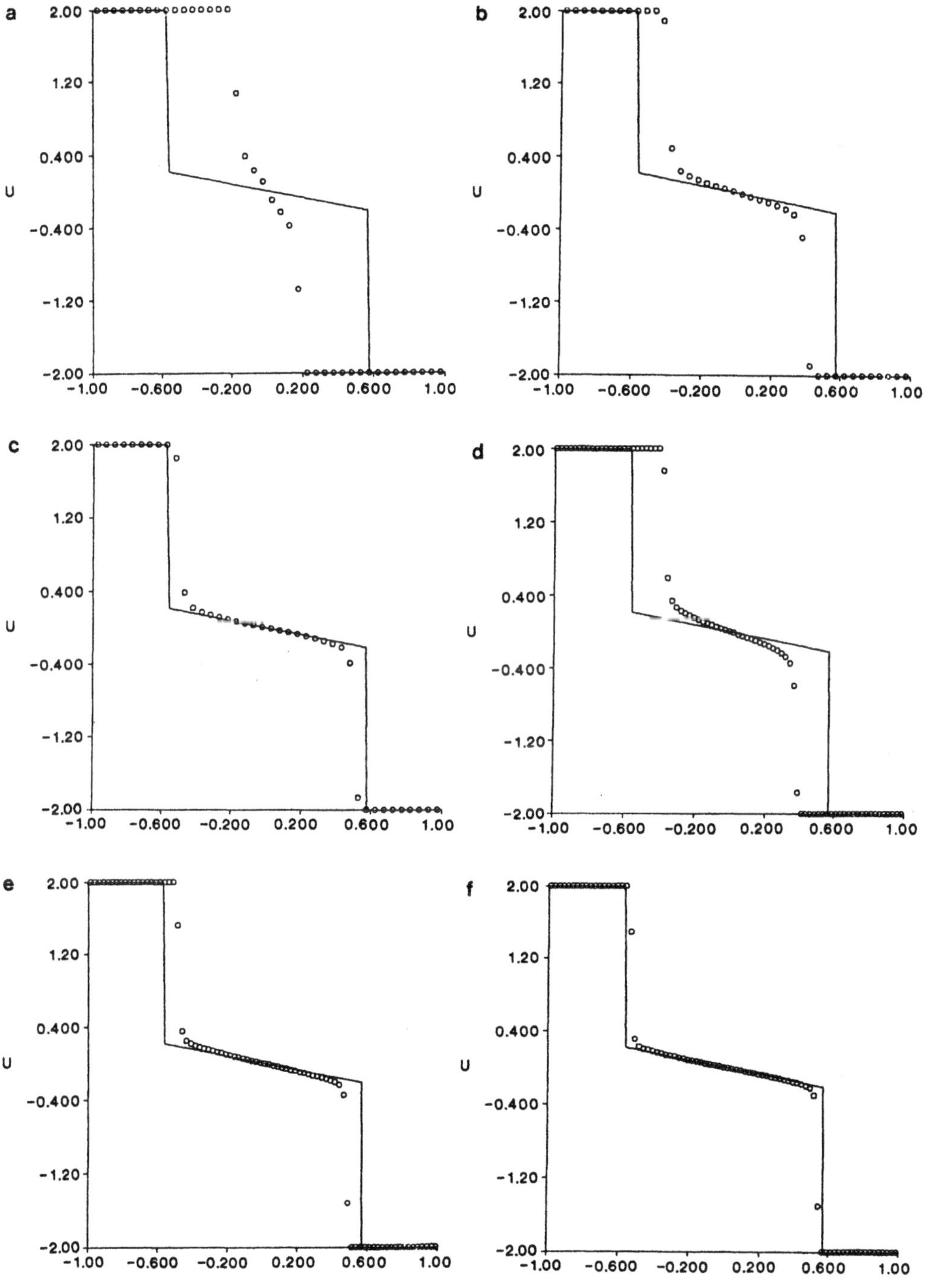

FIGURE 6

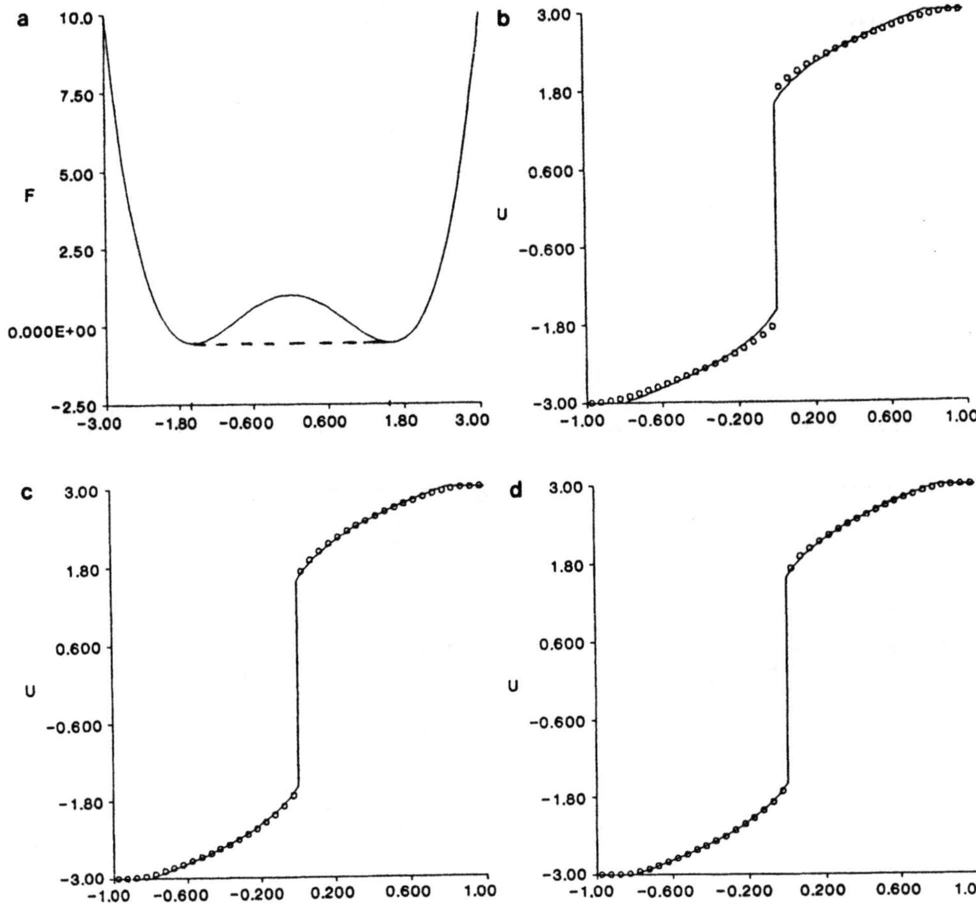

FIGURE 7

correct solutions; however, the quality of the numerical approximation depends strongly on the formal order of accuracy of the scheme.

We remark that an entropy correction to f^{ROE} which is more appropriate for the nonconvex case is obtained by using in (7.8a), $\varepsilon = \varepsilon(u_1, u_2)$ which is defined by

$$\varepsilon = \max[0, \bar{a}(u_1, u_2) - a_L, a_R - \bar{a}(u_1, u_2)], \tag{7.8c}$$

where

$$a_L = \min_{v \in [u_1, u_2]} \bar{a}(u_1, v), \qquad a_R = \max_{v \in [u_1, u_2]} \bar{a}(v, u_2); \tag{7.8d}$$

see [13]. In this case the modified f^{ROE} becomes computationally equivalent to the exact f^{R}.

Our purpose in presenting numerical experiments with the crude entropy correction (7.8b) rather than the more appropriate one (7.8c)–(7.8d) is to demonstrate that the importance of the Riemann solver in the formulation of the ENO schemes is decreasing with increasing order of accuracy. When $r=1$ $R(x; v^n)$ is piecewise constant and all the variation of the solution is contained in the discontinuities of the reconstruction. Consequently the Riemann solver is the only mechanism to describe time evolution. For $r>1$, the smooth polynomial variation in the cell (which is $O(h)$ in regions of smoothness) is generally larger than the variation in the discontinuities of the reconstruction (which is $O(h^r)$ in regions of smoothness)—Figs. 3, 14, and 16. Therefore the time evolution of the smooth polynomial part, namely the Cauchy–Kowalewski procedure is, in general, more important than the Riemann solver. The only exception is in the first few time-steps needed to introduce intermediate states in the solution to the Riemann IVP (7.7), where (u_L, u_R) is not a shock.

In all the calculations presented in this subsection we have used the ENO schemes with RD and CFL $=0.8$.

Case (i). $u_L=2$, $u_R=-2$. The exact solution in this case is (see Fig. 5a)

$$u(x,t)=\begin{cases} 2, & x/t<-0.5281529 \\ g(x/t), & |x|/t<0.5281529 \\ -2, & x/t>0.5281529; \end{cases} \tag{7.9a}$$

here $g(x/t)$ is a centered rarefaction wave: $g(y)$ is the solution of

$$y=f'(g)$$

in the concave part of f which is $|u|<\sqrt{5/6}$;

$$g(\pm 0.5281529)=\mp 0.2152504.$$

In Figs. 5b, c, d we show the results of the ENO schemes using the exact f^R as defined by (4.6a) for $r=1, 2, 4$, respectively; in these calculations we used $J=40$ in (7.4) and $N=80$ time steps. The exact solution is shown by the continuous line; the circles mark the values of $R(x_j; v^N)$. We observe that the structure of the solution in these calculations has developed at the correct rate; this is evident from the fact that the location of the computed shocks is accurate. In Fig. 5b we notice the "dog-leg" which is typical of Godunov's scheme.

In Figs. 6a, b, c we repeat the calculations in Figs. 5b, c, d but with f^R replaced by f^{ROE} (7.8a)–(7.8b). From these figures we see that the scheme develops the correct structure of the solution, but not at the correct rate. This is due to the fact that $\varepsilon=0.1/\lambda$ represents a fan which is much narrower than the initial fan in the exact solution. The location of the computed shocks lags behind the correct location by 8 cells for $r=1$, 3 cells for $r=2$, and only one cell for $r=4$. To verify that the numerical approximations converge to the entropy correct solution we refine the mesh by a factor of 2 and repeat the calculations of Figs. 6a, b, c with $J=80$, $N=160$; the results of these calculations are shown in Figs. 6d, e, f. Since the number of cells by which the computed shocks lags behind the correct location remains the same, we conclude that the numerical approximations indeed do converge to the entropy solution.

Taking into account the crudeness of the entropy correction (7.8b) we consider the performance of the 4th-order scheme in Fig. 6c and 6f to be surprisingly good.

Case (ii). $u_L = -3$, $u_R = 3$.
The exact solution in this case is (see Fig. 7a)

$$u(x,t) = \begin{cases} -3 & x/t \leqslant -19.5 \\ \tilde{g}(x/t) & -19.5 \leqslant x/t < 0 \\ -\tilde{g}(-x/t) & 0 < x/t \leqslant 19.5 \\ 3 & 19.5 \leqslant x/t; \end{cases} \tag{7.9b}$$

here $\tilde{g}(y)$ is the solution of

$$y = f'(\tilde{g})$$

in the convex part of f which is $|u| > \sqrt{5/6}$. Note that the solution (7.9b) is discontinuous at $x=0$; $\tilde{g}(0) = \sqrt{2.5}$.

In Figs. 7b, c, d we show the results of the ENO schemes using the exact f^R as defined by (4.6a) for $r=1, 2, 4$, respectively; in these calculations we used $J=40$ in (7.4) and $N=20$ time steps. We observe that the stationary shock at $x=0$ in these figures in perfectly resolved.

In Figs. 8a, b, c we repeat the calculations in Figs. 7b, c, d but with f^R replaced by f^{ROE} (7.8a)–(7.8b). Since the rarefaction fans in this case are not sonic, the quality of the numerical approximation of the rarefaction wave in Figs. 8a, b, c is similar to that of the corresponding one in Figs. 7b, c, d. We observe that the stationary shock at $x=0$ in Figs. 8a, b, c is somewhat smeared—this is due to the fact that the Riemann solver corresponding to (7.8b) places a fan of the size $|x/t| < \varepsilon$ around $x=0$. Nevertheless, if we compare the results of the 4th-order ENO schemes in the two experiments, we find that the results in Fig. 8c are only slightly inferior to those of Fig. 7d.

B. *Euler Equations of Gas Dynamics*

In this subsection we present numerical experiments with the ENO schemes for the Euler equations of gas dynamics for a polytropic gas with $\gamma = 1.4$ (see Sect. 6). In all these calculations we have used reconstruction via primitive function (RP) and f^{ROE} (5.11), (6.9) without any entropy corrections.

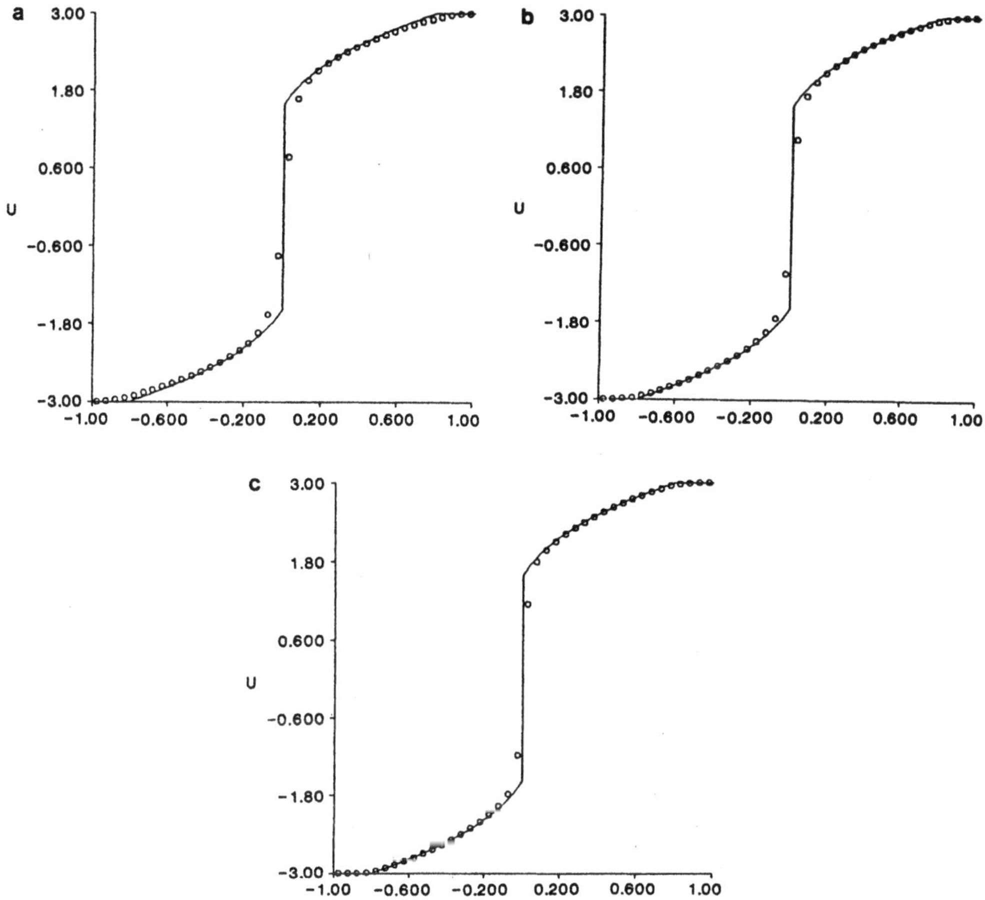

FIGURE 8

B1. *Riemann Problems*

In Figs. 9 and 10 we show the results of applying the ENO schemes with $r=2$ and $r=4$, respectively, to the Riemann problem (7.7a) with the initial data

$$(\rho_L, q_L, P_L) = (1, 0, 1); \qquad (\rho_R, q_R, P_R) = (0.125, 0, 0.10). \tag{7.10a}$$

In these calculations we have used the characteristic reconstruction (5.8), (6.3) with 100 cells, $h=0.1$, CFL $=0.8$, and 50 time steps.

In Fig. 11 we repeat the calculation of the "fourth-order" ENO scheme in Fig. 10 but with component-wise reconstruction (5.1). Comparing Fig. 10 with Fig. 11 we see that there is some "noise" in the component-wise reconstruction which is eliminated by using characteristic reconstruction. We note however that the level of "noise" in Fig. 11 may be considered acceptable for practical calculations.

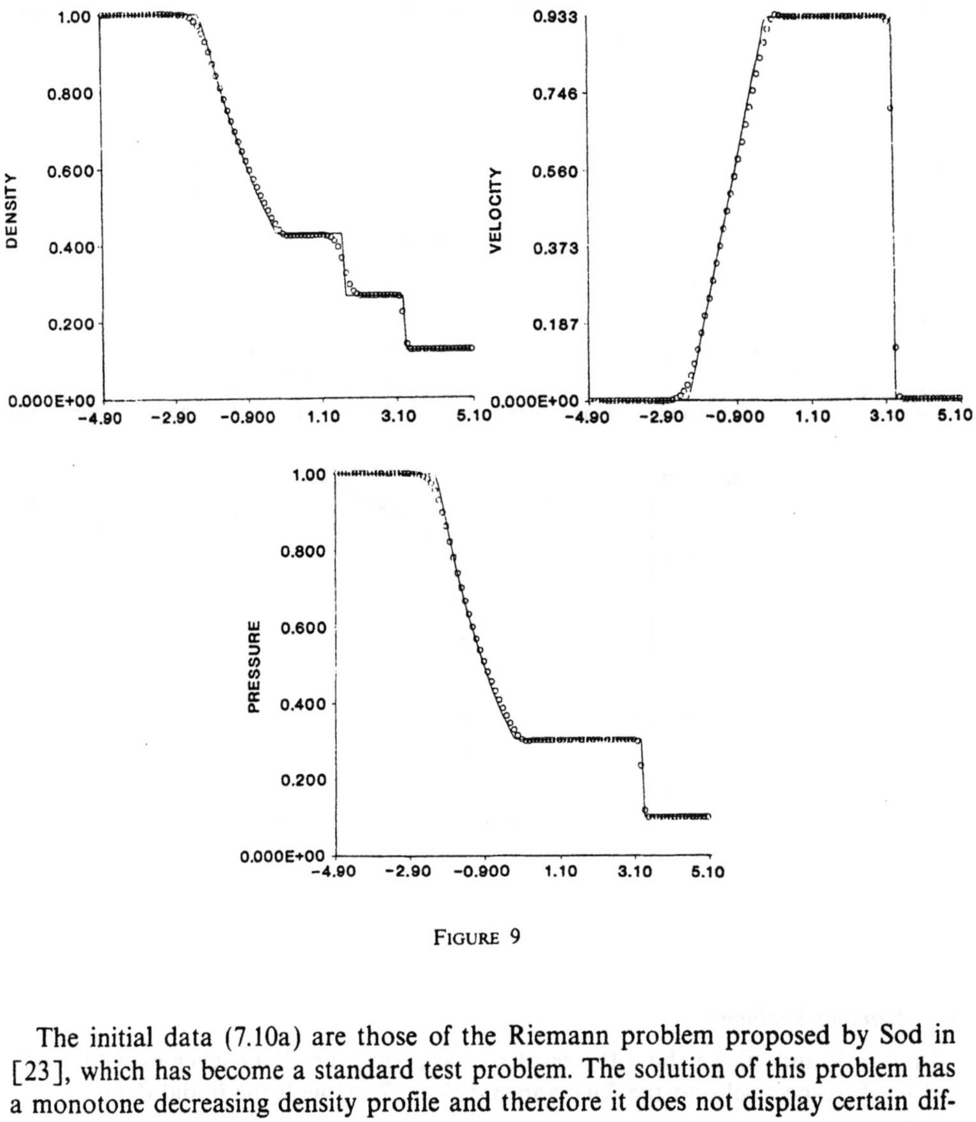

FIGURE 9

The initial data (7.10a) are those of the Riemann problem proposed by Sod in [23], which has become a standard test problem. The solution of this problem has a monotone decreasing density profile and therefore it does not display certain difficulties that may arise when the intermediate state has to be "built-up." In Figs. 12–16 we present calculations for the Riemann problem

$$(\rho_L, q_L, P_L) = (0.445, 0.698, 3.528); \qquad (\rho_R, q_R, P_R) = (0.5, 0, 0.571) \tag{7.10b}$$

used by Lax in [18]; see also [7, 9]. All these calculations were performed with 100 cells, $h = 0.1$, CFL $= 0.8$, and 85 time steps using a component-wise reconstruction (5.11). In Fig. 12 we show the results of the "4th-order" ENO scheme using a component-wise reconstruction (5.11). Comparing these results to Fig. 11 we see that the component-wise reconstruction here is much "noisier" than in Sod's problem. In

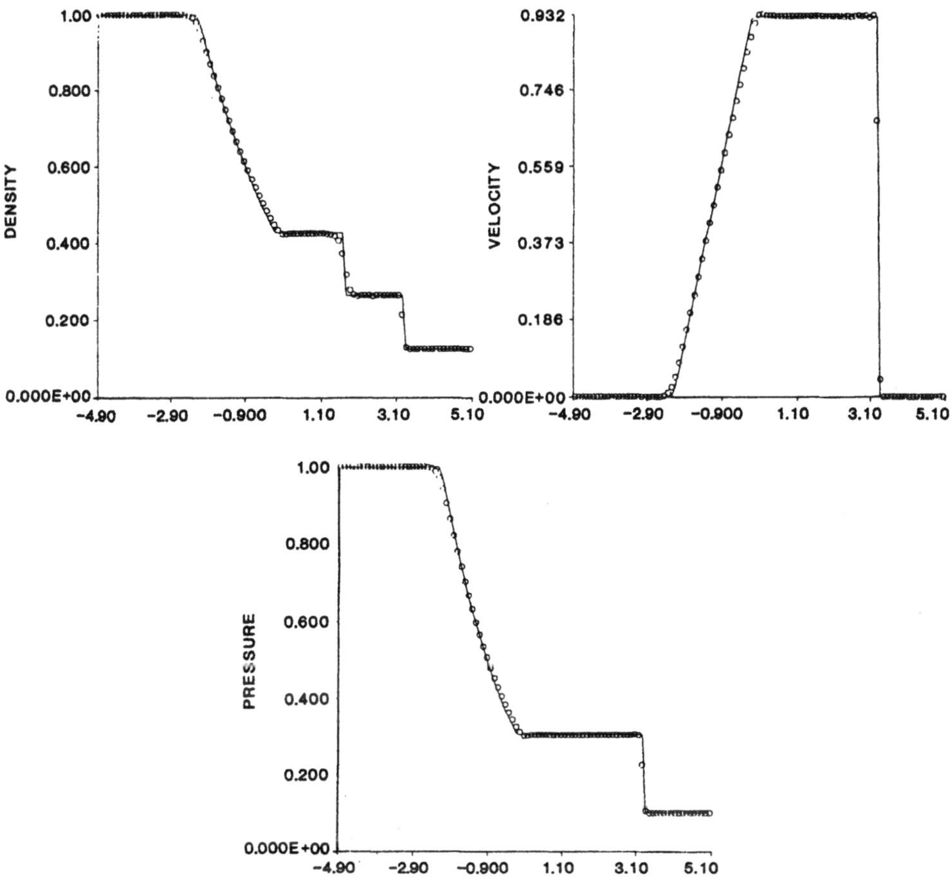

FIGURE 10

Figs. 13 and 15 we show the results of the ENO schemes using characteristic reconstruction (6.3) for $r=2$ and $r=4$, respectively; comparing Fig. 15 to Fig. 12 we see that most of the "noise" in Fig. 12 is eliminated.

In Figs. 14 and 16 we show the characteristic reconstruction $R(x; v^n)$ of the numerical solution in Figs. 13 and 15, respectively; $R(x; v^n)$ is piecewise-linear in Fig. 14 ($r=2$) and piecewise-cubic in Fig. 16 ($r=4$). The squares in these figures mark the values of $R(x_{j+1/2} \pm 0; v^n)$; thus the difference between the two squares at the same location shows the size of the discontinuity in the reconstruction there (we recall that the circles in Figs. 13 and 15 are the values of $R(x_j; v^n)$). We see that the discontinuities in the reconstruction of the rarefaction wave are small enough to be graphically imperceptible. Surprisingly the discontinuities in the reconstruction of the contact-discontinuity are also rather small. Comparing Fig. 16 to Fig. 14 we

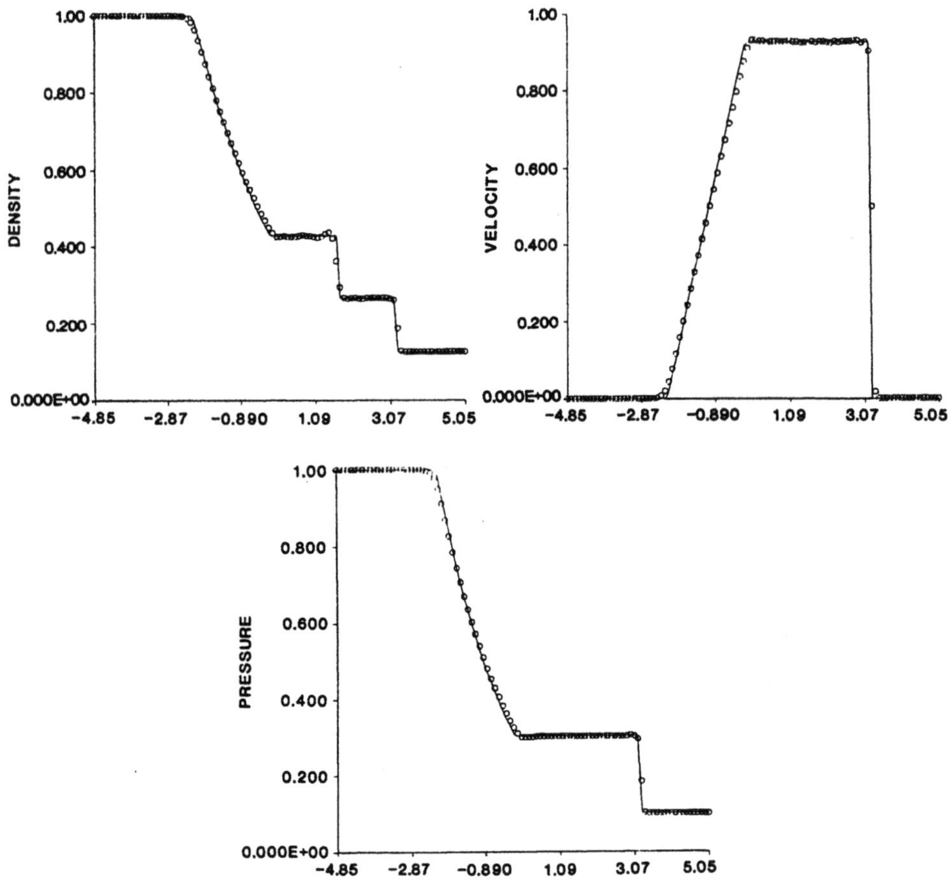

FIGURE 11

notice that the size of the discontinuities in the reconstruction for $r=4$ is always considerably smaller than that for $r=2$. It is interesting to note that even in the shock region in Fig. 16 ($r=4$), the sum of the jumps in the reconstruction is only about 35% of the size of the shock, while about 65% of the shock jump is described by the smooth polynomial part of the reconstruction.

We remark that because of the self-similar nature of the solution to the Riemann problem, the rate of convergence of any scheme is inherently limited to first order (see [27]). Comparing $r=4$ with $r=2$ in the solution of the above Riemann problems we notice a slight improvement in the smearing of the contact-discontinuity (we have not used artificial compression in these calculations) and the description of the rarefaction wave. Because of the self-similar nature of the solution it is better to compare the performances of two schemes by using x/t as the spatial

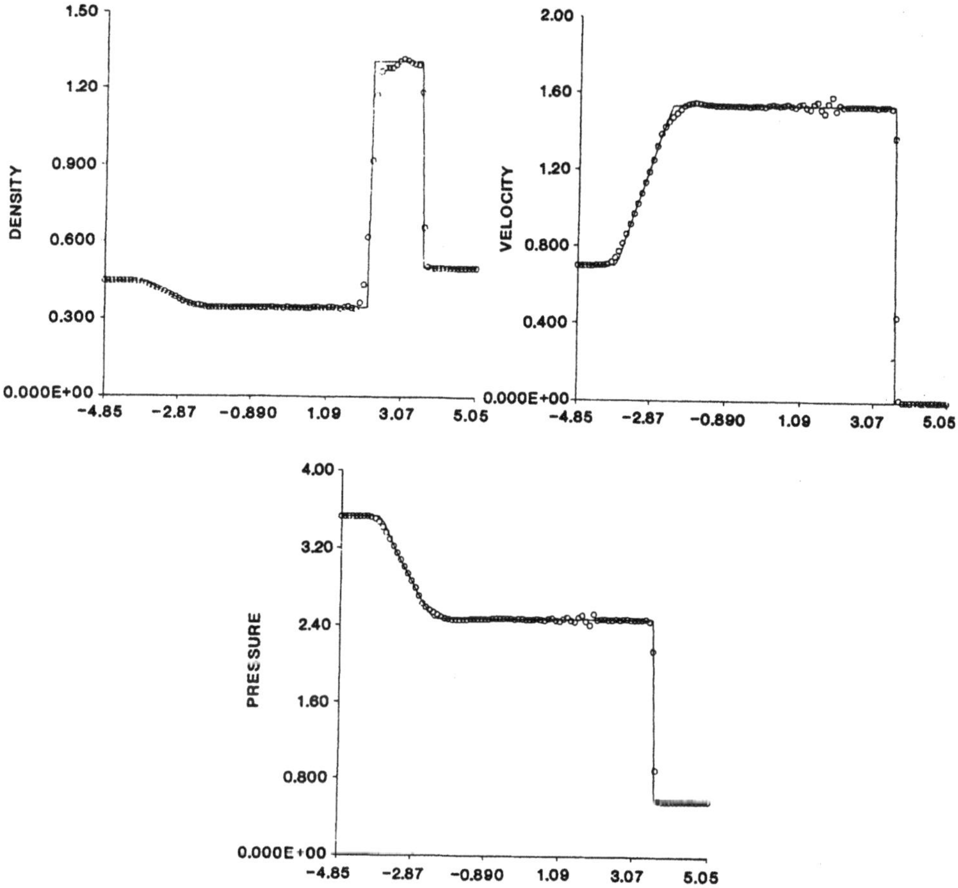

FIGURE 12

variable and to find how many time steps it takes to get well-resolved intermediate states. Doing so for the problem (7.10b) we find that $r=4$ with $N=35$ gives about the same result as $r=2$ with $N=70$.

B2. *Interaction of Blas Waves*

In this subsection we present numerical experiments with the ENO schemes for the problem of two interacting blast waves:

$$u(x,0)=\begin{cases} u_L & 0 \leqslant x<0.1 \\ u_M & 0.1 \leqslant x<0.9 \\ u_R & 0.9 \leqslant x<1, \end{cases} \tag{7.11a}$$

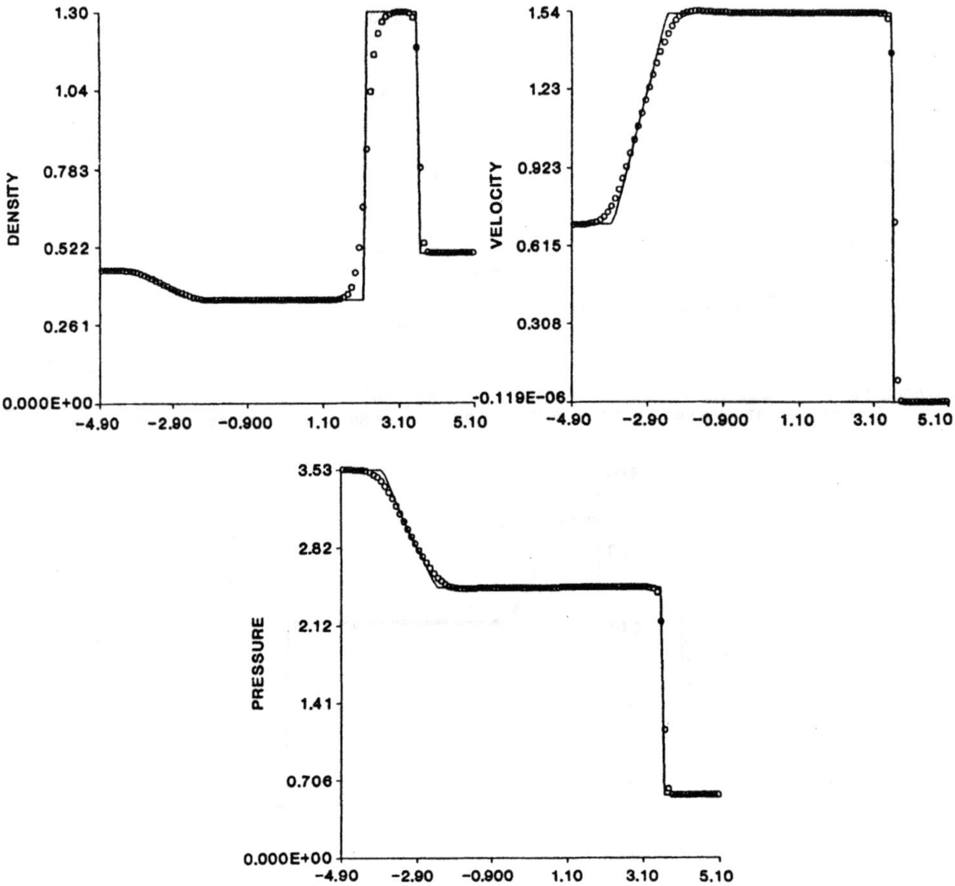

FIGURE 13

where

$$\rho_L = \rho_M = \rho_R = 1, \quad q_L = q_M = q_R = 0, \quad P_L = 10^3, \quad P_M = 10^{-2}, \quad P_R = 10^2; \quad (7.11b)$$

the boundaries at $x = 0$ and $x = 1$ are solid walls. This problem was suggested by Woodward and Colella as a test problem; we refer the reader to [31] where a comprehensive comparison of the performance of various schemes for this problem is presented.

In our calculations we divided the interval (0, 1) into J cells by

$$x_j = (j - \tfrac{1}{2})/J, \qquad j = 1, ..., J, \quad (7.12a)$$

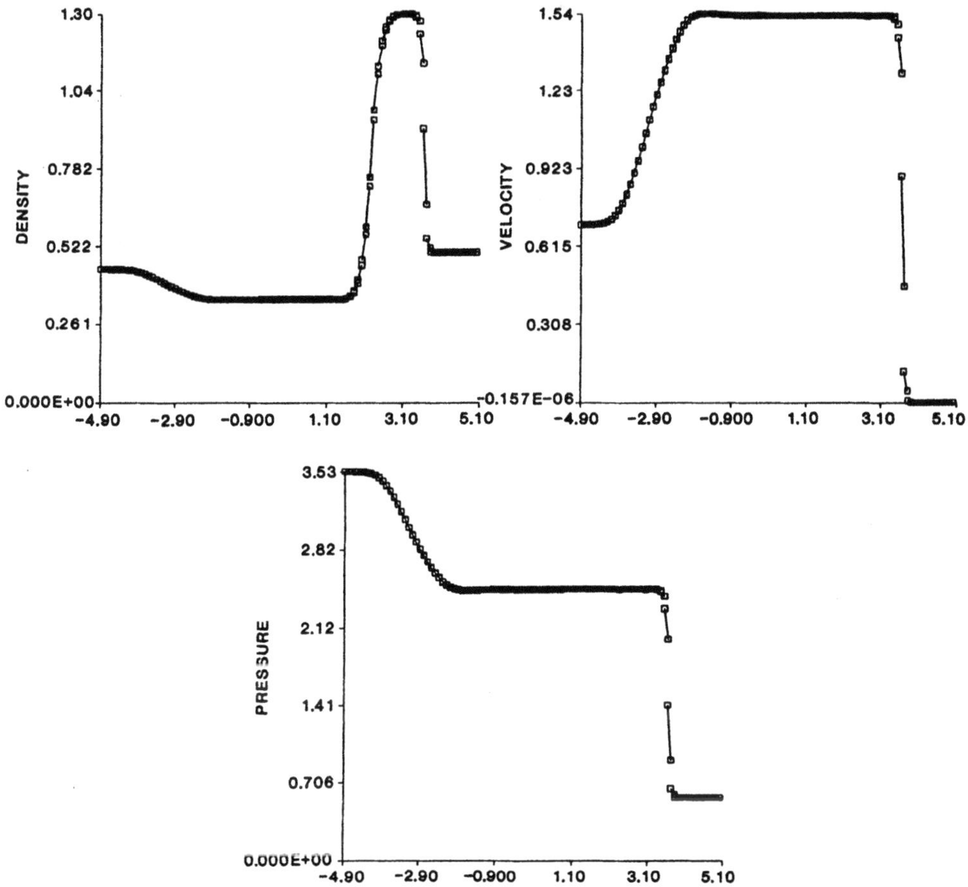

FIGURE 14

where x_j marks the center of the jth cell. The boundary conditions of a solid wall in $x=0$ and $x=1$ were treated by reflection, i.e., we defined auxiliary states $v_0^n, \ldots, v_{-r+1}^n$ for the left wall and $v_{J+1}^n, \ldots, v_{J+r}^n$ for the right wall by

$$\rho_{-j+1}^n = \rho_j^n, \qquad q_{-j+1}^n = -q_j^n, \qquad P_{-j+1}^n = P_j^n, \qquad j=1,\ldots,r \quad (7.12\text{b})$$

$$\rho_{J+j}^n = \rho_{J-j+1}^n, \qquad q_{J+j}^n = -q_{J-j+1}^n, \qquad P_{J+j}^n = P_{J-j+1}^n, \qquad j=1,\ldots,r. \quad (7.12\text{c})$$

We observe that representing the solid wall condition by the above reflection is very suitable for the characteristic reconstruction: A 3-wave approaching the right boundary is reflected as a 1-wave; consequently there is hardly any interaction between the waves in the characteristic variables (6.3a) and a situation of not having enough points of smoothness to choose from is thus avoided.

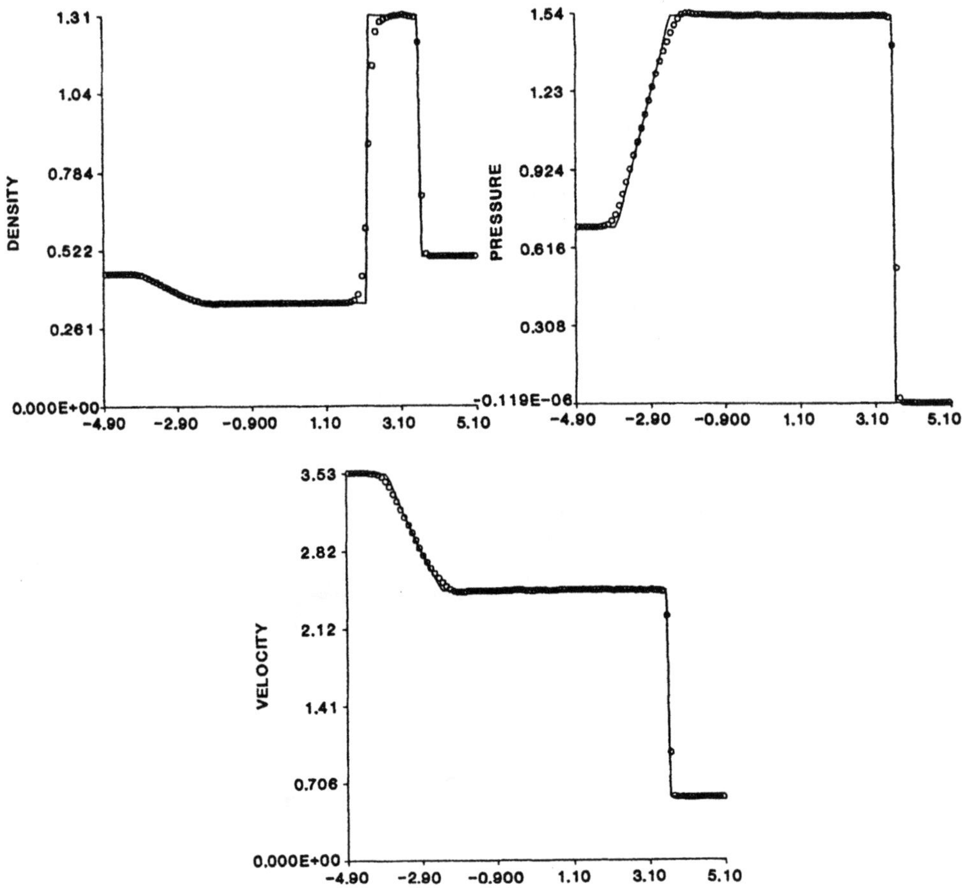

FIGURE 15

In Figs. 17a–h we show the solution of the "4th-order" ENO scheme at $t=0.010$, 0.016, 0.026, 0.028, 0.030, 0.032, 0.034, 0.038, respectively. We refer the reader to Fig. 2 in [31], where a highly accurate solution is displayed and a detailed description of the various interactions that occur at these instances is presented. The continuous line in Figs. 17a–h, 18, and 19 is the solution of the "4th-order" ENO scheme $J=800$ in (7.12a). Comparing this solution to the "exact" solution of Woodward and Colella in [31], we find that it shows all the important features of the various interactions and thus can be considered a "converged" solution. (The continuous line representing the solution with $J=800$ is the piecewise-linear interpolation of $\{R(x_j; v^n)\}$; consequently cusps in the solution, which do appear in $R(x; v^n)$, are chopped in the graphic representation). The circles in Figs. 17a–h show the values of $R(x_j; v^n)$ of the "4th-order" ENO scheme with

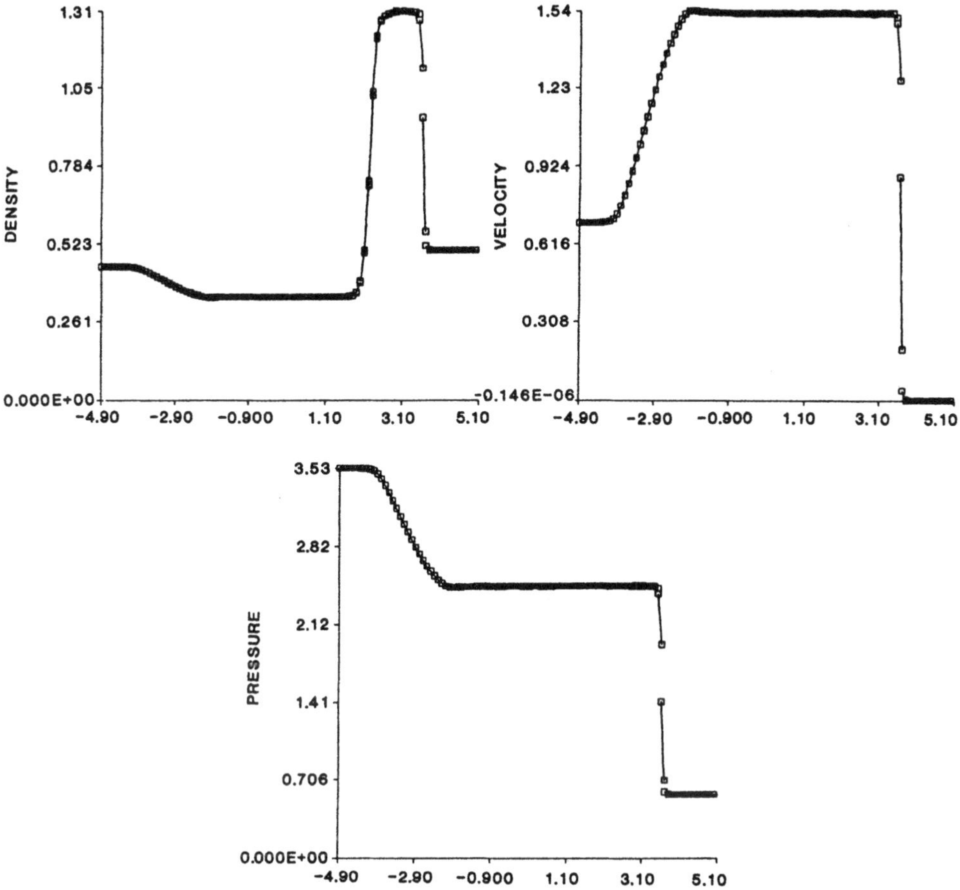

FIGURE 16

$J = 400$. Comparing the numerical solution for $J = 400$ to that of $J = 800$ we see that the velocity and pressure have already converged, while the density in Figs. 17g and 17h still deviates from the "converged" solution. This is due to the smearing of 3 contact discontinuities which are present in the solution at this time; the numerical results of Woodward and Colella demonstrate that the addition of "contact-discontinuity steepeners" improve the density profile considerably.

In Fig. 18 we show the solution of the "4th-order" ENO scheme with $J = 200$ at the final time $t = 0.038$; In Fig. 19 we repeat the calculation in Fig. 18 for the "2nd-order" ENO scheme. Comparing Figs. 18 and 19 we see that the "4th-order" scheme gives a much better resolution. We remark that the results of the "4th-order" scheme with $J = 100$ (not shown here) are of the same quality as those of the "2-order" scheme with $J = 200$.

We note that a parabola interpolating

$$P(x_{-1}) = 240, \qquad P(x_0) = 0.01, \qquad P(x_1) = 40$$

has an interval in which it is negative; the same is true for higher order interpolating polynomials that pass through these points. A situation of this type occurs in the calculation of the two interacting blast waves just before the interaction in Fig. 17d, when the low pressure region in Fig. 17c is shrinking to 1–2 computational cells. Since high order interpolating polynomials may produce negative values of pressure and density in such drastic situations, we have imposed a "positivity condition" on the reconstruction step of our programs for the Euler equations. To ensure that $\mathbf{R}(x; v^n)$ in the jth cell yields density and pressure that are positive, i.e.,

$$\begin{aligned} &P_j + \sum_{k=1}^{r-1} \left.\frac{\partial^k P}{\partial x^k}\right|_{x=x_j} \frac{(x-x_j)^k}{k!} > 0, \\ &\rho_j + \sum_{k=1}^{r-1} \left.\frac{\partial^k \rho}{\partial x^k}\right|_{x=x_j} \frac{(x-x_j)^k}{k!} > 0 \qquad \text{for} \quad |x - x_j| < h/2 \end{aligned} \tag{7.13a}$$

we check whether

$$\sum_{k=1}^{r-1} \left|\frac{\partial^k P_j}{\partial x^k}\right| \frac{(h/2)^k}{k!} < 0.8 P_j; \qquad \sum_{k=1}^{r-1} \left|\frac{\partial^k \rho_j}{\partial x^k}\right| \frac{(h/2)^k}{k!} < 0.8 \rho_j. \tag{7.13b}$$

If condition (7.13b) is not satisfied we reduce the order of the reconstruction *locally* at $x = x_j$ until positivity is ensured. We observe that the LHS of the inequalities in (7.13b) is $O(h)$ in smooth regions, hence this positivity condition does not reduce the asymptotic order of accuracy. Our computer program monitors does not reduce the asymptotic order of accuracy. Our computer program monitors occurrences of order reduction due to the positivity condition; we have found that the order in the calculations of the "4th-order" ENO scheme has been reduced during two time steps before the interaction in Fig. 16d, and only at the interaction zone itself; we have not encountered any order reduction in the solution to the Riemann problems (7.10).

C. *Variants and Extensions*

C1. *Characteristic Method for the Scalar Case*

In [15] we described an approximation to $v(x, t)$, the solution-in-the-small of (4.2), which is obtained by tracing approximate characteristics to the initial data. This approximation $\tilde{v}(x, t)$ can be extended to an arbitrary order of accuracy as follows: Let $\bar{a}^n_{j+1/2}$ denote

$$\bar{a}^n_{j+1/2} = \bar{a}(R(x_{j+1/2} - 0; v^n), \qquad R(x_{j+1/2} + 0; v^n)), \tag{7.14a}$$

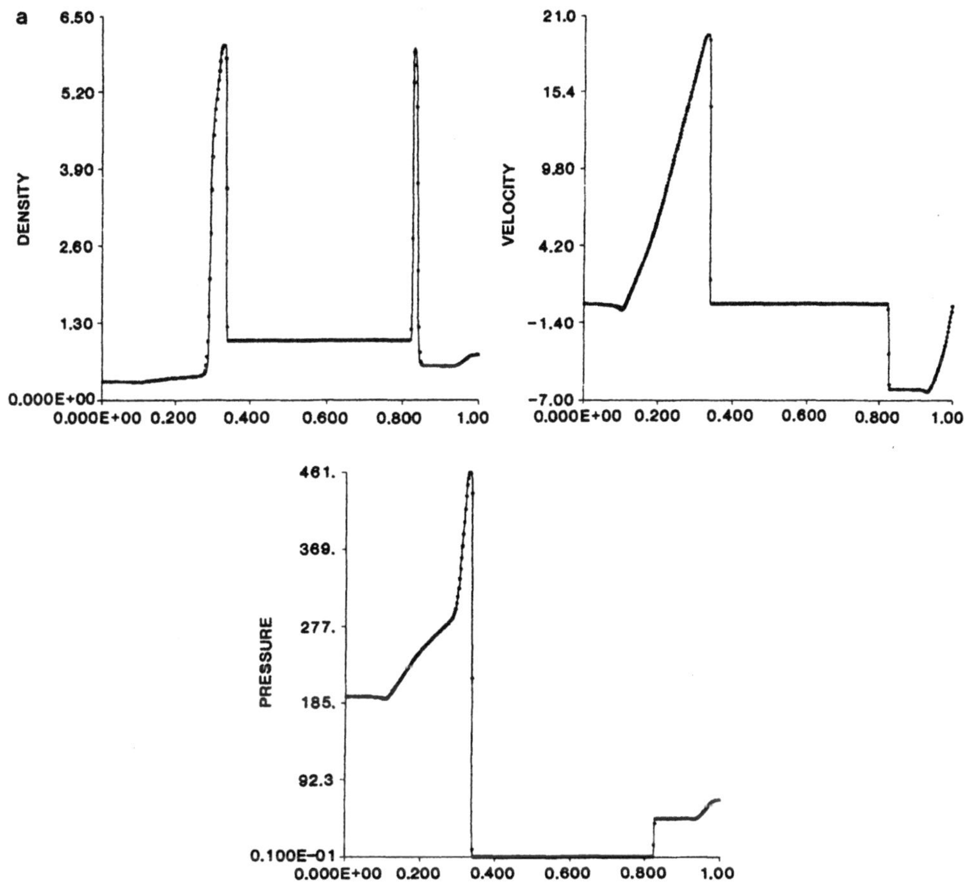

FIGURE 17

where $\bar{a}(u_1, u_2)$ is defined in (4.7), and let $\tilde{a}(x)$ denote the interpolation of $\tilde{a}^n_{j+1/2}$ by H_m (3.1) with $m = r - 1$, i.e.,

$$\tilde{a}(x_{j+1/2}) = \tilde{a}^n_{j+1/2} \tag{7.14b}$$

$$\tilde{a}(x) = H_m(x; \tilde{a}^n), \qquad m = r - 1. \tag{7.14c}$$

The approximation $\tilde{v}(x, t)$ is obtained by prescribing constancy of the solution along the approximate characteristic lines

$$x = x_0 + \tilde{a}(x_0)t, \tag{7.15a}$$

i.e.,

$$\tilde{v}(x_0 + \tilde{a}(x_0)t, t) = \tilde{v}(x_0, 0) = R(x_0; v^n); \tag{7.15b}$$

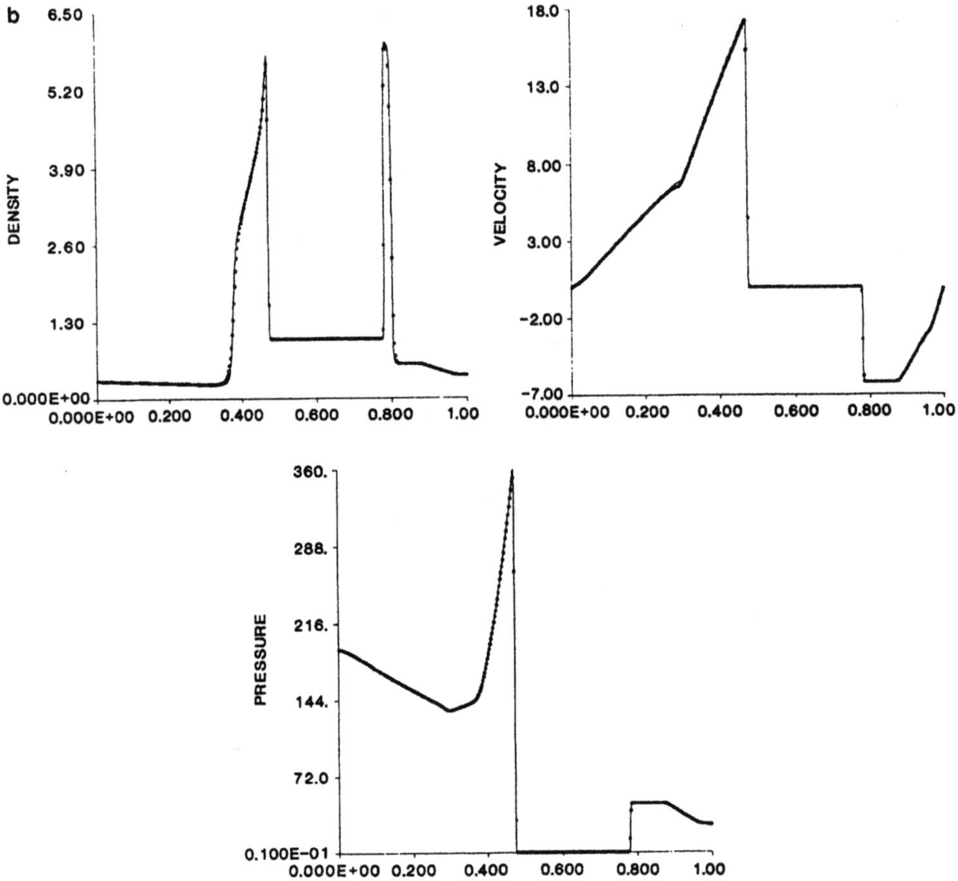

FIG. 17—*Continued.*

thus

$$\tilde{v}(x, t) = R(x_0(x, t); v^n), \tag{7.15c}$$

where $x_0(x, t)$ is the solution of the algebraic equation (7.15a). Let $x_0^1(x, t)$ denote the solution to (7.15a) for $m = 1$ in (7.14c); if x and t are such that

$$\tilde{x}_{j-1/2}(t) < x \leqslant \tilde{x}_{j+1/2}(t), \tag{7.16a}$$

where

$$\tilde{x}_{i+1/2}(t) = x_{i+1/2} + t\tilde{a}^n_{i+1/2} \tag{7.16b}$$

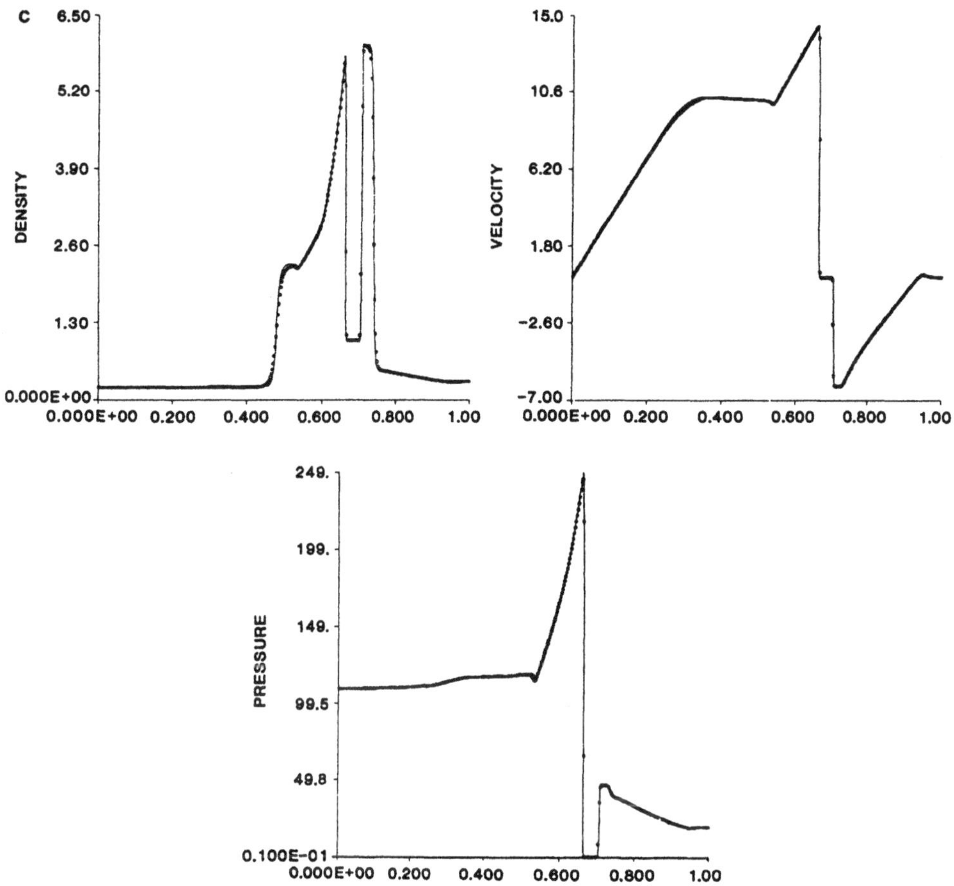

FIG. 17—*Continued.*

then

$$x_0^1(x, t) = x_{j-1/2} + \frac{x_{j+1/2} - x_{j-1/2}}{\tilde{x}_{j+1/2}(t) - \tilde{x}_{j-1/2}(t)} (x - \tilde{x}_{j-1/2}(t)). \tag{7.16c}$$

For $m > 1$ we obtain $x_0(x, t)$ by solving (7.15a) with Newton–Raphson iterations starting with the initial guess (7.16c).

Using $\tilde{v}(x, t)$ (7.15c) we define the following variant of (4.14):

$$v_j^{n+1} = v_j^n - \lambda(\bar{f}_{j+1/2} - \bar{f}_{j-1/2}) \tag{7.17a}$$

$$\bar{f}_{j+1/2} = \sum_{k=0}^{k} \alpha_k f(\tilde{v}(x_{j+1/2}, \beta_k \tau)). \tag{7.17b}$$

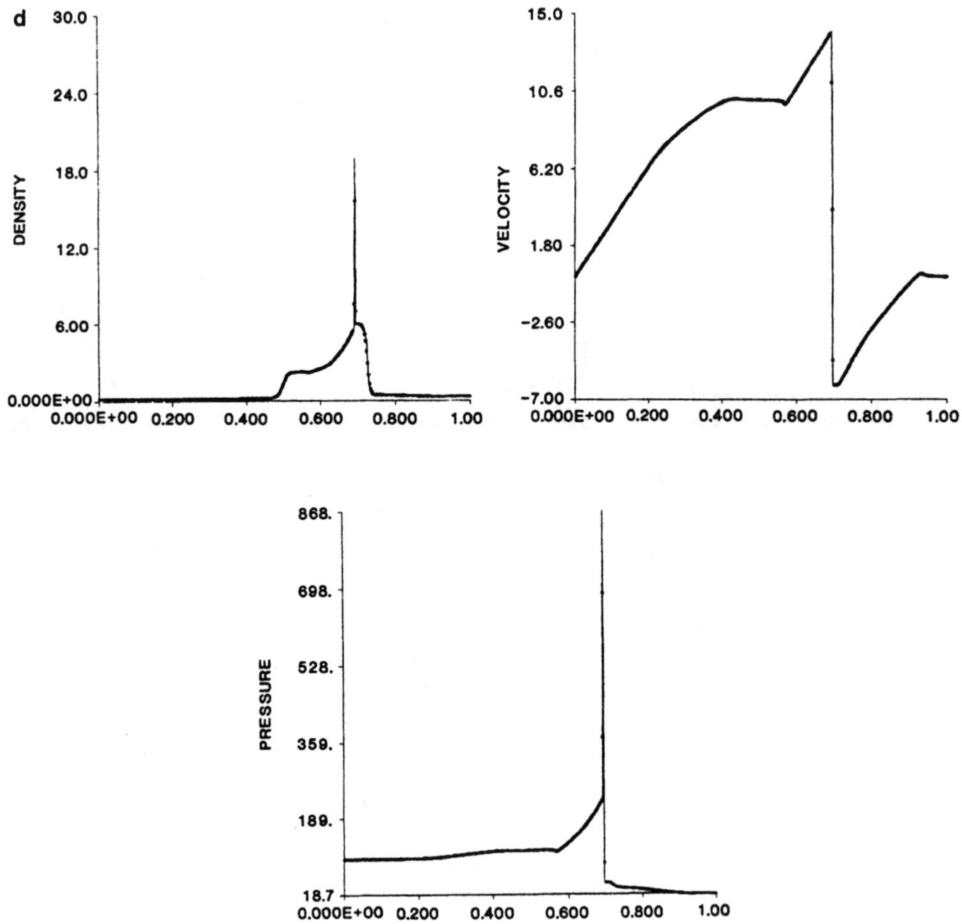

FIG. 17—*Continued.*

We have started the development of the ENOschemes with the version (7.17); later on we have replaced the characteristic method by the Cauchy–Kowalewski procedure which offers a unified approach in extending the scheme to include forcing terms and to systems of conservation laws. Our numerical experiments show that the two versions are computationally equivalent, although the version with the characteristic method (7.17) seems to be slightly more accurate than (4.14).

We remark that the scheme (7.17), as the scheme (4.14) with f^{ROE} in (4.21a), also admits any discontinuity with $\bar{a}(u_{\text{L}}, u_{\text{R}}) = 0$ as a stationary solution. This can be easily rectified by replacing the "shock curve" $\tilde{x}_{i+1/2}(t)$ in (7.16b) by an appropriate fan.

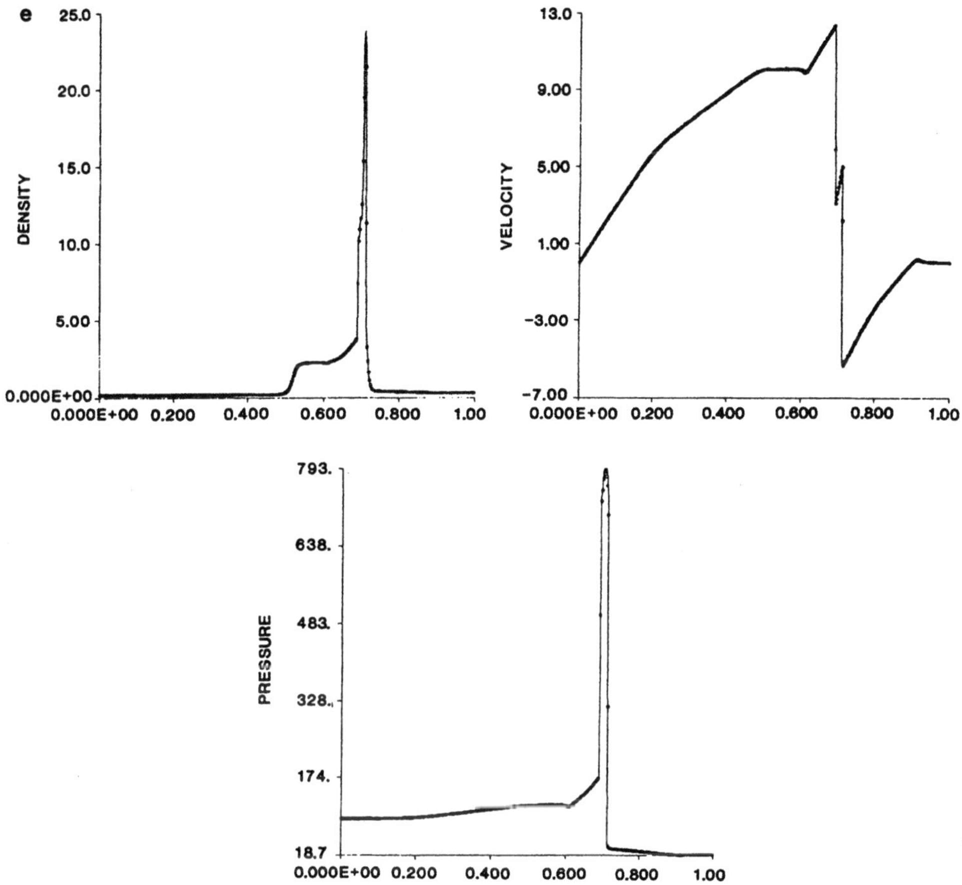

FIG. 17—*Continued*

C2. *Semi-discrete Formulation and Runge–Kutta Methods*

The semi-discrete version of the ENO schemes can be derived either directly from (1.4) or by letting $\tau \to 0$ in (4.14), (5.9). It takes the form

$$\frac{d}{dt} v_j(t) = -\frac{1}{h} [\bar{f}_{j+1/2}(t) - \bar{f}_{j+1/2}(t)] \equiv Q \cdot v_j(t), \tag{7.18a}$$

where

$$\bar{f}_{j+1/2}(t) = f^{\mathrm{R}}(R(x_{j+1/2} - 0; v(t)), R(x_{j+1/2} + 0; v(t))); \tag{7.18b}$$

here $v_j(t)$ is an approximation to $\bar{u}(x_j, t)$; $v(t) = \{v_j(t)\}$; $f^{\mathrm{R}}(u_1, u_2)$ is either the exact flux (4.5) or f^{ROE} (4.21), (5.11).

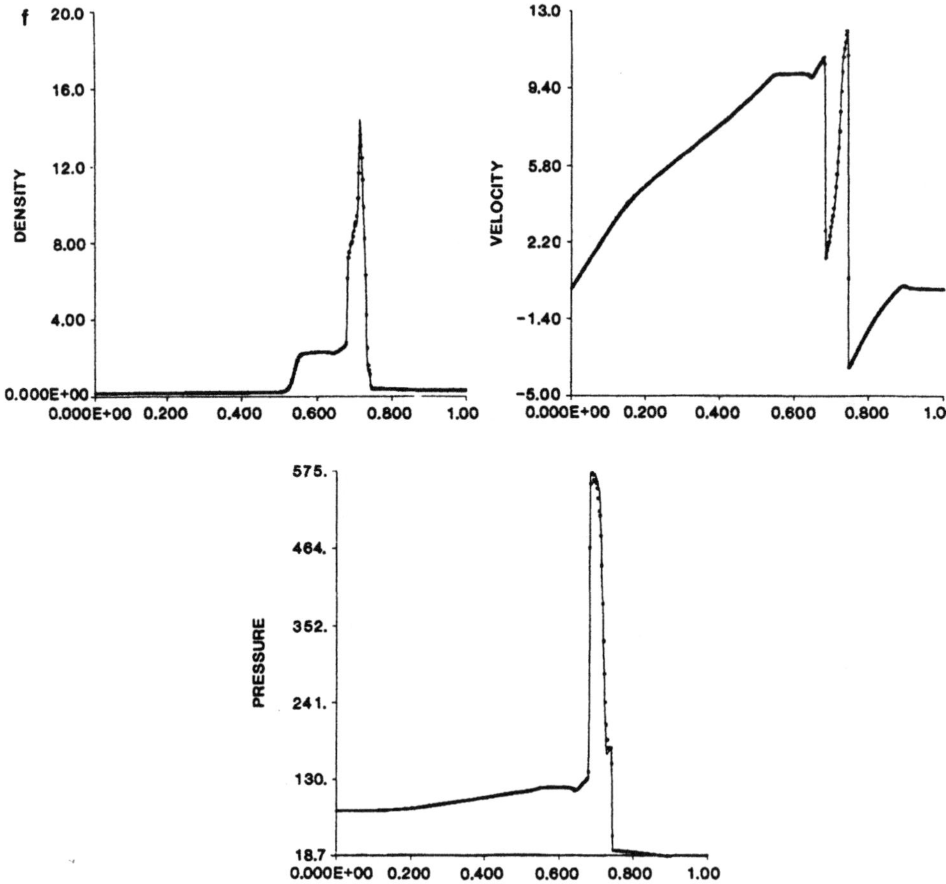

FIG. 17—*Continued.*

Considering (7.18) to be a system of ordinary differential equations in t for the vector $v(t) = \{v_j(t)\}$, we can solve the problem by using a numerical ODE solver. In [2] we present two sets of numerical experiments in which we use Runge–Kutta methods of appropriate order to approximate the solution of (7.18). In the first set of experiments we apply the scheme to the Riemann problem (7.10a) for $r = 1, 2, 3, 4, 5, 6$. In the second set of experiments we apply the scheme with $r = 2, 4$ to a Laval nozzle problem which involves the addition of a forcing term to the Euler equations (6.1). In these calculations we have used RP, f^{ROE} and CFL $= 0.5$. Comparing the results of the Riemann problem to those in the present paper we find them to be of similar quality. The numerical experiments of [2] indicate that the semi-discrete formulation (7.18) with Runge–Kutta temporal discretization does not generate spurious oscillations for CFL $\leqslant 0.5$; however when we increase the

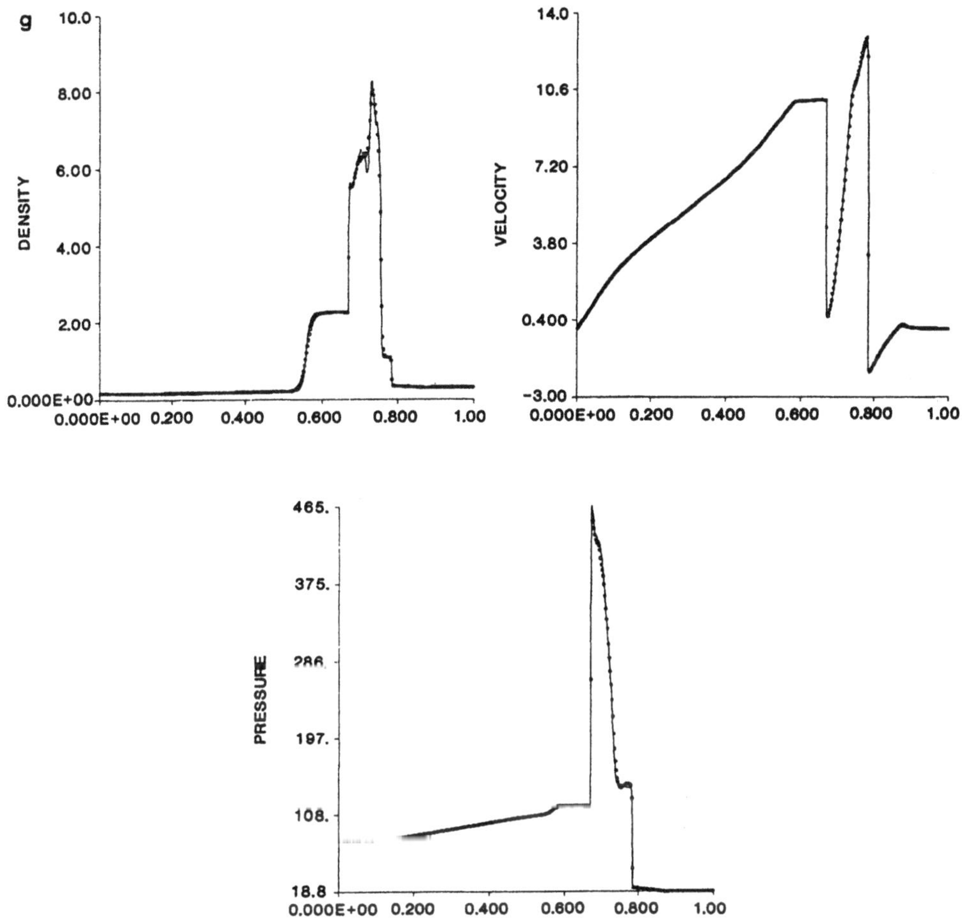

FIG. 17—*Continued.*

CFL number beyond 0.5 we start getting some oscillations and eventually the scheme becomes unstable.

The main advantage of using the Runge–Kutta temporal discretization is the ease of its programming; however it seems to be less efficient than the fully discrete formulation and also requires more storage.

C3. *Variable Grid and Front Tracking*

In Section 3 we have pointed out that the non-oscillatory interpolation H_m (3.1)–(3.5) and the reconstruction via primitive function (RP) (3.6)–(3.10) are well defined for non-uniform grids, see Appendix. Since the solution-in-the-small step

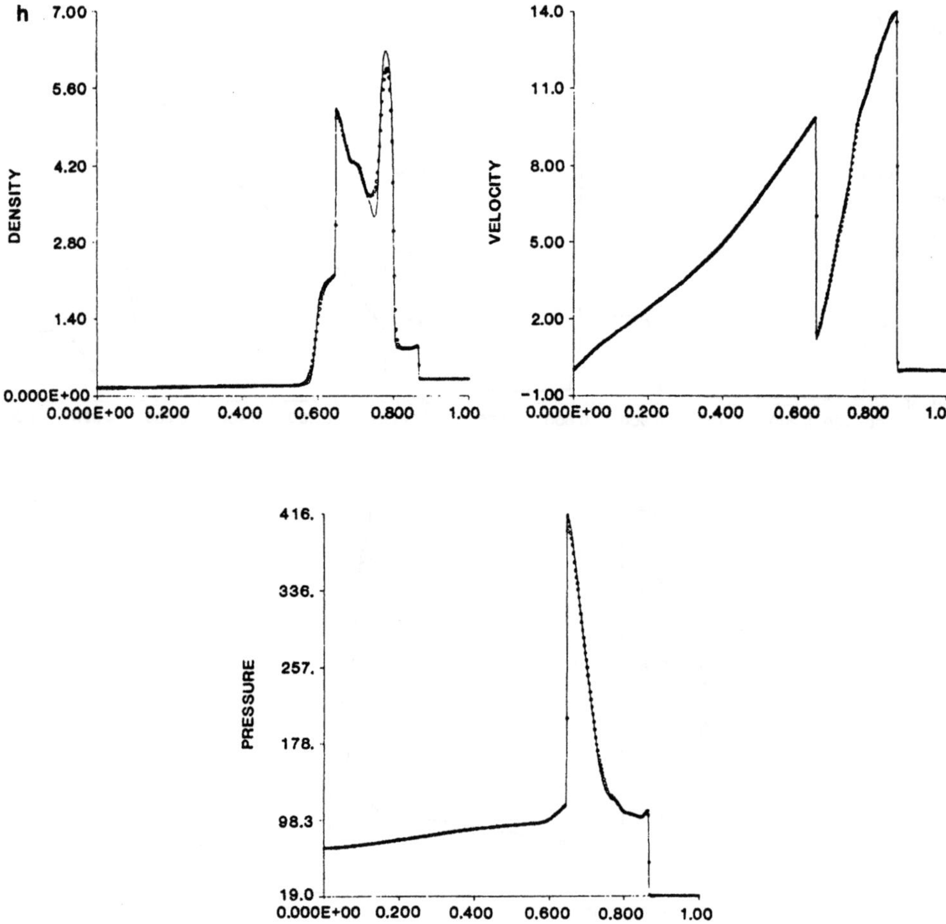

FIG. 17—*Continued.*

also does not require uniformity of the grid, we may compute new cell-averages v_j^{n+1} in (1.14c) on any choice of intervals $\{I_j^{n+1}\}$ by

$$v_j^{n+1} = \frac{1}{|I_j^{n+1}|} \int_{I_j^{n+1}} v_h(x, t_{n+1} - 0)\, dx; \tag{7.19}$$

here $I_k^l = (\xi_{k-1/2}^l, \xi_{k+1/2}^l)$ and $|I_k^l| = \xi_{k+1/2}^l - \xi_{k-1/2}^l$. Using the same rationale as before, our approximation to (7.19) becomes

$$|I_j^{n+1}|\, v_j^{n+1} = |I_j^n|\, v_j^n - \tau(\bar{f}_{j+1/2}^\sigma - \bar{f}_{j-1/2}^\sigma). \tag{7.20a}$$

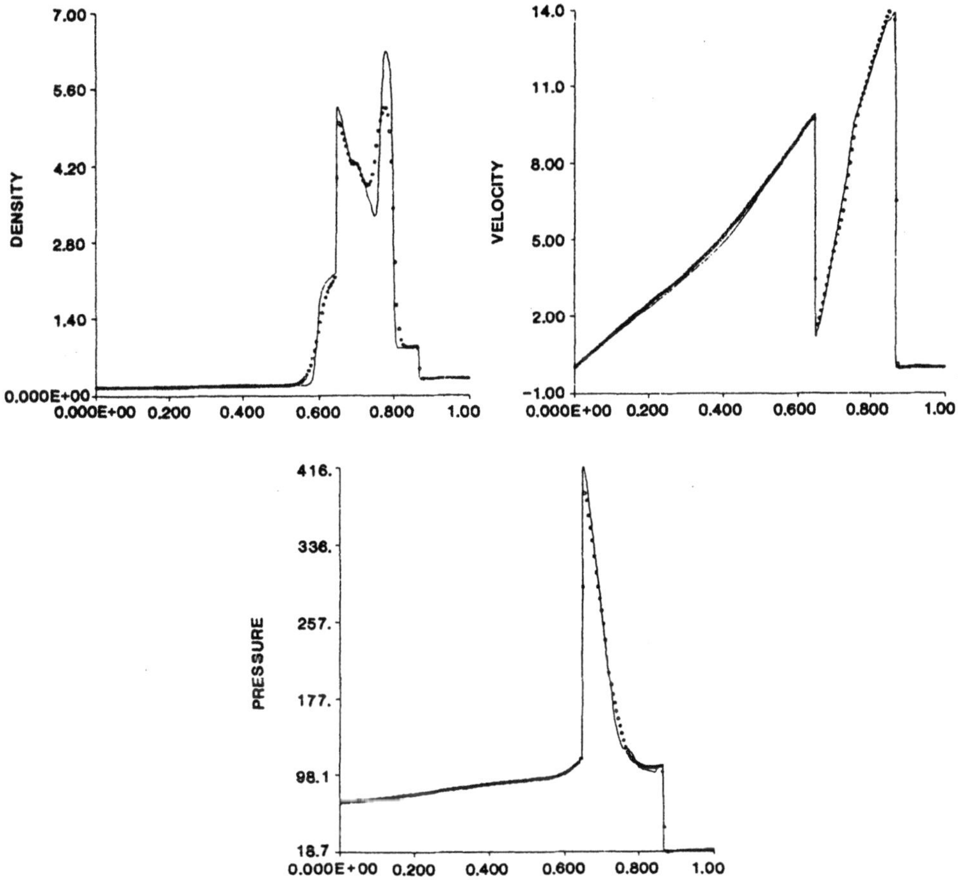

FIGURE 18

The numerical flux $\bar{f}^{\sigma}_{j+1/2}$ is consistent with $f(u)-\sigma_{j+1/2}u$

$$\sigma_{j+1/2}=(\xi^{n+1}_{j+1/2}-\xi^{n}_{j+1/2})/\tau, \tag{7.20b}$$

and can be expressed as

$$\bar{f}^{\sigma}_{j+1/2}=\sum_{k=0}^{k}\alpha_k \bar{f}^{\mathrm{R}}(\tilde{v}_j(x_{j+1/2}+\beta_k\tau\sigma_{j+1/2},\beta_k\tau),\tilde{v}_{j+1}(x_{j+1/2}+\beta_k\tau\sigma_{j+1/2},\beta_k\tau);\sigma_{j+1/2}) \tag{7.20c}$$

where

$$\bar{f}^{\mathrm{R}}(u_1,u_2;\sigma)=f(V(\sigma;u_1,u_2))-\sigma V(\sigma;u_1,u_2); \tag{7.20d}$$

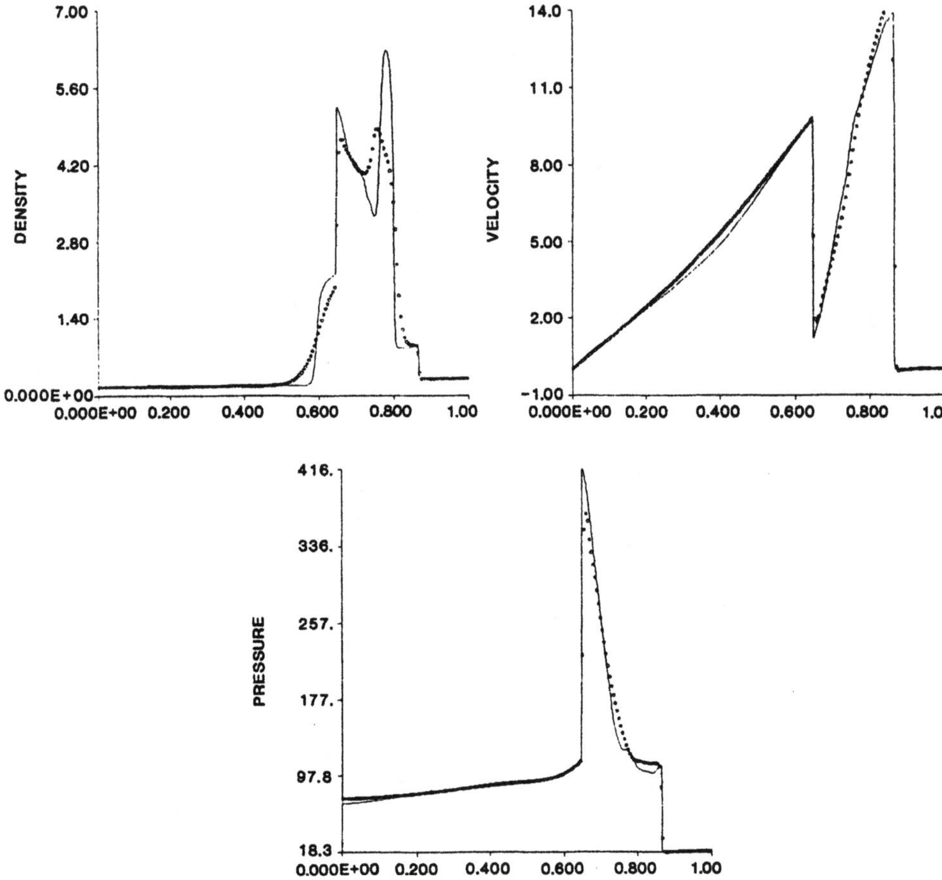

FIGURE 19

we recall that $V(\sigma; u_L, u_R)$ denotes the value of the solution to the Riemann problem (4.3) at $x/t = \sigma$. Roe's linearization (5.11) yields the following approximation

$$\bar{f}^{\text{ROE}}(u_1, u_2; \sigma) = \tfrac{1}{2}\Big[f(u_1) + f(u_2) - \sigma(u_1 + u_2) - \sum_{k=1}^{m} \delta_k(u_1, u_2) |a_k(\hat{u}) - \sigma| r_k(\hat{u}) \Big]. \tag{7.21}$$

In Fig. 20 we show the results of the scheme (7.20) with $\bar{f}^{\text{R}}$ approximated by $\bar{f}^{\text{ROE}}$ (7.21) for the Riemann problem (7.10b); the values of $\{\xi_j^{n+1}\}$ in this

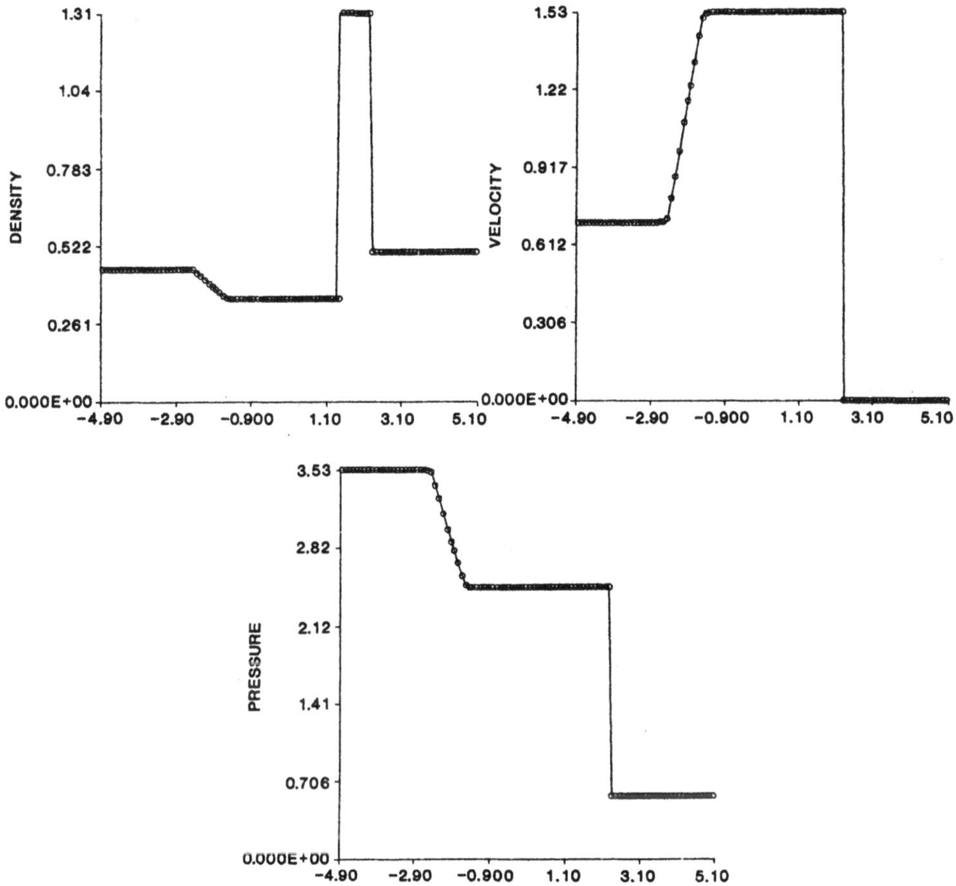

FIGURE 20

calculation were chosen by the self-adjusting grid algorithm of [13]. This algorithm provides and automatic way to place interval end points ξ_j^{n+1} at the location of significant discontinuities and thus avoid their smearing by the cell-averaging step (7.19). The calculation in Fig. 20 was initialized by taking the exact solution of the Riemann problem at $t=0.5$ (at which time there are 4 grid points between the contact-discontinuity and the shock). The results displayed in Fig. 20 show the numerical solution of the scheme with $r=4$ after 100 time steps with $\text{CFL}=0.5$. These results clearly demonstrate the adaptability of the ENO schemes to front tracking techniques.

We note that the use of irregular grids disallows the extra order of accuracy which was gained in (1.18) for a uniform grid. Numerical experiments with irregular grids (where ξ_j^{n+1} is randomly selected within a specified interval) show that the error in solving the scalar smooth problem (7.1) by the scheme (7.20) is $O(h^{r-1})$ in the L_1, L_2, and L_∞ norms. However, comparing Fig. 20 with Fig. 15 we observe a considerable gain in resolution in spite of the reduction in formal order of accuracy.

C4. *Extension to 2D*

In this subsection we outline the extension of the ENO schemes to the solution of the 2-dimensional IVP

$$u_t + f(u)_x + g(u)_y = 0, \qquad u(x, y, 0) = u_0(x, y). \tag{7.22}$$

We note that Strang-type dimensional splitting [29] is only second-order accurate in time, and therefore is unsuitable for extending the higher order accurate members of the ENO schemes to 2D.

Let $\bar{\bar{w}}$ denote the two-dimensional "sliding average" of w

$$\bar{\bar{w}}(x, y) = \frac{1}{\Delta x\, \Delta y} \int_{-\Delta x/2}^{\Delta x/2} \int_{-\Delta y/2}^{\Delta y/2} w(x+\xi, y+\eta)\, d\eta\, d\xi. \tag{7.23}$$

Integrating (7.22) over the computational cell $I_{ij} \times (t_n, t_{n+1}]$, $I_{ij} = [x_{i-1/2}, x_{i+1/2}] \times [y_{j-1/2}, y_{j+1/2}]$, we find that $\bar{\bar{u}}_{ij}^n = \bar{\bar{u}}(x_i, y_j, t_n)$ satisfies

$$\bar{\bar{u}}_{ij}^{n+1} = \bar{\bar{u}}_{ij}^n - \lambda_x(\hat{f}_{i+1/2,j} - \hat{f}_{i-1/2,j}) - \lambda_y(\hat{g}_{i,j+1/2} - \hat{g}_{i,j-1/2}), \tag{7.24a}$$

where $\lambda_x = \tau/\Delta x$, $\lambda_y = \tau/\Delta y$, and

$$\hat{f}_{i+1/2,j} = \frac{1}{\tau\, \Delta y} \int_0^\tau \int_{-\Delta y/2}^{\Delta y/2} f(u(x_{i+1/2}, y_j + \eta, t_n + t))\, d\eta\, dt, \tag{7.24b}$$

$$\hat{g}_{i,j+1/2} = \frac{1}{\tau\, \Delta x} \int_0^\tau \int_{-\Delta x/2}^{\Delta x/2} g(u(x_i + \xi, y_{j+1/2}, t_n + t))\, d\xi\, dt. \tag{7.24c}$$

The abstract form of the ENO schemes for the solution of (7.22) remains (1.10), i.e.,

$$v^{n+1} = A_h \cdot E(\tau) \cdot R(\cdot, \cdot; v^n), \qquad v^0 = \bar{\bar{u}}_0. \tag{7.25}$$

As before $E(t)$ is the exact evolution operator of (7.22); however, A_h is now the 2-dimensional cell-averaging (7.23) and $R(x, y; \bar{\bar{w}})$ is an appropriate 2-dimensional reconstruction of $w(x, y)$. In the scalar constant coefficient case

$$u_t + a u_x + b u_y = 0, \qquad u(x, y, 0) = u_0(x, y), \tag{7.26a}$$

the ENO scheme (7.25) becomes

$$v_{ij}^{n+1}=\bar{R}(x_i-a\tau,\ y_j-b\tau;\ v^n),\qquad v_{ij}^0=\bar{\bar{u}}_0(x_i,\ y_j). \tag{7.26b}$$

In [12] we present numerical experiments with the ENO scheme (7.26) for the scalar constant coefficient case, where the reconstruction $R(x, y; \bar{\bar{w}})$ is obtained via a two-dimensional deconvolution. Expanding $w(x+\xi,\ y+\eta)$ in (7.23) around $\xi=\eta=0$ we get as in (3.12)

$$\bar{\bar{w}}(x,\ y)=w(x,\ y)+\alpha_2[(\Delta x)^2 w_{xx}+(\Delta y)^2 w_{yy}]\ +\alpha_4[(\Delta_x)^4 w_{xxxx}$$
$$+2(\Delta x)^2(\Delta y)^2 w_{xxyy}+(\Delta y)^4 w_{yyyy}]+O(\Delta^6). \tag{7.27}$$

Multiplying both sides of (7.27) by $(\Delta x)^k(\Delta y)^{l-k}\ \partial^l/\partial x^k\ \partial y^{l-k}$ and truncating the expansion in the RHS at $O(\Delta^r)$, we get as in (3.13) an invertable system of linear equations which expresses $\bar{\bar{w}}$ and its deviatives in terms of w and its derivatives. We set

$$(\bar{\bar{D}}^{0,0})\ ij=\bar{\bar{w}}(x_i,\ y_j) \tag{7.28a}$$

and obtain approximation

$$(\bar{\bar{D}}^{k,l})_{ij}=(\Delta x)^k(\Delta y)^l\frac{\partial^{k+l}}{\partial x^k\ \partial y^l}\bar{\bar{w}}(x_i,\ y_j)+O(\Delta^r),\qquad 1\leqslant k+l\leqslant r-1; \tag{7.28b}$$

then, as in (3.17), we invert the system of linear equations to get the following approximations to w and its derivatives

$$(D^{k,l})_{ij}=(\Delta x)^k(\Delta y)^l\frac{\partial^{k+l}}{\partial x^k\ \partial y^l}w(x_i,\ y_j)+O(\Delta^r),\qquad 0\leqslant k+l\leqslant r-1. \tag{7.29a}$$

Using (7.29a) we define R in the cell I_{ij} by

$$R(x,\ y;\ \bar{\bar{w}})=\sum_{l=0}^{r-1}\frac{1}{l!}\sum_{k=0}^{l}(D^{k,l-k})_{ij}\binom{l}{k}\left(\frac{x-x_i}{\Delta x}\right)^k\left(\frac{y-y_j}{\Delta y}\right)^{l-k},\qquad (x,\ y)\in I_{ij}. \tag{7.29b}$$

The approximations $\bar{\bar{D}}^{kl}$ in (7.28b) are obtained by a sequence of applications of the one-dimensional operation (3.15b), which we rewrite now in the following operator form:

$$(G_m^l\bullet u)_j=M\left(\frac{d^l}{dz^l}H_m(z_j-0;u),\frac{d^l}{dz^l}H_m(z_j+0;u)\right); \tag{7.30}$$

here u denotes the one-dimensional vector $\{u(z_j)\}_{j=1}^{l}$. Using (7.30) and the notation convention $\bar{\bar{w}}_{\bullet,j}=\{\bar{\bar{w}}(x_i, y_j)\}_{i=1}^{l}$, we define

$$(\bar{\bar{D}}^{k,0})_{ij}=(\Delta x)^k(G_m^k \bullet \bar{\bar{w}}_{\bullet,j})_i, \qquad 1\leqslant k\leqslant r-1, \tag{7.31a}$$

$$(\bar{\bar{D}}^{0,l})_{ij}=(\Delta y)^l(G_m^l \bullet \bar{\bar{w}}_{i,\bullet})_j, \qquad 1\leqslant l\leqslant r-1. \tag{7.31b}$$

To obtain approximations to the mixed derivatives of $\bar{\bar{w}}$ we first evaluate

$$(\tilde{D}_1^{k,l})_{ij}=(\Delta y)^l[G_{m-k}^l \bullet(\bar{\bar{D}}^{k,0})_{i,\bullet}]_j, \qquad 1\leqslant l\leqslant r-1-k,$$

$$(\tilde{D}_2^{k,l})_{ij}=(\Delta x)^k[G_{m-l}^k \bullet(\bar{\bar{D}}^{0,l})_{\bullet,j}]_i, \qquad 1\leqslant k\leqslant r-1-l,$$

and then define

$$(\bar{\bar{D}}^{k,l})_{ij}=M((\tilde{D}_1^{kl})_{ij}, (\tilde{D}_2^{kl})_{ij}), \tag{7.31c}$$

where M is the min mod function (3.16).

We observe that the restriction of the two-dimensional reconstruction (7.29) to $y=y_j$, i.e., $R(x, y_j;\bar{\bar{w}})$ is identical to the one-dimensional reconstruction (3.18) applied to the restriction of $\bar{\bar{w}}$ to $y=y_j$, i.e., $R(x; \bar{\bar{w}}(\bullet\ y_j))$; the same observation applies to the restrictions to $x=x_i$.

We recall that the one-dimensional reconstruction is essentially non-oscillatory only if discontinuities are separated by at least $r+1$ points of smoothness. In the one-dimensional system case we had to overcome the problem of collision (in time) of discontinuities; in the two-dimensional case we also have to worry about intersections (in space) of curves of discontinuity. In order to study the severity of the problem we have experimented with the constant coefficient problem (7.26a) with the initial data

$$u_0(x, y)=\begin{cases}1 & (x, y)\in S;\\ 0 & (x, y)\in U-S.\end{cases} \tag{7.32}$$

Here $U=[-1, 1]\times[-1, 1]$ and S is a rotated square contained in U. In [12] we present numerical results which are obtained by applying the scheme (7.26) with $r=1, 2, 3, 4$ to the initial data (7.32) with periodic boundary conditions on ∂U. These results show that indeed small spurious oscillations are generated for $r\geqslant 2$ at the corners of S; however, it seems to us that they are small enough to be computationally acceptable.

APPENDIX: An Algorithm for Reconstruction

In this appendix we describe our algorithm for computing the coefficients $b_{j,k}$ in

$$R(x, \bar{w})=\sum_{k=0}^{r-1} b_{j,k}(x-x_j)^k, \tag{A.1}$$

where x_j is the center of the jth cell. To obtain these coefficients we start with Newton's form of interpolation

$$H_r(x; u) = \sum_{k=0}^{r} u[y_i, ..., y_{i+k}] \prod_{l=0}^{k-1} (x - y_{i+l}). \tag{A.2}$$

Here $i = i(j)$ is selected by the algorithm (3.4) with respect to the divided differences

$$d_{i,k} = u[y_i, ..., y_{i+k}]. \tag{A.3}$$

In the following we describe an algorithm to rewrite the polynomial on the RHS of (A.2) as a finite Taylor series around $x = x_c$:

$$q(x) = \sum_{k=0}^{r} d_{i,k} P_{i,k}(x) = \sum_{k=0}^{r} q^{(k)}(x_c)(x - x_c)^k / k!, \tag{A.4}$$

where

$$P_{i,k}(x) = \prod_{l=0}^{k-1} (x - y_{i+l}) \equiv \sum_{l=0}^{k} \mathscr{S}_{l,k}(x - x_c)^{k-l}. \tag{A.5}$$

Using the fact that the coefficients $\{\mathscr{S}_{l,k}\}$ satisfy a recursion relation we compute them as follows: We set

$$\begin{aligned} Z_l &= x_c - y_{i+l}, \qquad 0 \leqslant l \leqslant r-1; \\ \mathscr{S}_{0,k} &= 1, \qquad 0 \leqslant k \leqslant r; \end{aligned} \tag{A.6}$$

then we evaluate

$$\begin{aligned} &\textbf{do } l = 1, r \\ &\quad \mathscr{S}_{l,l} = \mathscr{S}_{l-1,l-1} Z_{l-1} \\ &\textbf{do } k = l+1, r \\ &\quad \mathscr{S}_{l,k} = \mathscr{S}_{l,k-1} + \mathscr{S}_{l-1,k-1} Z_{k-1}. \end{aligned} \tag{A.7}$$

It is easy to see that

$$q^{(k)}(x_c) = k! \sum_{l=k}^{r} \mathscr{S}_{l-k,l}\, d_{i,l}. \tag{A.8}$$

We note that the algorithm (A.6)–(A.8) is defined for a non-uniform grid. When the grid is uniform we can obtain (A.8) in two steps: First we take $x_c = y_i$ and observe that $Z_l = -lh$ in (A.6); consequently $\{\mathscr{S}_{l,k}\}$ are independent of i. Denoting

$$\bar{d}_k = h^k d_{i,k}, \qquad c_k = h^k q^{(k)}(y_i) \tag{A.9}$$

and using the convention $\bar{d}_k = 0$ for $k > r$, we get for $1 \leqslant r \leqslant 6$

$$\begin{aligned}
c_0 &= u(y_i) \\
c_1 &= \bar{d}_1 - \bar{d}_2/2 + \bar{d}_3/3 - \bar{d}_4/4 + \bar{d}_5/5 - \bar{d}_6/6 \\
c_2 &= \bar{d}_2 - \bar{d}_3 + 11\bar{d}_4/12 - 5\bar{d}_5/6 + 137\bar{d}_6/180 \\
c_3 &= \bar{d}_3 - 1.5\bar{d}_4 + 1.75\bar{d}_5 - 1.875\bar{d}_6 \\
c_4 &= \bar{d}_4 - 2\bar{d}_5 + 17\bar{d}_6/6 \\
c_5 &= \bar{d}_5 - 2.5\bar{d}_6 \\
c_6 &= \bar{d}_6
\end{aligned} \tag{A.10}$$

$$q(x) = \sum_{k=0}^{r} \frac{c_k}{k!} \left(\frac{x - y_i}{h} \right)^k.$$

Thus

$$h^k q^{(k)}(x_c) = \sum_{l=k}^{r} \frac{c_l}{(l-k)!} \left(\frac{x_c - y_i}{h} \right)^{l-k}. \tag{A.11}$$

Reconstruction via Primitive Function (RP)

In this case $I_k = (y_j, y_{j+1})$ is the jth cell and $x_c = x_j = \frac{1}{2}(y_j + y_{j+1})$ is its center. The given data is

$$\bar{w}_l = \frac{1}{y_{l+1} - y_l} \int_{y_l}^{y_{l+1}} w(y)\, dy, \tag{A.12}$$

from which we evaluate the point values of the primitive function

$$W(y_k) = \sum_{l=0}^{k-1} (y_{l+1} - y_l)\, \bar{w}_l. \tag{A.13}$$

Applying the algorithm (A.6)–(A.8) to

$$d_{i,k} = W[y_i, \ldots, y_{i+k}] \tag{A.14}$$

with $i = i(j)$ selected by (3.4), we obtain the values of $q^{(l)}(x_j)$. Using the definition (3.8) in (A.4) we get the coefficients of the Taylor expansion in (A.1) by

$$b_{j,k} = q^{(k+1)}(x_j)/k!. \tag{A.15}$$

We note that when the grid is uniform $y_k = x_{k-1/2}$ and we can also use the algorithm (A.9)–(A.11).

Reconstruction via Deconvolution (RD)

We recall that RD is used with a uniform grid so that the given data $\bar{w}_l$ can be thought of as point values $\bar{w}(x_l)$ of the sliding average function (1.3). Applying the algorithm (A.9)–(A.11) to

$$d_{i,k} = \bar{w}[x_i, ..., x_{i+k}] \tag{A.16}$$

with $x_c = x_j$ we get in (A.11) for $i = i(j-1)$ the values of

$$h^k \frac{d^k}{dx^k} H_r(x_j - 0; \bar{w}); \tag{A.17a}$$

when we apply this algorithm with $i = i(j)$ and $x_c = x_j$ we get in (A.11) the values of

$$h^k \frac{d^k}{dx^k} H_r(x_j + 0; \bar{w}). \tag{A.17b}$$

Next we evaluate $\bar{D}_{k,j}$ in (3.15) by taking the min mod of the appropriate values in (A.17a) and (A.17b). Finally we use the back-substitution (3.19) to obtain the coefficients of the Taylor expansion (A.1)

$$b_{j,k} = \frac{1}{k!} D_{k,j}/h^k. \tag{A.18}$$

We remark that the use of the algorithms (A.9)–(A.11) is preferable to that of (A.6)–(A.8) since it enables us to save computing time by rearranging the operations (A.16)–(A.17) as follows: First we set $i = i(j)$ in (A.16) and evaluate (*A*.9)–(A.10). Using the same coefficients c_k in (A.10) we now apply (A.11) to $x_c = x_j$ and $x_c = x_{j+1}$ to obtain $(d^k/dx^k) H_r(x_j + 0; \bar{w})$ and $(d^k/dx^k) H_r(x_{j+1} - 0; \bar{w})$, respectively; the min mod operation (3.15) is then performed in a following sweep.

Acknowledgment

The authors would like to thank P. L. Roe for his constructive criticisms of the first draft of this paper.

References

1. M. Ben-Artzi and J. Falcowitz, *SIAM J. Sci. Comput.*, in press.
2. S. R. Chakravarthy, A. Harten, and S. Osher, "Essentially non-oscillatory shock-capturing schemes of arbitrarily high accuracy," *AIAA* 86-0339, Reno, NA, 1986.
3. A. J. Chorin, *J. Comput. Phys.* **22**, 517 (1976).
4. P. Colella and P. R Woodward, *J. Comput. Phys.* **54**, 174 (1984).
5. S. K. Godunov, *Mat. Sb.* **47**, 271 (1959).
6. E. Harabetian, "A convergent series expansion for hyperbolic systems of conservation laws," NASA *Contractor Report 172557*, ICASE Report No. 85-13, 1985 (unpublished).

7. A. HARTEN, *Math. Comp.* **32**, 363 (1978).
8. A. HARTEN, *J. Comput. Phys.* **49**, 151 (1983).
9. A. HARTEN, *J. Comput. Phys.* **49**, 357 (1983).
10. A. HARTEN, *SINUM* **21**, 1 (1984).
11. A. HARTEN, "On high-order accurate interpolation for non-oscillatory shock capturing schemes," *MRC Technical Summary Report No. 2829*, University of Wisconsin, 1985 (unpublished).
12. A. HARTEN, "Preliminary results on the extension of ENO schemes to two-dimensional problems," Proceedings of the International Conference on Hyperbolic Problems, Saint-Etienne, January 1986 (in preparation).
13. A. HARTEN AND J. M. HYMAN, *J. Comput. Phys.* **50** 235 (1983).
14. A. HARTEN, P. D. LAX, AND B. VAN LEER, *SIAM Rev.* **25**, 35 (1983).
15. A. HARTEN AND S. OSHER, MRC Technical Summary Report No. 2823, May 1985; *SINUM*, in press.
16. A. HARTEN, B. ENGQUIST, S. OSHER, AND S. R. CHAKRAVARTHY, Uniformly high-order accurate non-oscillatory schemes, II, in preparation.
17. A. HARTEN, S. OSHER, B. ENGQUIST, AND S.R. CHAKRAVARTHY, *J. Appl. Num. Math.* **2**, 347 (1986).
18. P. D. LAX, *Comm. Pure Appl. Math.* **7**, 159 (1954).
19. P. D. LAX, "Hyperbolic systems of conservation laws and the mathematical theory of shock waves," in *Regional Conf. Series Lectures in Applied Math. Vol.* **11** (SIAM, Philadelphia, 1972).
20. P. D. LAX AND B. WENDROFF, *Commun. Pure Appl. Math.* **13**, 217 (1960).
21. P. D. LAX AND B. WENDROFF, *Commun. Pure Appl. Math.* **17** 381 (1964).
22. B. VAN LEER, *J. Comput. Phys.* **32** 101 (1979).
23. S. OSHER, SINUM **21**, 217 (1984).
24. S. OSHER AND E. TADMOR, *Math. Comp.*, in press.
25. P. L. ROE, *J. Comput. Phys.* **43**, 357 (1981).
26. P. L. ROE, Some contributions to the modeling of discontinuous flows, in *Lectures in Applied Mathematics Vol. 22* (Amer. Math. Soc., Providence, RI, 1985).
27. M. SEVER (MOCK), Order of dissipation near rarefaction centers, in *Progress and Supercomputing in Computational Fluid Dynamics, Proceedings of U. S.–Israel Workshop, 1984* (Birkhauser, Boston, 1985), p. 395.
28. G. A. SOD, *J. Comput. Phys.* **27**, 1 (1978).
29. G. STRANG, *SINUM* **5**, 506 (1968).
30. P. K. SWEBY, *SINUM* **21**, 995 (1984).
31. P. WOODWARD AND P. COLLELLA, *J. Comput. Phys.* **54** 115 (1984).
32. XIONG-HAU WU AND YOU-LAN ZHU, "A scheme of the singularity- separating method for the non-convex problem," Computer Center, Academia Sinica, 1984 (unpublished).

ENO Schemes with Subcell Resolution*

AMI HARTEN

School of Mathematical Sciences, Tel-Aviv University, Israel and Department of Mathematics, University of California, Los Angeles, California

Received July 9, 1987; revised February 29, 1988

DEDICATED TO EUGENE ISAACSON ON HIS 70TH BIRTHDAY

In this paper, we introduce the notion of subcell resolution, which is based on the observation that unlike point values, cell-averages of a discontinuous piecewise-smooth function contain information about the exact location of the discontinuity within the cell. Using this observation we design an essentially non-oscillatory (ENO) reconstruction technique which is exact for cell averages of discontinuous piecewise-polynomial functions of the appropriate degree. Later on we incorporate this new reconstruction technique into Godunov-type schemes in order to produce a modification of the ENO schemes which prevents the smearing of contact discontinuities.

1. INTRODUCTION

In [6–8] we have introduced a class of essentially nonoscillatory (ENO) schemes that generalizes Godunov's scheme [2] to a high order of accuracy.

In this paper we present a modification of the ENO schemes which is designed to prevent smearing of linear discontinuities. This is done by adding a correction term to the numerical flux of the ENO scheme. First, we shall derive a correction term to account for discontinuities in the scalar constant coefficient case. Later we shall apply the scalar correction to the linearly degenerate characteristic field in the Euler equations in order to improve the resolution of contact discontinuities.

Let $\{I_j \times [t_n, t_{n+1})\}$, where $I_j = [x_{j-1/2}, x_{j+1/2}]$, $x_\alpha = \alpha h$, $t_k = k\tau$, be a partition of $\mathbf{R} \times \mathbf{R}^+$. Let $\bar{u}_j^n$ be the "cell-average" of u at time t_n, i.e.,

$$\bar{u}_j^n = \frac{1}{h} \int_{I_j} u(x, t_n)\, dx. \tag{1.1}$$

* Research was supported under NSF Grant DMS85-03294, DARPA Grant in the ACMP Program, ONR Grant N00014-86-K-0691, NASA Ames Interchange NCA2-185, and NASA Langley Grant NAG1-270. Also research was partially supported under the National Aeronautics and Space Administration under NASA Contract NAS1-18107 while the author was in residence at the Institute for Computer Applications in Science and Engineering (ICASE), NASA Langley Research Center, Hampton, VA 23665-5225.

The cell-average of the solution to the initial value problem

$$u_t+f(u)_x=0, \qquad u(x,0)=u_0(x) \tag{1.2}$$

satisfies

$$\bar{u}_j^{n+1}=\bar{u}_j^n-\lambda[\bar{f}(x_{j+1/2},t_n;u)-\bar{f}(x_{j-1/2},t_n;u)], \tag{1.3a}$$

where $\lambda=\tau/h$ and

$$\bar{f}(x,t;u)=\frac{1}{\tau}\int_0^\tau f(u(x,t+\eta))\,d\eta. \tag{1.3b}$$

The ENO schemes can be written in the standard conservation form

$$v_j^{n+1}=v_j^n-\lambda(\bar{f}_{j+1/2}-\bar{f}_{j-1/2})\equiv[\bar{E}_h(\tau)\cdot v^n]_j; \tag{1.4a}$$

here $\bar{E}_h$ denotes the numerical solution operator and $\bar{f}_{j+1/2}$, the numerical flux, denotes a function of $2k$ variables

$$\bar{f}_{j+1/2}=\bar{f}(v_{j-k+1}^n,\ldots,v_{j+k}^n) \tag{1.4b}$$

which is consistent with the flux $f(u)$ in (1.2), in the sense that $\bar{f}(u,u,\ldots,u)=f(u)$. Unlike standard difference schemes, v_j^n in the ENO schemes is a high-order approximation to the cell-average $\bar{u}_j^n$, and not to the point value $u(x_j,t_n)$. Setting $v_j^n=\bar{u}_j^n$ in the numerical scheme (1.4) and comparing it to relation (1.3), we see that if the numerical flux $\bar{f}_{j+1/2}=\bar{f}(\bar{u}_{j-k+1}^n,\ldots,\bar{u}_{j+k}^n)$ can be expanded as

$$\bar{f}(\bar{u}_{j-k+1}^n,\ldots,\bar{u}_{j+k}^n)=\frac{1}{\tau}\int_0^\tau f(u(x_{j+1/2},t_n+\eta))\,d\eta+d(x_{j+1/2})h^r+O(h^{r+1}) \tag{1.5a}$$

then the truncation error

$$\bar{u}_j^{n+1}-[\bar{E}_h\cdot\bar{u}^n]_j=\lambda[d(x_{j+1/2})-d(x_{j-1/2})]h^r+O(h^{r+1}), \tag{1.5b}$$

is $O(h^{r+1})$ wherever $d(x)$ is Lipschitz continuous, i.e., the scheme (1.4) is rth order accurate in the sense of cell averages.

The most important ingredient in the ENO schemes is a preocedure to reconstruct a piecewise-smooth function $w(x)$ from its given cell-averages $\{\bar{w}_j\}$. This reconstruction, which we denote by $R(x;\bar{w})$, is a piecewise-polynomial function of x that has a uniform polynomial degree $(r-1)$ and satisfies:

(i) at all points x for which there is a neighborhood where w is smooth

$$R(x;\bar{w})=w(x)+e(x)h^r+O(h^{r+1}); \tag{1.6a}$$

(ii) conservation in the sense of

$$\frac{1}{h}\int_{x_{j-1/2}}^{x_{j+1/2}} R(\xi; \bar{w})\, d\xi = \bar{w}_j; \tag{1.6b}$$

(iii) it is essentially non-oscillatory

$$\mathrm{TV}(R(\cdot; \bar{w})) \leqslant \mathrm{TV}(\bar{w}) + O(h^{1+P}), \qquad P > 0, \tag{1.6c}$$

where TV denotes total variation in x.

Using such a reconstruction (1.6) we can express the abstract form of the ENO schemes by

$$\bar{E}_h(\tau) \cdot \bar{w} = A(I) \cdot E(\tau) \cdot R(\cdot; \bar{w}). \tag{1.7a}$$

Here $A(I)$ is the cell-averaging operator

$$A(I) \cdot w = \frac{1}{|I|} \int_I w(y)\, dy \tag{1.7b}$$

and $E(t)$ is the exact evolution operator of the IVP (1.2), i.e.,

$$u(\cdot, t) = E(t) u_0. \tag{1.7c}$$

We note that (1.7a) with the piecewise constant reconstruction

$$R(x; \bar{w}) = \bar{w}_j \qquad \text{for} \quad x_{j-1/2} \leqslant x \leqslant x_{j+1/2} \tag{1.8}$$

is exactly the first-order accurate Godunov's scheme [1]; (1.7a) with the piecewise linear reconstruction

$$R(x; \bar{w}) = \bar{w}_j + S_j(x - x_j) \qquad \text{for} \quad x_{j-1/2} \leqslant x < x_{j+1/2}, \tag{1.9a}$$

where

$$S_j = w_x(x_j) + O(h), \tag{1.9b}$$

is the abstract form of the second-order accurate extensions to Godunov's scheme described in [6–8]. We shall refer to these schemes as MUSCL-type schemes due to their similarity to van Leer's work in [9].

In the first-order case (1.8), the scheme (1.7a) can be expressed in the conservation form (1.4) with the numerical flux

$$\bar{f}_{j+1/2} = f^R(v_j^n, v_{j+1}^n); \tag{1.10}$$

here $f^R(u_1, u_2)$ is an approximation to the flux at the origin in a Riemann problem with u_1 to the left and u_2 to the right.

In the second-order case (1.9), the numerical flux of the abstract scheme cannot be expressed in a simple closed form, and we approximate it by

$$\bar{f}_{j+1/2} = \bar{f}^{\text{ENO}}_{j+1/2} = f^R(v^L_{j+1/2}, v^R_{j+1/2}), \tag{1.11a}$$

where

$$v^L_{j+1/2} = v^n_j + h(1 - \lambda a^n_j) S^n_j/2, \qquad v^R_{j+1/2} = v^n_{j+1} - h(1 + \lambda a^n_{j+1}) S^n_{j+1}/2; \tag{1.11b}$$

here $a^n_j = f'(v^n_j)$.

In this paper we pay special attention to the second-order accurate scheme (1.11), because at present this seems to be the state of the art. This class of second-order schemes (with various choices of S^n_j) performs rather well in smooth regions and shocks. However, it exhibits excessive smearing of linear discontinuities, i.e., contact discontinuitites. Usually such discontinuities are smeared more and more in time at the ate $O(n^{1/3})$, where n is the number of time-steps. To understand this smearing we note in (1.7) that whenever a discontinuity in the reconstruction R is propagated by the evolution operator E into the interior of the cell, then the cell-averaging operator $A(I)$ replaces this sharp discontinuity by a smeared transition. In the linear case there is nothing to stop this process and therefore it goes on forever. In the case of a shock wave, the fact that the characteristics converge into the shock counteracts the smearing, and a steady progressing profile is obtained.

The above observation is the basis for the artificial compression concept [3]. In order to prevent the excessive smearing of a linear discontinuity one can artificially induce convergence of the numerical characteristic field at each monotone strip of the solution. This can be accomplished by modifying the expression of the slopes S^n_j or by adding a corrective term to the numerical flux (1.11) (see [4]). The main advantage of artificial compression is that it is easy to use. The primary disadvantage is that one has to be extra careful (which also means to do a lot of checking...) not to generate unphysical discontinuities by applying it too strongly, where it need not be applied at all.

The piecewise-parabolic method (PPM) of Colella and Woodward [1] includes a mechanism to detect contact discontinuities and to correct the scheme by using a "steeper" reconstruction. The PPM proved itself to be a robust high resolution scheme in a large number of numerical tests [11]. In this paper we present a technique, which we call "subcell resolution," that is close in spirit to the PPM but is somewhat different in its methodology.

The present scheme is a "souped up" version of (1.11) in which the linear advection part is boosted to third-order accuracy (in L_1-sense) and is capable of propagating *linear* discontinuities perfectly (within third-order accuracy). The main ingredient in the new method is the observation that the information in cell-averages of a discontinuous function, unlike that of point values, contains the location of the discontinuity within the cell, e.g., the cell-averages $\bar{w}_j$,

$$\bar{w}_j = \begin{cases} u_L, & j \leq -1 \\ u_M, & j = 0 \\ u_R, & j \geq 1, \end{cases} \tag{1.12a}$$

with u_M between u_L and u_R, are identical to those of the step-function

$$w(x) = \begin{cases} u_L & x < (\theta - \frac{1}{2})h \\ u_R & x > (\theta - \frac{1}{2})h, \quad \theta = (u_R - u_M)/(u_R - u_L). \end{cases} \tag{1.12b}$$

Using this observation we can modify the ENO reconstruction of [7] to recover exactly any discontinuous quadratic function from its cell-averages.

In order to retain the relative simplicity of the numerical scheme (1.11) we use the new reconstruction to correct only the linear advection part. The new numerical flux is

$$\bar{f}_{j+1/2} = \bar{f}^{\text{ENO}}_{j+1/2} + \hat{g}_{j+1/2}; \tag{1.13}$$

here $\hat{g}_{j+1/2}$ is the flux through $x_{j+1/2}$ due to the linear advection of the *difference* between the modified reconstruction and the piecewise-linear one (1.9). In the constant coefficient case the scheme (1.13) is exact for discontinuous quadratic initial data.

Later on in this paper we present the extension of the "subcell resolution" concept to any finite order of accuracy and also extend the scheme to the Euler equations of gas dynamics.

2. ENO Reconstruction

In this section, we describe one of the techniques to obtain an ENO reconstruction. Given cell-averages $\{\bar{w}_j\}$ of a piecewise smooth function $w(x)$, we observe that

$$h\bar{w}_j = \int_{x_{j-1/2}}^{x_{j+1/2}} w(y)\, dy = W(x_{j+1/2}) - W(x_{j-1/2}), \tag{2.1a}$$

where

$$W(x) = \int_{x_0}^{x} w(y)\, dy \tag{2.1b}$$

is the primitive function of $w(x)$. Hence we can easily compute the *point values* $\{W(x_{i+1/2})\}$ by summation

$$W(x_{i+1/2}) = h \sum_{j=i_0}^{i} \bar{w}_j. \tag{2.2}$$

Let $H_m(x; u)$ be an interpolation of u at the points $\{y_j\}$, which is accurate to order m, i.e.,

$$H_m(y_j; u) = u(y_j), \tag{2.3a}$$

$$\frac{d^l}{dx^l} H_m(x; u) = \frac{d^l}{dx^l} u(x) + O(h^{m+1-l}), \qquad 0 \leqslant l \leqslant m. \tag{2.3b}$$

We obtain our "reconstruction via primitive function" technique by defining

$$R(x; \bar{w}) = \frac{d}{dx} H_r(x; W). \tag{2.4}$$

Relation (1.6a) follows immediately from (2.3b) with $l = 1$ and the definition (2.1); i.e.,

$$R(x; \bar{w}) = \frac{d}{dx} H_r(x; W) = \frac{d}{dx} W(x) + O(h^r)$$
$$= w(x) + O(h^r).$$

Relation (1.6b) is a direct consequence of (2.3a) and (2.2); i.e.,

$$A(I_j) R(\cdot; \bar{w}) = \frac{1}{h} \int_{x_{j-1/2}}^{x_{j+1/2}} \frac{d}{dx} H_r(x, W)\, dx$$
$$= \frac{1}{h} [H_r(x_{j+1/2}; W) - H_r(x_{j-1/2}; W]$$
$$= \frac{1}{h} [W(x_{j+1/2}) - W(x_{j-1/2})] = \bar{w}_j.$$

To obtain an ENO reconstruction, we take H_r in (2.4) to be the new ENO interpolation technique of the author [5]. In this case, $H_m(x; u)$ is a piecewise-polynomial function of x of degree m, which is defined (omitting the u dependence) by

$$H_m(x; u) = q_{j+1/2}(x) \qquad \text{for} \quad y_j \leqslant y \leqslant y_{j+1}, \tag{2.5a}$$

where $q_{j+1/2}$ is the unique polynomial of degree m that interpolates u at the $m+1$ points

$$S_m(i) \equiv \{y_{i+1}, \ldots, y_{i+m}\} \tag{2.5b}$$

for a particular choice of $i = i(j)$ (to be described in the following). To satisfy (2.3a), we need

$$q_{j+1/2}(y_j) = u(y_j), \qquad q_{j+1/2}(y_{j+1}) = u(y_{j+1});$$

therefore, we limit our choice of $i(j)$ to

$$j-m+1 \leqslant i(j) \leqslant j. \tag{2.5c}$$

The ENO interpolation technique is nonlinear: At each interval $[y_j, y_{j+1}]$, we consider the m possible choices of stencils (2.5b) subject to the restriction (2.5c) and assign to this interval the stencil in which u is "smoothest" in some sense; this is done by specifying $i(j)$ in (2.5b).

The information about the smoothness of u can be extracted from a table of divided differences. The kth divided difference of u,

$$u[y_i, y_{i+1}, ..., y_{i+k}] \equiv u[S_k(i)], \tag{2.6a}$$

is defined inductively by

$$u[S_0(i)] = u(y_i) \tag{2.6b}$$

and

$$u[S_k(i)] = (u[S_{k-1}(i+1)] - u[S_{k-1}(i)])/(y_{i+k} - y_i). \tag{2.6c}$$

If $u(x)$ is m times differentiable in $[y_i, y_{i+m}]$ then

$$u[S_m(i)] = \frac{1}{m!} u^{(m)}(\xi), \qquad \text{for some} \quad y_i \leqslant \xi \leqslant y_{i+m}. \tag{2.7a}$$

If $u^{(p)}(x)$ has a jump discontinuity in $[y_i, y_{i+m}]$ then

$$u[S_m(i)] = O(h^{-m+p}[u^{(p)}]), \qquad 0 \leqslant p \leqslant m-1 \tag{2.7b}$$

($[u^{(p)}]$ in the RHS of (2.7b) denotes the jump in the pth derivative).

Relations (2.7) show that $|u[S_m(i)]|$ is a measure of the smoothness of u in $S_m(i)$ and therefore can serve as a tool to compare the relative smoothness of u in various stencils. The simplest algorithm to assign $S_m(i(j))$ to the interval $[y_j, y_{j+1}]$ is the following:

ALGORITHM I. Choose $i(j)$ so that

$$|u[S_m(i(j))]| = \min_{j-m+1 \leqslant i \leqslant j} \{|u[S_m(i)]|\}. \tag{2.8}$$

Clearly (2.8) selects the "smoothest" stencil, provided that h is sufficiently small.

In order to make a sensible selection of stencil also when h is not sufficiently small, i.e., in the "pre-asymptotic" case, we prefer to use the following hierarchial algorithm:

ALGORITHM II. Let $i_k(j)$ be such that $S_k(i_k(j))$ is our choice of a $(k+1)$-point stencil for $[y_j, y_{j+1}]$. Obviously, we have to set

$$i_1(j) = j \tag{2.9a}$$

To choose $i_{k+1}(j)$, we consider as candidates the two stencils

$$S^L_{k+1} = S_{k+1}(i_k(j) - 1), \tag{2.9b}$$

$$S^R_{k+1} = S_{k+1}(i_k(j)), \tag{2.9c}$$

which are obtained by adding a point to the left of (or to the right of) $S_k(i_k(j))$, respectively. We select the one in which u is relatively smoother, i.e.,

$$i_{k+1}(j) = \begin{cases} i_k(j) - 1 & \text{if} \quad |u[S^L_{k+1}]| < |u[S^R_{k+1}]| \\ i_k(j) & \text{otherwise.} \end{cases} \tag{2.9d}$$

Finaly, we set $i(j) = i_m(j)$.

Using Newton's form of interpolation, we see that the polynomials $\{q_k(x)\}$, $1 \leqslant k \leqslant m$, corresponding to the stencils $S^k = S_k(i_k(j))$ selected by Algorithm II, satisfy the relation

$$q_{k+1}(x) = q_k(x) + u[S^{k+1}] \prod_{y \in S^k} (x - y). \tag{2.9e}$$

This shows that the choice made in (2.9d) selects q_{k+1} to be the one that deviates the least from q_k. It is this property that makes Algorithm II meaningful also for h in the pre-asymptotic range.

3. ENO Reconstruction with Subcell Resolution

In this section, we show how to modify the ENO reconstruction of the previous section so as to allow for the recovery of discontinuities in the interior of the cells. To illustrate the procedure, we first consider a discontinuous piecewise polynomial function $w(x)$ of the form

$$w(x) = \begin{cases} P_L(x), & x < x_d \\ P_R(x), & x > x_d, \end{cases} \tag{3.1a}$$

where $P_L(x)$ and $P_R(x)$ are polynomials of degree less than or equal to s,

$$\deg(P_L) \leqslant s, \qquad \deg(P_R) \leqslant s. \tag{3.1b}$$

We assume that $w(x)$ is actually discontinuous at x_d, i.e.,

$$P_L(x_d) \neq P_R(x_d) \tag{3.1c}$$

and that the discontinuity is located in the interior of the interval I_0

$$x_{-1/2} < x_d < x_{1/2}. \tag{3.1d}$$

(See Fig. 1a.)

Next we denote the cell-averages of $w(x)$ in (3.1) by $\{\bar{w}_j\}$ and consider the ENO reconstruction $R(x; \bar{w})$ applied to these data. To simplify our presentation let us denote the polynomial defining $R(x; \bar{w})$ in the cell I_j by $R_j(x; \bar{w})$. Clearly, provided h is sufficiently small, the stencils assigned to cells $\{I_k\}$, $j \neq 0$, are selected from the smooth part of the function. Therefore it follows from (2.3)–(2.5) that

$$\begin{aligned} R_j(x; \bar{w}) &= P_L(x) + O(h^r) \qquad \text{for} \quad j \leqslant -1 \\ R_j(x; \bar{w}) &= P_R(x) + O(h^r) \qquad \text{for} \quad j \geqslant 1. \end{aligned} \tag{3.2}$$

$R(x; \bar{w})$ in I_0 does not introduce spurious oscillations; however, it does not provide an accurte approximation to the discontinuous function $w(x)$ either (see Fig. 1b). Using (3.2) and the information contained in the cell-average $\bar{w}_0$, we can easily rectify this situation as follows: We extend $R_{-1}(x; \bar{w})$ to a point z in I_0 from the left, and extend $R_1(x; \bar{w})$ to z from the right and then approximate the location of the discontinuity in the cell I_0 by finding a value of z that will fit the cell average $\bar{w}_0$ (see Fig. 1c). This is done by finding a root of the algebraic equation $F_0(z) = 0$, where

$$F_0(z) = \frac{1}{h} \left\{ \int_{x_{-1/2}}^{z} R_{-1}(x; \bar{w})\, dx + \int_{z}^{x_{1/2}} R_1(x; \bar{w})\, dx \right\} - \bar{w}_0. \tag{3.3a}$$

When h is sufficiently small, the data near the cell I_0 approach those of a step-function. Therefore, as in (1.12) we expect to have

$$F_0(x_{-1/2}) \cdot F_0(x_{1/2}) \leqslant 0 \tag{3.3b}$$

and a single root of $F_0(z) = 0$ in I_0; we denote this root by θ_0,

$$F_0(\theta_0) = 0. \tag{3.3c}$$

It follows from (3.2) that

$$|\theta_0 - x_d| = O(h^r). \tag{3.3d}$$

What we mean by "ENO reconstruction with subcell resolution" is the modified ENO reconstruction $\hat{R}(x; \bar{w})$ which is defined in this case by

$$\hat{R}_j(x; \bar{w}) = R_j(x; \bar{w}) \qquad \text{for} \quad j \neq 0 \tag{3.4a}$$

$$\hat{R}(x; \bar{w}) = \begin{cases} R_{-1}(x; \bar{w} & \text{for} \quad x_{-1/2} < x < \theta_0 \\ R_1(x; \bar{w}) & \text{for} \quad \theta_0 < x < x_{1/2}. \end{cases} \tag{3.4b}$$

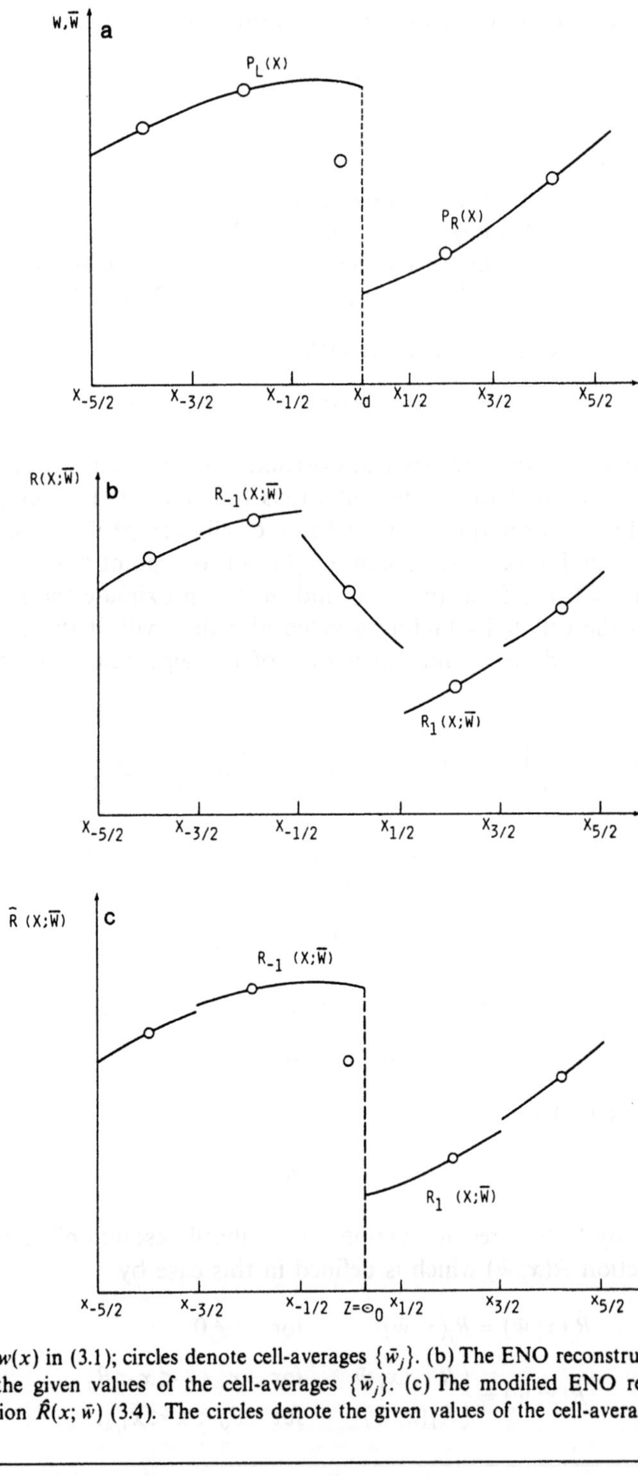

FIG. 1. (a) $w(x)$ in (3.1); circles denote cell-averages $\{\bar{w}_j\}$. (b) The ENO reconstruction $R(x;\bar{w})$. The circles denote the given values of the cell-averages $\{\bar{w}_j\}$. (c) The modified ENO reconstruction with subcelle resolution $\hat{R}(x;\bar{w})$ (3.4). The circles denote the given values of the cell-averages $\{\bar{w}_j\}$.

Clearly it follows from (3.2) and (3.3d) that $\hat{R}(x; \bar{w})$ is an $O(h^r)$ approximation to $w(x)$ in the L_1 sense, i.e., for any a and b,

$$\int_a^b |\hat{R}(x; \bar{w}) - w(x)| \, dx = O(h^r). \tag{3.5}$$

We observe that if the polynomial degree s in (3.1b) is less or equal $(r-1)$ then the primitive functions of P_L and P_R are polynomials of degree les or equal r, and therefore $H_r(x; w)$ in (2.5) is exact except at I_0. Hence,

$$R_j(x; \bar{w}) = P_L(x) \quad \text{for} \quad j \leqslant -1 \tag{3.6a}$$

$$R_j(x; \bar{w}) = P_R(x) \quad \text{for} \quad j \geqslant 1, \tag{3.6b}$$

and consequently $\theta_0 = x_d$ in (3.3). Thus we have shown

$$s \leqslant r-1 \Rightarrow \hat{R}(x; \bar{w}) \equiv \bar{w}(x). \tag{3.6c}$$

We turn now to describe the algorithm defining $\hat{R}(x; \bar{w})$ for a general piecewise-smooth function $w(x)$. As in the previous example we take $\hat{R}(x; \bar{w})$ in I_j to be $R_j(x; \bar{w})$, unless I_j is suspected of having a discontinuity of $w(x)$ in its interior. In the latter case we check whether

$$F_j(x_{j-1/2}) \cdot F_j(x_{j+1/2}) \leqslant 0, \tag{3.7a}$$

where

$$F_j(z) = \frac{1}{h} \left\{ \int_{x_{j-1/2}}^{z} R_{j-1}(x; \bar{w}) \, dx + \int_z^{x_{j+1/2}} R_{j+1}(x; \bar{w}) \, dx \right\} - \bar{w}_j. \tag{3.7b}$$

If (3.7a) holds, then there is a root $z = \theta_j$,

$$F_j(\theta_j) = 0, \qquad x_{j-1/2} \leqslant \theta_j \leqslant x_{j+1/2} \tag{3.7c}$$

in the cell I_j, and we define $\hat{R}(x; \bar{w})$ in this cell to be

$$\hat{R}_j(x; \bar{w}) = \begin{cases} R_{j-1}(x; \bar{w}) & \text{for} \quad x_{j-1/2} < x < \theta_j \\ R_{j+1}(x; \bar{w}) & \text{for} \quad \theta_j < x < x_{j+1/2}. \end{cases} \tag{3.7d}$$

If (3.7a) does not hold (which means that either there is no root in I_j or that there is an even number of roots in I_j), we take $\hat{R}_j(x; \bar{w})$ to be $R_j(x; \bar{w})$.

Let σ_j be some measure of the non-smoothness of the reconstruction $R(x; \bar{w})$ in I_j, e.g.,

$$\sigma_j = \left| \frac{d^k}{dx^k} R(x_j; \bar{w}) \right|, \qquad 1 \leqslant k \leqslant r-1, \tag{3.8a}$$

or a combination of such derivatives. Our algorithm identifies cells which are suspect of harboring a discontinuity of $w(x)$, as those which attain a local maximum of "non-smoothness" of the reconstruction, i.e.,

$$\sigma_j > \sigma_{j-1} \quad \text{and} \quad \sigma_j \geqslant \sigma_{j+1}. \tag{3.8b}$$

We summarize the algorithm defining $\hat{R}(x; \bar{w})$ by: If

$$\sigma_j > \sigma_{j-1}, \quad \sigma_j \geqslant \sigma_{j+1} \quad \text{and} \quad F_j(x_{j-1/2}) \cdot F_j(x_{j+1/2}) \leqslant 0 \tag{3.9a}$$

then

$$\hat{R}_j(x; \bar{w}) = \begin{cases} R_{j-1}(x; \bar{w}) & \text{for } x_{j-1/2} < x < \theta_j \\ R_{j+1}(x; \bar{w}) & \text{for } \theta_j < x < x_{j+1/2}; \end{cases} \tag{3.9b}$$

otherwise,

$$\hat{R}_j(x; \bar{w}) = R_j(x; \bar{w}). \tag{3.9c}$$

In an Appendix, we present analysis which motivates the choice of condition (3.8b). In the following we make several remarks and observations about $\hat{R}(x; w)$, the "ENO reconstruction with subcell resolution."

(1) $\hat{R}(x; \bar{w})$ is indeed essentially non-oscillatory (ENO). This follows from the fact that local maxima are isolated; i.e., if (3.8b) holds for I_j, it cannot hold either for I_{j-1} or for I_{j+1}. Consequently, if $\hat{R}(x; \bar{w})$ is defined in I_j by the discontinuous (3.9b), then in I_{j-1} and I_{j+1} it is defined by (3.9c); i.e.,

$$\hat{R}_{j-1}(x; \bar{w}) = R_{j-1}(x; \bar{w}), \qquad \hat{R}_{j+1}(x; \bar{w}) = R_{j+1}(x; \bar{w}). \tag{3.10}$$

(2) If, as in the example (3.1), there is a discontinuity of $w(x)$ in the interior of I_0, then

$$\sigma_0 = O(h^{-k}), \qquad \sigma_{-1} = O(1), \qquad \sigma_1 = O(1), \tag{3.11a}$$

where k is the order of the derivative used in (3.8a).

Therefore, provided h is sufficiently small, we get that

$$\sigma_0 > \sigma_{-1}, \qquad \sigma_0 > \sigma_1, \tag{3.11b}$$

and also as in (3.3b),

$$F_0(x_{-1/2}) \cdot F_0(x_{1/2}) < 0. \tag{3.11c}$$

This shows that $\hat{R}(x; \bar{w})$, as defined by the general algorithm (3.9), will recover any real discontinuity of $w(x)$.

(3) We observe that condition (3.9a) may hold also in the smooth part of $w(x)$ near a local maximum of $|d^k w/dx^k|$. In this case the algorithm places a discon-

tinuity at θ_j in the interior of I_j. However, because of the smoothness of $w(x)$ there and (1.6a), we have

$$R_{j\pm1}(\theta_j; \bar{w}) = w(\theta_j) + O(h^r). \tag{3.12}$$

Consequently the jump is of the size of the reconstruction error $O(h^r)$. We recall that the original ENO reconstruction is discontinuous at $\{x_{j+1/2}\}$. Therefore, the effect of the algorithm (3.9) is to replace the two discontinuities at $x_{j-1/2}$ and $x_{j+1/2}$ by a single one at θ_j.

(4) We observe that in order to evaluate $\hat{R}_j(x; \bar{w})$ in a cell I_j which contains a discontinuity at θ_j, we have to find out whether $x > \theta_j$ or $x < \theta_j$. Assuming θ_j to be the only root of $F_j(\theta_j) = 0$ in I_j, as is the case for a real discontinuity, we can use the logic of the interval-halving technique to evaluate $\hat{R}_j(x; \bar{w})$ without actually computing θ_j. To do so we calculate $F_j(x)$ and compare its sign with that of $F_j(x_{j-1/2})$ or $F_j(x_{j+1/2})$:

$$\hat{R}_j(x; \bar{w}) = \begin{cases} R_{j-1}(x; \bar{w}) & \text{if} \quad F_j(x) \cdot F_j(x_{j-1/2}) > 0 \\ R_{j+1}(x; \bar{w}) & \text{otherwise.} \end{cases} \tag{3.13}$$

4. A Second-Order Accurate ENO Scheme with Subcell Resolution

In the following we describe how to incorporate the reconstruction with subcell resolution into the ENO schemes, so as to improve their resolution of linear discontinuities. In this section, we present the derivation of (1.13), which is an improved version of the second-order accurate MUSCL-type scheme (1.11). In the next section, we shall generalize these ideas to any order of accuracy.

We start with the piecewise-parabolic reconstruction $R(x; \bar{w})$ which is defined by (2.4) with $r = 3$, i.e.,

$$R(x; \bar{w}) = \frac{d}{dx} H_3(x; W), \tag{4.1a}$$

where W is the primitive function (2.1b) of $w(x)$, and $H_3(x; W)$ is a piecewise-cubic ENO interpolation. Let $i = i(j)$ be the left endpoint of the stencil $(x_{i-1/2}, x_{i+1/2}, x_{i+3/2}, x_{i+5/2})$ assigned to the interval $(x_{j-1/2}, x_{j+1/2})$ by (2.8) or (2.9) or some other ENO technique; $j-2 \leqslant i \leqslant j$. The parabola describing $R(x; \bar{w})$ in $(x_{j-1/2}, x_{j+1/2})$ is given by

$$C_j = \frac{d^3 H_3(x; W)}{dx^3} = (\bar{w}_{i+2} - 2\bar{w}_{i+1} + \bar{w}_i)/h^2 \tag{4.1b}$$

$$S_j = \left.\frac{d^2 H_3(x; w)}{dx^2}\right|_{x = x_j} = (\bar{w}_{i+1} - \bar{w}_i)/h + \left(j - i - \frac{1}{2}\right) hC_j \tag{4.1c}$$

$$R(x; \bar{w}) = \left(\bar{w}_j - \frac{h^2}{24} C_j\right) + S_j(x - x_j) + \frac{1}{2} C_j(x - x_j)^2. \tag{4.1d}$$

Our basic scheme remains

$$v_j^{n+1} = A(I_j)E(\tau)L(\cdot; v^n) \tag{4.2a}$$

with the piecewise linear reconstruction

$$L(x; \bar{w}) = \bar{w}_j + S_j(x - x_j), \qquad x \in I_j, \tag{4.2b}$$

but we use the parabolic reconstruction in order to define the slope S_j by (4.1c); as in (1.11) we approximate its numerical flux by

$$\bar{f}_{j+1/2} = \bar{f}^{\mathrm{ENO}}_{j+1/2} = f^R(v^L_{j+1/2}, v^R_{j+1/2}), \tag{4.3a}$$

where

$$v^L_{j+1/2} = v_j^n + h(1 - \lambda a_j^n)S_j^n/2, \qquad v^R_{j+1/2} = v^n_{j+1} - h(1 + \lambda a^n_{j+1})S_{j+1}/2. \tag{4.3b}$$

We observe that by (2.3b) choosing S_j(in (4.4b) to be (4.1c) results in

$$S_j = w_x(x_j) + O(h^2), \tag{4.3c}$$

which is one order higher than (1.9b). Consequently, as the UNO scheme of Harten and Osher [6], this scheme is truly second order in all L_p norms (unlike TVD schemes which are first order in L_∞ and second order only in L_1).

Using the algorithm described in (3.9) we now define $\hat{R}(x; \bar{w})$, which modifies $R(x; \bar{w})$ in (4.1) so that it includes a discontinuity in the interior of each cell I_j which meets condition (3.9a). Ideally, we would like to use the scheme

$$v_j^{n+1} = A(I_j)E(\tau)\hat{R}(\cdot; v^n), \tag{4.4}$$

which is third-order accurate in L_1-sense (but only second-order accurate in the maximum norm). However, the proper approximation of the numerial flux of (4.2) is more complicated than (1.11), since it is one order more accurate in time. Bearing in mind that the main fault in the MUSCL-type scheme (1.11) that we want to correct is the smearing of linear discontinuities, we introduce subcell resolution into the scheme by correcting its numerical flux to account for the difference between using $\hat{R}$ and L in the constant coefficient case. Thus we use the scheme

$$v_j^{n+1} = v_j^n - \lambda(\bar{f}_{j+1/2} - \bar{f}_{j-1/2}) \tag{4.5a}$$

$$\bar{f}_{j+1/2} = \bar{f}^{\mathrm{ENO}}_{j+1/2} + \hat{g}_{j+1/2}, \tag{4.5b}$$

and define $\hat{g}_{j+1/2}$ by the requirement that the scheme (4.5) becomes identical to (4.4) in the constant coefficient case

$$u_t + au_x = 0, \qquad a = \text{const.} \tag{4.6}$$

This determines $\hat{g}_{j+1/2}$ to be

$$\hat{g}_{j+1/2}=\frac{a}{\tau}\int_0^\tau [\hat{R}(x_{j+1/2}-at;v^n)-L(x_{j+1/2}-at;v^n)]\,dt. \tag{4.7a}$$

Since in the constant coefficient case

$$\bar{f}^{\text{ENO}}_{j+1/2}=\frac{a}{\tau}\int_0^\tau L(x_{j+1/2}-at;v^n)\,dt, \tag{4.7b}$$

we get that

$$\bar{f}_{j+1/2}=\frac{a}{\tau}\int_0^\tau \hat{R}(x_{j+1/2}-at;v^n)\,dt. \tag{4.7c}$$

which shows that the scheme (4.5) is indeed identical to (4.4) in the constant coefficient case (4.6). Consequently, it is exact for discontinuous parabolic data of the form (3.1).

Next we derive an expression for $\hat{g}_{j+1/2}$ in the constant coefficient case (4.7a); this will later be generalized to the nonlinear case by "freezing" the characteristic speed within the cell.

In the constant coefficient case, (4.7a) can be rewritten as

$$\hat{g}_{j+1/2}=\frac{1}{\tau}\int_{x_{j+1/2}-a\tau}^{x_{j+1/2}} [\hat{R}(y;v^n)-L(y;v^n)]\,dy. \tag{4.8}$$

First let us assume $a>0$. When

$$\hat{R}(x;v^n)=R_j(x;v^n) \qquad \text{in } I_j, \tag{4.9a}$$

then

$$\hat{g}_{j+1/2}=\frac{1}{\tau}\int_{x_{j+1/2}-a\tau}^{x_{j+1/2}} [R_j(y;v^n)-L(y;v^n)]\,dy=\frac{a}{12}(\nu-1)(2\nu-1)h^2C_j; \tag{4.9b}$$

here $\nu=\lambda a$.

When there is a discontinuity in the interior of I_j, $\hat{R}_j(x;v^n)$ is given by (3.9b); i.e.,

$$\hat{R}(x;v^n)=\begin{cases} R_{j-1}(x;v^n), & x_{j-1/2}\leqslant x<\theta_j \\ R_{j+1}(x;v^n), & \theta_j\leqslant x\leqslant x_{j+1/2}, \end{cases} \tag{4.10a}$$

and we have to find out whether $(x_{j+1/2}-a\tau)$ is larger or smaller than θ_j. We recall that for (4.10a) to hold the relations

$$F_j(\theta_j)=0, \qquad F_j(x_{j-1/2})\cdot F_j(x_{j+1/2})\leqslant 0 \tag{4.10b}$$

have to be satisfied, where $F_j(z)$ is defined by (3.7b). Using the basic idea of the interval-halving method for calculating a root of algebraic equations, we can find out in which of the two cases we are without actually calculating θ_j (see Remark (4) at the end of the previous section). All we have to do is to compute $F_j(x_{j+1/2}-a\tau)$ and compare its sign to that of $F_j(x_{j-1/2})$, i.e.,

$$F_j(x_{j+1/2}-a\tau)\cdot F_j(x_{j-1/2})>0 \Rightarrow x_{j+1/2}-a\tau<\theta_j \tag{4.10c}$$

$$F_j(x_{j+1/2}-a\tau)\cdot F_j(x_{j-1/2})\leqslant 0 \Rightarrow x_{j+1/2}-a\tau\geqslant\theta_j. \tag{4.10d}$$

To express the integral in (4.8) let us introduce the notation

$$b_m(y_1, y_2)=\int_{y_1}^{y_2} R_m(x; v^n)\, dx=\left[\left(v_m^n-\frac{h^2}{24}C_m\right)y+\frac{1}{2}S_m y^2+\frac{1}{6}C_m y^3\right]_{y_1-x_m}^{y_2-x_m}. \tag{4.11}$$

In case $x_{j+1/2}-a\tau\geqslant\theta_j$, we get from (4.8) and (4.10a) that

$$\hat{g}_{j+1/2}=\frac{1}{\tau}\left\{b_{j+1}(x_{j+1/2}-a\tau,\, x_{j+1/2})-a\tau\left[v_j^n+\frac{1}{2}(h-a\tau)S_j^n\right]\right\}. \tag{4.12a}$$

When $x_{j+1/2}-a\tau<\theta_j$, we use the fact that

$$\frac{1}{h}\int_{x_{j-1/2}}^{x_{j+1/2}} \hat{R}(x; v^n)\, dx=v_j^n$$

to express the integral in (4.8) by

$$\int_{x_{j+1/2}-a\tau}^{x_{j+1/2}} \hat{R}(x; v^n)\, dx=hv_j^n-\int_{x_{j-1/2}}^{x_{j+1/2}-a\tau} \hat{R}(x; v^n)\, dx$$

$$=hv_j^n-\int_{x_{j-1/2}}^{x_{j+1/2}-a\tau} R_{j-1}(x; v^n)\, dx.$$

Rearranging terms we get in this case

$$\hat{g}_{j+1/2}=\frac{1}{\tau}\left\{(h-a\tau)\left(v_j^n-\frac{a\tau}{2}S_j^n\right)-b_{j-1}(x_{j-1/2},\, x_{j+1/2}-a\tau)\right\}. \tag{4.12b}$$

To summarize, the definition of $\hat{g}_{j+1/2}$ in the case $a>0$ is

$$\tau\hat{g}_{j+1/2}=\frac{a\tau}{12}(\nu-1)(2\nu-1)h^2C_j, \tag{4.13a$^+$}$$

unless the discontinuity condition (3.9a) holds for I_j; in the latter case we define

$$\tau\hat{g}_{j+1/2}=\begin{cases}(h-a\tau)\left(v_j^n-\dfrac{a\tau}{2}S_j\right)-b_j(x_{j-1/2},\, x_{j+1/2}-a\tau) & \\ \qquad\text{if}\quad F_j(x_{j+1/2}-a\tau)\cdot F_j(x_{j-1/2})>0 & \\ b_{j+1}(x_{j+1/2}-a\tau,\, x_{j+1/2})-a\tau[v_j^n+\tfrac{1}{2}(h-a\tau)S_j] & \\ \qquad\text{otherwise.} & \end{cases} \tag{4.13b$^+$}$$

Similarly for $a<0$ we get

$$\tau\hat{g}_{j-1/2}=\frac{a\tau}{12}(\nu+1)(2\nu+1)h^2C_j, \tag{4.13a$^-$}$$

unless the discontinuity condition (3.9a) holds for I_j; in the latter case we define

$$\tau\hat{g}_{j-1/2}=\begin{cases} -b_{j-1}(x_{j-1/2}, x_{j-1/2}-a\tau)-a\tau[v_j^n-\frac{1}{2}(h+a\tau)]S_j \\ \qquad\qquad \text{if} \quad F_j(x_{j-1/2})\cdot F_j(x_{j-1/2}-a\tau)>0 \\ b_{j+1}(x_{j-1/2}-a\tau, x_{j+1/2})-(h+a\tau)\left(v_j^n-\dfrac{a\tau}{2}S_j\right) \\ \qquad\qquad \text{otherwise.} \end{cases} \tag{4.13b$^-$}$$

Note that the expressions (4.13) are formulated as the contribution of the cell I_j to the numerical flux $\hat{g}$. Thus, if $a>0$, the contribution of from the I_j cell goes to $\hat{g}_{j+1/2}$, while if $a<0$ this contribution goes to $\hat{g}_{j-1/2}$.

In Section 6, we extend the subcell resolution ideas to the Euler equations of gas dynamics. Since shocks are highly resolved by the original ENO scheme, we apply subcell resolution only to the linearly degenerate characteristic field in order to improve the resolution of contact discontinuities. In this case the characteristic speed, which is the velocity of the flow, is not a constant but a function of the solution itself. Nevertheless, we use the same expressions as in (4.13), except that a in I_j is replaced by a_j, and $\nu=\lambda a_j$. We compute the corrective flux $\hat{g}$ in the following way: First we preset $\hat{g}=0$, and then we sweep over the mesh and collect contributions to $\hat{g}$ from each cell: If $a_j>0$ we add the RHS of $(4.13)^+$ to $\hat{g}_{j+1/2}$; if $a_j<0$ we add the RHS of $(4.13)^-$ to $\hat{g}_{j-1/2}$. Note that if $a_{j+1}<0$ and $a_j>0$, then $\hat{g}_{j+1/2}$ gets contributions from both I_j and I_{j+1}.

5. Extension to High Order of Accuracy

In this section, we describe the extension of the ENO scheme with subcell resolution to arbitrarily high order of accuracy. As in the second-order case (4.5), we introduce subcell resolution to the high order accurate ENO schemes via a corrective flux $\hat{g}_{j+1/2}$; i.e., we consider the modified scheme

$$v_j^{n+1}=v_j^n-\lambda(\bar{f}_{j+1/2}-\bar{f}_{j-1/2}) \tag{5.1a}$$

$$\bar{f}_{j+1/2}=\bar{f}^{\text{ENO}}_{j+1/2}+\hat{g}_{j+1/2}. \tag{5.1b}$$

First, we describe briefly the derivation of $\bar{f}^{\text{ENO}}_{j+1/2}$. We refer the reader to [7] for more details. Let $L(x;\bar{w})$ be an rth order accurate reconstruction of $w(x)$, such that

$$\frac{1}{h}\int_{I_j} L(x;\bar{w})\,dx=\bar{w}_j. \tag{5.2a}$$

As before we denote the $(r-1)$th degree polynomial describing $L(x; \bar{w})$ in I_j by $L_j(x; \bar{w})$, and express it in the following Taylor expansion

$$L_j(x; \bar{w}) = \sum_{k=0}^{r-1} \frac{\tilde{D}_k}{k!} (x - x_j)^k, \qquad \tilde{D}_k = \frac{d^k}{dx^k} L(x_j; \bar{w}). \tag{5.2b}$$

Let $u_j(x, t)$ be the solution to the initial value problem

$$u_t + f(u)_x = 0, \qquad u(x, 0) = L_j(x; \bar{w}). \tag{5.3a}$$

Since the initial data in (5.3a) are analytic, we know from the Cauchy–Kowalewski theorem that $u_j(x, t)$ exists uniquely and it is analytic for some time $0 \leqslant t \leqslant t_c$. Therefore, its Taylor expansion around $x = x_j$ and $t = 0$

$$u_j(x, t) = \sum_{k=0}^{\infty} \frac{1}{k!} \sum_{m=0}^{k} \binom{k}{m} \tilde{D}_{m,k-m} t^m (x - x_j)^{k-m}, \tag{5.3b}$$

$$\tilde{D}_{m,k-m} = \frac{\partial^k}{\partial t^m \partial x^{k-m}} u_j(x_j, 0), \tag{5.3c}$$

is valid for $0 \leqslant t \leqslant t_c$ and x sufficiently close to x_j. Using (5.3) we define $v_j(x, t)$ to be the truncated Taylor expansion

$$v_j(x, t) = \sum_{k=0}^{r-1} \frac{1}{k!} \sum_{m=0}^{k} \binom{k}{m} \tilde{D}_{m,k-m} t^m (x - x_j)^{k-m}, \tag{5.4a}$$

$$|v_j(x, t) - u_j(x, t)| = O(h^r). \tag{5.4b}$$

The coefficients $\tilde{D}_{m,k-m}$ in (5.4a) can be computed directly from the known coefficients $\{\tilde{D}_k\}$ in (5.2b) (note that $\tilde{D}_{0,k} = \tilde{D}_k$) by successive differentiation of the partial differential equation and substitution—see [7] for details.

Finally, using an appropriate numerical quadrature to approximate the integral in (1.5a), we define $\bar{f}^{\text{ENO}}_{j+1/2}$ to be

$$\bar{f}^{\text{ENO}}_{j+1/2} = \sum_{k=1}^{K} \beta_k f^R(v_j(x_{j+1/2}, \gamma_k \tau), v_{j+1}(x_{j+1/2}, \gamma_k \tau)). \tag{5.5}$$

Here β_k and γ_k are the coefficients of the numerical quadrature. In the second-order case we use the midpoint rule: $K = 1$, $\beta_k = 1$, $\gamma_k = \frac{1}{2}$. In the third- and fourth-order case we use the Gaussian quadrature: $K = 2$, $\beta_1 = \beta_2 = \frac{1}{2}$, $\gamma_1 = \frac{1}{2}(1 - 1/\sqrt{3})$, $\gamma_2 = \frac{1}{2}(1 + 1/\sqrt{3})$. Note that the second-order accurate scheme (1.11) is identical to (5.5) with $r = 2$.

Next we describe the derivation of the corrective flux $\hat{g}_{j+1/2}$ in (5.1b). Let $R(x; \bar{w})$ be another reconstruction of $w(x)$ which is at least rth order accurate. Using the algorithm (3.9) we define $\hat{R}(x; \bar{w})$, its modified version with subcell resolution. As in the second-order case (4.7a) we define $\hat{g}_{j+1/2}$ in the constant coefficient case to be

$$\hat{g}_{j+1/2} = \frac{a}{\tau} \int_0^\tau [\hat{R}(x_{j+1/2} - at; v^n) - L(x_{j+1/2} - at; v^n)]\, dt$$

$$= \frac{1}{\tau} \int_{x_{j+1/2} - a\tau}^{x_{j+1/2}} [\hat{R}(y; v^n) - L(y; v^n)]\, dy. \tag{5.6}$$

We note that relations (4.7b)–(4.7c) hold for any r; therefore, we can state that the scheme (5.1) in the constant coefficient case is identical to (4.4) in general. Consequently, if the reconstruction $R(x; \bar{w})$ is exact for *smooth* polynomial data of degree r, then the ENO scheme with subcell resolution (5.1) is exact for all initial data of *discontinuous* piecewise-polynomial functions of degree less than or equal to r.

Let us denote

$$d_m(y_1, y_2) = \int_{y_1}^{y_2} [R_m(x; v^n) - L_m(x; v^n)]\, dx, \tag{5.7a}$$

$$C_{m+1/2}(y_1, y_2) = \int_{y_1}^{y_2} [R_{m+1}(x; v^n) - R_m(x; v^n)]\, dx. \tag{5.7b}$$

Using these notations we can evaluate $\hat{g}_{j+1/2}$ by the following expressions:

If $a > 0$, then

$$\tau \hat{g}_{j+1/2} = d_j(x_{j+1/2} - a\tau, x_{j+1/2}), \tag*{(5.8a)$^+$}$$

unless

$$\sigma_j > \sigma_{j-1}, \quad \sigma_j \geqslant \sigma_{j+1}, \qquad F_j(x_{j+1/2}) \cdot F_j(x_{j-1/2}) \leqslant 0, \tag*{(5.8b)$^+$}$$

in which case

$$\tau \hat{g}_{j+1/2} = \begin{cases} d_j(x_{j+1/2} - a\tau, x_{j+1/2}) + C_{j+1/2}(x_{j+1/2} - a\tau, x_{j+1/2}) \\ \qquad\qquad \text{if } F_j(x_{j+1/2}) \cdot F_j(x_{j+1/2} - a\tau) > 0 \\ d_j(x_{j+1/2} - a\tau, x_{j+1/2}) + C_{j-1/2}(x_{j-1/2}, x_{j+1/2} - a\tau), \\ \qquad\qquad \text{otherwise.} \end{cases} \tag*{(5.8c)$^+$}$$

If $a < 0$, then

$$\tau \hat{g}_{j-1/2} = -d_j(x_{j-1/2}, x_{j-1/2} - a\tau), \tag*{(5.8a)$^-$}$$

unless

$$\sigma_j > \sigma_{j-1}, \quad \sigma_j \geqslant \sigma_{j+1}, \qquad F_j(x_{j+1/2}) \cdot F_j(x_{j-1/2}) \leqslant 0, \tag*{(5.8b)$^-$}$$

in which case

$$\tau\hat{g}_{j-1/2} = \begin{cases} -d_j(x_{j-1/2}, x_{j-1/2} - a\tau) + C_{j+1/2}(x_{j-1/2} - a\tau, x_{j+1/2}), \\ \qquad \text{if} \quad F_j(x_{j+1/2}) \cdot F_j(x_{j-1/2} - a\tau) > 0 \\ -d_j(x_{j-1/2}, x_{j-1/2} - a\tau) + C_{j-1/2}(x_{j-1/2}, x_{j-1/2} - a\tau), \\ \qquad \text{otherwise.} \end{cases} \tag{5.8c$^-$}$$

We observe tht up to this point we have not specified $L(x; \bar{w})$ and $R(x; \bar{w})$. One possibility is to generalize the set up of the second-order accurate scheme in Section 3 as follows: Let r be the desired order of accuracy of the scheme. We start with a reconstruction via primitive function $R(x; \bar{w})$ which is one order higher, i.e.,

$$R(x; \bar{w}) = \frac{d}{dx} H_{r+1}(x; W); \tag{5.9a}$$

here W is the primitive function of w and H_{r+1} is the ENO interpolation of Section 2. As before we denote the polynomial of degree r defining $R(x; \bar{w})$ in $_j$ by $R_j(x; \bar{w})$ and rewrite (5.9a) as a finite Taylor expansion:

$$R_j(x; \bar{w}) = \sum_{k=0}^{r} \frac{D_k}{k!} (x - x_j)^k, \qquad D_k = \frac{d^{k+1}}{dx^{k+1}} H_{r+1}(x_j; W). \tag{5.9b}$$

Using (5.9b) we now define $L(x; \bar{w})$ to be

$$L_j(x; \bar{w}) = (D_0 + \alpha_r h^r D_r) + \sum_{k=0}^{r-1} \frac{D_k}{k!} (x - x_j)^k, \tag{5.10a}$$

where

$$\alpha_k = \begin{cases} 0 & \text{for } k \text{ odd} \\ 2^{-k}/(k+1)! & \text{for } k \text{ even.} \end{cases} \tag{5.10b}$$

$L_j(x; \bar{w})$ is a polynomial of degree $(r-1)$ which reconstructs $w(x)$ in I_j to $O(h^r)$. We observe that

$$\frac{d^{r-1}}{dx^{r-1}} L_j(x; \bar{w}) = D_{r-1} = \frac{d^{r-1}}{dx^{r-1}} w(x_j) + O(h^2); \tag{5.10c}$$

this, as (4.3) does in the second-order case, eliminates some of the non-smoothness in the reconstruction eror which is due to the adaptive choice of stencils. Consequently, the ENO scheme based on this reconstruction is rth-order accurate in all L_p norms, including the maximum norm. Note that the correction to the first term in the RHS of (5.10a) takes care of the conservation property (5.2a), i.e.,

$$\frac{1}{h} \int_{I_j} L_j(x; \bar{w})\, dx = \bar{w}_j. \tag{5.10d}$$

Remark. There are other reasonable choices of $L(x; \bar{w})$ and $R(x; \bar{w})$. We may choose

$$L(x; \bar{w}) = \frac{d}{dx} H_r(x; W), \qquad R(x; \bar{w}) = \frac{d}{dx} H_{r+1}(x; W), \tag{5.11}$$

or even

$$L(x; \bar{w}) \equiv R(x; \bar{w}) = \frac{d}{dx} H_r(x; \bar{w}); \tag{5.12}$$

note that the expression for $\hat{g}_{j+1/2}$ in the latter case is much simpler since $d_j(y_1, y_2) \equiv 0$ in (5.7a).

6. Euler Equations of Gas Dynamics

In this section, we describe how to apply the scheme (5.1) to the Euler equations of gas dynamics for a polytropic gas:

$$u_t + f(u)_x = 0 \tag{6.1a}$$

$$u = (\rho, m, E)^{\mathrm{T}} \tag{6.1b}$$

$$f(u) = qu + (0, P, qP)^{\mathrm{T}} \tag{6.1c}$$

$$P = (\gamma - 1)(E - \tfrac{1}{2}\rho q^2). \tag{6.1d}$$

Here ρ, q, P, and E are the density, velocity, pressure, and total energy, respectively; $m = \rho q$ is the momentum and γ is the ratio of specific heats.

The eigenvalues of the Jacobian matrix $A(u) = \partial f/\partial u$ are

$$a_1(u) = q - c, \qquad a_2(u) = q, \qquad a_3(u) = q + c, \tag{6.2a}$$

where $c = (\gamma P/\rho)^{1/2}$ is the sound speed.

The corresponding right-eigenvectors are

$$r_1(u) = \begin{pmatrix} 1 \\ q - c \\ H - qc \end{pmatrix}, \qquad r_2(u) = \begin{pmatrix} 1 \\ q \\ \frac{1}{2}q^2 \end{pmatrix}, \qquad r_3(u) = \begin{pmatrix} 1 \\ q + c \\ H + qc \end{pmatrix}; \tag{6.2b}$$

here

$$H = (E + P)/\rho = c^2/(\gamma - 1) + \tfrac{1}{2}q^2 \tag{6.2c}$$

is the enthalpy.

To compute a left-eigenvector system $\{l_k(u)\}$ which is bi-orthonormal to $\{r_k(u)\}$

in (6.2b), we first form the matrix $\mathbf{T}(u)$, the columns of which are the right-eigenvectors in (6.2b)

$$\mathbf{T}(u)=(r_1(u), r_2(u), r_3(u))$$

and then define $l_k(u)$ to be the kth row in $\mathbf{T}^{-1}(u)$, the inverse of $\mathbf{T}(u)$. We get

$$\begin{aligned} l_1(u)&=\tfrac{1}{2}(b_2+q/c,\ -b_1q-1/c,\ b_1)\\ l_2(u)&=(1-b_2,\ b_1q,\ -b_1)\\ l_3(u)&=\tfrac{1}{2}(b_2-q/c,\ -b_1q+1/c,\ b_1), \end{aligned} \tag{6.2d}$$

where

$$b_1=(\gamma-1)/c^2 \tag{6.2e}$$

$$b_2=\tfrac{1}{2}q^2b_1. \tag{6.2f}$$

Given $\{v_j^n\}$, approximation to $\{\bar{u}(x_j, t_n)\}$, we use (6.2d)–(6.2f) to evaluate the locally defined characteristic variables $\bar{w}_i^k(v_j^n)$,

$$\bar{w}_i^k=l_k(v_j^n)v_i^n \qquad \text{for} \quad i=j-r, \ldots, j+r \quad \text{and} \quad k=1, 2, 3. \tag{6.3a}$$

Note that j is fixed in (6.3a) while i varies over the points which are relevant to the selection of the stencil for the cell I_j. Thus, the eigenvector system $\{l_k(v_j^n)\}_{k=1}^3$ should be regarded in this context as a constant system of coordinates. Next we apply our scalar algorithm to each of the locally defined characteristic variables in (6.3a); i.e., we select a (possibly different) stencil for each of the characteristic variables and define $R_j(x; \bar{w}^k)$ by (5.9); then we combine these scalar reconstruction by

$$\mathbf{R}_j(x; v^n)=\sum_{k=1}^{3} R_j(x; \bar{w}^k(v_j^n))r_k(v_j^n). \tag{6.3b}$$

Note that the vector reconstruction (6.3) satisfies all the requirements listed in (1.6).

As in Section 5 we rewrite the rth degree polynomial (6.3b) as a finite Taylor expansion, except that now the coefficients $\{D_k\}$ in (5.9) are vectors. With this convention in mind we proceed to define the vector reconstruction $L_j(x; v^n)$ and the numerical flux $\bar{f}^{\mathrm{ENO}}_{j+1/2}$ by (5.10) and (5.5), respectively.

We turn now to describe the vector $\hat{g}_{j+1/2}$, which introduced the subcell resolution to the numerical flux (5.1). As we have mentioned earlier in this paper., we use subcell resolution only in the linearly degenerate field ($k=2$ in (6.2)) in order to improve the resolution of contact discontinuities. We do so by applying the algorithm (5.8) scalarly to the linearly degenerate characteristic field $k=2$ as follows:

We define

$$\sigma_j = \left| l_2(v_j^n) \frac{d^k}{dx^k} \mathbf{R}_j(x_j; v^n) \right| \qquad \text{for some } k,\ 1 \leqslant k \leqslant r-1, \tag{6.4}$$

$$F_j(x) = \frac{1}{h} \cdot l_2(v_j^n) \left\{ \int_{x_{j-1/2}}^{z} \mathbf{R}_{j-1}(x; v^n)\, dx + \int_z^{x_{j+1/2}} \mathbf{R}_{j+1}(x; v^n)\, dx - h v_j^n \right\} \tag{6.5}$$

and, similarly,

$$d_j(y_1, y_2) = \int_{y_1}^{y_2} l_2(v_j^n)[\mathbf{R}_j(x; v^n) - \mathbf{L}_j(x; v^n)]\, dx \tag{6.6a}$$

$$C_{m+1/2}(y_1, y_2) = \int_{y_1}^{y_2} l_2(v_j^n)[\mathbf{R}_{m+1}(x; v^n) - \mathbf{R}_m(x; v^n)]\, dx \quad \text{for } m = j-1, j. \tag{6.6b}$$

The characteristic speed of the linerly degenerate field is the flow velocity q; which can be of different sign in different regions. The definition of $\hat{g}_{j+1/2}$ in (5.8) is formulated as the contribution of the cell I_j to the numerical flux. Therefore, it is convenient to program the calculation of the numerical flux in two stages: First we evaluate

$$\bar{f}_{j+1/2} = \bar{f}_{j+1/2}^{\text{ENO}} \tag{6.7}$$

by (5.5) for all j. Then we sweep over the mesh again and collect the contribution of each cell to the numerical flux. Using FORTRAN conventions this can be described by:

If $q_j^n > 0$ then

$$\bar{f}_{j+1/2} = \bar{f}_{j+1/2} + \frac{1}{\tau} \cdot d_j(x_{j+1/2} - \tau q_j^n, x_{j+1/2}) r_2(v_j^n); \tag{6.8$^+$}$$

if $q_j^n \leqslant 0$ then

$$\bar{f}_{j-1/2} = \bar{f}_{j-1/2} - \frac{1}{\tau} d_j(x_{j-1/2}, x_{j-1/2} - \tau q_j^n) r_2(v_j^n). \tag{6.8$^-$}$$

Next we check whether the discontinuity condition

$$\sigma_j > \sigma_{j-1}, \quad \sigma_j \geqslant \sigma_{j+1}, \qquad F_j(x_{j-1/2}) \cdot F_j(x_{j+1/2}) \leqslant 0 \tag{6.9}$$

is satisfied. If one or more of the inequalities in (6.9) is not true, we move on to the next cell. If all the inequalities in (6.9) are true we proceed to calculate as follows:

If $q_j^n > 0$, then

$$\delta_j = \begin{cases} c_{j+1/2}(x_{j+1/2} - \tau q_j^n, x_{j+1/2}) & \text{if } F_j(x_{j+1/2}) \cdot F_j(x_{j+1/2} - \tau q_j^n) > 0 \\ c_{j-1/2}(x_{j-1/2}, x_{j+1/2} - \tau q_j^n) & \text{otherwise} \end{cases} \tag{6.10a$^+$}$$

and

$$\tilde{f}_{j+1/2}=\bar{f}_{j+1/2}+\delta_j r_2(v_j^n). \qquad (6.10\text{b})^+$$

If $q_j^n \leqslant 0$, then

$$\delta_j=\begin{cases} c_{j+1/2}(x_{j-1/2}-\tau q_j^n,\, x_{j+1/2}) & \text{if } \ F_j(x_{j+1/2})\cdot F_j(x_{j-1/2}-\tau q_j^n)>0 \\ c_{j-1/2}(x_{j-1/2},\, x_{j-1/2}-\tau q_j^n) & \text{otherwise} \end{cases} \qquad (6.10\text{a})^-$$

and

$$\tilde{f}_{j-1/2}=\bar{f}_{j-1/2}+\delta_j r_2(v_j^n). \qquad (6.10\text{b})^-$$

Once we have completed the calculations in (6.10) we move on to the next cell.

7. Numerical Experiments

In this section, we present results of several computer experiments with the ENO schemes (5.5) and their modified version with subcell resolution (5.1); we refer to the latter as ENO/SR. In all these experiments we have used

$$\sigma_j=\left|\frac{d}{dx}R(x_j;v^n)\right| \qquad (7.1)$$

and, similarly, $k=1$ in (6.4) for systems. In all the calculations reported in this section we have used a CFL number of 0.8. The continuous line in Figs. 2–8 represents the exact solution. The circles in all figures represent values of $R(x_j; v^n)$ at the time specified.

We start with the scalar constant coefficient problem

$$u_t+u_x=0, \qquad u(x,0)=u_0(x), \qquad -1\leqslant x\leqslant 1, \qquad (7.2)$$

with periodic boundary conditions at $x=\pm 1$. In this case, we take

$$f^R(u_1,u_2)=u_1. \qquad (7.3)$$

First, we present numerical experiments with the highly discontinuous initial data

$$u_0(x+0.5)=\begin{cases} -x\sin(\tfrac{3}{2}\pi x^2), & -1<x<-\tfrac{1}{3} \\ |\sin(2\pi x)| & |x|<\tfrac{1}{3} \\ 2x-1-\sin(3\pi x)/6 & \tfrac{1}{3}<x<1. \end{cases} \qquad (7.4\text{a})$$

Note that the RHS of (7.4a) is shifted by (-0.5) for purposes of display. We initialized the calculation by taking

$$v_j^0=\frac{1}{h}\int_{x_{j-1/2}}^{x_{j+1/2}} u_0(x)\,dx=\frac{1}{h}\left[U_0(x_{j+1/2})-U_0(x_{j-1/2})\right], \qquad (7.4\text{b})$$

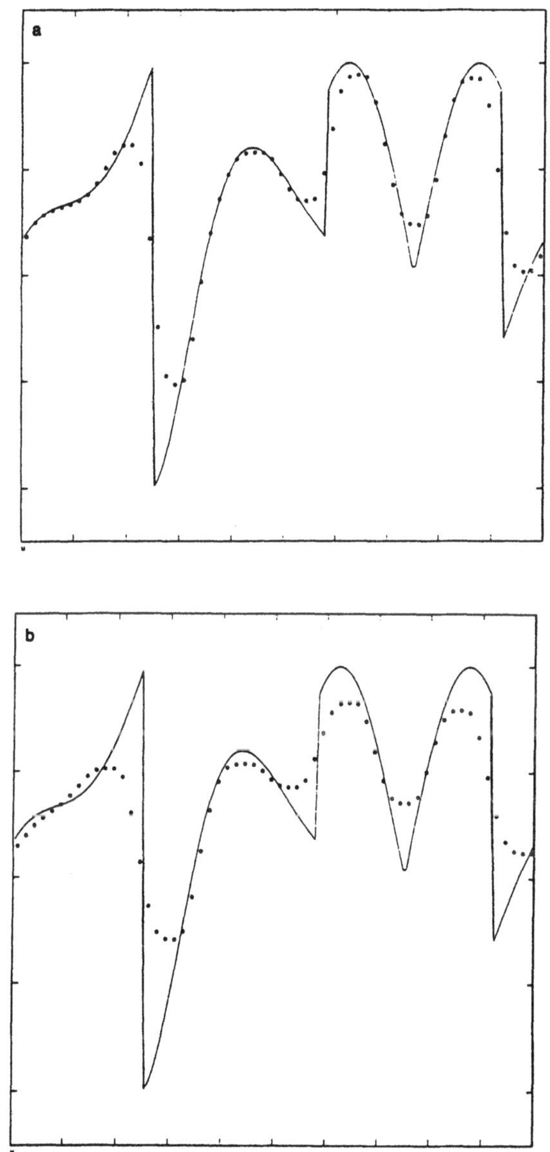

FIG. 2. MUSCL scheme, $-1 \leqslant x \leqslant 1$, $-1.25 \leqslant y \leqslant 1.25$: (a) $t=2$; (b) $t=8$.

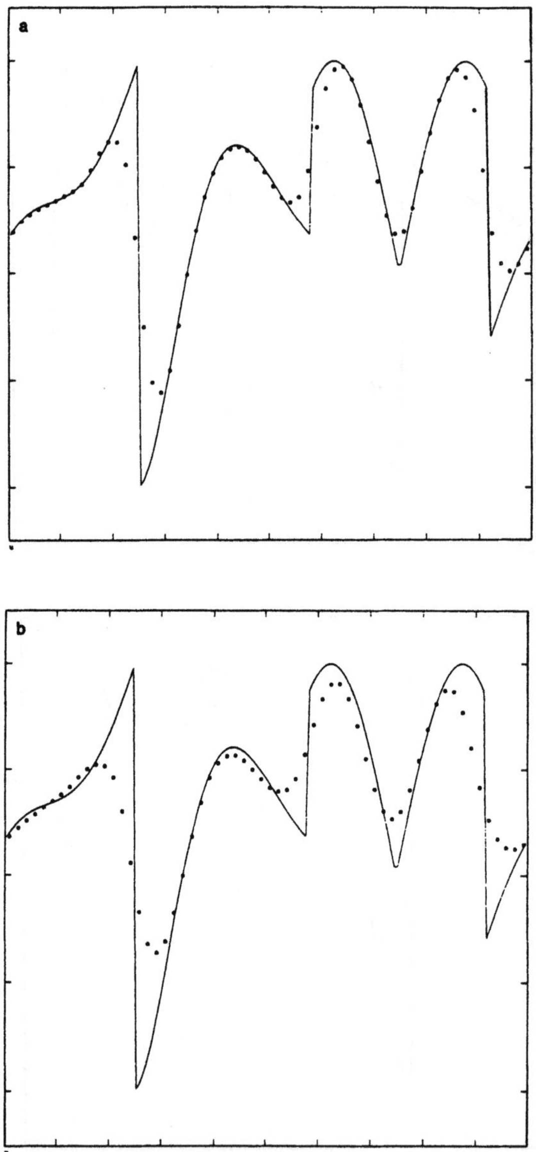

FIG. 3. Second-order ENO scheme, $-1 \leqslant x \leqslant 1$, $-1.25 \leqslant y \leqslant 1.25$: (a) $t=2$; (b) $t=8$.

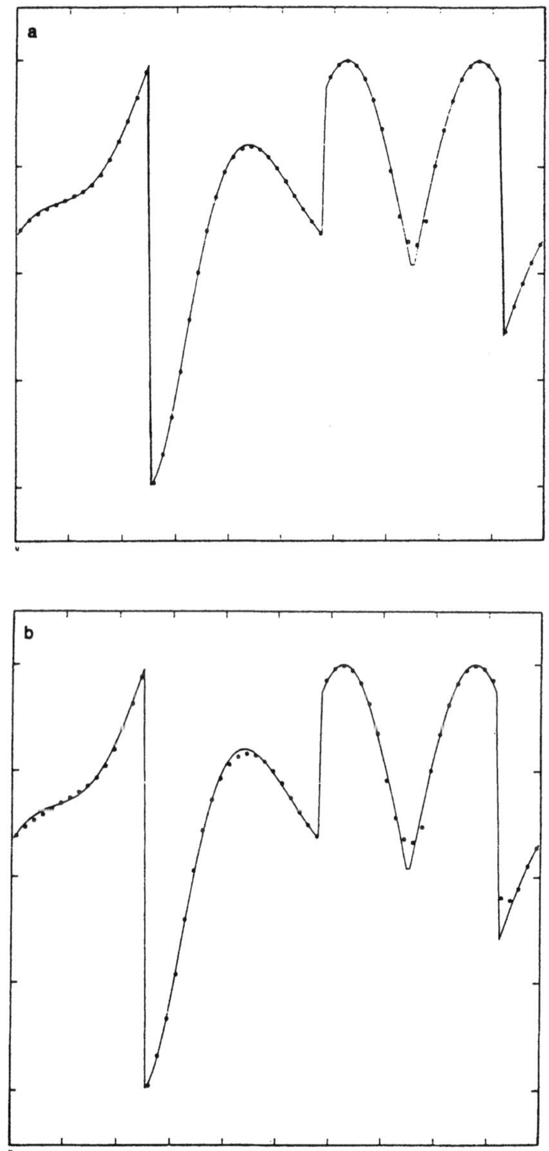

Fig. 4. Second-order ENO/SR, $-1 \leqslant x \leqslant 1$, $-1.25 \leqslant y \leqslant 1.25$: (a) $t=2$; (b) $t=8$.

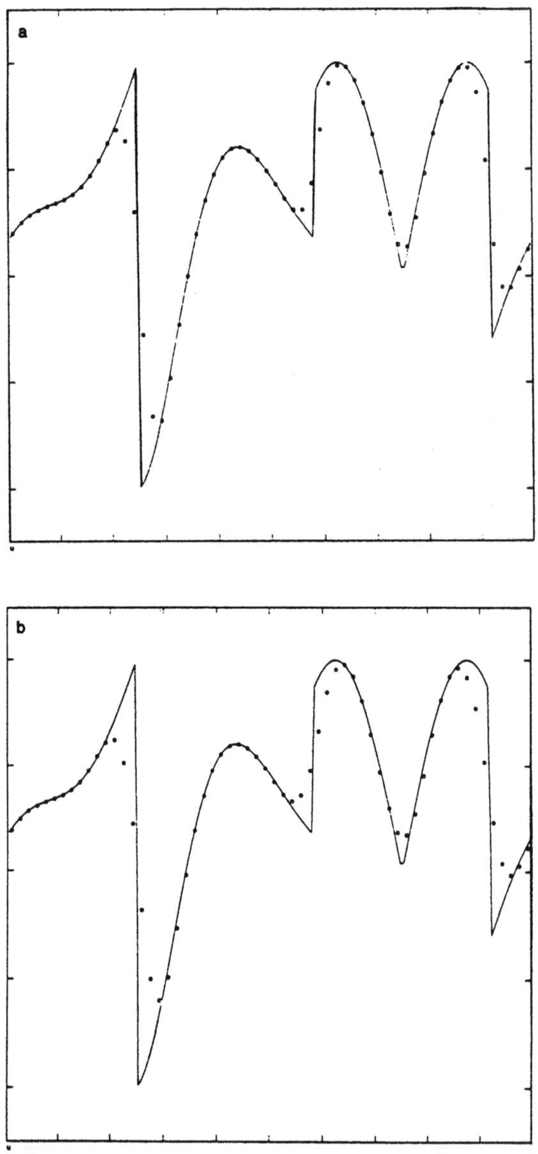

FIG. 5. Fourth-order ENO scheme, $-1 \leqslant x \leqslant 1$, $-1.25 \leqslant y \leqslant 1.25$: (a) $t=2$; (b) $t=8$.

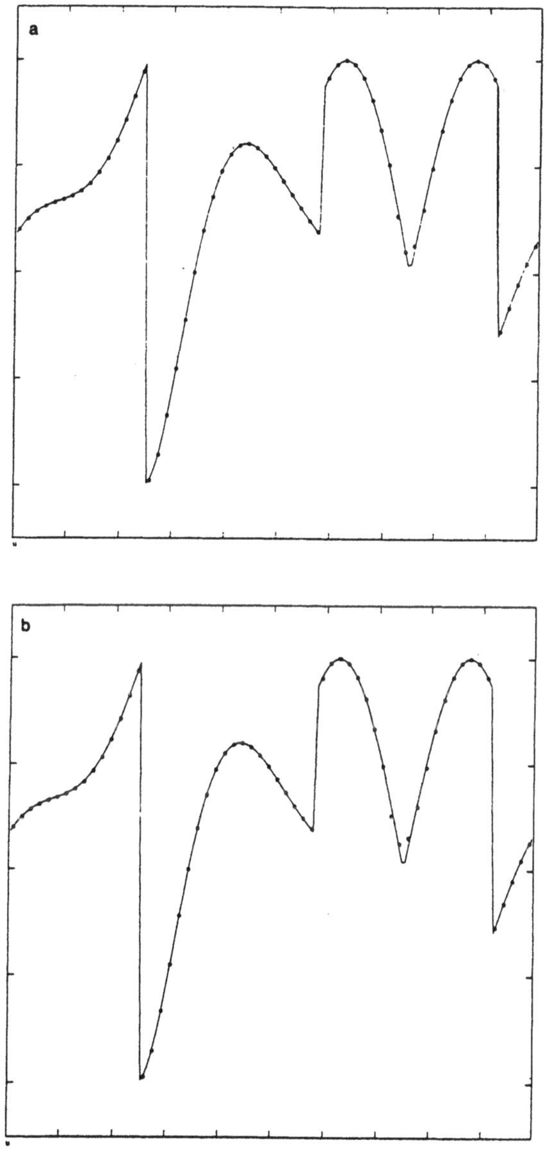

FIG. 6. Fourth-order ENO/SR, $-1 \leq x \leq 1$, $-1.25 \leq y \leq 1.25$; (a) $t=2$; (b) $t=8$.

where U_0 is the primitive function of $u_0(x)$. In Figs. 2 to 6, we show results with $h=\frac{1}{30}$ (i.e., 60 cells) at: (a) $t=2$ (after 1 period = 75 time-steps), (b) $t=8$ (after 4 periods = 300 time-steps). In Fig. 2, we show for sake of comparison the results of the MUSCL-scheme (1.11) with a slope S_j^n defined by

$$S_j^n = m(2(v_{j+1}^n - v_j^n), \tfrac{1}{2}(v_{j+1}^n - v_{j-1}^n), 2(v_j^n - v_{j-1}^n))/h; \tag{7.5a}$$

here $m(x, y, z)$ is the minmod function

$$m(x, y, z) = \begin{cases} s \cdot \min(|x|, |y|, |z|), & \text{if} \quad \operatorname{sgn}(x) = \operatorname{sgn}(y) = \operatorname{sgn}(z) = s \\ 0, & \text{otherwise.} \end{cases} \tag{7.5b}$$

In Fig. 3, we present results of the second-order accurate ENO scheme (4.3), and in Fig. 4 we show results of the corresponding second-order accurate ENO/SR (4.5) with $(4.13)^+$. In Fig. 5, we present the results of the fourth-order accurate ENO scheme (5.5) with $r=4$, and in Fig. 6 we show the corresponding results of the fourth-order accurate ENO/SR (5.1).

Next we demonstrate the kind of accuracy to be expected from these methods in smooth problems by calculating a refinement sequence for the periodic constant coefficient problem (7.2) with initial data

$$u_0(x) = \sin(\pi x).$$

In Table I, we show the results at $t=2$ with $h=\frac{1}{4}, \frac{1}{8}, \frac{1}{16}$ (i.e., 8, 16, and 32 cells, respectively). The quantity r_c in this table is the "computational order of accuracy" which is evaluated from two successive calculations by assuming the error to be a constant time h^{r_c}; clearly this definition is meaningful only for h sufficiently small.

TABLE I

Refinement Sequence for $u_t + u_x = 0$, $u(x, 0) = \sin \pi x$ with periodic boundary conditions

	No. of Cells	MUSCL	r_c	2nd Order ENO	r_c	2nd Order ENO/SR	r_c	4th Order ENO	r_c	4th Order ENO/SR	r_c
L_∞ error	8	1.454×10^{-1}		6.568×10^{-2}		4.532×10^{-2}		6.674×10^{-3}		8.592×10^{-3}	
			1.52		2.50		1.55		4.90		4.37
	16	5.506×10^{-2}		1.164×10^{-2}		1.546×10^{-2}		2.235×10^{-4}		4.146×10^{-4}	
			1.45		2.19		2.66		4.67		4.32
	32	2.019×10^{-2}		2.556×10^{-3}		2.445×10^{-3}		8.762×10^{-6}		2.082×10^{-5}	
L_1 error	8	7.040×10^{-2}		4.068×10^{-2}		3.098×10^{-2}		4.080×10^{-3}		5.670×10^{-3}	
			1.80		2.42		2.29		4.91		4.80
	16	2.028×10^{-2}		7.600×10^{-3}		6.340×10^{-3}		1.353×10^{-4}		2.042×10^{-4}	
			1.91		2.21		2.84		4.67		4.74
	32	5.380×10^{-3}		1.638×10^{-3}		8.800×10^{-4}		5.300×10^{-6}		7.630×10^{-6}	

Note that in the constant coefficient case the ENO/SR schemes are one-order higher in the L_1-sense than the corresponding ENO schemes.

We turn now to present numerical experiments with the Euler equation of gas dynamics (6.1). In these calculations we take $\gamma = 1.4$ and $f^R(u_1, u_2) = f^{\text{ROE}}(u_1, u_2)$, where we use Roe's linearization [10] to define

$$f^{\text{ROE}}(u_1, u_2) = \frac{1}{2}\left[f(u_1) + f(u_2) - \sum_{k=1}^{3} \delta_k \, |a_k(\hat{u})| \, r_k(\hat{u}) \right], \tag{7.6a}$$

with

$$\delta_k = l_k(\hat{u})(u_2 - u_1); \tag{7.6b}$$

here a_k, l_k, and r_k are the eigenvalues and the left- and right-eigenvectors, respectively. $\hat{u}$ is a particular average of u_1 and u_2 which is defined by:

$$\hat{q} = \langle q\sqrt{\rho}\rangle / \langle\sqrt{\rho}\rangle, \qquad \hat{H} = \langle H\sqrt{\rho}\rangle / \langle\sqrt{\rho}\rangle, \qquad \hat{c} = (\gamma - 1)^{1/2}\sqrt{\hat{H} - 1/2\hat{q}^2}; \tag{7.6c}$$

here $\langle\ \rangle$ denotes arithmetic average, i.e.,

$$\langle b \rangle = \tfrac{1}{2}(b_1 + b_2).$$

In Figs. 7 and 8, we show results of the Riemann initial value problem

$$u_t + f(u)_x = 0, \qquad u(x, 0) = \begin{cases} u_{\text{L}}, & x < 0 \\ u_{\text{R}}, & x > 0 \end{cases} \tag{7.7a}$$

with

$$(\rho_{\text{L}}, q_{\text{L}}, P_{\text{L}}) = (0.445, 0.698, 3.528); \qquad (\rho_{\text{R}}, q_{\text{R}}, P_{\text{R}}) = (0.5, 0, 0.571). \tag{7.7b}$$

These calculations were performed with 100 cells, $h = 0.1$, CFL $= 0.8$, and 85 time-steps. In Fig. 7, we show the density computed by the second-order ENO scheme and in Fig. 8 we show that of the corresponding ENO/SR.

Finally, we present numerical solutions to the problem of two interacting blast waves:

$$u(x, 0) = \begin{cases} u_{\text{L}}, & 0 \leqslant x < 0.1 \\ u_{\text{M}}, & 0.1 \leqslant x < 0.9 \\ u_{\text{R}}, & 0.9 \leqslant x < 1, \end{cases} \tag{7.8a}$$

where

$$\begin{aligned} &\rho_{\text{L}} = \rho_{\text{M}} = \rho_{\text{R}} = 1, \qquad q_{\text{L}} = q_{\text{M}} = q_{\text{R}} = 0, \\ &P_{\text{L}} = 10^3, \qquad P_{\text{M}} = 10^{-2}, \qquad P_{\text{R}} = 10^2; \end{aligned} \tag{7.8b}$$

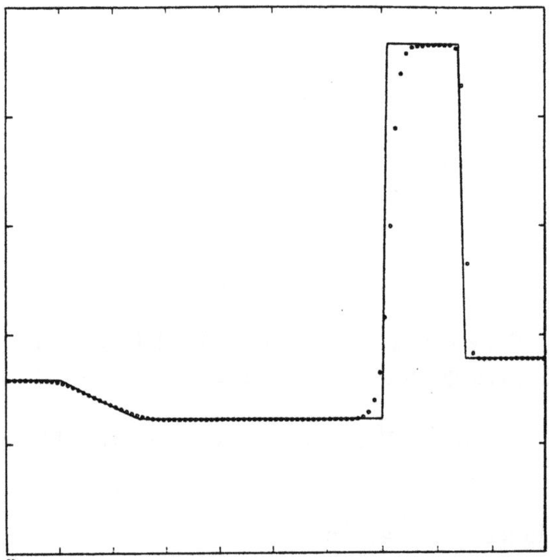

FIG. 7. Second-order ENO scheme, $0 \leqslant x \leqslant 1$, $0 \leqslant y \leqslant 1.4$: Density (85 time-steps).

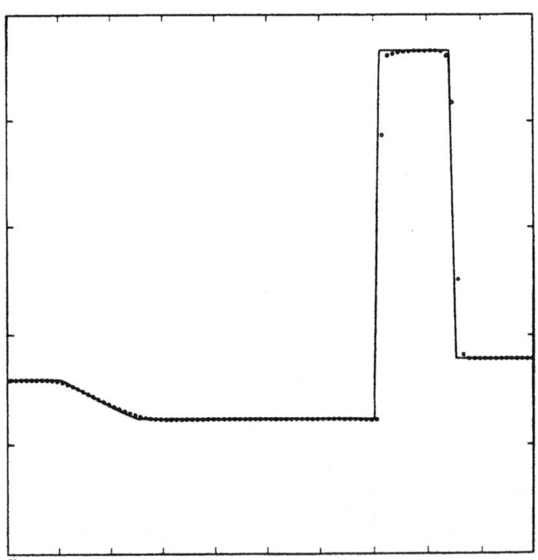

FIG. 8. Second-order ENO/SR, $0 \leqslant x \leqslant 1$, $0 \leqslant y \leqslant 1.4$: Density (85 time-steps).

the boundaries at $x=0$ and $x=1$ are solid walls. This problem was suggested by Woodward and Colella as a test problem; we refer the reader to [11] where a comprehensive comparison of the performance of various schemes for this problem is presented. We refer the reader to [7] for a detailed description of the implementation of the solid wall boundary condition in the ENO schemes.

In Fig. 9, we show the density at $t=0.038$ calculated by the second-order accurate ENO/SR with 800 cells and CFL $=0.8$. The circles in this figure represent values of $R(x_j, \rho^n)$; the continuous line is just the piecewise-linear interpolation of these values. Comparing these results to the solution presented by Woodward and Colella in [11], we find that it shows all the important features of the various interactions and thus can be considered a "converged" solution. We use this piecewise-linear interpolation of the calculation with 800 cells as the "exact solution" in Figs. 10 and 11. The circles in Figs. 10 and 11 are reconstructed values of density in a calculation with 200 cells. In Fig. 10, we show the calculation by the second order accurate ENO scheme; in Fig. 11 we show the results of the corresponding ENO/SR.

In the following, we make several remarks and observations concerning the numerical results presented in this section.

(1) In all our calculations we find that the subcell resolution technique is capable of producing perfctly resolved linear discontinuities. Observe that if $R(x; v^n)$ has a single intermediate value at a discontinuity then this discontinuity is perfectly resolved by $\hat{R}(x; v^n)$.

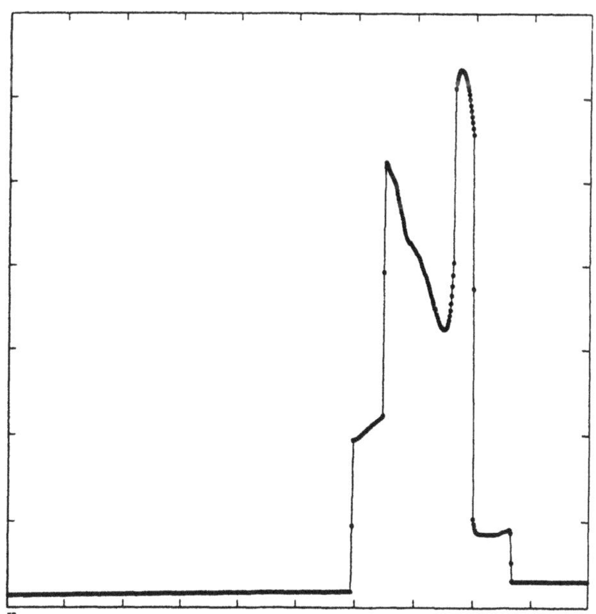

FIG. 9. Second-order ENO/SR, $0 \leqslant x \leqslant 1$, $0 \leqslant y \leqslant 7$: Density at $t=0.038$ with 800 cells.

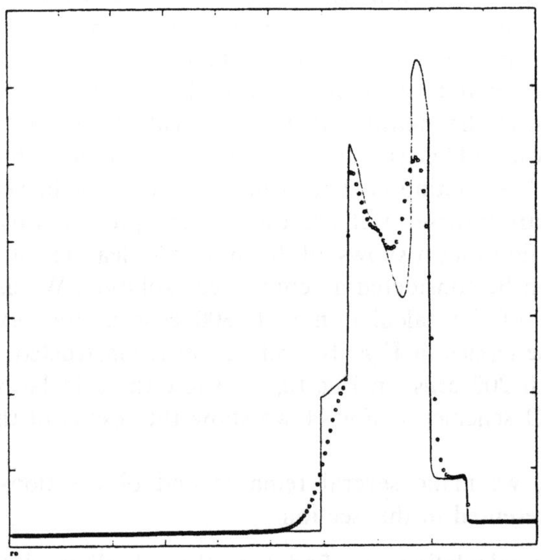

FIG. 10. Second-order ENO, $0 \leqslant x \leqslant 1$, $0 \leqslant y \leqslant 7$: Density at $t = 0.038$ with 200 cells.

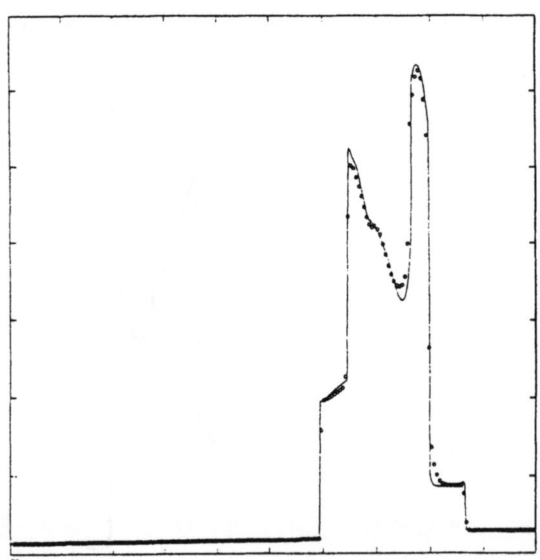

FIG. 11. Second-order ENO/SR, $0 \leqslant x \leqslant 1$, $0 \leqslant y \leqslant 7$: Density at $t = 0.038$ with 200 cells.

(2) When we study the effect of higher formal order of accuracy in the calculation of discontinuous data by the ENO schemes, we find that the most noticable improvement is due to the reduction in smearing of the linear discontinuitites. However, when we compare the second-order and the fourth-order ENO/SR schemes we see that the improvement is primarily due to higher accuracy in the smooth part of the solution. Consequently, there is no sense in going to higher order when solving a Riemann IVP. To justify the increased computational cost associated with higher order, one needs a lot of structure in the smooth part of the solution.

(3) Comparing the solution of the interacting blast waves (7.8) by the second-order ENO/SR to that of the PPM in [12], we find that the ENO/SR is more accurate. The ENO/SR highly resolves all three contact discontinuities in the problem, while for some reason the PPM well resolves two of the contact discontinuities but smears the one which results from the shock interaction. Another possible explanation for the difference in accuracy may be due to the fact that the ENO/SR is unifomly second-order accurate, while the PPM (because of its monotonicity constraints) reduces to first-order accuracy at points of local extremum.

(4) The numerical results for the Euler equations of gas dynamics clearly demonstrate that shocks are highly resolved by the original ENO schemes, and subcell resolution is not needed there. In any case, the expressions for $\bar{g}_{j+1/2}$ have to be modified before applying subcell resolution to a genuinely nonlinear field, as follows: (i) a_j^n should be replaced by the speed of the shock. (ii) Dissipation proportional to $(a_{j+1/2}-a_{j-1/2})$ should be added to a centered rarefaction wave. Fortunately, if the discontinuity condition (3.9a) is met in the cell I_j, then $\hat{R}(x; v^n)$ is continuous at $x_{j\pm 1/2}$ and no interaction terms need be added to the numerical flux. However, one has to account for the fact that the wave from the interior of I_j crosses its boundaries during the time-step.

APPENDIX: Derivation of the Discontinuity Condition

The reconstruction $R(x; \bar{w})$ is by definition discontinuous at $\{x_{j+1/2}\}$. In regions of smoothness of $w(x)$ the jump of the reconstruction at $x_{j+1/2}$ is of the order $O(h^r)$. When a discontinuity of $w(x)$ is located in the interior of I_j, then the discontinuities of the reconstruction at $x_{j\pm 1/2}$ are fragments of that of $w(x)$. (See Fig. 1b.)

In order to recover a possible discontinuity in the interior of each cell, we would like to associate the reconstruction with the boundaries of the cells $\{x_{j+1/2}\}$, rather than the cells themselves. Let $\tilde{R}_{j+1/2}(x; \bar{w})$ be the polynomial description of such a reconstruction which is valid in the neighborhood of $x_{j+1/2}$. Once this is done we consider reconstructing $w(x)$ in I_j by

$$\hat{R}_j(x; \bar{w}) = \begin{cases} \tilde{R}_{j-1/2}(x; \bar{w}) & \text{for } x_{j-1/2} \leqslant x < \theta_j \\ \tilde{R}_{j+1/2}(x; \bar{w}) & \text{for } \theta_j \leqslant x < x_{j+1/2} \end{cases} \tag{A.1}$$

if possible, i.e., if there is a θ_j such that

$$\tilde{F}_j(\theta_j) = 0, \qquad x_{j-1/2} \leqslant \theta_j < x_{j+1/2}, \tag{A.2a}$$

where

$$\tilde{F}_j(z) = \frac{1}{h}\left\{\int_{x_{j-1/2}}^{z} \tilde{R}_{j-1/2}(x; \bar{w})\, dx + \int_{z}^{x_{j+1/2}} \tilde{R}_{j+1/2}(x; \bar{w})\, dx\right\} - \bar{w}_j. \tag{A.2b}$$

If there is no such θ_j, we define

$$\hat{R}_j(x; \bar{w}) = R_j(x; \bar{w}). \tag{A.3}$$

The only thing which is left open at this point is the definition of $\tilde{R}_{j+1/2}(x; \bar{w})$. It is most natural within the framework of the ENO reconstruction to select the "smoother" of $R_j(x; \bar{w})$ and $R_{j+1}(x; \bar{w})$; i.e.,

$$\tilde{R}_{j+1}(x; \bar{w}) = \begin{cases} R_j(x; \bar{w}) & \text{if} \quad \sigma_j < \sigma_{j+1} \\ R_{j+1}(x; \bar{w}) & \text{if} \quad \sigma_j \geqslant \sigma_{j+1} \end{cases}. \tag{A.4}$$

Here σ_j is a monotone increasing function of the "non-smoothness," such as (3.8a). Note that since $R_i(x; \bar{w})$ is associated with a stencil of points, (A.4) is equivalent to assigning a stencil to $x_{j+1/2}$, the boundary of the cell.

We observe that there is a certain ambiguity in the definition of $\hat{R}_j(x; w)$ in (A.1), since θ_j need not be unique. We remove most of this ambiguity by adopting a policy of "no unnnecessary changes," and agree that

$$\tilde{R}_{j-1/2} = R_j \qquad \text{or} \qquad \tilde{R}_{j+1/2} = R_j \Rightarrow \hat{R}_j = R_j; \tag{A.5}$$

i.e., if one of the candidates for extension into I_j is no smoother than the original R_j, we just retain the original definition in $\hat{R}_j$. From the definition (A.4) of $\tilde{R}$ we can rewrite (A.5) in terms of σ_i as

$$\sigma_j \leqslant \sigma_{j-1} \qquad \text{or} \qquad \sigma_j < \sigma_{j+1} \Rightarrow \hat{R}_j = R_j. \tag{A.5$'$}$$

Hence we conclude that we define $\hat{R}_j(x; \bar{w})$ to be (A.1) only if

$$\sigma_j > \sigma_{j-1} \qquad \text{and} \qquad \sigma_j \geqslant \sigma_{j+1} \tag{A.6}$$

which is the discontinuity condition (3.8b). In this case $\tilde{R}_{j-1/2} = R_{j-1}$ and $\tilde{R}_{j+1/2} = R_{j+1}$ and the definition (3.9) follows.

REFERENCES

1. P. COLELLA AND P. R. WOODWARD, *J. Comput. Phys.* **54**, 174 (1984).
2. S. K. GODUNOV, *Math. Sb.* **47**, 271 (1959). [Russian]
3. A. HARTEN, *Math. Comput.* **32**, 363 (1978).
4. A. HARTEN, *J. Comput. Phys.* **49**, 357 (1983).
5. A. HARTEN, MRC Technical Symmary Report 2829, University of Wisconsin, 1985 (unpublished).
6. A. HARTEN AND S. OSHER, *SINUM* **24**, 279 (1987); MRC Technical Symmary Report 2823, May 1985 (unpublished).
7. A. HARTEN, B. ENGQUIST, S. OSHER, AND S. R. CHAKRAVARTHY, *J. Comput. Phys.* **71**, 231 (1987); ICASE Report 86–22, April 1986 (unpublished).
8. A. HARTEN, S. OSHER, B. ENGQUIST, AND S. R. CHAKRAVARTHY, *J. Appl. Numer. Math.* **2**, 347 (1986); in *Advances in Numerical and Applied Mathematics*, edited by J. C. South, Jr. and M. Y. Hussaini, ICASE Report 86–18, March 1986.
9. B. VAN LEER, *J. Comput. Phys.* **32**, 101 (1979).
10. P. L. ROE, *J. Comput. Phys.* **43**, 357 (1981).
12. P. WOODWARD AND P. COLELLA, *J. Comput. Phys.* **54**, 115 (1984).

Efficient Implementation of Essentially Non-oscillatory Shock-Capturing Schemes, II

CHI-WANG SHU*

Division of Applied Mathematics, Brown University, Providence, Rhode Island 02912

AND

STANLEY OSHER†

Department of Mathematics, University of California, Los Angeles, California 90024-1555

Received April 15, 1988; revised August 2, 1988

In this paper we extend our earlier work on the efficient implementation of ENO (essentially non-oscillatory) shock-capturing schemes. We provide a new simplified expression for the ENO construction procedure based again on numerical fluxes rather than cell-averages. We also consider two improvements which we label ENO-LLF (local Lax–Friedrichs) and ENO-Roe, which yield sharper shock transitions, improved overall accuracy, for lower computational cost than previous implementation of the ENO schemes. Two methods of sharpening contact discontinuities—the subcell resolution idea of Harten and the artificial compression idea of Yang, which those authors originally used in the cell average framework—are applied to the current ENO schemes using numerical fluxes and TVD Runge–Kutta time discretizations. The implementation for nonlinear systems and multidimensions is given. Finally, many numerical examples, including a compressible shock turbulence interaction flow calculation, are presented.

I. INTRODUCTION

In this paper we extend the construction we began in [12] of efficient implementation of ENO (essentially non-oscillatory) schemes approximating systems of hyperbolic conservation laws of the type:

$$\mathbf{u}_t + \sum_{i=1}^{d} \mathbf{f}_i(\mathbf{u})_{x_i} = 0 \qquad (\text{or} = \mathbf{g}(\mathbf{u}, \mathbf{x}, t), \text{ a forcing term}) \tag{11.1a}$$

$$\mathbf{u}(\mathbf{x}, 0) = \mathbf{u}^0(\mathbf{x}), \tag{1.1b}$$

* Research partically supported by NSF Grant DMS88-10150 and by the Institute for Mathematics and Its Applications.

† Research supported by NSF Grant DMS85-03294, DARPA grant in the ACMP Program, ONR Grant N00014-86-K-0691, NASA Langley Grant NAG-1-270.

where $\mathbf{u}=(u_1, \dots u_m)^T$, $\mathbf{x}=(x^1, x^2, \dots, x^d)$, and, for real $\xi=(\xi_1, \dots, \xi_d)$, the combination $\sum_{i=1}^d \xi_i(\partial \mathbf{f}_i/\partial \mathbf{u})$ always has m real eigenvalues and a complete set of eigenvectors. We use the notations $\mathbf{x_J}=\mathbf{J}\cdot\Delta\mathbf{x}$, $t_n=n\Delta t$, and use $\mathbf{u}_\mathbf{J}^n$ to denote the computed approximation to the exact solution $\mathbf{u}(\mathbf{x_J}, t)$ of (1.1). We use bold face letters for vectors and plain letters for scalars.

ENO schemes, constructed by Harten, Osher, Engquist, and Chakravarthy [2–5], use a local adaptive stencil to obtain information automatically from regions of smoothness when the solution develops discontinuities. As a result, approximations using these methods can obtain uniformly high-order accuracy right up to discontinuities, while keeping a sharp, essentially non-oscillatory shock transition. The original ENO schemes in [2–5] used a cell-average framework which involved a reconstruction procedure to recover accurate point values from cell averages and a Lax–Wendroff procedure (replacing time derivatives by space derivatives, using the PDE) for the time discretization. This can become a bit complicated for multi-dimensional problems [2]. Harten is currently investigating efficient and more local methods [7]. For ease of implementation we constructed [12] ENO schemes applying the adaptive stencil idea to the numerical fluxes and using TVD Runge–Kutta type time discretizations. The ENO schemes in [12] skip the reconstruction step and the Lax–Wendroff time discretization procedure, hence they are simpler to program, especially for multi-space-dimensional problems. We hope that [12] and this paper will encourage potential users to apply these high-order methods to real physical computations for systems of conservation laws in several space dimensions.

We shall use the same notation as in [12]: $\Delta_\pm$ are the usual difference operators $\Delta_\pm a_j = \pm(a_{j\pm 1}-a_j)$; Eq. (1.1a) is sometimes written in an abstract form:

$$\mathbf{u}_t=\mathscr{L}(\mathbf{u}). \tag{1.2}$$

The ENO spatial operator $\mathbf{L}(\mathbf{u})$ is supposed to approximate $\mathscr{L}(\mathbf{u})$ to rth order:

$$\mathbf{L}(\mathbf{u})=\mathscr{L}(\mathbf{u})+O(h^r) \tag{1.3}$$

for smooth $\mathbf{u}$, where h is the maximum mesh size, and the Euler forward version

$$\mathbf{w}=\mathbf{T}(\mathbf{u})=(\mathbf{I}+\Delta t\mathbf{L})(\mathbf{u}) \tag{1.4}$$

is *assumed* to be total variation stable, for scalar, 1D nonlinear problems, under a suitable CFL restriction

$$\lambda=\frac{\Delta t}{\Delta x}\leqslant\lambda_0 \tag{1.5}$$

where λ_0 is usually inversely proportional to max $|f'(u)|$. At present this stability cannot be proven for unmodified third-order or higher ENO schemes. There is, however, strong theorical and numerical evidence to indicate that the methods are indeed TV stable [3–5, 12].

An rth order TVD Runge–Kutta time discretization is then applied:

$$\mathbf{u}^{(i)} = \sum_{k=0}^{i-1} [\alpha_{ik}\mathbf{u}^{(k)} + \beta_{ik}\Delta t L(\mathbf{u}^{(k)})], \qquad i = 1, ..., \bar{r} \tag{1.6a}$$

$$\mathbf{u}^{(0)} = \mathbf{u}^n, \qquad u^{(\bar{r})} = \mathbf{u}^{n+1} \tag{1.6b}$$

with

$$\mathrm{TV}(\mathbf{u}^{n+1}) \leqslant \mathrm{TV}(T(\mathbf{u}^n)) \tag{1.7a}$$

under the CFL restriction:

$$\lambda = \frac{\Delta t}{\Delta x} \leqslant c_r \cdot \lambda_0. \tag{1.7b}$$

We call c_r the CFL coefficient. We constructed schemes [12] for $r \leqslant 4$ with $\bar{r} = r$ and $c_2 = c_3 = 1$, $c_4 = \frac{2}{3}$. For $r = 5$ we needed $\bar{r} = 6$ and $c_5 = \frac{7}{30}$. For $r \geqslant 4$ we also needed $\tilde{L}$ which approximates $\mathscr{L}$ in an "adjoint" equation $\mathbf{u}_t = -\mathscr{L}(\mathbf{u})$ because some of the β_{ik} become negative. The details are presented in [12]. Some of the schemes used in Section 5 are listed in Table I.

We shall always use conservative schemes, i.e., for the scalar 1D problem (1.1) we write

$$L(u)_j = -\frac{1}{\Delta x}(\hat{f}_{j+1/2} - \hat{f}_{j-1/2}) \tag{1.8a}$$

for a consistent numerical flux

$$\hat{f}_{j+1/2} = \hat{f}(u_{j-l}, ..., u_{j+k}); \qquad \hat{f}(u, ..., u) = f(u). \tag{1.8b}$$

TABLE I

TVD Runge–Kutta Schemes (1.6)

Order	α_i	β_i	CFL coefficients
2	1	1	1
	$\frac{1}{2}\ \ \frac{1}{2}$	$0\ \ \frac{1}{2}$	
3	1	1	
	$\frac{3}{4}\ \ \frac{1}{4}$	$0\ \ \frac{1}{4}$	1
	$\frac{1}{3}\ \ 0\ \ \frac{2}{3}$	$0\ \ 0\ \ \frac{2}{3}$	
4	1	$\frac{1}{2}$	
	$\frac{1}{2}\ \ \frac{1}{2}$	$-\frac{1}{4}\ \ \frac{1}{2}$	$\frac{2}{3}$
	$\frac{1}{9}\ \ \frac{2}{9}\ \ \frac{2}{3}$	$-\frac{1}{9}\ \ -\frac{1}{3}\ \ 1$	
	$0\ \ \frac{1}{3}\ \ \frac{1}{3}\ \ \frac{1}{3}$	$0\ \ \frac{1}{6}\ \ 0\ \ \frac{1}{6}$	

We proved in [12] the existence of constants $a_2, a_4, \ldots$ such that

$$\hat{f}_{j+1/2} = f_{j+1/2} + \sum_{k=1}^{[(r-1)/2]} a_{2k} \Delta x^{2k} \left(\frac{\partial^{2k}}{\partial x^{2k}} f \right)_{j+1/2} + O(\Delta x^{r+1}) \tag{1.9}$$

guarantees rth order accuracy (1.3). For example, $a_2 = -\frac{1}{24}$, $a_4 = \frac{7}{5760}, \ldots$.

The use of the a_{2k} may seem a bit unnatural. We need to evaluate all the even derivatives of the interpolating polynomials (except the last one if r is even), which increases the computational cost for large r. In Section 2 of this paper we provide a simplified (but equivalent) version of (1.9). We also consider in Section 2 two improvements labelled ENO-LLF (local Lax–Friedrichs) and ENO-Roe, which yield sharper shock transitions, better overall accuracy, and lower computational costs than the methods discribed in [12].

Shocks have a self-sharpening mechanism due to converging characteristics. This is not true for contact discontinuities, which are usually smeared in shock capturing calculations. Recently, Harten [6] introduced the notion of subcell resolution in order to sharpen contact discontinuities. He did this in the context of cell average based ENO schemes. Although in [6] the fact that cell averages contain information leading to the location of the shock is strongly used, this information is actually contained in any *conservative* scheme (1.8). In Section 3 we translate Harten's subcell resolution idea to the ENO schemes using fluxes and Runge–Kutta techniques. The result is simpler than [6] but less accurate in theoretical resolution. However, the numerical results (in Section 5) are quite close to those presented in [6]. The main drawback of our extension of Harten's approach is that it is not clear how to generalize it effectively to several space dimensions. A naive generalization gave unsatisfactory results in our 2D numerical tests. Currently Harten is considering truly 2D subcell resolution methods [7]. Also recently Yang [15] introduced a simple artificial compression technique applied to cell average based ENO schemes. He demonstrated that the technique preserves the TVD (or ENO) properties when applied to a TVD (or ENO) scheme. We translate his result to our present framework in Section 3. The numerical results presented in Section 5 indicate that this adaptation works well both for 1D and 2D problems.

In Section 4 we extend the present implementation to nonlinear systems and to multi-space dimensions. Section 5 contains numerical experiments, including scalar, 1D nonconvex Riemann problems; examples of sharpening contact discontinuities; 2D Riemann problems for Burger's equation; 1D and 2D Euler equations of gas dynamics including a compressible shock-turbulence interaction problem. Concluding remarks are presented in Section 6.

II. Simplifications and Improvements

We begin with the 1D scalar nonlinear problem, i.e., $d = m = 1$ in (1.1). To simplify (1.9) we use the following elementary result:

LEMMA 2.1. *If a function $h(x)$ satisfies*

$$f(u(x)) = \frac{1}{\Delta x} \int_{x-\Delta x/2}^{x+\Delta x/2} h(\xi)\, d\xi, \tag{2.1}$$

then

$$f(u(x))_x = \left(h\left(x + \frac{\Delta x}{2}\right) - h\left(x - \frac{\Delta x}{2}\right)\right) \Big/ \Delta x$$

The proof is trivial.

Lemma 2.1 tells us that the numerical flux $\hat{f}_{j+1/2}$ should approximate $h(x_{j+1/2})$ to a high order. It is not easy to obtain $(h(x)$ directly from (2.1). However, the "reconstruction via primitive function" technique in [4] can be applied to obtain the primitive function of $h(x)$

$$H(x) = \int_{-\infty}^{x} h(\xi)\, d\xi \tag{2.2}$$

at $x_{j+1/2}$ by

$$H(x_{j+1/2}) = \int_{-\infty}^{x_{j+1/2}} h(\xi)\, d\xi = \sum_{k=-\infty}^{j} \int_{x_{k-1/2}}^{x_{k+1/2}} h(\xi)\, d\xi = \Delta x \sum_{k=-\infty}^{j} f(u_k). \tag{2.3}$$

Noticed that the lower limit $-\infty$ is irrelevant. It can be changed to any fixed grid point $x_{j0+1/2}$.

Given the values of $H(x_{j+1/2})$, we can construct interpolating polynomials in an ENO fashion, i.e., by obtaining a locally "smoothest" stencil starting from one or two points, then adding one point to the stencil at each stage by comparing two divided differences and choosing the one which is smaller in absolute value (see [4]), and then using its derivative at $x_{j+1/2}$ as the numerical flux $\hat{f}_{j+1/2}$. We remark here that by (2.3), we have $(H(x_{j+1/2}) - H(x_{j-1/2}))/\Delta x = f(u_j)$, hence the $(k+1)$th-order divided differences of H can be easily obtained from the kth divided differences of f. Since we shall never need the 0th-order divided differences, we do not perform the summation in (2.3). We simply use the divided difference tables for f.

Lemma 2.1 also explains why the ENO schemes constructed here and in [12] are equivalent (for linear equations in one space dimension only) to the cell-average ENO schemes constructed in [4] via the primitive function for u.

The starting point in the choice of stencil process is very important. "Upwinding" is achieved by this initial choice, and this also is crucial for the evident stability of these methods. We have experimented with using x_{j+1} as the starting point in the choice of stencil process for evaluating $\hat{f}_{j+1/2}$, to solve Burgers' equation (i.e., upwinding in the wrong direction initially) with positive $f'(u)$. We observed instability even if the stencil is chosen in an ENO fashion up to third order. In [12], f is decomposed into $f = f^+ + f^-$ with $(f^+)' \geqslant 0$, $(f^-)' \leqslant 0$, then different starting points are assigned to $f^{\pm}$ according to the direction of the wind. This simple procedure may smear shocks and affect the overall accuracy (although the smearing is very mild comparing with lower order non-ENO schemes). It was also pointed out in [12], that instead of decomposing f into f^+ and f^- we could also use any E-flux $h_{j+1/2}$ [10], with $df^+_{j+1/2} = f_{j+1} - h_{j+1/2}$, $df^-_{j+1/2} = h_{j+1/2} - f_j$ replacing the first (undivided) differences $f^{\pm}_{j+1} - f^{\pm}_j$; we shall later refer to this procedure as ENO schemes with $h_{j+1/2}$ building blocks. However, that procedure requires smoothness of $h_{j+1/2}$ at sonic points to keep the high accuracy there. It is known that most E-fluxes are *not* smooth at sonic points, they are typically only Lipschitz continuous or perhaps C^1 there. Numerical results do indeed indicate a loss of accuracy at sonic points when such building blocks are used. Although we may modify any E-flux to make it smooth at sonic points, this procedure is too complicated to be computationally pleasing.

We now use a different approach to overcome this difficulty. We choose the first point in the stencil according to the *local* sign of $f'(u)$ at $x_{j+1/2}$. We may use, e.g., the "Roe" speed

$$\bar{a}_{j+1/2} = \frac{f(u_{j+1}) - f(u_j)}{u_{j+1} - u_j} \tag{2.4}$$

to determine the sign of $f'(u_{j+1/2})$. We then have the following algorithm:

ALGORITHM 2.1. (ENO–Roe).

(1) Compute the divided difference table of f, and identify

$$H[x_{l-1/2}, x_{l+1/2}] = f[u(x_l)] \tag{2.5a}$$

$$H[x_{l-1/2}, x_{l+1/2}, ..., x_{l+k+1/2}] = \frac{1}{k+1} f[u(x_l), ..., u(x_{l+k})], \qquad k = 1, 2, ..., r. \tag{2.5b}$$

(2) If $\bar{a}_{j+1/2} = (f(u_{j+1}) - f(u_j))/u_{j+1} - u_j \geqslant 0$, then

$$k^{(1)}_{\min} = j \tag{2.6a}$$

else

$$k^{(1)}_{\min} = j + 1. \tag{2.6b}$$

(3) $Q^{(1)}(x) = H[x_{k^{(1)}_{\min} - 1/2}, x_{k^{(1)}_{\min} + 1/2}](x - x_{k^{(1)}_{\min} - 1/2})$.

(4) Inductively, if $k^{(l-1)}_{\min}$ and $Q^{(l-1)}(x)$ are both defined, then let

$$a^{(l)} = H[x_{k^{(l-1)}_{\min} - 1/2}, ..., x_{k^{(l-1)}_{\min} + l - 1/2}] \tag{2.7a}$$

$$b^{(l)} = H[x_{k^{(l-1)}_{\min} - 1 - 1/2}, ..., x_{k^{(l-1)}_{\min} + l - 1 - 1/2}] \tag{2.7b}$$

and

(i) if $|a^{(l)}| \geqslant |b^{(l)}|$, then

$$c^{(l)} = b^{(l)}, \qquad k_{\min}^{(l)} = k_{\min}^{(l-1)} - 1 \tag{2.8a}$$

otherwise

$$c^{(l)} = a^{(l)}, \qquad k_{\min}^{(l)} = k_{\min}^{(l-1)} \tag{2.8b}$$

(ii) form

$$Q^{(l)}(x) = Q^{(l-1)}(x) + c^{(l)} \prod_{k=k_{\min}^{(l-1)}}^{k_{\min}^{(l-1)}+l-1} (x - x_{k-1/2}). \tag{2.9}$$

(5) $Q_{j+1/2}(x) = Q^{(r+1)}(x)$.

(6) We then take

$$\hat{f}_{j+1/2} = \frac{d}{dx} Q_{j+1/2}(x)|_{x=x_{j+1/2}}. \tag{2.10}$$

Remark 2.1. (a) Notice that unlike the ENO scheme which uses the Roe flux as a building block, Algorithm 2.1 gives a uniformly high-order flux $\hat{f}_{j+1/2}$, even at sonic points. The scheme described above will be $(r+1)$th-order accurate except perhaps at isolated zeros of derivatives of the flux $f(u(x))$, where it may degenerate to rth order.

(b) Since $x = x_{j+1/2}$ is a node of $Q_{j+1/2}$ in (2.10), the evaluation of the derivative in (2.10) costs very little.

Algorithm 2.1 yields a scheme admitting a stationary entropy-violating expansion shock since it is based on the first-order Roe scheme. However, the "entropy fix" in this framework turns out to be very simple. Before stating this remedy, let us present another algorithm:

ALGORITHM 2.2 (ENO-LLF).

(1) Compute the divided difference table of f and u, and identify

$$H^{\pm}[x_{l-1/2}, x_{l+1/2}] = \tfrac{1}{2}(f[u(x_l)] \pm \alpha_{j+1/2} u[x_l]),$$
$$l = j-r, ..., j+r \text{ for } H^+, \qquad l = j-r+1, ..., j+r+1 \text{ for } H^-; \tag{2.11a}$$

$$H^{\pm}[x_{l-1/2}, x_{l+1/2}, ..., x_{l+k+1/2}] = \frac{1}{k+1} \cdot \frac{1}{2} (f[u(x_l), ..., u(x_{l+k})] \pm \alpha_{j+1/2} u[x_l, ..., x_{l+k}])$$
$$l = j-r, ..., j+r-k \text{ for } H^+,$$
$$l = j-r+1, ..., j+r-k+1 \text{ for } H^-, k = 1, 2, ..., r, \tag{2.11b}$$

where

$$\alpha_{j+1/2} = \max_{u_j \leqslant u \leqslant u_{j+1}} |f'(u)|. \tag{2.12}$$

(2) For H^+, $k_{\min}^{(1)} = j$, then repeat steps (3)–(4) in Algorithm 2.1 to get

$$Q_+^{(r+1)}(x). \text{ Let } Q_{j+1/2}^+(x) = Q_+^{(r+1)}(x);$$

(3) For H^-, $k_{\min}^{(1)} = j+1$, then repeat steps (3)–(4) in Algorithm 2.1 to get

$$Q_-^{(r+1)}(x). \text{ Let } Q_{j+1/2}^- = Q_-^{(r+1)}(x);$$

(4) We then take

$$\hat{f}_{j+1/2} = \frac{d}{dx} Q_{j+1/2}^+(x)|_{x=x_{j+1/2}} + \frac{d}{dx} Q_{j+1/2}^-(x)|_{x=x_{j+1/2}}. \tag{2.13}$$

Remark 2.2. (a) When f is convex, $f'' \geqslant 0$, (2.12) simplifies to

$$\alpha_{j+1/2} = \max(|f'(u_j)|, |f'(u_{j+1})|). \tag{2.14}$$

(b) It can be verified that the local Lax–Friedrichs flux

$$h_{j+1/2}^{\text{LLF}} = \tfrac{1}{2}[f(u_j) + f(u_{j+1}) - \alpha_{j+1/2}(u_{j+1} - u_j)], \tag{2.15}$$

where $\alpha_{j+1/2}$ is defined by (2.12), is a monotone flux (see [1] for definitions). We omit the details. However, Algorithm 2.2 is not equivalent to the ENO scheme obtained by using $h_{j+1/2}^{\text{LLF}}$ as a building block [12]. The advantage of Algorithm 2.2 is again its uniformly high order of accuracy.

(c) In terms of cost, Algorithm 2.2 is equivalent to ENO schemes using f^+ in [12] (we shall denote it by ENO-LF), i.e., evaluating two divided difference tables, while Algorithm 2.1 is half as expensive because it only needs one divided difference table.

(d) Roughly speaking the viscosity of ENO-LF is larger than that of ENO-LLF, which is in turn larger than that of ENO-Roe. Hence in going from ENO-LF to ENO-LLF to ENO-Roe we can expect increasingly less shock smearing and better overall accuracy; this is verified by the numerical experiments of Section 5.

(e) As was true for the schemes described in [12], when $r = 1$, Algorithm 2.1 yields a familiar TVD second-order accurate scheme [9], when the minimum of the absolute value function is changed to the minmod function:

$$m(a_1, ..., a_n) = \begin{cases} s \cdot \min\limits_{1 \leqslant i \leqslant n} |a_i|, & \text{if } s = \text{sign}(a_1) = \cdots = \text{sign}(a_n) \\ 0, & \text{otherwise.} \end{cases} \tag{2.16}$$

We omit the details of the derivation here.

We now describe the "entropy fix" version of Algorithm 2.1.

ALGORITHM 2.3 (ENO-RF). Let $\hat{f}_{j+1/2}$ be defined by (2.10) if $f'(u)$ does not change sign between u_j and u_{j+1}; otherwise, let $\hat{f}_{j+1/2}$ be defined by (2.13).

Remark 2.3. If f is convex, $f''(u) \geqslant 0$, we can use (2.10) unless $f'(u_j) < 0 < f'(u_{j+1})$; i.e., we use (2.13) only in the cells which contain "expansion shocks." This is the case for, e.g., the Euler equations of fluid gas dynamics. See Section 5.

Notice that the cost of Algorithm 2.3 is very close to that of Algorithm 2.1 because sonic points are isolated and the divided differences of u need only be computed locally near sonic points.

We may associate Algorithm 2.3 with the following E-flux.

$$h^{RF}_{j+1/2} = \begin{cases} f(u_j), & \text{if } f'(u) \geqslant 0 \text{ between } u_j \text{ and } u_{j+1} \\ f(u_{j+1}), & \text{if } f'(u) \leqslant 0 \text{ between } u_j \text{ and } u_{j+1} \\ h^{\mathrm{LLF}}_{j+1/2}, & \text{otherwise;} \end{cases} \tag{2.17}$$

however, as mentioned above, this algorithm is not equivalent to the ENO scheme obtained by using $h^{RF}_{j+1/2}$ as a building block.

Numerically we observe that Algorithm 2.3 always yields the correct entropy solution even for nonconvex f. See Section 5 for some examples.

III. Sharpening of Contact Discontinuities

We first consider the 1D, scalar, linear version of (1.1), i.e., $d = m = 1$, $f(u) = au$. Discontinuous solutions to this problem (called contact discontinuities) are usually smeared more severely than shocks, when computed by shock capturing schemes.

Recently Harten [6] introduced the concept of subcell resolution which he used to sharpen contact discontinuities. This notion uses the observation that the cell averages carry information about shock locations. In fact, this information is also contained in any *conservative* scheme (1.8), whether cell average or numerical flux based. We can thus translate Harten's subcell resolution techniques to our ENO schemes which use fluxes and Runge–Kutta time discretizations. Notice that for a linear, constant coefficient problem $f(u) = au$, ENO-LF, ENO-LLF, and ENO-Roe are identical. We write out the following algorithm for the case $a > 0$.

ALGORITHM 3.1. For $a > 0$, at the beginning of every Runge–Kutta cycle:

(1) Let $s_j = |m(\Delta_+ u_j^n, \Delta_- u_j^n)|$, where m is the minmod function (2.16). Define the "critical intervals" (intervals containing discontinuities) $I_j = (x_{j-1/2}, x_{j+1/2})$ by $s_j \geqslant s_{j+1}$, $s_j > s_{j-1}$;

(2) For any "critical interval" I_j, let $\theta_j = (u_j - u_{j-1})/(u_{j+1} - u_{j-1})$ or more accurately let θ_j be the solution in $[0, 1]$ of the quadratic equation

$$(u_{j+1} + u_{j-1} - u_{j+2} - u_{j-2})\theta_j^2 + (3u_{j+2} + 3u_{j-1} - 5u_{j+1} - u_{j-2})\theta_j$$
$$+ (4u_{j+1} - 2u_{j+2} - 2u_j) = 0.$$

We use $x_{j-1/2}+\theta_j\Delta x$ as an approximation to the discontinuity location inside the cell I_j; we then perform each step of the Runge–Kutta cycle as follows:

(3) Let $\hat{f}_{j+1/2}$ be defined as usual (Algorithm 2.1) *unless* I_j or (for the second, third, etc. steps in the Runge–Kutta cycle) I_{j-1} is a "critical interval." If I_j is a "critical interval," we define

$$\hat{f}_{j+1/2}=(1-\xi_j)\hat{f}^{(L)}_{j+1/2}+\xi_j\hat{f}^{(R)}_{j+1/2} \qquad \text{with} \quad \xi_j=\min\left(\frac{(1-\theta_j)\Delta x}{a\Delta t},1\right), \tag{3.1}$$

where $\hat{f}^{(L)}_{j+1/2}$ and $\hat{f}^{(R)}_{j+1/2}$ are again computed by Algorithm 2.1 with the modification on step (2) only, i.e., the first step in the choice of stencil. We use $k^{(1)}_{\min}=j-1$ for $\hat{f}^{(L)}_{j+1/2}$ and $k^{(1)}_{\min}=j+1$ for $\hat{f}^{(R)}_{j+1/2}$. To be safe, for the second, third, etc. steps in the Runge–Kutta cycle, we may choose $k^{(1)}_{\min}=j+2$ for $\hat{f}^{(R)}_{j+1/2}$ if $\xi_j<1$, and, when I_{j-1} is a "critical interval" and $\xi_{j-1}<1$, we should choose $k^{(1)}_{\min}=j+1$ to evaluate $\hat{f}_{j+1/2}$.

Remark 3.1 (a) The case $a<0$ is easily obtained by symmetry.

(b) The discontinuity detector in step (1) is somewhat different than that of [6]. The purpose of our choice is to avoid identifying an interval containing a smooth extremum as "critical interval."

(c) As in [6], the modification in (3.1) does not affect accuracy if u happens to be smooth in I_j.

(d) The whole procedure here is simpler than in [6], especially if the first expression for θ_j is used in step (2). The price we pay is that it is less accurate in theoretical resolution, because it is based on piecewise constant (first choice of θ_j) or piecewise linear (second choice of θ_j) interpolants. Numerically, we observe results very close to those found in [6]. See Section 5 for details.

(e) The naive generalization to 2D, namely writing $u+\Delta tL_x u+\Delta tL_y u$ as $\frac{1}{2}(u+2\Delta tL_x u)+\frac{1}{2}(u+2\Delta tL_y u)$ and applying the algorithm separately to the two parts, does not work satisfactorily, according to our numerical tests (see Section 5). A truly 2D subcell resolution seems to be needed. Currently Harten [7] is investigating this within the cell average framework.

(f) For details of the analysis of Algorithm 3.1, we refer the readers to [6]. Although the two algorithms are different, we have essentially borrowed the philosophy of [6] and applied it to the present case.

Another interesting method we can use to sharpen contact discontinuities is an artificial compression method introduced by Yang [15], again in a cell average context, i.e., for MUSCL type schemes. The idea is to sharpen the slopes in the discontinuous cells while still preserving monotonicity. In the MUSCL framework one simply increases the slopes using the jumps in the interpolant across interfaces, while still keeping the ENO property. Translated to our current framework, it gives the following algorithm (again for $a>0$).

ALGORITHM 3.2. Let $\hat{f}^A_{j+1/2} = \hat{f}_{j+1/2} + c_{j+1/2}$, where

$$c_{j+1/2} = m\left[\frac{\alpha}{2} m(\hat{f}^{(R)}_{j+1/2} - \hat{f}_{j+1/2}, \hat{f}^{(R)}_{j-1/2} - \hat{f}_{j-1/2}), f_{j+1} - \hat{f}_{j+1/2}, \hat{f}^{(R)}_{j-1/2} - f_{j-1}\right]. \quad (3.2)$$

Here, as in (3.1), $\hat{f}^{(R)}_{j+1/2}$ corresponds to $k^{(1)}_{\min} = j+1$ in (2.6), step (2) of Algorithm 2.1. m is again the minmod function (2.16).

Remark 3.2. (a) α in (3.2) is a positive parameter. It can be tuned for each individual problem to optimize the results. The bigger the α, the more the compression. Monotonicity is preserved for any α, due to the last two arguments in the outer minmod function in (3.2). We usually use $2 \leqslant \alpha \leqslant 5$, but have not yet found a general rule for determining α.

(b) The approximation for $a < 0$ is constructed symmetrically.

(c) Other versions of (3.2) are possible. For example, we may use

$$\alpha = \alpha_0 \left(\frac{|u_{j+1} - 2u_j + u_{j-1}|}{|u_{j+1} - u_j| + |u_j - u_{j-1}|} \right)^\beta \quad (3.3)$$

in (3.2), where $\alpha_0 > 0$ and $\beta = \text{integer} \geqslant 0$ are two parameters. The heuristic reason to use (3.3) is that α is then very small ($\sim O(h^\beta)$) except near a discontinuity in either u or its derivative, hence over-compression in smooth regions should be avoided. Computationally this improves the performance, especially with some tuning of β. We have not yet been able to find a general rule in choosing the optimal β.

(d) Again, the modification (3.2) does not affect the accuracy in smooth regions.

Both algorithms in this section are easily generalizable to variable coefficient or nonlinear problems. However special caution is needed when one tries to sharpen a (nonlinear) shock, to avoid obtaining a nonphysical, entropy condition violating solution. In the computation for the Euler equations of compressible gas dynamics (Section 5), we only use the compression in the linearly degenerate fields.

IV. IMPLEMENTATIONS IN MULTI-DIMENSIONS AND SYSTEMS

A special advantage of ENO schemes using fluxes and Runge–Kutta methods is their relative simplicity in multi-dimensions. The algorithms in Section 2 are applied to each of the terms $f_i(u)_{x_i}$ in (1.1a), keeping all other variables fixed. The Runge–Kutta method (1.6) is then applied. A typical CFL restriction $(\Delta t/\Delta x) \max_u |f'(u)| \leqslant \lambda_0$ will be replaced by $\Delta t \max_u \sum_{i=1}^{d} (1/\Delta x_i)|f_i'(u)| \leqslant \lambda_0$.

For nonlinear systems, we simply apply the algorithms in Section 2 in each (local) characteristic field. We take an 1D system to exemplify this process. Let $A_{j+1/2}$ be some "average" Jacobian at $x_{j+1/2}$. Examples include $A_{j+1/2}=(\partial\mathbf{f}/\partial\mathbf{u})|_{\mathbf{u}=(1/2)(\mathbf{u}_j+\mathbf{u}_{j+1})}$ or, in the case of Euler equations of gas dynamics $A_{j+1/2}=(\partial\mathbf{f}/\partial\mathbf{u})|_{\mathbf{u}=\mathbf{u}^{(\mathrm{Roe})}_{j+1/2}}$, where $\mathbf{u}^{(\mathrm{Roe})}_{j+1/2}=\mathbf{R}(\mathbf{u}_j, \mathbf{u}_{j+1})$ is the Roe average of $\mathbf{u}_j$ and $\mathbf{u}_{j+1}$ [10]. We denote the eigenvalues and left and right eigenvectors of $A_{j+1/2}$ by $\lambda^{(p)}_{j+1/2}, l^{(p)}_{j+1/2}, \mathbf{r}^{(p)}_{j+1/2}, p=1, ..., m$, normalized so that

$$l^{(p)}_{j+1/2}\cdot\mathbf{r}^{(q)}_{j+1/2}=\delta_{pq}=\begin{cases}1, & \text{if } p=q\\ 0, & \text{if } p\neq q.\end{cases}$$

For any vector $\mathbf{a}$,

$$a^{(p)}=l^{(p)}_{j+1/2}\cdot\mathbf{a} \tag{4.1}$$

is the component of $\mathbf{a}$ in the pth (local) characteristic field, because

$$\mathbf{a}=\sum_{p=1}^{m} a^{(p)}\mathbf{r}^{(p)}_{j+1/2}. \tag{4.2}$$

Algorithm 2.1 now becomes:

ALGORITHM 4.1.

(1) Same as step (1) in Algorithm 2.1, changing vectors to bold face letters.

(2) For each j, compute $H^{(p)}[x_{l-1/2}, ..., x_{l+k+1/2}]$ for $l=j-\gamma, ..., j+1$ and $k=1, 2, ..., r$, by using the divided difference tables of H (derived from those of f) and (4.1). Here $1\leqslant p\leqslant m$.

(3) Apply steps (2)–(6) of Algorithm 2.1 to $H^{(p)}$, using $\bar{a}_{j+1/2}=\lambda^{(p)}_{j+1/2}$, to get $\hat{f}^{(p)}_{j+1/2}$.

(4) Use (4.2) to get $\hat{f}_{j+1/2}$.

We may similarly generalize Algorithms 2.2 and 2.3. Note that (2.12) becomes, for the pth field,

$$\alpha^{(p)}_{j+1/2}=\max_{\mathbf{u}\in L(\mathbf{u}_j, \mathbf{u}_{j+1})} |\lambda^{(p)}(\mathbf{u})|, \tag{4.3}$$

where $L(\mathbf{u}_j, \mathbf{u}_{j+1})$ is some curve (e.g., a straight line) in phase space connecting $\mathbf{u}_j$ and $\mathbf{u}_{j+1}$. For Euler equations of gas dynamics, the fields are either genuinely nonlinear or linearly degenerate, hence we may use

$$\alpha^{(p)}_{j+1/2}=\max(|\lambda^{(p)}(\mathbf{u}_j)|, |\lambda^{(p)}(\mathbf{u}_{j+1})|), \tag{4.4}$$

which is similar to (2.14).

In terms of computational cost ENO-Roe or ENO-RF is still about half as expensive as ENO-LF, because of the number of divided difference tables needed. The characteristic decompositions (4.1)–(4.2) do involve more vector multiplications for higher orders, e.g., third-order schemes need roughly $\frac{16}{9}$ times more vector multiplications than second-order ones. One way to reduce this cost is to apply the characteristic decomposition only in rapid transition regions. See [15] for an interesting approach to this. We shall not pursue this idea in this paper.

V. Numerical Results

We use the notation $\text{ENO}-X-(Y)-r$, where $X=$ Roe, LLF, RF, or LF refers to Algorithms 2.1, 2.2, 2.3 or the $f^{\pm}$ version in [11]; $Y=S$ or A refers to Algorithms 3.1 or 3.2; r is the order of the scheme. For the TVD Runge–Kutta methods used please see Table I. We have run most examples using second-, third-, and fourth-order schemes, but here we usually include only third-order results as representatives.

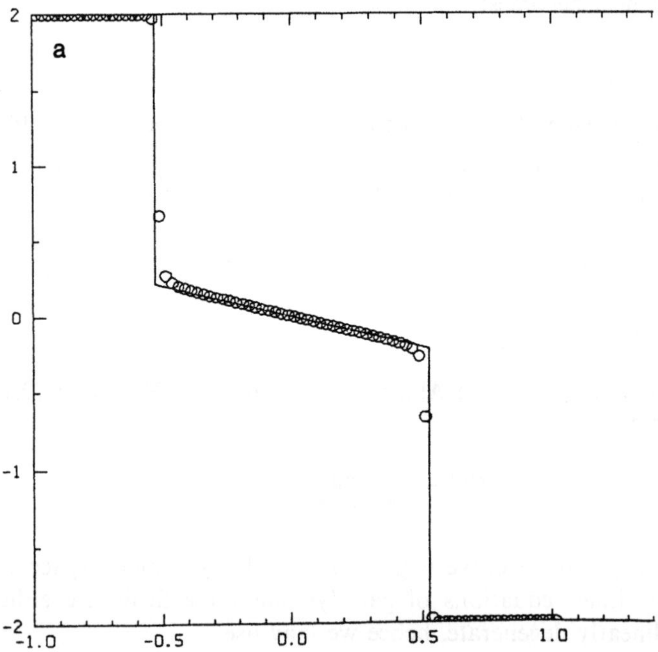

Fig. 1. ENO-RF-3, non-convex fluxes (solid lines are exact solutions; circles are numerical solutions): (a) (5.1) with $u_{\text{left}}=2, u_{\text{right}}=-2$; (b) (5.1) with $u_{\text{left}}=-3, u_{\text{right}}=3$; (c) (5.2) with $u^0(x)=-1$ in $[-\frac{1}{2}, 0]$; $u^0(x)=0$ elsewhere.

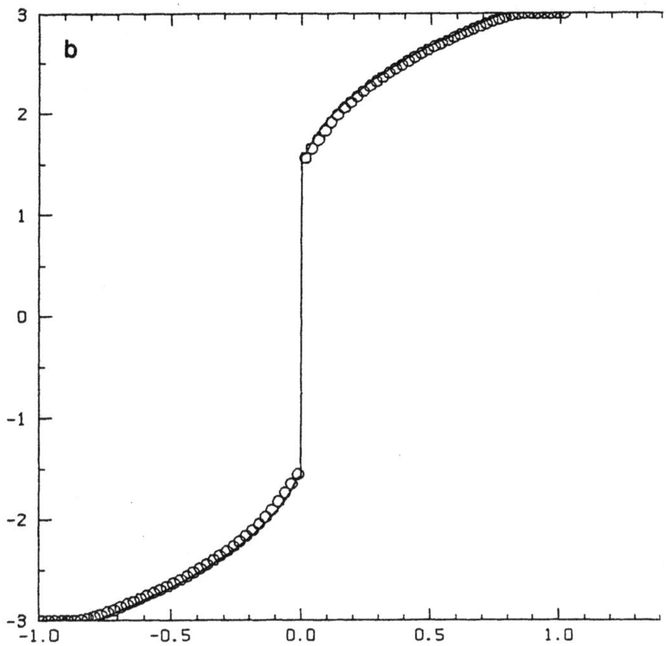

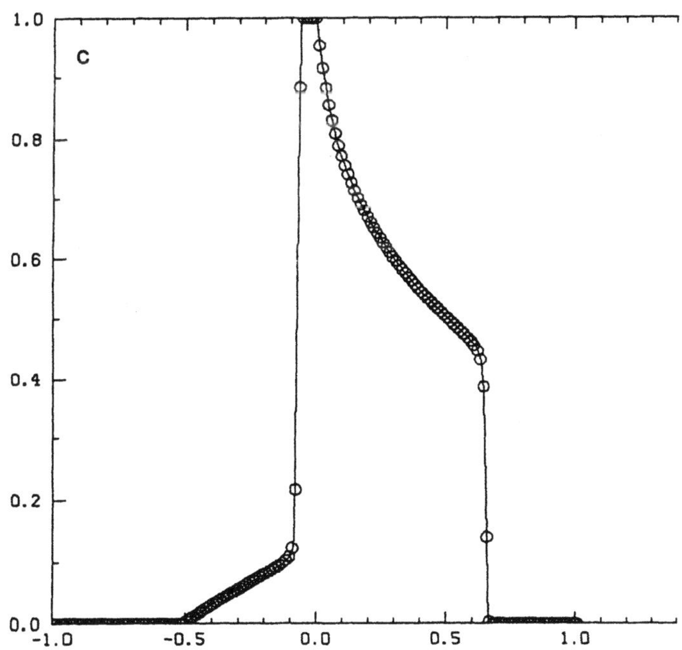

Fig. 1. *(Continued)*

EXAMPLE 1. This is the Example 4 in [12, Section IV] revisited. We solve the scalar 1D Riemann problems of (1.1) with two nonconvex functions f,

$$f(u)=\tfrac{1}{4}(u^2-1)(u^2-4) \tag{5.1}$$

$$f(u)=\frac{4u^2}{4u^2+(1-u)^2}. \tag{5.2}$$

ENO-RF-3 is used. The results are displayed in Fig. 1. We observe considerable improvements in shock transition and overall resolution compared to the results shown in Fig. 14, 18, and 22 in [12]. ENO-RF-3 and ENO-RF-4 were also tested on several other nonconvex problems. We always observed convergence to the correct entropy solution and a well-resolved solution. When applied to Burgers' equation (Example 1 in [12, Section IV]) we observed improvements in accuracy and shock transition of ENO-RF over ENO-LF. We omit the details here.

EXAMPLE 2. This example uses the two contact-discontinuity-sharpening algorithms in Section 3 applied to scalar 1D linear problems. We solve the model equation

$$u_t+u_x=0, \qquad -1\leqslant x<1 \tag{5.3a}$$

$$u(x,0)=u^0(x), \qquad u^0(x) \text{ periodic with period 2.} \tag{5.3b}$$

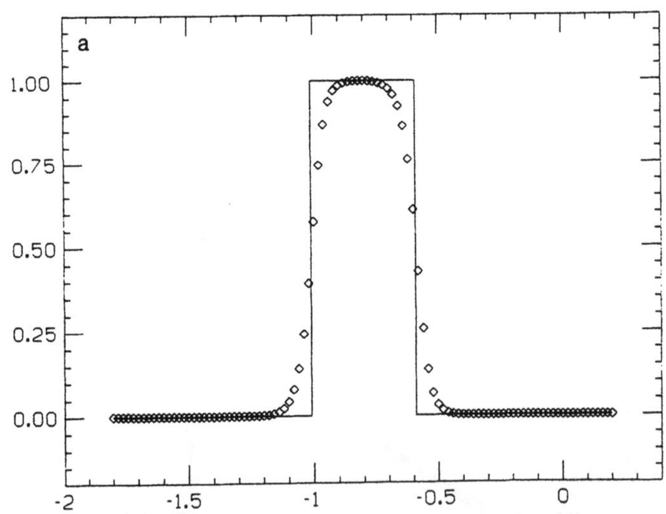

FIG. 2. (5.3)–(5.4a) (solid lines are exact solutions; diamonds are numerical solutions): (a) ENO-Roe-3; (b) ENO-Roe-S-3; (c) ENO-Roe-A-3.

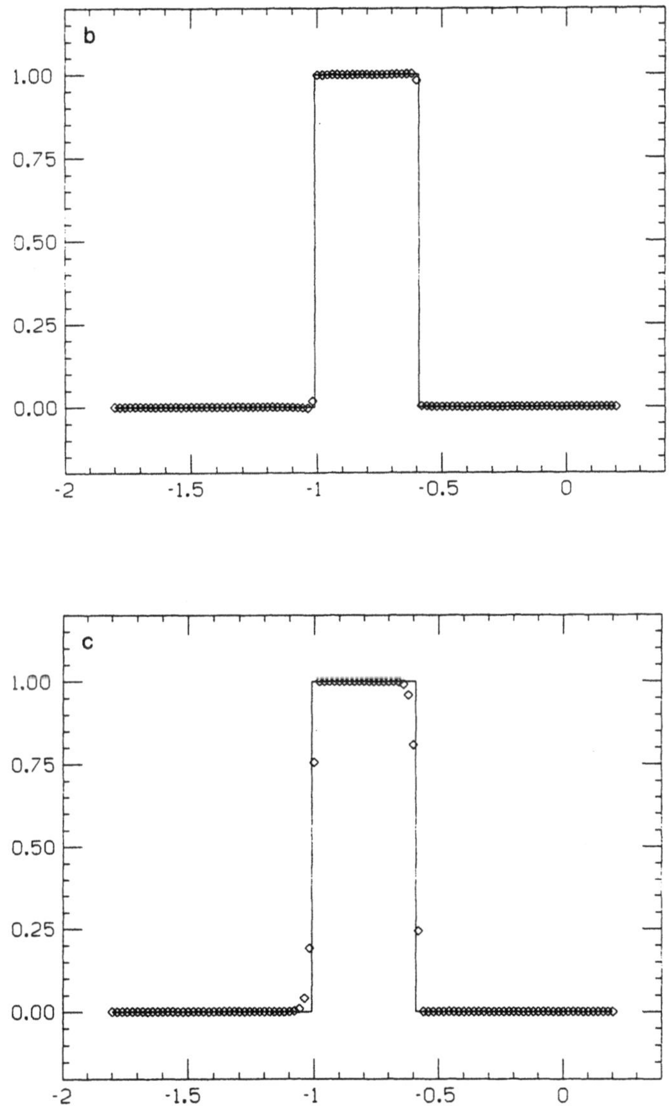

Fig. 2. (*Continued*)

Notice that in this linear case all the three algorithms ENO-Roe, ENO-LLF, and ENO-RF in Section 2 are identical.

Four initial conditions $u^0(x)$ are used. The first three are used by Zalesak [16], using 100 equally spaced grid points in $[-1, 1)$ with

$$u^0(x)=\begin{cases}1, & -\frac{1}{5}\leqslant x\leqslant\frac{1}{5},\\ 0, & \text{otherwise,}\end{cases}\quad \text{800 time steps, CFL}=0.2 \tag{5.4a}$$

$$u^0(x)=e^{-300x^2},\quad \text{600 time steps, CFL}=0.1 \tag{5.4b}$$

$$u^0(x)=\begin{cases}(1-(\frac{10}{3}x)^2)^{1/2}, & |x|<\frac{3}{10},\\ 0, & \text{otherwise,}\end{cases}\quad \text{600 time steps, CFL}=0.1 \tag{5.4c}$$

(since $u^0(x)$ is periodic we only give the definitions in $[-1, 1)$). The last is used by Harten *et al.* [4, 6]:

$$u^0(x+0.5)=\begin{cases}-x\sin(\frac{3}{2}\pi x^2), & -1\leqslant x<-\frac{1}{3},\\ |\sin(2\pi x)|, & |x|<\frac{1}{3},\quad t=8\\ 2x=1-\sin(3\pi x)/6, & \frac{1}{3}<x<1,\end{cases} \tag{5.4d}$$

The results are displayed in Fig. 2–5. We observe apparent improvements of both algorithms in all cases. For comparisons we refer the readers to [6, 16].

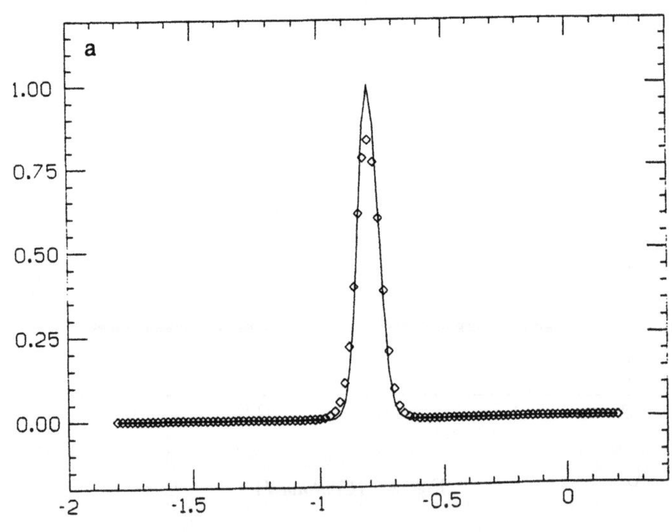

FIG. 3. (5.3)–(5.4b) (solid lines are exact solutions; diamonds are numerical solutions): (a) ENO-Roe-3; (b) ENO-Roe-S-3; (c) ENO-Roe-A-3.

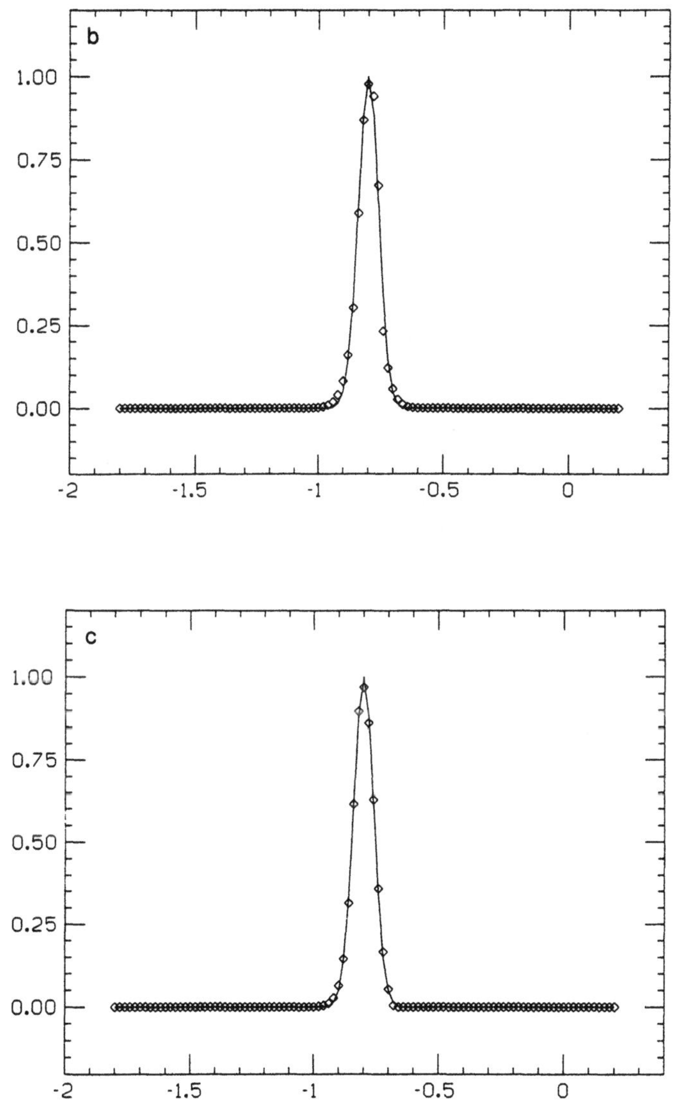

Fig. 3. (*Continued*)

EXAMPLE 3. We solve a Riemann problem for the 2D Burgers' equation

$$u_t + \left(\frac{u^2}{2}\right)_x + \left(\frac{u^2}{2}\right)_y = 0 \tag{5.5a}$$

of the type:

$$u(x, y, 0) = \begin{cases} u_1, & x > 0, y > 0 \\ u_2, & x < 0, y > 0 \\ u_3, & x < 0, y < 0 \\ u_4, & x > 0, y < 0. \end{cases} \tag{5.5b}$$

Depending on the orders of the u_i's, there are eight essentially different solution types. See [13] for details. We used ENO-RF and observed convergence to the correct entropy solution with good resolution for all cases. The results are displayed in Fig. 6–7. For one case (Fig. 6) we present the results of ENO-RF-3 for 20×20 and 80×80 grid points, and the result using first-order Engquist–Osher scheme for 80×80 grid points. We observe that ENO-RF-3 using 20×20 points has roughly the same resolution as does the first-order montone scheme with 80×80 points. For the remaining seven cases (Fig. 7) we only show the results of ENO-RF-3 with 80×80 points.

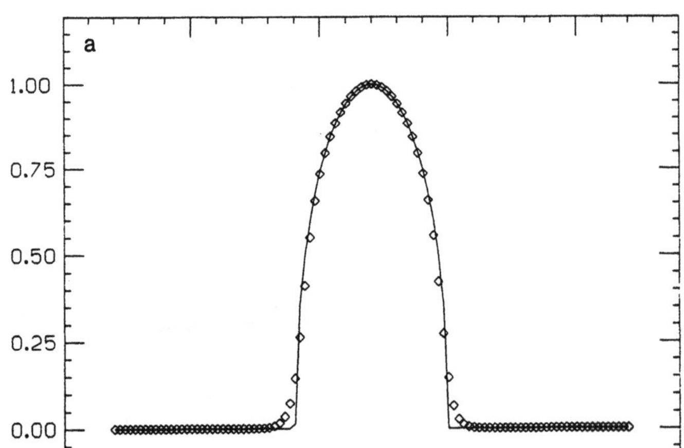

FIG. 4. (5.3)–(5.4c) (solid lines are exact solutions; diamonds are numerical solutions): (a) ENO-Roe-3; (b) ENO-Roe-S-3; (c) ENO-Roe-A-3.

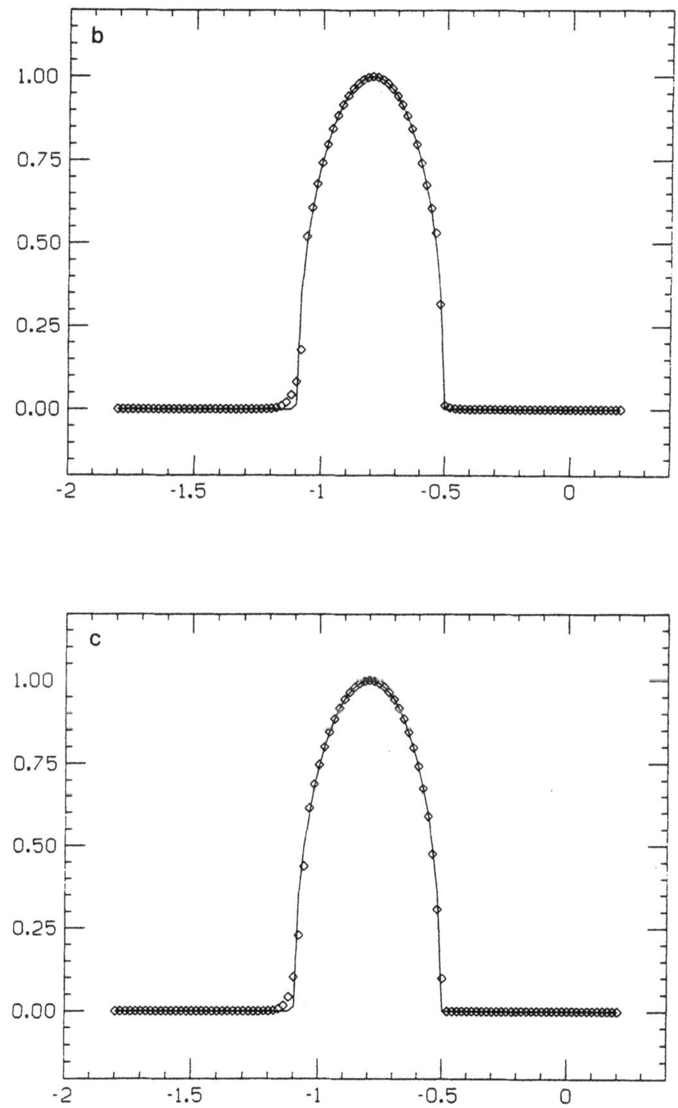

Fig. 4. *(Continued)*

We also remark here that Yang's artificial compression Algorithm 3.2 improves the resolution in this example according to our numerical experiments. Since we are mainly interested in applying Algorithm 3.1 and 3.2 to linear problems, we omit the details here.

EXAMPLE 4. This example is suggested by Professor D. Gottlieb (private communication). It simulates a 2D boundary layer problem. We solve the boundary value problem

$$u_t + \left(\frac{u^2}{2}\right)_x + u_y = 0, \qquad 0 \leqslant x < 2\pi, \quad 0 \leqslant y \leqslant 1 \tag{5.6a}$$

$$u(x, 0, t) = \alpha + \beta \sin x, \qquad u \text{ is periodic in } x \text{ with period } 2\pi \tag{5.6b}$$

to steady state. In steady state it resembles the 1D Burger's equation with $\alpha + \beta \sin x$ as "initial" condition, if y is identified as time t. Hence the exact solution to the steady state can be obtained. This experiment simulates a 2D boundary layer problem because the shock "dissolves" near the boundary $y = 0$. We can adjust the thickness of the boundary layer (i.e., the smooth region near $y = 0$) by adjusting β. We used ENO-RF-3 to compute (5.6), running it to steady state, with initial condition $u(x, y, 0) = \alpha + \beta \sin x$. We imposed periodic boundary conditions

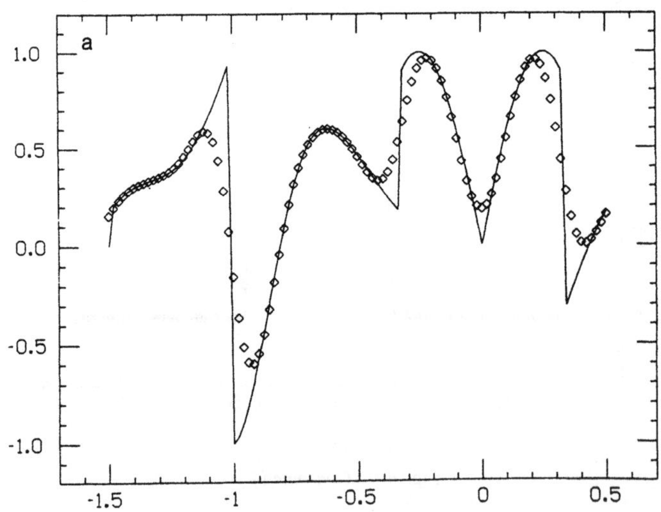

FIG. 5. (5.3)–(5.4d) (solid lines are exact solutions; diamonds are numerical solutions): (a) ENO-Roe-3; (b) ENO-Roe-S-3; (c) ENO-Roe-A-3.

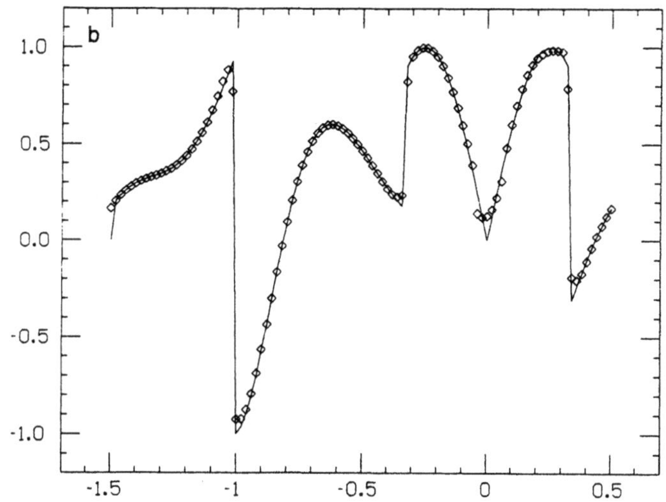

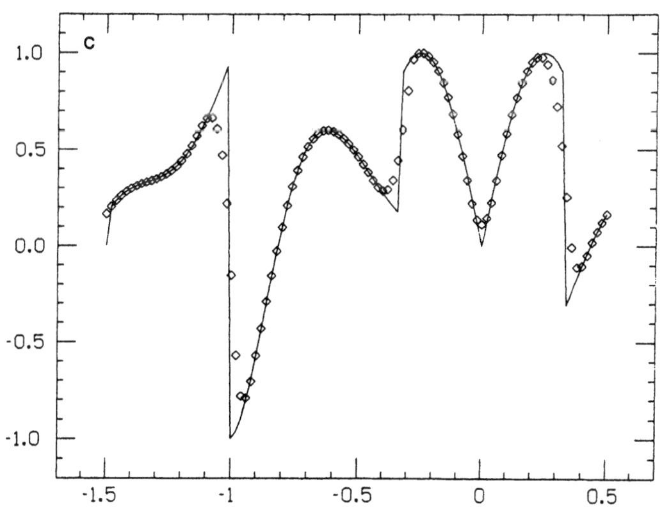

Fig. 5. (*Continued*)

in x, enforced (5.6b) at $y=0$, and imposed no boundary condition at the outflow boundary $y=1$. (For a detailed discussion of the implementation of boundary conditions for ENO schemes we refer the readers to [4].) Figure 8 contains the level curves for $\alpha=0$, $\beta=5$ and $\alpha=2$, $\beta=5$. 30×30 spatial points are used.

EXAMPLE 5. This example was suggested by Professor R. Sanders (private communication), to study the smearing of contact discontinuities in different directions when $\Delta x \neq \Delta y$. We solve

$$u_t + u_x + u_y = 0, \qquad u \text{ is periodic in } x \text{ and } y \text{ with periods } 1 \tag{5.7a}$$

$$u(x, y, 0) = \begin{cases} 1, & \text{if } (x-\frac{1}{2})^2 + (y-\frac{1}{2})^2 \leqslant \frac{1}{8} \\ 0, & \text{otherwise in } 0 \leqslant x, y < 1 \end{cases} \tag{5.7b}$$

with $\Delta y = 2\Delta x = \frac{1}{20}$, and ran it for $t=2$ (two periods). In Fig. 9 the "+" signs are for the cross section $y=\frac{1}{2}$, and the "*" signs are for the cross section $x=\frac{1}{2}$. We observe that ENO-Roe-3 does smear differently in the x and y directions with $\Delta x \neq \Delta y$, but ENO-Roe-A-3 does a much better job. On the other hand, ENO-

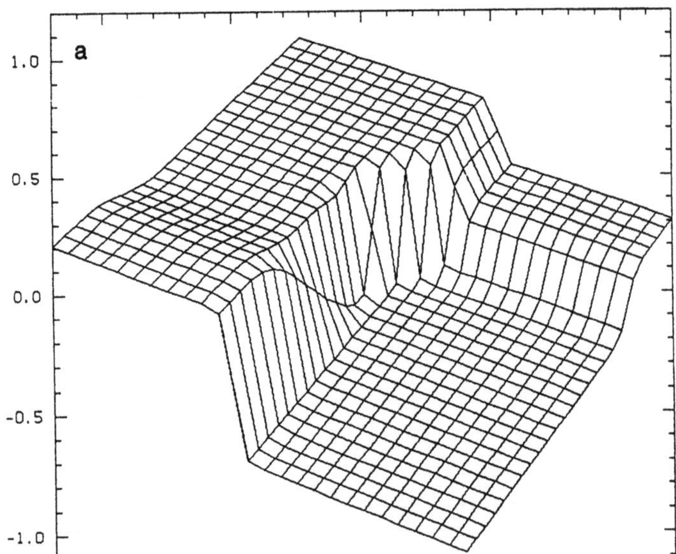

FIG. 6. (5.5 with $(u_1, u_2, u_3, u_4) = (-1, -0.2, 0.5, 0.8)$, $t=1$ (surfaces of numerical solutions): (a) ENO-RF-3, 20×20 points; (b) ENO-RF-3, 80×80 points; (c) First-order Engquist–Osher scheme, 80×80 points.

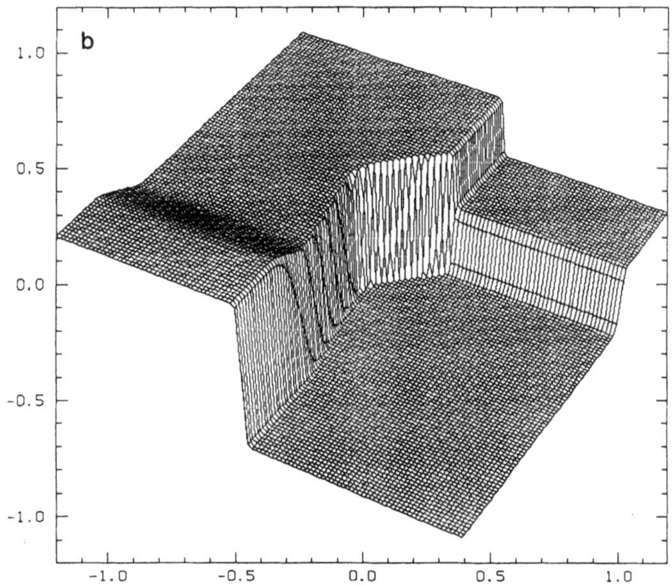

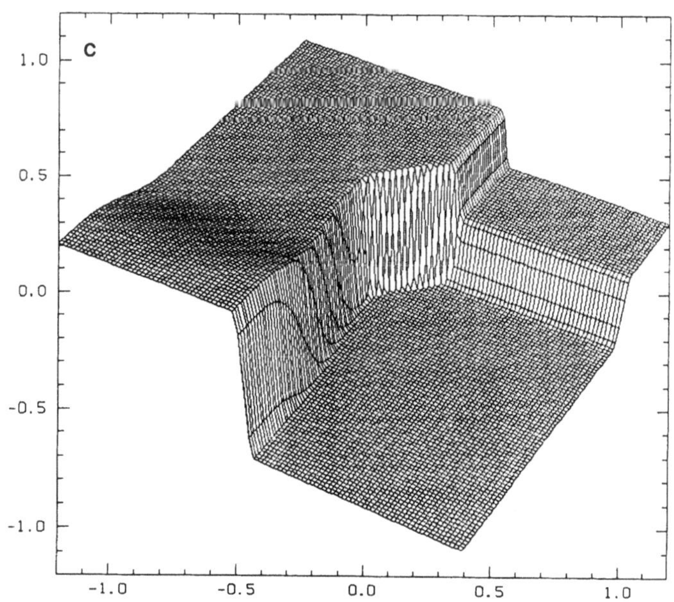

Fig. 6. (*Continued*)

Roe-S-2 (naive generalization to 2D) does not work well. It seems that some truly 2D subcell resolution techniques are needed.

EXAMPLE 6. We consider the Riemann problems for the Euler equations of gas dynamics for a polytropic gas, i.e., (1.1) with $d=1, m=3$, and

$$\mathbf{u}=(\rho, M, E)^{\mathrm{T}}, \qquad \mathbf{f}(\mathbf{u})=q\mathbf{u}+(0, P, qP)^{\mathrm{T}}, \tag{5.8a}$$

where

$$P=(\gamma-1)(E-\tfrac{1}{2}\rho q^2), \qquad M=\rho q \tag{5.8b}$$

for the initial condition

$$\mathbf{u}(x,0)=\begin{cases}\mathbf{u}_{\mathrm{L}}, & x<0\\ \mathbf{u}_{\mathrm{R}}, & x\geqslant 0.\end{cases} \tag{5.8c}$$

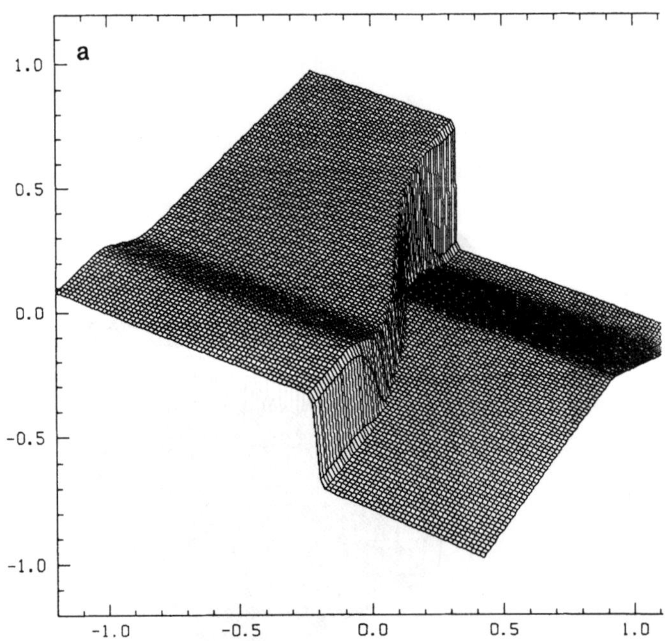

FIG. 7. ENO-RF-3, (5.5), $t=1$, 80×80 points (surfaces of numerical solutions): (a) $(u_1, u_2, u_3, u_4)=(-0.2, -1, 0.5, 0.8)$; (b) $(u_1, u_2, u_3, u_4)=(-1, 0.5, -0.2, 0.8)$; (c) $(u_1, u_2, u_3, u_4)=(-1, 0.5, -0.2, 0.8)$; (d) $(u_1, u_2, u_3, u_4)=(-1, -0.2, 0.8, 0.5)$; (e) $(u_1, u_2, u_3, u_4)=(0.8, -1, 0.2, -0.5)$; (f) $(u_1, u_2, u_3, u_4)=(0.8, -1, 0.5, -0.2)$; (g) $(u_1, u_2, u_3, u_4)=(0.8, -0.2, -1, 0.5)$.

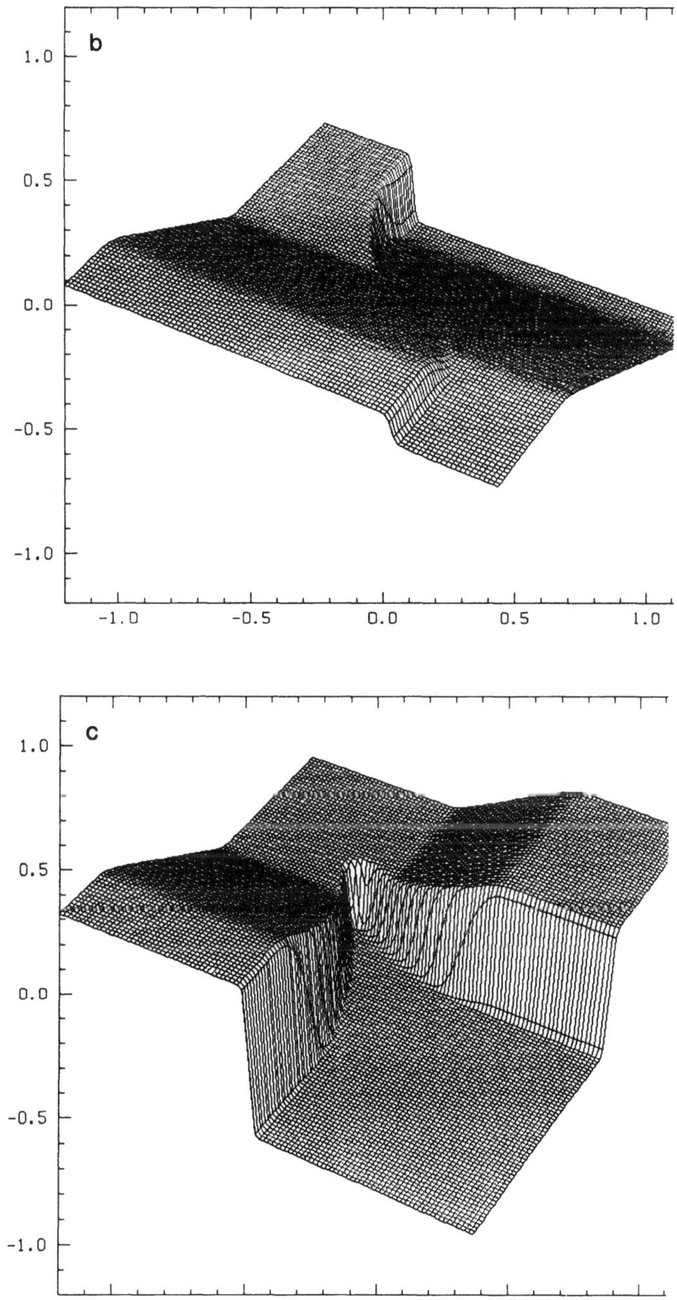

Fig. 7. *(Continued)*

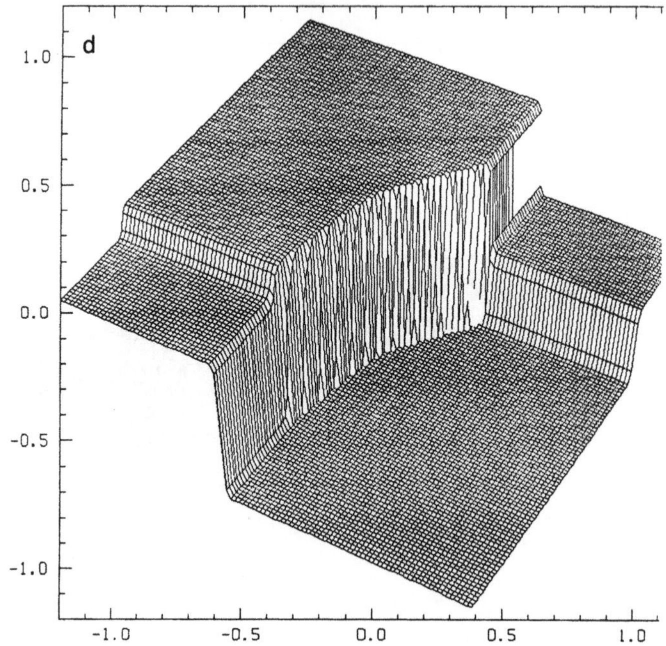

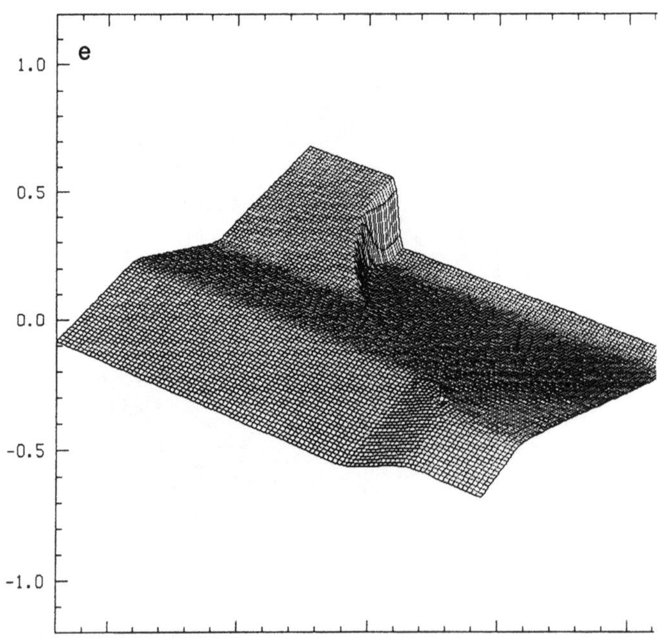

Fig. 7. (*Continued*)

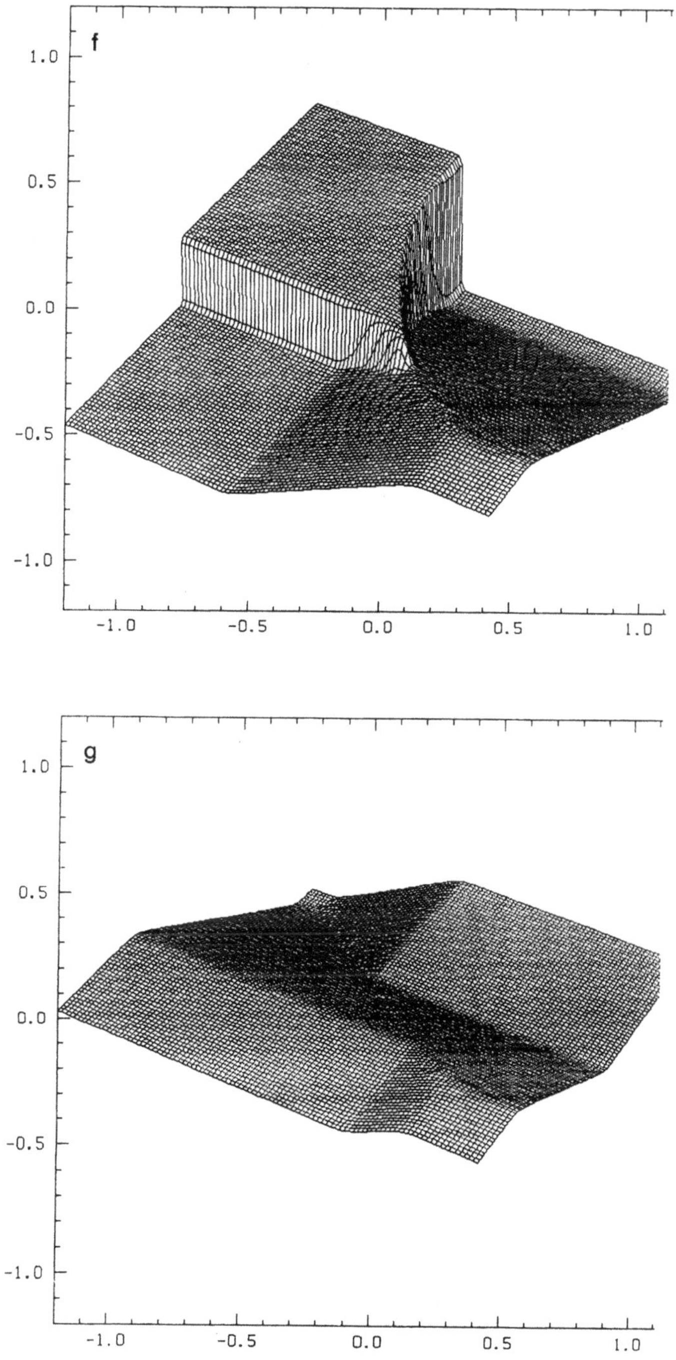

Fig. 7. (*Continued*)

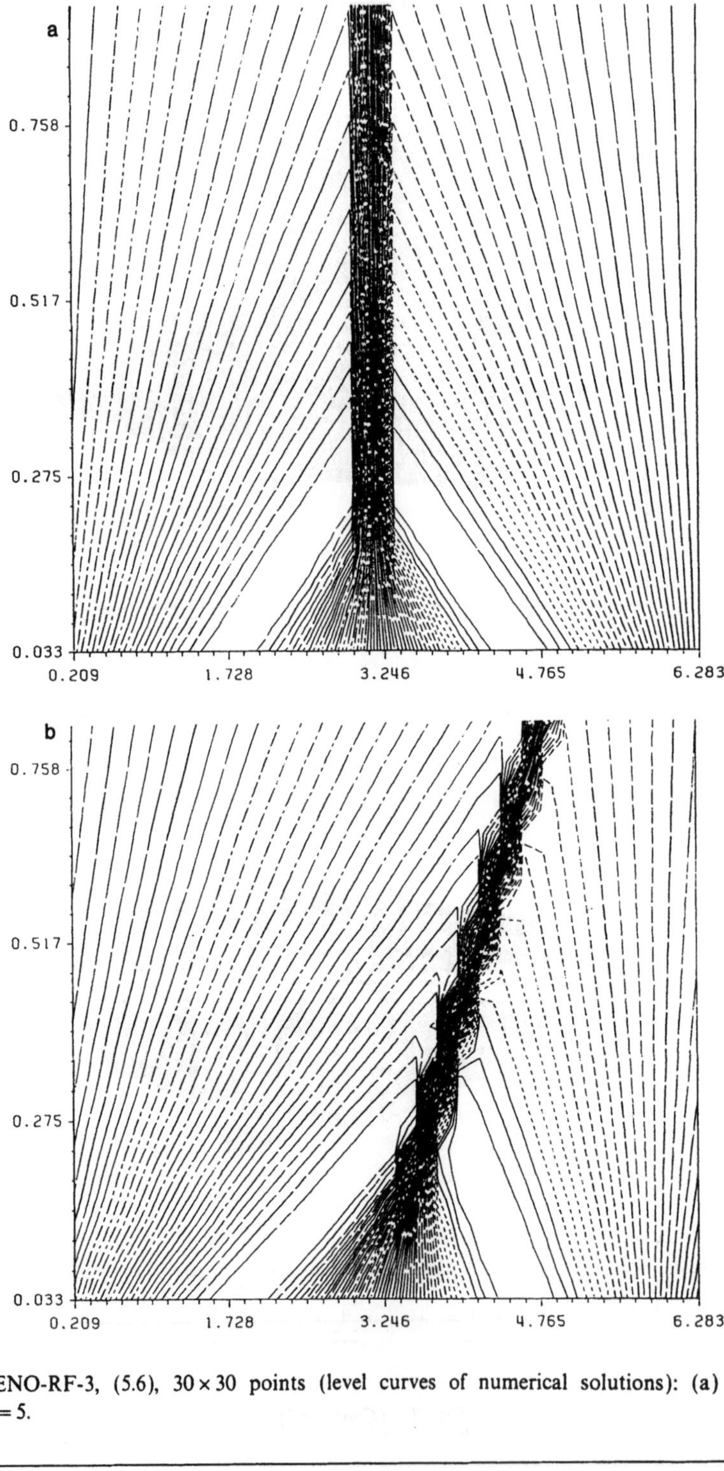

FIG. 8. ENO-RF-3, (5.6), 30 × 30 points (level curves of numerical solutions): (a) $\alpha = 0, \beta = 5$; (b) $\alpha = 2, \beta = 5$.

$\gamma = 1.4$ is used. For details of the Jacobian, its eigenvalues, and eigenvectors, see [4, 11] or Example 9 below.

We use the same two sets of initial conditions as in [4]:

$$(\rho_L, q_L, P_L) = (1, 0, 1); \qquad (\rho_R, q_R, P_R) = 0.125, 0, 0.10) \tag{5.9a}$$

$$(\rho_L, q_L, P_L) = (0.445, 0.698, 3.528); \qquad (\rho_R, q_R, P_R) = (0.5, 0, 0.571). \tag{5.9b}$$

The results are in Fig. 10–11. Notice the improved treatment of contact discontinuities after using the two contact sharpening agorithms of Section 3. Also notice that the corners of rarefaction waves (discontinuities in derivatives) are not resolved as well. Currently this phenomenon is still under investigation.

EXAMPLE 7. This is the same equation as in Example 6 with the initial condition

$$\mathbf{u}(x, 0) = \begin{cases} \mathbf{u}_L, & 0 \leqslant x < 0.1 \\ \mathbf{u}_M, & 0.1 \leqslant x < 0.9 \\ \mathbf{u}_R, & 0.9 \leqslant x < 1, \end{cases} \tag{5.10a}$$

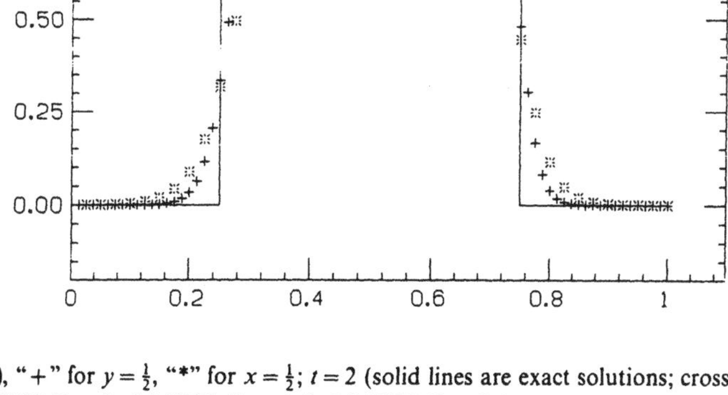

FIG. 9. (5.7), "+" for $y = \frac{1}{2}$, "*" for $x = \frac{1}{2}$; $t = 2$ (solid lines are exact solutions; crosses are numerical solutions): (a) ENO-Roe-3; (b) ENO-Roe-A-3; (c) ENO-Roe-S-2.

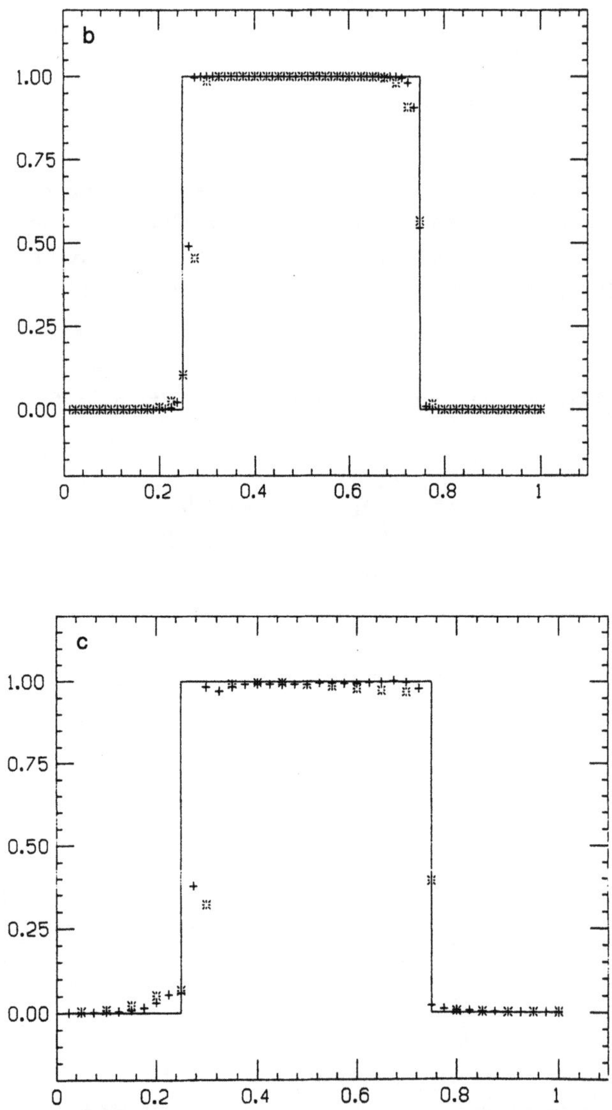

Fig. 9. (*Continued*)

where

$$\rho_L = \rho_M = \rho_R = 1, \qquad q_L = q_M = q_R = 0, \qquad P_L = 10^3, \qquad P_M = 10^{-2}, \qquad P_R = 10^2. \tag{5.10b}$$

A solid wall boundary condition is applied to both ends. See [4, 14] for details.

The results at the final time $t = 0.038$ are in Fig. 12–13. The solid lines are the numerical solutions using ENO-RF-S-3 with 800 points. It can be regarded as an exact solution. Notice that ENO-RF-3 has essentially converged for the velocity and the pressure, but not for the density, due to the smearing of contact discontinuities (Fig. 12). Also notice that ENO-RF-S-3 and ENO-RF-A-3 with 200 points have better resolution for the density than does ENO-RF-3 with 400 points (Fig. 13).

EXAMPLE 8. In the above problems we only show pictures for third-order schemes. If we compare with second- and fourth-order schemes we see the expected improvements in resolution with higher orders. These improvements are usually not very significant for problems lacking structure in smooth regions, as in the case for most of the above examples. This is particularly true when the two contact discontinuity sharpening algorithms in Section 3 are used. However, the advantage of higher order schemes becomes significant, when the problem involved has some structure. To exemplify this, we solve the Euler's equation in

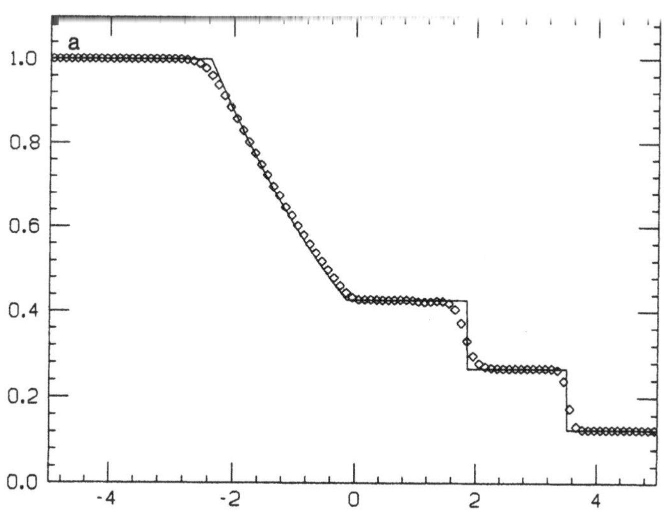

FIG. 10. ENO-RF-3, (5.8)–(5.9a), 100 points, $t = 2$ (solid lines are exact solutions; diamonds are numerical solutions): (a) density; (b) velocity; (c) pressure.

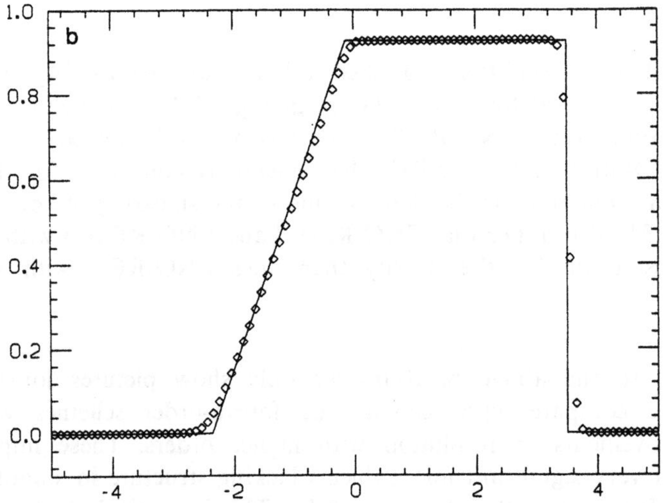

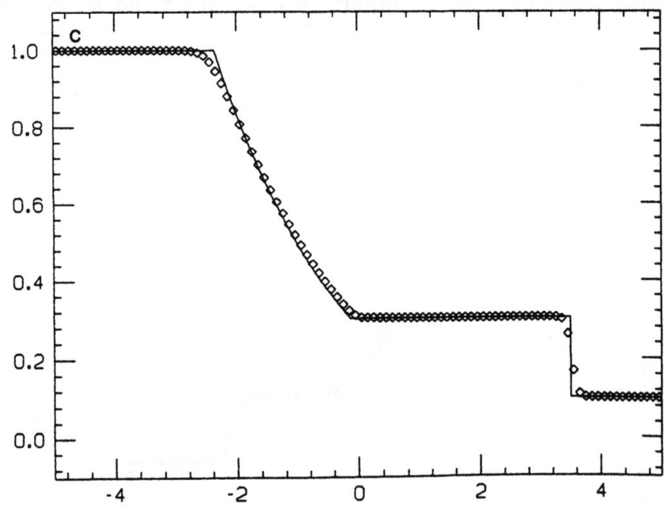

Fig. 10. (*Continued*)

Example 6 with a moving Mach = 3 shock interacting with sine waves in density, i.e., initially

$$\begin{array}{llll} \rho=3.857143; & q=2.629369; & P=10.33333 & \text{when } x<-4 \\ \rho=1+\varepsilon\sin 5x; & q=0; & P=1 & \text{when } x\geqslant -4. \end{array} \tag{5.11}$$

If $\varepsilon=0$, this is a pure Mach = 3 shock moving to the right.

We take $\varepsilon=0.2$. For a linearized analysis of this problem see [8]. The results are in Fig. 14–15. The solid lines are numerical solutions of ENO-RF-3 with 1600 grid points. It can be regarded as the exact solution. We observe that the fine structure in the density profile makes the higher order schemes perform much better than the lower order methods. ENO-RF-3 with 200 points has roughly the same resolution as the second-order MUSCL type TVD scheme [9] with 800 points (Fig. 14). On the other hand, the improvement of ENO-RF-3 over the second-order TVD scheme is not so significant for the velocity and pressure profiles, because they both lack any detailed structure (Fig. 15).

EXAMPLE 9. We apply ENO schemes to 2D Euler's equation of gas dynamics, i.e., (1.1) with $d=2$, $m=4$, and (we use $\mathbf{f}$, $\mathbf{g}$, x, y instead of $\mathbf{f}_1$, $\mathbf{f}_2$, x_1, x_2):

$$\mathbf{u}=(\rho, M_x, M_y, \mathrm{E})^{\mathrm{T}}, \qquad \mathbf{f}(\mathbf{u})=q_x\mathbf{u}+(0, P, 0, q_xP), \qquad \mathbf{g}(\mathbf{u})=q_y\mathbf{u}+(0, 0, P, q_yP), \tag{5.12a}$$

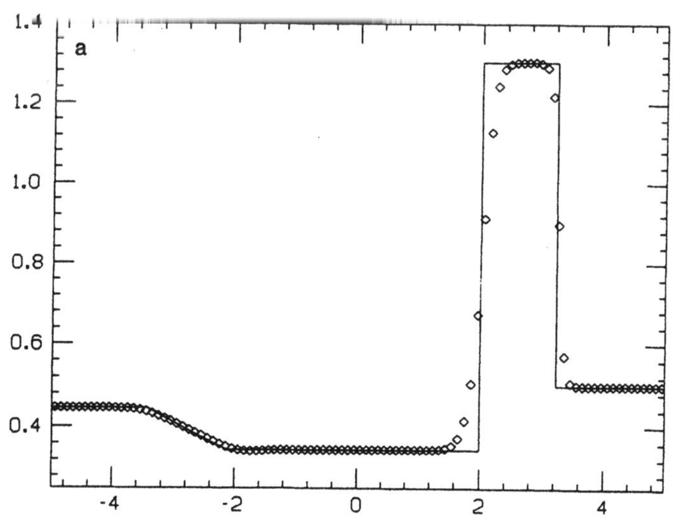

FIG. 11. (5.8)–(5.9b), 100 points, $t=1.3$, density (solid lines are exact solutions; diamonds are numerical solutions): (a) ENO-RF-3; (b) ENO-RF-S-3; (c) ENO-RF-A-3.

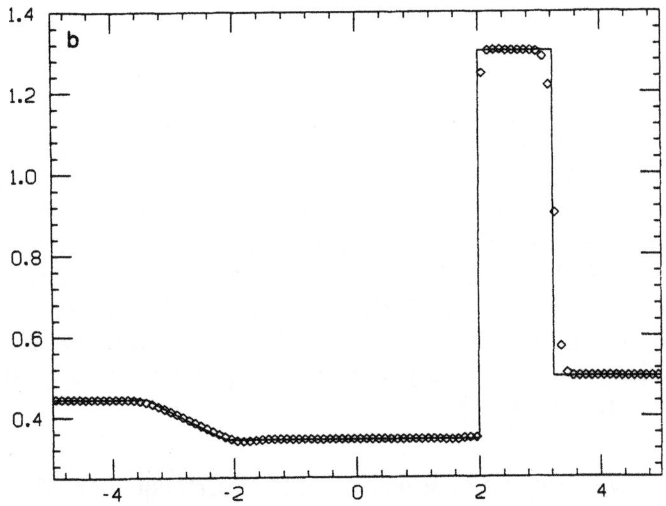

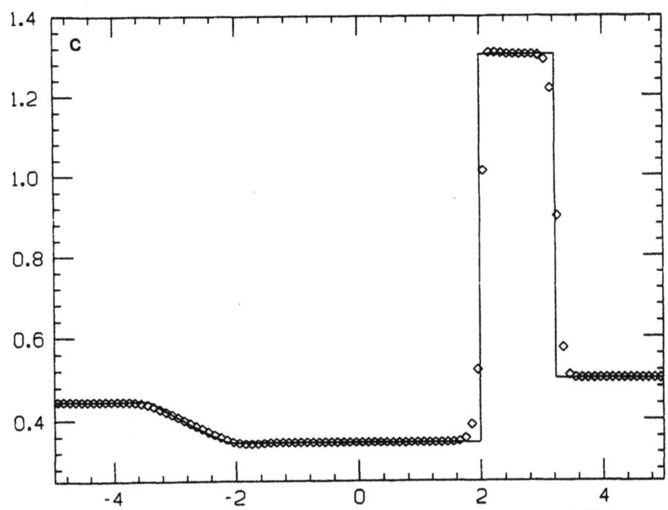

Fig. 11. (*Continued*)

where

$$P=(\gamma-1)(E-\tfrac{1}{2}\rho q^2), \qquad q^2=q_x^2+q_y^2, M_x=\rho q_x, M_y=\rho q_y. \tag{5.12b}$$

Also

$$c^2=(\gamma-1)(H-\tfrac{1}{2}q^2), \qquad H=\frac{E+P}{\rho}. \tag{5.12c}$$

For $\partial \mathbf{f}/\partial \mathbf{u}$, the eigenvalues are

$$\gamma_1=q_x-c, \qquad \lambda_2=\lambda_3=q_x, \qquad \lambda_4=q_x+c, \tag{5.13a}$$

the right eigenvectors are

$$\mathbf{r}_1=\begin{pmatrix}1\\ q_x-c\\ q_y\\ H-q_xc\end{pmatrix}, \qquad \mathbf{r}_2=\begin{pmatrix}0\\ 0\\ 1\\ q_y\end{pmatrix}, \qquad \mathbf{r}_3=\begin{pmatrix}1\\ q_x\\ q_y\\ \frac{1}{2}q^2\end{pmatrix}, \qquad \mathbf{r}_4=\begin{pmatrix}1\\ q_x+c\\ q_y\\ H+q_xc\end{pmatrix} \tag{5.13b}$$

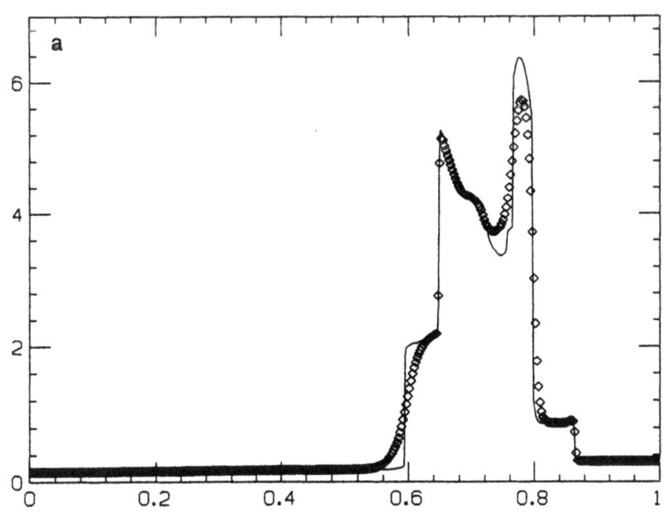

FIG. 12. ENO-RF-3, (5.8a)–(5.10), 400 points, $t=0.038$ (solid lines are the numerical solution of ENO-RF-3, 800 points): (a) density (diamonds); (b) velocity (diamonds); (c) pressure (diamonds).

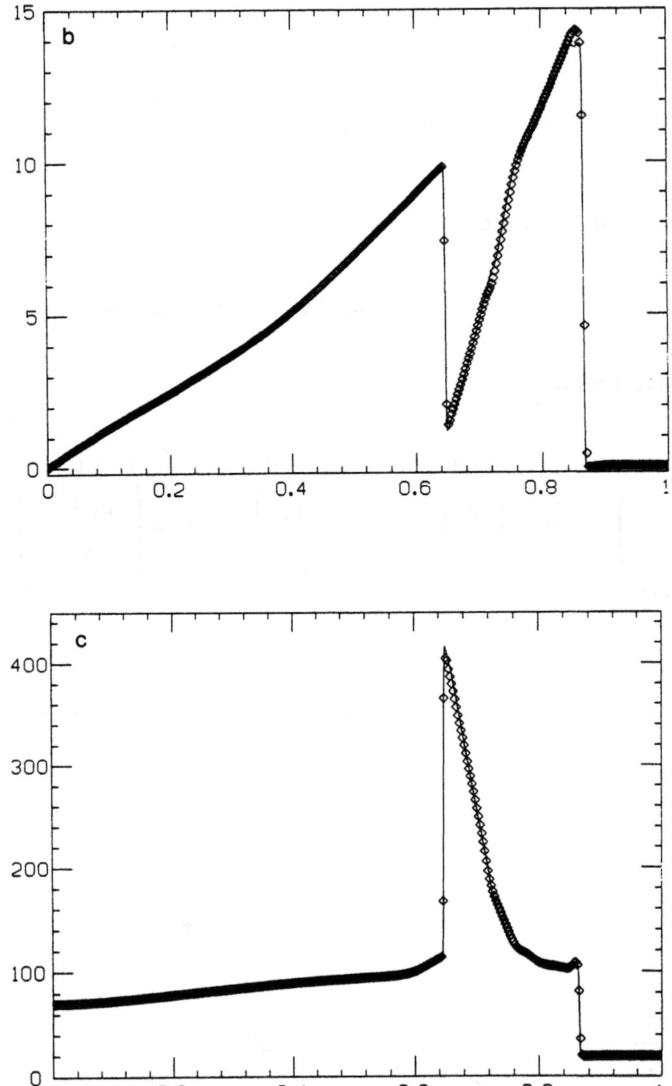

Fig. 12. (*Continued*)

and the left eigenvectors are

$$l_1=\frac{1}{2}\left(b_2+\frac{q_x}{2},\ -\frac{1}{c}-b_1 q_x,\ -b_1 q_y,\ b_1\right),$$

$$l_2=(-q_y, 0, 1, 0), \qquad l_3=(1-b_2, b_1 q_x, b_1 q_y, -b_1), \tag{5.13c}$$

$$l_4=\frac{1}{2}\left(b_2-\frac{q_x}{c}, \frac{1}{c}-b_1 q_x,\ -b_1 q_y,\ b_1\right),$$

where

$$b_1=(\gamma-1)/c^2 \tag{5.13d}$$

$$b_2=\tfrac{1}{2}q^2 b_1. \tag{5.13e}$$

We can get the results for $\partial \mathbf{g}/\partial \mathbf{u}$ by symmetry.

The test problem we choose is a moving shock interacting with compressible turbulence [17, 18]. At $t=0$, a Mach 8 shock at $x=-1.0$ is moving to the right into a state with $P_R=1, \rho_R=1$, and $q_x=-(c_R/P_R)\sin\theta_R\cos(xk_R\cos\theta_R+yk_R\sin\theta_R)$, $q_y=(c_R/P_R)\cos\theta_R\cos(xk_R\cos\theta_R+yk_R\sin\theta_R)$, where $k_R=2\pi$, $\theta_R=\pi/6$. We display the results at $t=0.20$ in Fig. 156. Notice that in [17, 18] similar results were obtained using a shock-fitting rather than a shock capturing method.

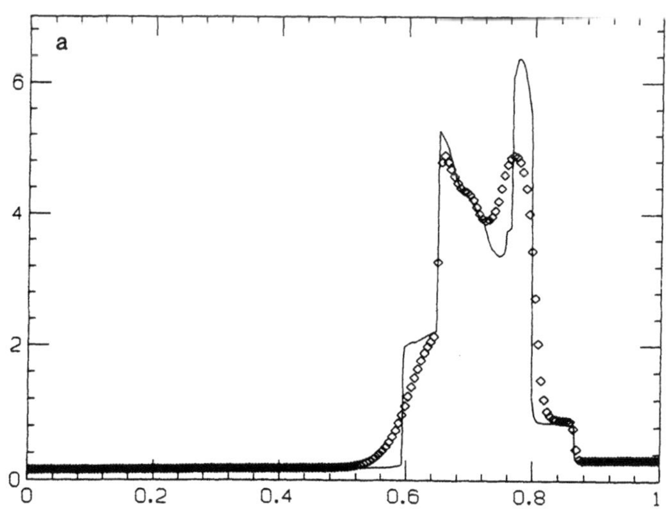

FIG. 13. (5.8a)–(5.10), 200 points, $t=0.038$, density (solid lines are the numerical solution of ENO-RF-S-3, 800 points): (a) ENO-RF-3 (diamonds); (b) ENO-RF-S-3 (diamonds); (c) ENO-RF-A-3 (diamonds).

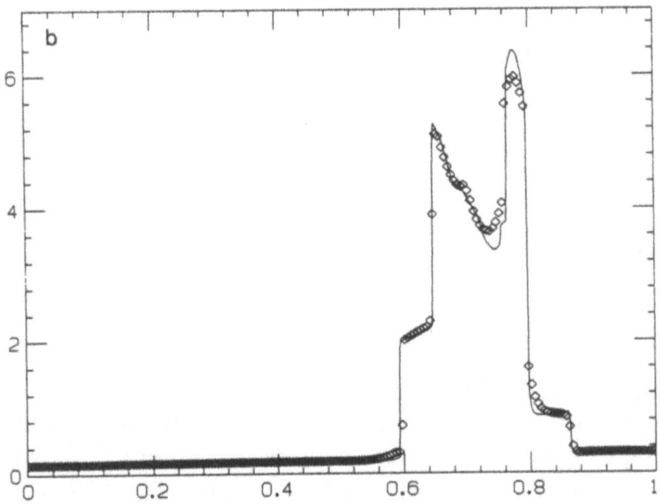

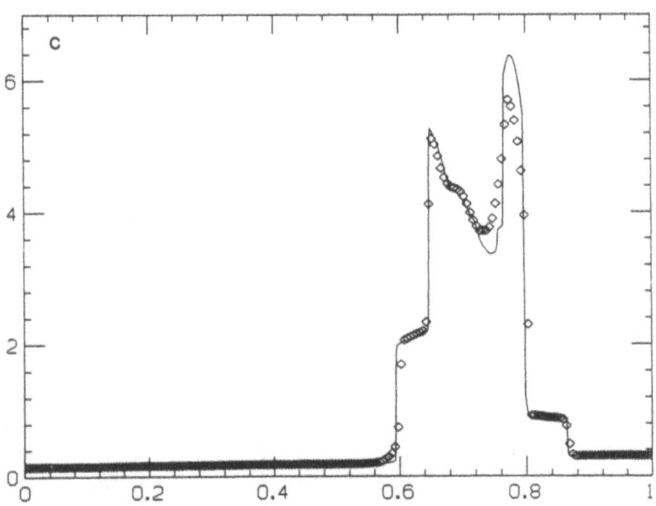

Fig. 13. *(Continued)*

This problem is difficult for shock capturing methods because both properties: high-order accuracy (to resolve the fine structure) and an oscillation-free shock transition, are needed. A small-spurious oscillation will be enough to kill any resolution of the specified small fluctuations. This example shows that ENO schemes have excellent possibilities for shock-turbulence computations.

VI. Concluding Remarks

ENO schemes based on fluxes and TVD Runge–Kutta type time discretizations seem to work very well in our numerical tests which include 1D and 2D scalar and systems problems. ENO-RF (Algorithm 2.3) seems preferred, since this method is half as expensive as ENO-LLF or ENO-LF, gives better resolution, and seems always to converge to the correct entropy solution. To sharpen contact discontinuities, Algorithm 3.1 (for 1D) or Algorithm 3.2 (for 1D or 2D) can be used. High-order ENO schemes show their special advantages for problems which have both discontinuities and detailed fine structure in smooth regions, e.g., Example 8 and Example 9 in Section V. Among the practical things needing further investigation are the resolution of rarefaction corners (discontinuities in derivatives). Finally, theoretical justification for the evident stability of these methods would be quite welcome.

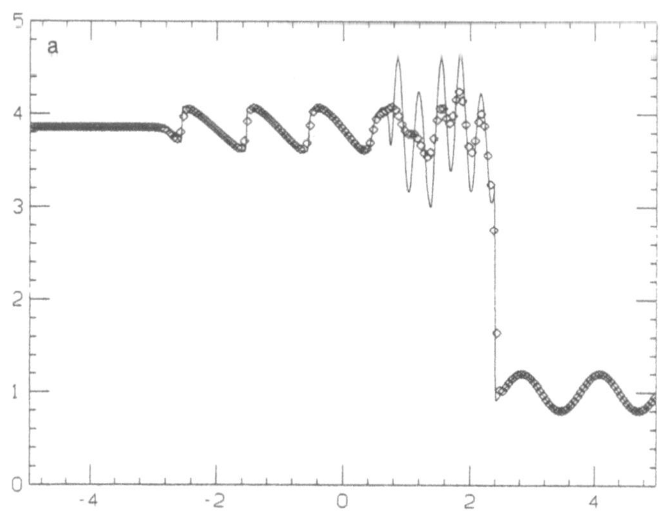

FIG. 14. (5.8a)–(5.11), $t = 1.8$, density (solid lines are the numerical solution of ENO-RF-3, 1600 points): (a) ENO-RF-3, 200 points (diamonds); (b) ENO-RF-3, 400 points (diamonds); (c) Second-order MUSCL-type TVD scheme, 800 points (diamonds).

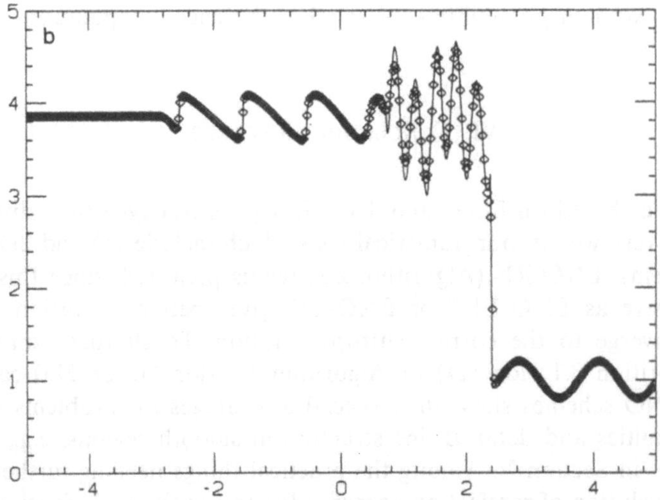

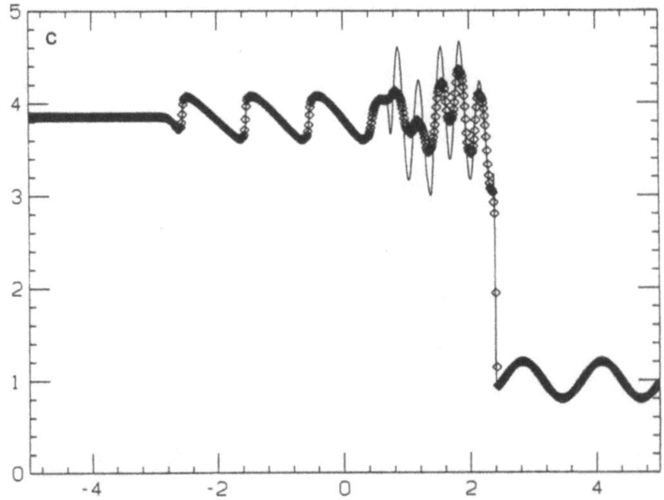

Fig. 14. (*Continued*)

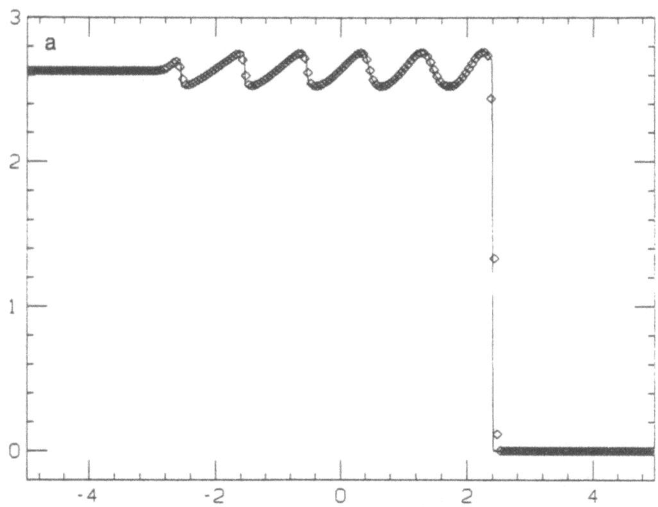

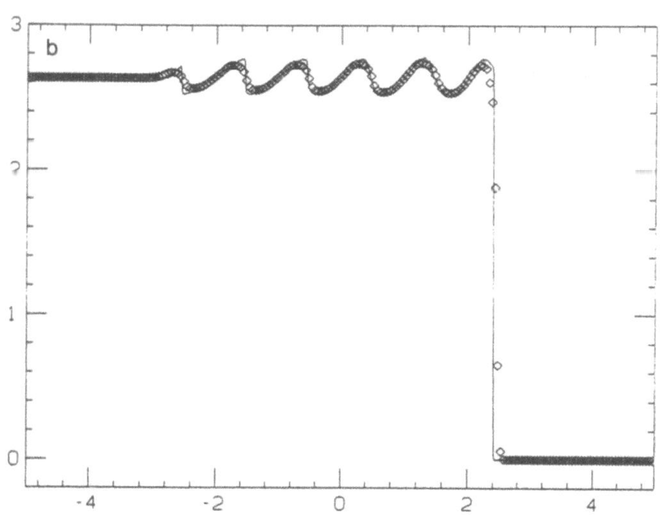

FIG. 15. (5.8a)–(5.11), 200 points, $t = 1.8$ (solid lines are the numerical solution of ENO-RF-3, 1600 points): (a) ENO-RF-3, velocity (diamonds); (b) Second-order MUSCL-type TVD scheme, velocity (diamonds); (c) ENO-RF-3, pressure (diamonds); (d) Second-order MUSCL-type TVD schemes, pressure (diamonds).

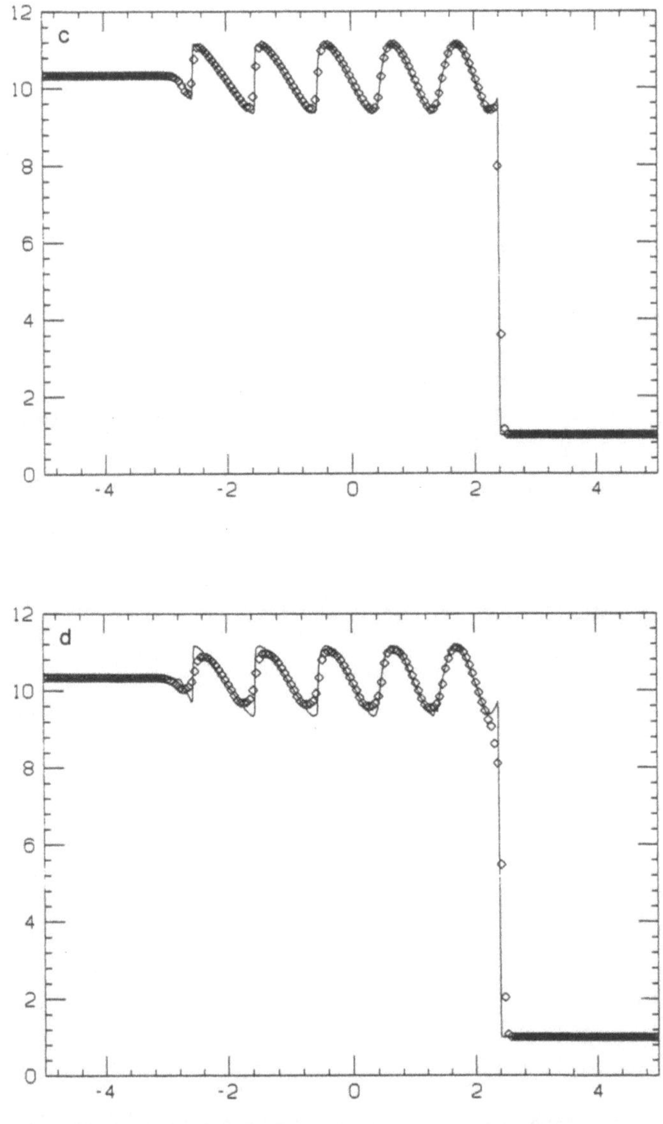

Fig. 15. *(Continued)*

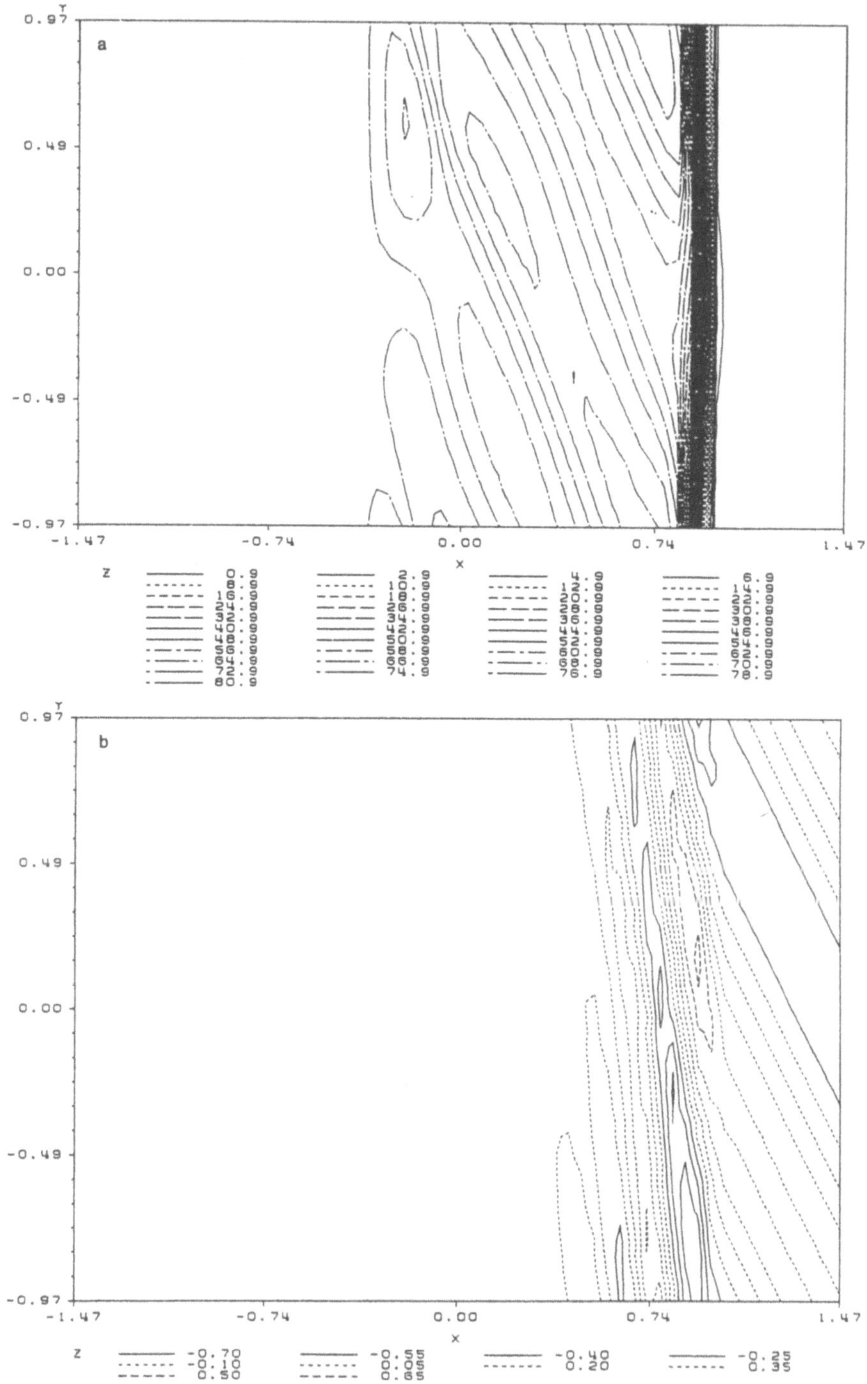

FIG. 16. ENO-RF-3, Example 9, $t = 0.2$ (level curves of numerical solutions): (a) 60×40 points, pressure; (b) 60×40 points, vorticity; (c) 60×40 points, entropy; (d) 120×80 points, pressure; (e) 120×80 points, vorticity; (f) 120×80 points, entropy.

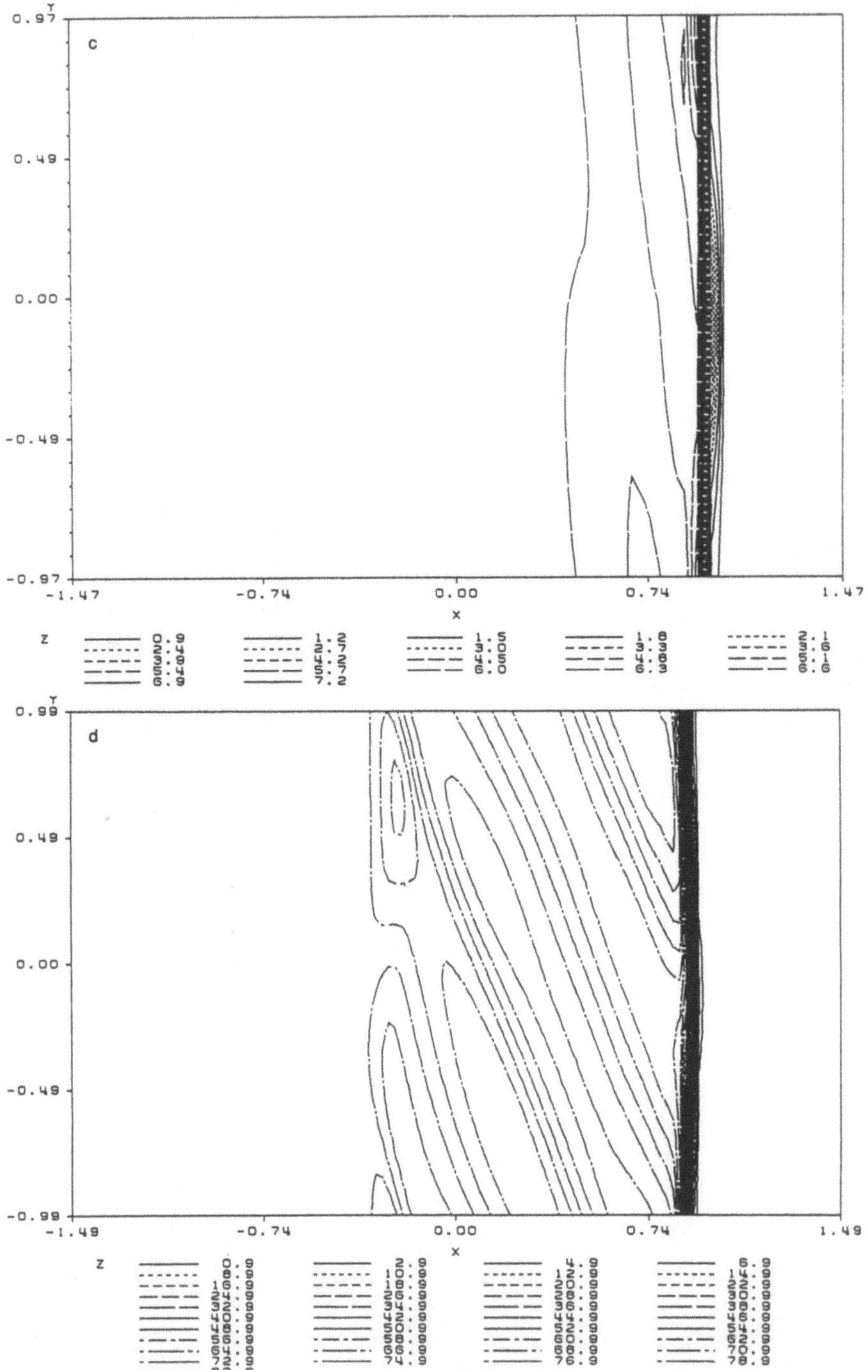

Fig. 16. (*Continued*)

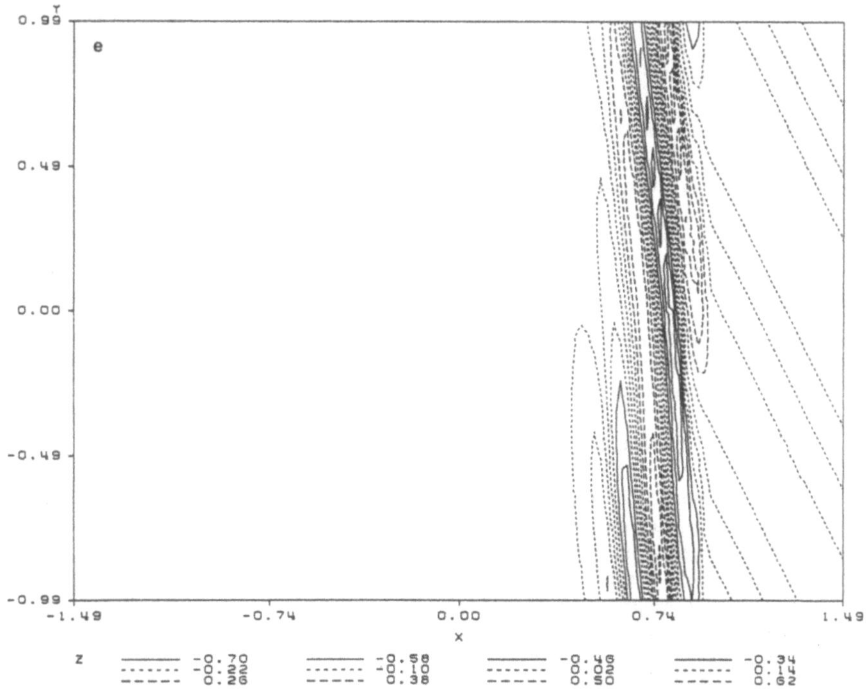

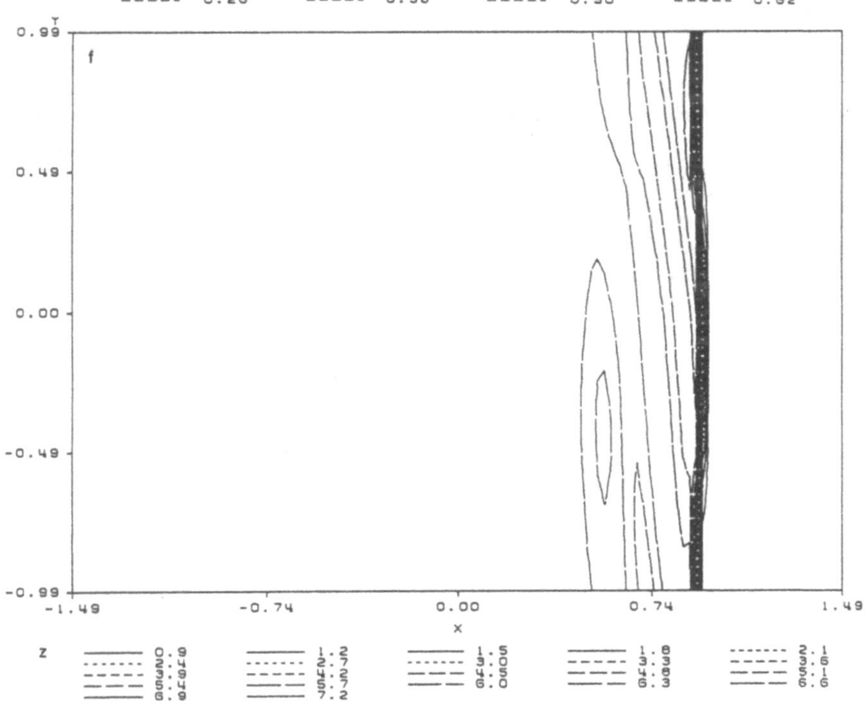

Fig. 16. (*Continued*)

77

Acknowledgments

We thank David Gottlieb, Ami Harten, Lawrence Sirovich, Bjorn Sjogreen, Huanan Yang, and Thomas Zang for helpful discussions and/or for suggesting numerical examples. We also thank the referees for some constructive comments on the first draft of this paper.

References

1. M. Crandall and A. Majda, *Math. Comput.* **34**, 1 (1980).
2. A. Harten, "Preliminary Results on the Extension of ENO Schemes to Two-Dimensional Problems," in *Proceedings of the International Conference on Hyperbolic Problems*, Saint-Etienne, January 1986.
3. A. Harten and S. Osher, *SIAM J. Numer. Anal.* **24**, 279 (1987).
4. A. Harten, B. Engquist, S. Osher, and S. Chakravarthy, *J. Comput. Phys* **71**, 231 (1987).
5. A. Harten, S. Osher, B. Engquist, and S. Chakravarthy, *J. Appl. Numer. Math.* **2**, 347 (1986).
6. A. Harten, ICASE Report 87-56, August 1987 (unpublished).
7. A. Harten, in preparation.
8. J. McKenzie and K. Westphal, *Phys. Fluids* **11**, 2350 (1968).
9. S. Osher, *SIAM J. Numer. Anal.* **22**, 947 (1985).
10. S. Osher and S. Chakravarthy, *SIAM J. Numer. Anal.* **21**, 955 (1984).
11. P. Roe, *J. Comput. Phys.* **43**, 357 (1981).
12. C.-W. Shu and S. Osher, ICASE Report 87-33; *J. Comput. Phys.* **77**, 439 (1988).
13. D. Wagner, *SIAM J. Math. Anal.* **14**, 534 (1983).
14. P. Woodward and P. Colella, *J. Comput. Phys.* **54**, 115 (1984).
15. H. Yang, Ph. D. thesis, UCLA Math. Dept., 1988; *J. Comput. Phys.* (1990) to appear.
16. S. Zalesak, "A Preliminary Comparison of Modern Shock-Capturing Schemes: Linear Advection," in *Advances in Computer Methods for Partial Differential Equations, VI*, edited by R. Vichnevetsky and R. Stepleman (IMACS, New Brunswick, NJ, 1987).
17. T. Zang, M. Hussaini, and D. Bushnell, *AIAA J.* **22**, 13 (1984).
18. T. Zang, D. Kopriva, and M. Hussaini, "Pseudospectral Calculation of Shock Turbulence Interaction," in *Proceedings of Numerical Methods Conference*, edited by C. Taylor (Pineridge Press, Swansea, U.K., 1984), pp. 210–222.

Essentially Nonoscillatory Spectral Fourier Methods for Shock Wave Calculations*

By Wei Cai, David Gottlieb, and Chi-Wang Shu

Dedicated to Professor Eugene Isaacson on the occasion of his 70th birthday

Abstract. In this paper, we present an essentially nonoscillatory spectral Fourier method for the solution of hyperbolic partial differential equations. The method is based on adding a nonsmooth function to the trigonometric polynomials which are the usual basis functions for the Fourier method. The high accuracy away from the shock is enhanced by using filters. Numerical results confirm that essentially no oscillations develop in the solution. Also, the accuracy of the spectral solution of the inviscid Burgers equation is shown to be higher than a fixed order.

1. Introduction. In this paper we discuss shock-capturing techniques using spectral methods. In particular, we would like to present an essentially nonoscillatory version of the spectral Fourier method when applied to a nonlinear hyperbolic equation. The main difficulty in applying spectral methods to discontinuous problems is of course the Gibbs phenomenon. In fact, this problem exists even on the approximation level. It is well known that if a discontinuous function $f(x)$ is approximated by its finite Fourier series $P_N f$,

$$P_N f = \sum_{k=-N}^{N} \hat{f}_k e^{ikx}, \tag{1.1a}$$

$$\hat{f}_k = \frac{1}{2\pi} \int_0^{2\pi} f(x) e^{-ikx}\, dx, \tag{1.1b}$$

then the order of convergence of $P_N f$ to f is only $O(1/N)$ for each fixed point. Moreover, $P_N f$ has oscillations of order 1 in a neighborhood of $O(1/N)$ of the discontinuity.

In the applications, we usually have piecewise C^∞ functions, and in this paper we will consider only those functions. It is known that it is possible to improve the accuracy of the approximation away from the shocks. There are currently two methods (see [7], [9]) that are being used. The first [9] amounts to modifying the Fourier coefficients by multiplying them by a decreasing function $\tau(k)$. Some of the

Received May 31, 1988.
1980 *Mathematics Subject Classification* (1985 *Revision*). Primary 65M99.
*This research was supported by the National Aeronautics and Space Administration under NASA Contract No. NAS1-18107 while the second author was in residence at the Institute for Computer Applications in Science and Engineering (ICASE), NASA Langley Research Center, Hampton, VA 23665. Additional support was provided by the Air Force Office of Scientific Research under Grant No. 85-0303 and DARPA URI contract N00014-86-K0754.

commonly used filters are

$$\tau_k = e^{-\alpha((k-k_0)/N)^{2m}}, \qquad |k| \geq k_0, \tag{1.2}$$
$$\tau_k = 1, \qquad |k| < k_0.$$

The second method [1] is based on convolving the approximation with an appropriate C^∞ function $\psi(x,y)$ such that

$$P_N f * \psi(x,y) \sim f(y). \tag{1.3}$$

While both (1.2) and (1.3) are effective away from the discontinuity, they do not eliminate the Gibbs phenomenon in the neighborhood of the shock. This is very important for the stability of the spectral method when applied to partial differential equations. In fact, in Section 2 we show that the total variation of $P_N f$ grows like $\log N$. It is easily shown that this is the case also for the filters in (1.2).

In Section 2, we show that by adding a sawtooth function to the basis functions e^{ikx} one can control the Gibbs phenomenon. This, in conjunction with the filters (1.2)–(1.3), yields a higher-order essentially nonoscillatory approximation to a piecewise C^∞ function. In Theorem 2, we prove that the total variation of the new approximation converges to that of the approximated function. We also prove that the convergence for the new approximation in the L^1 norm is one order higher than that of the usual spectral approximation.

Many modern nonlinear schemes for the solution of the conservation equation

$$u_t + f(u)_x = 0 \tag{1.4}$$

are based on two distinct steps, namely reconstruction and time marching. We use the cell averaging formulation to rewrite (1.4) as

$$\frac{\partial \overline{u}_j}{\partial t} + \frac{1}{\Delta x_j}(f(u_{j+1/2}) - f(u_{j-1/2})) = 0, \tag{1.5}$$

where

$$\Delta x_j = x_{j+1/2} - x_{j-1/2}, \qquad \overline{u}_j = \frac{1}{\Delta x_j}\int_{x_{j-1/2}}^{x_{j+1/2}} u\,dx,$$
$$f(u_{j+1/2}) = f(u(x_{j+1/2})).$$

The first step, then, is to reconstruct the function $u(x)$ from $\overline{u}(x)$. It is here that we use the essentially nonoscillatory technique developed in Section 2. For the second step, the time marching, we use the third-order Runge-Kutta scheme developed in [13]. We try to avoid any modification technique, such as the application of limiters, in order to avoid deterioration of the overall accuracy.

We demonstrate in the last section that the procedure applied to several model problems yields indeed essentially nonoscillatory results with an order of accuracy which is higher than algebraic away from the discontinuity.

2. Essentially Nonoscillatory Approximation. In this section, we suggest a method to reconstruct an essentially nonoscillatory approximation to a piecewise C^∞ periodic function from its first N Fourier coefficients. The approximation is essentially nonoscillatory in the sense that the total variation of the approximation converges to the total variation of the approximated function. Moreover, the approximation converges in the maximum norm outside a small interval around the point of discontinuity. Applying the filters (1.2)–(1.3) will increase the order of convergence away from the discontinuity, thus providing an essentially nonoscillatory spectral approximation.

For simplicity, assume that $u(x)$, $0 \le x \le 2\pi$, is a periodic piecewise C^∞ function with only one point of discontinuity at x_s, and denote by $[u]$ the value of the jump of $u(x)$ at x_s, namely

$$[u] = \frac{u(x_s^+) - u(x_s^-)}{2\pi}. \tag{2.1}$$

We assume also that the first $2N+1$ Fourier coefficients $\hat{u}_l$ of $u(x)$ are known:

$$\hat{u}_l = \frac{1}{2\pi}\int_0^{2\pi} u(x)e^{-ilx}\,dx, \qquad -N \le l \le N. \tag{2.2}$$

The objective is to construct an essentially nonoscillatory spectrally accurate approximation to $u(x)$ from the $\hat{u}_l$'s. We start by noting that the Fourier coefficients $\hat{u}_l$'s contain information about the shock position x_s and the magnitude $[u]$ of the shock. In fact we can state

LEMMA 1. *Let $u(x)$ be a periodic piecewise C^∞ function with one point of discontinuity x_s; then for $|l| \ge 1$ and for any $n > 0$,*

$$\hat{u}_l = e^{-ilx_s}\sum_{k=0}^{n-1}\frac{[u^{(k)}]}{(il)^{k+1}} + \frac{1}{2\pi}\int_0^{2\pi}\frac{[u^{(n)}]}{(il)^n}e^{-ilx}\,dx. \tag{2.3}$$

Proof. Since

$$\hat{u}_l = \frac{1}{2\pi}\int_0^{2\pi} u(x)e^{-ilx}\,dx = \frac{1}{2\pi}\int_0^{x_s} u(x)e^{-ilx}\,dx + \frac{1}{2\pi}\int_{x_s}^{2\pi} u(x)e^{-ilx}\,dx,$$

we can integrate by parts to get

$$\hat{u}_l = e^{-ilx_s}\frac{u(x_s^+) - u(x_s^-)}{2\pi il} + \frac{1}{2\pi}\int_0^{2\pi}\frac{u'(x)e^{-ilx}}{il}\,dx; \tag{2.4}$$

the rest is obtained by induction. This completes the proof. □

As an example, consider the sawtooth function $F(x, x_s, A)$ defined by

$$F(x, x_s, A) = A\begin{cases} -x, & x \le x_s, \\ 2\pi - x, & x > x_s. \end{cases} \tag{2.5}$$

Note that the jump of the function, $[F]$, is A and all the derivatives are continuous: $[d^kF/dx^k] = 0$ for $k \ge 1$. That means that the expansion (2.3) can be terminated after the first term, yielding the following results for $\hat{f}_k$, the Fourier coefficients of $F(x, x_s, A)$:

$$\begin{aligned} \hat{f}_k(x_s, A) &= A\frac{e^{-ikx_s}}{ik}, \qquad |k| \ge 1, \\ \hat{f}_0(x_s, A) &= A(\pi - x_s). \end{aligned} \tag{2.6}$$

This example suggests that we can rewrite (2.3) as

$$\hat{u}_l = \hat{f}_l(x_s, [u]) + e^{-ilx_s}\sum_{k=1}^{n-1}\frac{[u^{(k)}]}{(il)^{k+1}} + \frac{1}{2\pi}\int_0^{2\pi}\frac{u^{(n)}(x)e^{-ilx}}{(il)^n}\,dx, \qquad |l| \ge 1. \tag{2.7}$$

The order 1 oscillations in approximating $u(x)$ by its finite Fourier sum

$$P_N u = \sum_{l=-N}^{N} \hat{u}_l e^{ilx} \tag{2.8}$$

are caused by the slow convergence of

$$F_N(x, x_s, [u]) = \sum_{l=-N}^{N} \hat{f}_l(x_s, [u]) e^{ilx} \tag{2.9}$$

to the sawtooth function $F(x, x_s, [u])$. Therefore, those oscillations can be eliminated by adding a sawtooth function to the basis of the space to which $u(x)$ is projected. To be specific, we seek an expansion of the form

$$v_N(x) = \sum_{|l|\le N} a_l e^{ilx} + \sum_{|l|>N} \frac{A}{il} e^{-ily} e^{ilx} \tag{2.10}$$

to approximate $u(x)$. The $2N+3$ unknowns a_l ($|l| \le N$), A and y are determined by the orthogonality condition

$$\int_0^{2\pi} (u - v_N) e^{-ikx}\, dx = 0, \qquad |k| \le N+2. \tag{2.11}$$

The system of equations (2.11) leads to the following conditions:

$$a_l = \hat{u}_l, \qquad |l| \le N \tag{2.12}$$

(where $\hat{u}_l$ are the usual Fourier coefficients of $u(x)$, see (2.2)) and

$$\frac{A}{i(N+1)} e^{-i(N+1)y} = \hat{u}_{N+1}, \tag{2.13a}$$

$$\frac{A}{i(N+2)} e^{-i(N+2)y} = \hat{u}_{N+2}. \tag{2.13b}$$

Solving (2.13) for A and y one gets

$$e^{iy} = \frac{(N+1)\hat{u}_{N+1}}{(N+2)\hat{u}_{N+2}}, \tag{2.14a}$$

$$|A| = (N+1)|\hat{u}_{N+1}|. \tag{2.14b}$$

The sign of A is determined by (2.13).

Note that in the expansion presented in (2.10) the second sum starts at $|l| = N+1$. This is due to the fact that we make the additional basis function $F(x, y, A)$ orthogonal to e^{ikx}, thus we use $F(x, y, A) - F_N(x, y, A)$ in the expansion (2.10). The procedure described in (2.14) is second-order accurate in the location and jump of the shock. In fact, we can state

THEOREM 1. *Let $u(x)$ be a piecewise C^∞ function with one discontinuity at x_s. Let y and A be defined in* (2.14); *then*

$$|y - x_s| = O(1/N^2), \qquad |A - [u]| = O(1/N^2). \tag{2.15}$$

Proof. From (2.3) we get

$$e^{iy} = \frac{(N+1)\hat{u}_{N+1}}{(N+2)\hat{u}_{N+2}} = \frac{e^{-i(N+1)x_s}\left[[u] + \frac{[u']}{i(N+1)} + O\left(\frac{1}{(N+1)^2}\right)\right]}{e^{-i(N+2)x_s}\left[[u] + \frac{[u']}{i(N+2)} + O\left(\frac{1}{(N+2)^2}\right)\right]}$$
$$= e^{ix_s}\left[1 + O\left(\frac{1}{N^2}\right)\right].$$

By the same token,

$$|A| = (N+1)|\hat{u}_{N+1}| = \left[\left\{[u] - \frac{[u'']}{(N+1)^2}\right\}^2 + \frac{[u']^2}{(N+1)^2}\right]^{1/2}$$
$$= |[u]|\left[1 + O\left(\frac{1}{N^2}\right)\right]. \quad \square$$

It should be noted that a better approximation to the shock location x_s and its magnitude $[u]$ can be obtained if we add to the basis functions a function of the form

$$\sum_{|l|>N}\left[\frac{A}{il} + \frac{B}{(il)^2}\right]e^{-ily}e^{ilx} \tag{2.16}$$

and extend (2.11) to $|k| \leq N+3$. In practice, however, (2.10) is enough to get an essentially nonoscillatory scheme.

In order to demonstrate that the procedure described in (2.10), (2.12), and (2.14) is indeed essentially nonoscillatory, we recall the definition of the total variation of a function.

Definition. The total variation of u over $[0, 2\pi]$—$\mathrm{TV}[u]$—is defined as

$$\mathrm{TV}[u] = \sup \sum_{i=1}^{n} |u(x_i) - u(x_{i-1})| \tag{2.17}$$

where $0 \leq x_0 < x_1 < \cdots < x_n \leq 2\pi$ is a partition of $[0, 2\pi]$. The supremum is taken over all partitions.

It is clear that if $u'(x) \in L^1$ then

$$\mathrm{TV}[u] = \int_0^{2\pi} |u'(\xi)|\, d\xi. \tag{2.18}$$

If we approximate the function $u(x)$ by its finite Fourier series $P_N u$ defined in (2.8), then it is well known that the total variation of $P_N u$ need not approximate that of u. In fact we can state

LEMMA 2. *Let the sawtooth function $F(x,0,1)$ and its Nth Fourier approximation $F_N(x,0,1)$ be defined by* (2.5) *and* (2.9); *then*

$$\mathrm{TV}[F] = 4\pi, \tag{2.19}$$
$$\mathrm{TV}[F_N] = O(\log N). \tag{2.20}$$

Proof. Equation (2.19) follows directly from the definition of total variation. As for (2.20), we note that

$$\begin{aligned} \mathrm{TV}[F_N(x,0,1)] &= \int_0^{2\pi} |F_N'(x,0,1)|\,dx = \int_0^{2\pi} \left| \sum_{\substack{l=-N \\ l\neq 0}}^{N} e^{ilx} \right| dx \\ &= \int_0^{2\pi} \left| \frac{\sin(N+\frac{1}{2})x}{\sin\frac{1}{2}x} - 1 \right| dx \\ &= \int_0^{2\pi} \left| \frac{\sin(N+\frac{1}{2})x}{\sin\frac{1}{2}x} \right| dx + O(1). \end{aligned} \tag{2.21}$$

The first term on the right-hand side of (2.21) is the Lebesgue constant. It is known [14, p. 67] that it grows like $\log N$. Hence (2.20). □

We can therefore conclude that $\mathrm{TV}(P_N u)$ does not converge to $\mathrm{TV}[u]$. This reflects the existence of large oscillations in the neighborhood of the discontinuities.

The situation is different for v_N defined in (2.10). In fact we can state

THEOREM 2. *Let $u(x)$ be a piecewise C^∞ periodic function with one point of discontinuity x_s, and a jump of $[u]$. Let A and y be such that*

$$|y - x_s| = \Delta_1, \qquad |A - [u]| = \Delta_2. \tag{2.22}$$

Let v_N be defined in (2.10); *then*

$$\mathrm{TV}[v_N] \le \mathrm{TV}[u] + L_0\frac{\log N}{N} + L_1\Delta_1 N\log N + L_2\Delta_2\log N, \tag{2.23}$$

$$\|v_N - u\|_{L^1} \le C_0\frac{\log N}{N^2} + C_1\Delta_1\log N + C_2\Delta_2. \tag{2.24}$$

We present the proof in a series of lemmas in order to clarify the role of each one of the terms on the right-hand sides of (2.23) and (2.24).

LEMMA 3. *Let $F_N(x,\alpha,1)$ and $F_N(x,\beta,1)$ be defined by* (2.6)–(2.9) *and $\Delta = \alpha-\beta > 0$. Then, if $\Delta \le 1/N$,*

$$\mathrm{TV}[F_N(x,\alpha,1) - F_N(x,\beta,1)] = O(\Delta N\log N), \tag{2.25}$$

$$\|F_N(x,\alpha,1) - F_N(x,\beta,1)\|_{L^1} = O(\Delta\log N). \tag{2.26}$$

Proof. Since $F_N(x,\alpha,1), F_N(x,\beta,1)$ are trigonometric polynomials, they are C^∞ functions. Therefore

$$\begin{aligned} \mathrm{TV}[F_N(x,\alpha,1) - F_N(x,\beta,1)] &= \int_0^{2\pi} |F_N'(x,\alpha,1) - F_N'(x,\beta,1)|\,dx \\ &= \int_0^{2\pi} \left| \sum_{|l|\le N} [e^{il(x-\alpha)} - e^{il(x-\beta)}] \right| dx \\ &= 4\int_0^{2\pi} \left| \sum_{l=1}^{N} \sin l\left(x - \frac{\alpha+\beta}{2}\right)\sin l\frac{\alpha-\beta}{2} \right| dx. \end{aligned} \tag{2.27}$$

Upon defining $\sigma_l = \sin l\frac{\alpha-\beta}{2}$, we can rewrite (2.27) as

$$\mathrm{TV}[F_N(x,\alpha,1) - F_N(x,\beta,1)] = 4\int_0^{2\pi} \left| \sum_{l=1}^{N} \sigma_l \sin l\xi \right| d\xi; \tag{2.28}$$

we note that the σ_l are positive and monotone in l, $\sigma_{l-1} - \sigma_l < 0$. Define now

$$B_l(\xi) = \sum_{k=0}^{l} \sin k\xi \tag{2.29}$$

to get

$$\begin{aligned}\sum_{l=1}^{N} \sigma_l \sin l\xi &= \sum_{l=1}^{N} \sigma_l (B_l(\xi) - B_{l-1}(\xi)) \\ &= \sum_{l=1}^{N} (\sigma_{l-1} - \sigma_l) B_{l-1} + \sigma_N B_N.\end{aligned} \tag{2.30}$$

Therefore,

$$\begin{aligned}\int_0^{2\pi} \left| \sum_{l=1}^{N} \sigma_l \sin l\xi \right| d\xi &\le \sigma_N \int_0^{2\pi} |B_N(\xi)|\, d\xi \\ &\quad + \sum_{l=1}^{N} (\sigma_l - \sigma_{l-1}) \int_0^{2\pi} |B_{l-1}(\xi)|\, d\xi.\end{aligned} \tag{2.31}$$

Denote now

$$\mu = \max_{1 \le l \le N} \int_0^{2\pi} |B_l(\xi)|\, d\xi; \tag{2.32}$$

from (2.28) and (2.31) we obtain

$$\mathrm{TV}[F_N(x, \alpha, 1) - F_N(x, \beta, 1)] \le 8\mu\sigma_N. \tag{2.33}$$

In order to estimate μ, we first note that

$$B_l(\xi) = \sin \frac{(l+1)\xi}{2} \cdot \frac{\sin l\xi/2}{\sin \xi/2}. \tag{2.34}$$

Therefore,

$$\int_0^{2\pi} |B_l(\xi)|\, d\xi \le \int_0^{2\pi} \frac{|\sin l\xi/2|}{\sin \xi/2}\, d\xi = \mu_l. \tag{2.35}$$

But μ_l is exactly the Lebesgue constant; therefore,

$$\mu_l = O(\log l). \tag{2.36}$$

Since $\sigma_N \le N\Delta$, we get

$$\mathrm{TV}[F_N(x, \alpha, 1) - F_N(x, \beta, 1)] = O(\Delta N \log N).$$

To obtain (2.26), we follow the same arguments as above. Similar to (2.27) and (2.28) we have

$$\int_0^{2\pi} |F_N(x, \alpha, 1) - F_N(x, \beta, 1)|\, dx \le 2\pi\Delta + 4 \int_0^{2\pi} \left| \sum_{l=1}^{N} \hat{\sigma}_l \cos l\xi \right| d\xi, \tag{2.37}$$

where $\hat{\sigma}_l = \sigma_l / l$, $1 \le l \le N$. The $\hat{\sigma}_l$'s are positive and monotone in l, $\sigma_l - \sigma_{l-1} < 0$. If we define

$$\begin{aligned}\hat{B}_l(\xi) &= \sum_{k=1}^{l} \cos k\xi = \cos \frac{(l+1)\xi}{2} \cdot \frac{\sin l\xi/2}{\sin \xi/2}, \\ \hat{B}_0(\xi) &= 0,\end{aligned} \tag{2.38}$$

then similar to (2.33) we have

$$\int_0^{2\pi} |F_N(x,\alpha,1) - F_N(x,\beta,1)|\,dx \le 2\pi\Delta + 8\mu\hat{\sigma}_1, \tag{2.39}$$

where μ is defined in (2.32) with B_l replaced by $\hat{B}_l$ of (2.38). Notice that (2.35) also holds for $\hat{B}_l(x)$ and $|\hat{\sigma}_1| \le \Delta$. (2.26) now follows from (2.35), (2.37) and (2.39). □

LEMMA 4. *Let $F_N(x,\alpha,A)$ and $F_N(x,\beta,B)$ be defined in (2.6)–(2.9). Denote*

$$|\alpha - \beta| = \Delta_1, \qquad |A - B| = \Delta_2. \tag{2.40}$$

Then

$$\mathrm{TV}[F_N(x,\alpha,A) - F_N(x,\beta,B)] \le K_1\Delta_1 N\log N + K_2\Delta_2\log N, \tag{2.41}$$

$$\|F_N(x,\alpha,A) - F_N(x,\beta,B)\|_{L^1} \le C_1\Delta_1\log N + C_2\Delta_2 \tag{2.42}$$

for K_1, K_2, C_1, C_2 independent of N.

Proof. We have

$$\begin{aligned}\mathrm{TV}[F_N(x,\alpha,A) - F_N(x,\beta,B)] \le{}& \mathrm{TV}[F_N(x,\alpha,A) - F_N(x,\beta,A)]\\ &+ \mathrm{TV}[F_N(x,\beta,A) - F_N(x,\beta,B)].\end{aligned} \tag{2.43}$$

The first term in the right-hand side of (2.43) is bounded by (2.25); for the second term,

$$\begin{aligned}\mathrm{TV}[F_N(x,\beta,A) - F_N(x,\beta,B)] &= \mathrm{TV}\left[(A-B)\sum_{\substack{l=-N\\ l\ne 0}}^{N} \frac{e^{il(x-\beta)}}{il}\right]\\ &\le |A-B|\int_0^{2\pi}\left|\sum_{\substack{l=-N\\ l\ne 0}}^{N} e^{il(x-\beta)}\right| dx\\ &\le K_2\Delta_2\log N.\end{aligned} \tag{2.44}$$

Similarly, we have

$$\begin{aligned}&\|F_N(x,\alpha,A) - F_N(x,\beta,B)\|_{L^1}\\ &\quad\le \|F_N(x,\alpha,A) - F_N(x,\beta,A)\|_{L^1} + \|F_N(x,\beta,A) - F_N(x,\beta,B)\|_{L^1}\\ &\quad= \|F_N(x,\alpha,A) - F_N(x,\beta,A)\|_{L^1} + |A-B|\|F_N(x,\beta,1)\|_{L^1}.\end{aligned} \tag{2.45}$$

The first term on the right side of (2.45) is bounded by (2.26). For the second term,

$$|A-B|\|F_N(x,\beta,1)\|_{L^1} = \Delta_2\|F_N(x,\beta,1)\|_{L^1} \le C_2\Delta_2. \tag{2.46}$$

The assertion (2.42) now follows from (2.45), (2.46). □

LEMMA 5. *Let $S(x,\alpha,A)$ and $S_N(x,\alpha,A)$ be defined by*

$$S(x,\alpha,A) = A\sum_{\substack{l=-\infty\\ l\ne 0}}^{\infty} \frac{e^{il(x-\alpha)}}{(il)^2}, \qquad S_N(x,\alpha,A) = A\sum_{\substack{l=-N\\ l\ne 0}}^{N} \frac{e^{il(x-\alpha)}}{(il)^2}. \tag{2.47}$$

Then

$$\text{TV}[S(x,\alpha,A) - S_N(x,\alpha,A)] \le K_3 \frac{\log N}{N}, \tag{2.48}$$

$$\|S(x,\alpha,A) - S_N(x,\alpha,A)\|_{L^1} \le K_4 \frac{\log N}{N^2} \tag{2.49}$$

for K_3, K_4 *independent of* N.

Proof. It is clear that

$$\text{TV}(S(x,\alpha,A) - S_N(x,\alpha,A)) = \int_0^{2\pi} |F(x,\alpha,A) - F_N(x,\alpha,A)|\, dx.$$

The estimates (2.48)–(2.49) then follow from [14, p. 185]. □

We are now ready to prove Theorem 2.

Proof of Theorem 2. First we prove (2.23). In view of (2.3) we can write

$$u = F(x, x_s, [u]) + S(x, x_s, [u']) + g(x) \tag{2.50}$$

and therefore

$$P_N u = F_N(x, x_s, [u]) + S_N(x, x_s, [u']) + P_N g(x). \tag{2.51}$$

We can also rewrite (2.10) as

$$v_N(x) = P_N u + F(x,y,A) - F_N(x,y,A) \tag{2.52}$$

hence

$$v_N(x) = F_N(x, x_s, [u]) + S_N(x, x_s, [u']) + P_N g(x) + F(x,y,A) - F_N(x,y,A)$$

or

$$\begin{aligned} v_N(x) = {} & [F_N(x, x_s, [u]) - F_N(x,y,A)] \\ & + [F(x,y,[u]) + S(x,y,[u']) + g(x - y + x_s)] \\ & + [S_N(x, x_s, [u']) - S(x,y,[u'])] + [P_N g(x) - g(x - y + x_s)] \\ & + [F(x,y,A) - F(x,y,[u])]. \end{aligned} \tag{2.53}$$

The second term on the right-hand side is just the original function u shifted,

$$F(x,y,[u]) + S(x,y,[u']) + g(x - y + x_s) = u(x - y + x_s); \tag{2.54}$$

also from (2.48),

$$\begin{aligned} \text{TV}[S_N(x, x_s, [u']) - S(x,y,[u'])] &\le \text{TV}[S_N(x, x_s, [u']) - S_N(x,y,[u'])] \\ &\quad + \text{TV}[S_N(x,y,[u']) - S(x,y,[u'])] \\ &\le K \frac{\log N}{N} \end{aligned} \tag{2.55}$$

and finally, since $g(x)$ is smooth enough,

$$\text{TV}[P_N g - g(x - y + x_s)] \le \frac{K}{N}. \tag{2.56}$$

Therefore from (2.53) and Lemmas 4 and 5,

$$\text{TV}[v_N] \le \text{TV}[u] + L_0 \frac{\log N}{N} + L_1 \Delta_1 N \log N + L_2 \Delta_2 \log N. \tag{2.57}$$

Next we prove (2.24) following the same argument above. From (2.50), (2.51) and (2.52),

$$
\begin{aligned}
v_N(x) - u(x) = {} & [F_N(x, x_s, [u]) - F_N(x, y, A)] \\
& + [S_N(x, x_s, [u']) - S(x, x_s, [u'])] \\
& + [F(x, y, A) - F(x, x_s, [u])] + [P_N g(x) - g(x)].
\end{aligned} \tag{2.58}
$$

The first term will be bounded by (2.42), the second term by (2.49); for the third term,

$$
\int_0^{2\pi} |F(x, y, A) - F(x, x_s, [u])|\, dx \le C_1 \Delta_1 + C_2 \Delta_2. \tag{2.59}
$$

Now since $g(x)$ is smooth enough, we have

$$
\|P_N g - g\|_{L^1} = O\left(\frac{1}{N^2}\right). \tag{2.60}
$$

Therefore from Lemma 4, Lemma 5 and (2.59)–(2.60),

$$
\|v_N - u\|_{L^1} \le C_0 \frac{\log N}{N^2} + C_1 \Delta_1 \log N + C_2 \Delta_2,
$$

and the proof is completed. □

COROLLARY. *The method suggested in* (2.15) *yields*

$$
|A - [u]| = O\left(\frac{1}{N^2}\right) \quad \textit{and} \quad |y - x_s| = O\left(\frac{1}{N^2}\right),
$$

and therefore

$$
\mathrm{TV}[v_N] \le \mathrm{TV}[u] + K \frac{\log N}{N}, \tag{2.61}
$$

$$
\|v_N - u\|_{L^1} \le C \frac{\log N}{N^2}. \tag{2.62}
$$

Thus, the total variation of v_N converges to that of u. Convergence of v_N to u in the L^1 norm is one order higher than in the case of $P_N u$, for which the rate of convergence in L^1 is $O(\log N/N)$. The method therefore yields a reconstruction technique which is total variation bounded.

We conclude this section by pointing out that a similar result for collocation method and/or for Chebyshev expansions can be developed along the same lines. Computationally, we observe similar results for Galerkin and collocation methods (see Section 4). In practice, collocation is used more often than Galerkin, especially when solving a nonlinear PDE (Section 3).

3. Essentially Nonoscillatory Spectral Schemes. In this section we apply the techniques discussed in Section 2 to solve the PDE (1.4):

$$
u_t + f(u)_x = 0, \tag{3.1a}
$$

$$
u(x, 0) = u^0(x). \tag{3.1b}
$$

If the cell average of u is defined by

$$
\overline{u}(x, t) = \frac{1}{\Delta x} \int_{x - \Delta x/2}^{x + \Delta x/2} u(\xi, t)\, d\xi, \tag{3.2}
$$

then (3.1) can be rewritten as

$$\frac{\partial}{\partial t}\overline{u}(x,t) + \frac{1}{\Delta x}\left[f\left(u\left(x+\frac{\Delta x}{2},t\right)\right) - f\left(u\left(x-\frac{\Delta x}{2},t\right)\right)\right] = 0, \quad (3.3a)$$

$$\overline{u}(x,0) = \overline{u}^0(x). \quad (3.3b)$$

Hence a semidiscrete conservative scheme

$$\frac{d}{dt}\overline{u}_j = L(\overline{u})_j = -\frac{1}{\Delta x}(\hat{f}_{j+1/2} - \hat{f}_{j-1/2}) \quad (3.4)$$

will be of high order if the numerical flux $\hat{f}_{j+1/2}$ approximates $f(u(x_j + \Delta x/2, t))$ to high order. This is the approach used in the MUSCL type semidiscrete finite difference TVD and ENO schemes [11], [4]. Notice that (3.4) is a scheme for the cell averages $\overline{u}_j$. However, in evaluating $\hat{f}_{j+1/2}$, which should approximate $f(u(x_j + \Delta x/2, t))$, we also need accurate point values $u_{j+1/2} = u(x_j + \Delta x/2, t)$. For finite difference schemes the reconstruction from cell averages to point values is a major issue and causes difficulties, especially in several space dimensions [4], [5]. For spectral methods, this is very simple because $\overline{u}$ is just the convolution of u with the characteristic function of $(x_{j-1/2}, x_{j+1/2})$. To be precise, if

$$u(x) = \sum_{l=-N}^{N} a_l e^{ilx} \quad (3.5)$$

(we have suppressed the time variable t), then

$$\overline{u}(x) = \sum_{l=-N}^{N} \overline{a}_l e^{ilx} \quad (3.6)$$

with

$$\overline{a}_l = \sigma_l a_l, \qquad \sigma_l = \frac{\sin(l\Delta x/2)}{l\Delta x/2} \quad \text{for } 0 < |l| \le N,\ \sigma_0 = 1. \quad (3.7)$$

Notice that for collocation or Galerkin with $\Delta x = 2\pi/2N$, we have $|l\Delta x/2| \le \pi/2$ for $|l| \le N$, hence $2/\pi \le \sigma_l \le 1$. The division or multiplication by σ_l thus causes no stability difficulty. We point out that σ_l resembles the Lanczos filter [8, p. 65], which in our notations is $\sin(l\Delta x)/l\Delta x$, and approaches zero when $|l| \to N$.

The easy transform between u and $\overline{u}$ is also valid in several space dimensions and for other spectral expansions (e.g., Chebyshev expansions). We omit the details.

We now state our scheme as (3.4) with

$$\hat{f}_{j+1/2} = f(v_N(x_{j+1/2}, t)), \quad (3.8)$$

where v_N is defined by (2.10). We obtain the Fourier coefficients $\overline{a}_l$ of $\overline{u}$ from $\{\overline{u}_j\}$ by collocation, and obtain a_l of u needed in (2.10) by (3.7). The main difference between the conventional spectral method and the current approach is that we use the essentially nonoscillatory reconstruction v_N instead of the oscillatory $P_N u$ in (3.8).

The scheme, as it stands, can only treat a solution of not more than one discontinuity. However, it can be easily generalized.

We remark that if u is smooth, (2.10) keeps spectral accuracy because A determined by (2.14) will be spectrally small.

To discretize (3.4) in time, we use the high-order TVD Runge-Kutta methods in [13]:

$$\tag{3.9} \begin{aligned} &\overline{u}^{(i)} = \sum_{k=0}^{i-1} [\alpha_{ik}\overline{u}^{(k)} + \beta_{ik}\Delta t L(\overline{u}^{(k)})], \qquad i = 1, \ldots, r, \\ &\overline{u}^{(0)} = \overline{u}^n, \qquad \overline{u}^{(r)} = \overline{u}^{n+1}. \end{aligned}$$

In Section 4, we use a third-order scheme $r = 3$ with $\alpha_{10} = \beta_{10} = 1$, $\alpha_{20} = \frac{3}{4}$, $\beta_{20} = 0$, $\alpha_{21} = \beta_{21} = \frac{1}{4}$, $\alpha_{30} = \frac{1}{3}$, $\beta_{30} = \alpha_{31} = \beta_{31} = 0$, $\alpha_{32} = \beta_{32} = \frac{2}{3}$. We use a small Δt so that the temporal error can be neglected. These methods are TVD (or TVB) if the Euler forward version of (3.4) is TVD (or TVB). In light of (2.61) we expect the total variation of (3.4)–(3.8)–(3.9) to grow at most at the rate of $O(\ln N)$. In practice, we observe stable results (Section 4).

In summary, a suggested algorithm can be:

(1) Starting with $\{\overline{u}_j\}$, compute its collocation Fourier coefficients $\{\overline{a}_l\}$ and the Fourier coefficients $\{a_l\}$ of u by (3.7).

(2) Compute the shock location y and the shock strength A by (2.14).

(3) Compute $v_N(x)$ by

$$v_N(x) = a_0 + \sum_{\substack{l=-N \\ l \neq 0}}^{N} \tau_l \left(a_l - \frac{A}{il} e^{-ily} \right) e^{ilx} + F(x, y, A)$$

and a filtered $\{\overline{u}_j^*\}$ by

$$\overline{u}_j^* = \overline{a}_0 + \sum_{\substack{l=-N \\ l \neq 0}}^{N} \tau_l \left(\overline{a}_l - \sigma_l \frac{A}{il} e^{-ily} \right) e^{ilx_j} + \overline{F}(x_j, y, A),$$

where τ_l is some filter, e.g. (1.2).

(4) Use $\hat{f}_{j+1/2} = f(v_N(x_{j+1/2}, t))$ in (3.4), and use

$$\overline{u}^{(i)} = \sum_{k=0}^{i-1} [\alpha_{ik}\overline{u}^{*(k)} + \beta_{ik}\Delta t L(\overline{u}^{(k)})], \qquad i = 1, \ldots, r,$$

in (3.9).

As in the finite difference case [11], [12], we may also apply limiters to obtain provable TVB schemes while still keeping spectral accuracy. Let

$$\tag{3.10} \tilde{u}_j = u_{j+1/2} - \overline{u}_j, \qquad \tilde{\tilde{u}}_{j+1} = \overline{u}_{j+1} - u_{j+1/2},$$

where $u_{j+1/2} = v_N(x_{j+1/2}, t)$ in (3.8). We limit the increments by

$$\tag{3.11} \tilde{u}_j^{(\text{mod})} = m(\tilde{u}_j, \Delta_+\overline{u}_j, \Delta_-\overline{u}_j), \qquad \tilde{\tilde{u}}_{j+1}^{(\text{mod})} = m(\tilde{\tilde{u}}_{j+1}, \Delta_+\overline{u}_j, \Delta_-\overline{u}_j),$$

where m is the minmod function with TVB correction:

$$\tag{3.12} m(a_1, \ldots, a_k) = \begin{cases} a_1, & \text{if } |a_1| \leq M\Delta x^2, \\ s \cdot \min_{1 \leq i \leq k} |a_i|, & \text{if } |a_1| > M\Delta x^2 \text{ and } \operatorname{sign}(a_i) = s \ \forall i, \\ 0, & \text{otherwise,} \end{cases}$$

with $M = \frac{2}{3}M_2$ or $M = M_j = \frac{2}{9}(3+10M_2)M_2 \cdot \Delta x^2/(\Delta x^2 + |\Delta_+\overline{u}_j| + |\Delta_-\overline{u}_j|)$. Here, M_2 is the maximum of $|u^0_{xx}|$ in some region around the smooth critical points of $u^0(x)$. See [12], [2].

The flux (3.8) is modified to

$$\hat{f}_{j+1/2} = h(\overline{u}_j + \tilde{u}_j^{(\mathrm{mod})}, \overline{u}_{j+1} - \tilde{\tilde{u}}_{j+1}^{(\mathrm{mod})}), \tag{3.13}$$

where h is any monotone flux [3]. We then have the following lemma.

LEMMA 6. *Scheme* (3.9)–(3.13) *is TVB and formally spectrally accurate in space (i.e., the spatial local truncation error in smooth regions is spectrally small), if the filtering* (1.3) *is used.*

Proof. The proof for TVB can be found in [11], [12]. By [1], the local truncation error is spectrally small in smooth regions if the limiter (3.11) returns the first argument. The proof that (3.11) always returns the first argument in smooth regions, including at critical points, can be found in [2]. □

We remark that the scheme (3.4)–(3.8)–(3.9), with or without the TVB limiting (3.11), yields almost identical results in our numerical examples (Section 4). This indicates the good stability property of the scheme (3.4)–(3.8)–(3.9). We also remark that (3.13) yields a TVB scheme regardless of the underlying method (3.4). However, accuracy in smooth regions may be lost if the underlying method (3.4) is globally oscillatory, because the limiters (3.11) may be enacted in *smooth* regions to counterbalance these spurious oscillations. Numerical examples in Section 4 verify these remarks. In [10], McDonald also used some limiters to obtain a TVD spectral scheme. However, the accuracy in smooth regions is questionable in view of the above remarks.

4. Numerical Results. We use several numerical examples to illustrate the methods introduced in the previous sections.

Example 1. We use the approximation (2.10)–(2.12)–(2.14) on the following function

$$u(x) = \begin{cases} \sin\frac{x}{2}, & 0 \le x \le 0.9, \\ -\sin\frac{x}{2}, & 0.9 < x < 2\pi. \end{cases} \tag{4.1}$$

Notice that $[u^{(k)}] \neq 0$ for all $k \ge 0$. Both Galerkin and collocation methods are tested. Exponential filters (1.2) with $m = 4$, $k_0 = 0$ are used.

In Table 1, we list the errors of the shock location and shock strength determined by (2.14). Notice that the second-order accuracy (2.15) is verified.

Figure 1 displays the numerical solution of the Galerkin approximation (2.10)–(2.12)–(2.14) with $N = 64$. Figure 2 is the error of the approximation on a logarithm scale. We have found the same kind of results for collocation approximation. In Table 2 we list the L^1 error and numerical order in a smooth region (in this case, we define the smooth region to be 0.8 away from the discontinuity). We can see Galerkin and collocation have the same order of accuracy. There is no $O(1)$ error near the discontinuity, overall we achieve $O(\log N/N^2)$ for L^1 convergence, verifying (2.62). For comparison we refer the reader to [7].

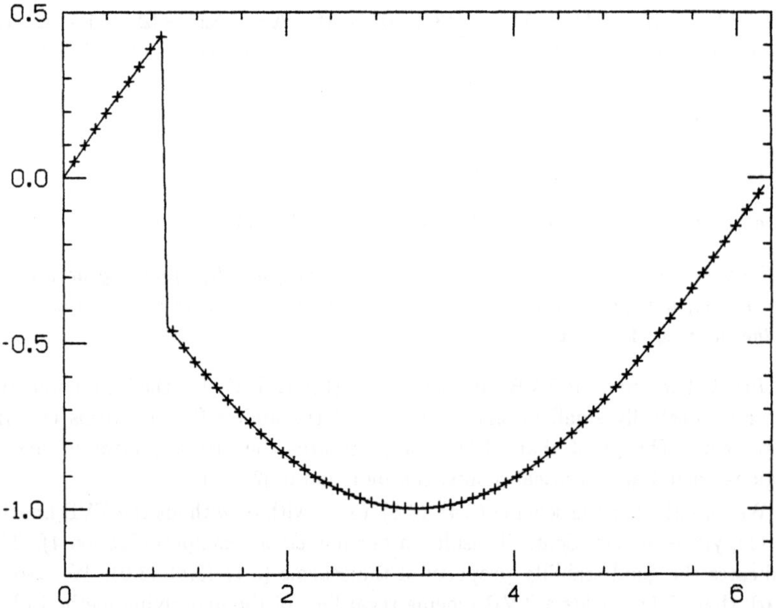

FIGURE 1

Example 1, Galerkin approximation. Solid line is the exact solution, the pluses the numerical solution, $N = 64$.

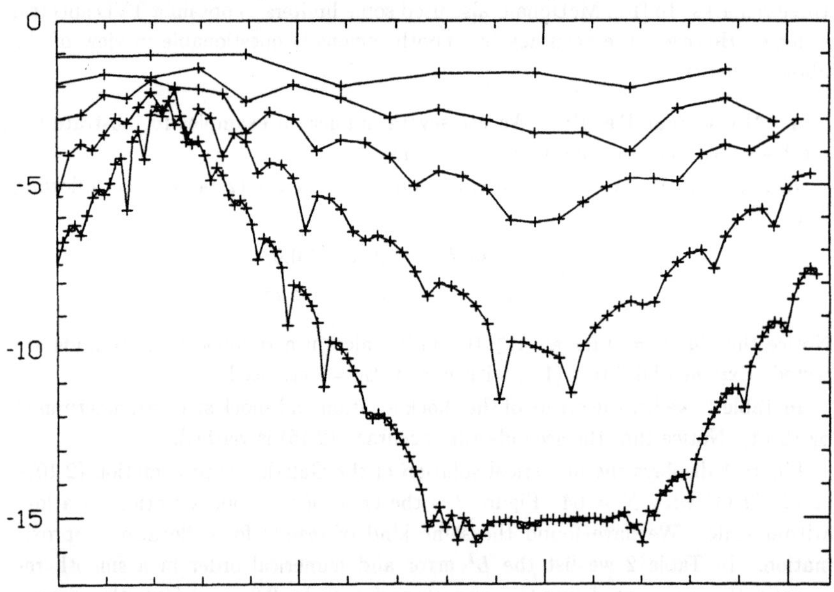

FIGURE 2

Example 1, error of the Galerkin approximation on logarithm scale for $N = 8, 16, 32, 64, 128$.

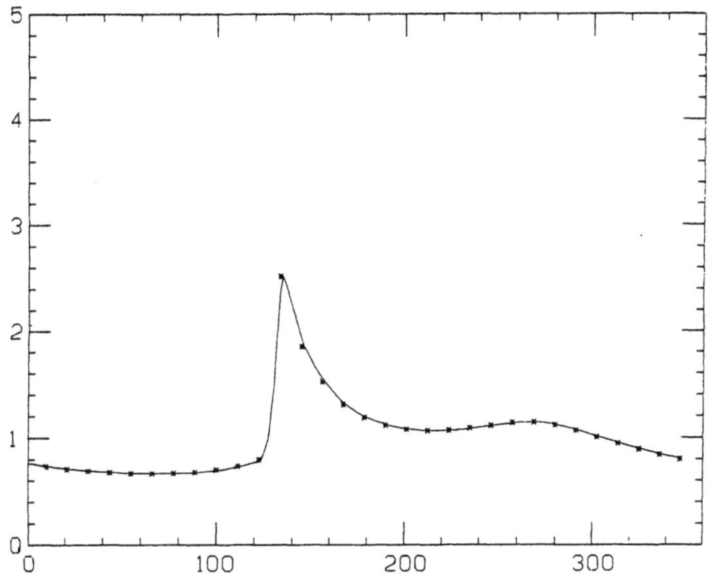

FIGURE 3

Galerkin approximation of (2.11) with N = 32 for the steady state solution of the astrophysics problem [6].

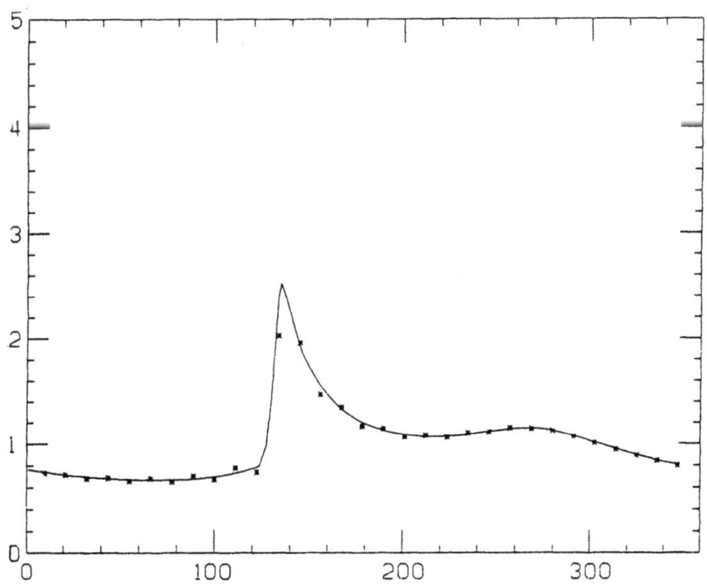

FIGURE 4

Usual Galerkin approximation for the steady state solution of the astrophysics problem, N = 32.

Example 2. We apply (2.10)–(2.12)–(2.14) on a discontinuous function which is the steady state solution of an astrophysics problem [6]. Figure 3 is v_N in (2.10) with $N = 32$. For comparison, Figure 4 is the usual Galerkin approximation $P_N u$ with $N = 32$. The improvement is apparent.

Example 3. We solve the Burgers' equation

$$u_t + \left(\frac{u^2}{2}\right)_x = 0, \quad u(x,0) = 0.3 + 0.7\sin x, \tag{4.2}$$

using scheme (3.4)–(3.8)–(3.9) and (3.4)–(3.9)–(3.13). We find the shock location and strength with (2.14). In our computation, we find that the coefficients of modes in the range of $\sqrt{N} \sim N^{3/4}$ give us the best results to detect shock location and strength. It can be proven that in this range of modes (2.14) will not fail in the presence of possible transition points in the numerical solutions. The errors of (3.4)–(3.8)–(3.9) in smooth regions (1.6 away from shock when it appears), at $t = 0.8$ (before shock), $t = 1.42$ (when the shock just develops), and $t = 2.00$ (after shock) are listed in Table 3. The numerical solutions are displayed in Figures 5–6. The error at $t = 2.00$, in logarithm scale, is displayed in Figure 7.

We seem to observe higher than algebraic order in smooth regions both before and after the shock develops. This might be the first time superalgebraic accuracy is observed in a shock-capturing spectral scheme solving a nonlinear PDE with shocks. The usual $O(1)$ Gibbs oscillation near the shock is also absent in all of our calculations. We also notice that the TVB limiter (3.11) does not change the numerical results significantly in the smooth region (see Table 4). Actually, we observe the same order of accuracy in the smooth region, comparing Table 4 with Table 3. This indicates that the scheme (3.4)–(3.8)–(3.9) is by itself very stable.

Finally, we run the usual spectral scheme (i.e., with v_N in (3.8) replaced by $P_N u$) with the TVB limiter (3.11). The errors in smooth regions (1.6 away from shock) are listed in Table 5 (compare with Table 4). Clearly we only get first-order accuracy in smooth regions after the shock develops. This indicates that TVB limiting can make a scheme stable but may not preserve the accuracy.

Example 4. 2-D Steady State. We solve a 2-dimensional scalar conservation law

$$\begin{cases} u_t + (\frac{u^2}{2})_x + u_y = 0, & (x,y) \in [0,2\pi] \times [-1,1], \\ u(x,0,t) = \sin x, & \\ u(0,y,t) = u(2\pi,y,t), & y \in [-1,1],\ t \geq 0. \end{cases} \tag{4.3}$$

We know that (4.3) has a steady state solution $u_\infty(x,y)$. $u_\infty(x,y)$ actually will be the solution to (4.2) if we replace t by y and set $u(x,0) = \sin x$ in (4.2).

As mentioned in Section 3, (3.4)–(3.8)–(3.9) can be extended to 2-dimensional cases and we can use either the Fourier or Chebyshev method in each of the spatial directions. To solve for the steady state of (4.3), we use the Fourier method in the x-direction and the Chebyshev method in the y-direction. The criterion we set for the steady state is that the relative L^1 residue between two consecutive time stages be less than 10^{-6}, i.e.,

$$\frac{\|u^{n+1} - u^n\|_{L^1}}{\|u^n\|_{L^1}} \leq 10^{-6}. \tag{4.4}$$

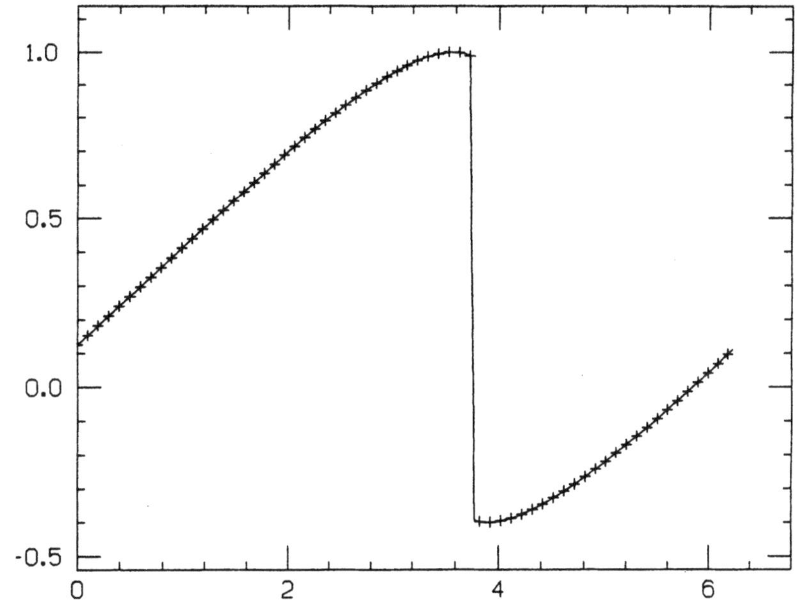

FIGURE 5

Example 3, inviscid Burgers' equation with initial data $u(x,0) = 0.3+0.7*\sin(x)$, *time* $t = 2.0, N = 64$. *Solid line is the exact solution, the pluses the numerical solution.*

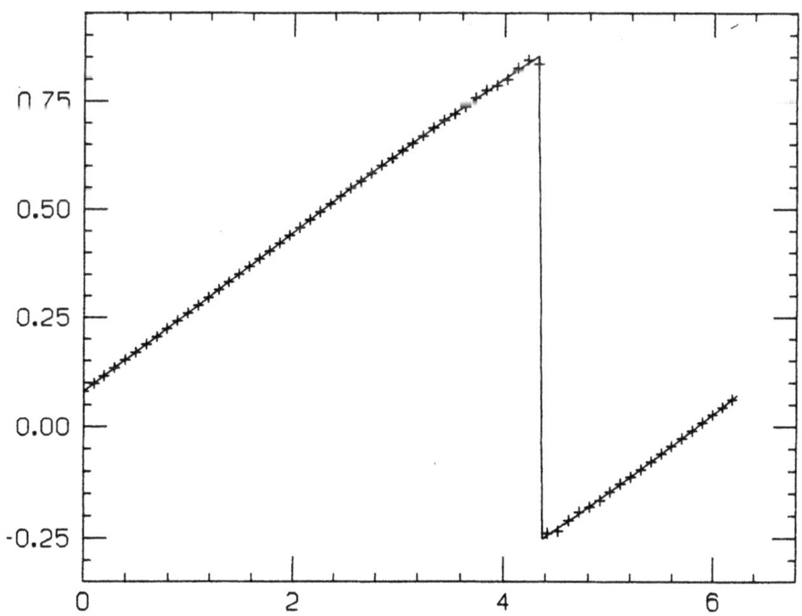

FIGURE 6

Example 3, same as Figure 5, except time $t = 4.0$.

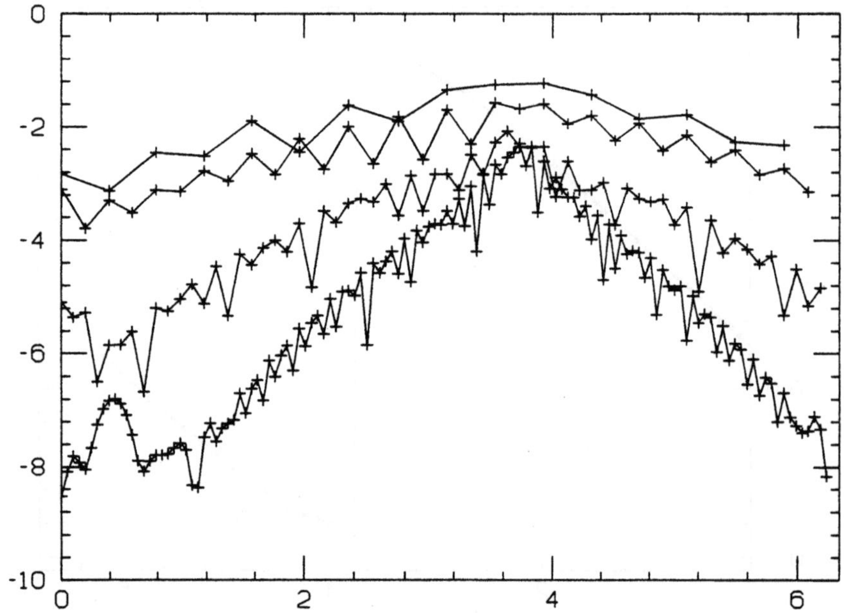

FIGURE 7

Example 3, inviscid Burgers' equation, $u(x,0) = 0.3+0.7*\sin(x)$. Errors of numerical solutions at time $t = 2.00$ in the logarithm scale for $N = 16, 32, 64, 128$.

Figure 8 displays the profile of the steady state at $y = 0.38$ and $y = 1.00$. We used 32 points in the x-direction and 8 points in the y-direction. Figure 9 is the contour plot for the numerical steady state solution.

TABLE 1

Errors of shock location and strength, Example 1.

	Galerkin				Collocation			
	Location		Strength		Location		Strength	
N	Error	Order	Error	Order	Error	Order	Error	Order
8	0.15(0)		0.12(−1)		0.36(0)		0.20(−1)	
16	0.24(−1)	2.6	0.22(−2)	2.4	−0.21(0)	0.8	0.12(−1)	0.7
32	0.49(−2)	2.3	0.48(−3)	2.2	−0.14(−1)	3.8	0.38(−2)	1.7
64	0.11(−2)	2.1	0.11(−3)	2.1	−0.32(−2)	2.2	0.11(−2)	1.8
128	0.26(−3)	2.1	0.27(−4)	2.1	−0.77(−3)	2.0	0.28(−3)	1.9

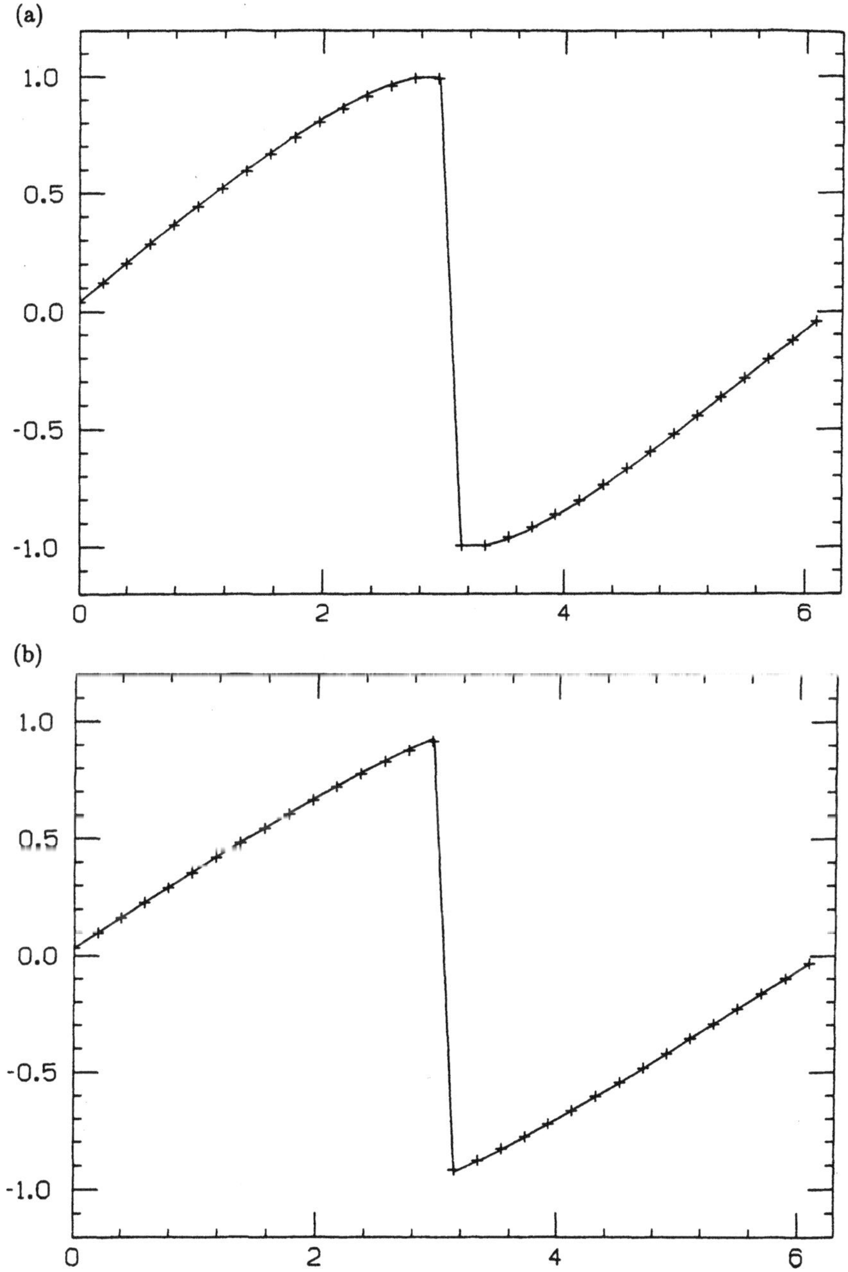

FIGURE 8

Example 4, steady state solution at (a) $y = 0.38$ (b) $y = 1.0$. *Solid lines are the exact solution, the pluses the numerical solution.*

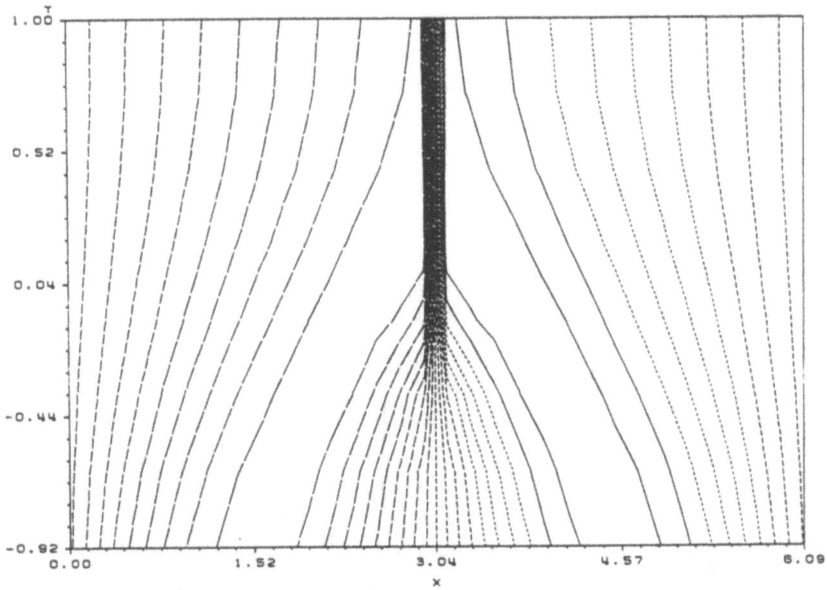

FIGURE 9

Example 4, contour plot of the steady state solution.

TABLE 2

L^1 Error in Region $\mathrm{I} = \{x \in [0, 2\pi], |x - x_s| > 0.8\}$ and Region $\mathrm{II} = [0, 2\pi]$, Example 1.

	Galerkin				Collocation			
	Region I		Region II		Region I		Region II	
N	Error	Order	Error	Order	Error	Order	Error	Order
8	0.32(−1)		0.14(0)		0.23(−1)		0.31(−1)	
16	0.32(−2)	3.30	0.75(−2)	4.27	−0.21(−2)	3.46	0.61(−2)	2.34
32	0.24(−3)	3.75	0.17(−2)	2.11	0.23(−3)	3.20	0.17(−2)	1.79
64	0.51(−5)	5.55	0.39(−3)	2.14	0.54(−5)	5.40	0.49(−3)	1.86
128	0.12(−7)	8.67	0.96(−4)	2.04	0.12(−7)	8.82	0.13(−3)	1.92

TABLE 3

Errors in smooth region for (4.2). *At* $t = 0.8$, *the smooth region is* $[0, 2\pi]$. *At* $t = 1.42, 2.0$, *the smooth region is* 1.6 *away from the shock.*

	$t = 0.8$		$t = 1.42$		$t = 2.0$	
N	L^1 Error	Order	L^1 Error	Order	L^1 Error	Order
16	0.94(−2)		0.39(−2)		0.44(−2)	
32	0.79(−3)	3.57	0.13(−2)	1.66	0.16(−2)	1.40
64	0.25(−4)	5.00	0.35(−4)	5.17	0.42(−4)	5.28
128	0.13(−6)	7.58	0.16(−6)	7.71	0.44(−6)	6.58

TABLE 4

Errors in smooth regions for (4.2) *of new spectral scheme with TVB limiting* (3.11). *For both* $t = 1.42$ *and* 2.0, *the* L^1 *errors are taken in the region* 1.6 *away from the shock.*

	$t = 1.42$		$t = 2.0$	
N	L^1 Error	Order	L^1 Error	Order
16				
	0.64(−2)		0.63(−2)	
32		1.90		1.67
	0.17(−2)		0.20(−2)	
64		5.55		5.29
	0.36(−4)		0.50(−4)	
128		7.75		7.11
	0.17(−6)		0.36(−6)	

TABLE 5

Errors in smooth regions for (4.2) *of the usual spectral scheme with TVB limiting* (3.11). *For both* $t = 1.42$ *and* 2.0, *the* L^1 *errors are taken in the region* 1.6 *away from the shock.*

	$t = 1.42$		$t = 2.0$	
N	L^1 Error	Order	L^1 Error	Order
16				
	0.25(−1)		0.16(−1)	
30		0.08		[illegible]
	0.98(−2)		0.17(−1)	
64		1.50		1.17
	0.34(−2)		0.79(−2)	
128		0.84		0.76
	0.19(−2)		0.47(−2)	

Division of Applied Mathematics
Brown University
Providence, Rhode Island 02912
E-mail (Gottlieb): am420000@brownvm.bitnet

1. S. ABARBANEL, D. GOTTLIEB & E. TADMOR, "Spectral methods for discontinuous problems," in *Numerical Methods for Fluid Dynamics*, II (K. Morton and M. Baines, eds.), Oxford University Press, New York, 1986, pp. 129–153.

2. B. COCKBURN & C.-W. SHU, *TVB Runge-Kutta Local Projection Discontinuous Galerkin Finite Element Methods for Conservation Laws* II: *General Framework*, IMA Preprint Series #392, University of Minnesota, 1988; *Math. Comp.*, this issue, pp. 411–435.

3. M. CRANDALL & A. MAJDA, "Monotone difference approximations for scalar conservation laws," *Math. Comp.*, v. 34, 1980, pp. 1-21.

4. A. HARTEN, B. ENGQUIST, S. OSHER & S. CHAKRAVARTHY, "Uniformly high order accurate non-oscillatory schemes, III," *J. Comput. Phys.*, v. 71, 1987, pp. 231-303.

5. A. HARTEN, *Preliminary Results on the Extension of ENO Schemes to Two-Dimensional Problems*, Proc. Internat. Conf. on Hyperbolic Problems, Saint-Etienne, January 1986.

6. M. HUSSAINI, D. KOPRIVA, M. SALAS & T. ZANG, "Spectral method for Euler equation, Part 1: Fourier method and shock capturing," *AIAA J.*, v. 23, 1985, pp. 234–240.

7. D. KOPRIVA, "A practical assessment of spectral accuracy for hyperbolic problems with discontinuity," *J. Sci. Comput.*, v. 2, 1987, pp. 249–262.

8. C. LANCZOS, *Discourse on Fourier Series*, Oliver & Boyd, Edinburgh, 1966.

9. A. MAJDA, J. MCDONOUGH & S. OSHER, "The Fourier method for nonsmooth initial data," *Math. Comp.*, v. 32, 1978, pp. 1041–1081.

10. B. MCDONALD, "Flux corrected pseudospectral methods for scalar hyperbolic conservation laws," *J. Comput. Phys.* (To appear.)

11. S. OSHER, "Convergence of generalized MUSCL schemes," *SIAM J. Numer. Anal.*, v. 22, 1985, pp. 947–961.

12. C.-W. SHU, "TVB uniformly high-order schemes for conservation laws," *Math. Comp.*, v. 49, 1987, pp. 105–121.

13. C.-W. SHU & S. OSHER, "Efficient implementation of essentially non-oscillatory shock-capturing schemes," *J. Comput. Phys.*, v. 77, 1988, pp. 439–471.

14. A. ZYGMUND, *Trigonometric Series*, v. 1, Cambridge Univ. Press, New York, 1959.

CELL AVERAGING CHEBYSHEV METHODS FOR HYPERBOLIC PROBLEMS[1]

Wei Cai
Department of Mathematics
University of North Carolina at Charlotte
Charlotte, NC 28223

David Gottlieb[2]
Division of Applied Mathematics
Brown University
Providence, RI 02912

Ami Harten [2]
Tel-Aviv University
Tel-Aviv

ABSTRACT

This paper describes a cell averaging method for the Chebyshev approximations of first order hyperbolic equations in conservation form. We present formulas for transforming between pointwise data at the collocation points and cell averaged quantities, and vice-versa. This step, trivial for the finite difference and Fourier methods, is nontrivial for the global polynomials used in spectral methods. We then prove that the cell averaging methods presented are stable for linear scalar hyperbolic equations and present numerical simulations of shock-density wave interaction using the new cell averaging Chebyshev methods.

[1]This work has been supported in part by AFOSR grant 85-0303 and by NSF grant DMS-8810150. The second author was partially supported by DARPA-URI contract N00014-86-k0754.

[2]Research was supported by the National Aeronautics and Space Administration under NASA Contract No. NAS1-18605 while the authors were in residence at the Institute for Computer Applications in Science and Engineering (ICASE), NASA Langley Research Center, Hampton, VA 23665.

i

1 Introduction

In this paper we introduce a new spectral technique for the numerical solution of nonlinear hyperbolic equations. This technique, as almost every modern finite difference scheme for shock computations, is based on the cell averaged form of the equations. This is essential for finite difference shock capturing techniques and it is our experience that it plays an essential role in a successful spectral simulation of problems that involve shock waves [1].

Consider the nonlinear hyperbolic equation

$$\begin{cases} U_t + F_x(U) = 0, & x \in [-1,1] \\ U(x,0) = U_0(x), & \end{cases} \tag{1.1}$$

with appropriate boundary conditions.

The cell averaged form of (1.1) is obtained by integrating (1.1) between any two points $-1 \leq a < b \leq 1$ to get

$$\frac{\partial}{\partial t}\bar{U} \equiv \frac{\partial}{\partial t}\frac{1}{b-a}\int_a^b U(x)\,dx = -\frac{1}{b-a}(F(U(b)) - F(U(a))). \tag{1.2}$$

Let $\bar{u}(x,t)$ be an approximation to $\bar{U}(x,t)$ at time t. Following Harten [4] we express the approximation to $\bar{U}(x,t+\tau)$ by

$$\bar{u}(x,t+\tau) = \mathcal{A}E(t)\mathcal{R}(\cdot\,;\bar{u}), \tag{1.3}$$

where $\mathcal{A}$ is the cell averaging operator and $E(t)$ is the exact time evolution operator corresponding to (1.1). Throughout the paper we will not distinguish between $\mathcal{R}(\cdot;\bar{u})$ and $\mathcal{R}\bar{u}$. The operator $\mathcal{R}(\cdot;\bar{u})$ is of extreme importance, it represents the way we reconstruct u from its given cell average values $\bar{u}_{j-\frac{1}{2}} = \frac{1}{x_{j-1}-x_j}\int_{x_j}^{x_{j-1}} u(x)\,dx$, where $\{x_j\}_{j=1}^N$ are the grid points. For finite difference schemes u is a piecewise polynomial of low degree, so that the reconstruction itself is simple. It becomes complicated only if one imposes also the requirement that the reconstruction should be essentially nonoscillatory. In [1] we have presented an essentially non-oscillatory Fourier method based on the cell averaging formulation (1.2). In that case

1

the transformations between the cell averages and the point values are simple and can be carried out efficiently by the Fast Fourier Transformations (FFT). This can be attributed to the fact that the boundary conditions are periodic and that the cell average of a trigonometric function is proportional to the function itself. However, for Chebyshev methods, the cell averaging operation (denoted by the operator $\mathcal{A}$) is not simple nor is $\mathcal{R}(\cdot\,;\bar{u})$. As a matter of fact not only the formulation but also the implementation of $\mathcal{A}$ and $\mathcal{R}$ is not straightforward.

In this paper we formulate the cell averaging technique for the Chebyshev method. We will discuss its stability for linear problems and show an example of its applicability to nonlinear systems of equations by simulating the problem of shock-density wave interaction. The cell averaging formulation is an essential part of the numerical code.

The outline of the paper is as follows: In Section 2 we show how to reconstruct efficiently point values of a polynomial from its cell averages and vice versa.In Section 3 we introduce the new numerical technique and show its stability for linear problems. Section 4 is devoted to numerical results obtained by using the new method.

2 Cell Averages and Point values

In this section we will discuss the cell averaging operator $\mathcal{A}$ and the reconstruction operator $\mathcal{R}$ in the context of the Chebyshev methods. In these methods the approximations are taken from the space of polynomials of degree N. It is therefore clear that $\mathcal{A}$ and $\mathcal{R}$, when restricted to the polynomial space, can be expressed as matrices $\mathbf{A_N}$ and $\mathbf{R_N}$. We will give the explicit formulas for these matrices. We start by discussing the operator $\mathcal{A}$.

Assume that $f(x)$ is in $C^r[-1,1]$, $r \geq 0$. Let $x_j = \cos\left(\frac{j\pi}{N}\right)$, $0 \leq j \leq N$ be the Chebyshev-Lobatto points in [-1,1]. For later use, define $x_{j-\frac{1}{2}} = \cos\left(\frac{(j-\frac{1}{2})\pi}{N}\right)$, $1 \leq j \leq N$. The cell averaged function $\bar{f}(x)$ of $f(x)$ is defined as

$$\bar{f}(x) = \mathcal{A}f =: \frac{1}{h_2(x) - h_1(x)} \int_{h_1(x)}^{h_2(x)} f(x)\,dx \qquad \text{for } -1 < x < 1, \tag{2.1}$$

2

where

$$\begin{aligned} h_1(x) &= \cos(\cos^{-1} x - \frac{\Delta\theta}{2}), \\ h_2(x) &= \cos(\cos^{-1} x + \frac{\Delta\theta}{2}), \\ \Delta\theta &= \frac{\pi}{N}. \end{aligned} \tag{2.2}$$

The reason for the definition in (2.1) is that

$$\bar{f}(x_{j-\frac{1}{2}}) = \frac{1}{x_{j-1} - x_j} \int_{x_j}^{x_{j-1}} f(x)\, dx, \qquad \text{for } 1 \le j \le N. \tag{2.3}$$

As stated before, we are interested in $\mathcal{A}$ operating in a polynomial space. In Lemma 1, we show that the result of $\mathcal{A}$ on a polynomial is still a polynomial of the same degree.

Lemma 1 *Let $U_k(x) = \frac{1}{k+1} T'_{k+1}(x)$, $k \ge 0$, be the second kind of the Chebyshev polynomials. Then*

$$\bar{U}_k(x) = \mathcal{A} U_k = \sigma_k U_k(x), \tag{2.4}$$

where

$$\sigma_k = \frac{\sin(k+1)\frac{\Delta\theta}{2}}{(k+1)\sin\frac{\Delta\theta}{2}}. \tag{2.5}$$

Proof: Substituting $U_k(x)$ in the right hand side of (2.1) and making the transformation $x = \cos\theta$, $0 \le \theta \le \pi$, we have

$$\bar{U}_k(x) = \frac{1}{\cos(\cos^{-1} x + \frac{\Delta\theta}{2}) - \cos(\cos^{-1} x - \frac{\Delta\theta}{2})} \int_{\cos(\cos^{-1} x - \frac{\Delta\theta}{2})}^{\cos(\cos^{-1} x + \frac{\Delta\theta}{2})} U_k(x)\, dx.$$

Since $T_k(x) = \cos(k\theta)$, then $U_k(x) = \frac{\sin(k+1)\theta}{(k+1)\sin\theta}$ and therefore

$$\begin{aligned} \bar{U}_k(x) &= \frac{1}{2(k+1)\sin\frac{\Delta\theta}{2}\sin\theta} \int_{\theta-\frac{\Delta\theta}{2}}^{\theta+\frac{\Delta\theta}{2}} \sin(k+1)\theta\, d\theta \\ &= \frac{1}{2(k+1)\sin\frac{\Delta\theta}{2}\sin\theta}(-1)\frac{\cos(k+1)\theta}{k+1}\Big|_{\theta-\frac{\Delta\theta}{2}}^{\theta+\frac{\Delta\theta}{2}} \\ &= \frac{\sin(k+1)\frac{\Delta\theta}{2}}{(k+1)\sin\frac{\Delta\theta}{2}} \left(\frac{1}{k+1} \frac{\sin(k+1)\theta}{\sin\theta} \right), \end{aligned}$$

3

i.e.

$$\bar{U}_k(x) = \sigma_k U_k(x).$$

Q.E.D.

¿From Lemma 1 one gets

Corollary 1 *The cell averaged function of any polynomial of order N is a polynomial of the same order.*

Proof: any polynomial of degree N has the expression

$$f(x) = \sum_{k=0}^{N} a_k U_k(x), \tag{2.6}$$

where a_k are constants.

Therefore, by Lemma 1

$$\bar{f}(x) = \sum_{k=0}^{N} \sigma_k a_k U_k(x). \tag{2.7}$$

Q.E.D.

Thus, Lemma 1 gives explicitly the eigenvalues σ_k of the matrix $\mathbf{A_N}$ (the restriction of $\mathcal{A}$ to polynomials of degree N). These eigenvalues are uniformly bounded from above and below, in fact

$$\frac{2}{\pi} \leq \sigma_k \leq \frac{\pi}{2}, \qquad 0 \leq k \leq N. \tag{2.8}$$

If $f(x)$ is a polynomial of degree N, then it is uniquely determined by its values $f(x_j)$, $j = 0, \cdots, N$. So theoretically $\bar{f}(x_{j-\frac{1}{2}})$, $j = 1, \cdots, N$ can be determined. Therefore the transformation from $f(x_j)$, $j = 0, \cdots, N$ to $\bar{f}(x_{j-\frac{1}{2}})$, $j = 1, \cdots, N$ is well defined and we only need to address the issue of its efficient implementation.

In general it is known that

$$f(x) = \sum_{k=0}^{N} a_k T_k(x), \tag{2.9}$$

where

$$\begin{aligned} a_k &= \frac{2}{N\bar{c}_k}\sum_{j=0}^{N} f(x_j)T_k(x_j), \\ \bar{c}_0 &= \bar{c}_N = 2, \qquad \bar{c}_k = 1 \text{ if } k \neq 0 \text{ or } N. \end{aligned} \tag{2.10}$$

Alternatively

$$f(x) = \sum_{k=0}^{N} b_k U_k(x), \tag{2.11}$$

$$\begin{aligned} b_k &= \frac{a_k}{2}, \qquad k = N-1 \text{ or } N, \\ b_k &= \frac{1}{2}(\bar{c}_k a_k - a_{k+2}), \qquad 0 \leq k \leq N-2 \end{aligned} \tag{2.12}$$

and therefore by corollary 1

$$\bar{f}(x) = \sum_{k=0}^{N} \sigma_k b_k U_k(x). \tag{2.13}$$

Now $\{\bar{f}(x_{j-\frac{1}{2}})\}_{j=1}^{N}$ are obtained by substituting $x_{j-\frac{1}{2}}$ in (2.13) (this can be carried out using the FFT).

We note that equations (2.9) - (2.13) describe how to get the vector $\bar{f}(x_{j-\frac{1}{2}})$, $j = 1, \cdots, N$ from the vector $f(x_i)$, $i = 0, \cdots, N$. Denote by $\mathbf{A_N}$ the $N \times (N+1)$ matrix defined by this transformation. We note that $\mathbf{A_N}$ can be written explicitly. In fact the polynomial $f(x)$ has an unique representation as

$$f(x) = \sum_{j=0}^{N} f(x_j) g_j(x), \tag{2.14}$$

where

$$g_j(x) = \frac{(1-x^2)T_N'(x)(-1)^{j+1}}{\bar{c}_j N^2 (x - x_j)} = \frac{2}{\bar{c}_j N}\sum_{k=0}^{N} \frac{T_k(x_j)T_k(x)}{\bar{c}_k}$$

with $\bar{c}_k$ defined in (2.10) .

It follows upon substituting (2.14) in (2.1) and using the fact that $\bar{T}_0(x) = U_0(x)$, $\bar{T}_1(x) = 2\sigma_1 U_1(x)$ and $\bar{T}_k(x) = \frac{1}{2}(\sigma_k U_k(x) - \sigma_{k-2}U_{k-2}(x))$ for $k \geq 2$,

$$\bar{f}(x) = \sum_{\ell=0}^{N} f(x_\ell)\bar{g}_\ell(x), \tag{2.15}$$

where

$$\bar{g}_\ell(x) = \frac{1}{2} + 2\sigma_1 T_1(x_\ell)U_1(x) + \frac{2}{\bar{c}_\ell N}\sum_{k=2}^{N} \frac{T_k(x_\ell)(\sigma_k U_k(x) - \sigma_{k-2}U_{k-2}(x))}{2\bar{c}_k} \tag{2.16}$$

with σ_k defined in (2.5).

Setting $x = x_{j-\frac{1}{2}}$ in (2.16)

$$(\mathbf{A_N})_{j\ell} = \bar{g}_\ell(x_{j-\frac{1}{2}}), \qquad 0 \le \ell \le N,\ 1 \le j \le N. \tag{2.17}$$

Thus we have outlined two procedures to get $\bar{f}(x_{j-\frac{1}{2}})$ from $f(x_j)$, one uses the FFT and another uses matrix vector multiplications.

We are now ready to discuss the reconstruction operator $\mathcal{R}(\cdot\,;\bar{f})$. Note that in the beginning of the solution process (1.3), we only have the values $\bar{f}_{j-\frac{1}{2}}$, $j = 1, \cdots, N$, thus we need another piece of information in order to define uniquely the N-th degree polynomial $f(x)$. This piece of information is provided by the boundary condition. For simplicity, we assume that the boundary condition is of the form

$$f(1) = f(x_0) = f_0. \tag{2.18}$$

The reconstruction is done in two steps. Define first a $(N-1)$-th polynomial $\bar{f}_{N-1}(x)$ which collocates $\bar{f}(x)$ at $\{x_{j-\frac{1}{2}}\}_{j=1}^{N}$, i.e. $\bar{f}_{N-1}(x_{j-\frac{1}{2}}) = \bar{f}(x_{j-\frac{1}{2}})$, $1 \le j \le N$, it is readily verified that

$$\bar{f}_{N-1}(x) = \sum_{k=0}^{N-1} c_k T_k(x), \tag{2.19}$$

where

$$c_k = \frac{2}{\bar{c}_k N}\sum_{j=1}^{N} \bar{f}(x_{j-\frac{1}{2}})T_k(x_{j-\frac{1}{2}}), \qquad \bar{c}_k \text{ is same as in (2.10).} \tag{2.20}$$

Alternatively we have

$$\bar{f}_{N-1}(x) = \sum_{k=0}^{N-1} b_k U_k(x), \tag{2.21}$$

where

$$\begin{aligned} b_k &= \frac{c_k}{2}, \qquad k = N-2,\ N-1, \\ b_k &= \frac{1}{2}(\bar{c}_k c_k - c_{k+2}), \qquad 0 \le k \le N-3. \end{aligned} \tag{2.22}$$

6

Now, by Corollary 1

$$f_{N-1}(x) = \sum_{k=0}^{N-1} \frac{b_k}{\sigma_k} U_k(x) \tag{2.23}$$

is a polynomial of degree $N-1$ such that $\mathcal{A} f_{N-1} = \bar{f}_{N-1}$.

Generally, $f_{N-1}(x)$ does not satisfy the boundary condition (2.18). There are two ways to modify $f_{N-1}(x)$ so that the boundary condition (2.18) is satisfied. In the first way we can add to $f_{N-1}(x)$ an N-th degree polynomial $Q(x)$ such that

$$\bar{Q}(x_{j-\frac{1}{2}}) = 0, \qquad 1 \leq j \leq N, \tag{2.24}$$

and

$$f_{N-1}(x) + Q(1) = f_0. \tag{2.25}$$

It can be verified that in order to satisfy (2.24)

$$Q(x) = c\left((1-x^2)T_N'(x)\right)'. \tag{2.26}$$

Let $f(x)$ be the sum of $f_{N-1}(x)$ and $Q(x)$,

$$\begin{aligned} f(x) &= f_{N-1}(x) + c\left((1-x^2)T_N'(x)\right)' \\ &= f_{N-1}(x) - c[xT_N'(x) + N^2 T_N(x)]. \end{aligned} \tag{2.27}$$

The last equality follows from the Chebyshev equation, and the constant c is now determined by the condition $f(x_0) = f_0$, i.e.

$$c = -\frac{1}{2N^2}\left(f_0 - f_{N-1}(1)\right). \tag{2.28}$$

Finally given $\bar{f}(x_{j-\frac{1}{2}})$, $j = 1, \cdots, N$ and f_0 we can get

$$f(x_i) = f_{N-1}(x_i) - c\left[x_i T_N'(x_i) + N^2 T_N(x_i)\right], \qquad i = 0, \cdots, N \tag{2.29}$$

where $f_{N-1}(x_i)$ can be evaluated from (2.23) by using the FFT.

Note that in this procedure we change the values of $f_{N-1}(x_i)$ at all the grid points.

Denote by $\mathbf{R_N}$ the $(N+1)\times(N+1)$ matrix transforming f_0 and $\bar{f}(x_{j-\frac{1}{2}})$, $1\le j\le N$ to $f(x_j)$, $0\le j\le N$, i.e.

$$(f(x_0),\cdots,f(x_N))^{\mathsf{T}}=\mathbf{R_N}\left(f(x_0),\bar{f}(x_{\frac{1}{2}}),\cdots,\bar{f}(x_{N-\frac{1}{2}})\right)^{\mathsf{T}}. \tag{2.30}$$

As before we can write $\mathbf{R_N}$ explicitly. Equation (2.19) can be rewritten as

$$\bar{f}_{N-1}(x)=\sum_{j=1}^{N}\bar{f}(x_{j-\frac{1}{2}})\bar{h}_j(x), \tag{2.31}$$

where $\bar{h}_j(x_{k-\frac{1}{2}})=\delta_{jk}$ and $h_j(x)$ are polynomials of degree $N-1$; explicitly

$$\bar{h}_j(x)=(-1)^{j+1}N\sin\left((j-\frac{1}{2})\frac{\pi}{N}\right)\frac{T_N(x)}{x-x_{j-\frac{1}{2}}}. \tag{2.32}$$

It can be shown that $\bar{h}_j(x)$ is the cell averaged polynomial of

$$h_j(x)=\sum_{k=0}^{N-1}\lambda_k U_k(x) \tag{2.33}$$

with λ_k defined as

$$\begin{aligned}\lambda_k &= \frac{1}{N\sigma_k}T_k(x_{j-\frac{1}{2}}), && \text{if } k=N-2,N-1\ ,\\ \lambda_k &= \frac{1}{N\sigma_k}(T_k(x_{j-\frac{1}{2}})-T_{k+2}(x_{j-\frac{1}{2}})), && \text{if } 0\le k\le N-3\ .\end{aligned} \tag{2.34}$$

As a result of (2.31) and (2.33) polynomial $f_{N-1}(x)$ takes the form

$$f_{N-1}(x)=\sum_{j=1}^{N-1}\bar{f}(x_{j-\frac{1}{2}})h_j(x), \tag{2.35}$$

and by (2.27)

$$f(x)=f_{N-1}(x)-\frac{1}{2N^2}(f_0-f_{N-1}(1))\left((1-x^2)T_N'(x)\right)'.$$

Using (2.31) we get

$$\begin{aligned}f(x) &= \sum_{j=1}^{N}\bar{f}(x_{j-\frac{1}{2}})h_j(x)-\frac{1}{2N^2}\left(f_0-\sum_{j=1}^{N}\bar{f}(x_{j-\frac{1}{2}})h_j(1)\right)\left((1-x^2)T_N'(x)\right)' \\ &= \sum_{j=1}^{N}\bar{f}(x_{j-\frac{1}{2}})\left[h_j(x)+\frac{h_j(1)}{2N^2}\left((1-x^2)T_N'(x)\right)'\right]-\frac{f_0}{2N^2}\left((1-x^2)T_N'(x)\right)'\end{aligned} \tag{2.36}$$

8

Substituting $x = x_i$ into (2.36) gives

$$(\mathbf{R_N})_{ij} = \begin{cases} \delta_{1j}, & \text{if } i = 0, \\ -\frac{1}{2N^2}\left((1-x_i^2)T_N'(x_i)\right)', & \text{if } j = 0, \\ h_j(x_i) + \frac{h_j(1)}{2N^2}\left((1-x_i^2)T_N'(x_i)\right)', & 1 \le i,j \le N. \end{cases} \tag{2.37}$$

To summarize we state

Lemma 2 *Let $f(x) = \mathcal{R}(\cdot\,;\bar{f})$ be the N-th polynomial defined in (2.27) then*

$$(\mathcal{A}f)(x_{j-\frac{1}{2}}) = \bar{f}(x_{j-\frac{1}{2}}) \qquad \text{for } 1 \le j \le N. \tag{2.38}$$

A different way to modify $f_{N-1}(x)$ in (2.23), in order to satisfy the boundary condition (2.18), is to add to it a polynomial $Q_1(x)$ of degree N such that the point values $f_{N-1}(x_i)$ remain unchanged and the new polynomial satisfies the boundary condition. Thus instead of (2.27) we define

$$f(x) = f_{N-1}(x) + (f_0 - f_{N-1}(1))\frac{(1+x)T_N'(x)}{2N^2}. \tag{2.39}$$

Computationally (2.39) is simpler than (2.23) (2.27). However, note that in this case

$$(\mathcal{A}f)(x_{j-\frac{1}{2}}) \ne \bar{f}(x_{j-\frac{1}{2}}), \tag{2.40}$$

which is in contrast to (2.38). The matrix corresponding to (2.40) can be formed similarly as in (2.37).

3 Cell Averaging Chebyshev (CAC) Method and Linear Stability

In this section we will establish the stability of the Cell Averaging Methods, presented in Section 2, when applied to a first order scalar hyperbolic equation. It is tempting to try to obtain stability in the L^1 norm because of the way the method (1.3) is presented. However we will only give the stability estimate based on a weighted Chebyshev norm.

Consider the initial boundary value problem of the scalar hyperbolic equation

$$\begin{cases} U_t = U_x, & x \in [-1,1], \\ U(x,0) = U_0(x), \\ U(1,t) = 0. \end{cases} \tag{3.1}$$

The cell averaged form of (3.1) is

$$\frac{\partial}{\partial t}\bar{U}(x,t) + \frac{1}{h_2(x) - h_1(x)}\left[U(h_2(x),t) - U(h_1(x),t)\right] = 0, \tag{3.2}$$

where $h_1(x)$ and $h_2(x)$ are defined in (2.2).

We follow the notation (1.3). Let the N-th degree polynomial $\bar{u}(x,t)$ be the approximation to (3.2). Our aim is to find the error equation for $\mathcal{R}(\cdot;\bar{u})$ defined either by (2.27) or (2.39). ¿From (1.3)

$$\bar{u}(x,t+\tau) = \mathcal{A}E\mathcal{R}\bar{u}(x,t). \tag{3.3}$$

Applying the reconstruction operator $\mathcal{R}$ we get

$$\mathcal{R}\bar{u}(x,t+\tau) = \mathcal{R}\mathcal{A}E\mathcal{R}\bar{u}(x,t) \tag{3.4}$$

¿From equation (3.4) it is clear that $\mathcal{R}\bar{u}$ satisfies exactly the equation

$$\frac{\partial \mathcal{R}\bar{u}}{\partial t} = \frac{\partial \mathcal{R}\bar{u}}{\partial x} + \tau\left[(1-x^2)T_N'\right]' \tag{3.5}$$

for the reconstruction procedure (2.27) and

$$\frac{\partial \mathcal{R}\bar{u}}{\partial t} = \frac{\partial \mathcal{R}\bar{u}}{\partial x} + \tau_1(1+x)T_N' \tag{3.6}$$

for (2.39). τ and τ_1 are quantities depending on time t.

Note that (3.6) is the error equation for the Chebyshev collocation method. For a scalar linear equation the CAC method corresponding to (3.6) is equivalent to the known collocation method [2]. It remains to investigate the stability of (3.5). For simplicity, in the remainder of this section we denote $\mathcal{R}\bar{u}$ by u. From the construction of $\mathcal{R}$ in section 2 we know that $u(x,t)$ satisfies the boundary condition in (3.1), i.e. $u(1,t) = 0$.

10

It is interesting to note from (3.6) that the CAC method with reconstruction operator (2.27) is equivalent - for constant coefficient case - to the collocation method where the grid points are the zeros of the polynomial

$$Q(x) = \left[(1-x^2)T_N'(x)\right]'. \tag{3.7}$$

Note that by using standard identities

$$Q(x) = -\left[xT_N'(x) + N^2 T_N(x)\right]. \tag{3.8}$$

The term τ in the right hand side of (3.5) is determined by the boundary condition $u(1,t) = \mathcal{R}\bar{u}(1,t)$. In fact substituting $x = 1$ in the equation (3.5) and noting that $\frac{\partial u}{\partial t}(1,t) = 0$ we get for τ

$$\tau = \frac{u_x(1,t)}{2N^2} \tag{3.9}$$

An alternative expression can be obtained by equating the highest coefficient on both sides of (3.5), thus if u is expanded as

$$u(x) = \sum_{k=0}^{N} \hat{u}_k T_k(x) \tag{3.10}$$

then

$$\tau = -\frac{1}{N(N+1)}\frac{\partial \hat{u}_N}{\partial t} \tag{3.11}$$

Before stating the main stability result of this paper, we state the following lemma

Lemma 3 *If $f(x) = \sum_{k=0}^{4N-1} a_k T_k(x)$, then*

$$\frac{\pi}{N}\sum_{j=0}^{N}\frac{1}{\bar{c}_j} f(x_j) = \int_{-1}^{1}\frac{f(x)}{\sqrt{1-x^2}}\,dx + \pi a_{2N} \tag{3.12}$$

where $\bar{c}_j$ is defined in (2.10); x_j $0 \le j \le N$ are the Chebyshev-Lobatto points.

Proof: see [3].

The main stability result follows from the following lemma

11

Lemma 4 *Let $U(x,t)$ be the solution of (3.2) and $u(x,t) = \mathcal{R}\bar{u}$ be the approximation to $U(x,t)$ by the CAC method (1.3) with the reconstruction operator (2.27) or (2.30), then*

$$\frac{\partial}{\partial t}\left[\frac{\pi}{2N}\sum_{j=1}^{N}\frac{1}{\bar{c}_j}\frac{1+x_j}{1-x_j}u^2(x_j,t) + \frac{\pi N}{2(N+1)}\hat{u}_N^2(t)\right] \leq 0. \tag{3.13}$$

Proof: We multiply (3.5) by $\frac{1}{\bar{c}_j}\frac{1+x_j}{1-x_j}u(x_j,t)$ and sum from the points $x_N = -1$ to x_1 to get

$$\begin{aligned}\frac{\partial}{\partial t}\frac{\pi}{2N}\sum_{j=1}^{N}\frac{1}{\bar{c}_j}\frac{1+x_j}{1-x_j}u^2(x_j,t) &= \frac{\pi}{N}\sum_{j=1}^{N}\frac{1}{\bar{c}_j}\frac{1+x_j}{1-x_j}u(x_j,t)u_x(x_j,t)\\ &- \tau N^2\frac{\pi}{N}\sum_{j=1}^{N}\frac{1}{\bar{c}_j}\frac{1+x_j}{1-x_j}u(x_j,t)T_N(x_j)\end{aligned} \tag{3.14}$$

We treat the two terms in the right hand side of (3.14) separately.

$$\begin{aligned} I &= \frac{\pi}{N}\sum_{j=1}^{N}\frac{1}{\bar{c}_j}\frac{1+x_j}{1-x_j}u(x_j,t)u_x(x_j,t)\\ &= \frac{\pi}{N}\sum_{j=0}^{N}\frac{1}{\bar{c}_j}\frac{1+x_j}{1-x_j}u(x_j,t)u_x(x_j,t) + \frac{\pi}{N}u_x^2(1,t).\end{aligned} \tag{3.15}$$

By noting the exactness of the Gauss Lobatto formula one gets

$$I = \int_{-1}^{1}\frac{1+x}{1-x}u(x,t)\frac{\partial}{\partial x}u(x,t)\frac{dx}{\sqrt{1-x^2}} + \frac{\pi}{N}u_x^2(1,t) \tag{3.16}$$

and integration by part yields

$$I = -\int_{-1}^{1}\frac{1+\frac{x}{2}}{\sqrt{1-x^2}}\frac{u^2(x,t)}{(1-x)^2}\,dx + \frac{\pi}{N}u_x^2(1,t). \tag{3.17}$$

We use again the Gauss Lobatto formula to get

$$I = -\frac{\pi}{N}\sum_{j=0}^{N}\frac{1}{\bar{c}_j}(1+\frac{x_j}{2})\frac{u^2(x_j,t)}{(1-x_j)^2} + \frac{\pi}{N}u_x^2(1,t) \tag{3.18}$$

and therefore

$$I \leq -\frac{3}{4}\frac{\pi}{N}u_x^2(1,t) + \frac{\pi}{N}u_x^2(1,t) = \frac{\pi}{4N}u_x^2(1,t). \tag{3.19}$$

We turn now to the second term in (3.14)

$$II = -\tau N^2\frac{\pi}{N}\sum_{j=1}^{N}\frac{1}{\bar{c}_j}\frac{1+x_j}{1-x_j}u(x_j,t)T_N(x_j) \tag{3.20}$$

12

or

$$II = -\tau N^2 \frac{\pi}{N} \sum_{j=0}^{N} \frac{1}{\bar{c}_j} \frac{1+x_j}{1-x_j} u(x_j,t) T_N(x_j) - \tau N^2 \frac{\pi}{N} u_x(1,t). \tag{3.21}$$

Using (3.9) and (3.11) for τ one gets

$$II = \frac{N}{N+1} \frac{\partial \hat{u}_N}{\partial t} \frac{\pi}{N} \sum_{j=0}^{N} \frac{1}{\bar{c}_j} \frac{1+x_j}{1-x_j} u(x_j,t) T_N(x_j) - \frac{\pi}{2N} u_x^2(1,t). \tag{3.22}$$

As $\frac{1+x}{1-x} u(x,t) T_N(x)$ is a polynomial of degree less than $4N-1$, one gets using lemma 3

$$\frac{\pi}{N} \sum_{j=0}^{N} \frac{1}{\bar{c}_j} \frac{1+x_j}{1-x_j} u(x_j,t) T_N(x_j) = \int_{-1}^{1} \frac{1+x}{1-x} \frac{u T_N}{\sqrt{1-x^2}} dx + \pi a_{2N}, \tag{3.23}$$

where a_{2N} is the $2N$ coefficient in the expansion of $\frac{1+x}{1-x} u(x,t) T_N(x)$

It can be easily verified that

$$\int_{-1}^{1} \frac{1+x}{1-x} \frac{u T_N}{\sqrt{1-x^2}} dx = -\frac{\pi}{2} \hat{u}_N, \tag{3.24}$$

$$a_{2N} = -\frac{1}{2} \hat{u}_N, \tag{3.25}$$

so

$$II = -\pi \frac{N}{N+1} \hat{u}_N \frac{\partial \hat{u}_N}{\partial t} - \frac{\pi}{2N} u_x^2(1,t). \tag{3.26}$$

Using (3.19) and (3.26) we get

$$I + II \leq -\pi \frac{N}{N+1} \hat{u}_N \frac{\partial \hat{u}_N}{\partial t} - \frac{\pi}{4N} u_x^2(1,t). \tag{3.27}$$

Substituting (3.27) into (3.14) yields (3.13). This completes the proof of the lemma.

Q.E.D.

We note that from Lemma 3

$$\frac{\pi}{2N} \sum_{j=1}^{N} \frac{1}{\bar{c}_j} \frac{1+x_j}{1-x_j} u^2(x_j,t) = \frac{1}{2} \int_{-1}^{1} \frac{1+x}{1-x} \frac{u^2(x,t)}{\sqrt{1-x^2}} dx - \frac{\pi}{4} \hat{u}_N^2(t), \tag{3.28}$$

so that we can finally state that

Theorem 1 *Let $u(x,t) = \mathcal{R}\bar{u}$ be the solution of the CAC method with the reconstruction operator (2.27) or (2.30), then*

$$\frac{1}{2} \int_{-1}^{1} \frac{1+x}{1-x} \frac{u^2(x,t)}{\sqrt{1-x^2}} dx + \frac{2N-1}{4(N+1)} \pi \hat{u}_N^2(t) \leq \frac{1}{2} \int_{-1}^{1} \frac{1+x}{1-x} \frac{u^2(x,0)}{\sqrt{1-x^2}} dx + \frac{2N-1}{4(N+1)} \pi \hat{u}_N^2(0). \tag{3.29}$$

13

4 Numerical Results

In this section we apply the CAC spectral method (1.3) to the one dimensional gas dynamics equations. The time evolution is done by the Runge-Kutta type method. Each step of the Runge-Kutta scheme is done as follows:

Fully Discretized CAC method

Step 1: Reconstruction:

given $\bar{u}^n_{j-\frac{1}{2}}$, $j = 1, \cdots, N$, we use the boundary condition and the matrix $\mathbf{R_N}$ to find the point values u_j, $j = 1, \cdots N$ as suggested in (2.27) or (2.30);

Step 2: Solution in time:

we update the values $\bar{u}^{n+1}_{j-\frac{1}{2}}$, $j = 1, \cdots, N$ using the forward scheme,

$$\bar{u}^{n+1}_{j-\frac{1}{2}} = \bar{u}^n_{j-\frac{1}{2}} - \frac{\Delta t}{x_{j-1} - x_j}\left[F(u^n_{j-1}) - F(u^n_j)\right], \; j = 1, \cdots, N, \tag{4.1}$$

where Δt is the time step.

The reconstruction operator $\mathbf{R_N}$ yields spectrally accurate point value approximations to the exact solutions if the exact solutions are smooth, thus $\bar{u}_{j-\frac{1}{2}}$, $j = 1, \cdots, N$ can be expected to approximate their cell averages accurately. However, if the exact solutions are discontinuous, the point values u_j, $j = 0, \cdots, N$ obtained by $\mathbf{R_N}$ will be oscillatory as the result of the Gibbs phenomenon. In [1] we proposed a practical way to obtain essentially nonoscillatory spectral reconstruction to a discontinuous function from its oscillatory Fourier approximations. The key idea there is to augment the Fourier space by adding simple discontinuous functions whose locations and magnitudes of discontinuities are approximations to those of the shock waves in the numerical solutions. In our computations of CAC method, we extend this idea along the same line to obtain essentially nonoscillatory reconstructions. The estimates on these reconstructions will be appearing in a separate work. We refer the reader to ([1]) for further details.

Now consider the Riemann problem for the Euler equations for a polytropic gas

14

$$\mathbf{U}_t(x,t)+\mathbf{f}_x(\mathbf{U})=0, \tag{4.2}$$

where $\mathbf{U(x,t)}$ and $\mathbf{f(U)}$ are defined as

$$\mathbf{U}=(p,M,E)^{\mathsf{T}}, \qquad -1\le x\le 1,$$
$$\mathbf{f}(\mathbf{U})=q\mathbf{U}+(0,P,qP)^{\mathsf{T}}, \tag{4.3}$$

where

$$P=(\gamma-1)(E-\frac{1}{2}\rho q^2), \qquad M=\rho q, \tag{4.4}$$

with $\gamma=1.4$ and the initial conditions are as follows

$$\begin{aligned}(\rho_L,q_L,P_L) &= (3.857143, 2.629369, 10.33333) \quad \text{when } x<-0.8,\\ (\rho_R,q_R,P_R) &= (1+\epsilon\sin 5\pi x, 0, 1) \quad \text{when } x>-0.8,\end{aligned} \tag{4.5}$$

where $\epsilon=0.2$.

The solutions to (4.2) - (4.5) model the interaction between a moving shock wave and disturbances. Note that in the right state of the density a sinuous perturbation of magnitude $\epsilon=0.2$ is superimposed upon a constant state. ¿From linear analysis it can be shown that the disturbances will interact with the shock wave. A density wave of different frequency and magnitude will emerge behind the shock wave. Also the disturbance in the density field will perturb the velocity and pressure fields behind the shock wave. The numerical solution of this Riemann problem mandates a high order scheme in order to capture the fine structures in the solutions for a correct interpretation of the physical process. We test this problem with second order MUSCL scheme [5] and third order point value version ENO finite difference scheme [6] and the CAC spectral method proposed in this paper. Our numerical results have shown clearly the advantage of a higher order numerical scheme for problems with complicated structures.

Figure 1 - 3 show the density profiles obtained by the three methods mentioned above. Figure 1 is the result using CAC spectral method. The solid lines are the solutions taken

15

from [6] using the third order ENO finite difference method with $N = 800$ which we take as the exact solutions . Figure 2 and Figure 3 are the results with the second order MUSCL scheme [5] and the third order ENO finite difference scheme respectively. In all three cases we use the same amount of mesh points $N = 220$. All the results are plotted at the same time $t = 0.3$.

ACKNOWLEDGMENT

We would like to thank Prof. Chi-wang Shu for many useful discussions.

References

[1] W. Cai, D. Gottlieb, C. W. Shu, "Essentially Non-Oscillatory Spectral Fourier method for Shock Wave Calculations" Math. Comput., Vol. 52, No. 186, April 1989, pp. 389-410.

[2] D. Gottlieb and E. Turkel, "Topics in Spectral Methods", Lecture Notes in Mathematics, No. 1127, edited by A. Dold and B. Eckman, 1983 , pp. 115 - 155

[3] D. Funaro and D. Gottlieb " A New Method of Imposing Boundary Conditions in Pseudospectral Approximation of Hyperbolic Equations" Math. Compt. December 1988

[4] A. Harten, "On the nonlinearity of modern shock capturing schemes" ICASE Report No. 86-69 (October 1986)

[5] S. Osher, "Convergence of Generalized MUSCL Schemes," SIAM J. Numer. Anal., v. 22, 1985, pp. 947-961.

[6] C.-W. Shu and S. Osher, "Efficient Implementation of Essentially Nonoscillatory Shock-Capturing Schemes, II," J. Compt. Phys., V. 83, No. 1, July 1989.

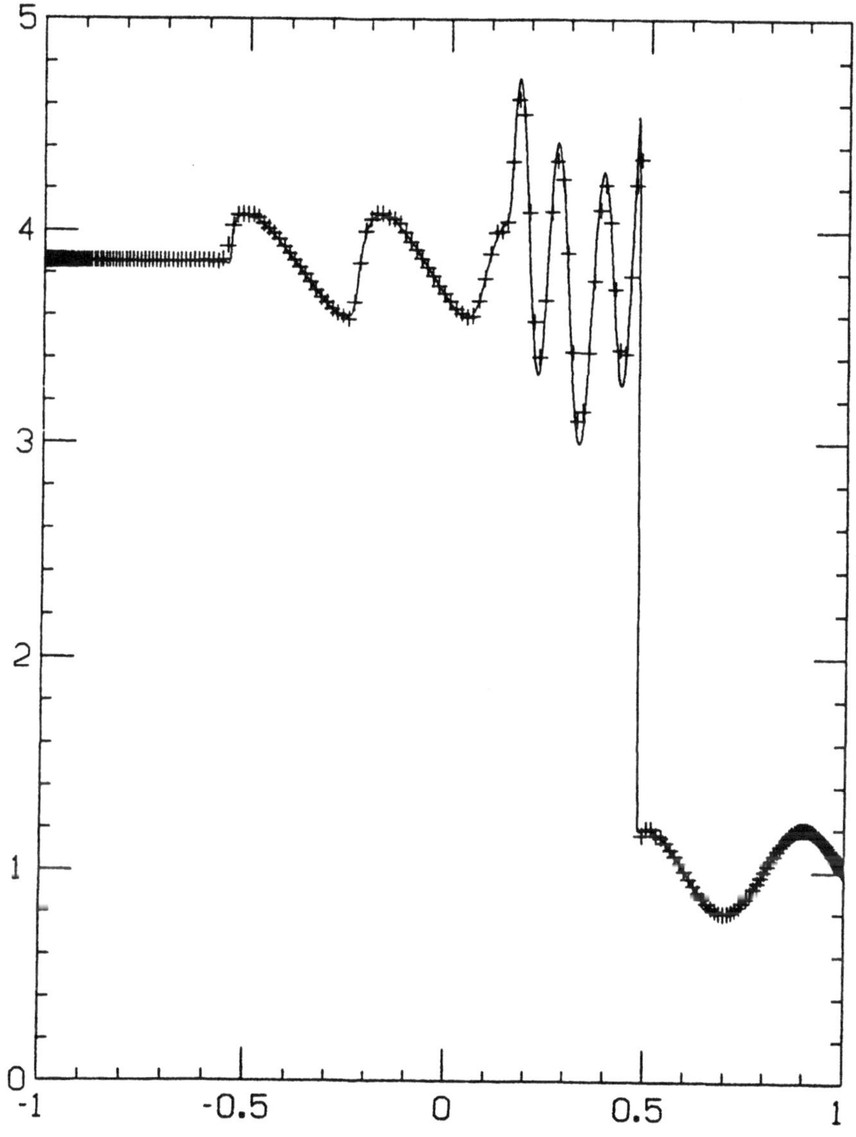

Figure 1: The CAC spectral method: density, N = 220, time t = 0.3. (+) - numerical solutions, solid line - exact solutions.

18

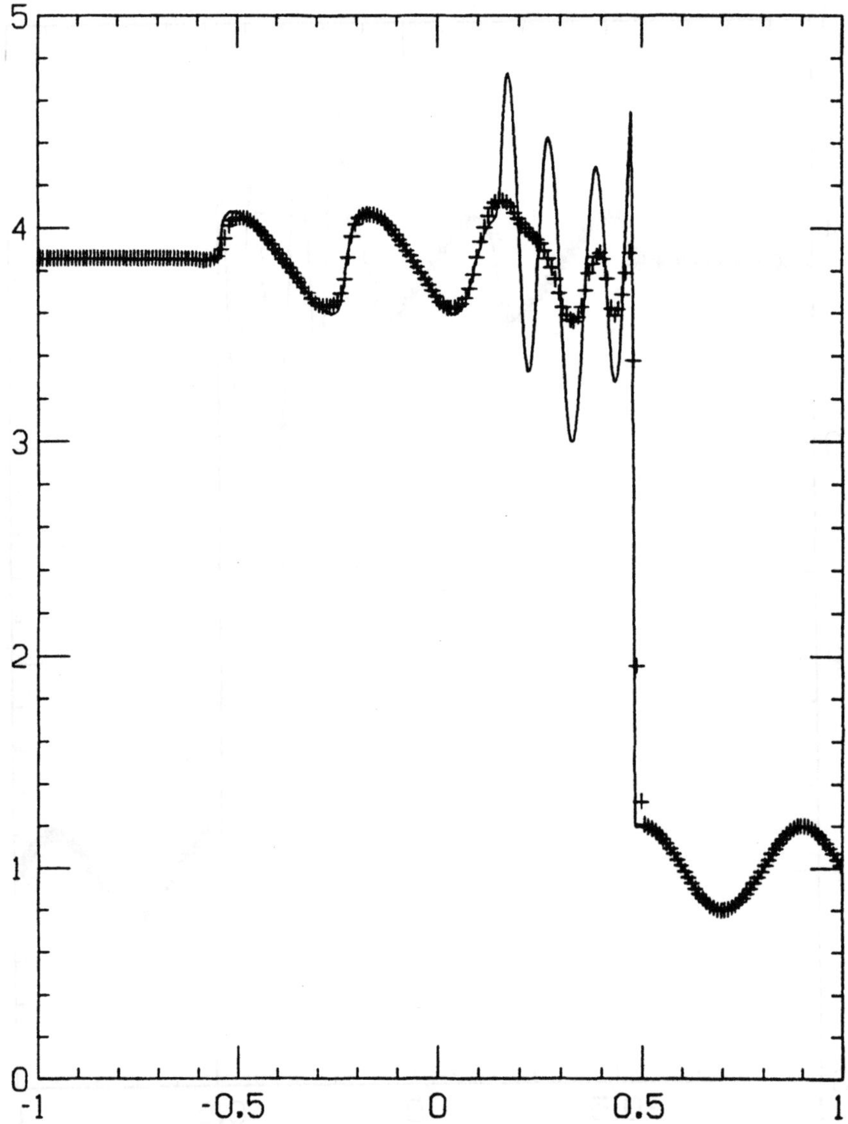

Figure 2: The second order MUSCL scheme: density, N = 220, time t = 0.3. (+) - numerical solutions, solid line - exact solutions.

19

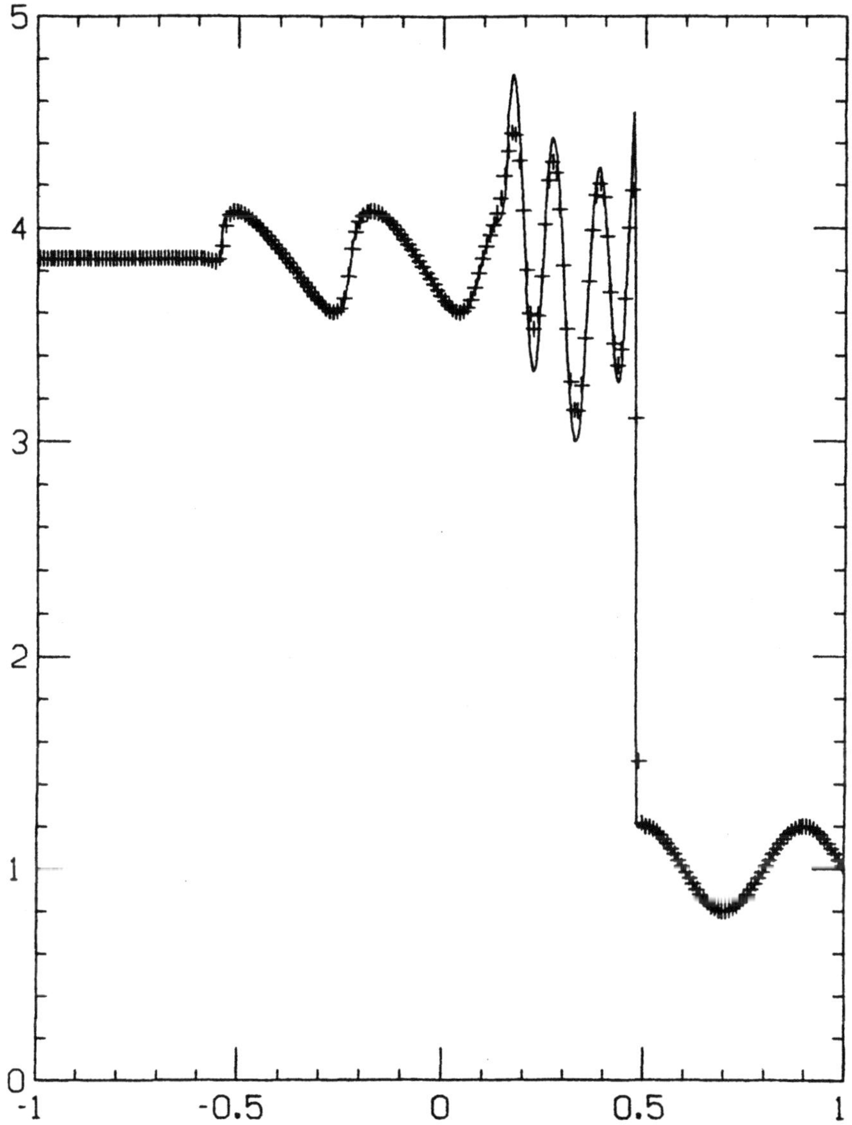

Figure 3: The third order ENO scheme: density, N = 220, time t = 0.3. (+) - numerical solutions, solid line - exact solutions.

20

Figure 3. The [illegible] order [illegible] density, $N = 220$, time $t = 0.8$ [illegible] numerical solution, solid line = exact solution.

Chapter IV

Multidimensional Schemes

Annotation

Once non-oscillatory interpolation and upwind differencing were routinely used as one-dimensional building blocks in constructing multi-dimensional codes, a small group of researchers started looking at genuinely multi-dimensional schemes. The papers in this last chapter deal with the construction of multi-dimensional algorithms that preserve all of the favorable properties of their one-dimensional predecessors. The multi-dimensional context raises a number of new issues, including questions of grid regularity, the choice of cell-centered *vs.* cell-vertex schemes, and the discrete representation of multi-dimensional fluid physics.

1. **A Rotationally Biased Upwind Difference Scheme for the Euler Equations** (S. F. Davis)
   This paper generated a small wave of papers on what may be called *multi-directional* flux functions. It describes the first attempt to use Riemann solvers in a rotated frame having more physical significance than the grid directions.

2. **Discrete Models for the Numerical Analysis of Time-Dependent Multidimensional Gas Dynamics** (P. L. Roe)
   This paper may be regarded as the birthplace of research in genuinely multi-dimensional methods for the Euler equations. It describes the use of plane traveling waves to model or "explain" the local residuals in a computational cell.

3. **Discontinuous Solutions to Hyperbolic Systems Under Operator Splitting** (P. L. Roe)
   Another key paper, showing that the error induced by operator splitting gets worse as one improves the one dimensional operators, thus motivating the development of multidimensional schemes.

4. **High Resolution Finite Volume Methods on Arbitrary Grids Via Wave Propagation** (R. J. LeVeque)
   Explains how, in a MUSCL-type code, by following waves as they cross more than one cell, a scheme can avoid the CFL time-step limit while remaining fully explicit. The author's use of this technique is in treating small irregular cells formed by a solid wall cutting through a non-aligned Cartesian grid.

5. **Design of an Essentially Nonoscillatory Reconstruction Procedure on Finite-Element-Type Meshes** (R. Abgrall)
   An attempt to ENO reconstruction on an arbitrary triangular grid; includes numerical examples.

6. **A Contribution to the Great Riemann Solver Debate** (J. J. Quirk)
   Details a number of failures of current-type Godunov schemes, and an array of repairs needed to make one-dimensional Riemann solvers work in multi-dimensional problems. This paper provides an excellent snap-shot of the state of the art in practical upwind schemes, and points out the need for continuing work in fully multi-dimensional schemes.

7. **An Approximate Riemann Solver for Magnetohydrodynamics (That Works in More Than One Dimension)** (K. G. Powell)
   By abandoning the conservation form of the MHD equations, a change in the eigenstructure is achieved that allows the formulation of a Riemann solver suited for multi-dimensional calculations, and prevents the accumulation of numerical errors in the value of $\nabla \cdot \mathbf{B}$ (= 0 nominally).

A Rotationally Biased Upwind Difference Scheme for the Euler Equations

STEPHEN F. DAVIS*

Institute for Computer Applications in Science and Engineering, NASA Langley Research Center, Hampton, Virginia 23665

Received September 27, 1983

The upwind difference schemes of Godunov, Osher, Roe and van Leer are able to resolve one-dimensional steady shocks for the Euler equations within one or two mesh intervals. Unfortunately, this resolution is lost in two dimensions when the shock crosses the computing grid at an oblique angle. To correct this problem, a numerical scheme is developed which automatically locates the angle at which a shock might be expected to cross the computing grid then constructs separate finite difference formulas for the flux components normal and tangential to this direction. Numerical results are presented which illustrate the ability of this new method to resolve steady oblique shocks.

1. INTRODUCTION

Beginning with the work of Godunov [2], considerable effort has been expended seeking numerical methods which can solve the equations of gas dynamics accurately and sharply resolve discontinuities.

In the case of one-dimensional flows containing steady discontinuities, this effort has paid off. Work by van Leer [14–16], Roe [12], Osher [1, 11], Harten [5] and others indicates that the mechanism of "shock capture" is well understood and that it is now possible to design second-order-accurate schemes which resolve steady discontinuities within one or two mesh intervals without wiggles. A recent paper by Harten and Hyman [3] indicates that these results can be extended to one-dimensional flows with moving discontinuities.

In the case of two-dimensional flows, some progress has been made but there is room for improvement. It is well known that the one-dimensional results cited above can be reproduced in two-dimensional calculations if the computing grid is chosen so that only one set of grid lines crosses a discontinuity in the flow. This would appear to be the optimum way to account for the two-dimensional aspects of the flow but in practice it is very difficult to choose an appropriate grid either a priori or adaptively during a calculation. We shall not examine this approach here.

* Research supported by the National Aeronautics and Space Administration under NASA Contract NASI-17070 while the author was in residence at ICASE, NASA Langley Research Center, Hampton, Virginia 23665.

An alternative approach is to use a locally one-dimensional or fractional step method (cf. Yanenko [18]) with a high-resolution, one-dimensional scheme in each one-dimensional step. This method ignores the two-dimensional orientation of discontinuities but appears to work surprisingly well (cf. Yee *et. al.* [19], Woodward and Colella [17]). Since this method is also simple to program, it is quite attractive for practical application.

Despite these attractions, we attempt to do better. In particular, in this paper, we derive a two-dimensional scheme which can sharply resolve weak shocks which are skew to the fixed computing grid. Locally one-dimensional methods have difficulty dealing with this situation. An outline of the paper follows.

In Section 2, we briefly discuss the ideas behind the Godunov [2] method and show heuristically why it resolves steady shocks so well. We then describe various simplifications of Godunov's method that retain its shock-resolving properties, including the flux vector splitting method of van Leer [16] which we later use.

In Section 3, we derive the numerical scheme. Since our scheme involves rotated differences we first review some early work by Jameson [6] in which he applies rotated differences to the transonic full potential equations. We then derive a rotated difference scheme for the Euler equations. Our scheme differs from Jameson's in that for our scheme both the finite difference formula and the computational stencil vary with angle whereas for Jameson's scheme only the finite difference formula varies.

In Section 4, we consider the choice of angle for the rotated differences. We present a choice which seems to be appropriate for shock resolution with the Euler equations and then we briefly discuss more general discontinuities and more general conservation laws.

Section 5 contains numerical computations using this scheme and comparisons with other schemes.

Section 6 summarizes the present work, discusses the results of the present work and presents an outline for future work.

2. Godunov-Type Methods and One-Dimensional Problems

In this section we study the Godunov method for one-dimensional systems of conservation laws

$$w_t + f(w)_x = w_t + A(w) w_x = 0 \tag{2.1a}$$

where

$$A(w) = \frac{df}{dw}(w). \tag{2.1b}$$

In particular we wish to determine which features of this scheme are responsible for its ability to resolve steady discontinuities. This discussion follows closely that of Harten *et al.* [4], so the reader is encouraged to consult their paper for additional details.

The construction of Godunov's scheme is as follows. At discrete time levels t_n, $n = 0, 1, \ldots$, the numerical approximation $v(x, t_n)$ to the solution $u(x, t_n)$ of (2.1) is taken to be a piecewise constant function of x, i.e.,

$$v(x, t_n) = v_j^n \qquad \text{for} \quad x \in I_j = ((j - \tfrac{1}{2})\Delta x, (j + \tfrac{1}{2})\Delta x). \tag{2.2}$$

To calculate the numerical approximation at the next time level $t_{n+1} = t_n + \Delta t$ we first solve exactly the problem of the breakup of the discontinuities in $v(x, t_n)$ at all interfaces. This means that at each discontinuity of $v(x, t_n)$ we solve a local Riemann problem. The solution to the Riemann problem due, for example, to the jump at the interface between I_j and I_{j+1} is a similarity solution which depends only on the states v_j^n and v_{j+1}^n and the ratio $(x - (j + \frac{1}{2})\Delta x)/(t - t_n)$. We denote this solution by $u\{(x - (j + \frac{1}{2})\Delta x)/(t - t_n); v_j^n, v_{j+1}^n\}$. Since signals propagate with finite velocity, there will be no interaction between neighboring Riemann problems if $(|a_{\max}|\Delta t)/\Delta x < \frac{1}{2}$, where $|a_{\max}|$ is the largest signal speed. In this case the exact solution of (2.1) with initial conditions (2.2), denoted by $u_n(x, t_n)$, is given by

$$u_n(x, t) = u\{(x - (j + \tfrac{1}{2})\Delta x)/(t - t_n); v_j^n, v_{j+1}^n\}$$

for

$$j\Delta x < x < (j + 1)\Delta x, \qquad t_n < t < t_{n+1}. \tag{2.3}$$

Godunov obtains a piecewise constant approximation $v(x, t_{n+1})$ for the next time step, by averaging $u_n(x, t_{n+1})$, i.e., by defining

$$v_j^{n+1} = 1/\Delta x \int_{I_j} u_n(x, t_{n+1})\, dx. \tag{2.4}$$

A three-point explicit finite difference scheme for (2.1) of the form

$$v_j^{n+1} = v_j^n - \frac{\Delta t}{\Delta x}\left[F(v_j^n, v_{j+1}^h) - F(v_{j-1}^h, v_j^n)\right] \tag{2.5}$$

is said to be in conservation form. The function F appearing in (2.5) is called the numerical flux function.

If we evaluate the integral in (2.4) by integrating (2.1) over $I_j x(t_n, t_{n+1})$ we can show that Godunov's method is in conservation form with a numerical flux function given by

$$F(v, \omega) = f(u(0; v, \omega)). \tag{2.6}$$

Details can be found in the paper by Harten, Lax, and van Leer [4].

Next we examine how this scheme resolves stationary shocks. For the sake of clarity we consider only a single conservation law. Qualitatively similar results can be derived for systems of conservation laws (see Lax [8]) but the details are far more complicated.

A shock is an exact discontinuous solution to the Riemann problem of the form

$$u_n(x,t) = \begin{cases} u_L, & x < st \\ u_R, & x > st \end{cases} \tag{2.7}$$

where s satisfies the Rankine–Hugoniot jump condition

$$s(u_R - u_L) = f(u_R) - f(u_L) \tag{2.8}$$

and the entropy condition

$$a_L > s > a_R \tag{2.9}$$

where

$$a_L = \frac{df}{du}(u_L), \qquad a_R = \frac{df}{du}(u_R). \tag{2.10}$$

A steady shock is a shock with $s = 0$.

If the initial conditions represent a steady shock located on the interface between cells I_{-1}, I_0, i.e., the initial data are

$$v_j^0 = u_0(x, t_0) = \begin{cases} u_L & \text{for } x \in I_j \text{ and } j < 0 \\ u_R & \text{for } x \in I_j \text{ and } j \geqslant 0 \end{cases} \tag{2.11}$$

then Eq. (2.9) will be satisfied exactly for all time levels and (2.11) is an exact solution to Godunov's method.

We now show that, if the initial conditions represent a steady shock located within a cell, say, I_0, then Eq. (2.4) will yield

$$v_j^0 = \begin{cases} u_L & \text{for } j < 0 \\ u_m & \text{for } j = 0 \\ u_R & \text{for } j > 0 \end{cases} \tag{2.12}$$

where u_m is some intermediate value between u_L and u_R.

Since Δt is chosen so that waves do not interact, at time t_1 the solution to the Riemann problem at the interface between I_{-1} and I_0 will consist of a shock moving at the speed

$$S_L = \frac{f(u_m) - f(u_L)}{u_m - u_L}. \tag{2.13}$$

If the shock is weak

$$f(u_m) \approx f(u_R) + \frac{df}{du}(u_R)(u_m - u_R) \approx f(u_R) + a_R(u_m - u_R). \tag{2.14}$$

Substitute this into (2.13) and note that $f(u_L)=f(u_R)$, so

$$S_L \approx a_R \frac{(u_m - u_R)}{(u_m - u_L)}. \tag{2.15}$$

Since u_m is between u_L and u_R, $(u_m - u_R)$ and $(u_m - u_L)$ will have opposite signs and their quotient will be negative. Equation (2.9) with $s=0$ shows that a_R is negative. Thus S_L is positive, the shock moves to the right and

$$u(0; u_L, u_m) = u_L. \tag{2.16}$$

By a similar argument we can show that on the interface between I_0 and I_1

$$u(0; u_m, u_R) = u_R. \tag{2.17}$$

Substitution of (2.16) and (2.17) into (2.6) and (2.5) yields

$$v_0^1 = v_0^0 - \frac{\Delta t}{\Delta x}\,[f(u_R) - f(u_L)] = v_0^0. \tag{2.18}$$

Thus (2.12) is an exact solution to Godunov's method.

We note in passing that the choice of numerical fluxes

$$F(u_L, u_m) = f(u_L)$$

and

$$F(u_m, u_R) = f(u_R)$$

which does not depend on u_m, confines the effect of the averaging which created u_m to a single cell. This would not be true of a conventional finite difference method whose numerical fluxes depended on all of its arguments. Subsequent time steps would propagate the effect of u_m through the entire domain and thus spread the shock over many cells.

The construction of solutions of Riemann problems for nonlinear systems is a complicated iterative procedure. In addition, Eq. (2.6) shows that although the entire Riemann solution is computed, only values at cell interfaces are actually used. For this reason much recent research has been devoted to the construction of numerical flux functions which retain the shock-capturing ability of Godunov's scheme but which are simpler to construct.

This is accomplished either by approximating the solution to the exact Riemann problem or by approximating certain features of the Godunov numerical flux. In the following we briefly describe some successful methods of both types. The reader is referred to the original papers for details.

Harten *et al.* [4] refer to methods which approximate Riemann solutions as Godunov-type methods. In their paper they construct two such methods by using the conservation laws (1.1) to lump together many of the features of the exact Riemann solution. These methods should be extremely simple to implement but, at present, no numerical results are available.

Another Godunov-type scheme is Roe's [12] method. Like Godunov, Roe begins with the piecewise constant approximation (2.2) but he replaces the solution to a Riemann problem for the exact equation due, for example, to the jump at the interface between I_j and I_{j+1} by the solution to a Riemann problem for the linearized equations

$$u_t + A(v_j^n, v_{j+1}^n)u_x = 0 \tag{2.19}$$

where the matrix A is chosen to satify

$$A(v_j^n, v_{j+1}^n) \cdot (v_{j+1}^n - v_j^n) = f(v_{j+1}^n) - f(v_j^n). \tag{2.20}$$

Conditions (2.20) assure that the Godunov method using the Roe approximate Riemann solution can be written in conservation form (2.5).

Osher and Solomon [11] construct a numerical flux which approximates the directional bias of the Godunov flux. To accomplish this, they write the flux difference as a path integral in state space as follows:

$$\Delta f := f(q) - f(p) = \int_p^q A(u)\, du. \tag{2.21}$$

The path of integration Γ is chosen to be piecewise parallel to the right eigenvectors R_k of A. Without going into detail, we point out that this choice of paths permits the right-hand side of (2.21) to be split as follows:

$$\int_p^q A(u)\, du = \int_p^q A^+(u)\, du + \int_p^q A^-(u)\, du. \tag{2.22}$$

Here, the integrals on the right of (2.22) are path integrals along those parts of the path Γ which are parallel to right eigenvectors R_k corresponding to the positive and negative eigenvalues of A, respectively. With this notation, the Osher–Solomon numerical flux can be written in the alternative forms:

$$F_{\text{os}}(p, q) = f(p) + \int_p^q A^-(u)\, du = f(q) - \int_p^q A^+(u)\, du. \tag{3.23}$$

The numerical flux corresponding to each method that we have dicussed satisfies

$$F(p, q) = f(p) \tag{2.24a}$$

if all eigenvalues of A are >0 and

$$F(p, q) = f(q) \tag{2.24b}$$

if all eigenvalues of A are <0. In [16] van Leer constructs a numerical flux functions for the Euler equations which satisfies (2.24a), (2.24b) and which permits steady shocks to be resolved within two cells.

To accomplish this van Leer splits the flux $f(w)$ of (2.1) for the Euler equations into a forward flux $f^+(w)$ and a backward flux $f^-(w)$ which satisfies

$$f(w) = f^+(w) + f^-(w) \tag{2.25}$$

where

$$f^+(w) = f(w), \qquad f^-(w) = 0 \tag{2.26a}$$

when the Mach number $M > 1$ and

$$f^+(w) = 0, \qquad f^-(w) = f(w) \tag{2.26b}$$

when the Mach number $M < -1$. The split fluxes are required to be continuously differentiable so that computed numerical solutions will be smooth. van Leer requires that the Jacobian matrices of the split fluxes $df^{\pm}/dw$ each have an eigenvalue vanish for $|M| < 1$ and he shows that this condition permits steady shocks to be resolved within two cells. Other conditions are imposed to determine the particular form of van Leer's split fluxes. These conditions and the reason for their choice are described in van Leer's paper.

For the one-dimensional Euler equations with ideal gas law considered as a function of density ρ, sound speed c and Mach number M, the resulting splittings for $|M| < 1$ are

(1) mass

$$\begin{aligned} \rho u &= f_{\text{mass}} = f^+_{\text{mass}} + f^-_{\text{mass}} \\ &= \rho c M = \rho c\{\tfrac{1}{2}(M+1)\}^2 - \rho c\{\tfrac{1}{2}(-M+1)\}^2 \end{aligned} \tag{2.27}$$

(2) momentum

$$\begin{aligned} \rho u^2 + p &= f_{\text{momentum}} = f^+_{\text{momentum}} + f^-_{\text{momentum}} \\ &= \rho c^2(M^2 + 1/\gamma) = \rho c^2 \left\{\frac{1}{2}(M+1)\right\}^2 \left(\frac{\gamma-1}{\gamma}M + 2/\gamma\right) \\ &\quad + \rho c^2 \left\{\frac{1}{2}(-M+1)\right\}^2 \left(-\frac{\gamma-1}{\gamma}M + 2/\gamma\right) \end{aligned} \tag{2.28}$$

(3) total energy

$$\begin{aligned} (e+p)u &= f_{\text{energy}} = f^+_{\text{energy}} + f^-_{\text{energy}} = \rho c^3 M(\tfrac{1}{2}M^2 + 1/(\gamma-1)) \\ &= \frac{\gamma^2}{2(\gamma^2-1)} \frac{(f^+_{\text{momentum}})^2}{(f^+_{\text{mass}})} + \frac{\gamma^2}{2(\gamma^2-1)} \frac{(f^-_{\text{momentum}})^2}{(f^-_{\text{mass}})}. \end{aligned} \tag{2.29}$$

The reader is referred to van Leer's paper for a detailed derivation of these expressions. The numerical flux based on this splitting is

$$F(u, v) = f^{+}(u) + f^{-}(v). \tag{2.30}$$

In the next section we incorporate this numerical flux into a two-dimensional rotated difference scheme.

3. Derivation of a Rotated Difference Scheme

In a supersonic region, the method of Murman and Cole [10] solves the transonic potential equation by replacing derivatives in the streamwise direction with upwind difference approximations and by replacing derivatives in the direction normal to the streamlines with central difference approximations. This is easy to do when the computing grid is approximately aligned with the streamwise and normal directions but is difficult otherwise.

Jameson [6] overcomes this problem in the following way. He writes the transonic potential equation in a coordinate system aligned with and normal to the local streamwise direction. He then expresses the derivatives in the streamwise-normal coordinate system in terms of derivatives in the coordinate system of his computing grid, making note of which terms come from streamwise derivatives and which terms come from normal derivatives. Finally, those terms which come from streamwise derivatives are approximated by upwind difference formulas and those terms which come from normal derivatives are approximated by central difference formulas. This creates a very effective method.

In the following we derive a method for the Euler equations which is based on this rotated difference idea. In particular, we attempt to choose a local coordinate system which permits us to apply the one-dimensional theory of Section 2. This means that our local coordinate directions must be normal and tangential to potential shock directions. In Section 4 we show how to determine these directions. Here we assume that the directions are known and derive difference formulas that are based on these directions. We note in passing that shocks are approximtely normal to the streamlines in transonic flows. Thus our choice of local coordinate system is equivalent to Jameson's in this case.

Consider a local cartesian coordinate system chosen as above. Such a coordinate system has coordinates (x', y'). The coordinate system of the computing grid, assumed for simplicity to be cartesian, has coordinates (x, y). Dependent variables measured in the local and global coordinate systems are primed and unprimed, respectively. This geometry is shown in Fig. 1.

Since the Euler equations are invariant under rotation, they can be written immediately in the local coordinate system as

$$\frac{\partial w'}{\partial t} + \frac{\partial f'(w')}{\partial x'} + \frac{\partial g'(w')}{\partial y'} = 0 \tag{3.1}$$

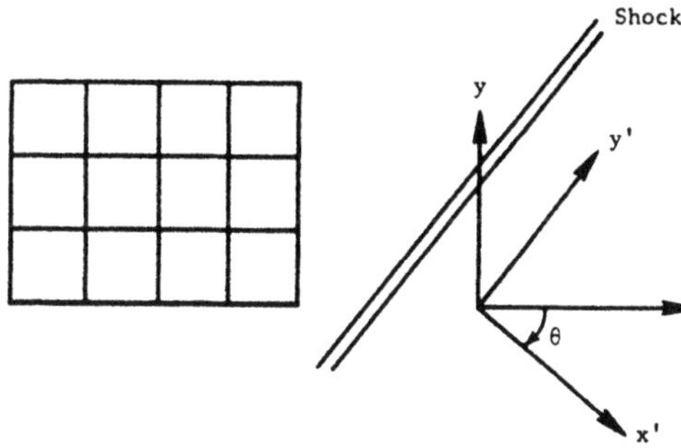

FIG. 1. Geometry of local and global coordinate systems.

where

$$w' = [\rho, \rho u', \rho v', e]^T$$

$$f' = [\rho u', \rho u'^2 + p, \rho u'v', (e+p)u']^T$$

$$g' = [\rho v', \rho u'v', \rho v'^2 + p, (e+p)v']^T \tag{3.2}$$

$$e = p/(\gamma - 1) + \tfrac{1}{2}\rho(u'^2 + v'^2) \tag{3.3}$$

and

$$u' = u \cos\theta + v \sin\theta$$

$$v' = -u \sin\theta + v \cos\theta. \tag{3.4}$$

If the coordinate system is chosen as described in the vicinity of a plane shock, the derivatives in the second term on the right of (3.1) exist and this term is equal to zero. This reduces the two-dimensional problem, locally, to the one-dimensional problem described in the previous section. We attempt to construct a numerical scheme which mimicks this behavior.

If we express the first term on the right of (3.1) in global coordinates we get

$$\frac{\partial f'}{\partial x'} = \cos\theta \frac{\partial f'}{\partial x} + \sin\theta \frac{\partial f'}{\partial y}. \tag{3.5}$$

This can be approximated by a finite difference expression

$$\frac{\cos\theta}{\Delta x}\left[F'(W'_{i+1,j}, W'_{i,j}) - F'(W'_{i,j}, W'_{i-1,j})\right]$$
$$+ \frac{\sin\theta}{\Delta y}\left[F'(W'_{i,j+1}, W'_{i,j}) - F'(W'_{i,j}, W'_{i,j-1})\right] \tag{3.6}$$

where F' is an appropriately chosen numerical flux and W' is a numerical approximation to w. The conditions that this expression be zero across a steady shock independent of Δx and Δy are

$$[F'(W'_{i+1,j}, W'_{i,j}) - F'(W'_{i,j}, W'_{i-1,j})] = 0$$

and

$$[F'(W'_{i,j+1}, W'_{i,j}) - F'(W'_{i,j}, W'_{i,j-1})] = 0. \tag{3.7}$$

The numerical flux functions corresponding to methods discussed in Section 2 admit solutions to (3.7) which spread steady shocks over only one or two mesh intervals. For this reason we choose F' to be one of these numerical flux functions. For this work we chose the van Leer flux vector splitting but any other method discussed would have sufficed.

The second term on the right of (3.1) says that g' is constant along lines which are parallel to a shock. To approximate this condition on a discrete grid we must make an assumption about the behavior of g' between grid points. For lack of any better information we assume that g' varies linearly between grid points.

Consider a local coordinate system defined by a rotation through an angle θ as shown in Fig. 2. Define

$$l = \Delta x/(-\tan\theta). \tag{3.8}$$

Then

$$l/\Delta y = \Delta x/(-\Delta y \tan\theta) = -\frac{\Delta x \cos\theta}{\Delta y \sin\theta}. \tag{3.9}$$

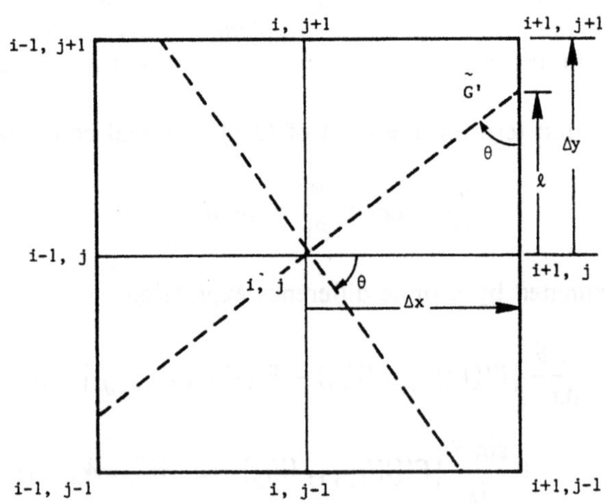

FIG. 2. Construction of tangential flux.

In the following we let $g'_{i,j}$ denote $g'(w'(x_i, y_j))$. The condition that g' varies linearly between grid points (x_{i+1}, y_j) and (x_{i+1}, y_{j+1}) can be written as

$$(g'_{i+1,j+1} - g'_{i+1,j})/\Delta y = (\tilde{g}' - g'_{i+1,j})/l$$

or

$$-(g'_{i+1,j} - \tilde{g}')\frac{\sin\theta}{\Delta x} + (g'_{i+1,j+1} - g'_{i+1,j})\frac{\cos\theta}{\Delta y} = 0. \tag{3.10}$$

Here $\tilde{g}'$ denotes $g'(w(x_{i+1}, y_j + l))$. The condition that g' be constant along a line parallel to a shock can be written

$$\tilde{g}' = g'_{i,j}. \tag{3.11}$$

If we express the second term on the right of (3.1) in global coordinates, we get

$$\frac{\partial g'}{\partial y'} = -\sin\theta\frac{\partial g'}{\partial x} + \cos\theta\frac{\partial g'}{\partial y}. \tag{3.12}$$

We allow the numerical flux on the cell boundary between the points (x_i, y_j) and (x_{i+1}, y_j) to depend on W' values at the points (x_i, y_j), $(x_i, y_{j\pm1})$, $(x_{i+1}\ y_j)$ and $(x_{i+1}, y_{j\pm1})$. We allow the numerical flux on the cell boundary between the points (x_i, y_j) and (x_i, y_{j+1}) to depend on W'' values at the points (x_i, y_j), $(x_{i\pm1}, y_j)$, (x_i, y_{j+1}) and $(x_{i\pm1}, y_{j+1})$. Then the numerical approximation to (3.12) takes the form

$$-\frac{\sin\theta}{\Delta x}[G'(W'_{i+1,j}, W'_{i+1,j\pm1}, W'_{i,j}, W'_{i,j\pm1}) - G'(W'_{i,j}, W'_{i,j\pm1}, W_{i-1,j}, W'_{i-1,j\pm1})]$$

$$+\frac{\cos\theta}{\Delta y}[G'(W'_{i,j}, W'_{i\pm1,j}, W'_{i,j+1}, W'_{i\pm1,j+1}) - G'(W'_{i,j-1}, W'_{i\pm1,j-1}, W'_{i,j}, W'_{i\pm1,j})]. \tag{3.13}$$

Since (3.10) and (3.11) represent a discrete approximation to (3.12) in the vicinity of a shock, we choose the numerical flux values in (3.13) to be equal to the corresponding physical flux values in (3.10) and (3.11). We note that by choosing lines parallel to a shock which pass through different grid points, we can construct other expressions which are similar to (3.10) and (3.11). Linear combinations of these formulas can be used to determine numerical flux functions for (3.13). Thus the formulas presented below are not unique but they have performed well in practice.

To make our G' numerical flux look similar to our F' numerical flux, we split it up as follows:

$$G'(W'_{i+1,j}, W'_{i+1,j\pm 1}, W'_{i,j}, W'_{i,j\pm 1})$$
$$= \tfrac{1}{2}[(1+Q)\,G'^{+}(W'_{i,j}, W'_{i,j\pm 1}) + (1-Q)\,G'^{-}(W'_{i+1,j}, W'_{i+1,j\pm 1})]. \quad (3.14)$$

Q is a parameter chosen to assure that the overall method remains stable. Numerical experiments indicate that if the method is stable, the computed results are not sensitive to the particular choice of Q but more study of these matters is needed. Further discussion on the choice of Q appears in Section 5. Table I specifies how G'^{+} and G'^{-} depend on the angle θ.

Until now we have discussed the derivation of numerical fluxes of local variables in local coordinate directions. To compute a global solution we need numerical fluxes of global variables in global coordinates in terms of the flux of local variables in local coordinates as shown below.

Case 1. Flux of a scalar variable:

$$f_1(w) = \rho u = \rho u' \cos\theta - \rho v' \sin\theta = f'_1(w')\cos\theta - g'_1(w')\sin\theta. \quad (3.15)$$

Case 2. Flux of a vector component:

$$f_2(w) = \rho u^2 + p = \rho(u'\cos\theta - v'\sin\theta)(u'\cos\theta - v'\sin\theta) + p$$
$$= \rho u'^2\cos^2\theta - 2\rho u'v'\cos\theta\sin\theta + \rho v'^2\sin^2\theta + p(\cos^2\theta + \sin^2\theta)$$
$$= f'_2(w')\cos^2\theta - g'_2(w')\cos\theta\sin\theta - g'_3(w')\cos\theta\sin\theta + f'_3(w')\sin^2\theta. \quad (3.16)$$

The numerical flux is computed by replacing the local flux by its corresponding local numerical flux. The two cases shown demonstrate that the flux of a scalar variable, such as ρ or e, transforms like a vector or first-order cartesian tensor and the flux of a vector component, such as a momentum component, transforms like a

TABLE I

Angle range	$G'^{+}(U'_{i,j}, U'_{i,j\pm 1})$	$G'^{-}(U'_{i+1,j}, U'_{i+1,j\pm 1})$	$G'^{+}(U'_{i,j}, U'_{i\pm 1,j})$	$G'^{-}(U'_{i,j+1}, U'_{i\pm 1,j+1})$
$-\infty < \text{Tan}\,\theta < -\frac{\Delta x}{\Delta y}$	$G'(U'_{i,j})$	$G'(U'_{i+1,j})$	$G'(U'_{i-1,j})$	$G'(U'_{i+1,j+1})$
$-\frac{\Delta x}{\Delta y} < \text{Tan}\,\theta < 0$	$G'(U'_{i,j-1})$	$G'(U'_{i+1,j+1})$	$G'(U'_{i,j})$	$G'(U'_{i,j+1})$
$0 < \text{Tan}\,\theta < \frac{\Delta x}{\Delta y}$	$G'(U'_{i,j+1})$	$G'(U'_{i+1,j-1})$	$G'(U'_{i,j})$	$G'(U'_{i,j+1})$
$\frac{\Delta x}{\Delta y} < \text{Tan}\,\theta < \infty$	$G'(U'_{i,j})$	$G'(U'_{i+1,j})$	$G'(U'_{i+1,j})$	$G'(U'_{i-1,j+1})$

second-order cartesian tensor. These transformation laws are well known (cf. Segal [13]), and are easily programmed.

To summarize, if we are given an angle θ, we compute a numerical flux as follows:

(1) Compute velocity components in the local coordinate system.

(2) Compute normal and tangential numerical flux components in the local coordinate system as described above.

(3) Use the cartesian tensor transformation laws to compute numerical flux components of the global variables in the global coordinate system.

We conclude this section with a brief discussion on what it means for a scheme to be conservative and consistent and thus show that the scheme described here possesses these characteristics.

A numerical scheme is called conservative if its numerical flux satisfies a discrete version of the divergence theorem. For a rectangular domain, this means that we must satisfy an identity of the form

$$\sum_{i=1}^{N}\sum_{j=1}^{N}\left[\frac{F_{i+1/2,j}-F_{i-1/2,j}}{\Delta x}+\frac{G_{i,j+1/2}-G_{i,j-1/2}}{\Delta y}\right]\Delta x\,\Delta y$$
$$=\sum_{i=1}^{N}\left[F_{N+1/2,j}-F_{1/2,j}\right]\Delta y+\sum_{i=1}^{N}\left[G_{i,N+1/2}-G_{i,1/2}\right]\Delta x. \tag{3.17}$$

This relation is trivially satisfied for the so-called finite-volume method. Therefore we formulate our scheme as a finite-volume method. That is, we integrate the Euler equations over the rectangular cell

$$x_{i-1/2}<x<x_{i+1/2}, \qquad y_{j-1/2}<y<y_{j+1/2} \tag{3.18}$$

where

$$x_{i\pm 1/2}=x_i\pm\frac{\Delta x}{2}, \qquad y_{j\pm 1/2}=y_j\pm\frac{\Delta y}{2}$$

to get

$$\frac{\partial}{\partial t}\int_{y_{j-1/2}}^{y_{j+1/2}}\int_{x_{i-1/2}}^{x_{i+1/2}} w(t,x,y)\,dx\,dy+\int_{y_{j-1/2}}^{y_{j+1/2}} f(w(t,x_{i+1/2},y))\,dy$$
$$-\int_{y_{j-1/2}}^{y_{j+1/2}} f(w(t,x_{i-1/2},y))\,dy+\int_{x_{i-1/2}}^{x_{i+1/2}} g(w(t,x,y_{j+1/2}))\,dx$$
$$-\int_{x_{i-1/2}}^{x_{i+1/2}} g(w(t,x,y_{j-1/2}))\,dx=0. \tag{3.19}$$

In the spirit of Godunov's method, our numerical method is an approximation to

(3.19). That is, the time derivative is approximated by a difference quotient, the numerical solution $W^n_{i,j}$ is interpreted as an approximation to

$$\frac{1}{\Delta x\,\Delta y}\int_{y_{j-1/2}}^{y_{j+1/2}}\int_{x_{i-1/2}}^{x_{i+1/2}} w(t_n, x, y)\,dx\,dy$$

the numerical flux $F^n_{i+1/2,j}$ is an approximation to

$$\frac{1}{\Delta y}\int_{y_{j-1/2}}^{y_{j+1/2}} f(w(t_n, x_{i+1/2}, y))\,dy$$

and the numerical flux $G^n_{i,j+1/2}$ is an approximation to

$$\frac{1}{\Delta x}\int_{x_{i-1/2}}^{x_{i+1/2}} g(w(t_n, x, y_{j+1/2}))\,dx.$$

The numerical fluxes are computed as described above; no time splitting is used.

Since each numerical flux value depends on an angle θ, we associate a value of θ with each cell boundary point. Thus there are four values of θ associated with each cell. The derivation of the tangential local flux G' assumed that θ was constant through a cell. This would indicate that we might prefer to associate the angle θ with cell centers rather than cell boundaries. Unfortunately we could not find a way to do this and maintain conservation.

Lax and Wendroff [7] have shown that a numerical scheme for hyperbolic conservation laws is consistent with the differential equations if the numerical flux reduces to the differential equation flux when all arguments of the numerical flux are set equal. That is,

$$F(u, u, \ldots, u) = f(u). \tag{3.20}$$

The scheme described here is derived so that (3.20) is satisfied by the numerical flux of local variables in local coordinates. Thus (3.20) is satisfied by the global numerical flux because both the global differential equation flux and the global numerical flux are computed from their respective local flux values by the same formulas.

4. Choice of Direction

In Section 3 we described a method designed to resolve shocks by using different numerical flux functions in directions normal and tangential to shocks. In this section we show how to find these directions.

It is important to note that the algorithm we describe does not determine whether or not a shock exists. Thus it is not a shock-fitting algorithm. Instead the algorithm always assumes that on each cell boundary a steady shock exists and computes its normal direction. Our numerical tests indicate that this approach locates the proper direction when it is needed and causes no problems otherwise.

To determine the direction of a steady oblique shock, we note, as do Gasdynamics texts (cf. Liepmann and Roshko [9]), that a steady oblique shock can be studied as a normal shock with a superimposed uniform tangential velocity. Thus, if we are given two velocity vectors, we can locate a possible shock direction by finding a rotated coordinate frame in which both vectors have a common component.

A simple way to accomplish this was suggested by John Strikwerda. He pointed out that since the velocity in the tangential direction does not change across the shock, the shock must be normal to the velocity jump. That is, to compute the angle at which a shock would cross the cell interface that passes through the point $(x_{i-1/2}, y_j)$, we use the two velocity vectors

$$\mathbf{v}(i-1, j) = [u(i-1, j), v(i-1, j)]$$

and

$$\mathbf{v}(i, j) = [u(i, j), v(i, j)]$$

to compute the normal to the potential shock direction as the direction of the vector.

$$\delta_x \mathbf{v}(i, j) = [u(i, j) - u(i-1, j), v(i, j) - v(i-1, j)] = [\delta_x u, \delta_x v]. \tag{4.1}$$

Thus the angle θ_x used to compute rotated numerical fluxes on this interface is

$$\theta_x = \arctan(\delta_x v / \delta_x u). \tag{4.2}$$

In practice, the velocity components are computed using a first-order finite difference formula. In the smooth parts of the flow the error in the velocity differences can be of the same order as the velocity differences themselves. This can cause the angle θ_x to vary wildly in a part of the flow field where little is happening. Numerical experiments indicate that this degrades the performance of the method.

To prevent this, we replace the velocity differences in Eq. (4.2) by weighted averages of the velocity differences at a number of points. At present we use the averaging

$$\delta_x u(i, j) = [\Delta_x u(i-1, j) + 4\Delta_x u(i, j) + \Delta_x u(i+1, j)]/6 \tag{4.3}$$

where

$$\Delta_x u(i, j) = \{[u(i, j-1) - u(i-1, j-1)] + 4[u(i, j) - u(i-1, j)] + [u(i, j+1) - u(i-1, j+1)]\}/6. \tag{4.4}$$

Similarly, the angle θ_y used to commute rotated numerical fluxes on the cell interface which passes through the point $(x_i, y_{j-1/2})$ is

$$\theta_y = \arctan(\delta_y v / \delta_y u) \tag{4.5}$$

where

$$\delta_y u(i,j) = [\Delta_y u(i,j-1) + 4\Delta_y u(i,j) + \Delta_y u(i,j+1)]/6 \tag{4.6}$$

and

$$\Delta_y u(i,j) = \{[u(i-1,j) - u(i-1,j-1)] + 4[u(i,j) - u(i,j-1)] + [u(i+1,j) - u(i+1,j-1)]\}/6. \tag{4.7}$$

Numerical tests indicate that this procedure locates the proper shock angles and provides smooth angle variations.

The disadvantage of this approach is that it locates steady shocks for the Euler equations but does not locate steady contact discontinuities for the Euler equations or steady discontinuities for other systems of conservation laws. This limitation is also true of the van Leer formulas that we used to compute numerical flux components in the normal direction. In the future we plan to construct a method which can locate these more general discontinuities and to use the method of Roe [12] to resolve them.

One way that we might do this is to note that for the scalar equation

$$u_t + f(u)_x + g(u)_y = 0 \tag{4.8}$$

the normal flux is continuous across a steady discontinuity but the tangential flux is not. Therefore the discontinuity, if it exists, is located in the direction of the vector

$$\delta_x \bar{f} = [f(i,j) - f(i-1,j), g(i,j) - g(i-1,j)] = [\delta_x f, \delta_x g]. \tag{4.9}$$

For systems we might take linear combinations of vectors like (4.9). One possibility for a system of m equations would be to seek potential discontinuities in the direction of the vector

$$\delta \bar{F} = \left[\sum_{j=1}^{m} \frac{(\delta f_j)(\delta u_j)}{(\delta u_j)^2}, \sum_{j=1}^{m} \frac{(\delta g_j)(\delta u_j)}{(\delta u_j)^2} \right]. \tag{4.10}$$

Thus far no numerical experiments have been performed using Eq. (4.10).

Baines[1] has proposed another promising approach to the problem of locating general discontinuities. He chooses the angle which satisfies a discrete approximation to the equation

$$\frac{\partial G'}{\partial y'} = -\sin\theta \frac{\partial G'}{\partial x} + \cos\theta \frac{\partial G'}{\partial y} = 0. \tag{4.11}$$

At this time, some details of this procedure need to be worked out and no numerical results are available.

[1] M. J. Baines, 1982, University of Reading, Reading, England, personal communiation.

5. Numerical Results

In this section we present some numerical computations which demonstrate the ability of the present first-order method to resolve steady, oblique shocks. The results are compared with the results of computations using state-of-the-art first- and second-order upwind methods.

Consider the problem of supersonic flow over a wedge which is illustrated in Fig. 3. The solution to this problem is a single oblique shock wave. We consider only cases where the flow is supersonic everywhere.

To solve this problem we construct a uniform computing grid aligned with the wedge as shown in Fig. 4. We specify all variables at the left and top boundaries and we extrapolate all variables at the right boundary. These boundary conditions are correct at the left and right boundaries but the top boundary is overspecified since the normal velocity at this boundary is subsonic. Fortunately, for the cases considered here, no signals reach this boundary from inside the computational domain and this overspecification causes no difficulties.

At the lower boundary we specify that the velocity normal to the wall be zero and use some numerical procedure to specify the remaining variables at the wall. We have experimented with a number of numerical boundary condition procedures and have found that the following gives the best results, at least for the methods and problems considered here.

To compute the pressure and density at the wall, we follow Chakravarthy and Osher [1] and assume that locally, only simple plane waves leave the boundary. In this case the relations

$$c - \frac{(\gamma - 1)}{?} v = \text{const}$$

$$p/\rho^{\gamma} = \text{const}$$

$$c = \sqrt{\frac{\gamma p}{\rho}} \tag{5.1}$$

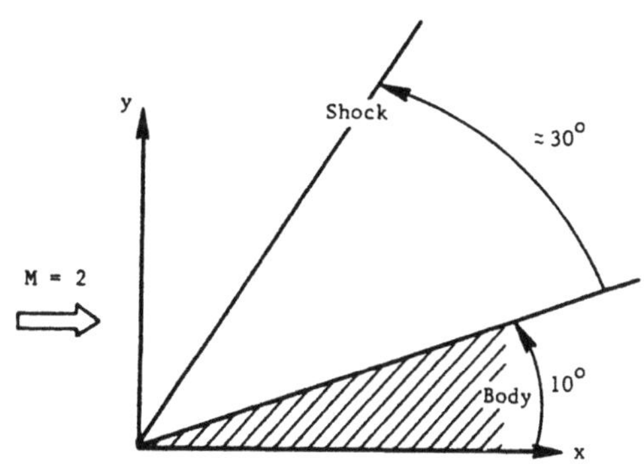

Fig. 3. Oblique shock problem in physical domain.

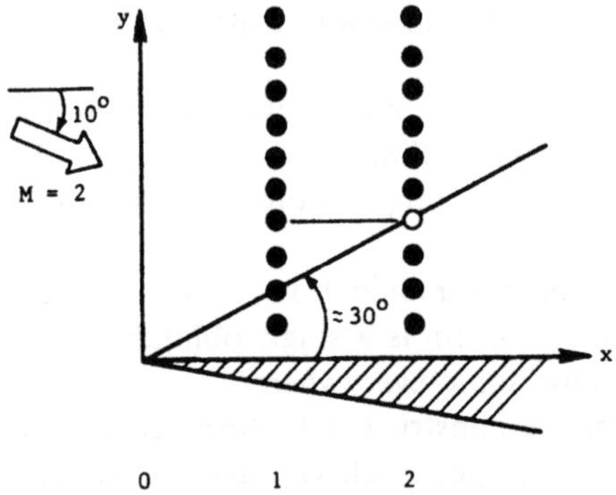

FIG. 4. Oblique shock problem in computational domain.

which are satisfied along the characteristics that reach the wall from the interior, and the wall boundary condition, $v = 0$, determine the pressure and density at the wall.

Various schemes have been used to determine the tangential velocity at the wall. Here we use the condition that, in the steady state, the total enthalpy along the wall should be constant. This gives a relation

$$\frac{e+p}{\rho} = \frac{\gamma}{\gamma - 1}\, p/\rho + u^2/2 = \text{const} \tag{5.2}$$

along the wall, which permits us to determine u once p and ρ are known.

Our initial conditions consist of the shock jump of the exact solution oriented at an angle of 45° to the computing grid.

Figure 5a is a three-dimensional plot of the density over a 10° wedge at Mach 2 computed using the first-order method of Osher and Solomon [11]. Figure 5b is a density profile along the line indicated by an arrow in Fig. 5a. The shock angle is approximately 30°. The 30° shock was very difficult for any method to resolve as can be seen from the fact that the shock is spread over many grid points. This is not an acceptable solution.

To be fair we note that the results shown for the Osher–Solomon method are typical of results for first-order upwind schemes. All first-order upwind schemes have difficulty with weak, oblique shocks and the Osher–Solomon method has no more difficulty than any other.

Figures 6a and b show the results of computations of the flow over a 10° wedge at Mach 2 using the second-order upwind method of van Leer described in [14] and [15]. The numerical flux for this method is based on the flux vector splitting formulas, described in Section 2, applied to a second-order approximation to the dependent variables. A slope-limiting function is used to prevent the oscillations which usually occur when shocks are computed using second-order methods. These results show that the scheme is monotonic and that it spreads the shock over approximately five grid points in this case.

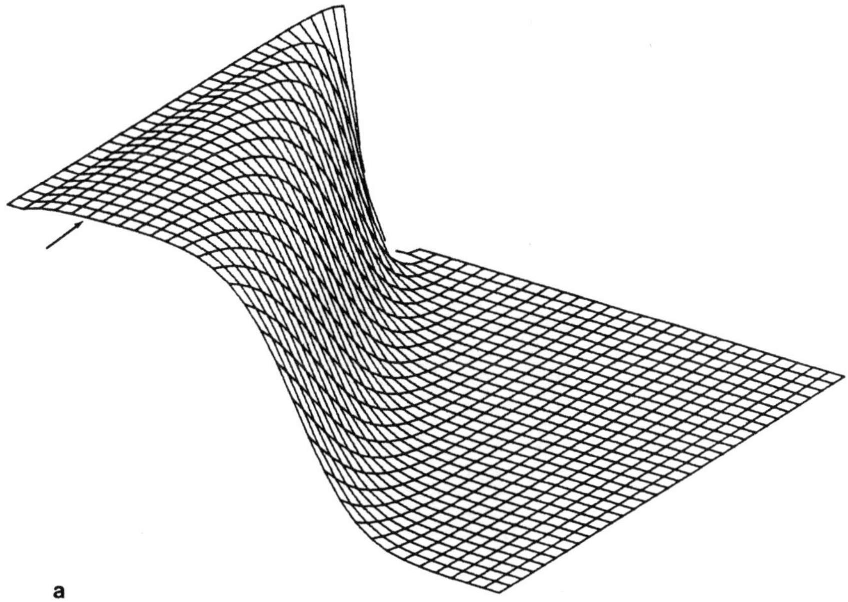

a

FIG. 5a. Three-dimensional plot of density for oblique shock problem computed using the method of Osher and Solomon.

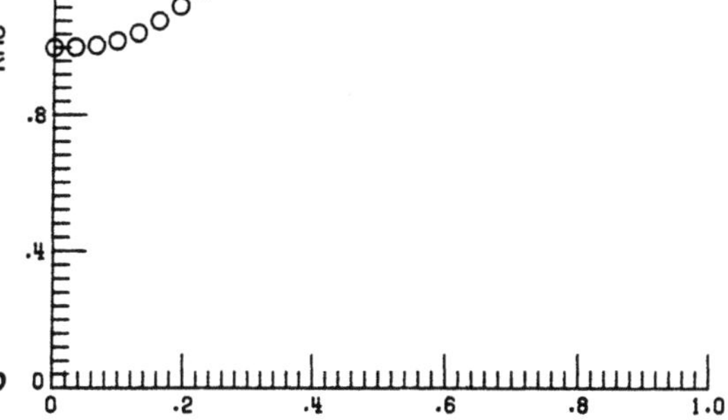

b

FIG. 5b. Density profile six grid points from the wall for the oblique shock problem computed using the method of Osher and Solomon.

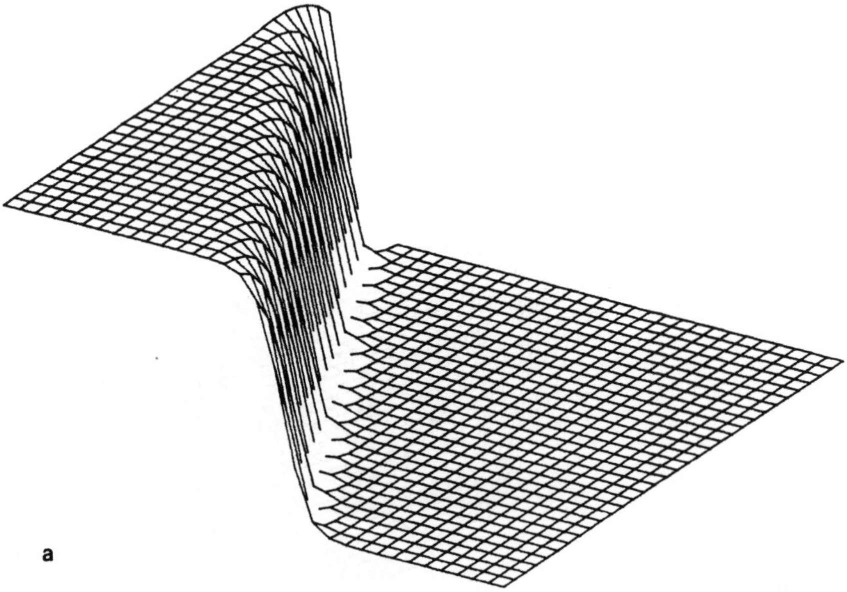

FIG. 6a. Three-dimensional plot of density for oblique shock problem computed using the second-order method of van Leer.

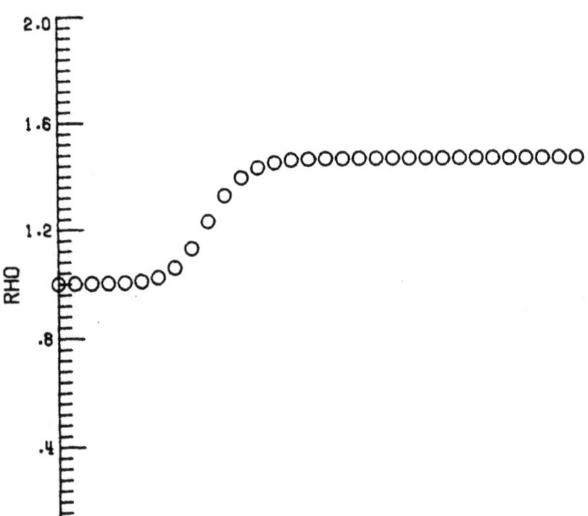

FIG. 6b. Density profile six grid points from the wall for the oblique shock problem computed using the second-order method of van Leer.

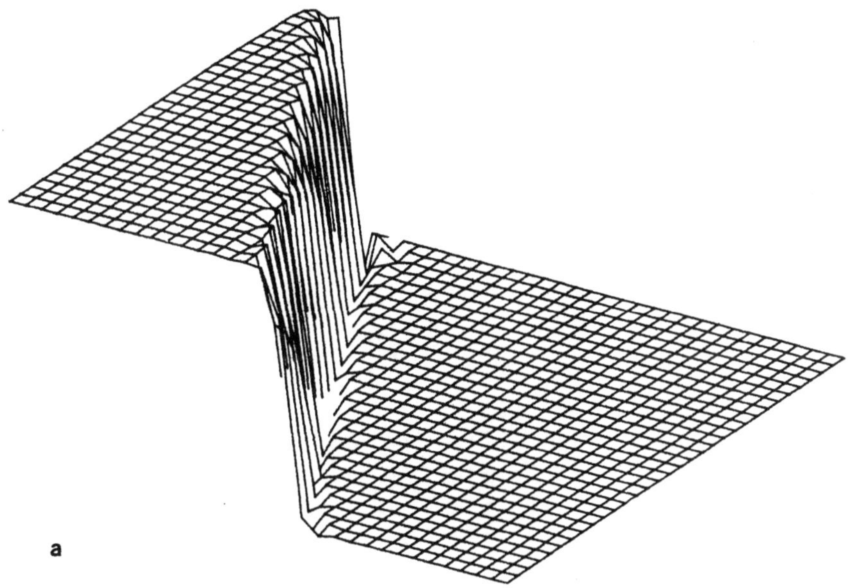

a

FIG. 7a. Three-dimensional plot of density for oblique shock problem computed using rotationally biased differences with exact shock angle.

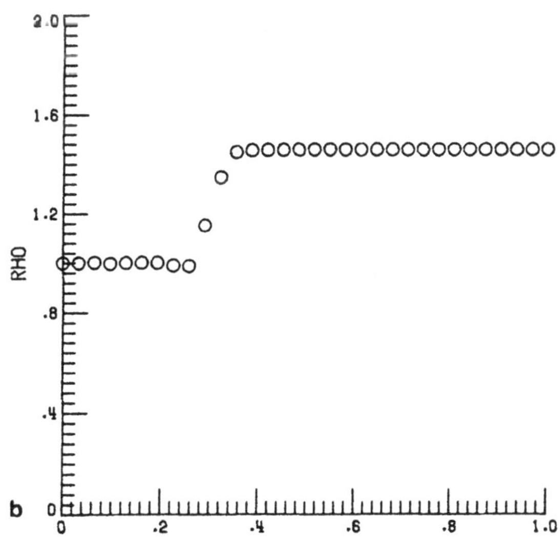

FIG. 7b. Density profile six grid points from the wall computed using rotationally biased differences with exact shock angle.

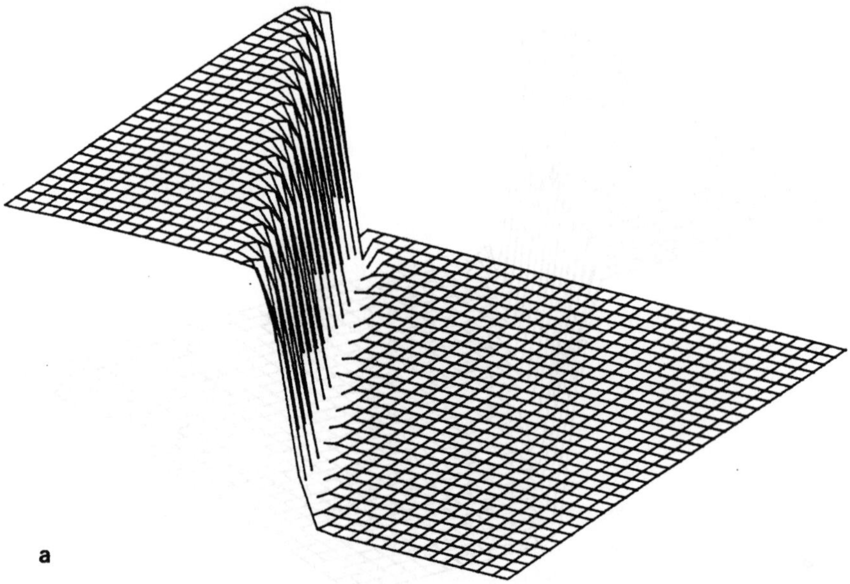

FIG. 8a. Three-dimensional plot of density for oblique shock problem computed using rotationally biased differences with automatic angle algorithm.

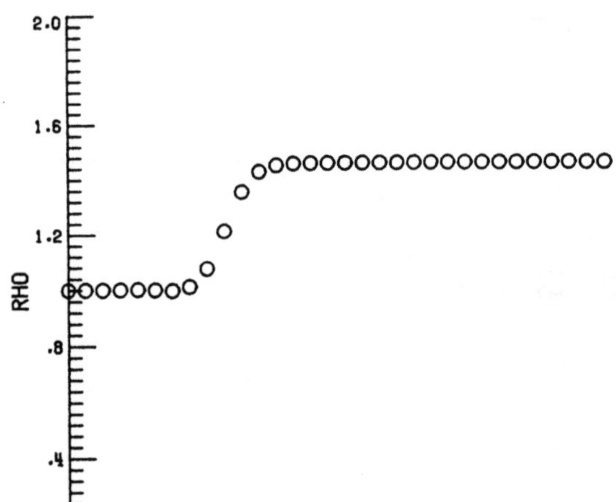

FIG. 8b. Density profile six grid points from the wall computed using rotationally biased differences with automatic angle algorithm.

Figures 7a and b illustrate an ideal situation. These results were computed using the rotated finite difference scheme of Section 3 with the exact shock angle specified and the parameter Q set to zero. In this case the shock is confined to between two and three grid points. In [16] van Leer shows that flux vector splitting requires two grid points to resolve steady one-dimensional shocks. This leads us to believe that the results shown are the best that can be expected for a two-dimensional method based on these numerical flux formulas. These results also indicate that the assumptions used to derive this method are reasonable.

Unfortunately, when we select the shock angle by the algorithm of Section 4, the method is unstable when $Q = 0$. We suspect that this is due to the spatial variations of the computed shock angle since these were not taken into account when the method was derived. We are presently trying to account for this.

At this time we can draw no specific conclusions as to the proper choice for the parameter Q. Extensive numerical experimentation indicates that the choice

$$Q_{i+1/2,j} = \frac{v'_{i,j} + v'_{i+1,j}}{2} \frac{\Delta t}{\Delta x} \tag{5.3}$$

is stable. Figures 8a and b demonstrate the performance of this choice combined with the automatic angle method of Section 4. The shock is confined to approximately three grid points.

The second problem that we study is the regular reflection of a shock from a plane wall. The physical situation is shown in Fig. 9.

The computations shown below were made on a 61×21 uniform grid. Exact values of all variables were specified on the left and top boundaries. All variables were extrapolated at the right boundary and the wall boundary conditions described above were imposed on the lower boundary.

The initial conditions imposed consisted of two shocks which had the same jumps as the exact solution but were located in the wrong place.

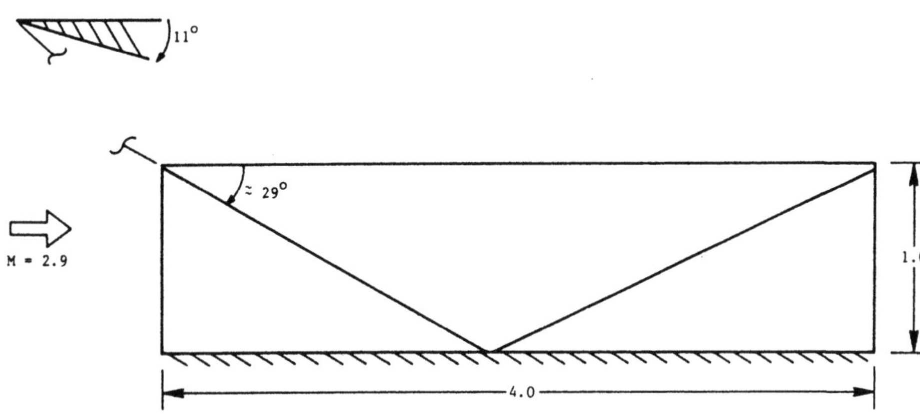

FIG. 9. Shock reflection problem.

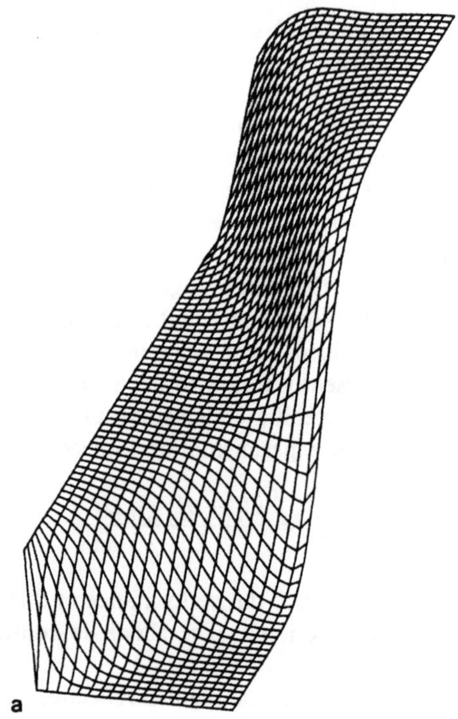

FIG. 10a. Three-dimensional plot of density for shock reflection problem computed using the method of Osher and Solomon.

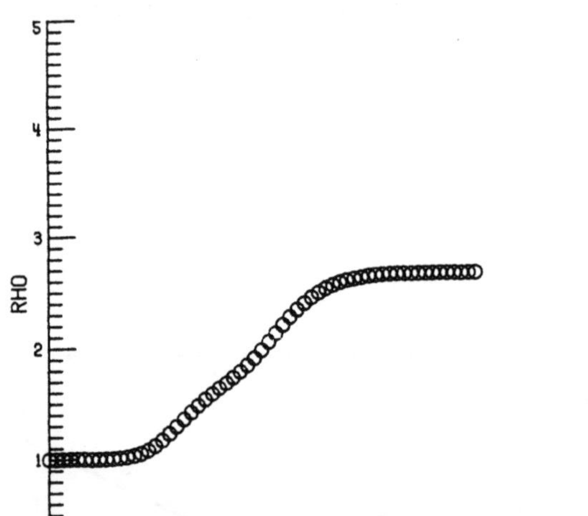

FIG. 10b. Desnity profile at $y = 0.25$ for shock reflection problem computed using the method of Osher and Solomon.

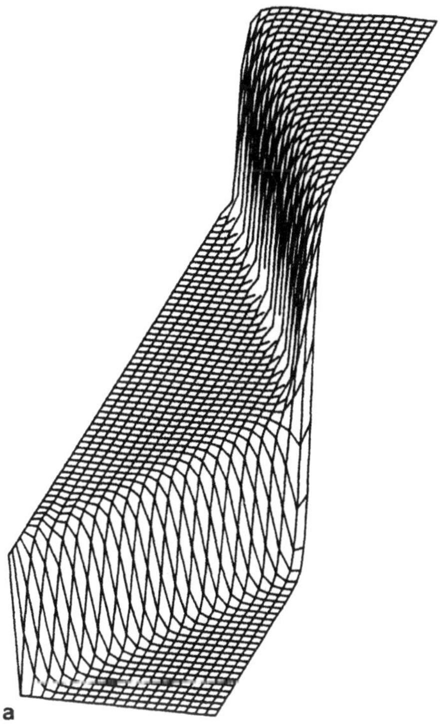

a

FIG. 11a. Three-dimensional plot of density for shock reflection problem computed using the second-order method of van Leer.

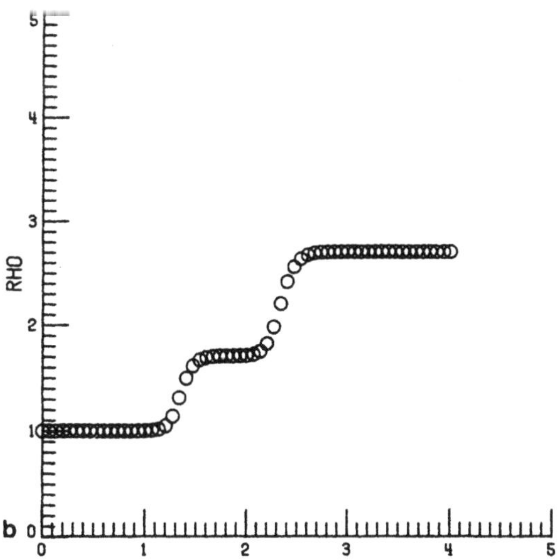

b

FIG. 11b. Density profile at $y = 0.25$ for shock reflection problem computed using the second-order method of van Leer.

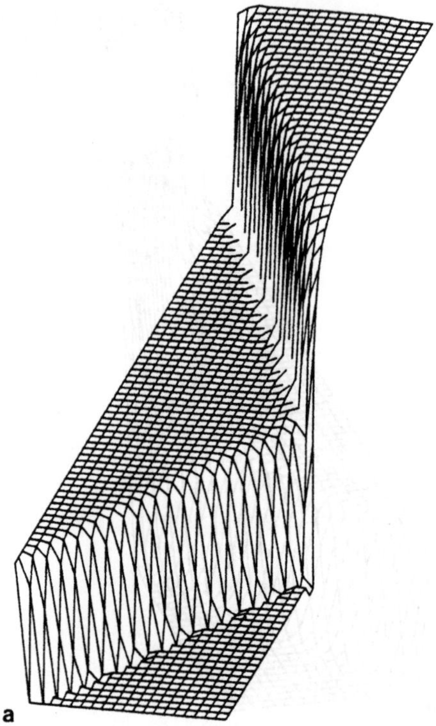

FIG. 12a. Three-dimensional plot of density for shock reflection problem computed using rotationally biased differences with automatic angle algorithm.

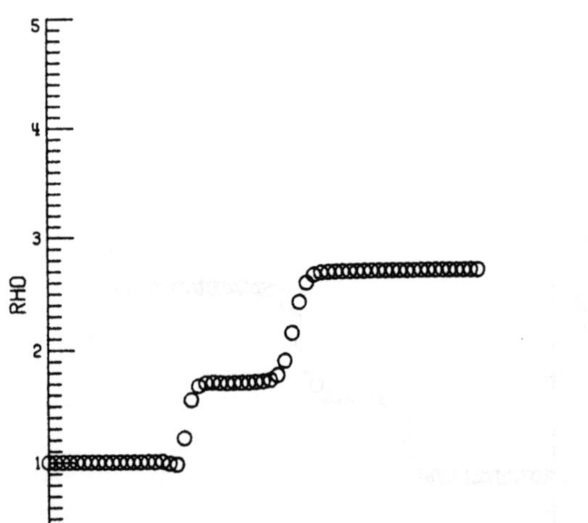

FIG. 12b. Density profile at $y = 0.25$ for shock reflection problem computed using rotationally biased differences with automatic angle algorithm.

Figure 10a is a carpet plot of the density for this flow as computed using the first-order method of Osher and Solomon. Figure 10b is a section of this plot taken at $y = 0.25$. These pictures show that this method smears the shocks so badly that, at $y = 0.25$, they cannot be distinguished.

Figures 11a and b are the corresponding density plots for this flow as computed using the second-order upwind method of van Leer. These results are much better than the previous ones. Here we can distinguish two distinct shocks. The first one is spread over between four and five mesh points and the second one is spread over between six and seven mesh points.

Finally, in Figs. 12a and b, we show density plots for this flow computed using the first-order rotated method. This method spreads the first shock over between two and three mesh points and the second shock over between five and six mesh points. There is a slight undershoot in front of the first shock.

At first sight, these results might appear to be less than dramatic but we should note that we have been comparing a first-order method with one of the more sophisticated of second-order methods. Under these circumstances, we are encouraged by the fact that the first-order method consistently resolved steady oblique shocks within fewer mesh intervals than did the second-order method. This confirms that our oblique shock model works. With additional effort we can make it work better.

6. Summary and Discussion

In this paper we describe a method which determines the orientation of possible shock solutions to the Euler equations. In addition we show how this information can be used to construct a first-order upwind method which computes solutions with greatly improved steady shock resolution. Indeed, we have shown that our first-order method resolves steady shocks within fewer mesh intervals than a sophisticated second-order upwind method.

This work has shown that the shock-resolving ability of numerical methods for hyperbolic equations can be improved if the methods take the orientation of possible shocks into account. Although the results thus far have been encouraging, we wish to improve the present method. Before this can be done some questions need to be answered. In particular, we wish to understand how the variation in angle affects the performance of the method and how we should select the parameter Q in the tangential flux. At present this parameter is selected in an ad hoc manner. It is also important that we construct a careful stability analysis for this method and that we study appropriate choices for boundary conditions. It might also be helpful to study other choices of upwind formulas for the normal flux.

On a more practical level, we realize that first-order methods are usually not accurate enough in regions of smoothly varying flow to be useful in applications. Therefore we are presently developing a second-order-accurate version of our method. This work will be described in a future paper.

Acknowledgments

I would like to thank Stanley Osher and Bram van Leer for many stimulating discussions on upwind schemes and for their encouragement during this work. I would also like to thank John Strikwerda for providing the simple derivation of Eq. (4.2) which appears in this paper.

References

1. S. Chakravarthy and S. Osher, "Numerical Experiments with the Osher Upwind Scheme for the Euler Equations," AIAA Paper 82-0975, AIAA/ASME 3rd Joint Thermophysics, Plasma and Heat Transfer Conference, June, 1982.
2. S. K. Godunov, A finite difference method for the numerical computation of discontinuous solutions of the equations of fluid dynamics (in Russian), *Mat. Sb.* **47** (1959); also Cornell Aeronautical Lab (CALSPAN) translation.
3. A. Harten and J. M. Hyman, "A Self-Adjusting Grid for the Computation of Weak Solutions Of Hyperbolic Conservation Laws," Report LA9105, Center for Nonlinear Studies, Theoretical Division, Los Alamos National Lab, Los Alamos, N. M., 1981.
4. A. Harten, P. D. Lax, and B. van Leer, On upstream differencing and Godunov-type schemes for hyperbolic conservation laws, *SIAM Rev.* **25** (1983), 35–62.
5. A. Harten, High resolution schemes for hyperbolic conservation laws, *J. Comput. Phys.* **49** (1983), 357–393.
6. A. Jameson, Iterative solution of transonic flows over airfoils and wings, including flows at Mach 1, *Comm. Pure Appl. Math.* **27** (1974), 283–309.
7. P. Lax and B. Wendroff, Systems of conservation laws, *Comm. Pure Appl. Math.* **13** (1960), 217–237.
8. P. D. Lax, "Hyperbolic Systems of Conservation Laws and the Mathematical Theory of Shock Waves," SIAM, Philadelphia, 1972.
9. H. W. Liepmann and A. Roshko, "Elements of Gasynamics," Wiley, New York, 1957.
10. E. M. Murman, Analysis of embedded shock waves calculated by relaxation methods, *AIAA J.* **12** (1974), 626–633.
11. S. Osher and F. Solomon, Upwind difference schemes for hyperbolic systems of conservation laws, *Math. Comp.* **38** (1982), 339–374.
12. P. L. Roe, Approximate Riemann solvers, parameter vectors and difference schemes, *J. Comput. Phys.* **43** (1981), 357–372.
13. L. A. Segal, "Mathematics Applied to Continuum Mechanics," Macmillan Co., New York, 1977.
14. G. D. van Albada, B. van Leer, and W. W. Roberts, "A Comparative Study of Computational Methods in Cosmic Gas Dynamics," ICASE Report 81-24, 1981.
15. B. van Leer, "On the Relation Between the Upwind-Differencing Schemes of Godunov, Enquist–Osher and Roe," ICASE Report 81-11, 1981.
16. B. van Leer, "Flux Vector Splitting for the Euler Equations," ICASE Report 82-30, 1982.
17. P. Woodward and P. Colella, *in* "Proceedings, 7th International Conference on Numerical Methods in Fluid Dynamics," Stanford/NASA Ames, 1980.
18. N. N. Yanenko, "The Method of Fractional Steps," Springer-Verlag, New York/Berlin, 1971.
19. H. C. Yee, R. F. Warning, and A. Harten, "On the Application and Extension of Harten's High Resolution Scheme," NASA TM 84256, 1982.

Discrete Models for the Numerical Analysis of Time-Dependent Multidimensional Gas Dynamics*

PHILIP L. ROE

College of Aeronautics, Cranfield Institute of Technology, Cranfield MK43 0AL, United Kingdom

Received February 7, 1985; revised May 31, 1985

This paper explores a possible technique for extending to multidimensional flows some of the upwind-differencing methods that have proved highly successful in the one-dimensional case. Attention here is concentrated on the two-dimensional case, and the flow domain is supposed to be divided into polygonal computational elements. Inside each element the flow is represented by a local superposition of elementary solutions consisting of plane waves not necessarily aligned with the element boundaries.

1. INTRODUCTION

The recent survey of Woodward and Colella [16] shows that for one-dimensional gas dynamics there is an order of magnitude difference in effectiveness between sophisticated codes physically based on correct transfer of information, and simpler codes combining central differences with artificial viscosity. The sophisticated codes need much more computational work to update the solution at each mesh point, but this is far outweighed by their ability to capture discontinuities on a coarser mesh. For two-dimensional problems the difference in efficiency is far less marked, and for less violent flows than the ones they consider the advantage is likely to be reversed.

The explanation is probably that the physics of one-dimensional flow is especially simple and well understood, and easy to imitate by numerical processes. Two-dimensional flows are more complex; in particular, acoustic waves can propagate in infinitely many directions rather than just two, and vorticity exists as a new phenomenon. Most extensions of upwind codes to two or more dimensions ignore these issues and advance the solution by "splitting," that is to say, through a sequence of one-dimensional operators. For examples, see the survey by Woodward and Colella, also Sells [13] and Chakravarthy and Osher [1]. There are also what may be called "one-and-a-half-dimensional" methods, in which the one-dimensional operators are interwoven, but the underlying physical model is still one involving

* Research was supported in part by the National Aeronautics and Space Administration under Contract NAS1-17070 while the author was in residence at the Institute for Computer Applications in Science and Engineering, NASA Langley Research Center, Hampton, Va. 23665.

wave propagation along the coordinate directions. This approach seems to yield some modest gains, as shown by Lytton [7] and Colella [2]. However, an observation is made in Section 3 which casts doubt on its real value.

If full advantage can be taken of upwinding techniques in two or more dimensions it is probably necessary to devise methods which take account of the actual directions in which information is propagated. The only results so far available for a method of this kind are those of Davis [4]. He assumes that the flow is locally dominated by a single shock wave whose unknown orientation may be deduced from the velocity field, or, in a later version of the code, from the pressure field (Davis, private communication). His method works very well on test problems where the flow is divided by shock waves into piecewise uniform regions. This is encouraging because it shows that a well chosen model of the flow can be used to numerical advantage.

It has been conjectured that the way forward into two dimensions is blocked by the complexity of a "two-dimensional Riemann solver," by which is meant an algorithm for computing the breakdown of initial conditions which are piecewise constant in two-dimensional cells. The solution of this problem close to the edge of a cell is straightforward, but secondary interactions near the corners are extremely difficult to compute. Even if a Riemann solver of this kind were computationally feasible, however, it would not be a satisfactory building block for two-dimensional calculations. It would, like the operator-splitting methods mentioned above, force the principal wave motions to take place normal to the cell boundaries.

In the present work we avoid this difficulty by thinking of the data as piecewise linear rather than piecewise constant, and in Section 2 we interpret one-dimensional upwinding schemes in that light. The gradients in the data are used to construct a "model flow" consisting of simple waves within each mesh interval. In Section 3 the corresponding simple wave solutions, propagating in arbitrary directions, are derived for the two-dimensional equations. In Section 4 we propose model flows which can be fitted to any data which varies linearly in two dimensions, and in Section 5 we describe a strategy for constructing conservative differencing schemes by fitting such models to the data given at vertices of an irregular two-dimensional mesh. Section 6 contains observations on the possibility of extending the work to three-dimensional flow, and Section 7 comments on the type of advection scheme needed to complete the algorithm.

2. Upwinding in One Dimension

We begin by observing that one way to derive upwind schemes for the Euler equations in one dimension is to suppose that the flow in each mesh interval $(i, i+1)$ is a locally linear superposition of simple waves having the form

$$\mathbf{w}(x, t)=\sum_k (\alpha_k \mathbf{r}_k)\cdot(x-\lambda_k t). \tag{2.1}$$

Here, $\mathbf{w}$ is the vector of unknowns, $\mathbf{r}_k$ is an eigenvector showing how the gradients due to the kth wave are distributed over the components of $\mathbf{w}$, α_k is the amplitude of the kth wave, and λ_k its speed. Any independent set of unknown variables $\mathbf{w}$ may be chosen, and the choice will not affect the values of α_k, λ_k, but $\mathbf{r}_k$ will be different for each choice. The values of $\lambda_{1,2,3}$ are

$$u-a, u, u+a \tag{2.2}$$

and the values of $\alpha_{1,2,3}$ are

$$\frac{1}{2a^2}[\Delta p-\rho a\Delta u],\ \Delta\rho-\frac{1}{a^2}\Delta p,\qquad \frac{1}{2a^2}[\Delta p+\rho a\Delta u] \tag{2.3}$$

where $\Delta(\cdot)=(\cdot)_{i+1}-(\cdot)_i$ and any local average values of ρ, a, u are valid. To achieve a conservative algorithm, two conditions are necessary. The eigenvectors $\mathbf{r}_k$ must show the effects of the waves upon the conserved variables (ρ, ρu, ρe). In these variables $\mathbf{r}_1, \mathbf{r}_2, \mathbf{r}_3$ are given by

$$\mathbf{r}_1=\begin{vmatrix}1\\ u-a\\ h-ua\end{vmatrix},\qquad \mathbf{r}_2=\begin{vmatrix}1\\ u\\ \tfrac{1}{2}u^2\end{vmatrix},\qquad \mathbf{r}_3=\begin{vmatrix}1\\ u+a\\ h+ua\end{vmatrix}, \tag{2.4}$$

where $h=a^2/(\gamma-1)+\frac{1}{2}u^2$ is the specific enthalpy. Also, the average values of ρ, a, u must now be chosen so that

$$\sum \alpha_k\lambda_k\mathbf{r}_k=\Delta\mathbf{F}/\Delta x, \tag{2.5}$$

where $\mathbf{F}$ is the vector of flux quantities. For details, see [9, 12].

This may be throught of as constructing, within each interval $(i, i+1)$ a local model of the flow. The model consists of elementary solutions of the Euler equations, linearized about a particular local average state. The model matches the observed data with respect to the spatial derivatives (or to be precise, with respect to the mesh differences). The time evolution of the model flow is readily predicted, and provides the information which is used to advance the global solution through one time step.

For example, a first-order scheme to update the solution [12, 15] involves subtracting a quantity

$$\frac{\Delta t}{\Delta x}\alpha_k\lambda_k\mathbf{r}_k$$

from the value of $\mathbf{w}$ at whichever end of the interval the kth wave is moving towards, i.e., from $\mathbf{w}_i$ if λ_k is negative in $[i, i+1]$ or from $\mathbf{w}_{i+1}$ if λ_k is positive. This algorithm causes selected information to be transmitted in directions which are determined by the model flow.

The present paper is motivated by the wish to create algorithms having similar properties with regard to multidimensional wave propagation, but its content is limited almost entirely to the devising of appropriate multidimensional models. The author hopes to develop the corresponding time-marching algorithms in subsequent publications.

3. Elementary Solutions in Two Dimensions

In this section, as a necessary preliminary to the construction of two-dimensional models, we investigate the elementary solutions from which they may be built. Consider the Euler equations in primitive variables $\mathbf{w}_p = (p, u, v, \rho)$.

$$\begin{aligned} p_t + up_x + vp_y + \rho a^2(u_x + v_y) &= 0, \\ u_t + uu_x + vu_y + \frac{1}{\rho} p_x &= 0, \\ v_t + uv_x + vv_y + \frac{1}{\rho} p_y &= 0, \\ \rho_t + u\rho_x + v\rho_y + \rho(u_x + v_y) &= 0. \end{aligned} \tag{3.1}$$

Corresponding to Eq. (2.1) there are solutions to (3.1) of the form

$$\mathbf{w} = \alpha(\theta)\, \mathbf{r}(\theta)(x \cos\theta + y \sin\theta - \lambda(\theta)\, t), \tag{3.2}$$

where θ is an arbitrary angle, and $\mathbf{r}(\theta)$ is an eigenvector. For acoustic waves and primitive variables it can be shown that

$$\mathbf{r}(\theta) = \begin{vmatrix} \rho a^2 \\ a \cos\theta \\ a \sin\theta \\ \rho \end{vmatrix} \tag{3.3}$$

and that the wave speed is

$$\lambda(\theta) = u \cos\theta + v \sin\theta + a. \tag{3.4}$$

Figure 1 shows that this speed corresponds to a wave front tangential to the Mach cone. We repeat here an observation from Roe [10]. Consider two Cartesian points having the same value of y, in a flow given by (3.2)–(3.4). An operator-split

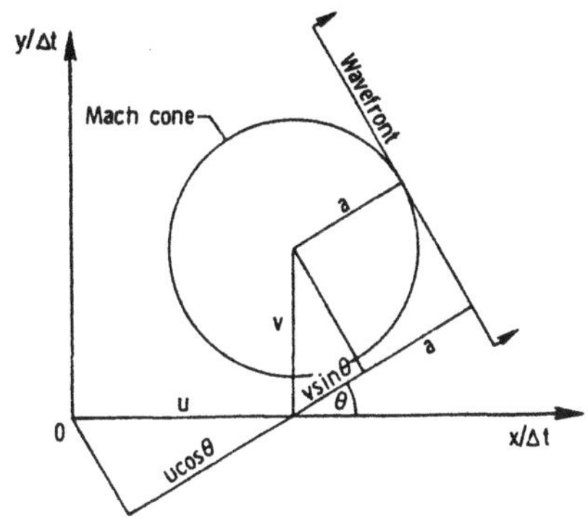

FIG. 1. An acoustic wave front.

method will attempt to explain the difference in states as due to waves passing in the x direction; it will compute

$$\Delta \begin{vmatrix} p \\ u \\ v \\ \rho \end{vmatrix} \propto \begin{vmatrix} \rho a^2 \\ a\cos\theta \\ a\sin\theta \\ \rho \end{vmatrix} = \tfrac{1}{2}(1+\cos\theta) \begin{vmatrix} \rho a^2 \\ a \\ 0 \\ \rho \end{vmatrix} + \sin\theta \begin{vmatrix} 0 \\ 0 \\ a \\ 0 \end{vmatrix} + \tfrac{1}{2}(1-\cos\theta) \begin{vmatrix} \rho a^2 \\ -a \\ 0 \\ \rho \end{vmatrix}, \tag{3.5}$$

where the r.h.s. shows the eigenvectors of two one-dimensional acoustic waves, and a slip line. These spurious waves may not even travel in the proper direction and their inclusion in a numerical method can hardly be realistic. This criticism applies even to the "unsplit" algorithms of Colella [2] and Lytton [7]. Our goal in the next section is to construct local models of the flow by superposing simple waves whose orientation is not assumed in advance.

Such a model cannot, however, be constructed purely out of acoustic waves since these are irrotational and the data may not be. There are two other fundamental flows which can be incorporated neatly into the model. One is a shear flow, which again has the general form of (3.2) but with

$$\mathbf{r}(\theta) = \begin{vmatrix} 0 \\ -\sin\theta \\ \cos\theta \\ 0 \end{vmatrix}, \tag{3.6}$$

$$\lambda(\theta) = u\cos\theta + v\sin\theta. \tag{3.7}$$

Another is solid-body rotation, or vorticity

$$u = -\tfrac{1}{2}\omega y, \qquad v = \tfrac{1}{2}\omega x. \tag{3.8}$$

There still remains an effect which is missing from the model, for all the flows above are isentropic. An entropy wave, across which pressure and velocity do not change, reveals itself in the primitive variables as a change of density. The general form is again (3.2) with $\lambda(\theta)$ given by (3.7), but with

$$\mathbf{r} = \begin{vmatrix} 0 \\ 0 \\ 0 \\ \rho \end{vmatrix} \tag{3.9}$$

independent of θ.

Another interesting fundamental solution (not directly used below) is obtained by superposing acoustic waves of the same strength with all possible propagation directions, i.e., by integrating (3.2) with respect to θ from 0 to 2π with $\alpha(\theta) = \alpha_0$. The result yields

$$\begin{aligned} p_x &= p_y = 0, \\ u_x &= v_y = \tfrac{1}{2}\alpha_0, \\ v_x &= u_y = 0, \\ \rho_x &= \rho_y = 0. \end{aligned} \tag{3.10}$$

This solution would appear in the data as a region of uniform (isotropic) velocity divergence. However, the same data could be explained equally well by the passage of four plane waves

$$\mathbf{w}(0) + \mathbf{w}(\pi/2) + \mathbf{w}(\pi) + \mathbf{w}(3\pi/2), \tag{3.11}$$

where $\mathbf{w}(\theta)$ is given by (3.2). For numerical purposes the discrete representation by four plane waves is more amenable than the representation by one circular wave, and this is how a uniformly diverging flow would be dealt with in the model we develop below. However, it may be worth noting that any three equal waves separated by angles of $2\pi/3$ would also produce (locally) the same effect.

4. The Discrete Models

It is not obvious how the model flows of Section 2 should be generalized from one dimension to two. The chief difficulty is that whereas in one dimension there are just three types of elementary wave, in two dimensions there are infinitely many if we count all the possible orientations as distinct. In one dimension there is only one model that can be constructed, and it has three parameters which are the unknown wave strengths. Matching the model to the spatial gradients of the three data quantities p, u, ρ gives three simple linear equations whose solution is (2.3). In two dimensions the data will allow us to estimate gradients in two directions of four quantities, yielding eight items of information. Whatever model we choose must have eight free parameters, some of which may be wave amplitudes, and the remainder will be orientation angles. If all the orientations are supposed to be known (aligned, e.g., with the grid directions) we will again find easily solved linear equations for the amplitudes. However, because of the observation made concerning Eq. (3.5), we reject this approach, and require that at least some of the orientations be left unspecified. However, the equations which must be solved for the parameters then become nonlinear. If the free parameters are not judiciously chosen, no closed form solution may be possible, or the solution may not always be real-valued, or the solution may be computationally expensive. In such cases, the model will be useless.

Two models, however, have been found whose parameters are given by simple real-valued expressions for all data. Each has, as its representation of the acoustic disturbances, a set of four orthogonal waves (Fig. 2). One of the four will have an orientation angle in the range $[\pm\pi/4]$ and we take this as reference. Its orientation is θ, and its amplitude α_1. The strength of the wave which moves in the opposite direction will be α_2, and the waves which travel at right angles to these two have strengths α_3, α_4. To this model we add an entropy wave with strength β and inclination ϕ, so that the model now contains seven unknown parameters.

To close the model we must introduce a fundamental solution incorporating vorticity, and it is only in this respect that the two models differ. In Model A we introduce a uniform vorticity ω, and in Model B we introduce a shear flow such that

$$\begin{aligned} u &= u_0(1 + k(v_0 x - u_0 y)), \\ v &= v_0(1 + k(v_0 x - u_0 y)). \end{aligned} \tag{4.1}$$

This is a special case of (3.6), (3.7) with $\tan\theta = -(u/v)$. We will first show the algebra for Model A, which is slightly simpler.

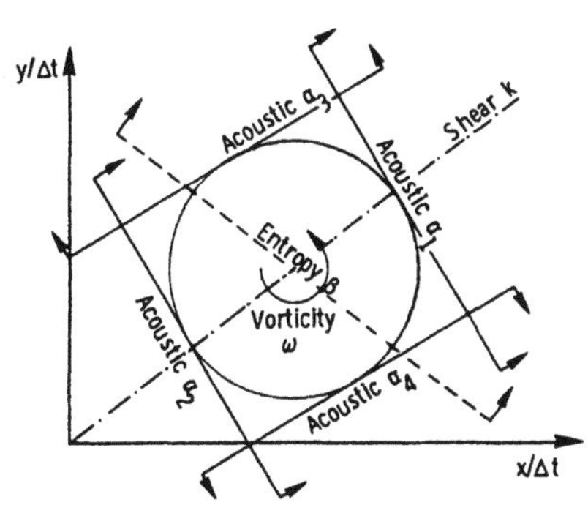

FIG. 2. The waves comprising the discrete models.

To tidy up the equations we write dimensionless derivatives

$$\begin{aligned} P_x &= p_x/\rho a^2, & P_y &= p_y/\rho a^2, \\ U_x &= u_x/a, & U_y &= u_y/a, \\ V_x &= v_x/a, & V_y &= v_y/a, \\ R_x &= \rho_x/\rho, & R_y &= \rho_y/\rho. \end{aligned} \tag{4.2}$$

By equating these to the sum of contributions produced by each component of the model, we find

$$P_x = \alpha_1 \cos\theta + \alpha_2 \cos\theta - \alpha_3 \sin\theta - \alpha_4 \sin\theta, \tag{4.3a}$$

$$P_y = \alpha_1 \sin\theta + \alpha_2 \sin\theta + \alpha_3 \cos\theta + \alpha_4 \cos\theta, \tag{4.3b}$$

$$U_x = \alpha_1 \cos^2\theta - \alpha_2 \cos^2\theta + \alpha_3 \sin^2\theta - \alpha_4 \sin^2\theta, \tag{4.3c}$$

$$\begin{aligned} U_y = {} & \alpha_1 \sin\theta\cos\theta - \alpha_2 \sin\theta\cos\theta - \alpha_3 \sin\theta\cos\theta \\ & + \alpha_4 \sin\theta\cos\theta - \tfrac{1}{2}\omega/a, \end{aligned} \tag{4.3d}$$

$$\begin{aligned} V_x = {} & \alpha_1 \sin\theta\cos\theta - \alpha_2 \sin\theta\cos\theta - \alpha_3 \sin\theta\cos\theta \\ & + \alpha_4 \sin\theta\cos\theta + \tfrac{1}{2}\omega/a \end{aligned} \tag{4.3e}$$

$$V_y = \alpha_1 \sin^2\theta - \alpha_2 \sin^2\theta + \alpha_3 \cos^2\theta - \alpha_4 \cos^2\theta, \tag{4.3f}$$

$$R_x = \alpha_1 \cos\theta + \alpha_2 \cos\theta - \alpha_3 \sin\theta - \alpha_4 \sin\theta + \beta\cos\phi, \tag{4.3g}$$

$$R_y = \alpha_1 \sin\theta + \alpha_2 \sin\theta + \alpha_3 \cos\theta + \alpha_4 \cos\theta + \beta\sin\phi. \tag{4.3h}$$

In these equations, the convention which distinguishes the contributions of α_1, α_2 is that the same angle θ is used, but the sign of a is reversed. The eight equations can be solved quite easily. From (4.3d) and (4.3e) we obtain at once

$$\begin{aligned} \omega &= a(V_x - U_y) \\ &= (v_x - u_y). \end{aligned} \tag{4.4}$$

Also

$$\begin{aligned} R_x - P_x &= \beta\cos\phi, \\ R_y - P_y &= \beta\sin\phi, \end{aligned} \tag{4.5}$$

whence β, ϕ. Next add (4.3d) and (4.3e) to obtain

$$U_y + V_x = 2\sin\theta\cos\theta(\alpha_1 - \alpha_2 - \alpha_3 + \alpha_4) \tag{4.6}$$

and subtract (4.3f) from (4.3c)

$$U_x - V_y = (\cos^2\theta - \sin^2\theta)(\alpha_1 - \alpha_2 - \alpha_3 + \alpha_4). \tag{4.7}$$

Dividing (4.6) by (4.7) yields

$$\tan 2\theta = \frac{U_y + U_x}{U_x - V_y} = \frac{N}{D}. \tag{4.8}$$

Since we have defined $|\theta| \leqslant \pi/4$ this result defines a unique orientation which is always real, coinciding, in fact, with the principal axis of the strain tensor. With θ known, the remaining equations are linear. We write (4.3c) and (4.3f) as

$$U_x = (\alpha_1 - \alpha_2)\cos^2\theta + (\alpha_3 - \alpha_4)\sin^2\theta,$$

$$V_y = (\alpha_1 - \alpha_2)\sin^2\theta + (\alpha_3 - \alpha_4)\cos^2\theta,$$

and combine them to give

$$\alpha_1 - \alpha_2 = \frac{U_x\cos^2\theta - V_y\sin^2\theta}{\cos^2\theta - \sin^2\theta}. \tag{4.9}$$

This expression must be rewritten to avoid the possible singularity. Noting that

$$\cos^2\theta = \tfrac{1}{2}(1 + \cos 2\theta), \qquad \sin^2\theta = \tfrac{1}{2}(1 - \cos 2\theta),$$

and that

$$\cos 2\theta = D/R, \tag{4.10}$$

where

$$R^2 = N^2 + D^2, \tag{4.11}$$

we find

$$\alpha_1 - \alpha_2 = \tfrac{1}{2}(U_x + V_y + R), \tag{4.12}$$

which is clearly always finite. By the same process we find

$$\alpha_3 - \alpha_4 = \tfrac{1}{2}(U_x + V_y - R). \tag{4.13}$$

It can be shown that these expressions (4.12), (4.13) are proportional to the greatest and least straining rates experienced by the fluid. In these results, R must have the same sign as D, since $|\theta| \leqslant \pi/4$ and so the r.h.s. of Eq. (4.10) must be positive. For locally one-dimensional flow in the x (resp. y) direction, R will equal U_x (resp. $-V_y$) and V_y (resp. U_x) will be zero. Equations (4.12), (4.13) will give the correct one-dimensional results. That is, $\alpha_1 - \alpha_2 = U_x$ (resp. 0), and $\alpha_3 - \alpha_4 = 0$ (resp. V_y). It is interesting that

$$\alpha_1 - \alpha_2 + \alpha_3 - \alpha_4 = U_x + V_y. \tag{4.14}$$

The l.h.s. is the total strength of the acoustic waves (the minus signs appear because of our conventions about a and θ) and the r.h.s. is the velocity divergence. Compare the result in Eq. (3.10).

The last step is to combine (4.3a), (4.3b) to give

$$\alpha_1+\alpha_2=P_x\cos\theta+P_y\sin\theta, \tag{4.15}$$

$$\alpha_3+\alpha_4=P_y\cos\theta-P_x\sin\theta, \tag{4.16}$$

and then the α's follow from (4.12), (4.13).

A remarkable identity concerning the wave strengths is the following.

$$\begin{aligned}
&\alpha_1^2+\alpha_2^2+\alpha_3^2+\alpha_4^2+\frac{1}{4}\omega^2/a^2\\
&\quad=\frac{1}{2}(\alpha_1+\alpha_2)^2+\frac{1}{2}(\alpha_1-\alpha_2)^2+\frac{1}{2}(\alpha_3+\alpha_4)^2+\frac{1}{2}(\alpha_3-\alpha_4)^2+\frac{\omega^2}{4a^2}\\
&\quad=\frac{1}{2}(P_x\cos\theta+P_y\sin\theta)^2+\frac{1}{8}(U_x+V_y+R)^2\\
&\qquad+\frac{1}{2}(P_y\cos\theta-P_x\sin\theta)^2+\frac{1}{8}(U_x+V_y-R)^2+\frac{\omega^2}{4a^2}\\
&\quad=\frac{1}{2}(P_x^2+P_y^2)+\frac{1}{4}(U_x+V_y)^2+\frac{1}{4}R^2+\frac{\omega^2}{4a^2}\\
&\quad=\frac{1}{2}(P_x^2+P_y^2)+\frac{1}{4}(U_x+V_y)^2+\frac{1}{4}(U_x-V_y)^2+\frac{1}{4}(U_y+V_x)^2+\frac{1}{4}(V_x-U_y)^2\\
&\quad=\frac{1}{2}[P_x^2+P_y^2+U_x^2+U_y^2+V_x^2+V_y^2].
\end{aligned} \tag{4.17}$$

Both ends of this chain are expressions representing some overall strength of the disturbance (excluding entropy effects which add another simple term).

The analysis of Model B is almost identical. The equations are altered by replacing the vorticity terms with the shear terms from (4.1) thus

$$P_x=\alpha_1\cos\theta+\alpha_2\cos\theta-\alpha_3\sin\theta-\alpha_4\sin\theta, \tag{4.18a}$$

$$P_y=\alpha_1\sin\theta+\alpha_2\sin\theta+\alpha_3\cos\theta+\alpha_4\cos\theta, \tag{4.18b}$$

$$U_x=\alpha_1\cos^2\theta-\alpha_2\cos^2\theta+\alpha_3\sin^2\theta-\alpha_4\sin^2\theta+ku_0v_0, \tag{4.18c}$$

$$U_y=\alpha_1\sin\theta\cos\theta-\alpha_2\sin\theta\cos\theta-\alpha_3\sin\theta\cos\theta+\alpha_4\sin\theta\cos\theta-ku_0^2, \tag{4.18d}$$

$$V_x=\alpha_1\sin\theta\cos\theta-\alpha_2\sin\theta\cos\theta-\alpha_3\sin\theta\cos\theta+\alpha_4\sin\theta\cos\theta+kv_0^2, \tag{4.18e}$$

$$V_y=\alpha_1\sin^2\theta-\alpha_2\sin^2\theta+\alpha_3\cos^2\theta-\alpha_4\cos^2\theta-ku_0v_0, \tag{4.18f}$$

$$R_x=\alpha_1\cos\theta+\alpha_2\cos\theta-\alpha_3\sin\theta-\alpha_4\sin\theta+\beta\cos\phi, \tag{4.18g}$$

$$R_y=\alpha_1\sin\theta+\alpha_2\sin\theta+\alpha_3\cos\theta+\alpha_4\cos\theta+\beta\sin\phi. \tag{4.18h}$$

The solution for β, ϕ is identical. Equations (4.18d), (4.18e) give

$$k=\frac{V_x-U_y}{u_0^2+v_0^2}. \tag{4.19}$$

The expression for θ in this case is

$$\tan 2\theta=\frac{U_y+V_x+k(u_0^2-v_0^2)}{U_x-V_y-2ku_0v_0} \tag{4.20}$$

which can be rewritten, using (4.19), and setting $v/u=\tan\delta$, as

$$\tan 2\theta=\frac{U_y+V_x+(V_x-U_y)\cos 2\delta}{U_x-V_y-(V_x-U_y)\sin 2\delta}=\frac{N}{D}. \tag{4.21}$$

Note that for irrotational flow, (4.21) agrees with (4.8). Again we introduce R, such that $R^2=N^2+D^2$, and having the same sign as D. In terms of this new R, we still have

$$\alpha_1-\alpha_2=\tfrac{1}{2}(U_x+V_y+R), \tag{4.22}$$

$$\alpha_3-\alpha_4=\tfrac{1}{2}(U_x+V_y-R), \tag{4.23}$$

and Eqs. (4.15), (4.16) are unaffected.

For Model B there seems to be no simple analogue of Eq. (4.17). Otherwise, the difference between the two models is that Model B involves computing slightly more expensive expressions for N and D, but may be able to fit itself to a greater variety of flows. Both models have the property that if the data is locally one-dimensional in any direction then waves will be predicted which are exactly those predicted by a one-dimensional linear Riemann solver aligned in that direction (rather than with the coordinate axis). However, Model B can simultaneously recognize a shear flow in some other direction. Neither model, however, could correctly recognize both shock waves of a colliding pair, unless these happened to be perpendicular. It would appear that any model flow must be a compromise between simplicity and generality. Simple models will generally be invalid at isolated points, and reliance must then be placed on conservation. It is to this aspect that the next section is devoted.

5. Conservation Properties

To create an algorithm capable of capturing shock waves, we must ensure that it is conservative. For present purposes, the most convenient definition of a conservative algorithm is that when it operates for one time step, the conserved quantities (mass, momentum, and energy) present within the computational domain are only changed because of events occurring on the boundaries of the domain. We will first set out a strategy which guarantees this. Then we will relate the results of previous sections to that strategy.

Suppose that the computational domain is tessellated into arbitrary polygons (see Fig. 3). Usually these would be quadrilaterals or triangles, and the formulae given below will then be very simple. However, we treat the general case to show that exceptional meshes create no difficulty, at least with regard to conservation. Consider, then, an arbitrary cell with vertices V_1, V_2,..., V_n and note that the area of the cell may be written

$$4A = \sum \mathbf{r}_i \times (\mathbf{r}_{i+1} - \mathbf{r}_{i-1}), \tag{5.1}$$

where $\mathbf{r}_i$ is the position vector of the ith vertex, and the counting is cyclic and anticlockwise. Equation (5.1) is proved by observing that the terms in the summation occur in equal pairs, and that every term $\mathbf{r}_i \times \mathbf{r}_{i+1}$ is twice the area of a triangle $V_i 0 V_{i+1}$, where 0 is an arbitrary origin. Rearrangement of terms in (5.1) leads to two alternative expressions

$$2A = \sum x_i (y_{i+1} - y_{i-1}) \tag{5.2}$$

$$= -\sum y_i (x_{i+1} - x_{i-1}). \tag{5.3}$$

Simple alterations of these formulae allow us to estimate the gradients within a cell of any quantity q which is defined at the vertices. Thus

$$2A \frac{\partial q}{\partial x} = \sum q_i (y_{i+1} - y_{i-1}), \tag{5.4}$$

$$2A \frac{\partial q}{\partial y} = -\sum q_i (x_{i+1} - x_{i-1}), \tag{5.5}$$

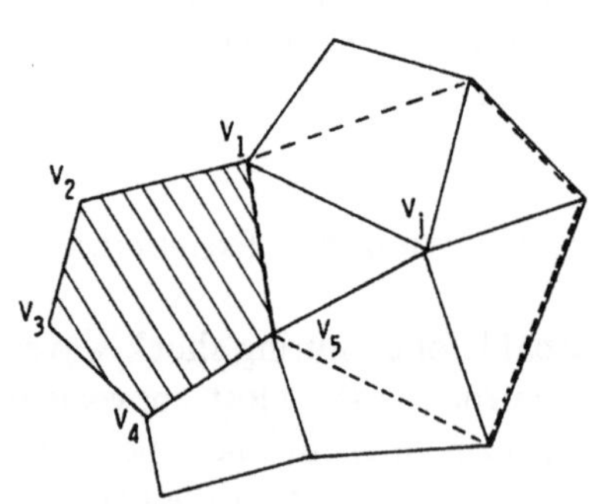

FIG. 3. Part of an irregular mesh.

and it can be seen that these estimates are exact whenever q is a linear function ($q=mx+ny$).

Now suppose that the quantities stored at the vertices are the variables defining flow of an ideal gas according to the Euler equations, written in conservation form as

$$\mathbf{w}_t+\mathbf{F}_x+\mathbf{G}_y=0. \tag{5.6}$$

Then an estimate for $\mathbf{w}_t$, averaged over the cell, is

$$2A\mathbf{w}_t=-\sum\left[\mathbf{F}_i(y_{i+1}-y_{i-1})-\mathbf{G}_i(x_{i+1}-x_{i-1})\right]. \tag{5.7}$$

An alternative way to obtain this formula is to integrate the passage of flux across the cell boundary, using the trapezium rule. We have followed this present derivation because the formulas (5.4), (5.5) are also useful for estimating the gradients from which, in Section 4, the local flow model was deduced.

The quantity $\mathbf{w}_t$, multiplied by a finite time step Δt reepresents the local accumulation of the conserved quantities. The solution can be advanced one time step by adding this change to the quantities stored at the vertices. The increments may be distributed equally or unequally to the vertices concerned. An equal distribution would, if applied on a regular rectangular mesh, reduce to a central-differencing scheme of the kind that can be allied with Runge-Kutta schemes [5]. An unequal distribution of increments, where the weights are obtained from the Jacobian matrices $\partial\mathbf{F}/\partial\mathbf{w}$ and $\partial\mathbf{G}/\partial\mathbf{w}$, has been used by Ni [8] to obtain an integration which is equivalent to Lax–Wendroff. The present work is intended for use with a scheme in which the increments are distributed with more regard to the "upwind" direction of each wave. By analogy with the algorithm described in Section 2 for one-dimensional flow, we propose to place the changes due to a given wave only on those vertices towards which the wave is moving. The selection of those vertices, and the weighting factors by which the effect of each wave is distributed over them will depend on the orientation of the wave relative to the cell. The details of this updating procedure are currently being studied. Meanwhile we prove that any distribution will lead to a conservative algorithm.

The total change of conserved quantities, within the computational domain, is obtained by summing (5.7) over all cells. A typical vertex V_j in the interior of the domain, contributes to this sum through all the cells which meet there. Its total contribution is, in fact

$$\Delta\mathbf{w}_j=\left[-\mathbf{F}_j\sum\Delta y+\mathbf{G}_j\sum\Delta x\right], \tag{5.8}$$

where the Δx, Δy are the adjacent chords of each cell meeting at V_j (see Fig. 3). But since the union of these chords is a closed polygon $\Delta\mathbf{w}_j=0$. Since this argument applies equally to all interior vertices, the sum of conserved quantities changes only due to events on the boundary, and this is what we require.

Next we demonstrate how the estimated total increment (5.7) may be decomposed into contributions due to each wave system. It has not been found possible to do this by any direct extension of the analysis in Section 4. When the spatial changes are large, there seems to be no simple choice of mean values which allows a tidy analysis of the flux gradients. Instead, we directly analyze the temporal changes inside each cell to produce a decomposition which is conservative but not unique. Uniqueness is imposed by incorporating results from Section 4.

First, observe that the time derivative of $\mathbf{w}$ due to the passage of a plane wave is the product of the amplitude and wave speed multiplied by an eigenvector which describes the effect of that wave on the conserved variables. Such eigenvectors are easily derived as projections onto the conserved variables of the eigenvectors shown for the primitive variables in Section 3. For an acoustic wave inclined at an angle θ, the eigenvector is

$$\mathbf{r}_a = \begin{vmatrix} \rho \\ \rho u + \rho a \cos\theta \\ \rho v + \rho a \sin\theta \\ \rho h + \rho a(u\cos\theta + v\sin\theta) \end{vmatrix} \tag{5.9}$$

and for an entropy wave at any angle it is

$$\mathbf{r}_e = \begin{vmatrix} \rho \\ \rho u \\ \rho v \\ \frac{1}{2}\rho(u^2+v^2) \end{vmatrix}. \tag{5.10}$$

For a shear wave it is

$$\mathbf{r}_s = \begin{vmatrix} 0 \\ -\rho a \sin\theta \\ \rho a \cos\theta \\ \rho a(v\cos\theta - u\sin\theta) \end{vmatrix}. \tag{5.11}$$

However, the shear wave included in Model B has zero speed (i.e., it is a steady solution of the Euler equations) so that the term involving $\mathbf{r}_s$ makes no contribution to $\mathbf{w}_t$. In this respect Model B is somewhat simpler than Model A, because the uniform vorticity does contribute to $\mathbf{w}_t$, in a way which is derived below. Introduce the notation

$$A_1 = \alpha_1\lambda_1 = \alpha_1(u\cos\theta + v\sin\theta + a), \tag{5.12a}$$

$$A_2 = \alpha_2\lambda_2 = \alpha_2(u\cos\theta + v\sin\theta - a), \tag{5.12b}$$

$$A_3 = \alpha_3\lambda_3 = \alpha_3(-u\sin\theta + v\cos\theta + a), \tag{5.12c}$$

$$A_4 = \alpha_4\lambda_4 = \alpha_4(-u\sin\theta + v\cos\theta - a), \tag{5.12d}$$

$$A_5 = \beta\lambda_e = \beta(u\cos\phi + v\sin\phi). \tag{5.12e}$$

Our strategy is to compute the $\{A_i\}$ within each cell in such a way that the total effect of all the disturbances in that cell will produce the correct conservative value of $\mathbf{w}_t$. First, though, it must be checked that the model does contain all the effects contributing to $\mathbf{w}_t$. Therefore, we evaluate

$$\sum_{i=1}^{i=4} A_i \mathbf{r}_{ai} + A_5 \mathbf{r}_e. \tag{5.13}$$

an expression which contains the contributions to $\mathbf{w}_t$ from the plane waves used in the models, but not any contribution from the vorticity contained in Model A.

As an example, the terms contributing to ρ_t are

$$\begin{aligned}\rho[A_1+A_2+A_3+A_4+A_5] &= \rho\alpha_1(u\cos\theta+v\sin\theta a+a)\\ &\quad+\rho\alpha_2(u\cos\theta+v\sin\theta-a)+\rho\alpha_3(v\cos\theta-u\sin\theta+a)\\ &\quad+\rho\alpha_4(v\cos\theta-u\sin\theta-a)+\rho\beta(u\cos\phi+v\sin\phi)\\ &=\rho(\alpha_1+\alpha_2)(u\cos\theta+v\sin\theta)+\rho(\alpha_1-\alpha_2)\,a\\ &\quad+\rho(\alpha_3+\alpha_4)(v\cos\theta-u\sin\theta)+\rho(\alpha_3-\alpha_4)\,a\\ &\quad+\rho\beta(u\cos\phi+v\sin\phi).\end{aligned}$$

Substituting the results of Section 4 into this expression, we obtain

$$\begin{aligned}\rho[A_1+A_2+A_3+A_4+A_5] &= \rho[P_x\cos\theta+P_y\sin\theta](u\cos\theta+v\sin\theta)\\ &\quad+\rho[P_y\cos\theta-P_x\sin\theta](v\cos\theta-u\sin\theta)\\ &\quad+\tfrac{1}{2}\rho[U_x+V_y+R]\,a+\tfrac{1}{2}\rho[U_x+V_y-R]\,a\\ &\quad+\rho u[R_x-P_x]+\rho v[R_y-P_y]\\ &=\rho a[U_x+V_y]+\rho uR_x+\rho vR_y\end{aligned}$$

or, in terms of the dimensional gradients,

$$\rho[A_1+A_2+A_3+A_4+A_5]=\rho[u_x+v_y]+u\rho_x+v\rho_y=-\rho_t. \tag{5.14}$$

This calculation, which is valid for Model A or B, checks the algebra and confirms the completeness of the model. Checking the other components of $\mathbf{w}_t$ is tedious, but necessary. It reveals that Model B supplies all the terms of $\mathbf{w}_t$ from the expression (5.13), but that when this expression is used to calculate the effects of the acoustic and entropy waves in Model A, there is a shortfall in the expression for $(\rho u)_t$ amounting to $\frac{1}{2}\rho v\omega$, and a surplus in the expression for $(\rho v)_t$ of $\frac{1}{2}\rho u\omega$. These terms represent the effects of convected vorticity. The expression for $(\rho e)_t$ turns out to be correct. Therefore, we write

$$-\rho_t = \rho(A_1 + A_2 + A_3 + A_4 + A_5), \tag{515a}$$

$$-(\rho u)_t = \rho A_1(u + a\cos\theta) + \rho A_2(u - a\cos\theta) + \rho A_3(y - a\sin\theta) + \rho A_4(u + a\sin\theta) + \rho A_5 u - \tfrac{1}{2}\rho v\omega, \tag{5.15b}$$

$$-(\rho v)_t = \rho A_1(v + a\sin\theta) + \rho A_2(v - a\sin\theta) + \rho A_3(v + a\cos\theta) + \rho A_4(v - a\cos\theta) + \rho A_5 v + \tfrac{1}{2}\rho u\omega, \tag{5.15c}$$

$$-(\rho e)_t = \rho A_1(h + au\cos\theta + av\sin\theta) + \rho A_2(h - au\cos\theta - av\sin\theta) + \rho A_3(h - au\sin\theta + av\cos\theta) + \rho A_4(h + au\sin\theta - av\cos\theta) + \tfrac{1}{2}\rho A_5(u^2 + v^2). \tag{5.15d}$$

where $\omega = 0$ for Model B.

These equations are the two-dimensional analogue of Eqs. (2.5). In each case we try to ensure that the changes of conserved variables predicted by the model are correct. Here, we assume that the *LHS* of each equation is obtained from the conservative formula (5.7) for each cell. Then we treat (5.15) as a set of conditions to be identically (not just approximately) satisfied by the $\{A_i\}$ and by θ, ϕ, ω. Since there are only four conditions for eight unknowns, the remaining information must be supplied from elsewhere. It seems natural to take the values of θ, ϕ, ω from Section 4. Conditions (5.15) are then an incomplete set of linear equations for the $\{A_i\}$, which may be partially analyzed as follows. We obtain at once

$$-(\rho u)_t + u\rho_t + \rho v\omega = \rho a\cos\theta(A_1 - A_2) - \rho a\sin\theta(A_3 - A_4), \tag{5.16a}$$

$$-(\rho v)_t + v\rho_t - \rho u\omega = \rho a\sin\theta(A_1 - A_2) + \rho a\cos\theta(A_3 - A_4), \tag{5.16b}$$

and hence $(A_1 - A_2)$, $(A_3 - A_4)$. Substituting these results into (5.15d) yields

$$\frac{\rho a^2}{\gamma - 1} A_5 = (h - u^2 - v^2)\,\rho_t + u(\rho u)_t + v(\rho v)_t - (\rho e)_t. \tag{5.17}$$

If the changes are so small that $(\cdot)_t$ may be treated as a derivative, rather than as a numerical estimate of a derivative, these equations simplify considerably, offering more insight into the models.

$$(A_2 - A_1) = u_t\cos\theta + v_t\sin\theta - \tfrac{1}{2}(v\cos\theta - u\sin\theta)\,\omega, \tag{5.18}$$

$$(A_4 - A_3) = v_t\cos\theta - u_t\sin\theta + \tfrac{1}{2}(u\cos\theta + v\sin\theta)\,\omega, \tag{5.19}$$

$$A_5 = (a^2\rho_t - p_t)/\rho a^2, \tag{5.20}$$

$$A_1 + A_2 + A_3 + A_4 = p_t. \tag{5.21}$$

Nonconservative schemes could use these simpler conditions; a fully conservative scheme would have to satisfy (5.15).

One more condition is needed on the $\{A_i\}$. In view of the symmetry of the results so far, we would like an expression for $A_1+A_2-A_3-A_4$ (which need not derive from the conservation form). By changing some signs in the analysis leading to (5.14) we find

$$A_1+A_2-A_3-A_4=P_x(u\cos 2\theta+v\sin 2\theta)$$
$$+P_y(u\sin 2\theta-v\cos 2\theta)+aR. \qquad (5.22)$$

It may be shown that the r.h.s. does not, in general, vanish when the data are taken from a steady flow. One might suppose that it should, since then the $\{A_i\}$ would all be zero, and either the strength or the speed of every wave would be zero. Instead of this, the models represent steady flow by a state of equilibrium between finite waves, such that $A_1=A_2=-A_3=-A_4$, and $A_5=0$.

We have now generated a conservative model of the flow, in which the effects of the various components are given by (5.15). The parameters of this decomposition ($\{A_i\}$, θ, ϕ, ω) are found from (4.4), (4.5), (4.8), or (4.21), (5.16), (5.17), and (5.22). Any consistent choice of local average values for ρ, u, v, a, h in these equations will be valid, and will not affect the conservation property. It may be asked, though, whether there are particular average values, similar to those which appear in the one-dimensional theory [9, 12], bestowing special "shock-recognition" properties. However, this question raises unsolved problems about the sort of "captured shock structure" that is possible in two-dimensional flow, and will not be discussed here.

6. Extension to Three Dimensions

No detailed formulae have been worked out for the three-dimensional case. However, merely counting the degrees of freedom makes it plausible that analogous models could be constructed. Data for the three-dimensional unsteady Euler equations would consist of five variables, so there would be fifteen gradients to be accounted for by the model. If the acoustic disturbances are again to be represented by a set of orthogonal plane waves (like an expanding cube) there will be six wave amplitudes and three angles involved (two angles to orient one wave, one angle to orient its neighbors). An entropy wave with one amplitude and two angles will bring the number of parameters up to twelve. The remaining three are available to represent rotational effects. The analogue of Model A would contain three independent vorticity components. The analogue of Model B could contain a shear flow

$$\mathbf{q}=\mathbf{q}_0[1+k_1(v_0x-u_0y)+k_2(w_0y-v_0z)+k_3(u_0z-w_0x)] \qquad (6.1)$$

which is, like (5.1), a steady solution to the Euler equations. However, the three shear components which it contains are not all independent, since all take place in a parallel flow, and one of the k_i can be dropped with no loss of generality. To complete the set of fifteen parameters one might add the flow

$$\begin{aligned} u &= u_0 = \omega(w_0 y - v_0 z), \\ v &= v_0 + \omega(u_0 z - w_0 x), \\ w &= w_0 + \omega(v_0 x - u_0 y). \end{aligned} \tag{6.2}$$

This is also a steady solution of the Euler equations and represents a swirling flow in which the vorticity is parallel to the streamline (u_0, v_0, w_0). Again there is a computational advantage to Model B in that some of its components are steady flows whose contribution to the time-marching process are identically zero. In fact, an analogue of Model B can be worked out for any number of space dimensions d, and the description of arbitrary data is reduced to the description of $(2d+1)$ non-linear scalar advection problems.

There are, however, geometrical difficulties which appear in three dimensions, if the partition of space is made into volumes whose facets have more than three sides. Most finite-volume schemes employ computational cells which are hexahedral, with quadrilateral faces specified by four vertices not normally lying in one plane. The boundary surfaces of the cells must be unambiguously defined so that the cells fill the computational space without overlaps or voids. This can be done by folding each surface into two triangles [6], or by choosing a particular doubly ruled surface to cover each face [3]. Once this has been done, any consistent formula for the volume can also be used to estimate the flux divergence, as in (5.7), but the consistent formulae are disconcertingly lengthy [6, 3]. The effect of simpler formulae on the computational accuracy awaits investigation.

7. Advection Schemes

The eventual goal of this work is to create an algorithm for multidimensional gas dynamics which will enjoy the same degree of success already obtained by upwind schemes in one dimension. In this paper we have addressed (at most) half the problem, showing how arbitrary disturbances in the data can be replaced by locally equivalent sets of plane waves and/or vorticity. To march forward in time, we need to apply to each wave some numerical advection scheme. We may hope that such schemes can be based on schemes for scalar problems, as has happened in the one-dimensional case. Also, we may anticipate that such schemes will show many of the typical features of successful one-dimensional schemes, such as asymmetric support and non-linear limiters [15]. However, the theory even of scalar advection algorithms in many dimensions is only in its infancy. Roe and Baines [11] present a criterion designed to avoid overshoots and describe a scheme which meets it. Smolarkiewicz [14] describes another distinctive, but related, approach. The next (rather large) step in the investigation reported here will be to experiment with these and other algorithms in the present context.

8. Conclusions

We have pointed out that the extension of upwind differencing schemes to more than one space dimension cannot be accomplished by operator splitting methods without losing the desirable property of recognizing data due to a simple wave. To construct "genuinely two-dimensional" schemes we propose model flows, composed of elementary solutions to the two-dimensional equations. These model flows are such that they can be matched to arbitrary data which varies linearly in some small region. The acoustic part of the flow is modelled by four orthogonal plane waves whose orientation is matched to the gradients in the data. Variation of entropy is represented by a single plane wave, and rotational effects either by uniform vorticity or by a parallel shearing motion. We show that the parameters of the model can be evaluated in such a way that a time-marching algorithm can be made exactly conservative.

References

1. S. R. Chakravarthy and S. Osher, *in* "Large-Scale Computations in Fluid Mechanics," Lectures in Applied Mathematics, Vol. 22, AMS 1985.
2. P. Colella, Multidimensional upwind methods for hyperbolic conservation law, Lawrence Berkeley Laboratory, Berkeley, Calif., preprint, LBL-17023, 1984.
3. D. E. Davies and Deborah Salmond, "Calculation of the volume of a general hexahedron for flow predictions," *AIAA J.* **23** (6) (1985), 954.
4. S. F. Davis, *J. Comput. Phys.* (1984), 65.
5. A. Jameson, W. Schmidt, and E. Turkel, "Numerical solutions of the Euler equations by finite-volume methods using Runge-Kutta time-stepping schemes," AIAA Paper 81-1259, 1981.
6. W. Kordulla and M. Vinokur, *AIAA J.* **21** (1983), 917.
7. C. C. Lytton (Formerly C. C. L. Sells), "Solution of the Euler Equations for Transonic Flow Past a Lifting Airfoil—The Bernoulli Formulation," Royal Aircraft Establishment, TR 84080, 1984.
8. R-H. Ni, *AIAA J.* **20** (11) (1982), 1565.
9. P. L. Roe, *J. Comput. Phys.* (1981), 357.
10. P. L. Roe, *in* "Proceedings, INRIA Workshop on Numerical Methods for the Euler Equations," SIAM, 1985, to appear.
11. P. L. Roe and M. J. Baines, *in* "Proceedings 5th GAMM Conference on Numerical Methods in Fluid Mechanics" (M. Pandolfi and R. Piva, Eds), Vieweg, Brunswick, 1984.
12. P. L. Roe and J. Pike, *in* "Computing Methods in Applied Sciences and Engineering VI" (R. Glowinski and J-L. Lions, Eds.), pp. 499–518, North-Holland, Amsterdam, 1984.
13. C. C. L. Sells, "Solution of the Euler Equations for Transonic Flow Past a Lifting Airfoil," Royal Aircraft Establishment, TR 80065, 1980.
14. P. K. Smolarkiewicz, *J. Comput. Phys.* **54** (1984), 325.
15. P. K. Sweby, *SINUM* **25** (5) (1984), 995.
16. P. Woodward and P. Colella, *J. Comput. Phys.* (1984), 115.

Discontinuous Solutions to Hyperbolic Systems Under Operator Splitting

P. L. Roe
*Cranfield Institute of Technology, Cranfield, MK43 0AL, United Kingdom**

Two-dimensional systems of linear hyperbolic equations are studied with regard to their behavior under a solution strategy that in alternate time-steps exactly solves the component one-dimensional operators. The initial data is a step function across an oblique discontinuity. The manner in which this discontinuity breaks up under repeated applications of the split operator is analyzed, and it is shown that the split solution will fail to match the true solution in any case where the two operators do not share all their eigenvectors. The special case of the fluid flow equations is analyzed in more detail, and it is shown that arbitrary initial data gives rise to "pseudo acoustic waves" and a nonphysical stationary wave. The implications of these findings for the design of high-resolution computing schemes are discussed.

I. INTRODUCTION

This article is concerned with two-dimensional systems of time-dependent partial differential equations. We write these in their matrix form

$$\mathbf{u}_t + A\mathbf{u}_x + B\mathbf{u}_y = 0, \tag{1.1}$$

and restrict our attention to the hyperbolic case, characterized by the fact that the $n \times n$ matrix $(A \cos \phi + B \sin \phi)$ possesses a full set of independent eigenvectors for all angles ϕ. Physically, this corresponds to the fact that for any direction ϕ it is possible to construct n different families of plane-wave solutions traveling in that direction.

A common technique for solving Eq. (1.1) numerically is that of *operator splitting* [1]. The operation of advancing Eq. (1.1) through a time interval Δt is replaced, in the simplest case, by the following two operations. First, the data is advanced by $\frac{1}{2}\Delta t$ by means of the one-dimensional operator

$$\mathbf{u}_t + 2A\mathbf{u}_x = 0, \tag{1.2}$$

and then the new solution is advanced through $\frac{1}{2}\Delta t$ by means of

$$\mathbf{u}_t + 2B\mathbf{u}_y = 0. \tag{1.3}$$

Repetition of this pair of operators N times is assumed to be equivalent to advancing Eq. (1.1) by $N\Delta t$. The attraction of the strategy is that it allows use to be made of very accurate and sophisticated one-dimensional numerical schemes. An impressive compendium of aerodynamic calculations accomplished by this and other means is available in Ref. 2.

*Present address: Dept. of Aerospace Engineering, University of Michigan, Ann Arbor, MI 48109.

The split operators are equivalent to the true operator in smooth regions of the flow. By this it is meant that the discrepancy between the two resulting solutions can be made vanishingly small by a suitable choice of Δt. In the above example, the splitting error is $O(\Delta t)$; better sequencing of the operators reduces this to $O(\Delta t^2)$ [1].

Despite this, most users of operator-split codes for computational fluid dynamics feel some unease concerning the correctness of the procedure. This is particularly the case when the one-dimensional methods employed are based on characteristic analysis [3] and aim to achieve highly resolved shock waves. It is well known (e.g., Ref. 4) that operator splitting will create errors in the treatment of flow discontinuities. These errors are of two kinds.

Crandall and Majda [5] have pointed out errors that arise when the problem is nonlinear, albeit scalar. Singularities of the solution may be shocks, expansions, or contacts, corresponding to converging, diverging, or parallel characteristics. The errors occur when a singularity of one type is misread as being of another type by either of the one-dimensional operators.

The error discussed in this article is also a misreading, arising even at the linear level if there is more than one unknown. A and B are then constant matrices, and the problem is found whenever the eigenvectors of A are different from the eigenvectors of B, that is to say, whenever A and B do not commute. In that case, we will show the consequences of operator splitting when the initial data consists of a single oblique discontinuity and each of the one-dimensional split problems is solved exactly. The aim is to identify, at a linearized level of analysis, the detailed nature of the error committed when operator splitting is employed. Since this error appears in an analytical solution, we see the nature of the numerical solution that would appear if the numerical operations were made increasingly exact. Therefore, the results do not relate to any particular numerical scheme, but can be used to explain certain anomalies that are observed in practice, and to suggest the kind of scheme that can be employed in given problems. A qualitative version of the analysis was given by Colella [4], but the present treatment reveals additional detail.

We will study two specific cases. One of these is the 2 × 2 system

$$\begin{vmatrix} u \\ v \end{vmatrix}_t + \begin{vmatrix} 1 & 0 \\ 0 & -1 \end{vmatrix} \begin{vmatrix} u \\ v \end{vmatrix}_x + \begin{vmatrix} 0 & 1 \\ 1 & 0 \end{vmatrix} \begin{vmatrix} u \\ v \end{vmatrix}_y = 0, \tag{1.4}$$

and the other is the 4 × 4 system

$$\begin{vmatrix} p \\ u \\ v \\ s \end{vmatrix}_t + \begin{vmatrix} u & \rho a^2 & 0 & 0 \\ 1/\rho & u & 0 & 0 \\ 0 & 0 & u & 0 \\ 0 & 0 & 0 & u \end{vmatrix} \begin{vmatrix} p \\ u \\ v \\ s \end{vmatrix}_x + \begin{vmatrix} v & 0 & \rho a^2 & 0 \\ 0 & v & 0 & 0 \\ 1/\rho & 0 & v & 0 \\ 0 & 0 & 0 & v \end{vmatrix} \begin{vmatrix} p \\ u \\ v \\ s \end{vmatrix}_y = 0. \tag{1.5}$$

The first of these is an abstract system that serves to introduce the methodology; the second is a linear version of the compressible flow (Euler) equations, using pressure, velocity, and entropy as unknowns.

In Secs. II and III we will briefly recall the time evolution of an oblique initial discontinuity in the data (Riemann problem). In Secs. IV and V we calculate how such a discontinuity evolves under the split operators when the governing equations are, respectively, (1.4) and (1.5). Section VI contains numerical examples, and Section VII discusses the implications.

II. THE EXACT SOLUTION

Consider initial data in which states $\mathbf{u}_L, \mathbf{u}_R$ occupy semi-infinite regions separated by a straight line inclined at an angle ϕ [Fig. 1(a)]. Introducing coordinates x', y' as shown (1.1) becomes

$$\mathbf{u}_t + (A\cos\phi + B\sin\phi)\mathbf{u}_{x'} + (A\sin\phi - B\cos\phi)\mathbf{u}_{y'} = 0. \quad (2.1)$$

Solutions not depending on y' having the form

$$\mathbf{u} = f(x' - \lambda t)\mathbf{r}$$

can be found if λ is an eigenvalue of $(A\cos\phi + B\sin\phi)$, and if $\mathbf{r}$ is the corresponding eigenvector. By a limiting process we can justify taking f to be the Heaviside function. To solve the proposed problem, project the given jump onto the eigenvectors

$$\mathbf{u}_R - \mathbf{u}_L = \sum_{i=1}^{i=n} \alpha_i(\phi)\mathbf{r}_i(\phi). \quad (2.2)$$

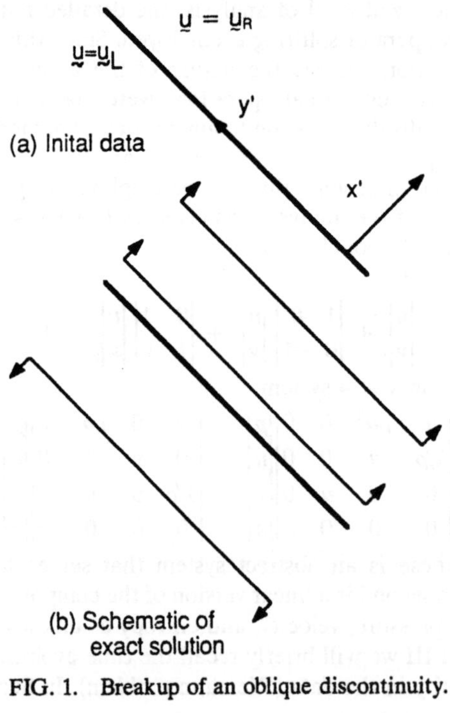

FIG. 1. Breakup of an oblique discontinuity.

In terms of Heaviside functions, the data is

$$\mathbf{u} = \mathbf{u}_L + \sum_{i=1}^{i=n} \alpha_i(\phi) H(x') \mathbf{r}_i(\phi), \tag{2.3}$$

and the solution at time t is [see Fig. 1(b)]

$$\mathbf{u} = \mathbf{u}_L + \sum_{i=1}^{i=n} \alpha_i(\phi) H(x' - \lambda_i t) \mathbf{r}_i(\phi). \tag{2.4}$$

Our object is to discover how well the split operators reproduce this exact solution.

III. THE OPERATOR-SPLIT SOLUTION

During those periods when the equation being solved is (1.2), any jumps in the solution are eigenvectors of A. If this equation is used to evolve the initial data, the result is, since all waves are being displaced horizontally as in Fig. 2(a),

$$\mathbf{u} = \mathbf{u}_L + \sum_{i=1}^{i=n} \alpha_i(0) H(x' - 2t\lambda_i(0) \cos \phi) \mathbf{r}_i(0). \tag{3.1}$$

Similarly, if we use Eq. (1.3), we obtain

$$\mathbf{u} = \mathbf{u}_L + \sum_{i=1}^{i=n} \alpha_i\left(\frac{1}{2}\pi\right) H(x' - 2t\lambda_i(\pi/2) \sin \phi) r_i\left(\frac{1}{2}\pi\right). \tag{3.2}$$

Now suppose that the initial discontinuity evolves for $\frac{1}{2}\Delta t$ according to (1.2). It will generate n distinct wavefronts [Fig. 2(a) shows the case $n = 2$]. Each of these generates new data to be evolved for a further $\frac{1}{2}\Delta t$ according to Eq. (1.3). At the end of the second interval we will have n^2 waves in the system. After time $k\Delta t$ we would have $n^{(2k)}$ waves.

IV. THE ABSTRACT SYSTEM

Here, the problem

$$\mathbf{u}_t + A\mathbf{u}_x + B\mathbf{u}_y = 0. \tag{4.1}$$

is defined by

$$\mathbf{u} = \begin{vmatrix} u \\ v \end{vmatrix}, \quad A = \begin{vmatrix} 1 & 0 \\ 0 & -1 \end{vmatrix}, \quad B = \begin{vmatrix} 0 & 1 \\ 1 & 0 \end{vmatrix}. \tag{4.2}$$

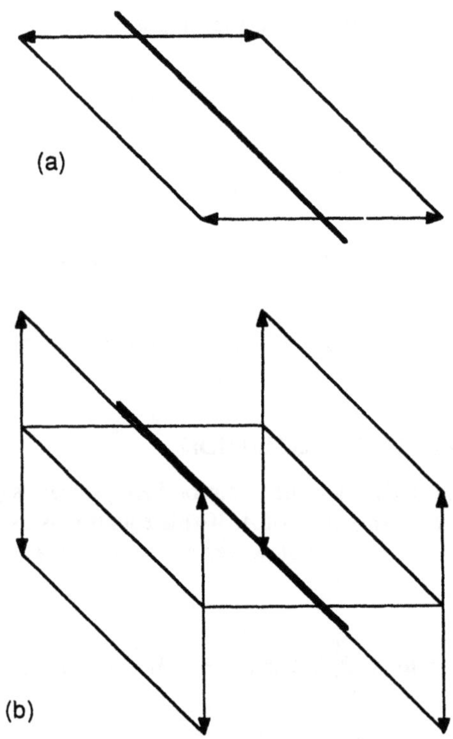

FIG. 2. Bifurcating wave patterns in the operator-split solution.

The eigenvalues of $A\cos\phi + B\sin\phi$ satisfy

$$\det\begin{vmatrix}\cos\phi-\lambda & \sin\phi\\ \sin\phi & -(\cos\phi+\lambda)\end{vmatrix} = 0,$$

i.e.,

$$\lambda^2 = 1. \tag{4.3}$$

Denoting by $\mathbf{r}^{\pm}(\phi)$ the eigenvectors for $\lambda = \pm 1$, we obtain

$$\mathbf{r}^-(\phi) = \begin{vmatrix}-\sin\frac{1}{2}\phi\\ \cos\frac{1}{2}\phi\end{vmatrix}, \qquad \mathbf{r}^+(\phi) = \begin{vmatrix}\cos\frac{1}{2}\phi\\ \sin\frac{1}{2}\phi\end{vmatrix}. \tag{4.4}$$

In particular,

$$r^-(0) = \begin{vmatrix}0\\1\end{vmatrix}, \qquad r^+(0) = \begin{vmatrix}1\\0\end{vmatrix}, \tag{4.5}$$

and

$$r^-(\tfrac{1}{2}\pi) = \begin{vmatrix}-1/\sqrt{2}\\ 1/\sqrt{2}\end{vmatrix}, \qquad r^+(\tfrac{1}{2}\pi) = \begin{vmatrix}1/\sqrt{2}\\ 1/\sqrt{2}\end{vmatrix}. \tag{4.6}$$

If these eigenvectors are denoted by $\mathbf{r}_x^-, \mathbf{r}_x^+, \mathbf{r}_y^-, \mathbf{r}_y^+$, we have the identities

$$\begin{aligned}
\mathbf{r}_x^+ &= \frac{1}{\sqrt{2}}(\mathbf{r}_y^+ - \mathbf{r}_y^-),\\
\mathbf{r}_x^- &= \frac{1}{\sqrt{2}}(\mathbf{r}_y^- + \mathbf{r}_y^+),\\
\mathbf{r}_y^+ &= \frac{1}{\sqrt{2}}(\mathbf{r}_x^- + \mathbf{r}_x^+),\\
\mathbf{r}_y^- &= \frac{1}{\sqrt{2}}(r_x^- - r_x^+).
\end{aligned} \tag{4.7}$$

For example, we see that at the end of an x step, each $\mathbf{r}_x^+$ wave will give rise to $\mathbf{r}_y^+$ and $\mathbf{r}_y^-$ waves of amplitudes, respectively, $1/\sqrt{2}$ and $-1\sqrt{2}$ times that of the source wave. To develop the notation further we introduce translation operators C and S, corresponding to translation of a wavefront in the x' direction by $\Delta t \cos\phi$ and $\Delta t \sin\phi$. We use notation ($\Rightarrow$) to mean "gives rise to" and derive from (4.7)

$$\begin{aligned}
\mathbf{r}_x^+ &\Rightarrow \frac{1}{\sqrt{2}}(S\mathbf{r}_y^+ - S^{-1}\mathbf{r}_y^-),\\
\mathbf{r}_x &\Rightarrow \frac{1}{\sqrt{2}}(S^{-1}\mathbf{r}_y^- + S\mathbf{r}_y^+),\\
\mathbf{r}_y^+ &\Rightarrow \frac{1}{\sqrt{2}}(C^{-1}\mathbf{r}_x^- + C\mathbf{r}_x^+),\\
\mathbf{r}_y^- &\Rightarrow \frac{1}{\sqrt{2}}(C^{-1}\mathbf{r}_x^- - C\mathbf{r}_x^+).
\end{aligned} \tag{4.8}$$

This produces a calculus for tracing the history of a wave through any number of split operations. Thus, an r_x^+ wave, operated on first in the x and then the y direction, evolves according to

$$\begin{aligned}
r_x^+ &\Rightarrow \frac{1}{\sqrt{2}}(Sr_y^+ - S^{-1}\mathbf{r}_y^-)\\
&\Rightarrow \tfrac{1}{2}[S(C^{-1}\mathbf{r}_x^- + C\mathbf{r}_x^+) - S^{-1}(C^{-1}r_x^- - Cr_x^+)]\\
&= \tfrac{1}{2}(SC + S^{-1}C)\mathbf{r}_x^+ + \tfrac{1}{2}(SC^{-1} - S^{-1}C^{-1})\mathbf{r}_x^-.
\end{aligned} \tag{4.9}$$

Similarly, under the same operations

$$r_x^- \Rightarrow \tfrac{1}{2}(SC - S^{-1}C)\mathbf{r}_x^+ + \tfrac{1}{2}(SC^{-1} + S^{-1}C^{-1})\mathbf{r}_x^-. \tag{4.10}$$

A powerful notation to describe the configuration of waves present after a certain number of steps is to introduce a vector $\mathbf{a} = (a_1, a_2)^T$ each of whose components is a sum of terms $k_{\alpha\beta}S^\alpha C^\beta$. Such a term describes a wave of amplitude $k_{\alpha\beta}$ at a location $x' = \alpha\sin\phi + \beta\cos\phi$. We use a_1 to describe the $\mathbf{r}_x^+$ waves and a_2 to describe the $\mathbf{r}_x^-$ waves. Evolution through a pair of split operations from $k\Delta t$ to $(k + 1)\Delta t$ can then be described as

$$\mathbf{a}_{k+1} = M\mathbf{a}_k . \tag{4.11}$$

where the matrix M can be obtained from Eqs. (4.9) and (4.10) as

$$M = \begin{vmatrix} \frac{1}{2}(SC + S^{-1}C) & \frac{1}{2}(SC - S^{-1}C) \\ \frac{1}{2}(SC^{-1} - S^{-1}C^{-1}) & \frac{1}{2}(SC^{-1} + S^{-1}C^{-1}) \end{vmatrix} . \tag{4.12}$$

The equation

$$\mathbf{a}_k = M^k\mathbf{a}_0 \tag{4.13}$$

is then a formal solution to the initial-value problem, if $\mathbf{a}_0 = (\alpha^+, \alpha^-)^T$ where α^+, α^- are the projections of the initial jump onto the eigenvectors $\mathbf{r}_\lambda^\pm$. However, computing high powers of M would be cumbersome, and it is more convenient to use the recurrence relationship

$$\mathbf{a}_k = \operatorname{tr}(M)\mathbf{a}_{k-1} - \det(M)\mathbf{a}_{k-2} . \tag{4.14}$$

This follows from applying to $\mathbf{a}_{k-2}$ the identity

$$M^2 - \operatorname{tr}(M)M + \det(M)I = 0,$$

which is the form taken for $n = 2$ by the Cayley–Hamilton theorem. Applying Eq. (4.14) to Eq. (4.12) gives

$$\mathbf{a}_k = \tfrac{1}{2}(S + S^{-1})(C + C^{-1})\mathbf{a}_{k-1} - \mathbf{a}_{k-2} . \tag{4.15}$$

The iteration need not start with an isolated discontinuity, but may take $\mathbf{a}_0$ arbitrary, $\mathbf{a}_1 = M\mathbf{a}_0$.

A particularly simple case that yields more general insights is to take $\phi = \frac{1}{4}\pi$, so that $S = C = T$ (e.g.) so that

$$\mathbf{a}_k = \tfrac{1}{2}(T^2 + 2 + T^{-2})\mathbf{a}_{k-1} - \mathbf{a}_{k-2} . \tag{4.16}$$

Let $\mathbf{c}_n^k$ be the coefficient of T^k in $\mathbf{a}_k$. Then Eq. (4.16) translates to

$$\mathbf{c}_n^{k+1} = \mathbf{c}_n^k + \tfrac{1}{2}(\mathbf{c}_{n-2}^k + \mathbf{c}_{n+2}^k) - \mathbf{c}_n^{k-1}, \tag{4.17}$$

or, rearranging,

$$\mathbf{c}_n^{k+1} - 2\mathbf{c}_n^k + \mathbf{c}_n^{k-1} = \tfrac{1}{2}(\mathbf{c}_{n-2}^k - 2\mathbf{c}_n^k + \mathbf{c}_{n+2}^k) . \tag{4.18}$$

This shows that $\mathbf{c}_n^k$ behaves like a leapfrog approximation to the wave equation

$$\mathbf{c}_{tt} = 2\mathbf{c}_{xx} = \mathbf{c}_{x'x'} . \tag{4.19}$$

The approximation has an effective Courant number of $1/\sqrt{2}$. In Eq. (4.19) we see waves propagating to left and right with unit velocity as in the exact solution, but in Eq. (4.18) we will expect the phase error and oscillations that typify leapfrog schemes.

The precise nature of the phase errors can be found by subjecting Eq. (4.18) to the usual Fourier analysis via the ansatz

$$\mathbf{c}_n^k = g^k \exp(i\theta x'/\Delta t), \tag{4.20}$$

where g is a complex amplification factor and θ is the Fourier frequency. Substituting Eq. (4.20) into Eq. (4.18) yields

$$g^2 - 2g \cos(\theta \sin \phi) \cos(\theta \cos \phi) + 1 = 0, \tag{4.21}$$

or

$$g = \cos(\theta \sin \phi) \cos(\phi \sin \phi) \pm i[1 - \cos^2(\theta \sin \phi) \cos^2(\theta \cos \phi)]^{1/2}. \tag{4.22}$$

Since $|g| = 1$ there is no dissipation. The phase shift per iteration is given by

$$\tan^{-1} \frac{[1 - \cos^2(\theta \sin \phi) \cos^2(\theta \cos \phi)]^{1/2}}{\cos(\theta \sin \phi) \cos(\theta \cos \phi)}, \tag{4.23}$$

which equals θ whenever $\phi = N\pi/2$. For other values of ϕ, Eq. (4.22) can be expanded in θ to give a phase velocity of

$$\frac{\sin^2 2\phi}{24} \theta^2 + 0(\theta^4). \tag{4.24}$$

Since the operator-splitting solution has emerged as a nondissipative second-order approximation to the exact solution (for plane waves), standard results for such schemes apply. In particular, an initial discontinuity will be dispersed over an area proportional to $t^{1/3}$, and the accompanying overshoots will reach a maximum amplitude of 27% of the initial amplitude [6].

V. THE LINEARIZED EULER EQUATIONS

In this section the problem

$$\mathbf{u}_t + A\mathbf{u}_x + B\mathbf{u}_y = 0 \tag{5.1}$$

is defined by

$$u = \begin{vmatrix} p \\ u \\ v \\ s \end{vmatrix}, A = \begin{vmatrix} u & \rho a^2 & 0 & 0 \\ 1/\rho & u & 0 & 0 \\ 0 & 0 & u & 0 \\ 0 & 0 & 0 & u \end{vmatrix}, B = \begin{vmatrix} v & 0 & \rho a^2 & 0 \\ 0 & v & 0 & 0 \\ 1/\rho & 0 & v & 0 \\ 0 & 0 & 0 & v \end{vmatrix}. \tag{5.2}$$

The eigenvalues of $(A \cos \phi + B \sin \phi)$ are the roots of

$$\det \begin{vmatrix} U - \lambda & \rho a \cos \phi & \rho a \sin \phi & 0 \\ \cos \phi/\rho & U - \lambda & 0 & 0 \\ \sin \phi/\rho & 0 & U - \lambda & 0 \\ 0 & 0 & 0 & U - \lambda \end{vmatrix} = 0, \tag{5.3}$$

where $U = u\cos\phi + v\sin\phi$, i.e.,

$$(U-\lambda)^2[(U-\lambda)^2 - a^2] = 0. \tag{5.4}$$

Corresponding to the two eigenvalues $\lambda = U$ are eigenvectors

$$\mathbf{r}_s = \begin{vmatrix} 0 \\ 0 \\ 0 \\ 1 \end{vmatrix}, \tag{5.5}$$

describing an *entropy wave,* and

$$\mathbf{r}_v = \begin{vmatrix} 0 \\ -\sin\phi \\ \cos\phi \\ 0 \end{vmatrix}, \tag{5.6}$$

called here a *vorticity wave,* across which the tangential component of velocity changes.

Corresponding to the eigenvalue $\lambda = U + a$ is an eigenvector

$$\mathbf{r}_a = \begin{vmatrix} \rho a \\ \cos\phi \\ \sin\phi \\ 0 \end{vmatrix} \tag{5.7}$$

which describes an *acoustic wave.* If θ ranges over $[-\pi, \pi]$ in Eq. (5.7), the eigenvalue $\lambda = U - a$ is accounted for as well.

With no real loss of generality we can assume that the linearization is about a state of rest (both the exact and operator-split solutions are invariant under steady translations). In that case the entropy and vorticity waves are stationary, and the acoustic waves can move with the sound speed a in any direction (ϕ).

To find the effect of the split operators we need to take $\phi = 0, \frac{1}{2}\pi$, yielding

$$\mathbf{r}_{sx} = \begin{vmatrix} 0 \\ 0 \\ 0 \\ 1 \end{vmatrix}, \mathbf{r}_{vx} = \begin{vmatrix} 0 \\ 0 \\ 1 \\ 0 \end{vmatrix}, \mathbf{r}_{ax}^{+} = \begin{vmatrix} \rho a \\ 1 \\ 0 \\ 0 \end{vmatrix}, \mathbf{r}_{ax}^{-} = \begin{vmatrix} \rho a \\ -1 \\ 0 \\ 0 \end{vmatrix}, \tag{5.8}$$

$$\mathbf{r}_{sy} = \begin{vmatrix} 0 \\ 0 \\ 0 \\ 1 \end{vmatrix}, \mathbf{r}_{vy} = \begin{vmatrix} 0 \\ -1 \\ 0 \\ 0 \end{vmatrix}, \mathbf{r}_{ay}^{+} = \begin{vmatrix} \rho a \\ 0 \\ 1 \\ 0 \end{vmatrix}, \mathbf{r}_{ay}^{-} = \begin{vmatrix} \rho a \\ 0 \\ -1 \\ 0 \end{vmatrix}. \tag{5.9}$$

The following identities determine how each time-split wave will fission into waves travelling in the other direction.

$$\begin{aligned}
\mathbf{r}_{sx} &= \mathbf{r}_{sy}, \\
\mathbf{r}_{vx} &= \tfrac{1}{2}[\mathbf{r}_{ay}^{+} - \mathbf{r}_{ay}^{-}], \\
\mathbf{r}_{ax}^{+} &= \tfrac{1}{2}[\mathbf{r}_{ay}^{+} + \mathbf{r}_{ay}^{-}] - \mathbf{r}_{vy}, \\
\mathbf{r}_{ax}^{-} &= \tfrac{1}{2}[\mathbf{r}_{ay}^{+} + \mathbf{r}_{ay}^{-}] + \mathbf{r}_{vy},
\end{aligned} \tag{5.10}$$

$$\begin{aligned}
\mathbf{r}_{sy} &= \mathbf{r}_{sx}, \\
\mathbf{r}_{vy} &= \tfrac{1}{2}[\mathbf{r}_{ax}^{-} - \mathbf{r}_{ax}^{+}], \\
\mathbf{r}_{ay}^{+} &= \tfrac{1}{2}[\mathbf{r}_{ax}^{-} + \mathbf{r}_{ax}^{+}] + \mathbf{r}_{vx}, \\
\mathbf{r}_{ay}^{-} &= \tfrac{1}{2}[\mathbf{r}_{ax}^{-} + \mathbf{r}_{ax}^{+}] - \mathbf{r}_{vx}.
\end{aligned} \tag{5.11}$$

From these results we can determine the evolution matrix for this problem. Let $\mathbf{a}$ be a four-vector whose components describe the configurations of waves $\mathbf{r}_x^+, \mathbf{r}_x^-, \mathbf{r}_v$, and $\mathbf{r}_s$, respectively. Then some algebra reveals that

$$\mathbf{a}_{k+1} = M\mathbf{a}_k, \tag{5.12}$$

where

$$M = \begin{vmatrix}
\tfrac{1}{4}(S + S^{-1} + 2I)C & \tfrac{1}{4}(S + S^{-1} - 2I)C & \tfrac{1}{4}(S - S^{-1})C & 0 \\
\tfrac{1}{4}(S + S^{-1} - 2I)C^{-1} & \tfrac{1}{4}(S + S^{-1} + 2I)C^{-1} & \tfrac{1}{4}(S - S^{-1})C^{-1} & 0 \\
\tfrac{1}{2}(S - S^{-1}) & \tfrac{1}{2}(S - S^{-1}) & \tfrac{1}{2}(S + S^{-1}) & 0 \\
0 & 0 & 0 & I
\end{vmatrix} \tag{5.13}$$

and S, C represent displacements through $a\Delta t \sin\phi$, $a\Delta t \cos\phi$.

The most obvious feature of Eq. (5.13) is that the entropy waves (at this linearized level) decouple completely from the rest of the waves, and are, in fact, represented exactly in the operator-split solution. It is easy to see that these statements are always true of any wave that is an eigenvector of both A and B (and hence of $A \cos\phi + B \sin\phi$).

The matrix (5.13) has another property not observed in the earlier problem (4.12). By treating S and C formally as real variables, we can attempt to find eigenvalues of M. It is quite clear that there will be one eigenvalue $\Lambda = I$, with eigenvector $(0, 0, 0, 1)^T$ which corresponds to a configuration that is a steady solution of the system. The cubic that gives the other eigenvalues is

$$(\Lambda^3 - I) - (\Lambda^2 - \Lambda)[\tfrac{1}{2}(S + S^{-1}) + \tfrac{1}{4}(S + S^{-1} + 2)(C + C^{-1})] = 0. \tag{5.14}$$

The existence of a second unit eigenvalue $\Lambda = I$ implies another steady-state configuration. The corresponding eigenvector

$$\mathbf{p} = \begin{vmatrix} C^{1/2}(S^{1/2} - S^{-1/2}) \\ -C^{-1/2}(S^{1/2} - S^{-1/2}) \\ (C^{-1/2} - C^{1/2})(S^{1/2} + S^{-1/2}) \\ 0 \end{vmatrix} \tag{5.15}$$

reproduces itself exactly and will be called the *persistent configuration.* (The other roots Λ_2, Λ_3 of (5.14) are pseudo-operators, having no simple interpretations, although we may remark that $\Lambda_2\Lambda_3 = I$. In this sense, they are inverse to each other and represent translations in opposing directions. We will describe the disturbances associated with them as "pseudo-acoustic" waves, and we shall see later that these waves obey a leapfrog-like recurrence relationship.)

If, as an example, we take $\phi = \tan^{-1}\frac{1}{2}$, so that $C = S^2$, the persistent configuration becomes

$$\mathbf{p} = \begin{vmatrix} S^{3/2} - S^{1/2} \\ S^{-3/2} - S^{-1/2} \\ S^{-3/2} + S^{-1/2} - S^{1/2} - S^{3/2} \\ 0 \end{vmatrix}, \tag{5.16}$$

and this is sketched in the upper part of Figure 3. All the waves that are radiated externally from this configuration will be self-cancelling. The profiles of pressure and velocity are shown in the lower part of Figure 3. Note that there is never any net change in any quantity across the persistent configuration (unless we add an entropy wave).

For other values of ϕ, the strengths of waves in the persistent configuration remain unaltered, but they are found in different locations. The waves A, B, C, D are displaced from the origin by amounts $\frac{1}{2}a\Delta t(\cos\phi + \sin\phi)$, $\frac{1}{2}a\Delta t(\cos\phi - \sin\phi)$, $\frac{1}{2}a\Delta t(\sin\phi - \cos\phi)$, $-\frac{1}{2}a\Delta t(\cos\phi + \sin\phi)$, and occupy a region of total breadth $a\Delta t(|\cos\phi| + \sin|\phi|)$. As ϕ approaches $N\pi/2$, however, the waves approach and cancel by pairs.

Note that the persistent configuration could be multiplied by any polynomial in the shift operators, to yield another persistent configuration.

We now turn to computing the evolution of arbitrary initial data. By the Cayley–Hamilton theorem we have

$$M^4 - (D+1)M^3 + 2DM^2 - (D+1)M + I = 0, \tag{5.17}$$

where

$$D = \tfrac{1}{2}(C + C^{-1}) + \tfrac{1}{2}(S + S^{-1}) + \tfrac{1}{4}(C + C^{-1})(S + S^{-1}) \tag{5.18}$$

so that

$$\mathbf{a}_k = (D+1)\mathbf{a}_{k-1} - 2D\mathbf{a}_{k-2} + (D+1)\mathbf{a}_{k-3} - \mathbf{a}_{k-4} \tag{5.19}$$

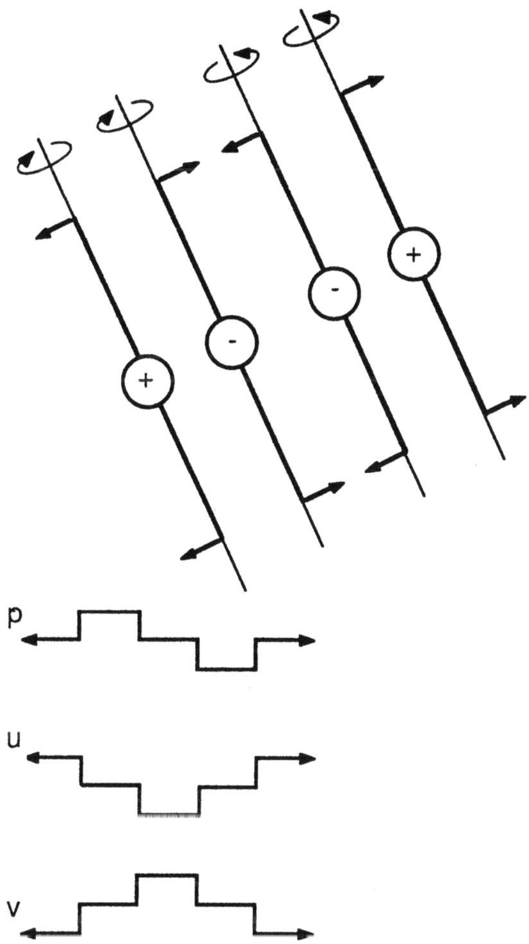

FIG. 3. The persistent wave configuration.

is a valid (but tedious) recurrence relationship. The trick is to separate **a** into its steady and unsteady parts, thus:

$$\mathbf{a} = \alpha\mathbf{p} + \beta\mathbf{r}_s + \mathbf{g},$$

where **g** lies in ξ, the complement of the space spanned by **p** and $\mathbf{r}_s$. The argument is formal, treating S and C as real numbers, but any interpretable statements reached in that way must be true.

Introduce the eigenvector factorization of M

$$M = R(\Lambda_p + \Lambda_q)L, \tag{5.20}$$

$$= P + Q, \tag{5.21}$$

where $\Lambda_p = \text{diag}(1,1,0,0)$ and $\Lambda_q = \text{diag}(0,0,\lambda_3,\lambda_4)$, where λ_3, λ_4 do not need to be explicitly computed. We then have that

$$\begin{aligned} Q\mathbf{p} = Q\mathbf{r}_s = P\mathbf{q} = 0, \\ P\mathbf{p} = \mathbf{p}, \quad P\mathbf{r}_s = \mathbf{r}_s. \end{aligned} \tag{5.22}$$

Inserting Eq. (5.21) into Eq. (5.17) and using Eq. (5.22) gives

$$(Q - I)^2[Q^2 - (D - 1)Q + I]\mathbf{q} = 0,$$

and since $Q - I$ is nonsingular, the part of the solution lying in ξ obeys

$$[Q^2 - (D - 1)Q + I]\mathbf{q} = 0 \tag{5.23}$$

or

$$\mathbf{q}_k = (D - 1)\mathbf{q}_{k-1} - \mathbf{q}_{k-2}. \tag{5.24}$$

Thus, the unsteady part of the operator-split Euler solution obeys a recurrence relationship very similar to that found for the abstract problem (cf. Eq. 415). In the special case $\phi = \pi/4$, we again find a leapfrog-type behavior for the discrete values

$$\mathbf{q}_k - 2\mathbf{q}_{k-1} + \mathbf{q}_{k-2} = (\tfrac{1}{4}T^2 + T - \tfrac{5}{2} + T^{-1} + \tfrac{1}{4}T^{-2})\mathbf{q}_{k-1}. \tag{5.25}$$

This shows that the unsteady part obeys a numerical approximation to $\mathbf{q}_{tt} = a^2\mathbf{q}_{x'x'}$. Fourier analysis reveals a nondissipative scheme; some lengthy algebra yields the phase speed in a general direction to be

$$a\left(1 + \frac{\sin^2 2\phi}{96}\theta^2\right) + O(\theta^4),$$

which is only slightly in error. The principal disparity between the exact and operator-split solutions is that, in both cases, an arbitrary initial discontinuity should project into a steady component (one that moves with the fluid) and an unsteady component. The chief failure of the split operators is that they do not correctly distinguish the two sorts of data (except for the entropy wave).

In the exact solution, that part of the data that projects onto the acoustic eigenvectors is treated as unsteady. In the split solution, it is not easy to identify the unsteady part of arbitrary data, but it generally differs from the exact projection. Here, we give a partial analysis of the decomposition

$$\boldsymbol{\alpha}_0 = \alpha\mathbf{p} + \beta\mathbf{r}_s + \mathbf{q}_0. \tag{5.26}$$

This is facilitated by the observation that M can be symmetrized

$$M = E^{-1}XE, \tag{5.27}$$

where

$$E = \frac{1}{\sqrt{2}}\begin{vmatrix} C^{-1/2} & C^{1/2} & 0 & 0 \\ C^{-1/2} & -C^{-1/2} & 0 & 0 \\ 0 & 0 & I & 0 \\ 0 & 0 & 0 & \sqrt{2I} \end{vmatrix}, \quad E^{-1} = \frac{1}{\sqrt{2}}\begin{vmatrix} C^{1/2} & C^{1/2} & 0 & 0 \\ C^{-1/2} & -C^{-1/2} & 0 & 0 \\ 0 & 0 & 2I & 0 \\ 0 & 0 & 0 & \sqrt{2I} \end{vmatrix} \tag{5.28}$$

and

$$X = \begin{vmatrix} \frac{1}{8}(C + C^{-1} + 2)(S + S^{-1} + 2) - I & \frac{1}{8}(C - C^{-1})(S + S^{-1} + 2I) & \frac{1}{4}(C^{1/2} + C^{-1/2})(S - S^{-1}) & 0 \\ \frac{1}{8}(C - C^{-1})(S + S^{-1} + 2I) & \frac{1}{8}(C + C^{-1} - 2)(S + S^{-1} + 2) + I & \frac{1}{4}(C^{1/2} - C^{-1/2})(S - S^{-1}) & 0 \\ \frac{1}{4}(C^{1/2} + C^{-1/2})(S - S^{-1}) & \frac{1}{4}(C^{1/2} - C^{-1/2})(S - S^{-1}) & \frac{1}{2}(S + S^{-1}) & 0 \\ 0 & 0 & 0 & I \end{vmatrix} \tag{5.29}$$

Since the eigenvectors of X will be orthogonal, a formal solution to the problem of determining α, β in Eq. (5.26) is

$$\alpha = \frac{\mathbf{p}^T E^T E \mathbf{a}_0}{\mathbf{p}^T E^T E \mathbf{p}}, \tag{5.30}$$

$$\beta = \frac{\mathbf{r}_s^T E^T E \mathbf{a}_0}{\mathbf{r}_s^T E^T E \mathbf{r}_s} = \mathbf{r}_s^T \mathbf{a}_0, \tag{5.31}$$

and it is only the evaluation of Eq. (5.30) that presents any difficulty. Formal manipulations yield

$$\alpha = \frac{-[C^{-1/2}(S^{1/2}-S^{-1/2}),\ -C^{1/2}(S^{1/2} - S^{-1/2}),\ \frac{1}{2}(C^{-1/2} - C^{1/2})(S^{1/2} + S^{-1/2}),\ 0]\mathbf{a}_0}{8\left[I - \left(\frac{C^{1/2} + C^{-1/2}}{2}\right)^2\left(\frac{S^{1/2} + S^{-1/2}}{2}\right)\right]^2}. \tag{5.30}$$

The inverse of the denominator of this expression can be expanded as

$$\frac{1}{8}\sum_{n=0}^{n=\infty}\left(\frac{C^{1/2} + C^{-1/2}}{2}\right)^{2n}\left(\frac{S^{1/2} + S^{-1/2}}{2}\right)^{2n}. \tag{5.31}$$

For Eq. (5.31) there is a coefficient of $C^P S^q$, e.g., given by a divergent infinite series. However, if Eq. (5.31) is multiplied by the numerator of Eq. (5.30), all coefficients are given by convergent, hence interpretable, series. For example, if the initial jump consists of an isolated $\mathbf{r}_x$ wave, [$\mathbf{a}_0 = (1,0,0,0)$] but inclined at $\phi = 45°$ we have $S = C = T$ and

$$\alpha = \frac{T^{-1} - I}{8}\sum_{n=0}^{n=\infty}\left(\frac{T^{1/2} + T^{-1/2}}{2}\right)^{4n}. \tag{5.32}$$

The largest coefficients are those of T^{-1} and T^0. We find

$$C_0 = -C_{-1} = \frac{1}{8}\sum_{n=0}^{n=\infty}\frac{(4n)!}{(2n)!\,(2n+1)!}\frac{1}{16^n}, \tag{5.33}$$

$$= \frac{1}{8}\sum_{n=0}^{n=\infty}\frac{\left(\frac{1}{4}\right)_n\left(\frac{2}{4}\right)_n\left(\frac{3}{4}\right)_n\left(\frac{4}{4}\right)_n}{\left(\frac{1}{2}\right)_n\left(\frac{2}{2}\right)_n\left(\frac{2}{2}\right)_n\left(\frac{3}{2}\right)_n}, \tag{5.34}$$

where $(x)_n = \Gamma(x + n)/\Gamma(n)$ is the Pochammer symbol. Cancellation of terms leads to

$$C_0 = -C_{-1} = \tfrac{1}{8}\,{}_2F_1[\tfrac{1}{4}, \tfrac{3}{4}; \tfrac{3}{2}; 1],$$

$$= \frac{1}{8}\frac{\Gamma(\frac{3}{2})\Gamma(\frac{1}{2})}{\Gamma(\frac{5}{4})\Gamma(\frac{3}{4})},$$

$$= \frac{1}{4}\frac{\Gamma(\frac{1}{2})\Gamma(\frac{1}{2})}{\Gamma(\frac{1}{4})\Gamma(\frac{3}{4})}.$$

This may be evaluated from the reflection formula

$$\Gamma(z)\Gamma(1 - z) = \pi/\sin(\pi z),$$

as

$$C_0 = -C_{-1} = \tfrac{1}{4}\sin(\pi/4). \tag{5.35}$$

The numerical value of these coefficients is 0.177, which could be regarded as the amplitude of the persistent wave configuration generated by the particular initial data. The same amplitude would result from an initial jump proportional to $\mathbf{r}_x^-$ and inclined at $\phi = 45°$.

If the initial jump is proportional to $\mathbf{r}_{vx}$ and inclined at 45°, then Eq. (5.30) yields

$$\alpha = \frac{T - T^{-1}}{16}\sum_{n=0}^{n=\infty}\left(\frac{T^{1/2} + T^{-1/2}}{2}\right)^{4n}. \tag{5.36}$$

In this expression the coefficient of T^0 vanishes, but the amplitude may be represented by

$$C_1 = -C_{-1} = \frac{1}{8}\sum_{n=0}^{n=\infty}\frac{(4n + 1)!}{(2n)!\,(2n + 2)!}\cdot\frac{1}{16^n},$$

$$= \frac{1}{16}\sum_{n=0}^{n=\infty}\frac{(\frac{2}{4})_n(\frac{3}{4})_n(\frac{4}{4})_n(\frac{5}{4})_n}{(\frac{1}{2})_n(\frac{2}{2})_n(\frac{3}{2})_n(\frac{4}{2})_n}, \tag{5.37}$$

$$= \tfrac{1}{16}\,{}_3F_2[\tfrac{3}{4}, \tfrac{5}{4}, 1; \tfrac{3}{2}, 2; 1].$$

The generalized hypergeometric function appearing here can be summed by Watson's theorem (Ref. 7, p. 54), giving

$$C_1 = -C_{-1} = \frac{1}{16}\left(\frac{\Gamma(\frac{1}{2})\Gamma(\frac{3}{2})}{\Gamma(\frac{7}{8})\Gamma(\frac{9}{8})}\right)^2,$$

$$= \left(\frac{\Gamma(\frac{1}{2})^2}{\Gamma(\frac{1}{8})\Gamma(\frac{7}{8})}\right)^2$$

which can be evaluated using the reflection formula as

$$C_1 = -C_{-1} = \sin^2\pi/8 = 0.147.$$

Results for arbitrary initial jumps can be found by superposing the special cases, but systematic conclusions would be hard to reach. The important point is that an appreciable fraction of the initial amplitude may be projected onto the persistent wave. This conclusion is reinforced by the numerical experiments in Sec. VI.

VI. NUMERICAL EXAMPLES

Figure 4 shows the distribution of pressure and vertical velocity through the wave system that results from 10 or 30 split time-steps being applied to data representing a shock inclined at $\phi = \pi/4$, i.e., $\mathbf{a}_0 = [\frac{1}{2}(1 + 1/\sqrt{2}), \frac{1}{2}(1 - 1/\sqrt{2}), 1/\sqrt{2}, 0]^T$. Note that a persistent wave establishes itself very quickly at the center of the domain, where the initial discontinuity was placed. A typical "leapfrog" wave moves to the right in a manner that imitates rather loosely the exact solution (also shown). Some very weak disturbances move to the left.

In Fig. 5 the initial data is a shear wave; $\mathbf{a}_0 = [0, 1/\sqrt{2}, -1/\sqrt{2}, 0]^T$. Here we see two pseudo-acoustic waves moving outward, leaving a pronounced persistent pressure defect. The region between is filled with low-amplitude noise, especially in the velocity.

Figure 6 demonstrates that the case $\phi = \pi/4$ is not untypical. Here we take $\phi = \tan^{-1}(\frac{1}{2})$, computing first the data corresponding to a shock $\mathbf{a}_0 = [\frac{1}{2}(1 + 2r), \frac{1}{2}(1 - 2r), r, 0]^T$ where $r^2 = 1/5$ and then due to a shear

$$\mathbf{a}_0 = (0, 2r, -r, 0)^T.$$

VII. RELEVANCE OF STUDY TO PRACTICAL COMPUTATIONS

We have seen that for discontinuous data, the solution of the split problem converges to the solution of the exact problem only in some weak sense, there being $O(1)$ errors even as $\Delta t \to 0$. That many practical calculations do successfully employ operator splitting is due to one fact: Two successive wavefronts in the split solution are separated by, at most, $a\Delta t \min[|\sin \phi|, |\cos \phi|] \leq a\Delta t$. For explicit marching schemes this is, in turn, less than the spatial mesh sizes Δx, Δy. In the frequency domain, the difference between the split solution and the true solution lies mostly above the folding frequency and thus cannot be resolved by the majority of numerical schemes.

Of course, this reasoning does not apply to implicit schemes, so any implicit method would have to contain "sufficient" dissipation. Nor does it apply to the explicit "large time-step" schemes developed by Harten [8] and LeVeque [9]. In fact, these latter schemes would be particularly inappropriate as they are "high-resolution" schemes that aim at preserving high frequencies.

Another scheme that performs very badly in split-operator applications is the random-choice method [4]. Again, this is because the one-dimensional operator is just too good. The states generated in solving the one-dimensional Riemann problems are not lost in any averaging process, but have a chance to appear in the solution. Although only some of them will appear, there is a striking similarity between the results of the last

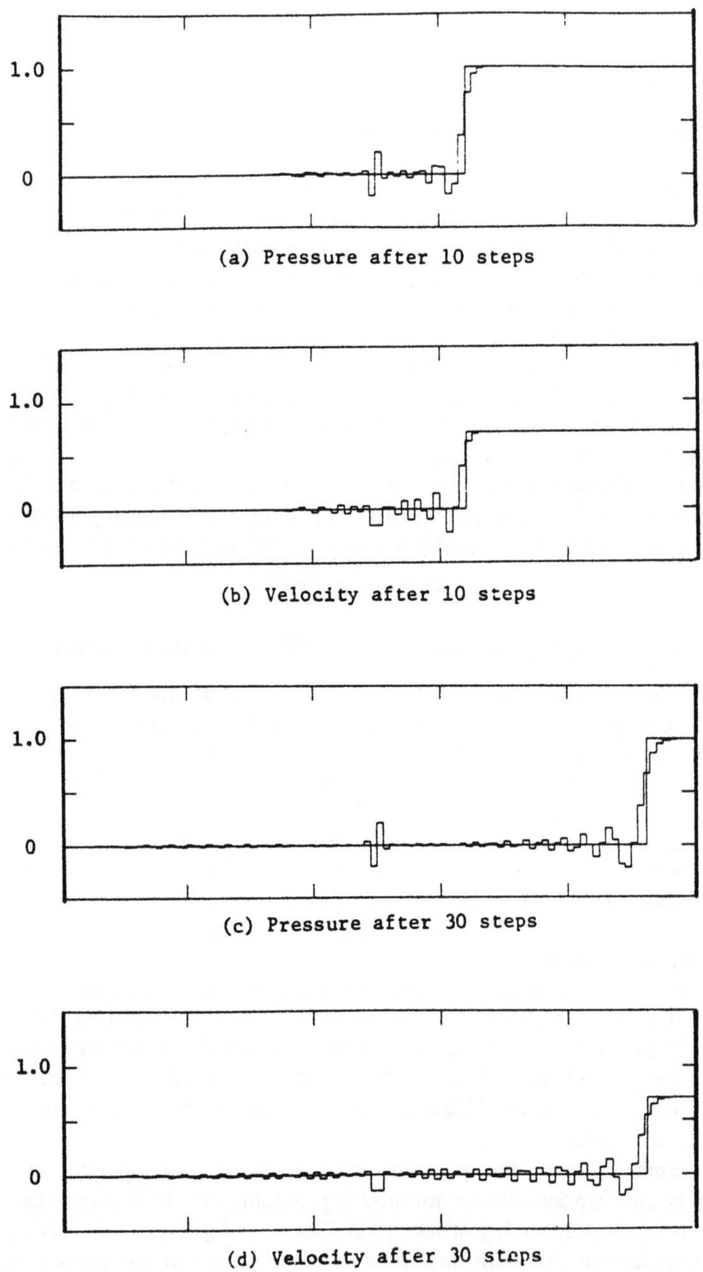

FIG. 4. Results from a shock at $\phi = \pi/4$. (a) Pressure after 10 steps. (b) Velocity after 10 steps. (c) Pressure after 30 steps. (d) Velocity after 30 steps.

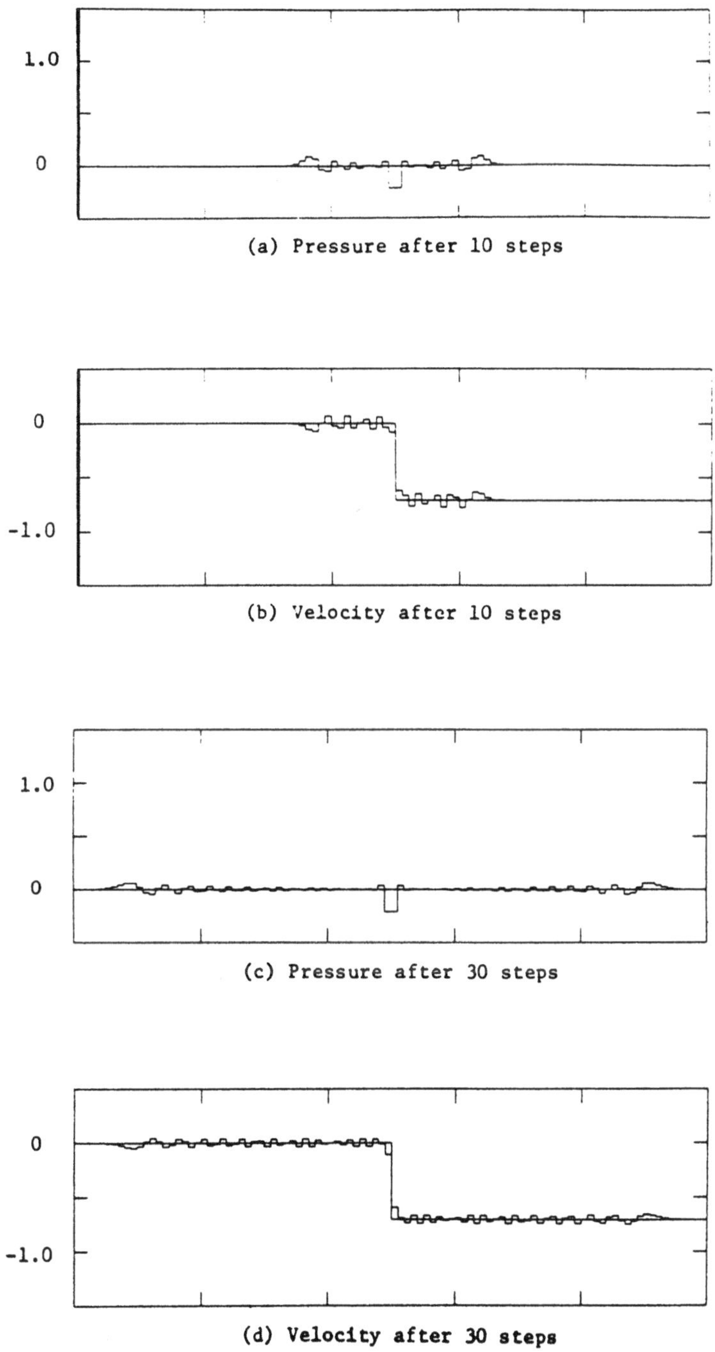

FIG. 5. Results for a shear wave at $\phi = \pi/4$. (a) Pressure after 10 steps. (b) Velocity after 10 steps. (c) Pressure after 30 steps. (d) Velocity after 30 steps.

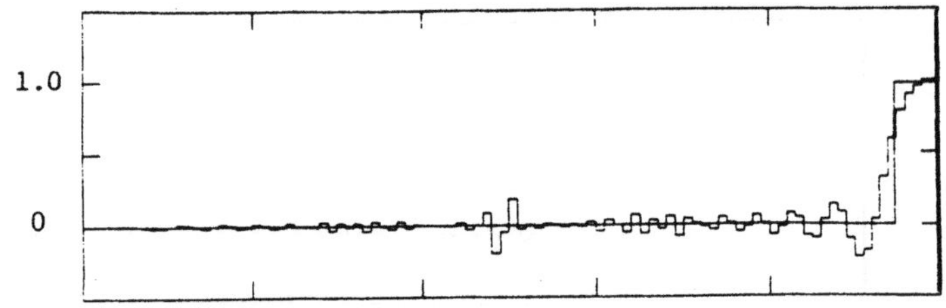

(a) Pressure due to a shock after 20 steps

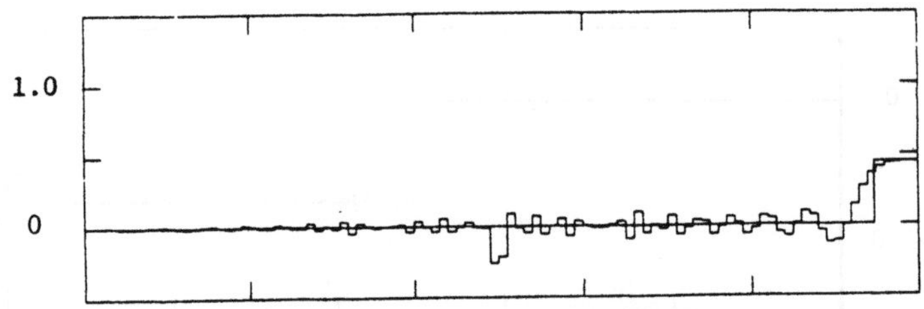

(b) Velocity due to a shock after 20 steps

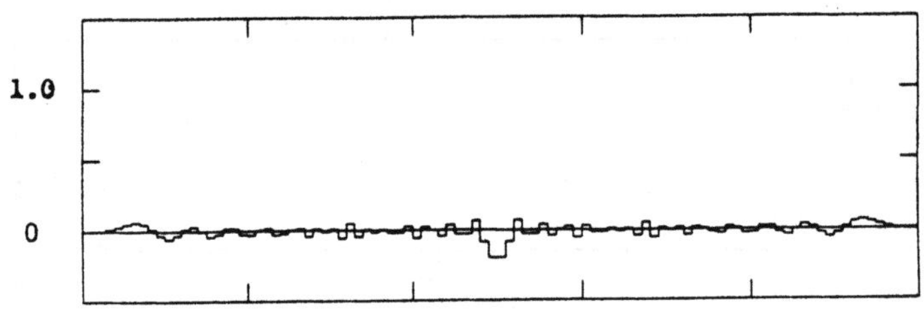

(c) Pressure due to a shear after 20 steps

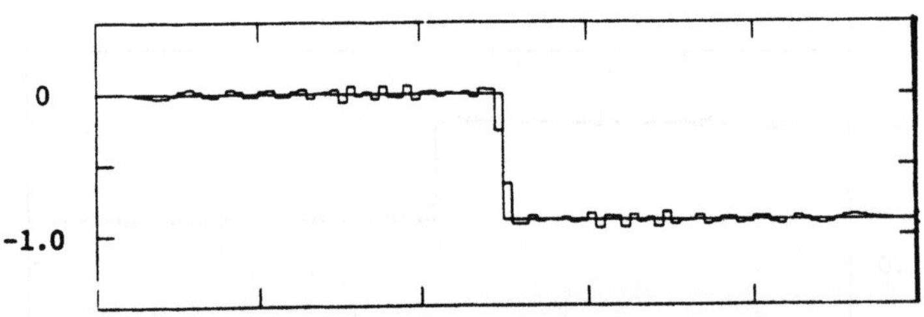

(d) Velocity due to a shear after 20 steps

FIG. 6. Results from waves at $\phi = \tan^{-1}\frac{1}{2}$. (a) Pressure due to a shock after 20 steps. (b) Velocity due to a shock after 20 steps. (c) Pressure due to a shear after 20 steps. (d) Velocity due to a shear after 20 steps.

section and the two-dimensional random choice results in Ref. 4. It is to be expected that other attempts to create "very-high-resolution" schemes, such as Harten's ACM [10] or Roe's Ultrabee [11] may experience similar difficulties.

More constructively, it is worth noting that any scheme based on decomposition of the data into different waves could use a very-high-resolution operator-split method on those eigenvectors common to both operators, with a slightly less ambitious method employed on the remaining data. In the context of gas dynamics, we might use moderate resolution on the acoustic waves. Shock waves would nevertheless appear sharp, because they are nonlinear, "self-steepening" phenomena, and expansions would be treated well where they became smooth. The difficult (because they are linear) entropy waves could be sharpened without creating any splitting problems. Even more valuable, perhaps, would be the application to chemically reactive or multiphase flows. These always introduce new equations (for the species concentrations) that add more linear wavefields having the same eigenvector in all directions [12]. The ability to use very-high-resolution schemes on these fields will help to reduce numerical dissipation.

The disturbance that seems most difficult to treat is the shear wave. Because it is linear it is prone to numerical diffusion, but a very-high-resolution splitting scheme would introduce into the solution the sort of noise displayed in Figs. 5(a)–5(d), 6(c) and 6(d). It may be necessary to control very carefully the degree of resolution applied to these waves, if they are not aligned with any coordinate direction.

This research was supported under the National Aeronautics and Space Administration under NASA Contract No. NASl-18107 while the author was in residence at the Institute for Computer Applications in Science and Engineering (ICASE), NASA Langley Research Center.

References

1. G. Strang, "On the Construction and Comparison of Difference Schemes," *SIAM J. Numer. Anal.*, **5**, 506 (1968).
2. H.C. Yee, "Upwind and Symmetric Shock-Capturing Schemes," *Proceedings Seminar on Computational Aerodynamics,* M.M. Hafez, Ed., University of California at Davis, AIAA Special Publication, Spring 1986.
3. P.L. Roe, "Characteristic-Based Schemes for the Euler Equations," *Ann. Rev. Fluid Mech.* 337 (1986).
4. P. Colella, "Glimm's Method for Gas Dynamics," *SIAM J. Sci. Statist. Comput.*, **3**, 77 (1982).
5. M. Crandall and A. Majda, "The Method of Fractional Steps for Conservation Laws," *Numer. Math.*, **34**, 285 (1980).
6. R.C.Y. Chin and G.W. Hedstrom, "A Dispersion Analysis for Difference Schemes: Tables of Generalized Airy Functions," *Math. Comp.*, **32**, 1163 (1978).
7. L.J. Slater, *Generalized Hypergeometric Functions,* Cambridge University Press, Cambridge, 1966.
8. A. Harten, "On a Large Time-Step High-Resolution Scheme," ICASE Report No. 82-34, 1982.
9. R. LeVeque, "A Large Time-Step Generalization of Godunov's Method for Systems of Conservation Laws," *SIAM J. Numer. Anal.*, **22**, 1051 (1985).

10. A. Harten, "The Artificial Compression Method for Computation of Shocks and Contact Discontinuities. III. Self-Adjusting Hybrid Schemes," *Math. Comp.,* **32**, 363 (1978).
11. P. L. Roe and M. J. Baines, "Asymptotic Behavior of Some Nonlinear Schemes for Linear Advection," in *Notes on Numerical Fluid Mechanics,* M. Pandolfi and R. Piva (Eds.), Vieweg, Braunsschweig, 1984. pp. 283–290.
12. H. C. Yee and J. L. Shinn, "Semi-Implicit and Fully Implicit Shock-Capturing Methods for Hyperbolic Conservation Laws with Stiff Source Terms," AIAA Report No. 87-1116, July 1987.

High Resolution Finite Volume Methods on Arbitrary Grids via Wave Propagation

RANDALL J. LEVEQUE*

Mathematics Department, University of Washington, Seattle, Washington 98195

Received April 30, 1987; revised August 24, 1987

In previous work by the author, a generalization of Godunov's method for systems of conservation laws has been developed and analyzed that can be applied with arbitrary time steps on arbitrary grids in one space dimension. Stability for arbitrary time steps is achieved by allowing waves to propagate through more than one mesh cell in a time step. In this paper the method is extended to second-order accuracy and to a finite volume method in two space dimensions. This latter method is based on solving one-dimensional normal and tangential Riemann problems at cell interfaces and again propagating waves through one or more mesh cells. By avoiding the usual time step restriction of explicit methods, it is possible to use reasonable time steps on irregular grids where the minimum cell area is much smaller than the average cell. Boundary conditions for the Euler equations are discussed and special attention is given to the case of a Cartesian grid cut by an irregular boundary. In this case small grid cells arise only near the boundary, and it is desirable to use a time step appropriate for the regular interior cells. Numerical results in two dimensions show that this can be achieved.

1. INTRODUCTION

Hyperbolic systems of conservation laws in two space dimensions, such as the Euler equations of gas dynamics, have the form

$$u_t + f(u)_x + g(u)_y = 0. \tag{1.1}$$

Here u is a vector of m variables and f and g are the flux functions. In order to solve these equations numerically, the domain is subdivided into computational grid cells (or intervals in one dimension) and an approximation to the solution over each cell is updated in each time step. With traditional explicit methods, the appropriate update is based on cell values in only a few neighboring cells. For example, with a three-point method in one dimension, only the nearest neighbor cells on either side can effect the solution. Most finite volume methods in two space dimensions also limit the domain of dependence to the nearest neighbors.

* Supported in part by NSF Grant DMS-8601363 and by the National Aeronautics and Space Administration under NASA Contracts NAS1-17070 and NAS1-18107 while the author was in residence at ICASE, NASA Langley Research Center, Hampton, VA 23665–5225.

This imposes an obvious limitation on the time step due to the CFL condition, which states that a necessary condition for convergence is that the domain of dependence of the numerical method include the domain of dependence of the true solution, at least in the limit as the grid is refined. For a three-point method in one dimension, this means that the time step k must be restricted by

$$\nu \equiv \frac{k}{h_{\min}} |s_{\max}| \leqslant 1, \tag{1.2}$$

where $h_{\min}$ is the length of the smallest cell and $s_{\max}$ is the maximum wave speed arising in the problem at hand. The quantity ν defined by (1.2) is called the Courant number.

This time step restriction may be reasonable on a uniform grid. An analysis of the truncation error will often show that (1.2) is consistent with the choice of time step required for the temporal accuracy to agree with the spatial accuracy. On the other hand, if the grid is not uniform we would like to replace $h_{\min}$ in (1.2) by h_{ave}, some average cell length. If $h_{\min} \ll h_{\text{ave}}$ then (1.2) imposes a severe time step restriction that may greatly reduce the efficiency of the method.

This situation arises, for example, in certain shock tracking procedures (e.g., [16]). In addition to a uniform grid, moving points are introduced that may fall arbitrarily close to the fixed points, creating small cells. Similar difficulties are encountered in the moving mesh method of Harten and Hyman [12], where they must take special care to ensure the proper separation of mesh points.

In more space dimensions, the problem of small mesh cells is difficult to avoid in several common situations. In particular, irregular boundaries cutting through a regular grid (as in Fig. 9.1 and 9.3) result in cells with arbitrarily small area. We would like to choose the time step based on the regular grid cells rather than being artificially restricted by the boundary cells. Small cells are often avoided by using body fitted coordinates, so that the grid cells have more uniform areas. However, this creates additional difficulties in first generating and then computing with the irregular grids. For complicated multicomponent geometries, particularly in three dimensions, the generation of such grids may be very difficult, requiring complicated grid patching. This has led several workers to reconsider the use of Cartesian grids (e.g., [6, 25, 28]).

Similar difficulties arise if a shock is tracked through a fixed grid, at the boundary of a region of mesh refinement or where independent grids are patched together. On general unstructured grids, such as triangulations of arbitrary regions, the cell areas may also vary widely.

In this paper a finite volume method is introduced that avoids these difficulties by dynamically extending the domain of dependence in such a way that the CFL condition is always satisfied. This is accomplished by fully exploiting the wave propagation that is inherent in hyperbolic equations. A one-dimensional Riemann problem is solved normal to each cell interface (meaning the initial data is piecewise constant with a jump discontinuity at the interface). Solving Riemann problems forms the basis of Godunov's classical method and numerous finite volume methods, but in most such methods the Riemann solution is used only to obtain a flux across the boundary. This flux is then used to update only the neighboring cell values.

In the method proposed here, the structure of the Riemann solution is used to a much greater extent. The Riemann solution consists of waves (shocks, contact discontinuities, or rarefaction waves) that propagate away from the interface, each carrying an increment in the flow variables. As a wave propagates, it modifies the flow variables in the cells it passes through. By allowing it to pass through more than one cell, the CFL condition remains satisfied with larger time steps. This gives a generalization of Godunov's method applicable for larger time steps on arbitrary grids. In one space dimension, this first-order accurate method has been developed and analyzed in a series of papers by the author [14–16]. This method is similar in spirit to the "transport collapse" method of Brenier [1–3]. In the present paper, the method is extended to second-order accuracy and arbitrary grids in two space dimensions.

In addition to propagating waves normally from cell interfaces, it is also possible to incorporate tangential wave propagation by solving auxiliary Riemann problems in the orthogonal direction. This leads to increased stability and enhanced modeling of two-dimensional phenomena and propagation of shock waves that are not aligned with the grid.

An additional advantage of the method proposed here is that boundary conditions can be incorporated into the algorithm relatively easily. Farfield nonreflecting boundary conditions and solid wall (or moving piston) boundary conditions for the Euler equations will be discussed here.

The basic idea of the method is described in Section 2 for a scalar equation in one space dimension. A new formulation of the Lax–Wendroff method is presented and then converted to a "high resolution" method (one with sharp nonoscillatory discontinuities and good nonlinear stability properties). The formulation of this method easily allows extension to arbitrary grids and time steps and incorporation of boundary conditions.

In Section 3 the method is extended to systems of conservation laws in one dimension. Boundary conditions for the Euler equations are discussed in Section 4 and then some one-dimensional numerical results are presented in Section 5.

In the remainder of the paper the method is extended to a finite volume method in two dimensions. The special case of a Cartesian grid with irregular boundaries receives particular attention. In Section 9, numerical results are presented for a shock tube problem oblique to the grid and single Mach reflection from a ramp.

2. One-Dimensional Scalar Equations

We first consider the one-dimensional scalar conservation law

$$u_t + f(u)_x = 0, \tag{2.1}$$

with $f(u)$ assumed to be a convex function. The computational grid is given by $x_0 < x_1 < \cdots < x_M$. The mesh spacing can be arbitrary and we set $h_i = x_{i+1} - x_i$. The approximate solution on the mesh cell (x_i, x_{i+1}) at time t_n is denoted by U_i^n. The time steps can also be arbitrary and varying. Since the methods are one-step methods and will be described over a single time step, we will denote the current time step simply by $k = t_{n+1} - t_n$ and use the shorthand abbreviations $U_i \equiv U_i^n$ and $\bar{U}_i \equiv U_i^{n+1}$.

For each value of j, set

$$s_j = \begin{cases} (f(U_{j+1}) - f(U_j))/(U_{j+1} - U_j) & \text{if } U_j \neq U_{j+1} \\ f'(U_j) & \text{if } U_j = U_{j+1}. \end{cases} \tag{2.2}$$

Then s_j is the speed at which the discontinuity between U_j and U_{j+1} propagates according to the Rankine–Hugoniot relation. For convenience at this point, suppose that $s_j > 0$ for all j. Then we set

$$u_j(x, t_n) = \begin{cases} U_j & \text{for } x < x_j \\ U_j + (x - x_{j+1/2})\sigma_j & \text{for } x_j \leqslant x < x_{j+1} \\ U_{j+1} & \text{for } x \geqslant x_{j+1}, \end{cases} \tag{2.3a}$$

where $x_{j+1/2} = \frac{1}{2}(x_j + x_{j+1})$ and σ_j is the slope of u_j over the jth interval, as yet unspecified (see Fig. 2.1). Based on these functions, we define an approximate solution $\tilde{u}(x, t)$ over the time interval $t_n \leqslant t \leqslant t_{n+1}$. At time t_n, $\tilde{u}(x, t_n)$ is simply the piecewise linear function that agrees with $u_j(x, t_n)$ on the jth cell:

$$\tilde{u}(x, t_n) = u_j(x, t_n) \qquad \text{for } x_j \leqslant x < x_{j+1}. \tag{2.4a}$$

For $t > t_n$ we let the wave form (2.3a) propagate with speed s_j, defining

$$u_j(x, t) = u_j(x - s_j(t - t_n), t_n). \tag{2.3b}$$

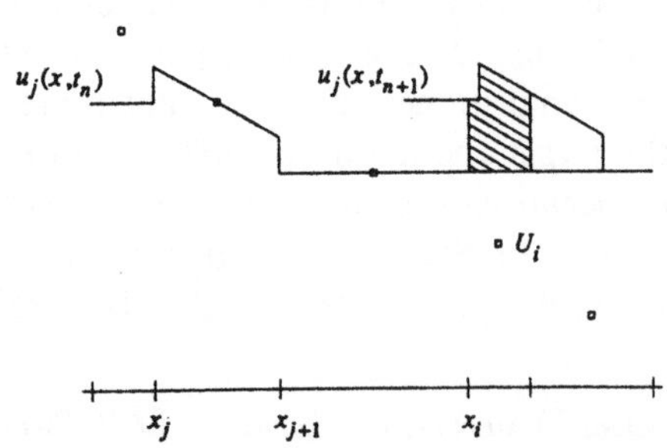

FIG. 2.1. The piecewise linear function u_j at time t_n and after propagation, at time t_{n+1}. Squares indicate the grid values U_{j-1}, U_j, ... The shaded region represents the contribution of this wave toward updating U_i.

These waves are then combined via linear superposition to obtain the approximate solution

$$\tilde{u}(x, t) = \tilde{u}(x, t_n) + \sum_j (u_j(x, t) - u_j(x, t_n)). \tag{2.4b}$$

The sum in (2.4b) is over all j but at any given point only a finite number of terms will be nonzero since $u_j(x, t_n)$ is constant outside of (x_j, x_{j+1}).

The approximation $\bar{U}_i$ at time t_{n+1} is obtained by projecting the approximate solution $\tilde{u}(x, t_{n+1})$ onto the grid, as in Godunov's method (i.e., by averaging over the mesh cell (x_i, x_{i+1})):

$$\begin{aligned}\bar{U}_i &= \frac{1}{h_i}\int_{x_i}^{x_{i+1}} \tilde{u}(x, t_{n+1})\, dx \\ &= U_i + \frac{1}{h_i}\sum_j \int_{x_i}^{x_{i+1}} (u_j(x, t_{n+1}) - u_j(x, t_n))\, dx. \end{aligned} \tag{2.5}$$

Note that $\bar{U}_i$ is obtained from U_i by adding to it contributions from several neighboring cells. Figure 2.1 shows the typical contribution to U_i, indicated by the shaded area.

For a linear problem $u_t + su_x = 0$ with constant speed s, the approximation $\tilde{u}(x, t)$ is in fact the true solution with initial data $\tilde{u}(x, t_n)$. For nonlinear problems the waveform (2.3a) should be distorted for $t > t_n$, and so (2.3b) represents a local linearization of the flux function. The superposition (2.4b) is a further linearization of the nonlinear problem. In spite of this, it is possible to achieve second-order accuracy with appropriate choices of σ_j.

There are various ways to specify the slope σ_j. Three possibilities are

$$\sigma_j^{\mathrm{G}} = 0 \qquad \text{(Godunov)} \tag{2.6a}$$

$$\sigma_j^{\mathrm{LW}} = (U_{j+1} - U_j)/(x_{j+3/2} - x_{j+1/2}) \qquad \text{(Lax–Wendroff)} \tag{2.6b}$$

$$\sigma_j^{\mathrm{M}} = \mathrm{minmod}(\sigma_j^{\mathrm{LW}}, \sigma_{j-1}^{\mathrm{LW}}) \qquad \text{(minmod)}, \tag{2.6c}$$

where the minmod function is defined by

$$\mathrm{minmod}(a, b) = \tfrac{1}{2}(\mathrm{sgn}(a) + \mathrm{sgn}(b)) \min(|a|, |b|).$$

The names assigned to the first two choices of σ are based on the fact that the method reduces to one of these standard methods on a uniform grid with Courant number $\nu < 1$. Then each U_i is influenced by at most two waves and the method can be rewritten as a three-point scheme. Setting $\nu_i = s_i k/h$ and computing the integrals in (2.5) explicitly, we obtain (still assuming $s_j > 0$ for all j)

$$\bar{U}_i = U_i - \nu_{i-1}(U_i - U_{i-1}) + \tfrac{1}{2}\nu_{i-1}(1 - \nu_{i-1})\sigma_{i-1}h - \tfrac{1}{2}\nu_i(1 - \nu_i)\sigma_i h. \tag{2.7}$$

This can be rewritten in conservation form as

$$\bar{U}_i = U_i - \frac{k}{h}(F_i - F_{i-1}), \tag{2.8}$$

where

$$F_j = f(U_j) + \tfrac{1}{2}(1 - \nu_j)\, s_j \sigma_j h. \tag{2.9}$$

If $\sigma_j \equiv 0$ this gives the first-order upwind scheme, which agrees with Godunov's method for the problem considered here. With slope (2.6b), this is easily seen to be the standard Lax–Wendroff method.

The Lax–Wendroff method is second-order accurate on smooth solutions, but it generally develops oscillations and possibly instabilities near discontinuities or sharp gradients. The minmod choice of slopes (2.6c) is designed to avoid this difficulty. Van Leer [27] used a similar geometric approach with limited slopes to define his second-order accurate MUSCL schemes. Later Goodman and Le Veque [10, 18] used the minmod slope (2.6c) to derive a total variation diminishing (TVD) version of van Leer's method. The method (2.7) with this choice of σ is closely related to these methods and to a variety of other "high resolution" schemes that have recently been proposed (e.g., [11, 19, 20, 29]). In particular, for a linear problem with $f(u) = au$ and $a > 0$, (2.7) with minmod slopes is precisely the minmod flux-limiter method introduced by Roe (see Sweby [26]) and hence is TVD.

It is interesting to note that the minmod method proposed here is not TVD on nonlinear problems. It is possible to produce examples where the variation increases slightly, although detrimental oscillations or instabilities have never been observed. Indeed, the lack of TVD may be an advantage since any TVD method must reduce to first order accuracy at extreme points [19]. In numerical experiments, the present method appears to perform better at such points.

The main advantage of this approach, however, is that we can easily go beyond formula (2.7) and apply the method as described by (2.5) with arbitrary grids and time steps.

Before discussing systems of conservation laws, we point out a slightly different formulation of the method that helps simplify its implementation and extension to two space dimensions. We can decouple the wave propagation into a two stage procedure. In the first stage we propagate the wave $u_j^{(1)}(x, t)$ given by (2.3) with $\sigma_j = 0$, corresponding to the first-order algorithm. In the second stage we propagate a second-order correction wave,

$$u_j^{(2)}(x, t) = u_j(x, t) - u_j^{(1)}(x, t),$$

with the same speed s_j. This wave has integral zero and initially has the form

$$u_j^{(2)}(x, t_n) = \begin{cases} (x - x_{j+1/2})\sigma_j & \text{for } x_j \leqslant x < x_{j+1} \\ 0 & \text{otherwise.} \end{cases}$$

When viewed in this way it is clear how the second-order correction relates to the flux corrected transport (FCT) algorithm of Boris and Book [3] and Zalesak [30] and to the ideas of Roe [22]. The results of the first-order accurate scheme are corrected by redistributing some of the flux between grid cells.

In the above discussion we assumed that $s_j > 0$. If $s_j < 0$ we must replace (2.3) by

$$u_j(x, t_n) = \begin{cases} U_j & \text{for} \quad x < x_{j+1} \\ U_j + (x - x_{j+3/2})\sigma_j & \text{for} \quad x_{j+1} \leqslant x < x_{j+2} \\ U_{j+1} & \text{for} \quad x \geqslant x_{j+2} \end{cases} \tag{2.10}$$

in order to obtain a method that reduces to the Lax–Wendroff or minmod method in the special cases described above. In the latter case we also redefine

$$\sigma_j^{\mathrm{M}} = \mathrm{minmod}(\sigma_j^{\mathrm{LW}}, \sigma_{j+1}^{\mathrm{LW}}).$$

Note that the linear segment in $u_j(x, t)$ is always behind the discontinuity as it propagates.

3. Systems of Conservation Laws in One Dimension

We now develop the method more generally for hyperbolic systems of conservation laws (2.1) with $u \in \mathbb{R}^m$. We assume that the Jacobian matrix $f'(u)$ has real eigenvalues $\lambda_j(u)$ for all u and that each characteristic field is either genuinely nonlinear or linearly degenerate. Then the Riemann problem, consisting of (2.1) with piecewise constant initial conditions

$$u(x, t_n) = \begin{cases} U_j & x < x_{j+1} \\ U_{j+1} & x \geqslant x_{j+1} \end{cases} \tag{3.1}$$

has a solution for $\|U_{j+1} - U_j\|$ sufficiently small. This similarity solution consists of $m+1$ constant states separated by shocks, contact discontinuities, or rarefaction waves. (See Lax [13] for a description of the theory.)

Although the method can be adapted to use this exact Riemann solution, we will base the method on an approximate Riemann solution of the type introduced by Roe [21]. The Riemann solution of (2.1) with data (3.1) is replaced by the solution of the linear problem

$$u_t + A_j u_x = 0, \tag{3.2}$$

where $A_j = A_j(U_j, U_{j+1})$ is an appropriate average value of $f'(u)$. This gives a good approximation, particularly for smooth solutions where $\|U_{j+1} - U_j\| = O(h)$. Denote the solution of this problem by $u_j^{(1)}(x, t)$ for $t \geqslant t_n$. Then $u_j^{(1)}(x, t)$ consists of $m+1$ constant states separated by discontinuities travelling with speeds

$$s_{j1} \leqslant s_{j2} \leqslant \cdots \leqslant s_{jm} \tag{3.3}$$

which are eigenvalues of A_j. The jump in $u_j^{(1)}$ across the pth such wave is an eigenvector R_{jp} of A_j, so that

$$A_j R_{jp} = s_{jp} R_{jp} \tag{3.4}$$

and $U_{j+1} - U_j$ is decomposed as

$$U_{j+1} - U_j = \sum_{p=1}^{m} R_{jp}. \tag{3.5}$$

To obtain high resolution we will also introduce a correction wave $u_j^{(2)}(x, t)$ by generalizing the scalar case discussed in Section 2. We break $u_j^{(2)}$ up into m correction waves $u_{jp}^{(2)}$, $p = 1, 2, \ldots, m$ each of which travels with speed s_{jp} and has an initial form that is piecewise linear and nonzero over a single mesh interval. The choice of interval again depends on the sign of s_{jp}: For $s_{jp} > 0$:

$$u_{jp}^{(2)}(x, t_n) = \begin{cases} (x - x_{j+1/2})\sigma_{jp} & \text{for} \quad x_j \leqslant x < x_{j+1} \\ 0 & \text{otherwise.} \end{cases} \tag{3.6a}$$

For $s_{jp} \leqslant 0$:

$$u_{jp}^{(2)}(x, t_n) = \begin{cases} (x - x_{j+3/2})\sigma_{jp} & \text{for} \quad x_{j+1} \leqslant x < x_{j+2} \\ 0 & \text{otherwise,} \end{cases} \tag{3.6b}$$

where σ_{jp} is now a slope vector with m components, to be defined below. We now set

$$u_j^{(2)}(x, t) = \sum_{p=1}^{m} u_{jp}^{(2)}(x, t) = \sum_{p=1}^{m} u_{jp}^{(2)}(x - s_{jp}(t - t_n), t_n) \tag{3.7}$$

and

$$u_j(x, t) = u_j^{(1)}(x, t) + u_j^{(2)}(x, t). \tag{3.8}$$

Based on these functions $u_j(x, t)$, we can define an approximate solution $\bar{u}(x, t)$ just as in the scalar case via (2.4). Averaging $\bar{u}(x, t_{n+1})$ over the ith interval as in (2.5) gives the numerical solution $\bar{U}_i$ at time t_{n+1}.

If $\sigma_{jp} \equiv 0$ then $u_j^{(2)}(x, t) \equiv 0$ for all j. In this case the method reduces to the large time step generalization of Godunov's method (with Roe's approximate Riemann solver) that has been previously studied [14–16]. This method is conservative for any mesh ratio as is shown in [14] where the method is rewritten in conservation form. Note that the method remains conservative with arbitrary slope vectors since $\int_{-\infty}^{\infty} u_j^{(2)}(x, t)\, dx = 0$ for any σ_{jp}. We can view the function $u_j(x, t)$ for $t \geqslant t_n$ as a new approximate Riemann solver which gives second-order accuracy for an appropriate choice of σ_{jp}. Because of this the high resolution method with nonzero σ_{jp} can also be written in conservation form following Section 4 of [14].

Corresponding to the three slope choices (2.6), we now have

$$\sigma_{jp}^{G}=0 \qquad \text{(Godunov–Roe)}, \tag{3.9a}$$

$$\sigma_{jp}^{LW}=R_{jp}/(x_{j+3/2}-x_{j+1/2}) \qquad \text{(Lax–Wendroff)}, \tag{3.9b}$$

$$\sigma_{jp}^{M}=\text{minmod}(\sigma_{jp}^{LW},\sigma_{Jp}^{LW}) \qquad \text{(minmod)}, \tag{3.9c}$$

where in (3.9c) $J=j-\text{sgn}(s_{jp})$ and the minmod function is applied to vectors componentwise. Either of the last two choices appears to give second-order accuracy on smooth solutions. In order to achieve stability and nonoscillatory behavior near discontinuities, the minmod slope (3.9c) is recommended and has been used in the numerical experiments reported below.

This method appears difficult to implement from the description just given. In fact, it is quite simple if we consider each wave arising from each Riemann problem in turn and compute its effect on the solution. Due to the linearization of wave interactions that is used here, each wave can be propagated and averaged onto the grid independently of all other waves. For the case $\sigma_{jp}\equiv 0$ this algorithm has already been described in some detail in [15, 16]. This gives propagation of the waves $u_{jp}^{(1)}$.

The correction waves $u_{jp}^{(2)}$ simply give a redistribution between mesh cells. In particular, if the mesh is uniform, then the nonzero portion of $u_{jp}^{(2)}$ overlaps at most two cells, independent of the Courant number, and the second-order correction is easily implemented.

For linear systems of equations, solution of the Riemann problems gives resolution into the characteristic variables and $\bar{u}(x,t)$ is again the exact solution with the piecewise linear data $\bar{u}(x,t_n)$, for all $t\geq t_n$. Consequently, the method gives second-order accuracy with any size time step in this case. For nonlinear problems with smooth solutions, the superposition (2.4) can be shown to be consistent with second-order accuracy. Towards this end, a semidiscrete version of this superposition (discretized only in time) is investigated in [17] and shown to produce an approximate solution with $O(k^3)$ error over a time step of length k.

4. Boundary Conditions for the Euler Equations

One advantage of the approach used here is the ease with which boundary conditions can be derived and implemented, based on the physics of the problem. This will be illustrated for the Euler equations at a free field (non-reflecting) boundary and at a solid wall (or moving piston) boundary.

The Euler equations of gas dynamics have the form

$$\begin{pmatrix}\rho\\ \rho v\\ E\end{pmatrix}+\begin{pmatrix}\rho v\\ \rho v^2+p\\ (E+p)v\end{pmatrix}=0, \tag{4.1}$$

where ρ, v, p, and E represent the density, velocity, pressure, and energy, respectively. These quantities are related by $E = p/(\gamma - 1) + \frac{1}{2}\rho v^2$ with $\gamma = 1.4$ for air.

Non-reflecting boundary conditions are easily achieved by simply allowing all waves which cross the boundary to pass out of the region, with no incoming waves. In a subsonic problem these boundary conditions are not strictly correct. For example, if two shocks propagate out of the domain at different times, they may interact (in the exterior) and give rise to a weak rarefaction wave which should reenter the domain at a later time. Correctly modeling such behavior requires boundary conditions which are nonlocal in time. Simply letting the waves pass out appears to be the best that can be achieved with local boundary conditions and is easily implemented with arbitrary grids and time steps.

At a solid wall boundary the correct boundary condition is $v = 0$. This gives rise to a boundary Riemann problem which can be solved for the incoming wave. Note that the boundary is characteristic in this case, so that there is only one incoming wave.

Another way to impose this boundary condition for the PDE is to extend the domain beyond the boundary and specify initial conditions obtained by suitably reflecting the values from the interior. For example, if we wish to compute over time k and $s_{\max}$ represents the maximum wave speed, then it suffices to extend the domain from $x = 0$, say, to $x^* = -ks_{\max}$ and set

$$\begin{aligned} \rho(x) &= \rho(-x) \\ v(x) &= -v(-x) \\ p(x) &= p(-x) \end{aligned} \tag{4.2}$$

for $x^* \leqslant x \leqslant 0$. Solving the problem over this extended domain (using, for example, non-reflecting boundary conditions at $x = x^*$), will give the desired solution for $x \geqslant 0$ with the boundary condition $v(0, t) = 0$ satisfied.

This latter approach is easy to implement numerically on arbitrary grids. New grid points are introduced exterior to the original domain by reflecting the interior points which fall within distance $ks_{\max}$ of the boundary. The values on these mesh cells are obtained by reflecting the interior values as in (4.2). Then any wave leaving the original domain during the timestep is matched by a "reflected" wave which enters at the same point in time. The jump in v in the entering wave is the negative of the jump in v in the exiting wave (due to the symmetry in solutions to the Riemann problem under the reflection (4.2)) and hence the boundary condition is always satisfied. This same symmetry exists in Roe's approximate Riemann solver and the slopes σ_{jp} will also be symmetric, so that the boundary condition remains satisfied for the method discussed here. Again the procedure is easily implemented for arbitrary time steps and meshes. Finally, it is clear that the method has the same order of accuracy near the boundary as in the interior, since after the reflection the boundary essentially disappears and we are simply using the interior scheme everywhere.

If the solid wall boundary is moving (as, for example, in a piston problem), then boundary conditions are also easily derived. If the piston is located at $x = z_n$ at time t_n and is moving with speed s_0 (assumed constant) for $t_n \leqslant t \leqslant t_{n+1}$, then the mesh cells for $z_n \leqslant x \leqslant z_n + s_{\max} k$ are reflected to the left of z_n with grid values

$$\begin{aligned} \rho(z_n - x) &= \rho(z_n + x) \\ v(z_n - x) &= 2s_0 - v(z_n + x) \\ p(z_n - x) &= p(z_n + x). \end{aligned} \tag{4.3}$$

The waves generated from these cell boundaries will interact with the interior waves in such a way that $v(z_n + s_0(t - t_n), t) = s_0$ will be satisfied for $t_n \leqslant t \leqslant t_{n+1}$.

5. Numerical Results in One Dimension

On a uniform grid with $v < 1$ the present minmod method performs much like the minmod flux limiter method discussed by Sweby [26]. Figure 5.1 shows some results for a shock tube problem with the Euler equations and initial data with

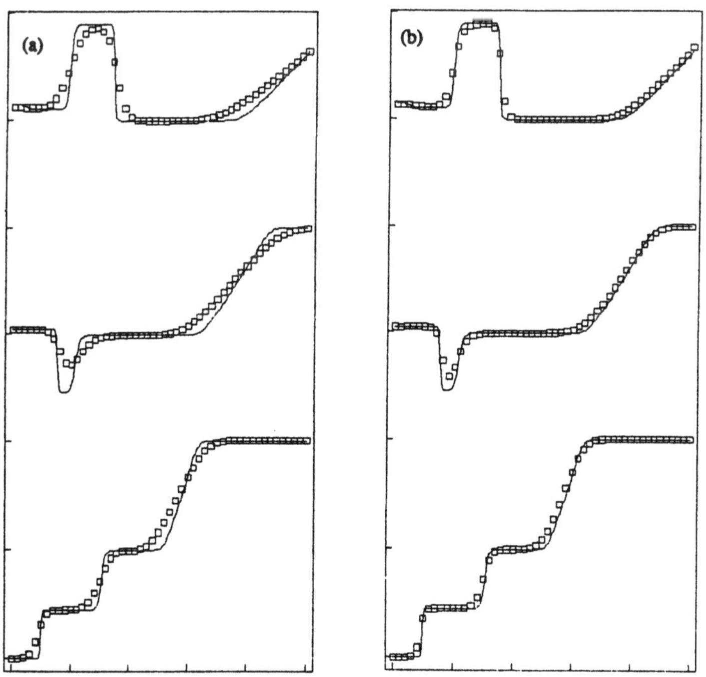

FIG. 5.1. The density ρ in the shock tube problem at three different times (reading upward). Figure (a) shows the values obtained with slopes $\sigma_{jp} = 0$. Figure (b) shows the values obtained with minmod slopes (3.9c).

$v=0$ everywhere and $\rho=p=1$ for $x<0.4$, $\rho=p=3$ for $x>0.4$. The solution at three different times is shown. Note that the shock reflects off a solid wall boundary at $x=0$ and then interacts with the contact discontinuity. Nonreflecting boundary conditions were used at $x=1$, where the rarefaction wave begins to leave the region.

Figure 5.1a shows the density ρ computed using the first-order accurate method with Godunov slopes $\sigma_{jp}=0$. Figure 5.1b shows the results with minmod slopes. In each case $h=0.02$ and $k=h/2$ corresponding to $\nu\cong 0.75$. The solid line in each figure is the solution computed with 201 mesh points using minmod slopes.

Figure 5.2 demonstrates the ability of this method to deal with nonuniform meshes. In this problem smooth initial conditions were specified:

$$\rho(x,0)=1.0+0.1\cos(2.5x),\quad v(x,0)=0.1\sin(5x),\quad p(x,0)=1.8+0.1\cos(2.5x),$$

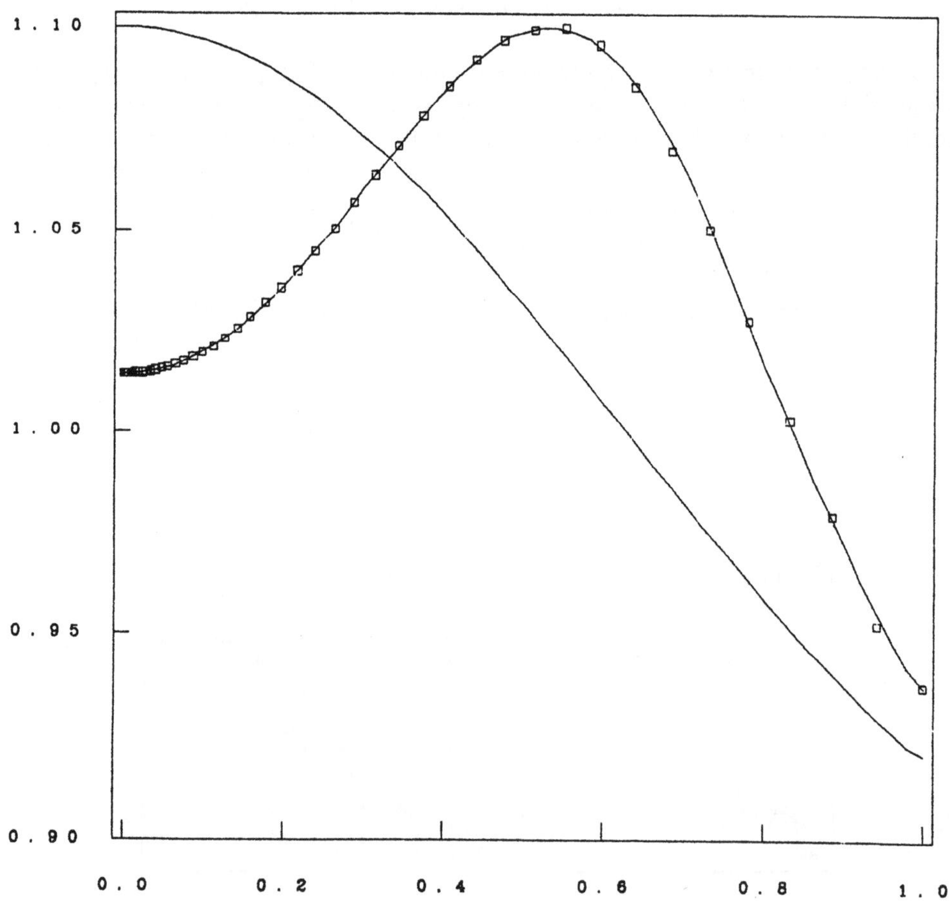

FIG. 5.2. Density in a smooth solution computed on a uniform grid (solid line) and on nonuniform grid (squares) with Courant number 2500. The initial conditions $\rho(x,0)$ are also shown.

again with a solid wall boundary at $x=0$ and nonreflecting boundary conditions at $x=1$. The solid line shows the resulting density at $t=0.1$ computed with 51 equally spaced mesh points and $k=h=0.02$, giving $\nu \cong 1$. The squares show the solution computed with 51 mesh points $x_j=((j-1)h)^3$ densely clustered near $x=0$. Again $k=0.02$ but now the Courant number, based on the smallest mesh width h_1, is roughly 2500. The numerical solution does not appear to be adversely affected by this.

6. Finite Volume Methods in Two Dimensions

In two space dimensions, the wave propagation approach can be used to rewrite a standard finite volume method in a way that can be easily extended to allow larger timesteps than are normally possible. As noted in the Introduction, this might be valuable with nonuniform grids where some of the grid cells are considerably smaller than average. Standard explicit methods have a stability restriction that limits the time step in terms of the smallest cell area. Of particular interest is a uniform Cartesian grid in which some of the grid cells are cut off by an irregular boundary, a tracked shock, or an area of mesh refinement. To some extent the discussion will be specialized to this situation, but the ideas can be applied on more general unstructured grids as well.

In one space dimension, we have seen that nonuniform grid cells are easily handled by allowing waves to propagate entirely through a small cell and out the other side. The same approach can be taken in two dimensions with waves arising at cell interfaces. The basic algorithm will be described on a general unstructured grid.

Associated with each grid cell C_j is an approximation U_j to the solution $u(x, y, t_n)$ over this grid cell. We also define A_j to be the area of C_j and h_{ij} to be the length of the face separating adjacent cells C_i and C_j. We begin by describing a Godunov-type scheme and will later introduce high resolution corrections based on piecewise linear approximations.

In a finite volume method, fluxes across the cell faces are defined and used to update the cell values on either side of the face. Consider, for example, the face separating C_i and C_j in Fig. 6.1. A flux can be defined by solving a one-dimensional Riemann problem normal to the face. Define new variables ξ (normal to the face) and η (tangential) by

$$\xi=\alpha x+\beta y, \qquad \eta=-\beta x+\alpha y \tag{6.1}$$

with $\alpha=\cos\theta$, $\beta=\sin\theta$, where θ is the angle of the cell interface. The conservation laws

$$u_t+f(u)_x+g(u)_y=0 \tag{6.2}$$

are transformed to

$$u_t+F(u)_\xi+G(u)_\eta=0, \tag{6.3}$$

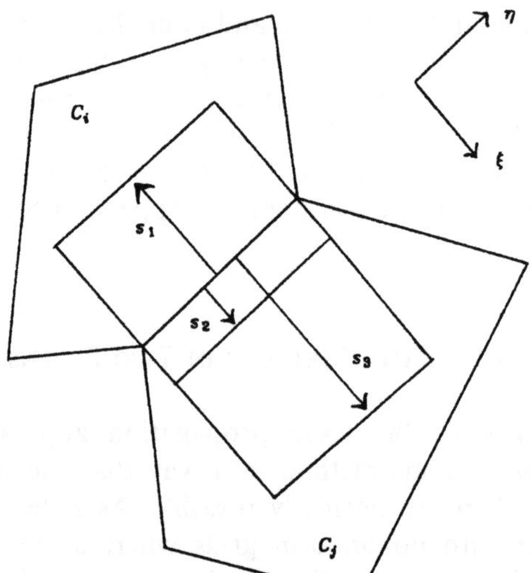

FIG. 6.1. Two grid cells C_i and C_j from an unstructured grid. Wave propagation from the cell interface is determined by solving a one dimensional Riemann problem in the ξ-direction, giving rise to waves propagating with velocities s_1, s_2, and s_3.

where

$$\begin{aligned} F(u) &= \alpha f(u) + \beta g(u) \\ G(u) &= -\beta f(u) + \alpha g(u). \end{aligned} \tag{6.4}$$

In the new coordinates u is constant in η and so (6.3) reduces to a one-dimensional Riemann problem

$$u_t + F(u)_\xi = 0 \tag{6.5}$$

with left and right states U_i and U_j. For rotationally invariant equations such as the Euler equations this is particularly simple since, after a change of dependent variables to rotate the velocity field, the form of $F(u)$ agrees with $f(u)$ and hence a single Riemann solver suffices for all angles θ.

As in the one-dimensional algorithm, we use Roe's approximate Riemann solver and obtain a set of waves traveling with speeds $s_1, s_2, ..., s_m$ in the ξ-direction. As before we denote the jumps in u across these waves by the vectors $R_1, ..., R_m$, so that

$$U_j - U_i = \sum_p R_p. \tag{6.6}$$

Figure 6.1 shows these waves propagating from the interface in a situation where the waves lie entirely within neighboring cells.

We think of u being incremented by $-\text{sgn}(s_p)R_p$ at each point swept out by the pth wave. Consequently, the average value U_i of u over the cell C_i in Fig. 6.1, for example, will be incremented by

$$\left(\frac{|s_1|\, kh_{ij}}{A_i}\right) R_1. \tag{6.7}$$

The factor multiplying R_1 is the ratio of the area swept out by the 1-wave to the total cell area. Similarly, U_j is incremented by

$$\left(\frac{s_2 kh_{ij}}{A_j}\right)(-R_2) + \left(\frac{s_3 kh_{ij}}{A_j}\right)(-R_3). \tag{6.8}$$

These increments to U_i and U_j can be interpreted more traditionally in terms of a flux-difference splitting. The total increment to u (which has been split between C_i and C_j and averaged over these cells) is the sum of A_i times (6.7) plus A_j times (6.8), giving

$$-kh_{ij}[s_1 R_1 + s_2 R_2 + s_3 R_3]. \tag{6.9}$$

A basic property that Roe's approximate Riemann solver shares with the exact Riemann solution is that

$$\sum s_p R_p = F(U_j) - F(U_i) \tag{6.10}$$

so that (6.9) becomes

$$-kh_{ij}(F(U_j) - F(U_i)). \tag{6.11}$$

This is the flux difference between C_i and C_j which has been split via (6.7) and (6.8).

The method as described so far is thus equivalent to traditional flux-difference splitting methods. By integrating the conservation laws around each cell, it is easily shown that such methods are conservative for any splitting of the flux difference. A variety of such methods have been proposed, most of which share the property that flux differences are split only between C_i and C_j with no contribution to other cells. This leads to a restriction on the time step, since clearly the factor multiplying R_1 in (6.7), for example, should be no greater than 1 for stability. If A_i is very small, this gives a severe restriction on k.

The advantage of the wave propagation interpretation of Fig. 6.1 is twofold. First, it makes it clear that in some circumstances other cells should also be affected. More importantly, it makes it easy to determine exactly what that effect should be. Figure 6.2 shows a typical wave propagating over a different grid. Each cell value should clearly be updated by $\text{sgn}(s_p)R_p$ times the ratio of the shaded area within the cell to the total cell area. Note that this ratio is always less than or equal to 1. Splitting each wave in this manner and adding up all increments to all cells

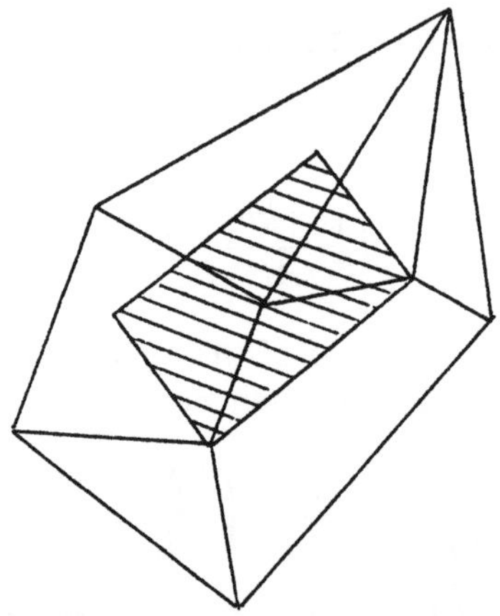

FIG. 6.2. Propagation of a wave over a general unstructured grid. Cell values are updated in every cell the wave intersects. The contribution to each cell is proportional to the ratio of the shaded area within that cell to the total cell area.

gives the same total increment (6.9) and hence the method remains conservative, while being much more stable in a situation such as Figure 6.2 than a method that increments only C_i and C_j.

Instabilities can still arise, however, as illustrated by the following simple example. Consider the scalar linear advection equation

$$u_t + u_x + u_y = 0 \tag{6.12}$$

on a uniform rectangular mesh with initial data consisting of a sawtooth discontinuity at 45° to the mesh:

$$U_{ij} = \begin{cases} 1 & \text{if} \quad i+j \leqslant M \\ 0 & \text{if} \quad i+j \geqslant M+1 \end{cases}$$

for some M. If $k \leqslant h/2$ then the method described above is stable and in fact for $k = h/2$ gives the exact solution, a wave propagating through one additional diagonal of mesh cells in each time step:

$$\bar{U}_{ij} = \begin{cases} 1 & \text{if} \quad i+j \leqslant M+1 \\ 0 & \text{if} \quad i+j \geqslant M+2. \end{cases}$$

For larger k, however, the method is unstable. For example, if $k = h$ then the wave should propagate through two additional diagonals in each step. Instead the waves propagating in the x- and y-directions overlap and give the value $\bar{U}_{ij} = 2$ for $i + j = M + 1$ with no propagation into the second diagonal. In later time steps this instability grows exponentially.

In order to avoid this difficulty, it appears to be necessary to include tangential wave propagation as well as normal propagation. For Eq. (6.12) it is easy to see how to include tangential propagation. All waves should propagate with tangential

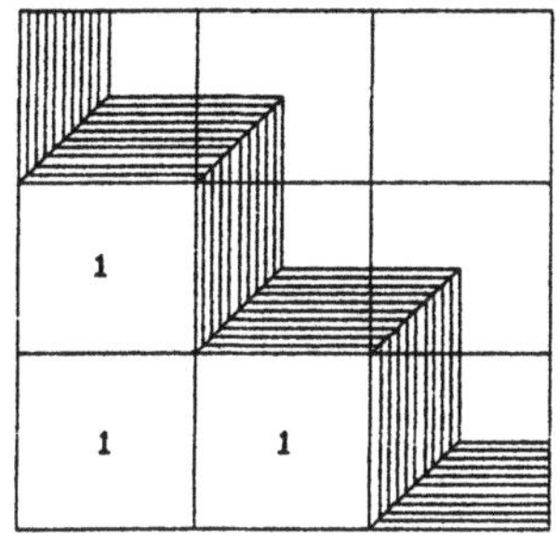

FIG. 6.3. Wave propagation for the scalar problem (6.12) when tangential wave propagation is included. Shown for a time step $k = h/2$.

velocity 1 as well as normal velocity 1. Figure 6.3 shows this wave propagation in the case $k = h/2$, giving

$$\bar{U}_{ij} = \begin{cases} 1 & \text{if} \quad i+j<M+1 \\ 0.75 & \text{if} \quad i+j=M+1 \\ 0.25 & \text{if} \quad i+j=M+2 \\ 0 & \text{if} \quad i+j>M+2 \end{cases}.$$

The method now gives exact propagation with $k = h$ and indeed, except for smearing introduced by the averaging process, this modified method gives exact propagaton for any time step k.

For nonlinear systems of equations the tangential propagation is introduced as follows. After solving the normal Riemann problem we have waves R_p traveling with speeds s_p normal to the cell face. Suppose now that we split up the wave R_p into subwaves $R_p^{(1)}, R_p^{(2)}, \ldots, R_p^{(m)}$ with

$$\sum_{q=1}^{m} R_p^{(q)} = R_p \tag{6.13}$$

and assign tangential velocities $s_p^{(1)}, \ldots, s_p^{(m)}$. The wave in Figure 6.2, for example, might be split into two waves $R_p^{(1)}$ and $R_p^{(2)}$ with tangential speeds $s_p^{(1)}$ and $s_p^{(2)}$ as shown in Fig. 6.4. Note that the area swept out by each wave is $|s_p|\, kh_{ij}$, the same as the area of the original wave in Fig. 6.2, independent of the tangential speeds $s_p^{(q)}$. If each of these waves is propagated as before to increment neighboring mesh cells, the total increment will be

$$-kh_{ij}[s_p R_p^{(1)} + s_p R_p^{(2)}] = -kh_{ij} s_p R_p \tag{6.14}$$

in view of (6.13). Hence the method remains conservative for an arbitrary splitting $R_p^{(q)}$ and arbitrary speeds $s_p^{(q)}$ provided only that (6.13) is satisfied.

A natural way to choose the splitting is by solving a tangential Riemann problem with initial data having a discontinuity of magnitude R_p. For the equation we take the tangential portion of (6.3),

$$u_t + G(u)_\eta = 0. \tag{6.15}$$

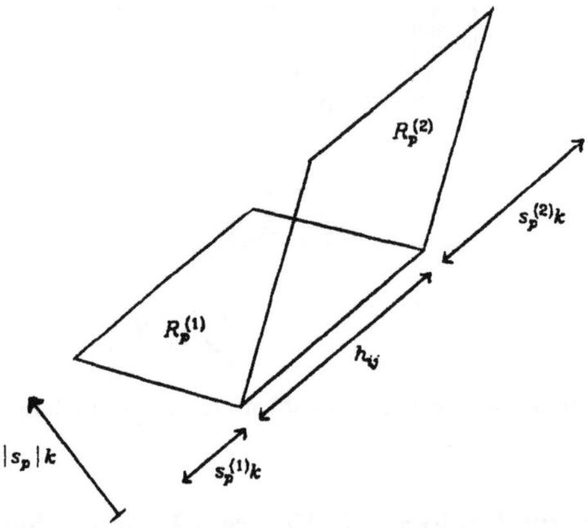

FIG. 6.4. The wave R^p is split tangentially into two waves $R_p^{(1)}$ and $R_p^{(2)}$ with tangential speeds $s_p^{(1)}$ and $s_p^{(2)}$.

Figure 6.1 suggests a possible set of initial data. The wave R_p propagates with the state

$$U_i + \sum_{q<p} R_q \tag{6.16a}$$

to one side and

$$U_j - \sum_{q>p} R_q \tag{6.16b}$$

to the other. Taking these as the left and right states for Eq. (6.15) gives a discontinuity of magnitude R_p which will then be resolved into waves $R_p^{(1)}, ..., R_p^{(m)}$ propagating in the η direction with speeds $s_p^{(1)}, ..., s_p^{(m)}$.

For the linear equation (6.12), $m=1$, $s_1=1$, and $s_1^{(1)}=1$ for all i, j. This gives the correct propagation as described above. More generally, consider a linear system

$$u_t + Au_x + Bu_y = 0.$$

First suppose that A and B have identical eigenvectors and hence are simultaneously diagonalizable. Then each wave R_p obtained by splitting $U_j - U_i$ into eigenvectors of A is already an eigenvector of B as well, so that $R_p^{(p)} = R_p$ and $R_p^{(q)} = 0$ for $q \neq p$. The normal and tangential speeds s_p and $s_p^{(p)}$ are the corresponding eigenvalues of A and B, respectively. In this case the method gives exact propagation (followed by smearing due to the averaging process) for any time step k and we are essentially using the method of characteristics.

If A and B are not simultaneously diagonalizable, or for nonlinear systems, each wave R_p will in general be split into m waves $R_p^{(q)}$. Except in special circumstances the method will not give exact propagation and the time step will need to be restricted in order to obtain reasonable results. Also note that when waves from different interfaces overlap, they are combined by linear superposition as in the one-dimensional algorithm. This incurs additional errors on nonlinear problems. A serious error analysis of the method has not yet been attempted, but the numerical results of Section 9 give some indication of its capabilities.

In the special case of a uniform Cartesian grid, with $h_{ij} \equiv h$ and $A_i \equiv h^2$, the wave propagation becomes relatively simple, particularly if the Courant number $kh\,|s_{\max}|/h^2$ is less than 1. Then each wave affects at most two cells, as in Figure 6.3, and the areas of intersection are easy to compute. Although we do not present the details, the method can be written as a nine-point finite difference method in conservation form. If tangential wave propagation is suppressed, the method reduces to a five-point scheme.

7. High Resolution in Two Dimensions

Greater accuracy for smooth flows and better resolution of discontinuities can be obtained by introducing piecewise linear approximations as in the one-dimensional method. Ideally one might like to use bilinear approximations matching slopes in the x- and y-directions simultaneously, but it turns out that a simpler approach is sufficient to reproduce the Lax–Wendroff method for scalar problems on uniform grids. Introducing minmod slope limiting gives a stable second-order correction that is easy to generalize and implement for nonlinear systems on arbitrary grids.

First consider a uniform Cartesian grid with Courant number less than one and suppress tangential propagation. Then we obtain a five-point flux-splitting method as discussed above. We can think of implementing this by applying the first-order one-dimensional algorithm of Section 3 simultaneously in the x- and y-directions. It is natural then to also apply the second-order correction wave introduced there. In handling the interface between U_{ij} and $U_{i+1,j}$, for example, we obtain waves $R_1, \ldots, R_m$ propagating in the x-direction. Suppose that R_p propagates with speed $s_p > 0$. Then $U_{i+1,j}$ is incremented by

$$-\left(\frac{s_p k}{h}\right) R_p$$

according to the first-order algorithm. For the second-order correction we propagate the wave $u^{(2)}$ from Section 3, viewed now as a two-dimensional wave that is constant in y over the grid cells in question and piecewise linear in x. Thus $U_{i+1,j}$ is incremented by $\frac{1}{2}(k/h)\,s_p(h-ks_p)\sigma_p$ while U_{ij} is incremented by $-\frac{1}{2}(k/h)\,s_p(h-ks_p)\sigma_p$, where σ_p is the slope vector in the x-direction, again defined by one of (3.9). Wave propagation in the y-direction is handled in an analogous manner.

If we choose the Lax–Wendroff slope (3.9b), $\sigma_p = R_p/h$, then for scalar equations the method just described reduces to the following:

$$U_{ij}^{n+1} = U_{ij} - \frac{k}{2h}\left[f(U_{i+1,j}) - f(U_{i-1,j}) + g(U_{i,j+1}) - g(U_{i,j-1})\right]$$
$$+\frac{k^2}{2h^2}\left[\frac{(f(U_{i+1,j})-f(U_{ij}))^2}{U_{i+1,j}-U_{ij}} - \frac{(f(U_{ij})-f(U_{i-1,j}))^2}{U_{ij}-U_{i-1,j}}\right.$$
$$\left.+\frac{(g(U_{i,j+1})-g(U_{ij}))^2}{U_{i,j+1}-U_{ij}} + \frac{(g(U_{ij})-g(U_{i,j-1}))^2}{U_{ij}-U_{i,j-1}}\right]. \tag{7.1}$$

This method is not second-order accurate as it stands, since it is missing the cross derivative terms $\frac{1}{2}k^2[(f'(u)\,g(u)_y)_x + (g'(u)f(u)_x)_y]$ that arise in a Taylor series expansion of $u(x_i, y_j, t_{n+1})$. Remarkably, if we now introduce tangential propagation of the first order waves, the algorithm is modified by terms modeling these cross derivatives and second order accuracy is achieved. The modification is one-sided and the exact form depends on the signs of $f'(u)$ and $g'(u)$.

Note that the second order correction waves are **not** split tangentially. Consequently, on uniform grids the second-order correction is very easy to implement and adds little cost to the algorithm. In practice the method is modified by using minmod slopes (3.9c) to eliminate oscillations while retaining second order accuracy on smooth flows.

On unstructured grids, a high resolution correction can be defined in various ways. Here is one untested possibility, motivated by the above treatment of uniform grids. Let ξ and η be the local coordinate directions defined by the interface between C_i and C_j. For the pth wave, define a slope vector σ_p in the ξ-direction (normal to the interface). The simplest choice would be the "Lax–Wendroff" slope

$$\sigma_p = R_p/h^*, \tag{7.2}$$

where h^* is some measure of the normal distance between cells C_i and C_j, such as the difference in ξ-coordinates of the centers of mass of these two cells. Then a piecewise linear wave with this slope in the ξ-direction, zero slope in the η-direction, and total integral zero can be propagated in the ξ-direction with velocity s_p. The shape of the region where the slope is σ_p is still undetermined. Following the one-dimensional method, we could choose the region to be the cell C_i if $s_p > 0$ and C_j if $s_p < 0$. To be specific, assume that $s_p > 0$ and let (ξ_c, η_c) be the center of mass of the cell C_i, in the local coordinates determined by the interface. Then the high resolution correction, which must be averaged onto the cells, is the piecewise linear function

$$u^{(2)}(\xi, \eta, t_{n+1}) = \begin{cases} (\xi - s_p k - \xi_c)\sigma_p & \text{if } (\xi - s_p k, \eta) \in C_i \\ 0 & \text{otherwise.} \end{cases}$$

In problems with shocks one would again like to introduce a slope limiter instead of using (7.2) directly. The best way to do this on unstructured grids has not yet been determined.

8. Boundary Conditions in Two Dimensions

Nonreflecting outflow boundary conditions are implemented in two dimensions in exactly the same way as in one dimension. Waves propagating out of the computational domain at such a boundary simply disappear and no new waves are propagated inward from boundary interfaces.

At a solid wall boundary, the proper boundary conditions for the Euler equations are obtained by specifying zero normal velocity. If the boundary is flat, this boundary condition can be implemented in the same way as was done in Section 4 for the one-dimensional Euler equations. We will consider this simple case first and see how to reinterpret it in a manner more suited to arbitrary boundaries. For flat boundaries, the grid cells near the boundary can be reflected across the boundary to form fictitious cells as in Fig. 8.1. We assign values of U in these exterior cells based on the values in the mirror-image interior cells so that the density, pressure, and tangential velocity in the exterior cell agree with those values in the interior while the normal velocity is negated. Then we can simply compute over the extended domain using the algorithm described above, ignoring the boundary. Each wave from the interior which crosses the boundary will be matched by a wave from the exterior in such a way that the normal velocity in mirror-image cells adjacent to the boundary will remain negatives of one another, simulating zero normal velocity on the wall itself.

Alternatively, rather than creating fictitious exterior cells and propagating waves from their interfaces to match interior waves, we can simply reflect the interior waves when they hit the boundary as suggested by Fig. 8.1. If the wave carries increments R_p then the reflected wave carries the same increments of density, pressure, and tangential velocity. The increment of normal velocity is negated.

The latter approach has the advantage that it extends more easily to general boundaries. If the boundary is not flat, creating globally well-defined fictitious cells

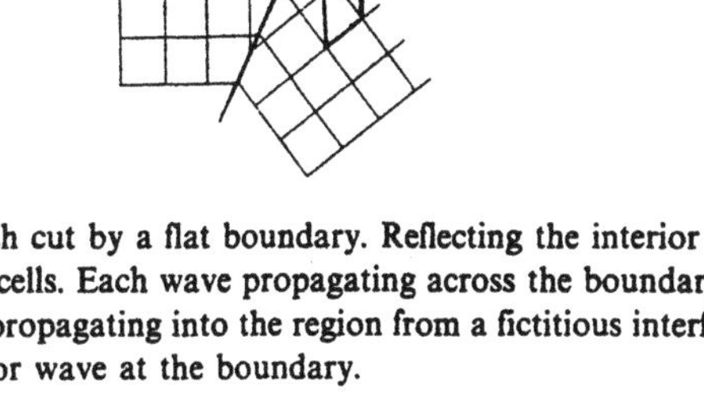

FIG. 8.1. Cartesian mesh cut by a flat boundary. Reflecting the interior cells through this boundary gives the fictitious exterior cells. Each wave propagating across the boundary from an interior cell interface is matched by a wave propagating into the region from a fictitious interface. This can also be viewed as a reflection of the interior wave at the boundary.

by reflecting the interior cells is not possible. It is, however, still possible to reflect waves at the boundary provided only that we assume the boundary is locally flat where it intersects the wave, as can be done to good approximation if the boundary is smooth.

In addition to handling the reflection of waves arising from interior interfaces, it is also necessary to generate a wave from the cell face defining the boundary itself. We solve a one-dimensional Riemann problem normal to the boundary. The left state comes from the interior cell. The right state is the same except that the normal velocity is negated, as motivated by Fig. 8.1 and the previous discussion. Because the two states are related in this manner, the solution of the Riemann problem will give only wave propagating into the computational domain, with zero normal velocity in its wake. Of the other three waves, two will have zero velocity and the third will propagate out of the computational domain and is ignored.

9. Numerical Results

To date the method has been implemented only on a Cartesian grid with "small" time steps, meaning the Courant number is less than one relative to the area h^2 of the regular grid cells. However, boundaries are allowed to cut through this grid and the area of the resulting boundary cells may be arbitrarily small. The boundary is approximated by piecewise linear segments connecting the points where the boundary crosses the grid lines. Consequently all cells are polygons with at most five sides, since we assume each cell contains at most one boundary segment.

Because of the Courant number restriction, wave propagation in the interior cells is easily implemented as discussed in Section 6. It is only near the boundaries that waves may intersect several cells in an unpredictable manner. Each wave is also a polygon which can be represented by a list of its vertices. This polygon can be intersected with the polygonal grid cells and the area of intersection calculated using standard algorithms of computational geometry. A wave that strikes the boundary is reflected by intersecting it with a half space which locally approximates the exterior. This gives a polygon representing the portion of the wave exterior to the domain which is then reflected through the boundary of the half space to obtain an interior polygon.

This basic algorithm has been fully implemented. The high resolution correction terms have been properly included only at the regular grid cells in the interior. At the boundary an approximation is used that gives reasonable results but can probably be improved upon.

The test cases presented here involve only flat boundaries. The handling of boundary conditions is then equivalent to reflecting the interior cells across the boundary. This minimizes the possible negative effects of the boundary conditions and gives a reasonable test of the basic algorithm in situations where there are irregular small grid cells.

Allowing arbitrarily small boundary cells poses some numerical difficulties not associated with the theoretical stability properties of the method. If the size of a grid cell is on the order of the machine rounding error, then the area of the intersection of that cell with a wave cannot be accurately computed. This can lead to gross errors that will eventually cause noticeable errors in other grid cells and possibly nonphysical flow quantities. In order to avoid such problems, we have chosen to ignore any boundary grid cells with area less than $10^{-2}h^2$. This amounts to introducing imperceptible holes in the otherwise solid boundary. The experiments reported here were performed on a Vaxstation II in single precision.

The first test case is a one-dimensional shock tube problem in a channel at 40° to the grid. We take the same initial data as in Section 5. Figure 9.1 shows the boundary cells of the computational grid and contour lines of constant density for the numerical solution at $t=0.3$, using the high resolution method with $h=\frac{1}{40}$ and timestep $k=0.0075$. This gives a Courant number of roughly 0.5 relative to the regular cells, though it is nearly 100 times larger for the smallest irregular cells that are not ignored. Figure 9.1a shows the results obtained with the basic first-order method. Figure 9.1b shows the results when the high resolution correction waves discussed in Section 7 are included. Both figures show contour lines of density as well as the irregular grid cells adjacent to the boundary. Figure 9.2 shows corresponding plots of the density along the lower wall of the channel. Values in each boundary cell are plotted vs distance along the wall. No attempt has been made to extrapolate to the actual boundary location, or to account for the nonuniformity of these boundary cells. Consequently some nonsmoothness is visible in this boundary data (and even more so in Fig. 9.4 below). Plots of grid values down the center of the channel would show much smoother results.

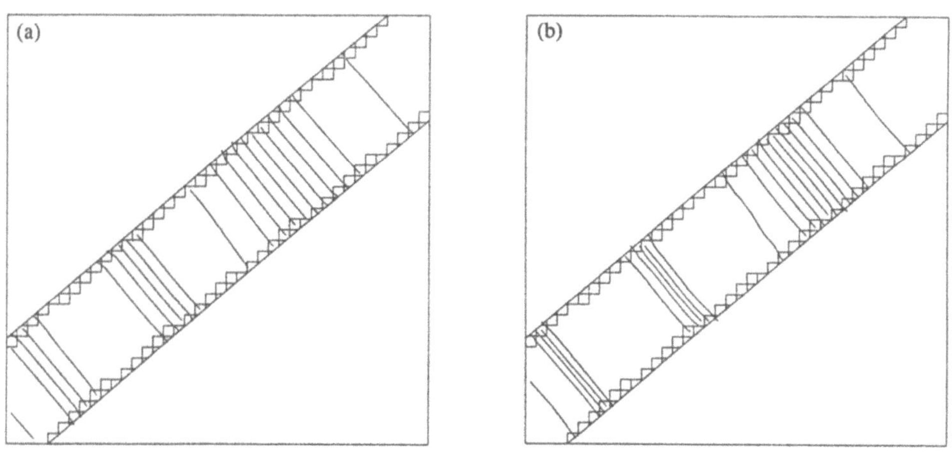

FIG. 9.1. Density contours for a shock tube at 40° to the grid. The irregular boundary cells of the computational grid are also shown. (a) First-order method. (b) With high resolution correction waves included.

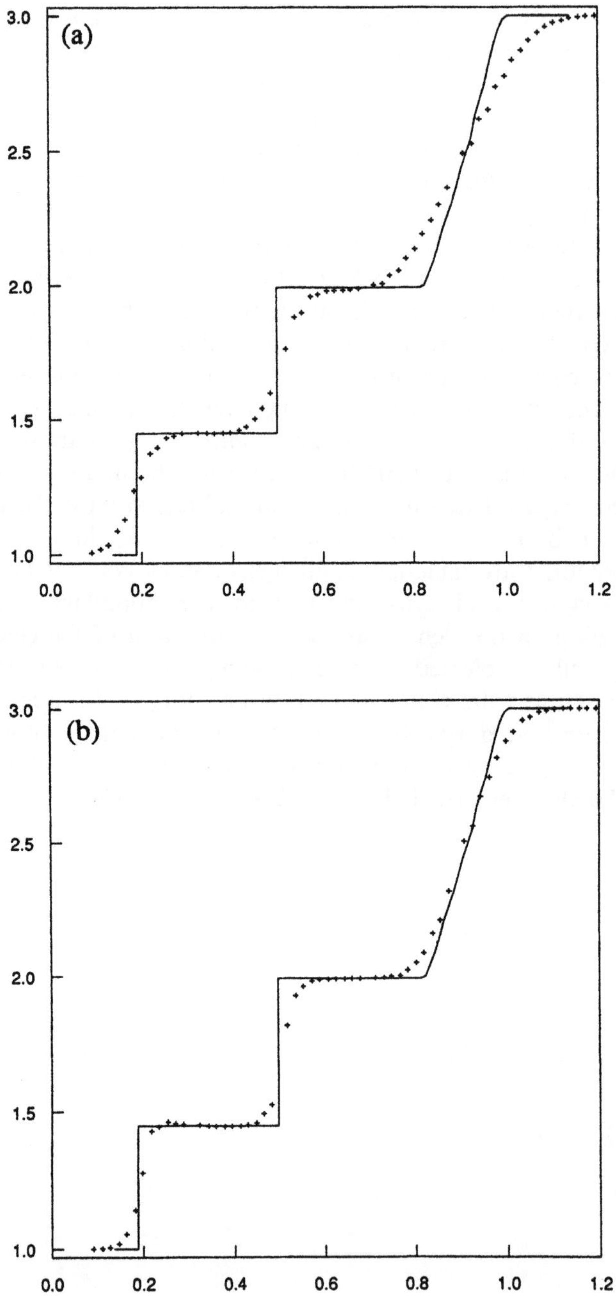

FIG. 9.2. Density in the shock tube along the lower wall. Values in the boundary cells are plotted versus arclength. The solid line is the true solution. (a) First-order method. (b) With high resolution correction waves included.

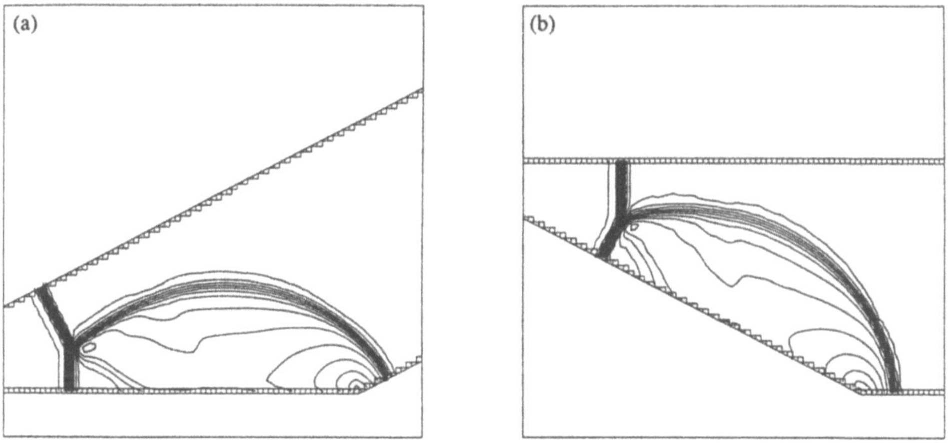

FIG. 9.3. Density contours for Mach 2.03 shock reflection off a 27° ramp. (a) Case I: grid aligned with ramp. (b) Case II: grid aligned with incident shock.

Smearing is clearly reduced by introducing the high resolution correction waves. In both cases the contour plots show that the method is capable of maintaining one-dimensionality of the solution, with essentially the same accuracy in the irregular cells as in the interior. The small boundary cells cause no stability problems.

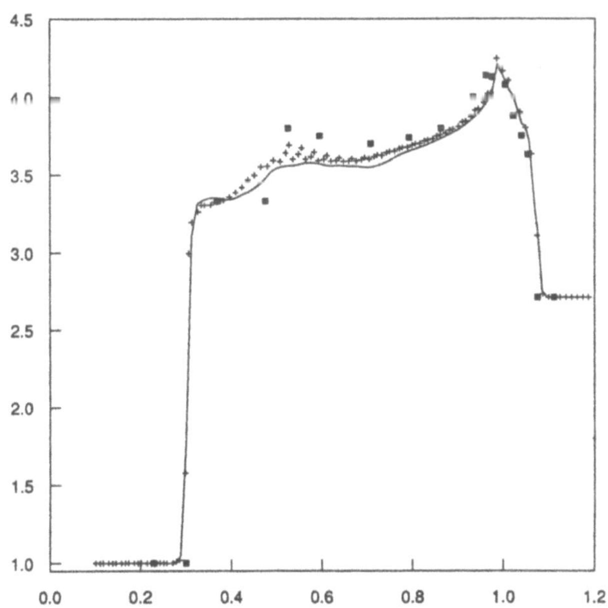

FIG. 9.4. Density along the lower boundary. The solid line shows results from Case I, the + symbols are results from Case II. The squares are experimental results taken from [9].

The second test example is an unsteady shock reflection from a 27° ramp at Mach 2.03, with single Mach reflection. Experimental as well as numerical results for this problem can be found in the paper of Glaz, Collela, Glass, and Deschambault (Case 2 in [9]). The initial flow conditions are $\rho = 0.387$, $v = 0$, $p = 33.3$ ahead of the shock, $\rho = 1.05$, $v = -14.06$, $p = 154.4$ behing the shock, where v if the velocity normal to the shock. The shock was initially distance 0.05 from the ramp corner and results are presented at time 0.275 with $h = \frac{1}{80}$ and $k = 2.5 \times 10^{-4}$ (110 time steps). The results presented here were computed with the high resolution method. Computations with the first-order version have also been done and give the same qualitative results with much more smearing.

For comparison purposes, two different orientations of the ramp are considered. In Case I (Fig. 9.3a), the grid is aligned with the ramp. This is the orientation often used in other codes to avoid irregular grid cells in the region of primary interest (e.g., [9, 29]). In Case II (Fig. 9.3b) the grid is aligned with the incident shock and the ramp cuts through the grid obliquely. In each case 30 equally spaced density contours are plotted. The results agree closely and have essentially the same resolution.

Figure 9.4 shows a plot of relative density (ρ/ρ_0, where $\rho_0 = 0.387$) along the lower wall. The solid line shows the results from Case I while the + symbols show results from Case II. The squares are experimental results taken from [9]. The results are in good agreement and also compare reasonably well with numerical results of [9]. The discontinuities obtained here are not as sharp, particularly the slip line connecting the triple point to the wall. In large part this is due to the coarseness of the grid (essentially 80×40 grid points compared with 350×130 in [9]). Figure 9.4 shows that even in Case II, where the ramp is oblique to the grid, there are at most two points in the shocks.

10. Conclusions

An approach for deriving high resolution numerical methods has been presented that allows generalization to stable methods on arbitrary grids. Numerical results in two space dimensions show that the accuracy in irregular grid cells is comparable to what is obtained with regular cells. In particular, for Cartesian grids cut by irregular boundaries, it is possible to obtain stable high resolution results with a time step chosen based on the interior regular cells.

Another application of these ideas is to mesh refinement using overlapping rotated Cartesian grids, as in the work of Berger [1, 2]. Again irregular cells are created at the grid interface and the present approach gives a way of deriving conservative interface conditions. This is currently being studied by the author and M. Berger.

A similar approach toward achieving stability has been taken by Chern and Colella [5]. They give an algorithm for shock tracking across a Cartesian grid in which fluxes are redistributed from small grid cells to neighboring grid cells to ensure stability. Their redistribution procedure is rather ad hoc compared to the wave propagation approach taken here but presumably has the advantage of being much cheaper. A major shortcoming of the present method is its expense. If tangential wave propagation is used in two dimensions, then at each cell interface $m+1$ Riemann problems must be solved and m^2 waves propagated. Even using Roe's approximate Riemann solver (which is easy to vectorize over multiple Riemann problems, particularly on a Cartesian grid), this may be prohibitive for a practical algorithm. On the other hand, it does not seem that so much information should be necessary, and efforts are currently under way to find simplifications of the algorithm by using a smaller number of approximate waves that carry the essential information. One possibility is to use a different kind of wave decomposition in place of the Riemann solutions. Some of the ideas of Roe [24] or Deconinck, Hirsch, and Peuteman [8] might be useful in this regard.

Another practical possibility is to use the algorithm developed here only at irregular boundaries of a Cartesian grid, coupling it to a more efficient standard method in the regular interior cells.

On the other hand, in some applications the approach discussed here may have advantages over standard finite volume or dimensionally split methods even on regular grids. The fact that waves are propagated obliquely to the grid (when tangential propagation is included) may be advantageous in modeling complex two-dimensional phenomena. Tests have shown that including tangential propagation, even when not required for stability, increases the accuracy with which waves oblique to the grid can be tracked.

In some special applications it could also prove useful to use Courant numbers larger than one even on a regular grid. For example, in problems with widely varying wave speeds where the phenomena of primary interest occurs on the slow scale, the fast waves may carry little information but will limit the time step with traditional methods. Using the wave propagation approach allows one to choose a time step appropriate for the slow scale phenomena. This could lead to a simultaneous increase in efficiency and reduction in the numerical dissipation of slow waves.

References

1. M. Berger, Stanford Computer Science Dept. Report STAN-CS-82-924, 1982 (unpublished).
2. M. Berger and J. Oliger, *J. Comput. Phys.* **53**, 484 (1984).
3. J. P. Boris and D. L. Book, *J. Comput. Phys.* **11**, 38 (1973).
4. Y. Brenier, *SIAM J. Numer. Anal.* **21**, 1013 (1984).
5. I. L. Chern and P. Colella, Lawrence Livermore National Laboratory Report UCRL-97200, 1987 (unpublished).
6. D. K. Clarke, M. D. Salas, and H. A. Hassan, *AIAA J.* **24**, 343 (1986).
7. S. F. Davis, ICASE NASA Contractor Report 172179, 1983 (unpublished).
8. H. Deconinck, C. Hirsch, and J. Peuteman, in *10th Int. Conf. on Num. Meth. in Fluid Dyn., Beijing*, 1986.

9. H. M. Glaz, P. Colella, I. I. Glass, and R. L. Deschambault, *Proc. R. Soc. London A* **398**, 117 (1985).
10. J. B. Goodman and R. J. LeVeque, *SIAM J. Numer. Anal.* **25**, 268 (1988).
11. A. Harten, *J. Comput. Phys.* **49**, 357 (1983).
12. A. Harten and J. M. Hyman, *J. Comput. Phys.* **50**, 235 (1983).
13. P. D. Lax, *Hyperbolic Systems of Conservation Laws and the Mathematical Theory of Shock Waves*, SIAM Regional Conference Series in Applied Mathematics Vol. 11 (SIAM, Philadelphia, 1973).
14. R. J. LeVeque, *Commun. Pure Appl. Math.* **37**, 463 (1984).
15. R. J. LeVeque, *SIAM J. Numer. Anal.* **22**, 1051 (1985).
16. R. J. LeVeque, in *Proceedings, 7th Int. Conf. Methods in Appl. Sci. and Eng., Versailles, France*, 1985.
17. R. J. LeVeque, *SIAM J. Numer. Anal.* **25**, 1 (1988).
18. R. J. LeVeque and J. B. Goodman, Lectures in Appl. Math. Vol. 22, p. 51 (Amer. Math. Soc., Providence, RI, 1985).
19. S. Osher and S. Chakravarthy, *SIAM J. Numer. Anal.* **21**, 955 (1984).
20. P. L. Roe, in *Proceedings, Seventh Intl. Conf. on Num. Meth. in Fluid Dynamics*, edited by W. C. Reynolds and R W. MacCormack, Lecture Notes in Physics Vol. 141 (Springer-Verlag, New York/Berlin, 1981).
21. P. L. Roe, *J. Comput. Phys.* **43**, 357 (1981).
22. P. L. Roe, in *Numerical Methods for Fluid Dynamics*, edited by K. W. Morton and M. J. Baines, (Academic Press, New York/London, 1982).
23. P. L. Roe, Lectures in Appl. Math. Vol. 22, p. 163 (Amer. Math. Soc., Providence, RI, 1985).
24. P. L. Roe, ICASE Report 85-18, 1985 (unpublished).
25. P. E. Rubbert *et al.*, in *Computational Mechanics—Advances and Trends*, AMD Vol. 75, edited by A. K. Noor, pp. 49–84, (1986).
26. P. K. Sweby, *SIAM J. Numer. Anal.* **21**, 995 (1984).
27. B. van Leer, *J. Comput. Phys.* **32**, 101 (1979).
28. B. Wedan and J. C. South, AIAA Paper 83-1889-CP, 6th AIAA Computational Fluid Dynamics Conf., 1983.
29. P. Woodward and P. Colella, *J. Comput. Phys.* **54**, 115 (1984).
30. S. T. Zalesak, *J. Comput. Phys.* **31**, 335 (1979).

DESIGN OF AN ESSENTIALLY NONOSCILLATORY RECONSTRUCTION PROCEDURE ON FINITE-ELEMENT-TYPE MESHES

RÉMI ABGRALL

ABSTRACT. In order to construct high-order finite volume schemes on triangular unstructured meshes, we propose in this article an Essentially Nonoscillatory Reconstruction on these meshes. Given a function u that may admit discontinuities on a locally regular curve, we consider its Lagrange interpolation and another polynomial approximation that is formally average preserving. We analyze the asymptotic behavior of the highest-degree coefficients of of these polynomial approximations of u. These results are then applied to derive a conservative essentially nonoscillatory approximation of this type of function. Some numerical examples are given. They demonstrate the efficiency of the reconstruction.

INTRODUCTION

During the past few years, a growing interest has emerged for constructing high-order accurate schemes for compressible flow simulations. This is not an obvious problem because it is well known that even for smooth initial conditions, these flows may develop discontinuities. At the beginning of the 80s, the class of Total Variation Diminishing (TVD) schemes appeared, and they have been successfully and widely used with many types of meshes (see, for example, [1] for a review and, among many others, [2] for simulations on finite element type meshes). Nevertheless, one of their main weaknesses is that their order of accuracy falls to first order in regions of discontinuity and at extrema, leading to excessive numerical dissipation.

Various methods have been proposed to overcome this difficulty (adaptation of the mesh, for example) but one promising way is the class of the Essentially Nonoscillatory schemes (ENO for short) introduced by Harten, Osher and others [3, 4, 5].

Briefly speaking, the ENO schemes are finite volume MUSCL type schemes where the linear reconstruction procedure is replaced by a higher-degree reconstruction in order to reduce numerical dissipation ; hence a wider stencil for the reconstruction is needed. If a fixed stencil were used, it is well known that the reconstruction

1991 *Mathematics Subject Classification*. Primary 65D05, 65N30.

Key words and phrases. Interpolation, finite element methods.

This work has been partly done under contract NAS-18605 while the author was in residence at ICASE, NASA Langley Research Center, Hampton, Virginia.

would become highly oscillatory near discontinuities, so that the resulting scheme would become unstable.

The idea of Harten et al. is to use an adaptive stencil instead, with the following principles. In order to simplify the discussion, we assume now that one wants to integrate a scalar hyperbolic equation with an initial condition in $\mathbb{R}$; the unknown is u. In any finite volume scheme, the numerical solution is described by the averaged values u_i of the (numerical) solution in "control volumes" $[x_{i-1/2}, x_{i+1/2}]$ that cover the domain of integration. The reconstruction is conservative, i.e., the average of the approximation is u_i in $[x_{i-1/2}, x_{i+1/2}]$. Harten et al. describe two procedures to define a function w whose values at certain points y_i of $\mathbb{R}$ are known from the averages of u in the $[x_{i-1/2}, x_{i+1/2}]$'s, namely the reconstruction via primitive function (RP) or via deconvolution (RD). To describe how this can be achieved, let us recall two well-known (or more precisely, well-admitted, see the counter-examples of §1.2) properties of the Lagrange interpolation : denote by P the Lagrange interpolation of degree $\leq n$ of w with the stencil $\{y_1 < \cdots < y_{n+1}\}$,

- if w is of class C^m, $m > n$, in $[y_1, y_{n+1}]$, then the coefficients of the Taylor expansion of P around any point of $]y_1, y_{n+1}[$ remain bounded,
- else, if w is of class C^{p-1}, $p > 1$, on $]y_1, y_{n+1}[$, and if there exists $x_0 \in]y_1, y_{n+1}[$ such that w is C^m, $m > n$ on $]y_1, y_{n+1}[-\{x_0\}$ and such that

$$\lim_{x \to x_0, x > x_0} w^{(l)}(x) \text{ and } \lim_{x \to x_0, x < x_0} w^{(l)}(x)$$

exist for all $p \leq l \leq n$ and are different at least for $l = p$, then the kth coefficient of the Taylor expansion of w, $k \geq p$, around any point of (y_1, y_{n+1}), behaves like $(y_{n+1} - y_1)^{-k+p}$: it is unbounded for $k > p$.

By using these two remarks, it is possible to localize the discontinuities of w and those of $w^{(p)}$, $p < n$, at least asymptotically, by comparing the highest coefficients of P, and then to select a stencil lying in the smooth part of w. Some attempts have been made to extend these ideas to multidimensional flows (e.g. [7]), but only for smoothly varying structured meshes : the reconstruction is done direction by direction.

In contrast, the reconstruction problem is much more complicated for general unstructured meshes because of the lack of privileged direction. It is also easy to see that a reconstruction method that would use a primitive function w like the RP procedure is unfeasible : this primitive function has to be known at some points of the plane ; this is equivalent to assuming that some rectangles $[a, b] \times [c, d]$ of the plane are exactly covered by control volumes, which is generally impossible.

In order to handle the reconstruction problem in several dimensions on general meshes, we have to define new reconstruction procedures. In order to achieve this goal, we consider two approximation methods in the first part of this paper : the Lagrange interpolation technique and a polynomial approximation that exactly conserves the mean on given control volumes. We study the asymptotic behavior of the coefficients of these polynomial approximations in the case where u, the function to be approximated, is smooth except on locally C^1 curves. The result for the Lagrange interpolation of a smooth function is of course well known, but the other results seem to be new.

Then we propose two algorithms for the ENO reconstruction on triangular conformal meshes. The first is an adaptation of the reconstruction via a deconvolution procedure, while the second uses the formally conservative approximation. A similar method has also been considered by Barth et al. [8] (but it ensures conservation *only* for polynomials), and the same method has independently been considered by Harten et al. [6]. What is new here is, first, that we can prove the asymptotic behavior of the polynomial coefficients and, second, that the set of stencils we propose is much easier to handle.

In conclusion, some numerical examples are given to illustrate the method.

Let us start with some notations. We denote by $P_n(\mathbb{R}^2)$ the vector space of polynomials of total degree $\leq n$, in two variables (x and y), with real coefficients. An element of $P_n(\mathbb{R}^2)$ may be described by its expansion in terms of powers of x and y :

$$P(x,y) = \sum_{l=0}^{n} \sum_{i+j=l; i,j\geq 0} a_{ij} x^i y^j .$$

If (x_0, y_0) is a point of $\mathbb{R}^2$, another expansion of P may be written in terms of the monomials $(x-x_0)^i(y-y_0)^j$ with the help of the Taylor formula. The total degree of P, i.e., the smallest integer l such that $a_{ij} = 0$ for all (i,j) with $l+1 \leq i+j \leq n$, does not depend on the point (x_0, y_0). The space $P_n(\mathbb{R}^2)$ has dimension $N(n) = \frac{(n+1)(n+2)}{2}$; a basis is the set of monomials $(x-x_0)^i(y-y_0)^j$ for $i+j \leq n$.

1. Two approximation problems in $\mathbb{R}^2$

Throughout this paper, a stencil S will be either a (finite) collection of points A_i or a (finite) collection of cells C_i. When the elements of S are points A_i whose coordinates are (x_i, y_i), we will often denote either by $u(A_i)$ or $u(x_i, y_i)$ the value of a function u at the point A_i. When the elements of S are cells C_i, we assume they fulfill the following properties (where N_0 is a given fixed integer) :

- they are polygonals of nonempty interior, their boundary has at most N_0 extremal points ;
- they are connected ;
- the intersection of two different cells is of empty interior.

The role of N_0 will be made clear in §1.2.

Let u be a L^1 real-valued function in $\mathbb{R}^2$ and let N be an integer. We consider the following linear forms on L^1 :

- if $S = (A_i)_{1\leq i\leq N}$ is a family of N points in $\mathbb{R}^2$, we set

(1) $$L_{A_i}(u) = u(A_i),\ 1 \leq i \leq N;$$

- if $S = (C_i)_{1\leq i\leq N}$ is a family of cells, we set

(2) $$L_{C_i}(u) = < u >_{C_i} := \frac{\int_{C_i} u(x,y)\, dxdy}{\text{area}(C_i)} ;$$

and the approximation problem $(\mathcal{I})$:

Given N, n, two integers, and the family $(L_{\sigma_i})_{1\leq i\leq N}$ of N linear forms associated with the stencil $\mathcal{S}=\{\sigma_i\}_{1\leq i\leq N}$, find an element P of $P_n(\mathbb{R}^2)$ such that, for any element of $\mathcal{S}$, one has $L_{\sigma_i}(P)=L_{\sigma_i}(u)$ for $1\leq i\leq N$.

For this problem to have a unique solution, two conditions must be fulfilled :

(1) one must have $N:=\frac{(n+1)(n+2)}{2}=N(n)$;
(2) the following generalized Vandermonde determinant must be nonzero :

(3)
$$\Delta_{\mathcal{S}}=\det\left[L_{\sigma_l}(x^i\,y^j)\right]\Big|_{\substack{i+j\leq n\\ 1\leq l\leq N}}=$$

$$\begin{vmatrix} 1 & L_{\sigma_1}(x) & L_{\sigma_1}(y) & \cdots & L_{\sigma_1}(x^n) & L_{\sigma_1}(x^{n-1}y) & \cdots L_{\sigma_1}(y^n)\\ \vdots & \vdots & \vdots & \vdots & \vdots & \vdots & \vdots\\ 1 & L_{\sigma_N}(x) & L_{\sigma_N}(y) & \cdots & L_{\sigma_N}(x^n) & L_{\sigma_N}(x^{n-1}y) & \cdots L_{\sigma_N}(y^n)\end{vmatrix}.$$

In what follows, the notation $\mathcal{S}^{(n)}$ will stand for a stencil with $N(n)$ elements that is admissible for degree n, i.e., $\Delta_{\mathcal{S}^{(n)}}\neq 0$. If a set $\mathcal{S}^{(n)}$ is admissible, for any function u of L^1, there exist $N(n)$ coefficients, (a_{ij}), such that the solution of the problem $(\mathcal{I})$ is :

$$P=\sum_{l=1}^{n}\sum_{i+j=l;i,j\geq 0}a_{ij}x^iy^j.$$

In the sequel, the polynomial P will often be denoted as $\Pi_L(u)$ in the case of Lagrange interpolation (i.e., $L_{A_i}(u)=u(A_i)$) and $\Pi_M(u)$ in the case of approximation in the mean (i.e., $L_{\sigma_i}(u)=<u>_{C_i}$). The linear operators Π_L and Π_M are projectors onto $P_n(\mathbb{R}^2)$.

The problem of characterizing the admissible sets for the Lagrange interpolation has been widely studied (e.g. [9] and the references therein). We do not know of any criteria for selecting admissible stencils for approximation in the mean.

The aim of this section is to obtain the asymptotic behavior of the highest-degree coefficients of $P=\Pi_L(u)$ (resp. $P=\Pi_M(u)$) when the size of the convex hull K of the points (resp. the collections of the cells) of $\mathcal{S}^{(n)}$ in the case of Π_L (resp. Π_M) becomes small. We first study the case when u admits a $(n+1)$st bounded derivative in K. Then, we proceed to the case when the pth derivative, D^pu, $0\leq p\leq n$, of u admits a jump $[D^pu]>\gamma>0$ on a locally C^1 curve contained in K.

Throughout this paper, we call "convex hull of $\mathcal{S}^{(n)}$" either the convex hull of the points of $\mathcal{S}^{(n)}$ for Lagrange interpolation, or the convex hull of the cells of $\mathcal{S}^{(n)}$ for approximation in the mean.

1.1. The case of a "smooth" stencil. This problem, for Lagrange and Hermite interpolation, has been studied for example, by Ciarlet and Raviart in [13]. As will be shown in §1.1.1, their error study of the Lagrange interpolation can easily be extended to the mean approximation. In the following theorem, K denotes the convex hull of $\mathcal{S}^{(n)}$.

Theorem 1.1. *Let $\mathcal{S}^{(n)}$ be an admissible (for degree n) stencil of $\mathbb{R}^2$, and let h and ρ be respectively the diameter of K and the supremum of the diameters of the circles contained in K. Let u be a function that admits everywhere in K a $(n+1)$st derivative $D^{n+1}u$ with*

$$M_{n+1} = \sup\{||D^{n+1}u(x)||; x \in K\} < +\infty.$$

If Π denotes either Π_L or Π_M, the projectors on the space of polynomials of degree less than or equal to n, we have for any integer m with $0 \leq m \leq n$,

$$\sup\{||D^m u(x) - D^m \Pi(u)(x)||; x \in K\} \leq C M_{n+1} \frac{h^{n+1}}{\rho^m},$$

for some constant

$$C = C(n, m, \mathcal{S}^{(n)}).$$

Moreover, if $\mathcal{S}^{(n)\prime}$ is obtained from $\mathcal{S}^{(n)}$ by an affine transformation, then $C(n, m, \mathcal{S}^{(n)}) = C(n, m, \mathcal{S}^{(n)\prime})$.

From this inequality, we see that the "flatter" K is, the poorer is the estimate.

1.1.1. *Proof of the error estimate for approximations in the mean .* The proof of this results is directly inspired by Ciarlet and Raviart almost without any change.

Lemma 1.2. *Let $n \geq 1$ be an integer, assume that $\mathcal{S}^{(n)} = (C_i)_{1 \leq i \leq N(n)}$ is admissible, and let $u \in C^p$ with $p > n$. Then, for any $x \in K$ and for any integer m, $0 \leq m \leq n$, we have :*

$$D^m \Pi_M(u)(x) = D^m u(x) + \frac{1}{m!} \sum_{i=1}^{N(n)} \left[D^{n+1}u(\nu_i(x)).(\alpha_i - x)^{n+1}\right] D^m p_i, \tag{4}$$

where the p_i's are the polynomials defined by $< p_i >_{C_j} = \delta_i^j$, the α_i's are points of C_i, and the ν_i's are $\nu_i(x) = \theta_i x + (1-\theta_i)\alpha_i$ for $0 < \theta_i < 1$.

Proof. From the definition of the p_i's, we have :

$$\Pi_M(u) = \sum_{i=1}^{N(n)} < u >_{C_i} p_i.$$

Since C_i is connected and u is continuous in C_i, there exists $\alpha_i \in C_i$ such that $< u >_{C_i} = u(\alpha_i)$, and consequently

$$\Pi_M(u) = \sum_{i=1}^{N(n)} u(\alpha_i) p_i, \quad D^m \Pi_M(u) = \sum_{i=1}^{N(n)} u(\alpha_i) D^m p_i.$$

The rest of the proof follows exactly that of Ciarlet and Raviart [13]. □

Lemma 1.3. *The property of admissibility remains invariant under affine transformations.*

Proof. We have to prove this for the approximation in the mean. Let $\widehat{\mathcal{S}^{(n)}} = \{\widehat{C}_i\}_{1\leq i\leq N(n)}$ be an admissible stencil, and $\mathcal{S}^{(n)} = \{C_i\}_{1\leq i\leq N(n)}$ its image by the affine transformation which translation and matrix are T and B. For any L^1 function v, we have

$$< u >_{C_i} = < u \circ T^{-1} \circ B^{-1} >_{\widehat{C}_i}$$

where T is the translation by the vector (x_0, y_0). Moreover, if $P \in P_n(\mathbb{R}^2)$ then, $P \circ T^{-1} \circ B^{-1} \in P_n(\mathbb{R}^2)$ because $T^{-1} \circ B^{-1}$ is an affine transformation. Hence, the stencil $\mathcal{S}^{(n)}$ is admissible if and only if $\widehat{\mathcal{S}^{(n)}}$ is admissible. □

Following Ciarlet and Raviart, we have also the lemma :

Lemma 1.4. *If $\mathcal{S}^{(n)}$ is obtained from $\widehat{\mathcal{S}^{(n)}}$ by an affine transformation of matrix B, then*

$$||B|| \leq \frac{h}{\widehat{\rho}}, \; ||B^{-1}|| \leq \frac{\widehat{h}}{\rho},$$

where h (res. $\widehat{h}$) and ρ (resp. $\widehat{\rho}$) are the diameter and the maximum diameter of the circles contained in the convex hull of $\mathcal{S}^{(n)}$ (resp. $\widehat{\mathcal{S}^{(n)}}$).

The combination of Lemmas 1.3 and 1.4 with Lemma 1.2 gives Theorem 1.1.

1.2. The case of a "nonsmooth" stencil . We begin with some notations. Let $\mathcal{S}^{(n)}$ be an admissible stencil. For the sake of simplicity, we assume that (x_0, y_0) is any of the points of $\mathcal{S}^{(n)}$, or the center of gravity of any element of $\mathcal{S}^{(n)}$, depending on the kind of approximation we are dealing with. Throughout this subsection, we adopt the lexicographic ordering for polynomials : if i and j are two indices such that $i + j = p \leq n$, we set $l = N(p-1) + i + 1$ (with the convention $N(-1) = 0$) and denote by P_l the monomial $(x - x_0)^i(y - y_0)^j$. We also set

$$R_l = \left(L_{\sigma_1}(P_l), \cdots, L_{\sigma_{N(n)}}(P_l)\right)^T.$$

Given a set of $N(n)$ real numbers u_i, the solution of problem $(\mathcal{I})$,

$$u_i = \sum_{1\leq j\leq N(n)} \alpha_j L_{\sigma_i}(P_j) \text{ for all } i,$$

may be seen as the solution of the linear system $\mathcal{M}_n \left(\alpha_1, \cdots, \alpha_{N(n)}\right)^T = U = \left(u_1, \cdots, u_{N(n)}\right)^T$, where the lth column of $\mathcal{M}_n$ is R_l. The solution of this system is given by

$$\alpha_l = \frac{\det(R_1, \cdots, \widehat{R_l}, \cdots R_{N(n)})}{\det(R_1, \cdots, R_l, \cdots, R_{N(n)})}, \; 1 \leq l \leq N(n), \tag{5}$$

where $\widehat{R_l} = U$.

We need to work with admissible stencils such that the polynomial P, solution of problem $(\mathcal{I})$ (for degree n) with data that are either 0 or taken among linear combination of terms belonging to $\{L_\sigma(P_l)\}_{\sigma\in\mathcal{S}^{(n)}}$ is *exactly* of degree n. More precisely, we define the set $\mathcal{P}^{\alpha,\beta}_{n,p}$ for α, $\beta > 0$ and $p \in \{1, \cdots n\}$ by

Definition 1.5. *Given any $p \in \{1, \cdots n\}$, and two real number $\alpha > 0$, $\beta > 0$ we define the set $\mathcal{P}^{\alpha,\beta}_{n,p}$ as follows : $(\lambda_{N(p-1)+1}, \cdots \lambda_{N(p)}, \mathcal{S}^{(n)}) \in \mathcal{P}^{\alpha,\beta}_{n,p}$ if and only if*

(1) *the diameter h of K, the convex hull of* $\mathcal{S}^{(n)}$, *is 1 and* $(0,0) \in K$;
(2) *the stencil* $\mathcal{S}^{(n)}$ *is admissible and*

$$\left|\det(R_1, \cdots, R_{N(n)})\right| \geq \beta \tag{6}$$

(3) $\max_{N(p-1)+1 \leq l \leq N(p)} |\lambda_l| = 1$;
(4) *for any partition* $\mathcal{S}_0$, $\mathcal{S}_1$ *of* $\mathcal{S}^{(n)}$, *with* $\#\mathcal{S}_0 > 0$ *and* $\#\mathcal{S}_1 > 0$, *the polynomial* $P \in P_n(\mathbb{R}^2)$ *defined by*

$$L_{\sigma_k}(P) = \begin{cases} 0 & \text{if } \sigma_k \in S_0 \\ \sum_{l=N(p-1)+1}^{N(p)} \lambda_p L_{\sigma_k}(P_l) & \text{if } \sigma_k \in S_1 \end{cases}$$

satisfies

$$\sum_{i+j=n} |a_{ij}| \geq \alpha. \tag{7}$$

One can find stencils so that (7) is not satisfied, whatever $\alpha > 0$. We give two counterexamples for the Lagrange interpolation with $n = 2$, one in $\mathbb{R}$ and the other in $\mathbb{R}^2$.

(1) Consider the stencil $\{x_0 = 0, x_1 = 1, x_2 = 2\}$, and P, polynomial of degree two such that $P(0) = 1$, $P(1) = 0$, $P(2) = 0$. We then have $P(3) = 1$. But the stencil $\{x_0 = 0, x_1 = 1, x_2 = 2, x_3 = 3\}$ is admissible for degree three, and hence does not satisfy the equivalent of Property 1.5 in $\mathbb{R}$.
(2) Let $Q(x, y) = x^2 + y^2 - 1 \in P_2(\mathbb{R})$. The sets $\{Q = 0\}$ and $\{Q = 1\}$ are two circles that have the same center. Then, one can find two distinct lines, D_1 and D_2, such that D_1 cuts $\mathcal{E} = \{Q(Q-1) = 0\}$ in four points, M_1, M_2, M_3, M_4, and D_2 cuts $\mathcal{E}$ in at least three points, M_5, M_6, M_7. Then one chooses M_8, M_9, M_{10}, three other points that do not belong to D_1 or D_2, that are not aligned ; we also assume that they are on $\{Q = 0\}$. The polynomial Q has value one for at least three of the elements of $\mathcal{S} = \{M_1, \cdots, M_{10}\}$, and zero for at least three among the other ones. We claim that $\mathcal{S}$ is admissible for the Lagrange interpolation in $\mathbb{R}$ with $n = 3$.

Proof. It is equivalent to show that the Vandermonde determinant (3) does not vanish or to show that the points of $\mathcal{S}$ do not lie on a cubic curve $\mathcal{C}$. Let $P_1 = 0$ be an equation of D_1, $P_2 = 0$ be an equation of D_2 and $H = 0$ be an equation of the cubic that goes through $M_1, \cdots, M_9$. The polynomials P_1 and P_2 are prime, and since $D_1 \neq D_2$, they are relatively prime. Since D_1 cuts $\mathcal{C}$ in more that three points, P_1 divides H (make the Euclidian division of H by P_1 in $\mathbb{R}(X)[Y]$, the polynomial ring over the field of fraction $\mathbb{R}(X)$ to see that). The same is true for P_2. Then $H = P_1R_1$ and $H = P_2R_2$, where R_1 and R_2 belong to $P_2(\mathbb{R}^2)$. Since $P_3(\mathbb{R}^2)$ is factorial and P_1 and P_2 are prime, we have $H = P_1P_2P_3$, where P_3 belongs to $P_1(\mathbb{R}^2)$. The set $\{P_3 = 0\}$ is a line that goes through M_8 and M_9, but not through M_{10} because these points are not aligned. This concludes the proof. □

Conversely, there exist stencils and real numbers $\lambda_{N(p-1)+1} \cdots \lambda_{N(p)}$ that satisfy Property (7). We show that for the Lagrange interpolation ; the same arguments can be used for the approximation in the mean because of the assumption we have made on the cells C_i, in particular, the boundary is a polygon. In the following argument, we look at the coefficients a_{ij} of the expansion of any polynomial as an abstract rational function depending on the λ_is and on the coordinates of the points of $\mathcal{S}^{(n)}$; we explain why these rational functions cannot be identically vanishing. We consider any integer p, $1 \leq p \leq n$, and a partition of $\mathcal{S}^{(n)} = S_1 \bigcup S_2$ as in property 1.5. If one multiplies a_{ij} by $\det(R_1, \cdots, R_{N(n)})$, one sees that the condition $a_{ij} = 0$ can be rewritten is the form

$$Q = \sum_{l=N(l-1)+1}^{N(l)} \lambda_l Q_l(M_1, \cdots, M_{N(n)}) = 0, \tag{8}$$

where the Q_l are polynomials of degree less or equal to $(N(n))!$: they correspond to the determinants one gets when all the λ_is but one which is 1 are zero. This Q is not identically zero : if it were identically vanishing, the polynomials Q_l would also be identically vanishing, which is absurd. There is a finite number of possible polynomials Q : one for each p and each partition of $\mathcal{S}^{(n)}$. Then the intersection of the algebraic varieties that they define is of empty interior. Thus there exist stencils satisfying property 1.5.

To end this paragraph, let us notice that for any stencil $\mathcal{S}^{(n)}$, if there exist some $\lambda_{N(p-1)+1} \cdots \lambda_{N(p)}$ so that $(\lambda_{N(p-1)+1} \cdots \lambda_{N(p)}, \mathcal{S}^{(n)}) \in \mathcal{P}^{\epsilon}_{n,p}$, then there may exist another choice $\mu_{N(p-1)+1} \cdots \mu_{N(p)}$ so that $(\mu_{N(p-1)+1} \cdots \mu_{N(p)}, \mathcal{S}^{(n)}) \notin \mathcal{P}^{\epsilon'}_{n,p}$, for any $\epsilon' > 0$ because (8) is linear in the λ_is.

When the stencils are made of cells, we have assumed the existence of an integer N_0 so that the maximum number of extrema points of the cell is less or equal to N_0. Because of that, we can consider the set of stencils, admissible or not, as a subset of $\mathbb{R}^{N_0 \times 2N(n)}$. When the stencils are made of points, we can consider the set of stencils as a subset of $\mathbb{R}^{2N(n)}$. In both cases, we can induce the topology of $\mathbb{R}^M$, for $M = N_0 \times 2N(n)$ or $2N(n)$, on the set of stencils. The same argument enable us to define a topology on $\mathcal{P}^{\alpha,\beta}_{n,p}$. Now we can state :

Lemma 1.6. *Given any n, p, α, β as in Definition 1.5, one has :*

(1) *the set of all stencils $\mathcal{S}^{(n)}$ with $h = 1$, $(0,0)$ belonging to the convex hull of $\mathcal{S}^{(n)}$ and $|\det(R_1, \cdots, R_{N(n)})| \geq \beta$ is compact ;*
(2) *there exists $\gamma > 0$ such that the maximum diameter ρ of all the circles contained in K, the convex hull of any stencil so that $h = 1$ and $|\det(R_1, \cdots, R_{N(n)})| \geq \beta$ satifies $\rho \geq \gamma$;*
(3) *the set $\mathcal{P}^{\alpha,\beta}_{n,p}$ is compact.*

Proof. (1) Let $\mathcal{M}$ be the set of stencils considered in the (1) of the lemma. All the extremal points of K are at a distance ≤ 1 of $(0,0)$. If $(\mathcal{S}^{(n)}_k)_{k \geq 0}$ is a sequence of stencils, there exists an accumulation point to that sequence. It satisfies (6), then it is an admissible stencil belonging to $\mathcal{M}$: this set is compact.

(2) we can consider the function ϕ, defined on $\mathcal{M}$, by $\phi(\mathcal{S}^{(n)}) =$ the maximum diameter of all the circles contained in K, the convex hull of $\mathcal{S}^{(n)}$. This function is continuous, reaches its minimum since $\mathcal{M}$ is compact. The minimal internal diameter is > 0 :

- Case of the Lagrange interpolation : since ρ is the internal diameter, $\rho = 0$ means that the interior of K is empty. Since K is convex, then K is a segment, and then $\det(R_1, \cdots, R_{N(n)}) = 0$ which is absurd.
- Case of the interpolation in the mean. All the rows $R_2, \cdots, R_{N(n)}$ of the determinant are zero because K is of empty interior : it is a set of zero area, and so are the cells.

(3) $\mathcal{P}_{n,p}^{\alpha,\beta}$ is closed and contained in $[0,1]^{N(l)-N(l-1)} \times \mathcal{M}$.

□

Consider now an admissible stencil $\mathcal{S}^{(n)}$ of cardinality $N(n)$ and two nonempty subsets $\mathcal{S}_0$, $\mathcal{S}_1$ of $\mathcal{S}^{(n)}$ having empty intersection with their union being $\mathcal{S}^{(n)}$. The set K is still the convex hull of $\mathcal{S}^{(n)}$. We have

Lemma 1.7. *Let p, n, α and β be integers and real numbers as in Definition 1.5. We also consider $\delta > 0$. Let (x_0, y_0) be any point of $\mathbb{R}^2$. Let $\mathcal{S}^{(n)}$ be an admissible stencil and $\lambda_{N(p-1)+1}, \cdots, \lambda_{N(p)}$ be real numbers such that*

(1) *there exists an affine transformation ϕ so that $(\lambda_{N(p-1)+1}, \cdots, \lambda_{N(p)}, \phi^{-1}(\mathcal{S}^{(n)})) \in \mathcal{P}_{n,p}^{\alpha,\beta}$,*

(2) *the ratio $\frac{h}{\rho}$ is bounded by δ.*

Let $\mathcal{S}_0 \bigcup \mathcal{S}_1 = \mathcal{S}^{(n)}$ be any partition of $\mathcal{S}^{(n)}$ as in Definition 1.5. Then there exist two constants $C_1(n,p,\alpha,\beta,\delta) > 0$ and $C_2(n,p,\beta,\delta) > 0$ such that if h, ρ are the diameter and the supremum of the diameters of the circles contained in K, the convex hull of $\mathcal{S}^{(n)}$ and if $(a_{ij})_{i\geq 0;\, j\geq 0;\, i+j\leq n}$ are the coefficients of the Taylor expansion of P, the polynomial defined by

$$L_{\sigma_k}(P) = \begin{cases} 0 & \text{if } \sigma_k \in \mathcal{S}_0 \\ \sum_{l=N(p-1)+1}^{N(p)} \lambda_l L_{\sigma_k}(P_l) & \text{if } \sigma_k \in \mathcal{S}_1 \end{cases}$$

satisfies

$$C_2(n,p,\beta,\delta) h^{-n+p} \geq \sum_{i+j=n} |a_{ij}| \geq C_1(n,p,\alpha,\beta,\delta) h^{-n+p}. \tag{9}$$

Furthermore, the constants C_1 and C_2 are invariant by affine transformation.

The notation $GL_2(\mathbb{R})$ denotes the set of all invertible 2×2 real matrices.

Proof. We set $\mathcal{S}^{(n)} = \{\sigma_i\}_{1\leq i\leq N(n)}$. The integer p satisfies $1 \leq p \leq n$. Let B be the matrix of the affine transformation ϕ as above and set $\widehat{\mathcal{S}^{(n)}} = \phi^{-1}(\mathcal{S}^{(n)}))$. We assume that $(\lambda_{N(p-1)+1}, \cdots, \lambda_{N(p)}, \phi^{-1}(\mathcal{S}^{(n)}) \in \mathcal{P}_{n,p}^{\alpha,\beta}$.

The sum $\sum_{i+j=n} |a_{ij}|$ can be seen, because of (5), as a function Ψ of $\lambda_{N(p-1)+1}$, $\cdots, \lambda_{N(p)}, \widehat{\sigma_1} = \phi^{-1}(\sigma_1), \cdots, \widehat{\sigma_{N(n)}} = \phi^{-1}(\sigma_{N(n)})$, that is

$$\Psi\left(\lambda_{N(p-1)+1}, \cdots, \lambda_{N(p)}, B, \widehat{\sigma_1}, \cdots, \widehat{\sigma_{N(n)}}\right) = \sum_{i+j=n} |a_{ij}|.$$

We do not mention the translation of the affine transformation because the a_{ij} are the highest degree coefficient of P : they are invariant by any translation. Since $\frac{h}{\rho} \leq \delta$, because of the inequalities of Lemmas 1.7 and 1.6, $||B^{-1}||$ is bounded by a constant C that depends on δ and γ (of Lemma 1.6). Clearly, Ψ is continuous on $\mathcal{P}_{n,p}^{\alpha,\beta} \times GL_2(\mathbb{R})$, Moreover, for all $\mu > 0$, we have

$$\Psi\left(\lambda_{N(p-1)+1}, \cdots, \lambda_{N(p)}, \mu' B, \sigma_1, \cdots, \mu'' \sigma_{N(n)}\right) = \mu^{p-n} \times$$

$$\Psi\left(\lambda_{N(p-1)+1}, \cdots, \lambda_{N(p)}, B, \widehat{\sigma_1}, \cdots, \widehat{\sigma_{N(n)}}\right).$$

Thus, we can consider the restriction of Ψ on

$$\mathcal{E} = \mathcal{P}_{n,p}^{\alpha,\beta} \times \left[GL_2(\mathbb{R}) \cap \{B \text{ such that } ||B|| = 1 \text{ and } ||B^{-1}|| \leq C\}\right]$$

which is compact, thanks to Lemma 1.6. Thus Ψ is bounded below by $C_2(n, p, \alpha, \beta, \delta)$ and above by $C_1(n, p, \alpha, \beta, \delta)$. The constant $C_2(n, p, \alpha, \beta, \delta)$ cannot be zero because it is reached, hence there is a polynomial P such that the corresponding sum $\sum_{i+j=n} |a_{ij}|$ is $C_2(n, p, \alpha, \beta, \delta)$. This polynomial is *exactly* of degree n because $(\lambda_{N(p-1)+1}, \cdots, \lambda_{N(p)}, \phi^{-1}(\mathcal{S}^{(n)}) \in \mathcal{P}_{n,p}^{\alpha,\beta}$ and because the degree of a polynomial is invariant by affine transformations. The constant $C_1(n, p, \alpha, \beta, \delta)$ can be made independent of α because $\mathcal{P}_{n,p}^{\alpha,\beta}$ can be replaced by $[0,1]^{N(p)-N(p-1)} \times \{\widehat{\mathcal{S}^{(n)}} \text{ such that } \hat{h} = 1\}$. Hence, we have got

$$C_2(n, p, \alpha, \beta, \delta) ||B||^{p-n} \leq \sum_{i+j=n} |a_{ij}| \leq C_1(n, p, \beta, \delta) ||B||^{p-n}. \tag{10}$$

The inequalities of Lemmas 1.4 and 1.6 combined with (10) conclude the proof.
□

Now, we may state our main result :

Theorem 1.8. *Let n, p, α, β and δ be integers and real numbers as in Lemma 1.7 and $\mathcal{S}^{(n)}$ be a admissible stencil. We consider u, a real function defined on an open subset of Ω in $\mathbb{R}^2$, that is C^p, $p \leq n$, except perhaps on a locally C^1 simple curve C. The intersection of the convex hull K of $\mathcal{S}^{(n)}$ and C is assumed to be nonempty. We also assume that the pth-order derivative of u may have a jump such that $\min_{(x,y) \in C} ||[D^p u](x, y)|| \geq \gamma > 0$. Furthermore, we also assume that h, the diameter of K is small enough for $([D^p u], \mathcal{S}^{(n)})$ to satisfy the assumption of Lemma 1.7. Then there exits a constant $C(n, p, \alpha, \beta, \delta) > 0$ such that the coefficients in the Taylor expansion around any point (x_0, y_0) satisfy*

$$\sum_{i+j=n} |a_{ij}| \geq C(n, p, \epsilon_0) \frac{\gamma}{h^{n-p}} \tag{11}$$

when h is small enough.

Proof. We first establish the proof of the theorem for Lagrange interpolation and then indicate how to adapt it for approximation in the mean.

The idea of the proof is the following :

(1) an appropriate Taylor expansion of u enables us to express any coefficient a_{ij}, $i + j = n$, of the Taylor expansion of P with terms that make use only of the behavior of $D^p u$ in the neighborhood of C ;
(2) the lemma 1.7 enables us to estimate all the remaining terms.

Let $\epsilon > 0$. We consider stencils of diameter $\leq \epsilon$. From the assumptions, the curve $\mathcal{C}$ divides K in two connected parts, K^+ and K^-, each being on one side of $\mathcal{C}$. Let (x_0, y_0) be a point of K, say in K^+. If $a = (a_x, a_y) \in \mathcal{S}^{(n)}$, we may assume that the line $\mathcal{L} = [(x_0, y_0), a]$ cuts $\mathcal{C}$ intersected with the disk of center (x_0, y_0) and radius ϵ only once at a_C for ϵ small enough, because $\mathcal{C}$ is locally C^1. If a is between (x_0, y_0) and a_C (i.e., in K^+), the Taylor formula gives

$$\begin{aligned} u(a_x, a_y) &= \sum_{l=0}^{p-1} \frac{1}{l!} \sum_{i+j=l} \frac{\partial^{i+j}}{\partial^i x \partial^j y} u(x_0, y_0)(a_x - x_0)^i (a_y - y_0)^j \\ &+ \frac{1}{p!} \sum_{i+j=p} A_{ij}^+(a_x, a_y)(a_x - x_0)^i (a_y - y_0)^j, \end{aligned} \tag{12}$$

where, for some $\nu \in [0,1]$, $A_{ij}^+(a_x, a_y) = \frac{\partial^{i+j}}{\partial^i x \partial^j y} u(\nu x_0 + (1-\nu)a_x, \nu y_0 + (1-\nu)a_y) = D_{ij}^+ + o(1)$. Here, D_{ij}^+ is the limit of $\frac{\partial^{i+j}}{\partial^i x \partial^j y} u$, when $(x, y) \in \mathcal{L}$ tends to a given reference point, $a_0 \in \mathcal{C} \cup K$, from the left.

Now, if $a \in \mathcal{S}^{(n)} \bigcap K^-$, we consider $v = u + \frac{1}{p!} \sum_{i+j=p} [\frac{\partial^{i+j} u}{\partial^i x \partial^j y}](x - x_0)^i (y - y_0)^j$ where $[\frac{\partial^{i+j} u}{\partial^i x \partial^j y}]$ represents the jump of $\frac{\partial^{i+j}}{\partial^i x \partial^j y} u$ at a_C, and we get

$$\begin{aligned} u(a_x, a_y) &= \sum_{l=0}^{p-1} \frac{1}{l!} \sum_{i+j=l} \frac{\partial^{i+j}}{\partial^i x \partial^j y} u(x_0, y_0)(a_x - x_0)^i (a_y - y_0)^j \\ &+ \frac{1}{p!} \sum_{i+j=p} A_{ij}^-(a_x, a_y)(a_x - x_0)^i (a_y - y_0)^j, \end{aligned} \tag{13}$$

where $A_{ij}^- \longrightarrow D_{ij}^-$, the limit, when (x, y) tends to a_0 from the right, of $\frac{\partial^{i+j}}{\partial^i x \partial^j y} u$. We may assume that (12) holds for the m first elements of $\mathcal{S}^{(n)}$ while (13) is true for the $N(n) - m + 1$ last ones.

We want to evaluate $a_{i_0 j_0}$ for (i_0, j_0) such that $i_0 + j_0 = n$. By the rules of evaluating determinants, if we set $k_0 = N(n-1) + i_0 + 1$, we get (using the lexicographic ordering),

$$\begin{aligned} a_{i_0 j_0} &= \frac{1}{p!} \sum_{i+j=p} \frac{\det(R_1, \cdots, \widehat{U_{k_0}}, \cdots, R_{N(n)})}{\det(R_1, \cdots, R_{N(n)})} \\ &+ \frac{1}{p!} \sum_{i+j=p} \frac{\det(R_1, \cdots, \widehat{V_{k_0}}, \cdots, R_{N(n)})}{\det(R_1, \cdots, R_{N(n)})} \end{aligned} \tag{14}$$

with

(15)
$$\widehat{U_{k_0}} = \begin{pmatrix} D_{ij}^+ L_{\sigma_1}[(x - x_0)^i (y - y_0)^j] \\ \vdots \\ D_{ij}^+ L_{\sigma_m}[(x - x_0)^i (y - y_0)^j] \\ D_{ij}^- L_{\sigma_{m+1}}[(x - x_0)^i (y - y_0)^j] \\ \vdots \\ D_{ij}^- L_{\sigma_{N(p)}}[(x - x_0)^i (y - y_0)^j] \end{pmatrix}, \quad \widehat{V_{k_0}} = \begin{pmatrix} o(1)\ D_{ij}^+ L_{\sigma_1}[(x - x_0)^i (y - y_0)^j] \\ \vdots \\ o(1)\ D_{ij}^+ L_{\sigma_m}[(x - x_0)^i (y - y_0)^j] \\ o(1)\ D_{ij}^- L_{\sigma_{m+1}}[(x - x_0)^i (y - y_0)^j] \\ \vdots \\ o(1)\ D_{ij}^- L_{\sigma_{N(p)}}[(x - x_0)^i (y - y_0)^j] \end{pmatrix}$$

In $\widehat{V_{k_0}}$, $|o(1)| \leq \epsilon$. Both vectors $\widehat{U_{k_0}}$ and $\widehat{V_{k_0}}$ stand in the k_0th column of the determinants.

For the sake of simplicity, we denote by $\mu^{ij}_{i_0j_0}$ and $\nu^{ij}_{i_0j_0}$, for any of the (i,j), $i+j=p$, the coefficients obtained from those of the first sum of the right-hand side of equation (14) by replacing D^+_{ij} by 1 and D^-_{ij} by 0 for μ, and vice versa for ν. For the same reason, we also denote by $\mu^{ijk}_{i_0j_0}$ (resp. $\nu^{ijk}_{i_0j_0}$), for the same indices (i,j), the coefficient obtained from the same sum by replacing only the $k \leq m$th term (resp. $k > m$th term) D^+_{ij} by 1 (resp. D^-_{ij} by 1) and all the other terms by 0). Hence, we can write

$$|a_{i_0j_0} - \frac{1}{p!}\sum_{i+j=p}\left(D^+_{ij}\mu^{ij}_{i_0j_0} + D^-_{ij}\nu^{ij}_{i_0j_0}\right)| \leq \epsilon \sum_{i+j=p}\sum_{k=1}^{k=N(n)}\left(|\mu^{ijk}_{i_0j_0}| + |\nu^{ijk}_{i_0j_0}|\right).$$

We notice that $\mu^{ij}_{i_0j_0} + \nu^{ij}_{i_0j_0} = 0$ and that $\sum_{i+j=n}|\mu^{ijk}_{i_0j_0}|$, $\sum_{i+j=n}|\nu^{ij}_{i_0j_0}|$, $\sum_{i+j=n}|\mu^{ij}_{i_0j_0}|$ and $\sum_{i+j=n}|\nu^{ij}_{i_0j_0}|$ fulfill the inequalities (9) because C cuts K, so that

(16)

$$\begin{aligned}\gamma\, h^{p-n}\textstyle\sum_{i_0+j_0=n} \quad & \frac{1}{p!}\textstyle\sum_{i+j=p}|\nu^{ij}_{i_0j_0}| \leq h^{p-n}\sum_{i_0+j_0=n}|a_{i_0j_0}| \\ + \quad & \epsilon h^{p-n}\frac{1}{p!}\textstyle\sum_{i_0+j_0=n}\sum_{i+j=p}\sum_{k=1}^{k=N(n)}\left(|\mu^{ijk}_{i_0j_0}| + |\nu^{ijk}_{i_0j_0}|\right).\end{aligned}$$

By Lemma 1.7, the left-hand side of (16) is bounded above by $C_1\frac{\gamma}{p!}$ and the right-hand side is bounded by $h^{p-n}\sum_{i_0+j_0=n}|a_{i_0j_0}| + \frac{N(n)C_2}{(p-1)!}\epsilon$. This gives the result for Lagrange interpolation when ϵ is small enough.

In the case of approximation in the mean, the same kind of arguments may be applied because :

- if $C_a \subset K^+$ or K^-, there exists $\alpha_a \in C_a$ such that
$$< u >_{C_i} = u(\alpha_a)$$
- if C_a cuts C, there exist $\alpha_a^+ \in C_a \cap K^+$ and $\alpha_a^- \in C_a \cap K^-$ such that
$$< u >_{C_i} = \frac{\text{area}(C_a \cap K^+)}{\text{area}(C_a)}u(\alpha_a^+) + \frac{\text{area}(C_a \cap K^-)}{\text{area}(C_a)}u(\alpha_a^-).$$

□

This result enables us to detect the regions of smoothness from those where a jump in one of the derivatives occurs.

1.3. Some remarks on the practical calculation of the reconstruction. We intend to study the condition number of the linear system that provides the reconstruction. We will consider two kinds of polynomial expansion :

(1) the "natural" expansion : for any point $(x_0, y_0) \in \mathbb{R}^2$,

$$P(x,y) = \sum_{i+j\leq n} a_{ij}(x-x_0)^i(y-y_0)^j; \tag{17}$$

(2) an expansion using "barycentric" coordinates as follows : let $\mathcal{S}^{(n)} = \{\sigma_1, \sigma_2, \sigma_3, \ldots, \sigma_{N(n)}\}$ be an admissible stencil. Hence, at least one subset of three elements of $\mathcal{S}^{(n)}$ is an admissible stencil for $n = 1$, for example the set $\{\sigma_1, \sigma_2, \sigma_3\}$. We consider the three polynomials Λ_i of degree 1 defined by

$$L_{\sigma_j}(\Lambda_i) = \delta_i^j, \ 1 \leq i \leq 3, \ 1 \leq j \leq 3. \tag{18}$$

The symbol δ_i^j is the Kronecker symbol. Clearly, we have $\Lambda_1 + \Lambda_2 + \Lambda_3 = 1$. These polynomials are the barycentric coordinates of the triangle constructed on the elements of $\mathcal{S}^{(1)}$ for the Lagrange interpolation and on the gravity centers of σ_1, σ_2, and σ_3 for the approximation in the mean. In order to get expansion (17), a strategy may be to look first for the expansion of the polynomial P in terms of powers of Λ_2 and Λ_3 :

$$P = \sum_{i+j \leq n} a_{ij} \Lambda_2^i \Lambda_3^j, \tag{19}$$

and then to get the Taylor expansion of P at a point (x_0, y_0) in the convex hull of $\mathcal{S}^{(n)}$. (Theorems 1.1 and 1.8 give the behavior of the leading coefficients of P_u in whatever way the point is chosen in the convex hull of $\mathcal{S}$). The use of these barycentric coordinates is of course borrowed from Finite Element Methods [12].

In order to get the expansions (17) or (19), we have to solve linear $N(n) \times N(n)$ systems :

$$\mathcal{M}_n(a_{00}, \cdots, a_{0n})^T = U^T, \tag{20}$$

where the elements of $\mathcal{M}_n$ are $L_\sigma((x - x_0)^i (y - y_0)^j)$ for (17), and $L_\sigma(\Lambda_2^i \Lambda_3^j)$ for (19). We now study the properties of these linear systems.

Before starting this study, we note that an element of $\mathcal{P}^{\alpha,\beta}_{n,n}$ writes $(\lambda, \mathcal{S}^{(n)})$, where λ is ± 1. Consequently, in the sequel, we talk about stencils $\mathcal{S}^{(n)} \in \mathcal{P}^{\alpha,\beta}_{n,n}$, omitting λ.

1.3.1. *The case of expansion (16).* A very easy consequence of the inequality (11) is that :

Proposition 1.9. *Assume that the conditions of Lemma 1.7 hold for $p = n$, and let h be the supremum of the diameters of the circles containing the convex hull of $\mathcal{S}^{(n)}$. Then the condition number of system (20) is at least $O(h^{-n})$ for h small enough.*

Proof. For the sake of simplicity, we consider the following norm on $P_n[\mathbb{R}^2]$: for $P = \sum_{i+j \leq n} a_{ij}(x - x_0)^i (y - y_0)^j$, $||P|| = \sum_{i+j \leq n} |a_{ij}|$. On $\mathbb{R}^{N(n)}$, we consider the L^1 norm. Let U be a set of data for the right-hand side of (20), and consider the perturbation δU,

$$\delta U = (0, \cdots, \epsilon, \cdots)^T,$$

where ϵ is in the lth position, $l \geq N(n-1) + 1$. All the other entries of δU are zero. If we consider the function u defined on $\cup C_i$ by

$$(x, y) \in C_i, \ u(x, y) = \delta U_i,$$

we may apply Theorem 1.8. Hence, the perturbation δP has a norm satisfying

$$||\delta P|| \geq \sum_{i+j=n} |\delta a_{ij}| \geq C\frac{\epsilon}{h^n},$$

since $||\delta U|| = \epsilon$. This completes the proof. □

This fact is well known for 1D Lagrange interpolation and has motivated the search for more efficient algorithms, such as Newton's algorithm. There exist algorithms that generalize it [10, 11]. They involve numerous solutions of linear systems, so that we have preferred a more classical approach where the coefficients of the linear systems are obtained from the "barycentric" coordinate expansion (19).

1.3.2. *The case of expansion (18).* In the case of expansion (19), we have the following result :

Proposition 1.10. *Assume that the conditions of Lemma 1.7 holds for $p = n$, then the condition number of the system (20) for the expansion (19) is bounded above and below by constants independent on h, the supremum of the diameters of the circles containing the convex hull of $S^{(n)}$.*

Proof. The proof is also based on that of Lemma 1.7. As in Proposition 1.9, the only thing we have to do is to study the effect on the a_{ij}'s of a perturbation δU. We denote by P the solution of the problem for δU. The proof can be achieved in two stages :

(1) Let B be any invertible matrix. Consider the stencil

$$\widehat{S^{(n)}} = \{B[\sigma_i] + (x_0, y_0)\}_{1\leq i \leq N(n)}$$

for any (x_0, y_0). It is clear, from the definition of the Λ_i's, that $\widehat{\Lambda}_i(\widehat{x}, \widehat{y}) = \Lambda_i(x, y)$ if $(\widehat{x}, \widehat{y}) = B(x, y) + (x_0, y_0)$. Hence, the sum $S(P)$ of the absolute values of the coefficients of P in the basis $\Lambda_2^i\Lambda_3^j$ is the same as that for the development of P in the basis $\widehat{\Lambda}_2^i\widehat{\Lambda}_3^j$. This is a homogeneity property.

(2) Following the arguments of Lemma 1.7, $S(P)$ is bounded below and above, independently of B, hence independently of h. There exist $C_1 > 0$ and $C_2 > 0$ such that

$$C_1 \geq S(P) \geq C_2 > 0.$$

The constant C_2 is larger than zero because $\delta U \neq 0$.

This completes the proof. □

1.3.3. *Some final remarks.* These results indicate that it is better to evaluate the reconstruction with the help of barycentric coordinates. In §2, we propose a hierarchical algorithm for selecting the "best" stencil in a cell C. A sequence of stencils is constructed such that $\{C\} \subset S^{(1)} \subset \cdots \subset S^{(n)}$. It is possible to use this property in order to reduce the cost of the method. This is what we explain now.

The idea is, instead of looking directly for the coefficients of $P^{(n)}$, the approximation for $S^{(n)}$, to get first those of all of the $P^{(k)}$'s, the approximations for $S^{(k)}$, $k \leq n$, and then to construct those of $P^{(n)}$. In the ENO algorithm described in §2, this involves *no* extra cost and simplifies the evaluation of the a_{ij}'s. This has also the advantage of reducing the size of the linear systems and also to improve their condition numbers.

Assume that $P^{(1)}, \dots, P^{(p)}$ are known. We first get the coefficients of $P^{(p+1)} - P^{(p)}$,

$$P^{(p+1)} - P^{(p)} = \sum_{i+j \le p+1} a'_{ij} \Lambda_2^i \Lambda_3^j$$

by solving the linear system

$$\mathcal{M}_{p+1} \begin{pmatrix} \underline{a_1} \\ \underline{a_2} \end{pmatrix} = \begin{pmatrix} \mathcal{M}_p & B_{p,p+1} \\ C_{p,p+1} & D_{p,p+1} \end{pmatrix} \begin{pmatrix} \underline{a_1} \\ \underline{a_2} \end{pmatrix} = \begin{pmatrix} \underline{u_1} \\ \underline{u_2} \end{pmatrix}. \tag{21}$$

In equation (21), $\underline{a_1}$ (respectively $\underline{a_2}$) stands for the coefficients $\{a_{ij}\}_{i+j \le p}$ (resp. $\{a_{ij}\}_{i+j=p+1}$). The block matrices $\mathcal{M}_p$, $B_{p,p+1}$, $C_{p,p+1}$ and $D_{p,p+1}$ are defined according to this decomposition. In particular, we notice from the hypothesis, that $\mathcal{M}_p$ is invertible.

By assumption, we have $\underline{u_1} = 0$, so that the system (21) can be decomposed as

$$\begin{aligned} &\underline{a_1} = -\mathcal{M}_p^{-1} B_{p,p+1} \underline{a_2}, \\ &\left[-C_{p,p+1}\mathcal{M}_p^{-1}B_{p,p+1} + D_{p,p+1}\right] \underline{a_2} = \underline{u_2}. \end{aligned} \tag{22}$$

Since $\mathcal{S}^{(p+1)}$ is admissible, $E_p = \left[-C_{p\ p+1}\mathcal{M}_p^{-1}B_{p,p+1} + D_{p,p+1}\right]$ is also invertible, so that we can get $\underline{a_2}$, then $\underline{a_1}$ and finally the coefficients of $P^{(p+1)}$.

Simple manipulations show that

$$\mathcal{M}_{p+1}^{-1} = \begin{pmatrix} \mathcal{M}_p^{-1}B_{p,p+1}E_p^{-1}C_{p,p+1}\mathcal{M}_p^{-1} + \mathcal{M}_p^{-1} & -\mathcal{M}_p^{-1}B_{p,p+1}E_p^{-1} \\ -E_p^{-1}C_{p,p+1}\mathcal{M}_p^{-1} & E_p^{-1} \end{pmatrix}$$

so that we can quite easily look for the next step.

Finally, one should notice that the condition number of this method is always better than that of the original one, because it depends only on the condition number of parts of the original system.

2. Two ENO Reconstruction Techniques

In this section, we show how to generalize the ENO reconstruction methods to *unstructured triangular meshes.* In all what follows, we also assume that there is a procedure to construct a cell around each node of the mesh such that there is a one-to-one correspondence between the nodes of the mesh and the cells. This point is essential because the heuristic method we propose to look for the "smoothest" stencil basically relies on the connectivity of the mesh. The construction of the cells (or control volumes) is done in such a way that the general assumptions of §1 are true. An example is the following. Let $\mathcal{T}$ be a triangulation of Ω, a domain of $\mathbb{R}^2$. Around each node A_i of $\mathcal{T}$, we construct the control volume C_i whose boundary is the segment joining the centroids G_{ijk} of the triangles (A_i, A_j, A_k) having A_i as a vertex and the midpoints of the segments of those triangles (see Figure 1).

As suggested by the results of §§1.1 and 1.2, we must deal with meshes where the stencils we need satisfy the property of Theorems 1.1 and 1.8. This will be the case for most of them (and in practice for all of them). We will always consider a locally regular function u that fulfills the assumptions of these theorems. The results we have obtained in §§1.1 and 1.2 can be summarized as follows : If the a_{ij}'s are the coefficients of the Taylor expansion of the approximation polynomial of total degree n defined on an admissible stencil $\mathcal{S}^{(n)}$, then

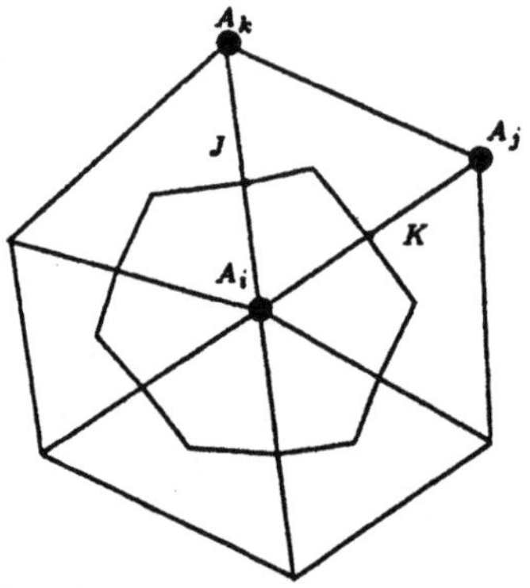

FIGURE 1. Control volume around the node A_i

- if u is smooth in the convex hull K of $\mathcal{S}^{(n)}$, $\sum_{i+j=n} |a_{ij}|$ remains finite;
- if u admits continuously differentiable derivatives only up to the order $k < n$ in the convex hull K, then $\sum_{i+j=n} |a_{ij}|$ becomes unbounded as the diameter of K tends to zero.

2.1. The selection of the "smoothest" stencil . The ENO reconstruction we propose utilizes a recursive selection algorithm to get the "smoothest" stencil, following ideas close to those of [3, 4]. The same method is applied to Lagrange interpolation as well as to the reconstruction in the mean. In the latter case, we construct a connectivity on the cells : we identify the cell C_{M_i} constructed around the node M_i, so that if $(M_{i_1}, M_{i_2}, M_{i_3})$ is a triangle of the original mesh, $(C_{M_{i_1}}, C_{M_{i_2}}, C_{M_{i_3}})$ is a "triangle" of the "mesh" formed with the cells. To simplify the description, since the correspondence between the nodes of $\mathcal{T}$ and the cells is one-to-one, we use the word "node" instead of "cell" when this algorithm is applied to the reconstruction in the mean.

For any node A_i, we consider the following inductive procedure :

(1) If $n = 1$, let $\{T_i\}$ be the set of triangles of $\mathcal{T}$ having A_i as a vertex. For each T_i, we consider the linear interpolation of u on this triangle. The triangle $T_{\min}$ is the one where the sum $\sum_{i+j=1} |a_{ij}|$ is minimal. The stencil $\mathcal{S}^{(1)}$ is the set of the nodes of $T_{\min}$.
(2) If $n > 1$, let $\mathcal{S}^{(n-1)}$ be the stencil defined by the previous step. Consider all the nodes surrounding $\mathcal{S}^{(n-1)}$ in $\mathcal{T}$ and consider all the stencils obtained from $\mathcal{S}^{(n-1)}$ by adding $n+1$ of the nodes surrounding $\mathcal{S}^{(n-1)}$. The stencil $\mathcal{S}^{(n)}$ is the stencil that minimizes $\sum_{i+j=n} |a_{ij}|$.

We have intentionally left the second point imprecise because it is obvious that the number of stencils to consider is in general huge. We have to define criteria for selecting the "good" from "bad" stencils. These criteria are essentially heuristic and a priori ones.

A first criterion that seems natural is that the stencil should be as compact as possible. Another one is to achieve the smallest possible number of stencils, but the potential stencils must not be confined to a particular angular area of the plane, in order not to favor any direction. Last, the node (or the cell) around which we want to reconstruct should belong to all stencils.

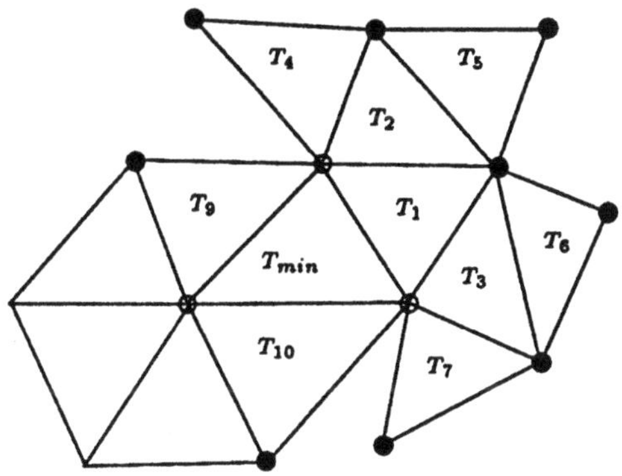

FIGURE 2. Some possible interpolation points. Circles : points of $T_{\min}$ (second-order interpolation), black circles : points that may be added to obtain a stencil for third-order interpolation.

With this in mind, two possible sets of stencils for third-order interpolation are (see Figure 2) :

- the nodes of the triangle $T_{\min}$ plus, for each of its edges, the three additional nodes of triangles T_1, T_2, T_3. This leads to a maximum of three stencils per triangle;
- the nodes of triangles consisting of $T_{\min}$ plus, for each of its edges, the three additional nodes of triangles T_1, T_2, T_3, or T_1, T_2 and T_4 or T_5, or T_1, T_3 and T_6 or T_7, or T_1, T_9 or T_{10}, and one of the six triangles $T_2, T_3, T_4, T_5, T_6, T_7$.

The second solution leads to a maximum number of 52 stencils once $T_{\min}$ has been found. We have made several tests to evaluate the "performance" of each type of stencil. They are given in §3 and indicate that the first kind of stencil is generally sufficient.

These experimental results suggest the following method for selecting stencils for higher order of accuracy (of course, in this case, the problem of choosing the smallest number of possible stencils is much more critical than for third-order interpolation !).

To begin with, we notice that the six points of $T_{\min}$, T_1, T_2 and T_3, denoted by $\mathcal{S}^{(3)}$ in the sequel, form a set isomorphic to the standard stencil for P_2 interpolation (see Figure 3). Now, this stencil can be embedded in the standard stencil for P_3 interpolation (Figure 3) where the additional points 7, 8, 9, 10 are the new points that one has to introduce.

Consider a "side" of $\mathcal{S}^{(3)}$, say the points 4, 5, 6. Since the mesh is conformal, the segment 4-5 is the edge of two triangles in general, T_1 and a new one, T_4. The same is true for the segment 5-6, and a new triangle T_5 is introduced. Then the sides of these new triangles are the edges of two triangles (except at the boundary) ; for T_4 they are T_6 and T_8 and for T_5, they are T_7 and T_9. At most four new triangles, T_6, T_7, T_8, T_9 are introduced, and the new stencils are made of $\mathcal{S}^{(3)}$ and the points of $\{T_4, T_6, T_8, T_5\}$ or $\{T_4, T_5, T_9, T_7\}$ or $\{T_4, T_8, T_9, T_5\}$ when $T_8 \neq T_9$. These two or three new stencils are isomorphic to the standard P_3 stencil as shown in Figure 3. This new procedure may be repeated for the three sides of $\mathcal{S}^{(3)}$, and so at most 12 stencils must be considered for $\mathcal{S}^{(3)}$.

This number can even be reduced if $\mathcal{T}$ is constructed from a "coarser" triangulation $\mathcal{T}'$ by adding to the nodes of $\mathcal{T}'$ the nodes needed for the classical Lagrange P_3 interpolation [12]. With this trick, only three stencils are generally constructed from a stencil $\mathcal{S}^{(2)}$. This idea can be generalized to higher degree and is used in the numerical tests.

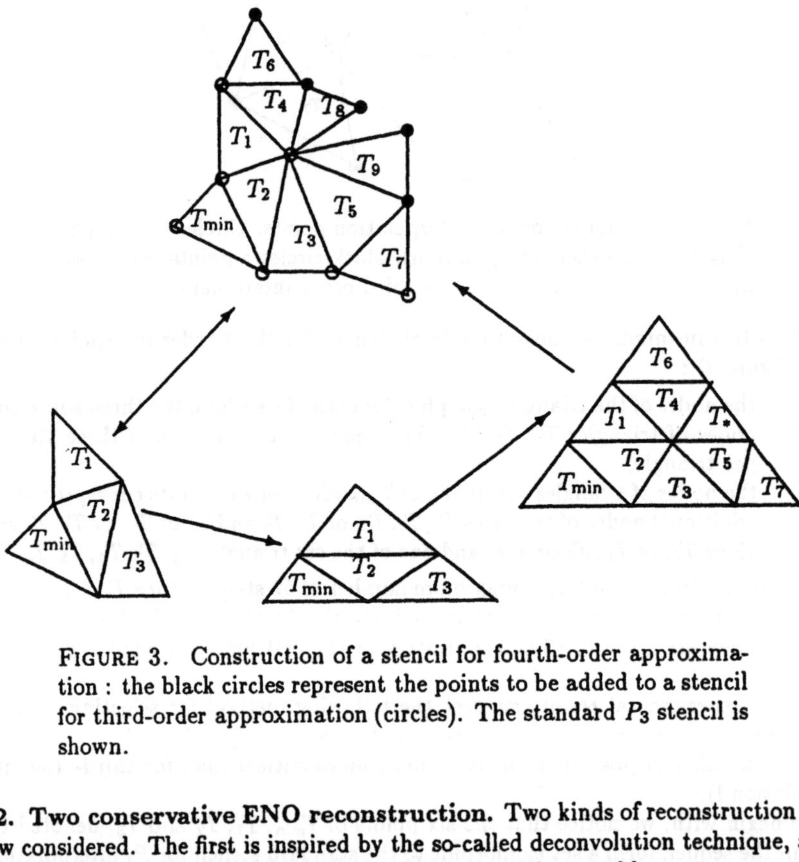

FIGURE 3. Construction of a stencil for fourth-order approximation : the black circles represent the points to be added to a stencil for third-order approximation (circles). The standard P_3 stencil is shown.

2.2. Two conservative ENO reconstruction. Two kinds of reconstruction are now considered. The first is inspired by the so-called deconvolution technique, and a second is based on approximation in the mean. They enable us to compute a function that is polynomial (of degree n) in each control volume. This function is generally discontinuous at the edges of each cell.

2.2.1. *Deconvolution technique revisited* . We start by briefly recalling the RD method in $\mathbb{R}$ (see [4] for more details). If $(x_i)_{i\in\mathbb{N}}$ is a regular mesh of $\mathbb{R}$ and u is a real-valued function on $\mathbb{R}$, the reconstruction by the deconvolution technique consists of applying the selection algorithm not to u but to

$$\bar{u}_i(y) = \frac{1}{x_{i+1} - x_i} \int_{x_{i-1/2}}^{x_{i+1/2}} u(x + y - x_i)dx, \tag{23}$$

where, as usual, the mesh size $\Delta x_{i+1/2} = x_{i+1} - x_i$ is constant and $x_{i+1/2} = (x_i + x_{i+1})/2$. In particular, we see that the function $\overline{u}_i$ does not depend on i and that $\overline{u}(x_i)$ is the average of u on $[x_{i-1/2}, x_{i+1/2}]$. Let $\Pi_L(\overline{u})$ be the $(n+1)$st order Lagrange interpolation of $\overline{u}$. The idea is to perform a Taylor expansion of $\overline{u}$ and its successive derivatives around x_i, to truncate each kth derivative at order $n-k$, then replace the values $\overline{u}, \cdots, \overline{u}^{(n)}$ by $\Pi_L(\overline{u})$ and its n successive derivatives. This gives a triangular linear system of order n whose solution provides an approximation of the successive derivatives of order $\leq n$ of u evaluated at x_i.

The fundamental reason why one achieves the expected order of accuracy is because all polynomials of degree $\leq$ n are left invariant by the method. They are left invariant by the construction because the shape of the control cells does not change from one point to another. If this were not the case, i.e., if $\Delta x_{i+1/2} = x_{i+1} - x_i$ were not constant, the formula (23) would indeed depend on the point x_i and this property would be lost. To show this, we simply consider $u(x) = x$ and a $(n+1)$st-order interpolation that has values $\overline{u}_i$ at points x_i. Assuming that $\{x_0, x_1, \cdots\}$ is the stencil selected by the ENO algorithm, we have

$$\Pi_L(\overline{u}) = \overline{u}_0 + K(x - x_0) + \cdots, \text{ where } K = \frac{1}{2} + \frac{\Delta x_{3/2} + \Delta x_{-1/2}}{\Delta x_{1/2}}.$$

For u to be invariant, we should have

$$\Pi_L(\overline{u})(x_0) = u(x_0) + \frac{1}{8}\frac{\Delta x_{1/2}^2 - \Delta x_{-1/2}^2}{\Delta x_{1/2}} u'(x_0), \quad \Pi_L(\overline{u})'(x_0) = K = u'(x_0).$$

The second equation indicates that one must have $K = 1$, which is, in general, not true.

To overcome this problem, we propose the following technique : to reconstruct in the cell around the node x_0, instead of considering the convolution product of u and the characteristic function of C_i divided by its area, apply the ENO to

$$\overline{u}(y) = \frac{1}{\text{area}(C_{\mathcal{S}^{(n)}})} \int_{C_{\mathcal{S}^{(n)}}} u(x + y - x_0) dx, \tag{24}$$

where $\mathcal{S}^{(n)}$ is any possible stencil around x_0 that we have to test, and $C_{\mathcal{S}^{(n)}}$ is the union of the control volumes of each node in $\mathcal{S}^{(n)}$:

$$C_{\mathcal{S}^{(n)}} = \bigcup_{C_M \in \mathcal{S}^{(n)}} C_M. \tag{25}$$

The problem is to evaluate the integral (24) from the average value of u. This cannot generally be achieved. Nevertheless, this is possible for the polynomials of $P_n(\mathbb{R}^2)$, at least in general. The condition is that the Vandermonde-type determinant associated with the approximation in the mean does not vanish. If this condition is satisfied, then we can find coefficients $\alpha_l(y)$, $1 \leq l \leq N(n)$, so that

$$\frac{1}{\text{area}(C_{\mathcal{S}^{(n)}})} \int_{C_{\mathcal{S}^{(n)}}} u(x + y - x_0)\, dy = \sum_{l=1}^{N(n)} \alpha_l(y) < u >_{C_l}, \tag{26}$$

when u belongs to $P_n(\mathbb{R}^2)$. If $u \notin P_n(\mathbb{R}^2)$, the equation (26) gives a $(n+1)$st-order quadrature formula. With all this, we get the following proposition whose proof is obvious.

Proposition 2.1. *The algorithm defined by the ENO technique with (24), (25),(2) and (26) leaves invariant the polynomials of $P_n(\mathbb{R}^2)$ and hence gives a $(n+1)$st-order approximation of smooth functions. Moreover, this approximation is conservative up to the quadrature formula (26).*

2.2.2. *A formally conservative reconstruction.* The second reconstruction technique is obtained by using the approximation in the mean on u directly instead of the Lagrange interpolation with a modified deconvolution procedure. The "smoothest" stencil is computed with the method of §2.1 when the word "node" is understood in the sense of control volume. This reconstruction procedure is conservative by construction, because for any cell C_M, this cell belongs to each stencil considered in the algorithm of §2.1. A proposition similar to Proposition 2.1 can be formulated.

This second method is less expensive (in term of CPU cost) than the previous one because it does not involve additional solution of linear systems for recovering conservativity. Its cost is the same as a Lagrange interpolation with random stencils.

3. Numerical examples

We have performed several tests on the second-, third- and fourth-order ENO interpolation and ENO reconstruction, but we only report the third- and fourth-order results since they are a priori more challenging. In particular, we intend to check numerically that the expected order of accuracy is in fact attained for smooth functions.

In all these examples, the control volumes are those given in Figure 1. The practical calculations of the averages in these control volumes have been performed with a 5th-order quadrature formula [12].

All error estimates have been obtained on irregular meshes, as the one shown on Figure 4. These meshes are obtained by random perturbations of regular structured meshes. The set of points that we obtain is triangulated to obtain a Delaunay triangulation. The main difference between such a mesh and the regular structured one is that the number of triangles each node belongs to, may change from one node to another. We also have done the same tests with regular meshes, and we have not seen any degradation of convergence.

The tests on smooth functions are performed on $u(x,y) = \cos(2\pi(x^2+y^2))$. The locally smooth function we have chosen is obtained by a modification of that used by Harten in [4] for example : if ϕ is any angle, let f_ϕ be

$$\left.\begin{array}{l} \text{if } r \le -\frac{1}{3},\ f_\phi(x,y) = -r\sin\left(\frac{\pi}{2}r^2\right) \\[2ex] \text{if } r \ge \frac{1}{3},\ f_\phi(x,y) = 2r - 1 + \frac{1}{6}\sin(3\pi r) \\[2ex] \text{if } |r| < \frac{1}{3},\ f_\phi(x,y) = |\sin(2\pi r)| \end{array}\right\} \text{ where } r = x + \tan(\phi)y, \tag{27}$$

and let u be

$$\text{(28)} \qquad \begin{aligned} &\text{if } x \leq \tfrac{1}{2}\cos(\pi y), \; u(x,y) = f_{\sqrt{\pi/2}}(x,y), \\ &\text{if } x > \tfrac{1}{2}\cos(\pi y), \; u(x,y) = f_{-\sqrt{\pi/2}}(x,y) + \cos(2\pi y). \end{aligned}$$

The function u defined by (27)-(28) shows discontinuities in the function itself and its first-order derivatives; some of the discontinuities are straight lines (never aligned to the mesh), one is a curved line where the jump changes from one point to another. Last, the behavior of u is basically one-dimensional on the left of the curve $x = \cos(\pi y)/2$ and trully two-dimensional on the right. A plot of this function is given in Figure 5.

No visible difference can be observed between the two kinds of reconstruction (directly from mean values and by deconvolution), except in the cost of each algorithm. Because of that, most of the results if not specified otherwise, are presented from the "reconstruction in the mean" technique.

For the third-order results, two types of stencils (as presented in §2.1) have been tested. The use of the second type of possible stencils results in a much more expensive approximation (in general, we must test 52 stencils per triangle versus only 3 in the simplest version) and the results have *never* been improved. All the results that are presented below have been obtained with the 3-stencil version of the method. More generally speaking, all the results presented here are obtained with the simpliest version of the selection algorithm.

Results on $u(x,y) = cos(2\pi(x^2 + y^2))$. We have displayed in Figure 6 the L^∞ error of the third-order ENO interpolation (solid line) and the third-order ENO conservative reconstruction (solid line with circles) The dashed line indicates the slope -3. The same tests have been performed for the fourth-order interpolation and reconstruction of the smooth function. Figure 7 shows the L^∞ error of the ENO reconstruction for random meshes. One can see that the expected order of accuracy is indeed attained.

Results on the nonsmooth function. In Figures 8(a) and 9(a), we have displayed the node values of the third-order ENO reconstruction for two meshes (1600 nodes and 6400 nodes). Figures 8(b) and 9(b) show pictures of the fourth-order ENO reconstruction. A better capture of the area surrounding the triple points is the only visible difference between third- and fourth- order reconstruction. There are no oscillations in the reconstruction. To better see the behavior of both approximation techniques, we also show the cross section on the line $y = 0$ (Fig. 10). It is obtained from the 1600-nodes mesh by the deconvolution technique, on 200 regularly spaced points. Where the function is smooth, we should recover the asymptotic order of convergence obtained for a smooth function. In order to check this, we have computed the error between the reconstruction and the exact function u at different points of the line $y = 0$, namely at $x = 0.4, 0.2, 0.1, 0, -0.2, -0.5, -0.75$: there is at least one point in each region playing a role in the definition of u. The results are presented in Figure 11 for fourth-order conservative reconstruction. We get what is expected. In particular, the point $x = 0$ is on the line where u is continuous but its first-order derivative has a jump, so that only a first-order approximation is obtained in any case. These results have been obtained in single precision.

To end this subsection, we must note that the algorithm for choosing the stencils may lead to some difficulties at the boundaries, as can be seen in Figure 5 on the left upper corner : the most left upper triangle of the mesh (Figure 4) does not admit any additional points of the type we consider to make a stencil.

4. Conclusion

We have developed and analyzed three methods for the reconstruction of a function admitting discontinuities only on regular planar curves from their node values or from their averages in control volumes that surround them. They are based on two polynomial approximations of real-valued functions, the Lagrange interpolation and what we have called the approximation in the mean. In order to give a firm basis to the Essentially Nonoscillatory Interpolation and Reconstruction, we have studied the behavior of the highest order coefficients of these two polynomial approximations of smooth functions and nonsmooth ones for which the discontinuities lie on regular curves. We have also given an adaptation of the so-called "reconstruction via deconvolution" to irregular triangulated meshes. The approximation in the mean leads to a very efficient ENO reconstruction that seems to be the cheapest one. We also propose a heuristic algorithm for selecting the smallest number of possible stencils.

These techniques have been shown to work very well on smooth and nonsmooth functions. In particular, we have shown on typical examples that the minimum number of possible stencils was sufficient for our purpose. We have not been able to make a rigorous study of the quality in term of convergence of these new approximations ; that study is clearly needed. In a future paper, we will give applications of these new approximations for deriving new algorithms for compressible Fluid Mechanics.

Acknowledgements : I would like to thanks José Grimm, from Projet Safir of INRIA who has suggested the second counterexample of §§1.2. I am also very much indebt of the numerous criticisms of Tamir Tassa, Pr. W. Gautchi and an unknown referee.

References

1. H.C. Yee, *Upwind and Symmetric Shock-Capturing Schemes.* Technical Report-TM-89464, NASA, May 1987.
2. L. Fezoui and B. Stoufflet, A Class of Implicit Upwind Schemes for Euler Simulations with Unstructured Meshes. *J. of Comput. Phys.*, 84(1989), 174-206.
3. A. Harten and S. Osher, Uniformly High-Order Accurate Nonoscillatory Schemes I, *SIAM J. of Numer. Anal.*, 24(1987), 279-309.
4. A. Harten, B. Engquist, S. Osher and S. R. Chakravarthy, Uniformly High Order Accurate Essentially Non-Oscillatory Schemes III, *J. of Comput. Phys.*, 71(1987), 231-303.
5. C.W. Shu and S. Osher, Efficient Implementation of Essentially Non-Oscillatory Shock-Capturing Schemes, *J. of Comput. Phys.*, 77(1988), 439-471.
6. A. Harten and S. R. Chakravarthy, Multi-dimensional ENO Schemes for General Geometries, CAM Report 91-16, UCLA, August 1991.
7. J. Casper and H.L. Atkins, Finite-Volume Application of High Order ENO Schemes to Two-Dimensional Boundary-Value Problems, Submitted to *J. of Comput. Phys.*.
8. T. J. Barth and P.O. Frederickson, High Order Solution of the Euler Equations on Unstructured Grids using Quadratic reconstruction, AIAA Paper No 90-0013, 28 th Aerospace Science Meeting, January 8-11, Reno, Nevada.

9. K.C. Chung and T. H. Yao, On Lattrices Admitting Unique Lagrange Interpolations, *SIAM J. of Numer. Anal.*, 14(1977) 735-743.
10. G. Muhlbach, The General Recurrence Relation for Divided Differences and the General Newton-Interpolation-Algorithm with Applications to Trigonometric Interpolation. *Numer. Math.*, 32(1979), 393-408.
11. G. Muhlbach, The General Neville-Aitken-Algorithm and Some Applications. *Numer. Math.*, 31(1978), 97-110.
12. G. Strang and G. J. Fix, An analysis of the finite element method. Prentice-Hall, Englewood Cliffs, N.J., 1973
13. P.G. Ciarlet and P. A. Raviart, General Lagrange and Hermite interpolation in R^n with application to Finite Element Methods. *Arch. Rational Mech. and Anal.*, 42(1972), 177-199.

INRIA, BP 93, 06902 SOPHIA ANTIPOLIS CEDEX, FRANCE
E-mail address: abgrall@sophia.inria.fr

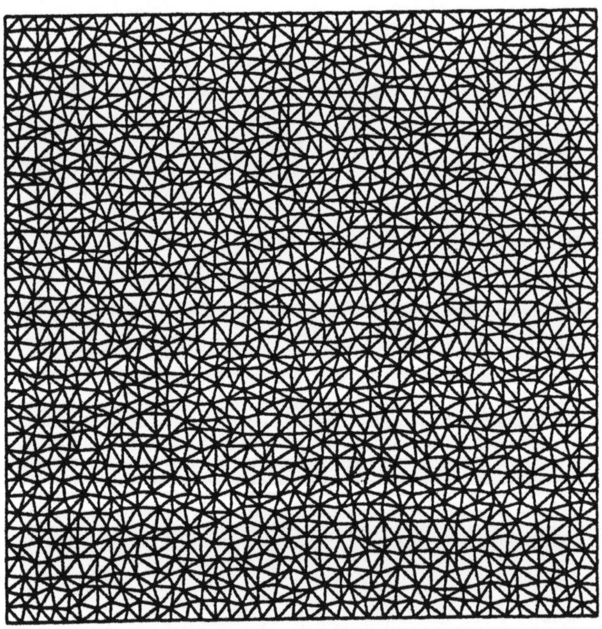

FIGURE 4. Typical mesh : 1600 nodes, 3042 triangles

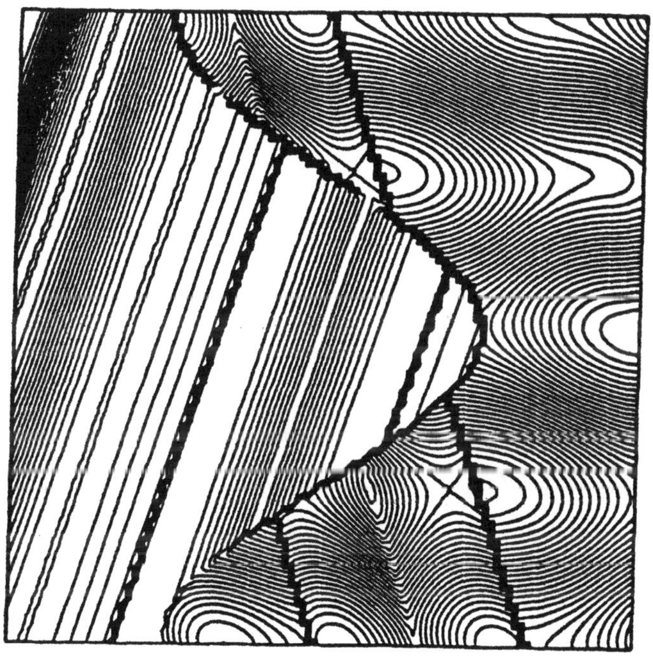

FIGURE 5. Exact function : Min=-1.331, Max=2.650, $\delta = 0.1$

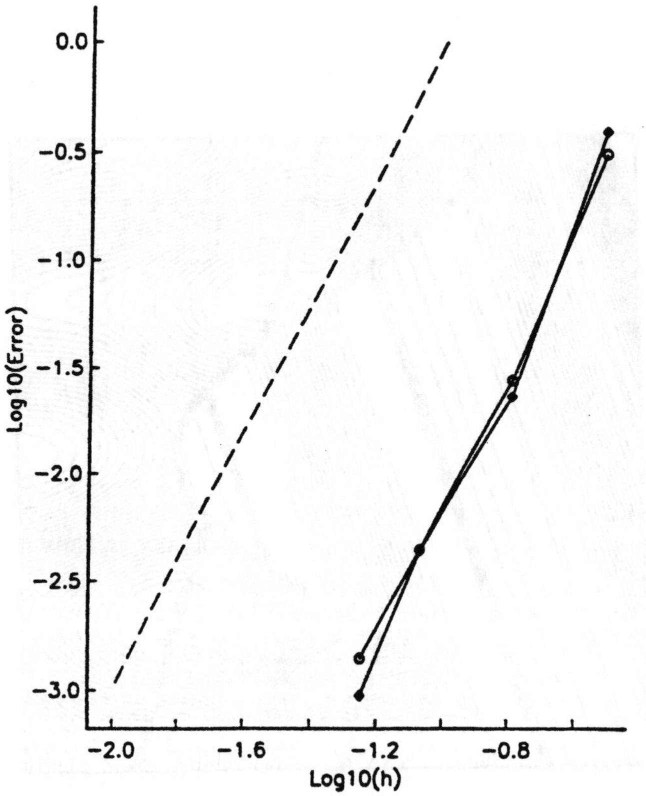

FIGURE 6. L^∞ error for $f(x,y) = cos(2\pi(x^2 + y^2))$. Squares : ENO interpolation only, Circles : ENO + reconstruction, Dashed line : slope -3

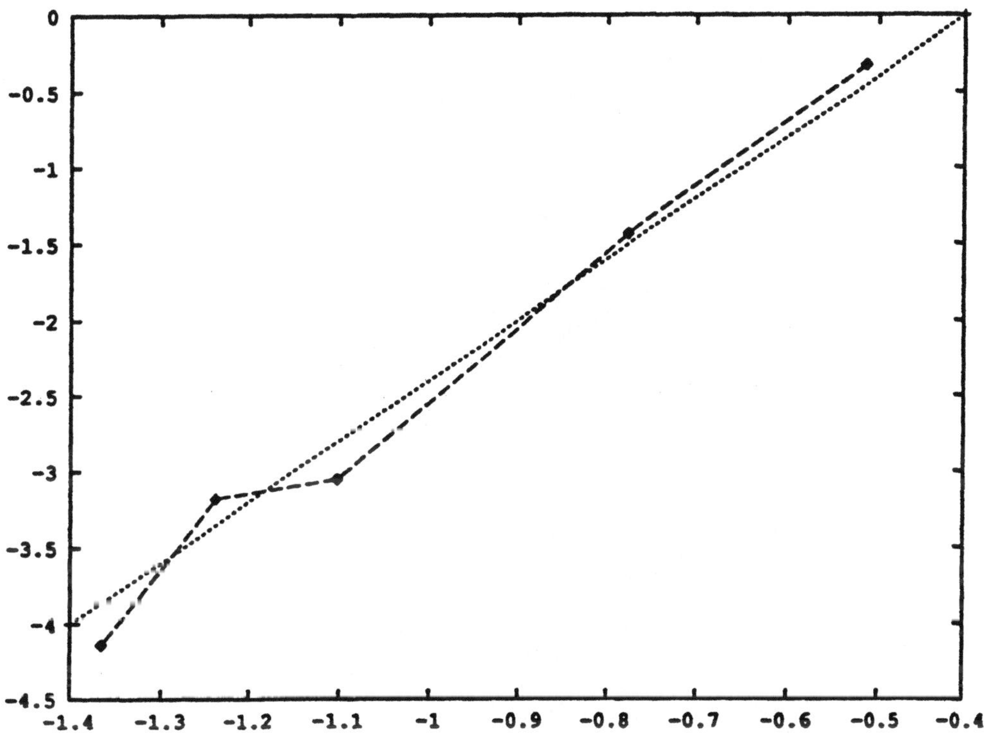

FIGURE 7. L^∞ error for $f(x,y) = cos(2\pi(x^2+y^2))$. Squares : Dashed line : slope -4, Dashed+squares : ENO + reconstruction

(a) Third-order

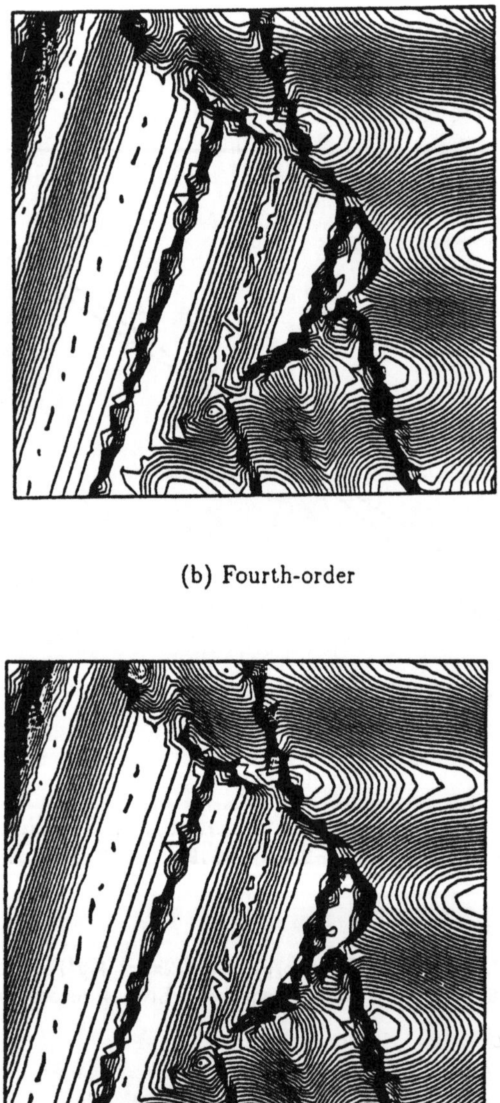

(b) Fourth-order

FIGURE 8. ENO reconstruction "in the mean" with the 1600-nodes mesh ; Min=-1.30, Max=2.6, $\delta = 0.1$

(a) Third order

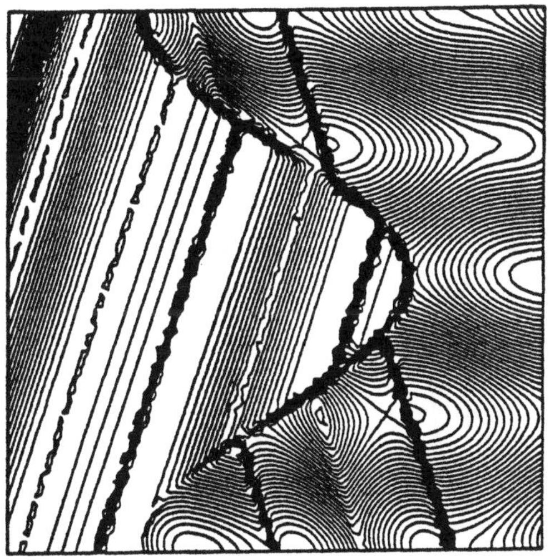

(b) Fourth order

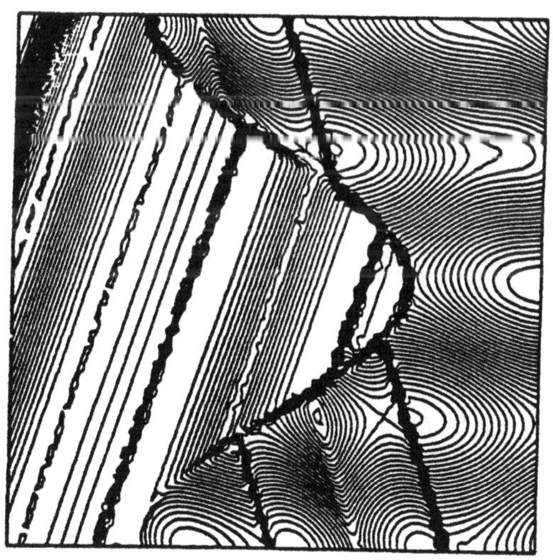

FIGURE 9. ENO reconstruction "in the mean" with the 6400-nodes mesh ; Min=-1.30, Max=2.6, $\delta = 0.1$

(a) E.N.O. interpolation.

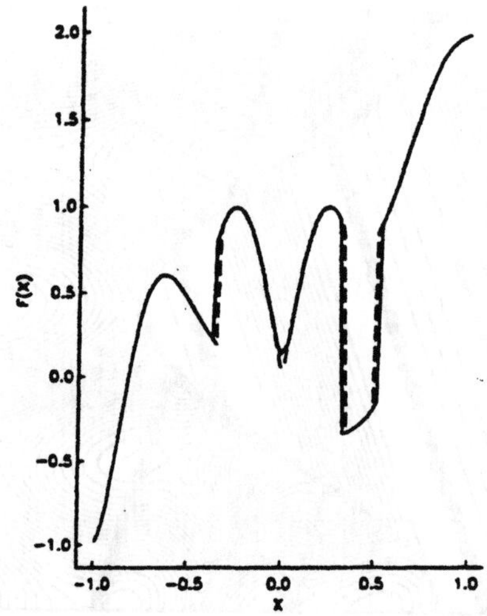

(b) E.N.O. reconstruction.

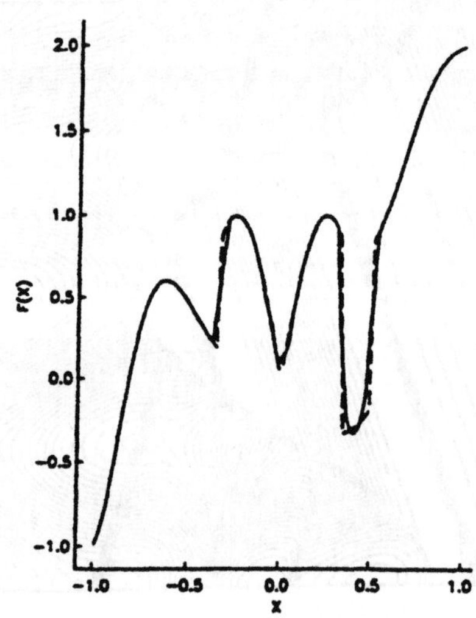

FIGURE 10. Cross section at $Y = 0$ of the ENO interpolation (a) and the ENO reconstruction (b) for the 1600-nodes mesh.

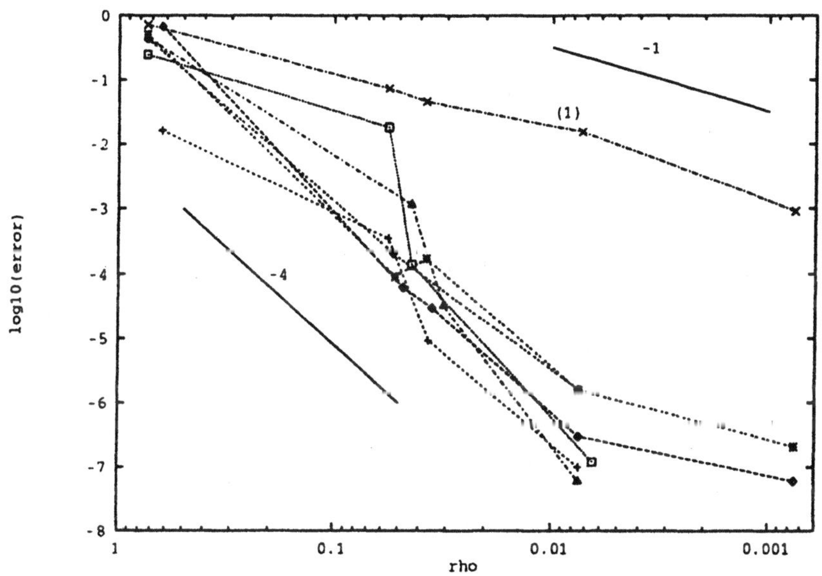

FIGURE 11. Convergence history for the points $x = 0.4, 0.2, 0.1, 0, -0.2, -0.5, -0.75$ and $y = 0$; the line corresponding to $x = 0$ is marked by (1)

A CONTRIBUTION TO THE GREAT RIEMANN SOLVER DEBATE

JAMES J. QUIRK

Institute for Computer Applications in Science and Engineering, NASA Langley Research Center, Hampton, VA 23681, U.S.A.

SUMMARY

The aims of this paper are threefold: to increase the level of awareness within the shock-capturing community of the fact that many Godunov-type methods contain subtle flaws that can cause spurious solutions to be computed; to identify one mechanism that might thwart attempts to produce very-high-resolution simulations; and to proffer a simple strategy for overcoming the specific failings of individual Riemann solvers.

KEY WORDS Riemann solvers Shock waves Numerical artifacts

1. INTRODUCTION

Over recent years a plethora of shock-capturing schemes have been developed for the Euler equations of gas dynamics. During this period it has emerged that one of the more successful strategies for designing a shock-capturing scheme is to follow Godunov's lead[1] and utilize a classic initial value problem known as a Riemann problem.[2] Godunov assumed that a flow solution could be represented by a series of piecewise constant states. Thus the numerical representation closely approximates the true solution near discontinuities, and regions of smooth flow are reasonably well approximated by a series of step functions. He evolved this discretized flow solution by considering the non-linear interactions between its component states. Viewed in isolation, each pair of neighbouring states constitutes a Riemann problem, the solution to which may be found exactly.[3] The results from these separate Riemann problems may then be averaged so as to advance the flow solution through some time increment. Because it mimics much of the relevant physics, Godunov's scheme results in an accurate and well-behaved treatment of shock waves.

Although it provides the bedrock upon which most modern schemes are built, in its original form Godunov's method is of limited use. Firstly, the scheme proves to be highly dissipative and so requires an inordinately fine mesh to resolve complex shock-on-shock interactions. Secondly, since a Riemann problem has no closed form solution and can only be solved by some iterative method, Godunov's scheme is significantly more expensive than schemes which employ ordinary finite difference operators.

One of the first people to address this second shortcoming was Roe.[1] He argued that since the Riemann problems associated with Godunov's method arise from an approximation of the data, it might be sufficient to find only approximate solutions to these Riemann problems, provided that they still describe important, non-linear behaviour—his motivation being that approximate solutions can be computed much more cheaply than exact solutions. Thus the

Received 2 November 1992

industry of designing approximate Riemann solvers was born.[4–7] Now, whilst Godunov-type schemes are often held up to be models of robustness, they can on occasions fail quite spectacularly. For example, when computing shock reflection problems, Roe's method can sometimes go awry by admitting solutions for which the Mach stem is inexplicably kinked. The existence of such failings partly explains why no consensus of opinion has been reached concerning the ideal Riemann solver. Whenever a new failing is unearthed, it adds fuel to the great Riemann solver debate: method X is better than method Y because of reasons A, B and C. It is our contention that for all the current crop of Riemann solvers at least one set of circumstances may be found for which any one solver is found wanting; some failings are just more visible than others.

In Section 2 we catalogue a number of situations in which anomalous behaviour is known to occur. This catalogue should serve to increase the general level of awareness within the shock-capturing community of the current limitations of Riemann solver technology. At present this awareness is not all that it should be: there are many instances in the literature where suspect numerical results are presented with either little or no adverse comment. We believe that one of the failings listed in our catalogue has hitherto gone unreported. In Section 3 we proffer a diagnosis of the mechanism which causes this new failing.

At this juncture it should be noted that any foibles that a specific Riemann solver might have may usually be controlled by the judicious use of a small amount of artificial dissipation. Indeed, it is worth pointing out that Colella and Woodward's PPM scheme,[8] which has proved itself to be more robust than most Godunov-type schemes, does in fact employ an elaborate artificial dissipation mechanism to supplement the dissipation provided via upwinding. As will be described in Section 4, we favour a strategy whereby the weaknesses of any one solver are overcome by combining it with one or more complementary solvers. The main advantages of this approach over that of adding artificial dissipation are twofold. Firstly, it does not degrade the resolution of the base Riemann solver; it is possible to control certain instabilities by changing the flavour of the dissipation mechanism rather than increasing the absolute level of dissipation. Secondly, it does not necessitate a host of tunable parameters and so this synergetic strategy does not negate the principal advantage of Godunov-type schemes over other shock-capturing methods. Of course, we are left with the difficulty of deciding when to use one Riemann solver in preference to another; however, we present a number of computations which suggest that this difficulty is not particularly bothersome.

Finally, in Section 5 we list the main conclusions that we have drawn from this work. Note that in this paper we do not address the first shortcoming of Godunov's method, namely its low resolution. Following van Leer,[9] it is assumed that a high-order extension to a first-order method can always be achieved by preprocessing the data supplied to the Riemann solver.

2. A CATALOGUE OF FAILINGS

We now present several instances where various Riemann solvers are known to give unreliable results. While most of these cases will be known to the aficionado, we believe that one of the cases which we are about to describe has hitherto gone unreported in the literature. Although our catalogue is not exhaustive, we hope that it might save some investigators from the harrowing experience of spending weeks or even months searching for coding errors that simply do not exist.

All the computational results shown in this section are for first-order schemes. Since such methods have low resolution, our calculations employed relatively fine meshes; for clarity, grids are drawn using every other fourth grid line.

2.1. *Expansion shocks*

By far the most widely investigated failing is that some Riemann solvers do not satisfy an 'entropy condition', such schemes can admit non-physical solutions such as expansion shocks. Osher[4] has found a general condition for a scheme to be entropy-satisfying when applied to scalar equations and he designates such schemes E-schemes. At present, however, any extension to a system of equations contains a large amount of empiricism and must therefore remain suspect. Indeed, Godunov's method is classified as an E-scheme but, as observed by Woodward and Colella,[10] it can give rise to nearly discontinuous expansion fans near sonic points. The density contours shown in Figure 1 illustrate this deficiency of Godunov's method quite clearly. These results are taken from the diffraction of a strong shock wave, $M_s = 5{\cdot}09$ with $\gamma = 1{\cdot}4$, around a 90° corner.

In its basic form Roe's scheme is another solver that admits expansion shocks; however, several fixes have been proffered which cure Roe's scheme of this particular affliction.[11–13] Such fixes are typical of the way in which Riemann solver deficiencies have been countered up to now. Whilst this strategy has proved reasonably successful, it has a number of drawbacks. Sometimes a fix uses a parameter which must be retuned between problems and hence one of the major advantages of Riemann-based schemes over, say, artificial dissipation schemes is lost. Alternatively, a scheme may require more than one fix and it may be unclear how the different fixes interact with one another.

2.2. *Negative internal energies*

Another situation in which some Riemann solvers are found wanting occurs whenever the dominant energy mode is kinetic rather than thermal. For such solvers the kinetic energy computed from a numerical approximation to the conservation laws of mass and momentum can exceed the total energy computed via an approximation to the conservation law of energy. Thus they can yield negative internal energies and hence negative pressures, which cause the scheme to fail. Einfeldt *et al.*[12] call any scheme which can be guaranteed not to yield negative

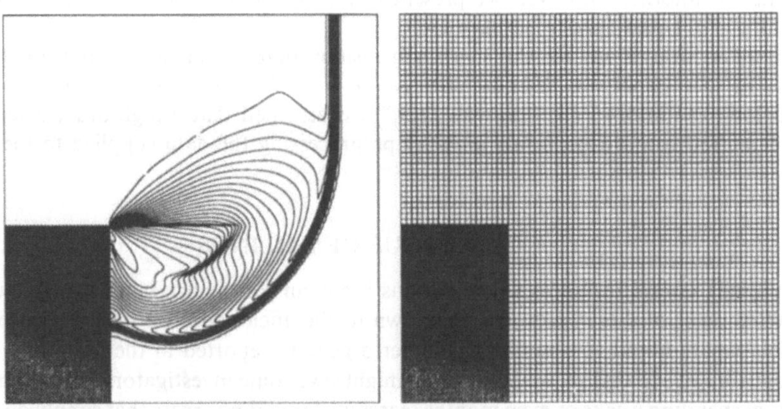

(a) (b)

Figure 1. A strong shock diffracting around a corner gives rise to an expansion shock: (a) density contours; (b) computational grid pressures 'positively conservative'. They have shown that while Godunov's scheme is 'positively conservative', the reverse is true for any Godunov-type scheme based on a linearized Riemann solver. Indeed, the basic form of Roe's scheme is unable to cope with the test problem shown in Figure 1; the strength of the diffracting shock is sufficient to cause a negative pressure to be computed near the apex of the corner. Roe's scheme may be made 'positively conservative' by modifying its wave speeds; in essence, the scheme is made more dissipative by increasing the spread in velocity between the two acoustic waves.[12]

2.3. *Slowly moving shocks*

Since shock-capturing schemes do not resolve the internal structure of a shock wave, no physical significance can be attached to the discrete shock structure produced by a numerical scheme. Methods are built upon the premise that shock profiles are monotone; the precise structure comes about as a matter of course and is not preordained. Unfortunately, Roberts has shown that the nature of the shock structure produced by a particular scheme can have a large bearing on how well the scheme copes with slowly moving shock waves.[14] Godunov-type methods fare quite badly in this respect: as the shock moves relative to the mesh, the shock profile flexes, perturbing the supposedly passive characteristic fields as it does so.

Figure 2 shows a snapshot of the shock profile produced by Einfeldt's HLLE scheme[6] taken from the simulation of a shock wave which is moving slowly from left to right; the pre-shock state (density, velocity, pressure) is (1, −3·44, 1) and the post-shock state is (3·86, −0·81, 10·33). Note that for a Courant number of one it takes 50 time steps for this shock to traverse one mesh cell. The low-frequency perturbations observed in this figure are produced to a greater or lesser extent by any scheme which attempts to 'recognize' a shock wave. For fast moving shocks the post-shock noise will be of a much shorter wavelength than is the case here and will be effectively damped by the dissipation of the scheme. Roberts reports that Osher's scheme[15] does not produce low-frequency noise for slowly moving shocks, since it never connects two adjacent states by a shock, and he concludes that there may be advantages to using flux formulae that do not recognize the analytic shock jump conditions.

Another situation where the perturbation of a shock from its preferred profile results in perturbations on the passive characteristic fields occurs whenever a shock crosses a discontinuity in mesh spacing.[16] However, in this case sizeable perturbations may occur whatever the speed of the shock.

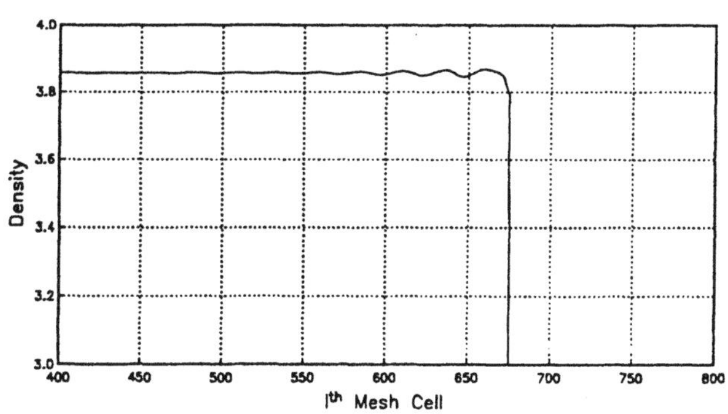

Figure 2. Low-frequency, post-shock oscillations occur for slowly moving shock waves

2.4. The carbuncle phenomenon

Several authors have by now reported a failing of Roe's scheme which has been dubbed the 'carbuncle phenomenon.'[17–19] For steady state, blunt body calculations Roe's scheme sometimes admits a spurious solution in which a protuberance grows ahead of the bow shock along the stagnation line. It appears that this effect is more pronounced the more closely the grid is aligned to the bow shock. Also, a carbuncle is more likely to appear for high-Mach-number flows than for low-Mach-number flows. Figure 3 shows such a spurious solution; here the freestream Mach number was taken to be 10. Note that along the stagnation line the bow shock is almost perfectly aligned with the grid. Consequently, parallel to the shock Roe's scheme will not add any dissipation via the contact and shear waves to counteract perturbations that appear through the acoustic waves; this appears to be a recurring theme whenever Roe's method fails. It is interesting to note that if Harten's entropy fix[13] is applied to the contact and shear waves, any shortcoming of Roe's scheme is invariably cured. However, there is no justification, either physical or mathematical, for applying this fix to these waves; it is just a convenient method for introducing an amount of artificial dissipation into the scheme.

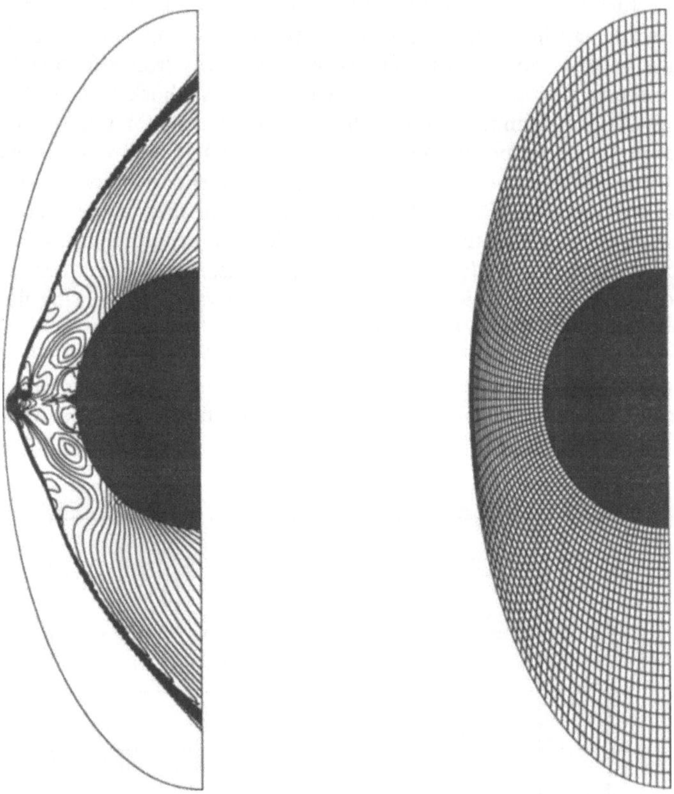

(a) (b)

Figure 3. The *carbuncle phenomenon*: (a) density contours; (b) computational grid

2.5. Kinked Mach stems

During the course of developing a mesh adaptation scheme, we encountered a failing of Roe's scheme which is not dissimilar to the 'carbuncle phenomenon'.[16] When the reflection of a plane shock wave from a ramp lies in the double-Mach reflection regime, the principal Mach stem is sometimes inexplicably kinked. Figure 4 shows a snapshot of the pressure contours taken during the reflection of a plane shock, $M_s = 5{\cdot}5$ with $\gamma = 1{\cdot}4$, from a 30° ramp. The principal Mach stem is so severely kinked that it has given rise to a spurious triple point. Similarly strange results have been produced by Sawada[20] and Itoh *et al.*[21] As before, because of the way the Mach stem is aligned with the grid, there is probably insufficient dissipation added via the contact and shear waves to counteract perturbations that appear via the acoustic waves.

2.6. Odd–Even decoupling

By far the most insidious failing that we have come across has, we believe, gone unreported in the literature. During the course of producing very-high-resolution simulations, we have noticed a tendency for odd–even decoupling to occur along the length of planar shocks which are aligned with the mesh (for an example see Figure 11). Of the Riemann solvers that we have at our disposal, this failing afflicts an exact solver,[22] Roe's solver and Toro's linearized solver.[7] We emphasize the fact that this phenomenon only becomes apparent for very-high-resolution simulations which suffer some systematic perturbation. However, as will be shown below, the required perturbation can arise quite innocuously, so we suspect that this failing will prove fairly widespread once very-high-resolution simulations become commonplace owing to increases in computer power.

Now since we obtain our high grid resolution by means of a fairly complex mesh adaption scheme,[16] it seemed reasonable to suppose that this odd–even decoupling was attributable to some coding error, but an exhaustive search for such an error proved fruitless. Subsequently we have managed to reproduce this failing in a more controlled manner, as has a colleague using an independent code,[23] so we have little doubt that this tendency for odd–even decoupling to occur constitutes a genuine failing rather than being the manifestation of some deficiency of our code. That said, our adaptive mesh scheme clearly exasperates this failing. In the next section

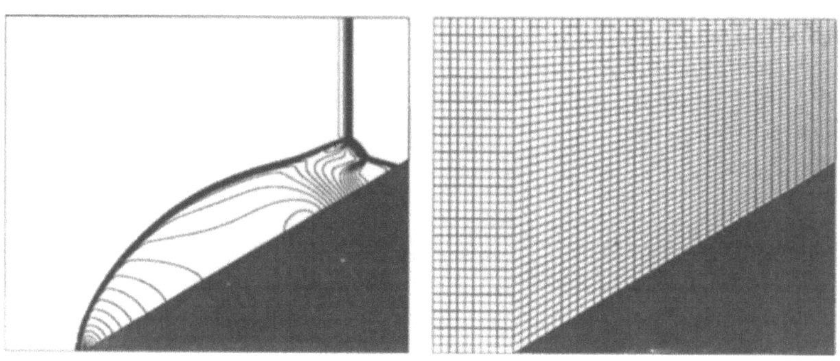

(a) (b)

Figure 4. The principal Mach stem arising from the reflection of a plane shock from a ramp is inexplicably kinked: (a) pressure contours; (b) computational grid we shall present a possible diagnosis of the mechanism which causes this mode of failure; here we merely present the evidence that it exists.

Figure 5 shows several snapshots of the density contours from the simulation of a plane shock wave, $M_s = 6$ with $\gamma = 1{\cdot}4$, propagating down a duct. For this calculation we have used Roe's scheme together with a nominally uniform grid of 20×800 cells with unit spacing, the centreline of which is perturbed from that of a perfectly uniform mesh in the following manner:

$$Y_{i,jmid} = \begin{cases} Y_{jmid} + 10^{-6} & \text{for } i \text{ even,} \\ Y_{jmid} - 10^{-6} & \text{for } i \text{ odd.} \end{cases}$$

This perturbation to the grid centreline promotes odd–even decoupling along the length of the shock. Note that the shock has propagated some 15 channel widths before the decoupling first becomes apparent; see frame (b). For this point in the calculation Figure 6 shows a series of slices across the duct, for both the density and pressure fields, as one moves from the head to the foot of the shock. Interestingly, within the shock the decoupling of the pressure field is out of phase with the decoupling of the density field. As the shock continues to propagate down the duct, so the decoupling becomes progressively worse, until the shock breaks down completely.

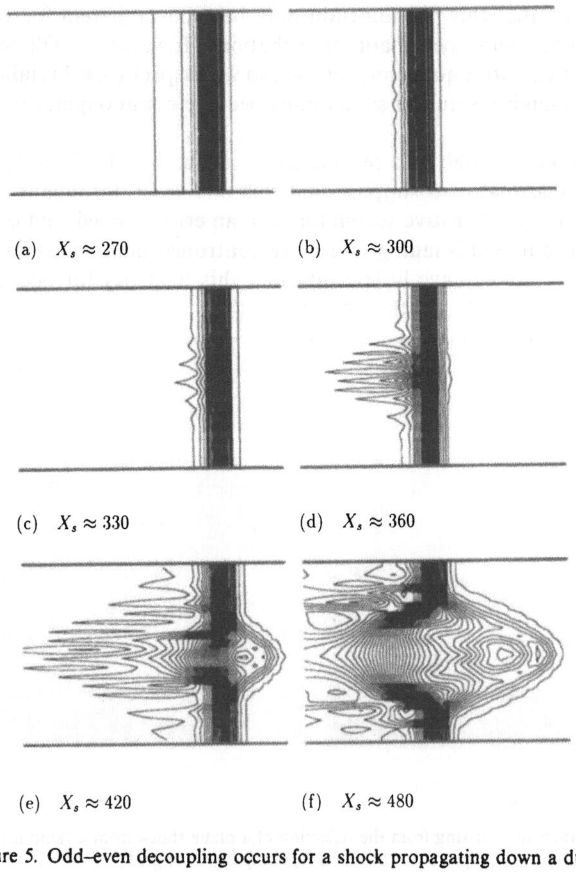

Figure 5. Odd–even decoupling occurs for a shock propagating down a duct

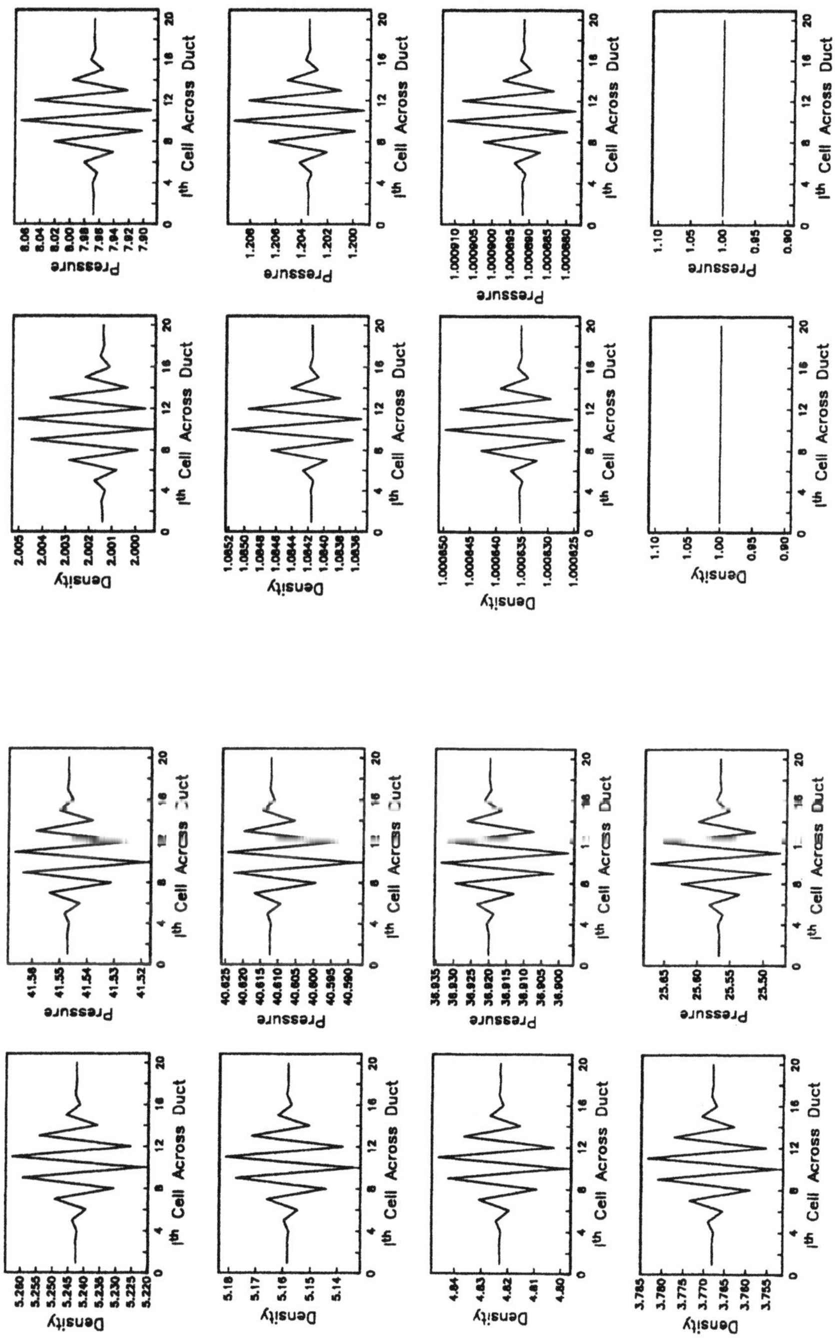

Figure 6. Transverse slices of the pressure and density fields for the shock shown in frame (b) of Figure 5

However, at no stage in the calculation does the code blow up in the sense that it generates a floating point exception; it simply goes astray.

3. ODD–EVEN DECOUPLING—A DIAGNOSIS?

As yet, the tools do not exist that would allow us to perform a rigorous non-linear stability analysis for some shock-capturing scheme applied to the Euler equations. However, it is possible to examine the way in which a scheme evolves certain sets of prescribed data so as to ascertain its likely stability characteristics. Since some Riemann solvers allow odd–even decoupling to develop along the length of a plane shock, it might prove fruitful to examine how different schemes evolve sawtooth-type initial data. To this end we consider the one-dimensional Euler equations with a passive component of shear velocity:

$$\frac{\partial}{\partial t}\begin{pmatrix}\rho\\ \rho u\\ \rho v\\ E\end{pmatrix}+\frac{\partial}{\partial y}\begin{pmatrix}\rho v\\ \rho uv\\ \rho v^2+p\\ (E+p)v\end{pmatrix}=0. \tag{1}$$

The quantities ρ, p, u, v and E are density, pressure, the passive shear component of velocity, the velocity in the y-direction and the total energy per unit volume respectively. For a perfect gas

$$E=\frac{p}{\gamma-1}+\tfrac{1}{2}\rho(u^2+v^2),$$

where γ is the ratio of specific heats. We assume that the computational mesh is uniform, with mesh spacing Δy, and that the discrete solution at time t^n is given by

$$\rho_j^n=\rho+\hat{\rho}^n,\qquad p_j^n=p+\hat{p}^n,\qquad u_j^n=u,\qquad v_j^n=0 \tag{2}$$

if j is even and by

$$\rho_j^n=\rho-\hat{\rho}^n,\qquad p_j^n=p-\hat{p}^n,\qquad u_j^n=u,\qquad v_j^n=0$$

if j is odd. Here $\hat{\rho}^n$ and $\hat{p}^n$ are the amplitudes of the sawtooth profiles for the density and pressure fields respectively. We shall consider two schemes which may be expressed in the form

$$\mathbf{W}_j^{n+1}=\mathbf{W}_j^n-\frac{\Delta t}{\Delta y}(\mathbf{G}_{j+1/2}^n-\mathbf{G}_{j-1/2}^n), \tag{3}$$

where $\mathbf{W}$ is the conserved variable vector $(\rho, \rho u, \rho v, E)^{\mathrm{T}}$ and $\mathbf{G}_{j+1/2}^n$ is a first-order flux function computed from the states $\mathbf{W}_j^n$ and $\mathbf{W}_{j+1}^n$.

3.1. *Roe's scheme*

Using Roe's scheme,[1] the interface flux for the system of equations (1) may be written as

$$\mathbf{G}_{j+1/2}^n=\tfrac{1}{2}(\mathbf{G}_j^n+\mathbf{G}_{j+1}^n)-\tfrac{1}{2}\sum_{k=1}^{k=4}\alpha_{j+1/2}^{(k)}|\lambda_{j+1/2}^{(k)}|\mathbf{e}_{j+1/2}^{(k)},$$

where the wave speeds $\{\lambda^{(k)}\}$, wave strengths $\{\alpha^{(k)}\}$ and eigenvectors $\{\mathbf{e}^{(k)}\}$ are given by

$$\lambda^{(1)} = \tilde{v} - \tilde{a}, \qquad \lambda^{(2)} = \tilde{v}, \qquad \lambda^{(3)} = \tilde{v}, \qquad \lambda^{(4)} = \tilde{v} + \tilde{a},$$

$$\alpha^{(1)} = \frac{\Delta p - \tilde{\rho}\tilde{a}\Delta v}{2\tilde{a}^2}, \qquad \alpha^{(2)} = \Delta\rho - \frac{\Delta p}{\tilde{a}^2}, \qquad \alpha^{(3)} = \rho\Delta u, \qquad \alpha^{(4)} = \frac{\Delta p + \tilde{\rho}\tilde{a}\Delta v}{2\tilde{a}^2},$$

$$\mathbf{e}^{(1)} = \begin{pmatrix} 1 \\ \tilde{u} \\ \tilde{v} - \tilde{a} \\ \tilde{h} - \tilde{a}\tilde{v} \end{pmatrix}, \qquad \mathbf{e}^{(2)} = \begin{pmatrix} 1 \\ \tilde{u} \\ \tilde{v} \\ \frac{1}{2}(\tilde{u}^2 + \tilde{v}^2) \end{pmatrix}, \qquad \mathbf{e}^{(3)} = \begin{pmatrix} 0 \\ 1 \\ 0 \\ \tilde{u} \end{pmatrix}, \qquad \mathbf{e}^{(4)} = \begin{pmatrix} 1 \\ \tilde{u} \\ \tilde{v} + \tilde{a} \\ \tilde{h} + \tilde{a}\tilde{v} \end{pmatrix}.$$

Here quantities written as $(\tilde{\cdot})$ are the so-called Roe-averaged quantities, $\tilde{a}$ and $\tilde{h}$ are the Roe-averaged sound speed and total enthalpy respectively and $\Delta(\cdot)$ represents the forward difference operator $(\cdot)_{j+1} - (\cdot)_j$. Now for our chosen data

$$(\tilde{\cdot})_{j+1/2} = (\tilde{\cdot})_{j-1/2}, \qquad \Delta(\cdot)_{j+1/2} = -\Delta(\cdot)_{j-1/2}.$$

Therefore

$$\sum_{k=1}^{k=4} \alpha^{(k)}_{j+1/2} |\lambda^{(k)}_{j+1/2}| \mathbf{e}^{(k)}_{j+1/2} = -\sum_{k=1}^{k=4} \alpha^{(k)}_{j-1/2} |\lambda^{(k)}_{j-1/2}| \mathbf{e}^{(k)}_{j-1/2}.$$

Also $\mathbf{G}^n_{j-1} = \mathbf{G}^n_{j+1}$, so the evolution scheme (3) may be written as

$$\mathbf{W}^{n+1}_j = \mathbf{W}^n_j + \frac{\Delta t}{\Delta y} \sum_{k=1}^{k=4} \alpha^{(k)}_{j+1/2} |\lambda^{(k)}_{j+1/2}| \mathbf{e}^{(k)}_{j+1/2},$$

which can be simplified to

$$\mathbf{W}^{n+1}_j = \mathbf{W}^n_j + \nu_y \frac{\Delta p}{\tilde{a}^2} \begin{pmatrix} 1 \\ \tilde{u} \\ 0 \\ \tilde{h} \end{pmatrix}, \qquad (4)$$

where ν_y is the Courant number $\tilde{a}\Delta t/\Delta y$. Recognizing that $\mathbf{W}^n_j$ may be expressed as

$$\begin{pmatrix} \rho \\ \rho u \\ 0 \\ E \end{pmatrix} \pm \begin{pmatrix} \hat{\rho}^n \\ u\hat{\rho}^n \\ 0 \\ \hat{p}^n/(\gamma - 1) + (u^2/2)\hat{\rho}^n \end{pmatrix}$$

and that by definition

$$\tilde{a}^2 = (\gamma - 1)\left(\tilde{h} - \frac{\tilde{u}^2}{2} - \frac{\tilde{v}^2}{2}\right),$$

equation (4) may be manipulated to give

$$\hat{\rho}^{n+1} = \hat{\rho}^n - \frac{2\nu_y}{\tilde{a}^2}\hat{p}^n, \qquad \hat{p}^{n+1} = \hat{p}^n(1 - 2\nu_y).$$

From this it can be seen that the initial perturbation to the pressure field is damped provided that the CFL condition is met, i.e.

$$\nu_y \leq 1.$$

However, the form of the evolution for the density perturbation exposes a flaw in Roe's scheme; the density perturbation is fed directly from the pressure perturbation. Making the loose approximation that $\tilde{a}$ remains constant, for a one-off disturbance $(\hat{\rho}^0, \hat{p}^0)$ we have

$$\hat{\rho}^n = \hat{\rho}^0 - \frac{2\nu_y}{\tilde{a}^2}\hat{p}^0[1 + (1 - 2\nu_y) + (1 - 2\nu_y)^2 + \cdots + (1 - 2\nu_y)^{n-1}]$$

Thus

$$\hat{\rho}^\infty = \hat{\rho}^0 - \frac{\hat{p}^0}{\tilde{a}^2}. \tag{5}$$

Therefore if $\hat{p}^0$ is of opposite sign to $\hat{\rho}^0$, then for a one-off disturbance $\hat{\rho}$ will grow but remain bounded; but if the pressure field is continuously perturbed in a systematic manner, no matter how small the pressure perturbations, $\hat{\rho}$ will grow without bound, albeit slowly. For two-dimensional calculations, although we cannot prove it, we suspect that a strong shock wave moving normal to the y-direction provides this systematic perturbation.

Firstly, it is interesting to note that the failing reported in Section 2.6 is only observed for strong shocks. For a strong shock wave it seems reasonable to suppose that $|\hat{p}^0/\tilde{a}^2|$ is more likely to be larger than $|\hat{\rho}^0|$ than would be the case for weak shocks. Thus even if $\hat{\rho}^0$ and $\hat{p}^0$ are initially of the same sign, they need not remain so; see (5). Now consider frame (b) of Figure 5 and the associated profiles shown in Figure 6. Within the shock the odd–even decouplings of the pressure and density fields are indeed out of phase with one another, which is consistent with the observations made above. Such behaviour will cause the local sound speed to vary along the length of the shock and its profile will exhibit a sawtooth perturbation which is in phase with that of the pressure field. Consequently, the individual segments of the shock will be moving alternately faster and slower than the nominal shock speed. Such movements will exaggerate the sawtooth perturbation to the pressure field along the length of the shock, but diminish that for the density field. The increased pressure perturbations will then promote an increase in the density perturbations as detailed above, and so the whole process repeats itself.

Since there are two competing processes that affect the density perturbations, namely the relative movements of the shock and the decoupling along the length of the shock, we cannot categorically state that Roe's method is bound to break down. However, the weight of numerical evidence suggests that at least for strong shocks Roe's scheme will break down in the manner described here. Given our arguments, it should come as no surprise that Godunov's method also exhibits a tendency to allow odd–even decoupling to occur along the length of a strong shock wave. Since it is the sweep parallel to the shock that primarily causes the instability, the differences between using an exact Riemann solver and Roe's linearized solver for data that are nominally uniform should have little bearing on the growth of the instability.

Finally, before moving on, it should be noted that none of the popular entropy fixes which are applied to Roe's scheme cures this particular failing, except for the case where Harten's fix is applied to the shear and contact waves.* Simply altering the acoustic wave speeds can have

* To reiterate the comment made in Section 2.4, applying Harten's entropy fix to the linearly degenerate wave fields has no mathematical or physical justification, but is merely a convenient way in which to add an amount of artificial dissipation.

no affect; because of the symmetry of the data, both waves will be changed by the same amount and so the problem will persist. Also, moving to a high-order version of Roe's scheme will not improve matters, because the odd–even decoupling will cause a high-order flux function to drop to the first-order function.

3.2. *Einfeldt's HLLE scheme*

For Einfeldt's HLLE scheme[6] the interface flux function is given by

$$\mathbf{G}^n_{j+1/2} = \frac{b^{(+)}\mathbf{G}^n_{j+1} - b^{(-)}\mathbf{G}^n_j}{b^{(+)} - b^{(-)}} + \frac{b^{(+)}b^{(-)}}{b^{(+)} - b^{(-)}}(\mathbf{W}^n_{j+1} - \mathbf{W}^n_j),$$

where

$$b^{(+)} = \max(0, \lambda^{(4)}_{j+1/2}, v_{j+1} + a_{j+1}), \qquad b^{(-)} = \min(0, \lambda^{(1)}_{j-1/2}, v_j - a_j)$$

and $\lambda^{(1)}$ and $\lambda^{(4)}$ are the two acoustic wave speeds from Roe's method. Now for our chosen data $(\tilde{\cdot})_{j+1/2}$ is equal to $(\tilde{\cdot})_{j-1/2}$, therefore

$$b^{(+)}_{j+1/2} = -b^{(-)}_{j-1/2} \approx \tilde{a}, \qquad b^{(-)}_{j+1/2} = -b^{(+)}_{j-1/2} \approx -\tilde{a}.$$

Using these signal weightings, it may be found that

$$\hat{\rho}^{n+1} = (1 - 2\nu_y)\hat{\rho}^n, \qquad \hat{p}^{n+1} = (1 - 2\nu_y)\hat{p}^n,$$

where

$$\nu_y \approx \frac{\tilde{a}\Delta t}{\Delta y}.$$

From this it can be seen that both the density and pressure perturbations are damped provided that the CFL condition is satisfied. Just as important, however, is the fact that the pressure perturbation does not feed into the density perturbation, so we would not expect the HLLE solver to exhibit the odd–even decoupling that afflicts both Roe's scheme and Godunov's scheme; numerical experimentation confirms this expectation.

It is our contention that any scheme for which it can be shown that the perturbation to the pressure field feeds the perturbation to the density field will be afflicted by the odd–even decoupling shown in Figure 5. Thus it comes as no surprise to find that Toro's linearized Riemann solver[7] is afflicted by this failing, but Liou and Steffen's AUSM scheme[19] is not. The way is now open for some mathematician to perform a more rigorous analysis than we are able, so as to shed additional light on the mechanism which causes this particular failing.

4. AN ADAPTIVE RIEMANN SOLVER

Having exposed many of the weaknesses of Riemann solvers, we now present a simple strategy that we have found useful for improving the robustness of Godunov-type schemes. In essence, we select the precise flavour of upwinding to match the local flow data such that a particular Riemann solver is only employed in those situations where it is known to give reliable results. By recognizing the limitations of any one solver, it is possible to reap its advantages without suffering its attendant failings.

Our synergetic strategy has a number of attractions, not least of which is that some favoured solver need not be jettisoned simply because it occasionally fails. However, it does introduce the difficulty of how to decide when to use one Riemann solver in preference to another. It has been our experience that this added difficulty is not particularly bothersome, since we tend to combine a single high-resolution Riemann solver with just one or two other solvers that prove more reliable under conditions which are fairly well defined, and so a set of *ad hoc* switching functions suffice. For example, some of the worst failings of Riemann solvers occur in the vicinity of strong shock waves. To overcome such failings, we employ Einfeldt's HLLE scheme. Now it makes little sense to chop and change the choice of Riemann solver used along the length of a shock wave, since to do so would inevitably perturb a planar shock front. Hence we apply this particular Riemann solver throughout the immediate vicinity of a strong shock. Thus the HLLE switching function need only locate the position of a shock wave, but such functions already exist in the guise of mesh refinement monitor functions.

A simple test that identifies those cell interfaces which are in the vicinity of a strong shock is to check whether or not

$$\frac{|p_r - p_l|}{\min(p_l, p_r)} > \alpha, \tag{6}$$

where α is some threshold parameter which is problem-dependent and p_r and p_l refer to the pressures which act on the interface. If this condition is met, the two cells separated by the interface are flagged as lying within a strong shock. Thus when it comes to computing cell–interface fluxes, if the cells either side of an interface are both flagged as lying within a strong shock, the flux is computed using the HLLE solver. Note that since numerical shocks are invariably smeared over several mesh cells, it is worth locating shocks using a projection of the flow solution on a grid which is coarser than that used for the calculation. On such a grid a shock will appear much less smeared and so the left-hand side of the above switching function will be a fair indication of its strength. Once a set of cells have been flagged on this coarse mesh, the flags may be prolongated to the actual computational mesh so as to find those cells which lie in the vicinity of a shock wave.

Before proceeding further, several observations should be made. Firstly, although the HLLE solver is adjudged to be a low-resolution scheme, it does in fact resolve *isolated* shocks as well as an exact Riemann solver does. Consequently, using this robust solver in the vicinity of strong shock waves does not necessarily pollute a scheme's resolution, as would be the case if artificial dissipation were used to augment the dissipation provided via upwinding. For many inviscid calculations the amount of pollution proves to be negligible, and whilst some degradation would be expected for the case of a strong shock interacting with a boundary layer, it may well be unnecessary to employ the HLLE solver in such a situation because of the extra dissipation provided by the real viscous terms. Secondly, although the HLLE switching function requires a tunable parameter α, the retuning of this parameter is less involved than the retuning of an artificial dissipation mechanism; in general, it is far simpler to determine where extra dissipation should be added than it is to determine how much extra dissipation to add. For many problems, assuming that shocks are located as described above, a sensible threshold on the shock strength can be specified *a priori*. Lastly, it should be noted that our strategy of switching Riemann solvers may not prove suitable for those implicit schemes which require that the numerical flux function be differentiable.

Figure 7 shows how the HLLE solver may be used to correct the tendency of Roe's scheme

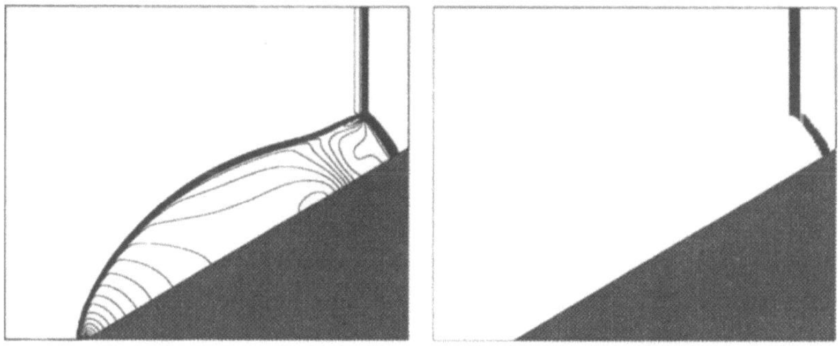

(a) (b)

Figure 7. The HLLE scheme can be used to circumvent the tendency of Roe's method to produce kinked Mach stems: (a) pressure contours; (b) HLLE switching function

to produce kinked Mach stems (cf. Figure 4). For this calculation the HLLE switching function was tuned such that it would only be activated by the incident shock and the principal Mach stem; α was simply set to half the strength of the incident shock. Note that apart from the region near the Mach stem, these new results are very similar to the old ones. This shows that the HLLE scheme has had no adverse affect on the resolution of Roe's scheme. Similarly, Figure 8 shows how the carbuncle phenomenon may be circumvented (cf. Figure 3). Here we have restricted the HLLE solver to cells near the stagnation line in order to demonstrate how localized the failing of Roe's scheme really is. In practice, however, we would advocate using the HLLE scheme along the whole length of the bow shock so as to maximize robustness without compromising resolution. Again a sensible value of α can be found *a priori* by using some large fraction of the shock strength along the stagnation line, which can be estimated, given the freestream Mach number, by assuming that the flow is locally one-dimensional. As shown in Figure 9, the HLLE solver may also be used to good effect to prevent Godunov's scheme from admitting expansion shocks (cf. Figure 1). Here we have employed the HLLE solver along the sonic line and in regions where the expansion waves are strong.

Having presented the gist of our strategy, we see little point in trying to sell a particular combination of solvers. Starting with some high-resolution Riemann solver, whose choice will inevitably be a matter of personal taste, the correct combination of solvers will depend both on that scheme's weaknesses and on the specific application in hand. In turn, the combination of Riemann solvers will dictate the choice of switching functions. Therefore we shall resist the temptation to recommend a specific course of action; instead, we present two simulations that show how an adaptive Riemann solver might be used to good effect. Briefly, both simulations were done using the two-dimensional analogue of the one-dimensional Euler equations given in Section 3. These equations were integrated using the two-step, finite volume scheme which is attributable to Hancock.[24] This scheme employs van Leer's MUSCL approach[9] to achieve a second-order extension to Godunov's method; hence different Riemann solvers may be slotted directly into the scheme so as to vary the flavour of the upwinding. Although the calculations were performed using an adaptive mesh algorithm,[16,25] the mesh refinement monitor function was such that the calculations employed a nominally uniform Cartesian mesh.

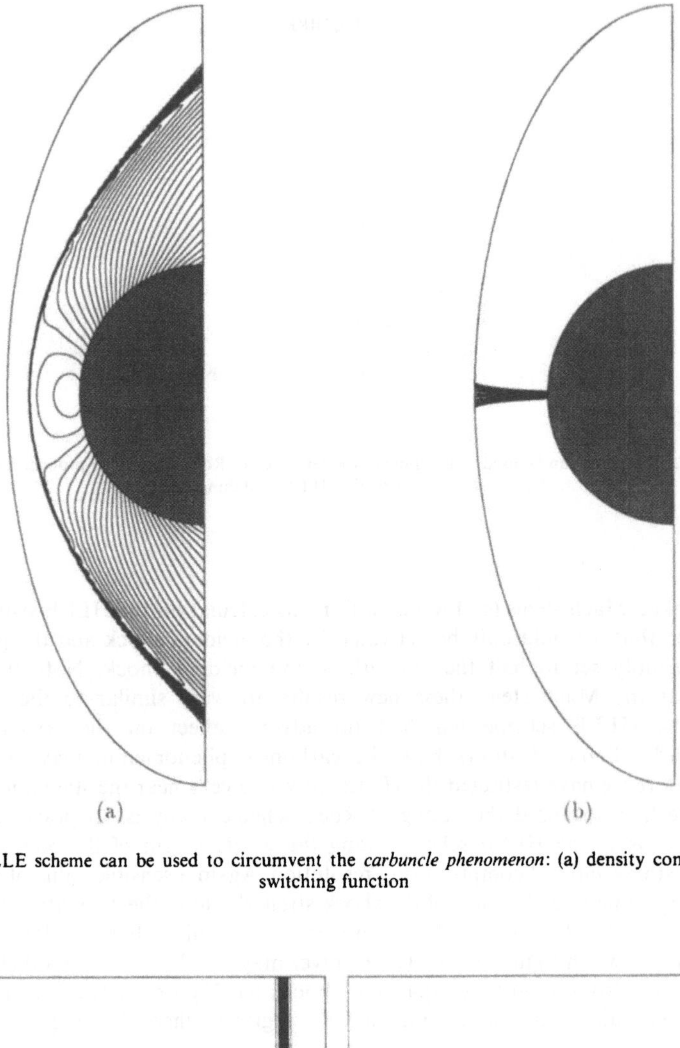

Figure 8. The HLLE scheme can be used to circumvent the *carbuncle phenomenon*: (a) density contours; (b) HLLE switching function

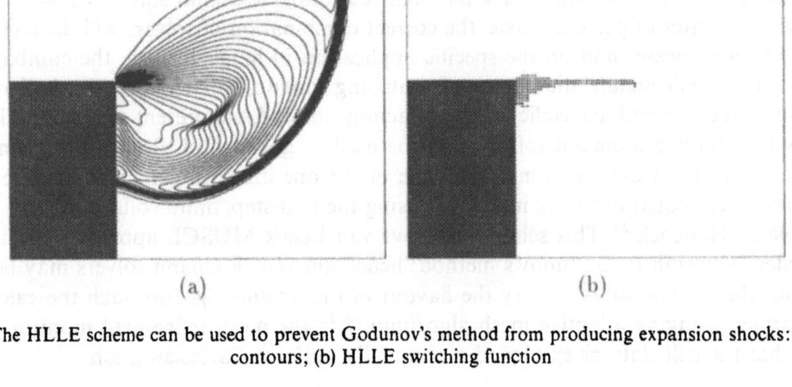

Figure 9. The HLLE scheme can be used to prevent Godunov's method from producing expansion shocks: (a) density contours; (b) HLLE switching function

Our first example concerns the simulation of a strong shock wave diffracting around a 90° corner; the shock Mach number and the ratio of specific heats are 5·09 and 1·4 respectively. We have computed this test problem using a combination of three different Riemann solvers: Toro's linearized Riemann solver was used to perform the MUSCL reconstruction step of Hancock's scheme as described by Quirk,[25] and the upwinding step was performed by adaptively selecting between Roe's solver and the HLLE solver. The parameter α used by the switching function (6) was set to unity so as to limit the HLLE solver to the incident and diffracted shock fronts and to a small region near the apex of the corner. Figure 10 shows a schlieren-type snapshot taken from this simulation. The different shades of grey depict the magnitude of the gradient of the density field; the darker the shade, the larger the magnitude; details of this shading procedure are given in Appendix A. Here it is not our intention to assess the accuracy of these results; the interested reader may do this using the experimental results of Bazhenova *et al.*[26] and the computational results of Hillier.[27] Instead, we wish to illustrate the fact that certain Riemann solver failings, if left unaddressed, can place an upper limit on the resolution of simulations that may be performed.

Consider the consequences of disabling the HLLE switching function so that Roe's solver alone is used for the upwinding stage of Hancock's scheme. The tendency of Roe's solver to allow odd–even decoupling to occur along a planar shock wave which is aligned with the grid will sooner or later cause this simulation to come to grief (see Figure 11), thus precluding the possibility of performing very detailed simulations. By way of comparing the resolution of these two sets of results, for Figure 10 there are 560 mesh cells from the apex of the corner to the point where the Mach stem meets the wall, while for Figure 11 there are only 120 cells. The question of whether or not our adaptive mesh algorithm contains some flaw which exasperates the odd–even decoupling is largely academic. The fact remains that Roe's solver is susceptible to this mode of failure whereas the HLLE solver is not. Whether the initial stimulus comes from a distorted mesh as in Section 2 or from some component of the mesh adaption scheme as seems likely here is immaterial.

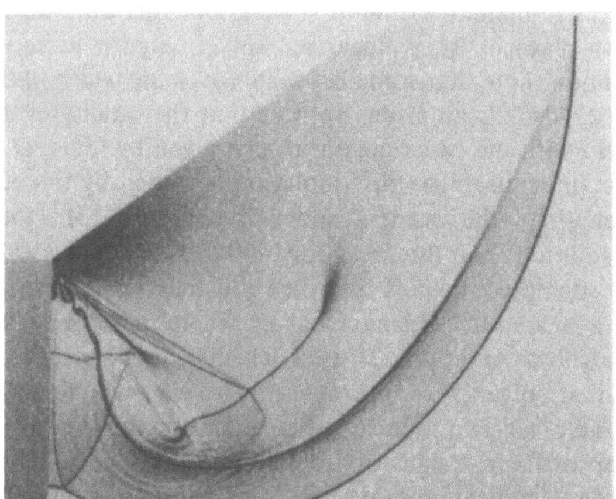

Figure 10. A schlieren- type snapshot from the diffraction of a strong shock wave around a 90° degree corner

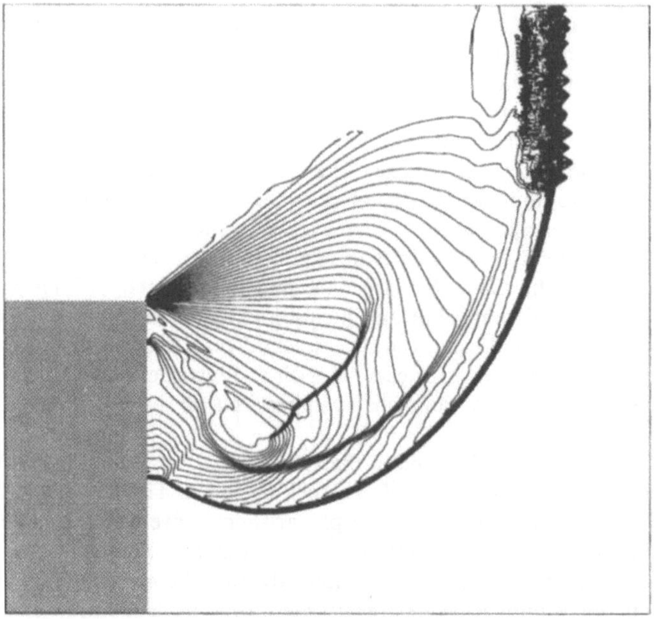

Figure 11. On its own, Roe's approximate Riemann solver cannot be used to reproduce the resolution of the simulation shown in Figure 10

So as not to leave the impression that the above shortcoming is somehow peculiar to Roe's method, we present a second set of results which are taken from the interaction of a planar shock wave with a half-diamond; the shock Mach number is 2·85, the ratio of specific heats is 1·4 and the angle at the apex of the diamond is 90°. As before, we have run this test problem using a combination of three different Riemann solvers, but this time we have substituted an exact Riemann solver[22] in place of Roe's linearized solver. Figure 12 shows a schlieren-type snapshot from this calculation; note that some 800 cells cover the width of the diamond, so this calculation is well resolved. Also, as an aside, we note that the quality of these results may be gauged by comparing them with the experimental results given by Glass *et al.*[28] Once again, if the HLLE switching function is disabled, the simulation is ruined by the odd–even decoupling that occurs along the length of the incident shock (see Figure 13). Note that this second calculation is of lower resolution than the first; only 400 cells cover the width of the diamond.

In this section we have attempted to show that the robustness of Godunov-type schemes may be improved by employing an adaptive Riemann solver, where the specific flavour of upwinding is altered to suit the local flow conditions. If used sensibly, this strategy can overcome most known failings of individual solvers. Despite our efforts, we recognize that the majority of shock-capturing practitioners will continue to use artificial dissipation as a band aid to fix a particular Riemann solver at the first signs of any failing, simply because it is expedient to do so. Whilst we find this approach disappointing, the principal aim of this paper is to emphasize the fact that most of the Riemann solvers that are in common use *must be augmented in some way* if they are to be used for the purpose of producing genuinely high-resolution simulations of shock hydrodynamic phenomena.

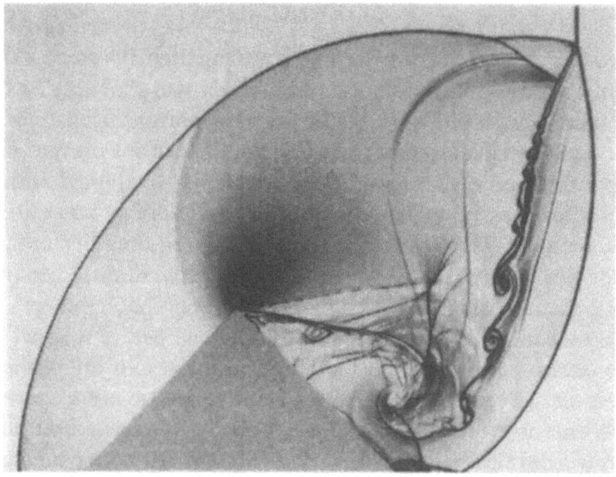

Figure 12. A schlieren-type snapshot from the interaction of a planar shock wave with a half-diamond

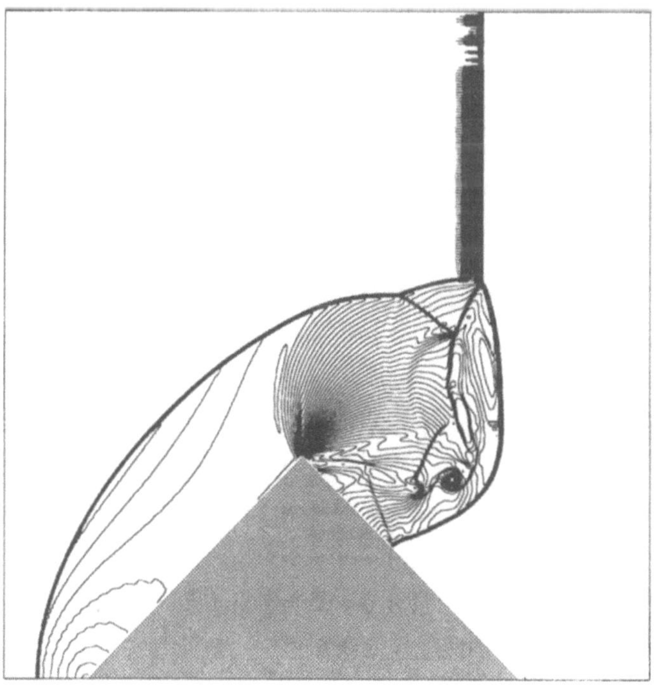

Figure 13. On its own, an exact Riemann solver cannot be used to reproduce the resolution of the simulation shown in Figure 12

5. CONCLUSIONS

Unless the dissipation provided via upwinding is augmented by some other mechanism, any Godunov-type scheme built upon a single Riemann solver will probably be flawed. For example, when subjected to some small but systematic form of perturbation, most popular Riemann solvers for the Euler equations, including the exact solver, cannot prevent odd–even decoupling occurring along the length of a strong shock wave which is aligned with the computational mesh. Thus far, this flaw has gone largely unnoticed simply because it is only exposed by very-high-resolution simulations. However, given that the required perturbations can arise quite innocuously, this mode of failure should prove fairly widespread once genuinely high-resolution simulations become commonplace owing to increases in computer power.

Although most flaws can be controlled by the judicious use of a small amount of artificial dissipation, to do so necessarily leads to a reduction in the resolution of the scheme. We favour an alternative approach whereby the failings of any one Riemann solver are circumvented by combining it with one or more complementary solvers. In essence, we advocate selecting the precise flavour of upwinding to suit the flow data. Admittedly, this synergistic strategy is not as aesthetically pleasing as having a single Riemann solver for all occasions, but we have shown that it can be made to work quite effectively; besides which, Riemann solvers are sometimes touted as being a solution-adaptive technique, so the concept of an adaptive Riemann solver is not that contrived.

Looking to the future, it is to be hoped that genuinely multidimensional Riemann solvers will overcome many of the shortcomings of today's dimensionally split schemes. However, given the way in which the present shortcomings have been stumbled across, these multidimensional schemes may themselves arrive complete with subtle failings with which to ensnare the unwary.

ACKNOWLEDGEMENTS

I am indebted to Dr. P. A. Jacobs for taking the time to reproduce the failing reported in Section 2.6 and I wish to thank Professor P. L. Roe for his comments concerning this work. This research was supported by the National Aeronautics and Space Administration under NASA Contract NAS1-18605 while the author was in residence at the Institute for Computer Applications in Science and Engineering (ICASE), NASA Langley Research Center, Hampton, VA 23681.

APPENDIX: SCHLIEREN-TYPE PLOTS

The plots shown in Figures 10 and 12 depict the magnitude of the gradient of the density field, namely

$$\sqrt{\left[\left(\frac{\partial \rho}{\partial x}\right)^2 + \left(\frac{\partial \rho}{\partial y}\right)^2\right]},$$

and hence they may be viewed as idealized schlieren images. So as to accentuate weak flow features, the following non-linear shading function has been used:

$$\mathit{shade} = \exp(-k\psi).$$

Here k is a constant and ψ is a weighting function given by

$$\frac{|\nabla\rho| - |\nabla\rho|_0}{|\nabla\rho|_1 - |\nabla\rho|_0},$$

where

$$|\nabla \rho|_0 = k_0 |\nabla \rho|_{\max}, \qquad |\nabla \rho|_1 = k_1 |\nabla \rho|_{\max},$$

k_0 and k_1 being constants. Note that *shade* is limited to values between 0 and 1, so for a 24-bit colour graphics system the grey level *shade* may be converted to an $\langle R, G, B\rangle$ triplet using

$$\langle 255 * shade, 255 * shade, 255 * shade \rangle.$$

For both figures the constants k, k_0 and k_1 were set to 5, 0·05 and -0·001 respectively.

REFERENCES

1. P. L. Roe, 'Characteristic-based schemes for the Euler equations', *Ann. Rev. Fluid Mech.*, **18**, 337–365 (1986).
2. R. Courant and K. O. Friedrichs, *Supersonic Flow and Shock Waves*, Interscience, New York, 1948; reprinted as *Applied Mathematical Physics*, Vol. 21, Springer, New York, 1976.
3. M. Holt, *Numerical Methods in Fluid Dynamics*, Springer, Berlin, 2nd edn, 1984.
4. S. Osher, 'Riemann solvers, the entropy condition, and difference approximations', *SIAM J. Numer. Anal.*, **21**, 289–315 (1984).
5. M. Pandolfi, 'A contribution to numerical prediction of unsteady flows', *AIAA J.*, **22**, 602–610 (1984).
6. B. Einfeldt, 'On Godunov-type methods for gas dynamics', *SIAM J. Numer. Anal.*, **25**, 294–318 (1988).
7. E. F. Toro, 'A linearised Riemann solver for the time-dependent Euler equations of gas dynamics', *Proc. R. Soc. Lond. A*, **434**, 683–693 (1991).
8. P. Colella and P. R. Woodward, 'The piecewise parabolic method (PPM) for gas-dynamical simulations', *J. Comput. Phys.*, **54**, 174–201 (1984).
9. B. van Leer, 'Towards the ultimate conservative difference scheme, II. Monotonicity and conservation combined in a second-order scheme', *J. Comput. Phys.*, **14**, 361–376 (1974).
10. P. R. Woodward and P. Colella, 'The numerical simulation of two-dimensional fluid flow with strong shocks', *J. Comput. Phys.*, **54**, 115–173 (1984).
11. P. L. Roe, 'Some contributions to the modelling of discontinuous flows', in *Lectures in Applied Mathematics*, Vol. 22, American Mathematical Society, Providence, RI, 1985.
12. B. Einfeldt, C. D. Munz, P. L. Roe and B. Sjögreen, 'On Godunov-type methods near low densities', *J. Comput. Phys.*, **92**, 273–295 (1991).
13. H. C. Yee, R. F. Warming and A. Harten, 'Implicit total variation diminishing (TVD) schemes for steady-state calculations', *J. Comput. Phys.*, **57**, 327–360 (1985).
14. T. W. Roberts, 'The behaviour of flux difference splitting schemes near slowly moving shock waves', *J. Comput. Phys.*, **90**, 141–160 (1990).
15. S. Osher and F. Solomon, 'Upwind difference schemes for hyperbolic systems of conservation laws', *Math. Comput.*, **38**, 339–374 (1982).
16. J. J. Quirk, 'An adaptive grid algorithm for computational shock hydrodynamics', *Ph.D. Thesis*, College of Aeronautics, Cranfield Institute of Technology, 1991.
17. K. M. Peery and S. T. Imlay, 'Blunt-body flow simulations', *AIAA Paper 88-2904*, 1988.
18. H. C. Lin, 'Dissipation additions to flux-difference splitting', *AIAA Paper 91-1544-CP*, 1991.
19. M.-S. Liou and C. J. Steffen, Jr., 'A new flux splitting scheme', *NASA TM 104404*, 1991.
20. K. Sawada, 'A multi-dimensional extension of the preprocessing approach for cell centred finite volume scheme', *AIAA Paper 91-1536*, 1991.
21. K. Itoh, K. Takayama and G. Ben-Dor, 'Numerical simulation of the reflection of a planar shock wave over a double wedge', *Int. J. Numer. Methods Fluids*, **13**, 1153–1170 (1991).
22. E. F. Toro, 'A fast Riemann solver with constant covolume applied to the random choice method', *Int. J. Numer. Methods Fluids*, **9**, 1145–1164 (1989).
23. P. A. Jacobs, University of Queensland, personal communication, 1991.
24. G. D. van Albada, B. van Leer and W. W. Roberts, 'A comparative study of computational methods in cosmic gas dynamics', *Astron. Astrophys.*, **108**, 76–84 (1982).
25. J. J. Quirk, 'An alternative to unstructured grids for computing gas dynamic flows around arbitrarily complex two-dimensional bodies', *ICASE Rep. 92-7*, 1992; *Comput. Fluids*, **23** (1), 125–142 (1994).
26. S. K. Bazhenova, L. G. Gvozdeva and M. G. Nettleton, 'Unsteady interactions of shock waves', *Prog. Aerospace Sci.*, **21**, 249–331 (1984).
27. R. Hillier, 'Computation of shock wave diffraction at a ninety degrees convex edge', *Shock Waves*, **1**, 89–98 (1991).
28. I. I. Glass, J. Kaca, D. L. Zhang, H. M. Glaz, J. B. Bell, J. A. Trangenstein and J. P. Collins, 'Diffraction of planar shock waves over half-diamond and semicircular cylinders: an experimental and numerical comparison', *AIP Conf. Proc.*, 1989.

AN APPROXIMATE RIEMANN SOLVER FOR MAGNETOHYDRODYNAMICS
(That Works in More than One Dimension)

*Kenneth G. Powell**
Department of Aerospace Engineering
The University of Michigan
Ann Arbor, MI 48109-2118

Abstract

An approximate Riemann solver is developed for the governing equations of ideal magnetohydrodynamics (MHD). The Riemann solver has an eight-wave structure, where seven of the waves are those used in previous work on upwind schemes for MHD, and the eighth wave is related to the divergence of the magnetic field. The structure of the eighth wave is not immediately obvious from the governing equations as they are usually written, but arises from a modification of the equations that is presented in this paper. The addition of the eighth wave allows multi-dimensional MHD problems to be solved without the use of staggered grids or a projection scheme, one or the other of which was necessary in previous work on upwind schemes for MHD. A test problem made up of a shock tube with rotated initial conditions is solved to show that the two-dimensional code yields answers consistent with the one-dimensional methods developed previously.

*This research was supported by the National Aeronautics and Space Administration under NASA Contract No. NAS1-19480 while the author was in residence at the Institute for Computer Applications in Science and Engineering (ICASE), NASA Langley Research Center, Hampton, VA 23681-0001.

1 Introduction

The governing equations of ideal magnetohydrodynamics (MHD) describe the physics of a conducting fluid in which the following assumptions hold:

$$\begin{aligned} \frac{\lambda}{L} &\ll 1 \\ \frac{\epsilon}{\tau\sigma} &\ll 1 \\ \left(\frac{V}{c}\right)^2 &\ll 1 \\ \frac{\mu}{\rho V L} &\ll 1 \end{aligned} \tag{1}$$

where ρ, V, τ and L are, respectively, characteristic density, speed, time and length scales for the problem, c is the speed of light, and ϵ and σ represent the dielectric constant and conductivity of the fluid. These equations, written in conservation-law form, are

$$\frac{\partial}{\partial t}\begin{pmatrix} \rho \\ \rho \mathbf{u} \\ \mathbf{B} \\ E \end{pmatrix} + \nabla \cdot \begin{pmatrix} \rho \mathbf{u} \\ \rho \mathbf{u}\mathbf{u} + \mathbf{I}\left(p + \frac{\mathbf{B}\cdot\mathbf{B}}{2}\right) - \mathbf{B}\mathbf{B} \\ \mathbf{u}\mathbf{B} - \mathbf{B}\mathbf{u} \\ \left(E + p + \frac{\mathbf{B}\cdot\mathbf{B}}{2}\right)\mathbf{u} - \mathbf{B}\left(\mathbf{u}\cdot\mathbf{B}\right) \end{pmatrix} = 0 \tag{2}$$

where $\mathbf{I}$ is a 3×3 identity matrix, ρ is the density, $\mathbf{u}$ is the velocity, p is the pressure, $\mathbf{B}$ is the magnetic field, and E is the energy, defined as

$$E = \frac{p}{\gamma - 1} + \rho\frac{\mathbf{u}\cdot\mathbf{u}}{2} + \frac{\mathbf{B}\cdot\mathbf{B}}{2} \,. \tag{3}$$

Solutions of these equations can yield insight into a number of problems governed by fluid-dynamic and electromagnetic effects.

Much of the past work in solving these equations has been based on Rusanov and Lax-Wendroff techniques. Only recently have authors begun to work on upwind schemes for solving these equations. In particular, Brio and Wu [2], Zachary and Colella [?], and Dai and Woodward [7] have done some of the early development of Riemann-solver-based schemes for the MHD equations. Their work has been based not on the system of eight conservation laws as written in Equation 2, but instead on the closely related system that comes from assuming B_x = constant and dropping the evolution equation for B_x. This yields a 7×7 system. The reason for their use of this modified system arises from the fact that one of the equations governing the magnetic field is

$$\nabla \cdot \mathbf{B} = 0 \,, \tag{4}$$

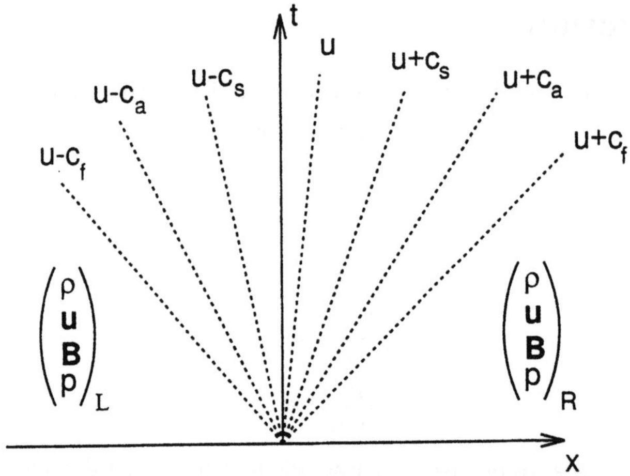

Figure 1: Waves in the One-Dimensional MHD Riemann Problem

which, in one dimension, becomes the constraint $B_x =$ constant.

The eigenvalues and eigenvectors of this 7×7 system are well known (see, for example, the book by Jeffrey and Taniuti [3]); they correspond to:

- one entropy wave traveling with speed u;
- two Alfvén waves traveling with speed $u \pm c_a$ where
$$c_a = \frac{B_x}{\sqrt{\rho}}$$
is the Alfvén speed;
- four magneto-acoustic waves, two "fast" and two "slow", traveling with speed $u \pm c_f$ and $u \pm c_s$ respectively, where
$$c_{f,s}^2 = \frac{1}{2}\left(\frac{\gamma p + \mathbf{B}\cdot\mathbf{B}}{\rho} \pm \sqrt{\left(\frac{\gamma p + \mathbf{B}\cdot\mathbf{B}}{\rho}\right)^2 - 4\frac{\gamma p B_x^2}{\rho^2}}\right).$$

An (x,t) diagram of the wave interactions at a cell interface is shown in Figure 1.

Given these seven eigenvalues and corresponding right and left eigenvectors, it is possible to develop a linear approximate Riemann solver ala Roe [2, ?], or a more nonlinear approximate Riemann solver [7]. Once some questions as to how to scale the left and right eigenvectors of the system are answered (for a very nice description of the problems of scaling in the MHD eigensystem, and an elegant solution to these problems, see the paper by Roe and Balsara [5]), a robust solver for one-dimensional unsteady problems in MHD can be developed.

Building a code capable of solving two- or three-dimensional problems from the one-dimensional Riemann solver building block is not, unfortunately, as straightforward as in the case of the Euler equations. In the one-dimensional problem, no evolution equation is necessary for the component of **B** normal to a cell interface, because the condition of Equation 4 implies that $B_{n_L} = B_{n_R}$. However, in a two-dimensional problem, this is no longer true. In two dimensions, the discrete constraint corresponding to Equation 4 is

$$\sum_{faces} B_n \, ds = 0 \; , \tag{5}$$

and so a jump in B_n *is* allowed across a face; it simply must be balanced by the jumps across the other faces of the cell. Thus, a separate procedure for updating this portion of the magnetic field must be implemented, and must be implemented in such a way as to satisfy the constraint implied in Equation 5. It should be noted that the $\nabla \cdot \mathbf{B}$ constraint is a headache not just for *upwind* schemes for MHD, but for solution of MHD problems in more than one dimension by *any* method. Typically, one of three approaches is taken to satisfy this constraint:

- a projection scheme, in which a Poisson equation must be solved to subtract off the portion of the magnetic field that leads to a non-zero divergence;
- non-collocated variables (e.g. a staggered-grid approach), so that the constraint is met identically,
- a vector-potential description of the magnetic field, so that the constraint is met identically.

A very different approach is taken in the work presented here. Instead of solving a seven-wave Riemann problem, with an added procedure to update the remaining B-field component which assures that Equation 4 is satisfied, an eight-wave Riemann solver, in which *all* of the magnetic field components are updated, is developed and tested.

2 Derivation of the Eight-Wave Riemann Solver

Given the primitive variables

$$\mathbf{W} = (\rho, u, v, w, B_x, B_y, B_z, p) \; , \tag{6}$$

3

Equation 2 may be rewritten in quasilinear form as

$$\frac{\partial \mathbf{W}}{\partial t} + A_p \frac{\partial \mathbf{W}}{\partial x} + B_p \frac{\partial \mathbf{W}}{\partial y} + C_p \frac{\partial \mathbf{W}}{\partial z} = 0\,, \tag{7}$$

where, for example

$$A_p = \begin{bmatrix} u & \rho & 0 & 0 & 0 & 0 & 0 & 0 \\ 0 & u & 0 & 0 & -\frac{B_x}{\rho} & \frac{B_y}{\rho} & \frac{B_z}{\rho} & \frac{1}{\rho} \\ 0 & 0 & u & 0 & -\frac{B_y}{\rho} & -\frac{B_x}{\rho} & 0 & 0 \\ 0 & 0 & 0 & u & -\frac{B_z}{\rho} & 0 & -\frac{B_x}{\rho} & 0 \\ 0 & 0 & 0 & 0 & 0 & 0 & 0 & 0 \\ 0 & B_y & -B_x & 0 & -v & u & 0 & 0 \\ 0 & B_z & 0 & -B_x & -w & 0 & u & 0 \\ 0 & \gamma p & 0 & 0 & (\gamma - 1)\,\mathbf{u}\cdot\mathbf{B} & 0 & 0 & u \end{bmatrix} . \tag{8}$$

The Riemann solver would normally be based on the eigensystem of A_p, but it is evident that this matrix is singular — the fifth row of the matrix is zero, leading to a zero eigenvalue. This zero eigenvalue is clearly non-physical — the eigenvalues should appear either singly as the x−component of the flow speed, u, or in pairs symmetric about u. It also does not bode well numerically — the mode corresponding to this eigenvalue will be undamped.

The approach taken here is to look for a way in which to modify the governing equations so as to make A_p non-singular. The criteria that should be met by the modified matrix A_p' are:

- The eigenvalues and eigenvectors corresponding to the seven waves in the one-dimensional ($B_x =$ constant) Riemann solver remain unchanged;
- The eigenvalue corresponding to the new eighth wave is equal to u (the only physical choice for a single eigenvalue);
- The left and right eigenvectors corresponding to the new eight wave "make sense";
- In the case $B_x =$ constant, the eight-wave Riemann problem reduces to the seven-wave Riemann problem.

With these criteria in mind, it becomes possible to find a modified version of A_p, given some patience and some facility with Maple's symbolic manipulation capabilities. The modified matrix that meets the above criteria is

$$A'_p = \begin{bmatrix} u & \rho & 0 & 0 & 0 & 0 & 0 & 0 \\ 0 & u & 0 & 0 & 0 & \frac{B_y}{\rho} & \frac{B_z}{\rho} & \frac{1}{\rho} \\ 0 & 0 & u & 0 & 0 & -\frac{B_x}{\rho} & 0 & 0 \\ 0 & 0 & 0 & u & 0 & 0 & -\frac{B_x}{\rho} & 0 \\ 0 & 0 & 0 & 0 & u & 0 & 0 & 0 \\ 0 & B_y & -B_x & 0 & 0 & u & 0 & 0 \\ 0 & B_z & 0 & -B_x & 0 & 0 & u & 0 \\ 0 & \gamma p & 0 & 0 & 0 & 0 & 0 & u \end{bmatrix} . \tag{9}$$

The eigensystem of this matrix is composed of the following eight waves, with their corresponding eigenvalues λ, left eigenvectors $\tilde{\ell}$ and right eigenvectors $\tilde{r}$:

One Entropy Wave

$$\begin{aligned} \lambda_e &= u \\ \tilde{\ell}_e &= \left(1,0,0,0,0,0,0,-\frac{1}{a^2}\right) \\ \tilde{r}_e &= (1,0,0,0,0,0,0,0)^{\mathrm{T}} \ ; \end{aligned} \tag{10}$$

Two Alfvén Waves

$$\begin{aligned} \lambda_a &= u \pm \frac{B_x}{\sqrt{\rho}} \\ \tilde{\ell}_a &= \left(0,0,-B_z,B_y,0,\pm\frac{B_z}{\sqrt{\rho}},\mp\frac{B_y}{\sqrt{\rho}},0\right) \\ \tilde{r}_a &= (0,0,-B_z,B_y,0,\pm\sqrt{\rho}B_z,\mp\sqrt{\rho}B_y,0)^{\mathrm{T}} \ ; \end{aligned} \tag{11}$$

Four Magneto-acoustic Waves

$$\begin{aligned} &\lambda_{f,s} = u \pm c_{f,s} \\ &\tilde{l}_{f,s} = \left(0, \pm\rho c_{f,s}, \mp\frac{B_x B_y \rho c_{f,s}}{\rho c_{f,s}^2 - B_x^2}, \mp\frac{B_x B_z \rho c_{f,s}}{\rho c_{f,s}^2 - B_x^2}, 0, \frac{B_y \rho c_{f,s}^2}{\rho c_{f,s}^2 - B_x^2}, \frac{B_z \rho c_{f,s}^2}{\rho c_{f,s}^2 - B_x^2}, 1\right) \\ &\tilde{r}_{f,s} = \left(\rho, \pm c_{f,s}, \mp\frac{B_x B_y c_{f,s}}{\rho c_{f,s}^2 - B_x^2}, \mp\frac{B_x B_z c_{f,s}}{\rho c_{f,s}^2 - B_x^2}, 0, \frac{B_y \rho c_{f,s}^2}{\rho c_{f,s}^2 - B_x^2}, \frac{B_z \rho c_{f,s}^2}{\rho c_{f,s}^2 - B_x^2}, \gamma p\right)^{\mathrm{T}} \ ; \end{aligned} \tag{12}$$

5

One "Divergence" Wave

$$\begin{aligned} \lambda_d &= u \\ \tilde{\ell}_d &= (0,0,0,0,1,0,0,0) \\ \tilde{r}_d &= (0,0,0,0,1,0,0,0)^{\mathrm{T}} \ . \end{aligned} \tag{13}$$

It is important to note that the first seven waves yield the same eigenvectors and eigenvalues as the seven-wave Riemann problem, with the additional information that none of them carries a change in B_x (the fifth entry of each right eigenvector is zero), and none of the wave strengths is proportional to a jump in B_x (the fifth entry of each left eigenvector is zero). The new eighth wave travels with the $x-$component of the flow speed (its eigenvalue is u), and it carries a jump in B_x (the only non-zero entry in the left eigenvector is the entry corresponding to B_x), and affects only the $x-$component of the magnetic field (the only non-zero entry in the right eigenvector is the entry corresponding to B_x).

It is clear that the eigensystem of this modified matrix has all of the desired properties. In the case $B_x =$ constant, the strength of the eighth wave is zero, and the model reverts to that of the seven-wave problem. The new wave simply gives a rational procedure for dealing with non-zero jumps in B_x across the cell faces, which will in general occur when problems in two or three dimensions are being solved. The question remains, however, of what the modification of the matrix A_p (and the corresponding changes to B_p and C_p) has done to the system of conservation laws.

This can be seen by collecting the source terms due to the modifications to A_p, B_p and C_p and transforming to conserved variables. The new equation set, which has the eight-wave eigensystem described above, is

$$\frac{\partial}{\partial t}\begin{pmatrix} \rho \\ \rho\mathbf{u} \\ \mathbf{B} \\ E \end{pmatrix} + \nabla\cdot\begin{pmatrix} \rho\mathbf{u} \\ \rho\mathbf{u}\mathbf{u} + \mathbf{I}\left(p + \frac{\mathbf{B}\cdot\mathbf{B}}{2}\right) - \mathbf{B}n\mathbf{B} \\ \mathbf{u}\mathbf{B} - \mathbf{B}\mathbf{u} \\ \left(E + p + \frac{\mathbf{B}\cdot\mathbf{B}}{2}\right)\mathbf{u} - \mathbf{B}\left(\mathbf{u}\cdot\mathbf{B}\right) \end{pmatrix} = -\begin{pmatrix} 0 \\ \mathbf{B} \\ \mathbf{u} \\ \mathbf{u}\cdot\mathbf{B} \end{pmatrix}\nabla\cdot\mathbf{B}\ . \tag{14}$$

This is a noteworthy result: *the source term that must be added to Equation 2 is proportional to* $\nabla\cdot\mathbf{B}$. At the partial differential equation level, only terms that are equal to zero have been added to the conservative form of the governing equations. So, while technically the equations are no longer in conservative form, the deviations from conservation will be very small. It is only by writing the equations in this slightly non-conservative form that the singularity related to $\nabla\cdot\mathbf{B}$ can be removed. It has been previously noted that solving the momentum equation in non-conservative form can remove instabilities related to non-zero

6

$\nabla \cdot \mathbf{B}$ [1]; the current work hopefully reinforces this earlier result, and sheds further light on the mechanism for stabilizing the equations, as well as applying the idea in a novel way to develop a Riemann solver for multi-dimensional MHD.

It is interesting to note another justification of this particular choice of source term. Rewriting Equation 2 slightly by expanding some of the terms, the following form of the equations

$$\begin{aligned}
\frac{\partial \rho}{\partial t} + \nabla \cdot (\rho \mathbf{u}) &= 0 \\
\frac{\partial}{\partial t}(\rho \mathbf{u}) + \nabla \cdot (\rho \mathbf{u}\mathbf{u}) + \nabla\left(p + \frac{\mathbf{B}\cdot\mathbf{B}}{2}\right) - \mathbf{B}\cdot\nabla\mathbf{B} - \underline{\mathbf{B}\nabla\cdot\mathbf{B}} &= 0 \\
\frac{\partial \mathbf{B}}{\partial t} + \mathbf{u}\cdot\nabla\mathbf{B} + \mathbf{B}\nabla\cdot\mathbf{u} - \mathbf{B}\cdot\nabla\mathbf{u} - \underline{\mathbf{u}\nabla\cdot\mathbf{B}} &= 0 \\
\frac{\partial E}{\partial t} + \mathbf{u}\cdot\nabla\left(E + p + \frac{\mathbf{B}\cdot\mathbf{B}}{2}\right) + \left(E + p + \frac{\mathbf{B}\cdot\mathbf{B}}{2}\right)\nabla\cdot\mathbf{u} - & \\
\mathbf{B}\cdot\nabla(\mathbf{u}\cdot\mathbf{B}) - \underline{(\mathbf{u}\cdot\mathbf{B})\nabla\cdot\mathbf{B}} &= 0 \qquad (15)
\end{aligned}$$

is obtained. The terms that are proportional to $\nabla \cdot \mathbf{B}$ have been underlined; they are exactly the same as the source term defined above. Thus it can be seen that the addition of the source term in Equation 14 simply acts to remove the terms proportional to $\nabla \cdot \mathbf{B}$ that appear in Equation 2.

Another interesting note is what the evolution equation for $\nabla \cdot \mathbf{B}$ is for the two forms of the governing equations. This may be seen by taking the divergence of the evolution equation for the magnetic field in Equations 2 and 14. For the original form of the equations, the evolution equation is

$$\begin{aligned}
\nabla \cdot \left(\frac{\partial \mathbf{B}}{\partial t} + \mathbf{u}\cdot\nabla\mathbf{B} + \mathbf{B}\nabla\cdot\mathbf{u} - \mathbf{B}\cdot\nabla\mathbf{u} - \mathbf{u}\nabla\cdot\mathbf{B}\right) &= 0 \\
\frac{\partial}{\partial t}(\nabla\cdot\mathbf{B}) &= 0\,. \qquad (16)
\end{aligned}$$

¿From the partial differential equation point of view, this might well seem the correct result; $\nabla \cdot \mathbf{B} = 0$ is an initial condition, and this equation guarantees that $\nabla \cdot \mathbf{B} = 0$ is maintained throughout the evolution. For the modified form of the equations, the evolution equation for the magnetic field is

$$\begin{aligned}
\nabla \cdot \left(\frac{\partial \mathbf{B}}{\partial t} + \mathbf{u}\cdot\nabla\mathbf{B} + \mathbf{B}\nabla\cdot\mathbf{u} - \mathbf{B}\cdot\nabla\mathbf{u}\right) &= 0 \\
\frac{\partial}{\partial t}(\nabla\cdot\mathbf{B}) + \nabla\cdot(\mathbf{u}\nabla\cdot\mathbf{B}) &= 0\,. \qquad (17)
\end{aligned}$$

Thus the addition of the source term has modified the evolution equation for $\nabla \cdot \mathbf{B}$ so that the quantity $\nabla \cdot \mathbf{B}/\rho$ is treated as a passive scalar. This is clearlv the more numerically stable of the two evolution equations; any local $\nabla \cdot \mathbf{B}$ that is created is convected away.

The above derivation gives all the pieces for building an ideal MHD solver that works for two-dimensional problems, without having to resort to non-collocated variables or a projection algorithm. Specifically, a Roe-type approximate Riemann solver has been implemented, where the wave strengths and speeds are derived from the above left eigenvectors and eigenvalues. The eigenvectors are properly normalized to avoid difficulties associated with coinciding wave speeds [5]. The average state needed at cell interfaces is computed by a simple average of left and right states (although a Roe average does exist for the ideal MHD equations [4]). The source term, though small, is calculated in each cell, and added to the residual. The resulting code is first order in space and time.

3 A Test of the Eight-Wave Riemann Solver

Brio and Wu [2] developed a test problem for one-dimensional MHD solvers based on the shock-tube problem of Sod [6]. Two stationary plasmas are separated by a membrane which is removed at $t = 0$, allowing the plasmas to interact. The test problem used here for the two-dimensional MHD solver is a rotated version of the Brio-Wu problem. The left and right input states, and the orientation of propagation of disturbances to the grid, is shown in Figure 2. In the Brio-Wu problem (the top figure), the boundary conditions are that the problem is periodic along a line $y =$ constant; in the current test problem (the bottom figure), the boundary conditions are that the problem is periodic along a line $x + y =$ constant.

Both the rotated and non-rotated problems were run on coarse (600 cells in x) and fine (1200 cells in x) grids. The time step was taken as $\Delta t/\Delta x = 0.2$, which corresponds to a CFL number of approximately 0.8 on the non-rotated problem. The ratio of specific heats, γ, was 2.0. The number of time steps taken on the coarse and fine grids were 100 and 200, respectively. The $x-$axis in the plotted results from the rotated problem was scaled by a factor of $\sqrt{2}$, to account for the fact that the CFL number is lower for the rotated problem than for the non-rotated problem.

Figures 3–7 show comparisons of the results on the fine grid of the non-rotated shock-tube problem with the (scaled) results of the rotated shock-tube problem for

3. density (ρ);

4. pressure (p);

5. velocity component normal to the original discontinuity (u_n);

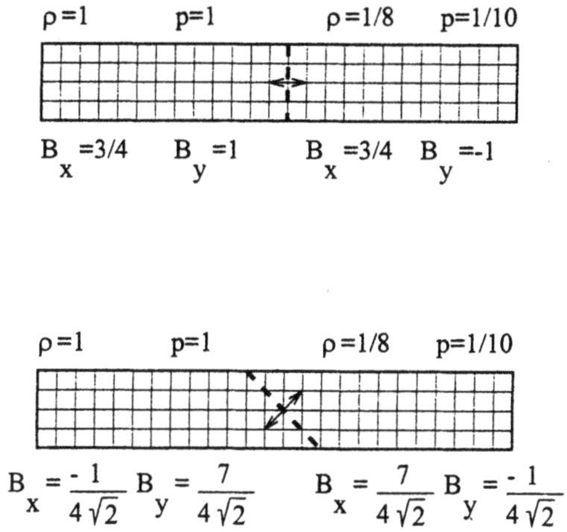

Figure 2: A Test Problem for Two-Dimensional MHD

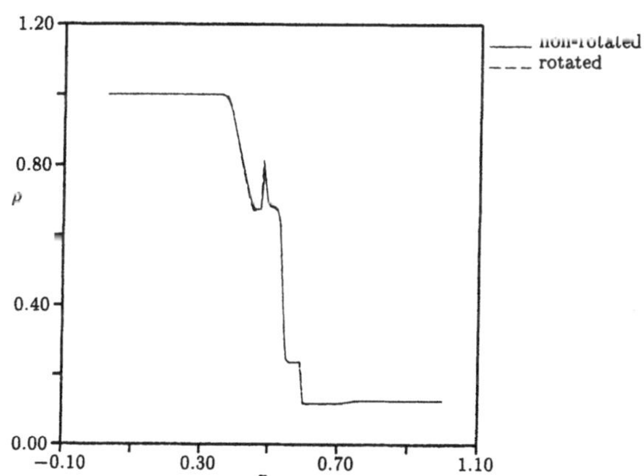

Figure 3: Density in the Rotated and Non-Rotated Shock Tubes

9

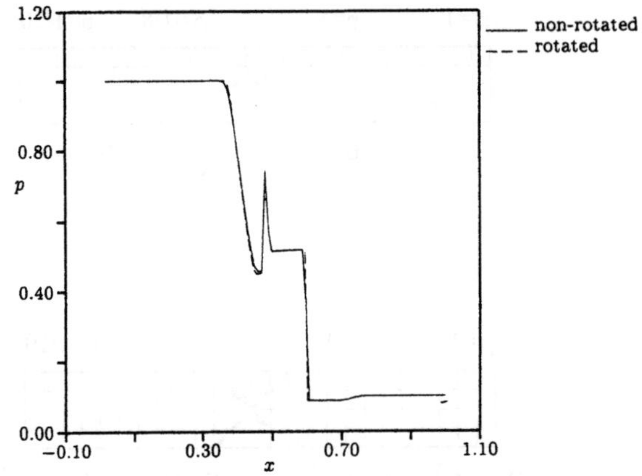

Figure 4: Pressure in the Rotated and Non-Rotated Shock Tubes

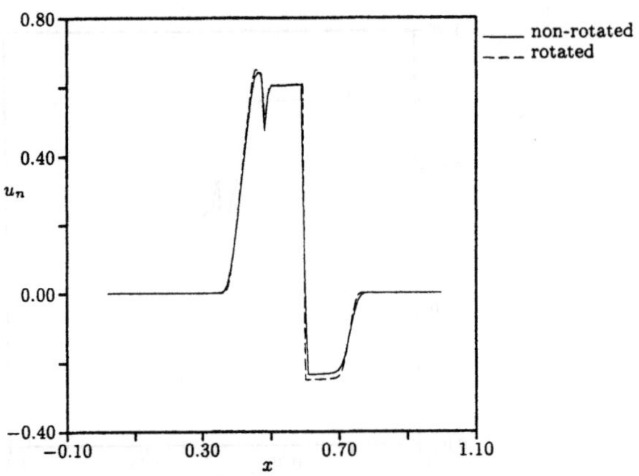

Figure 5: Normal Velocity in the Rotated and Non-Rotated Shock Tubes

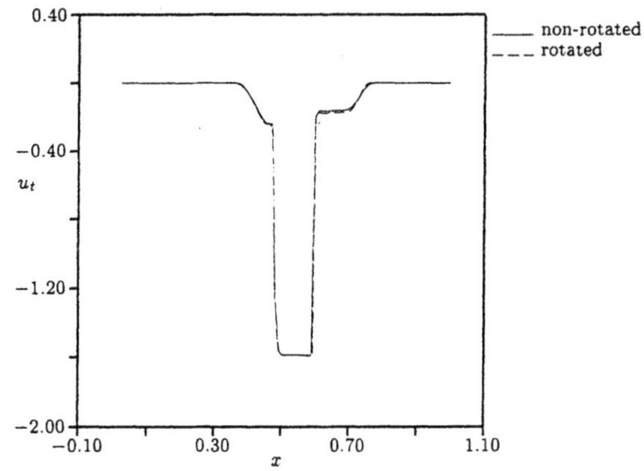

Figure 6: Tangential Velocity in the Rotated and Non-Rotated Shock Tubes

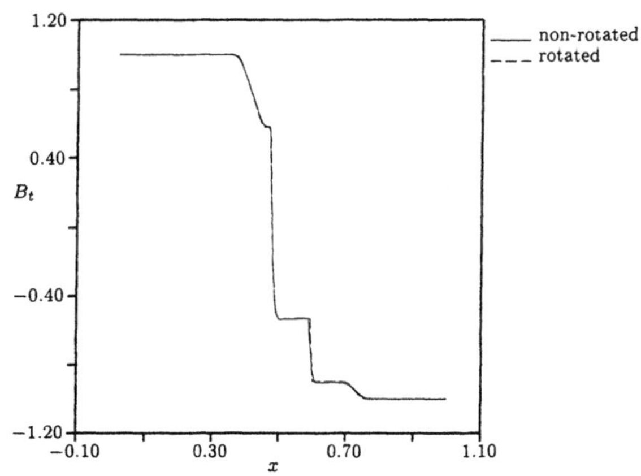

Figure 7: Tangential Magnetic Field in the Rotated and Non-Rotated Shock Tubes

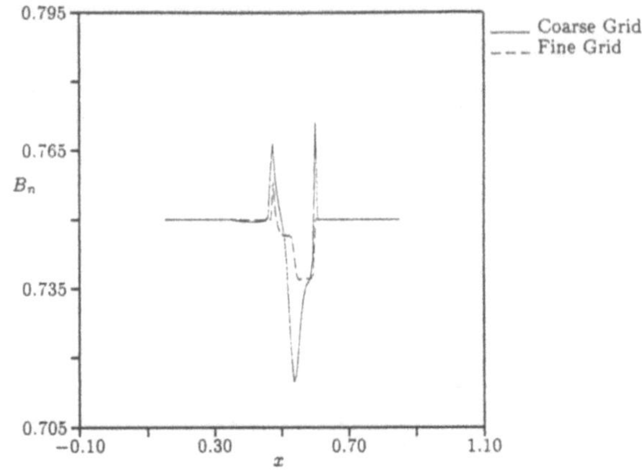

Figure 8: Normal Magnetic Field in the Rotated Shock Tube (Coarse and Fine)

6. velocity component tangential to the original discontinuity (u_t);
7. magnetic-field component tangential to the original discontinuity (B_t).

As can be seen, the agreement is quite good, with the results of the two cases nearly indistinguishable for all but the normal component of velocity. The errors in u_n are balanced by errors in the magnetic-field component normal to the original discontinuity (B_n). Figure 8 shows B_n for the rotated shock-tube problem on the coarse and fine grids. In the non-rotated problem, $B_n = 0.75$ throughout the tube. As can be seen, there are errors on the order of a few percent in B_n on the coarse grid, but the errors are reduced as the grid is refined.

4 Concluding Remarks

In some respects, this paper presents the development of only one-eighth of a Riemann solver. Seven of the eight waves of the Riemann solver are the same as those used in previous work on upwind methods for MHD. The deceptively simple eighth wave that arises from the analysis, however, is of a different character than the other seven — it arises only in multi-dimensional problems, and it is crucial for understanding and solving those problems. It plays the very important role of stabilizing the numerical method with respect to the small amounts of $\nabla \cdot \mathbf{B}$ generated in solving the discrete MHD equations.

Given the meteoric rise of Riemann solvers in the computation of compressible gas dynamics, it is not very risky to predict that schemes based on Riemann solvers will play an increasingly important role in the computation of compressible conducting flows. The ability of Riemann solvers to capture strong discontinuities robustly and with minimal dissipation, the framework that Riemann solvers provide for implementing stable boundary procedures, and the aesthetically appealing physical basis of Riemann solvers are all strong arguments for their use. The aim of this paper is to remove what is hopefully one of the last major obstacles to the use of Riemann solvers in large-scale codes for computing multi-dimensional conducting flows.

Acknowledgments

The work presented here was carried out at the Institute for Computer Applications in Science and Engineering (ICASE), and the Centrum voor Wiskunde en Informatica (CWI). The author thanks both ICASE and CWI for supporting his visits there, and particularly Dr. Barry Koren, who coordinated the author's visit to CWI. At ICASE, the author's work was sponsored through NASA Contract Number NAS1-19480. The appropriate addresses are: CWI, Kruislaan 413, 1098 SJ Amsterdam, NETHERLANDS; ICASE, NASA Langley Research Center, Hampton VA, 23681. In addition, the author thanks Phil Roe, who was, as always, an inspiration in this work.

References

[1] J. U. Brackbill and D. C. Barnes. The effect of nonzero $\nabla \cdot \mathbf{B}$ on the numerical solution of the magnetohydrodynamic equations. *Journal of Computational Physics*, 35:426–430, 1980.

[2] M. Brio and C. C. Wu. An upwind differencing scheme for the equations of ideal magnetohydrodynamics. *Journal of Computational Physics*, 75:400–422, 1988.

[3] A. Jeffrey and A. Taniuti. *Nonlinear Wave Propagation.* Academic Press, 1964.

[4] P. L. Roe, 1994. Private Communication.

[5] P. L. Roe and D. S. Balsara. Notes on the eigensystem of magnetohydrodynamics. *Journal of Computational Physics*, 1994. To appear.

[6] G. A. Sod. A survey of finite-difference methods for systems of nonlinear conservation laws. *Journal of Computational Physics*, 27(1):1–31, 1978.

[7] P. R. Woodward and W. Dai. A Riemann solver for ideal magnetohydrodynamics. *Journal of Computational Physics*, 1994. To appear.

Papers Reprinted

Chapter I. Riemann Solvers and Upwind Schemes

1. On the Relation Between the Upwind-Differencing Schemes of Godunov, Engquist–Osher and Roe
   B. van Leer
   SIAM J. Sci. Stat. Comput., vol. 5, no. 1, March 1984, pp. 1–20
   Originally published as ICASE Report No. 81-11
   Copyright 1984, Society for Industrial and Applied Mathematics
2. On Upstream Differencing and Godunov-Type Schemes for Hyperbolic Conservation Laws
   A. Harten, P.D. Lax, and B. van Leer
   SIAM Review, vol. 25, no. 1, January 1983, pp. 35–61
   Originally published as ICASE Report No. 82-5
   Copyright 1983, Society for Industrial and Applied Mathematics
3. Flux-Vector Splitting for the Euler Equation
   B. van Leer
   Proc. of the 8th Intl. Conf. on Numerical Methods in Fluid Dynamics, Aachen, Germany;
   Lecture Notes in Physics, no. 170, E. Krause (ed.), pp. 507–512, Springer-Verlag, 1982
   Originally published as ICASE Report No. 82-30

Chapter II. Second-Order Total-Variation Diminishing Schemes

1. A Comparative Study of Computational Methods in Cosmic Gas Dynamics
   G.D. van Albada, B. van Leer, and W.W. Roberts, Jr.
   Astron. Astrophys., vol. 108, no. 1, April 1982, pp. 76–84
   Originally published as ICASE Report No. 81-24
   Copyright 1982, Springer-Verlag
2. High Resolution Schemes and the Entropy Condition
   S. Osher and S. Chakravarthy
   SIAM J. Numer. Anal., vol. 21, no. 5, October 1984, pp. 955–984
   Originally published as ICASE Report No. 83-49
   Copyright 1984, Society for Industrial and Applied Mathematics

3. Convergence of Generalized MUSCL Schemes
   S. Osher
   SIAM J. Numer. Anal., vol. 22, no. 5, October 1985, pp. 947–961
   Originally published as ICASE Report No. 84-10
   Copyright 1985, Society for Industrial and Applied Mathematics
4. A Simplified TVD Finite Difference Scheme Via Artificial Viscosity
   S.F. Davis
   SISC, vol. 8, no. 1, January 1987, pp. 1–18
   Originally published as ICASE Report No. 84-20
   Copyright 1987, Society for Industrial and Applied Mathematics
5. On Central-Difference and Upwind Schemes
   R.C. Swanson and E. Turkel
   J. Comp. Physics, vol. 101, no. 2, August 1992, pp. 292–306
   Originally published as ICASE Report No. 90-44
   Copyright 1992, Academic Press, Inc.

Chapter III. Essentially Non-oscillatory Schemes

1. Uniformly High-Order Accurate Nonoscillatory Schemes. I
   A. Harten and S. Osher
   SIAM J. Numer. Anal., vol. 24, no. 2, April 1987, pp. 279–309
   Copyright 1987, Society for Industrial and Applied Mathematics
2. Uniformly High Order Accurate Essentially Non-oscillatory Schemes, III
   A. Harten, B. Engquist, S. Osher, and S.R. Chakravarthy
   J. Comp. Physics, vol. 17, no. 1, August 1987, pp. 231–303
   Originally published as ICASE Report No. 86-22
   Copyright 1987, Academic Press, Inc.
3. ENO Schemes with Subcell Resolution
   A. Harten
   J. Comp. Physics, vol. 83, no. 1, 1989, pp. 148–184
   Originally published as ICASE Report No. 87-56
   Copyright 1989, Academic Press, Inc.
4. Efficient Implementation of Essentially Non-oscillatory
   Shock-Capturing Schemes, II
   C.-W. Shu and S. Osher
   J. Comp. Physics, vol. 83, no. 1, July 1989, pp. 32–78
   Originally published as ICASE Report No. 88-24
   Copyright 1989, Academic Press, Inc.
5. Essentially Nonoscillatory Spectral Fourier Methods
   for Shock Wave Calculations
   W. Cai, D. Gottlieb, and C.-W. Shu
   Math. Comp., vol. 52, no. 186, 1989, pp. 389–410
   Originally published as ICASE Report No. 88-37
   Copyright 1989, American Mathematical Society

6. Cell Averaging Chebyshev Methods for Hyperbolic Problems
   W. Cai, D. Gottlieb, and A. Harten
   Computers and Mathematics with Applications, vol. 24, no. 5/6, September 1992
   Originally published as ICASE Report No. 90-27
   Copyright 1992, Pergamon Press

Chapter IV. Multidimensional Schemes

1. A Rotationally Biased Upwind Difference Scheme for the Euler Equations
   S.F. Davis
   J. Comp. Physics, vol. 56, no. 1, October 1984, pp. 65–92
   Originally published as ICASE Report No. 83-37
   Copyright 1984, Academic Press, Inc.
2. Discrete Models for the Numerical Analysis of Time-Dependent Multidimensional Gas Dynamics
   P.L. Roe
   J. Comp. Physics, vol. 63, no. 2, April 1986, pp. 458–476
   Originally published as ICASE Report No. 85-18
   Copyright 1986, Academic Press, Inc.
3. Discontinuous Solutions to Hyperbolic Systems Under Operator Splitting
   P.L. Roe
   Numerical Methods for Partial Differential Equations, vol. 7, 1991, pp. 277–297
   Originally published as ICASE Report No. 87-64
   Copyright 1991, John Wiley & Sons, Ltd.
4. High Resolution Finite Volume Methods on Arbitrary Grids Via Wave Propagation
   R.J. LeVeque
   J. Comp. Physics, vol. 78, 1988, pp. 36–63
   Originally published as ICASE Report No. 87-68
   Copyright 1988, Academic Press, Inc.
5. Design of an Essentially Nonoscillatory Reconstruction Procedure on Finite-Element-Type Meshes
   R. Abgrall
   International Conference on Spectral and High-Order Methods, Montpellier, France
   Originally published as ICASE Report No. 91-84
6. A Contribution to the Great Riemann Solver Debate
   J.J. Quirk
   Inter. J. for Num. Methods in Fluids, vol. 18, 1994, pp. 555–574
   Originally published as ICASE Report No. 92-64
   Copyright 1994, John Wiley & Sons, Ltd.

7. An Approximate Riemann Solver for Magnetohydrodynamics (That Works in More Than One Dimension)
K.G. Powell
ICASE Report No. 94-24

MIX
Papier aus verantwortungsvollen Quellen
Paper from responsible sources
FSC® C105338

If you have any concerns about our products,
you can contact us on
ProductSafety@springernature.com

In case Publisher is established outside the EU,
the EU authorized representative is:
Springer Nature Customer Service Center GmbH
Europaplatz 3, 69115 Heidelberg, Germany

Printed by Libri Plureos GmbH
in Hamburg, Germany